Advances in Composite Science and Technology 2018

IOP Conference Series: Materials Science and Engineering
Volume 683

Moscow, Russia
5 - 8 December 2018

ISBN: 978-1-7138-0935-7
ISSN: 1757-8981

Printed from e-media with permission by:

Curran Associates, Inc.
57 Morehouse Lane
Red Hook, NY 12571

Some format issues inherent in the e-media version may also appear in this print version.

Institute of Physics
Dirac House, Temple Back
Bristol BS1 6BE UK

Phone: 44 1 17 929 7481
Fax: 44 1 17 920 0979

techtracking@iop.org

Additional copies of this publication are available from:

Curran Associates, Inc.
57 Morehouse Lane
Red Hook, NY 12571 USA
Phone: 845-758-0400
Fax: 845-758-2633
Email: curran@proceedings.com
Web: www.proceedings.com

TABLE OF CONTENTS

IOP Conf. Series: Materials Science and Engineering **683** (2019) 011001 doi:10.1088/1757-899X/683/1/011001

Preface

The 1st international conference on composites **"Advances in Composite Science and Technologies"** was held in Bauman Moscow State Technical University from 5 to 8 December, 2018. The conference is meant to be a unified international platform for the knowledge and experience exchange between representatives of the scientific community from the field of composite materials. The following sections covered the most important topics in the composite sciences field.

Session 1. "Smart" composite materials and composites for biomedical application

Session 2. Advanced metal matrix composites

Session 3. Physical and technological maintenance of the quality of composite materials products

Session 4. Composite materials for rocket and space applications

Session 5. Mechanics of composite materials and structures

Session 6. Inorganic composite materials

Session 7. Fiber surface modification methods

Session 8. Modelling in composite science

Session 9. High Temperature composites

Session 10. Russian engineering school (educational session)

The success of the conference is supported with more than 150 of proceeding papers received by Organizing Committee and only 95 were selected for publication after reviewing process. We are grateful to all participants as well as to discussion leaders, members of Organizing and Scientific boards for taking part in the conference and looking forward to the next year event.

Acknowledgments
This work was supported by Russian Foundation of Basic Research (RFBR) grant 18-08-20139 \18.

IOP Conf. Series: Materials Science and Engineering **683** (2019) 011002 doi:10.1088/1757-899X/683/1/011002

Peer review statement

All papers published in this volume of *IOP Conference Series: Materials Science and Engineering* have been peer reviewed through processes administered by the proceedings Editors. Reviews were conducted by expert referees to the professional and scientific standards expected of a proceedings journal published by IOP Publishing.

Synthesis of cross-linked polyurethane with self-healing properties

E O Platonova[a], E Vlasov[a], A Kireynov[a], A V Polezhaev[a,b*]

[a]Bauman Moscow State Technical University, 105005, Russia, Moscow, 2nd Baumanskaya str., 5/1
[b]A.N. Nesmeyanov Institute of Organoelement Compounds, 119334, Russia, Moscow, Vavilova str., 28

E-mail: avp@emtc.ru

Abstract. Prepolymer synthesized from toluene-2,4-diisocyanate (TDI) (hard segment), and polypropylene glycol (PPG-2000) (soft segment) was end-capped with difurfurylamine groups. A new cross-linked self-healing polyurethane was obtained via Diels-Alder reaction of the en-capped prepolymer with N,N'-bismaleimido-4,4'-diphenylmethane (BMI). Thermal reversibility of the Diels-Alder reaction (DA-reaction) between terminal furanic rings of the polyurethane prepolymer and bismaleimide cross-linking agent designed to be responsible for self-healing. Polymers were characterized by NMR and IR spectroscopy and thermal analysis (DSC and TGA). Thermal reversibility of the DA-bonds in the cross-linked polymer was confirmed by DSC. The synthesized polymer has good processability and potential recyclability.

1. Introduction

Polyurethanes (PU) are widely used as elastomers, adhesives, covers, and foams due to their physical and mechanical properties [1, 2]. However, PUs are affected by the environment (mainly air moisture and light), mechanical and chemical influence, UV radiation and also with a combination of these factors causing microcracking and other structural defects [3]. Another disadvantage is that molded polyurethanes cannot be recycled as most thermosetting plastics and could pollute the environment. Self-healing and recyclable resins are desirable materials that can reduce the major disadvantages of PUs. Self-healing properties could be realized in many ways such as encapsulation [4], hollow fibers with healing resins [5], microvascular networks [6], the addition of thermoplastic particles [7, 8] and e.t.c. All methods mentioned above demand incorporation of a healing agent and catalyst [9-11], and the number of self-healing attempts is limited with one or few. Internal healing based on reversible chemical interaction between functional groups in a polymer are more attractive due to unlimited mending cycles, but external stimulus such as heat or light must be applied [12-14]. There is a wide variety of internal self-healing approaches based on photoactivated [2+2]-cycloaddition [15], reversible formation of disulfide bridges [16] and hydrogen bonds [17], radical reactions [18], pH-dependent processes [19], Diels-Alder reactions [1, 20]. Diels-Alder (DA) reaction between furans and maleimides is a thermally reversible process [21]. Formation of bicyclic DA-adduct usually occurs at 40-70 °C and at higher temperatures (>100 °C) it undergoes cleavage (reverse DA-reaction, re-DA) which results in regeneration of starting components [21, 22]. A few examples of linear polyurethanes, prepared via DA-based polymerization of difuranic oligomers and bismaleimides are known [1, 23-25]. We present here the synthesis of crosslinked PU based on tetrafuranic oligomer and bismaleimide. Tetrafuranic block (Fu2N-TDI-PPG2000) was prepared by the termination of NCO-ended oligomer (TDI-PPG2000) with difurfurylamine, and commercially available N,N'-bismaleimido-4,4'-diphenylmethane was used as maleimide linker. Difurfurylamine was synthesized by two-step route from furfural, an industrial byproduct of vegetable stock processing. Utilization of monomers obtainable from biorenewable sources allows to reduce using non-renewable stock sources as oil [26-28]; some furane-based resins are actively used as reactive diluents for epoxy and phenol-formaldehyde resins[28-30]. We present here the synthesis, spectroscopic characterization, and evaluation of self-healing abilities of these material.

2. Experimental

2.1. Materials and methods

Furfural (99.5%) was purchased from “Komponent-reactive” and purified by vacuum distillation. TDI (toluene-2,4-diisocyanate, 80%), sodium borohydride (98%) and BMI (1,10-(methylenedi-1,4-phenylene)bismaleimide, 95%) were purchased from Aldrich and used as received. PPG2000 (polypropylene glycol, Mn = 2000) was purchased from Aldrich and dried under vacuum at 110 °C prior to use. Hydrofuramide was prepared by the common method [1] from furfural and ammonia. DMF (N,N-dimethylformamide) was purchased from Acros dried over CaH2 and distilled prior to use.

All NMR spectra were recorded by a Bruker Avance 600 NMR Spectrometer (600.1 MHz), using residual proton signal of deuterated solvent as a reference, chemical shifts were reported as parts per million downfield from tetramethylsilane (TMS). ATR-FTIR was performed on Nicolet iS10 spectrometer in the range of 4000 to 650 cm-1 on a germanium crystal. The thermal behavior was examined by DSC, with a NETZH DSC 204 F1 Phoenix within a temperature range of -60 to 160 °C at heating/cooling rates of 10 K min-1 in an argon atmosphere. A sample weight of about 16 mg was used for the measurements. Thermogravimetric analysis (TGA) was performed on NETZH TG 209 F1 Libra within a temperature range of 30 to 550 °C at the heating/cooling rate of 10 K min-1 in an argon atmosphere.

2.2. Synthesis of Fu_2N-TDI-PPG.

Fu2N-prepolymer was prepared using a conventional two-step method. At the first step, TDI (10.0 g, 57.4 mmol) was reacted with PPG2000 (57.0 g, 28.5 mmol) in a 100 mL two necked round bottom flask, equipped with a magnetic stirrer and dropping funnel. The reaction was carried out at 60 °C under an argon atmosphere for three hours, and the reaction progress was monitored by IR-spectroscopy. At the second step reaction mixture was cooled in an ice-water bath, diluted with DMF (3 mL) and difurfuryl amine (3.0 g, 16.9 mmol) dissolved in DMF (3 mL) was added dropwise into the isocyanate end-capped prepolymer solution in about 15 min. The reaction mixture was stirred at room temperature for two hours. Yield: 21.0 g. ^{1}H NMR (400 MHz, DMSO-d_6): 9.52 (s, 2H, –**NH**–CO–O–), 8.05 (s, 2H, –HNC(O)**NH**–Ar), 7.62 (s, 2H, =C**H**O– of furan ring), 7.53–7.40 (m, 2H, –Ar), 7.17 (d, J_{HH} = 7.8 Hz,2H, –Ar), 7.07-6.98 (m, 2H, –Ar), 6.42 (s, 4H, =C**H**–CH= of the furan ring), 6.32 (s, 4H, =CH–C**H**= of the furan ring), 4.96-4.77 (m, 2H, –**CH**(CH$_3$)–OC(O)–NH–), 4.53 (s, 8H, –N-C**H$_2$**–), 3.52-3.91 (m, 64H,–**CH$_2$**–CH(CH$_3$)–O–), 3.38-3.28 (m, 32H,–CH$_2$–**CH**(CH$_3$)–O–), 1.27-1.14 (m, 6H, **CH$_3$**–Ar), 1.04 (d, J_{HH} = 5.7 Hz, –O–CH(**CH$_3$**)–CH$_2$–); ^{13}C NMR (101 MHz, DMSO-d_6): 155.45, 153.54, 151.84, 143.02, 138.27, 137.56, 130.33, 127.19, 127.09, 110.90, 108.42, 107.39, 75.14, 75.05, 74.94, 74.83, 74.70, 72.91, 72.72, 44.96, 44.88, 42.81, 17.76, 17.71, 17.36; IR (ATR, neat, cm^{-1}): 3304, 2970, 2930, 2895, 2869, 1727, 1673, 1600, 1531, 1506, 1453, 1373, 1225, 1102, 1012, 926, 868, 816, 751, 667.

2.3 Synthesis of cross-linked PU

BMI (0.88 g, 2.5 mmol) was added to Fu_2N-prepolymer (7.0 g, 2.5 mmol) in 6 mL DMF. Reaction mixture was heated to 60 °C for three hours, then carried onto a glass and cured in oven within 96 hours to afford transparent polymer film.

3. Results and discussion

A synthetic approach for the self-healing polyurethane PU-DA is shown in Figure 1. Briefly, TDI reacted with oligomeric glycol yielding isocyanate-terminated preform which was further treated with difurfurylamine with the formation of the furan-terminated oligomer.

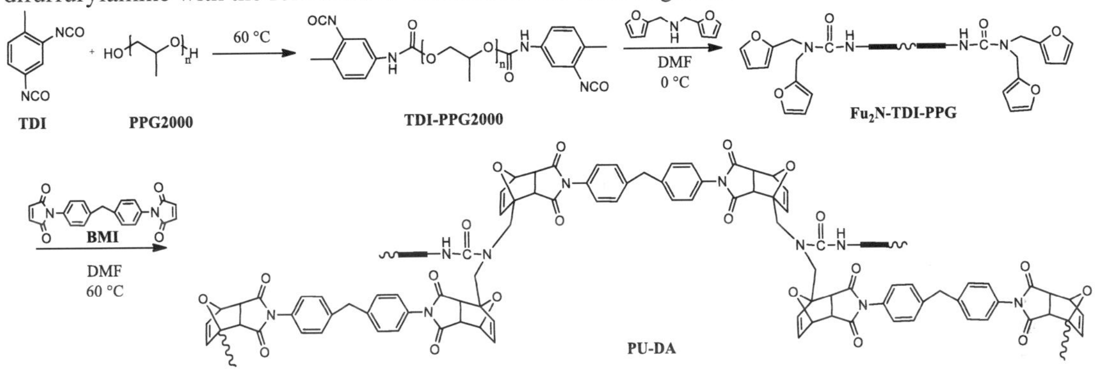

Figure 1. Synthesis of PU-DA.

3.1. Characterization of PU-DA

Structure of the obtained polyurethane was characterized by NMR and IR spectroscopy. Figure 2 represents ^{1}H NMR spectra of the initial monomer and resulting polymer.

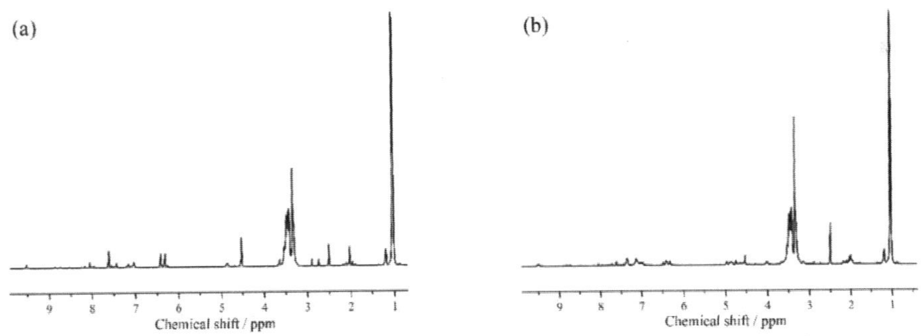

Fugure 2. ^{1}H NMR of Fu_2N-TDI-PPG (a) and PU-DA (b).

Both spectra demonstrate signals corresponding to hard segment: at 7.53-6.98 ppm corresponding to phenyl ring, and at 1.27-1.14 ppm corresponding to methyl groups of TDI [1] – and signals from protons of a soft segment: at 3.52-3.91 ppm corresponding to methylene groups, at 3.38-3.28 ppm – to methine protons of backbone chain, and a multiplet at 4.96-4.77 ppm complies to methine protons of the terminal PPG-2000 groups nearest to N-C(O)O-fragments. [2] In the polymer spectra (Fig. 2(b)) intensity of furanic ring signals (7.62, 6.42 and 6.32 ppm) is mainly smaller, which along with the appearance of multiplets in the range 6.52-6.26 ppm indicates completed reaction between monomer and bismaleimide and formation of the DA-adduct. [3]

All monomer and a resulting polyurethane were also characterized by IR-spectroscopy. An absence of the band at 2270 cm-1 in the Fu2N-TDI-PPG spectrum (Figure 3 (a)) corresponding to isocyanate groups indicates completion of difurfurylamine addition reaction. A small intensity band at 1775 cm-1 in the

PU-DA spectrum (Figure 3(b)) corresponds to DA-adduct. [4] Also, a decreasing of 751 and 1012 cm-1 bands intensity in the polymer spectrum indicates completion of Diels-Alder reaction.

The thermal behavior of PU-DA was characterized by TGA analysis. From the TGA curve, PU-DA shows a slow weight loss after 200 °C that is indicative of high thermal stability of PU-DA adduct (Fig. 4).

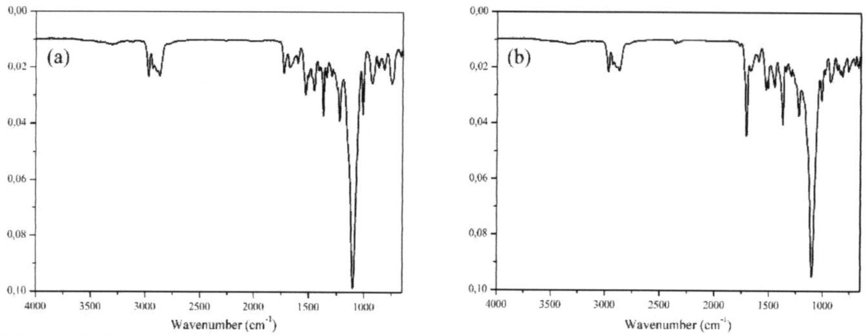

Figure 3. IR-spectra of Fu_2N-TDI-PPG (a) and PU-DA (b).

3.2 Self-healing properties of PU-DA.

Self-healing properties of the polyurethane were tested by DSC method using two-step heating-cooling cycle (Fig. 5). At the first heating curve, an endothermic peak at 98 °C appeared corresponding to reverse DA-reaction process (re-DA). An absence of this peak at the second heating suggests that material is lack of DA-bonds. Then the sample was placed to a thermostat at 60 °C for 48 hours and subsequent DSC-analysis observed pattern similar to the previous one: the first heating curve shows a peak of the reverse Diels-Alder reaction which is absent at the second heating curve.

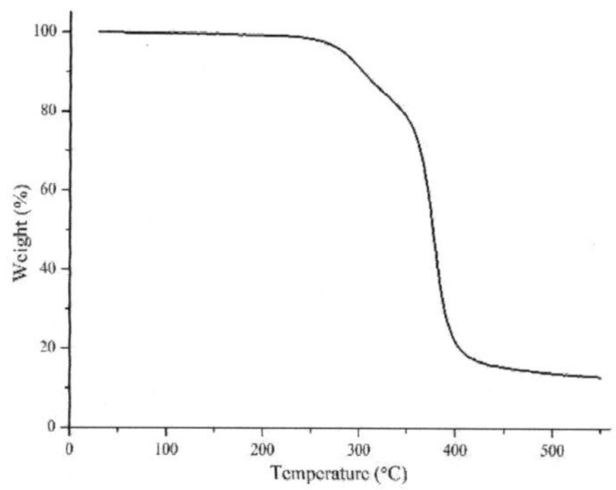

Figure 4. TGA curve of PU-DA.

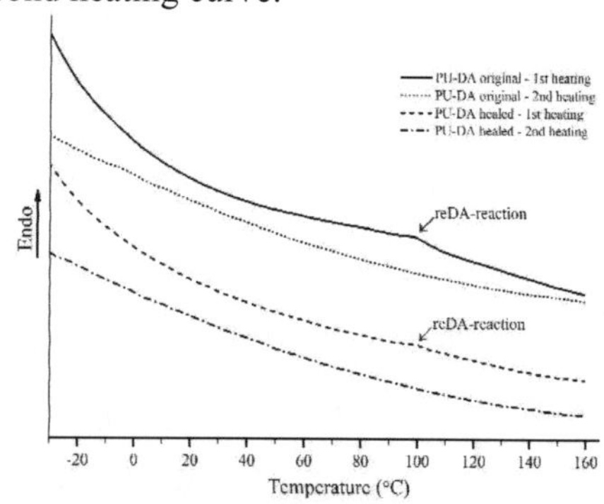

Figure 5. DSC curves of PU-DA.

4. Conclusion

A new self-healing cross-linked polyurethane was synthesized via Diels-Alder reaction between bismaleimide and furan end-capped prepolymer. Self-healing properties of the resulting material were examined by DSC analysis.

Acknowledgments

This work was supported by Russian Foundation of Basic Research (RFBR) grants 18-29-18036 and 18-29-18037. NMR experiments were performed with the financial support from Ministry of Science and Higher Education of the Russian Federation using the equipment of Center for molecular composition studies of INEOS RAS.

References

[1] Feng L, Yu Z, Bian Y, Lu J, Shi X and Chai C 2017 Self-healing behavior of polyurethanes based on dual actions of thermo-reversible Diels-Alder reaction and thermal movement of molecular chains *Polymer* **124** 48-59

[2] Yu X, Yang P, Zhang Z, Wang L, Liu L and Wang Y 2018 Self-healing polyurethane nanocomposite films with recoverable surface hydrophobicity *J. Appl. Polym. Sci.* **135** 46421

[3] Osswald T A and Menges G 2012 *Materials Science of Polymers for Engineers*: Carl Hanser Verlag GmbH & Co. KG) pp 423-87

[4] White S R, Sottos N R, Geubelle P H, Moore J S, Kessler M R, Sriram S R, Brown E N and Viswanathan S 2001 Autonomic healing of polymer composites *Nature* **409** 794

[5] Patrick J F, Sottos N R and White S R 2012 Microvascular based self-healing polymeric foam *Polymer* **53** 4231-40

[6] Toohey K S, Sottos N R, Lewis J A, Moore J S and White S R 2007 Self-healing materials with microvascular networks *Nature Mater.* **6** 581

[7] Nji J and Li G 2010 A biomimic shape memory polymer based self-healing particulate composite *Polymer* **51** 6021-9

[8] Borodulin A S, Kalinnikov, A.N., Bazheva, R.C., Beshtoev, B.Z. 2018 Synthesis and properties of aromatic polyethersulfones *Int.l J. Mech. Eng. Tech.* **9** 1109-16

[9] Jin H, Mangun C L, Stradley D S, Moore J S, Sottos N R and White S R 2012 Self-healing thermoset using encapsulated epoxy-amine healing chemistry *Polymer* **53** 581-7

[10] Zhang H, Wang P and Yang J 2014 Self-healing epoxy via epoxy–amine chemistry in dual hollow glass bubbles *Compos. Sci. Technol.* **94** 23-9

[11] Nelyub V A, Borodulin A S, Kobets L P and Malysheva G V 2016 Capillary hydrodynamics of oligomer binders *Polymer Science Series D* **9** 322-5

[12] Teixeira R F A, Hillewaere X K D, Billiet S and Du Prez F E 2013 Chemistry of Crosslinking Processes for Self-Healing Polymers *Self-Healing Polymers*

[13] Wang Y, Pham D T and Ji C 2015 Self-healing composites: A review *Cogent Engineering* **2** 1075686

[14] Hia I L, Vahedi V and Pasbakhsh P 2016 Self-Healing Polymer Composites: Prospects, Challenges, and Applications *Polymer Reviews* **56** 225-61

[15] Ling J, Rong M-z and Zhang M-q 2014 Effect of molecular weight of PEG soft segments on photo-stimulated self-healing performance of coumarin functionalized polyurethanes *Chin. J. Polym. Sci.* **32** 1286-97

[16] Gewei Z, Jianzhong P, Rui L, Pengbo L and Bochao Z 2018 The preparation and characterization of a novel self-healing based on the dynamic translocation of disulfide bonds *Materials Research Express* **5** 105301

[17] Grzelak A W, Boinard P and Liggat J J 2018 The influence of diol chain extender on morphology and properties of thermally-triggered UV-stable self-healing polyurethane coatings *Prog. Org. Coat.* **122** 1-9

[18] Zhang Z P, Rong M Z and Zhang M Q 2018 Mechanically Robust, Self-Healable, and Highly Stretchable "Living" Crosslinked Polyurethane Based on a Reversible C□C Bond *Adv. Funct. Mater.* **28** 1706050

[19] Wang S, Zhang Z, Chen B, Shao J and Guo Z 2018 Self-healing hydrogel of poly(vinyl alcohol)/graphite oxide with pH-sensitive and enhanced thermal properties *J. Appl. Polym. Sci.* **135** 46143

[20] Gu L and Wu Q-Y 2018 Recyclable bio-based crosslinked polyurethanes with self-healing ability *J. Appl. Polym. Sci.* **135** 46272
[21] Chen X, Dam M A, Ono K, Mal A, Shen H, Nutt S R, Sheran K and Wudl F 2002 A Thermally Re-mendable Cross-Linked Polymeric Material *Science* **295** 1698
[22] Du P, Wu M, Liu X, Zheng Z, Wang X, Joncheray T and Zhang Y 2013 Diels–Alder-based crosslinked self-healing polyurethane/urea from polymeric methylene diphenyl diisocyanate *J. Appl. Polym. Sci.* **131**
[23] Liu X, Du P, Liu L, Zheng Z, Wang X, Joncheray T and Zhang Y 2013 Kinetic study of Diels–Alder reaction involving in maleimide–furan compounds and linear polyurethane *Polym. Bull.* **70** 2319-35
[24] Du P, Liu X, Zheng Z, Wang X, Joncheray T and Zhang Y 2013 Synthesis and characterization of linear self-healing polyurethane based on thermally reversible Diels–Alder reaction *RSC Advances* **3** 15475-82
[25] Du P, Wu M, Liu X, Zheng Z, Wang X, Sun P, Joncheray T and Zhang Y 2014 Synthesis of linear polyurethane bearing pendant furan and cross-linked healable polyurethane containing Diels–Alder bonds *New Journal of Chemistry* **38** 770-6
[26] Gandini A, Carvalho A J F, Trovatti E, Kramer R K and Lacerda T M 2017 Macromolecular materials based on the application of the Diels–Alder reaction to natural polymers and plant oils *Eur. J. Lipid Sci. Technol.* **120** 1700091
[27] Bessonov I V, Kopitsyna M N, Polezhaev A V and Nelyub V A 2016 A mechanistic study of the reaction between furfural-acetone resins and polyamines *Polymer Science Series D* **9** 17-21
[28] Kopitsyna M N, Bessonov I V, Gusev S A, Kireinov A V, Kopitsyn D S, Solodilov V I and Kotomin S V 2018 The Properties of Epoxy Binders Modified by Furan Resin and Polysulfone *Polymer Science, Series B* **60** 299-306
[29] Ugryumov S A 2009 Furane oligomers in plywood and wood-board production *Polymer Science. Series D* **2** 106-8
[30] Ugryumov S A 2017 A study of the viscosity of phenol–formaldehyde resin modified with furfural–acetone monomer FA *Polymer Science, Series D* **10** 99-102
[31] Gol'dfarb Y L, Ostapenko é G, Vinogradova V G, Zverev A N, Polyakov A V, Yanovskii A I, Yufit D S and Struchkov Y T 1987 Synthesis, structures, and redox properties of some new dithiocarbamates that include a heteroaromatic ring *Chemistry of Heterocyclic Compounds* **23** 740-7
[32] Han H, Li S, Zhu X, Jiang X and Kong X Z 2014 One step preparation of porous polyurea by reaction of toluene diisocyanate with water and its characterization *RSC Advances* **4** 33520-9
[33] Jackson A T and Robertson D F 2008 *Comprehensive Analytical Chemistry,* ed J M Chalmers and R J Meier: Elsevier) pp 171-203
[34] Goussé C, Gandini A and Hodge P 1998 Application of the Diels−Alder Reaction to Polymers Bearing Furan Moieties. 2. Diels−Alder and Retro-Diels−Alder Reactions Involving Furan Rings in Some Styrene Copolymers *Macromolecules* **31** 314-21

IOP Conf. Series: Materials Science and Engineering **683** (2019) 012002 doi:10.1088/1757-899X/683/1/012002

Synthesis of self-healing polymers precursors from available bio-renewable raw materials.

D V Zakharova[1,2], A A Pavlov[2], A V Polezhaev[1,2]

[1] Bauman Moscow State Technical University, 105005, Russia, Moscow, 2nd Baumanskaya str., 5/1
[2] A.N.Nesmeyanov Institute of Organoelement Compounds of Russian Academy of Sciences (INEOS RAS), 119334, Russia, Moscow, Vavilova str., 28

E-mail: DZakharova@emtc.ru, avp@emtc.ru

Abstract. A series of a new diglycidyl derivatives of aromatic amines containing a furan fragment was obtained. The synthetic scheme consists of two stages: reductive amination and epoxidation. We used industrially produced inexpensive amines and furfural as starting material. Diepoxides obtained are key components for development of epoxy binders capable of intrinsic self-healing based on the reversible Diels-Alder reaction between furan and maleimide. To demonstrate the possibility of self-healing, the Diels-Alder reaction adduct of one of the products with N-phenyl maleimide was isolated and characterized by 2D NMR spectroscopy. The reversibility of the reaction was confirmed by DSC experiment.

1. Introduction

Epoxy binders are a crucial component of carbon and fiberglass composites that have a decisive influence on the performance of the composite. [1, 2] Polymer matrix bind fibers, protects them from environmental factors such as light, moisture, aggressive chemicals, e.t.c., and also contributes to the redistribution of mechanical stresses. [3, 4] The polymer matrix is always less durable than the reinforcing fiber. Defects in its structure, even microcracks, can significantly reduce the strength of a composite material. One of the problems of widespread use of a fiber-reinforced plastic is the complexity of prediction of defect formation and propagation in the epoxy binder layer.[4, 5] Development of self-healing materials is a possible solution to this problem. Two approaches, extrinsic and intrinsic self-healing were proposed. The first approach involves an introduction of microcapsules containing a healing agent and a catalyst into a polymer matrix. A healing agent releases from a microcapsule after cracking. [6] The main disadvantages of this approach are: the impossibility to predict the degree of cross-linking in the damaged area after the "healing" process, the limited amount of the "healing" agent and the irreversible nature of the epoxidation reaction. All the disadvantages mentioned above make the "healing" process possible only once. Moreover, to increase the self-healing ability, one should increase the number of microcapsules in the volume of the material, but at the same time, the mechanical properties might fall. [7-9] The "intrinsic" self-healing is based on the design of materials capable of independent internal self-healing that results from physical or chemical impacts, which provide the ability to rebound and repair damage due to the presence of specific functional groups in the material. [7] The reversible Diels-Alder reaction between furan and maleimide residues is a promising method for self-healing polymers design. [10]

Thus, we have proposed the synthesis of compounds via the Diels-Alder reaction between furan and maleimide residues. Such synthetic strategy leads to the formation of a highly cross-linked polymer structure and opens up new opportunities for its functionalization. Specific features of the polymer structure, the self-healing ability due to the sequential reverse and direct Diels-Alder reaction determines the improved mechanical properties of these compounds.

The creation of epoxy binders based on furan-containing compounds and maleimide cross-linking reagents was reported previously. Authors demonstrated self-healing properties of the compounds obtained as a result of the thermal cycling process. [7] We propose a synthetic strategy for getting new epoxy-containing monomers to create self-healing binders with improved thermomechanical properties. It is worth notice that one of the initial reagents of the synthetic scheme (Scheme 1) is furfural - a compound obtained from bio-renewable raw materials, in particular, from various agricultural wastes. [11-13] This reduces the cost of the target compound and makes the process of its production more environmentally friendly [14-16].

2. Results and discussion

Earlier, the Ming Qiu Zhang group demonstrated thermal remendability of the epoxy blend consist of N,N-diglycidylfurfurylamine, N,N'-(4,4-diphenylmethane)-bismaleimide and methyl hexahydrophthalic anhydride [1]. However, the described self-healing cycle was too long for practical application, and the mechanical characteristics were relatively poor. We suggest that these shortcomings can be corrected by introducing a hard aromatic scaffold into the structure of the epoxy component of the mixture and spacing the epoxy groups making their structure closer to well-known Bisphenol A and F - based epoxy resins.

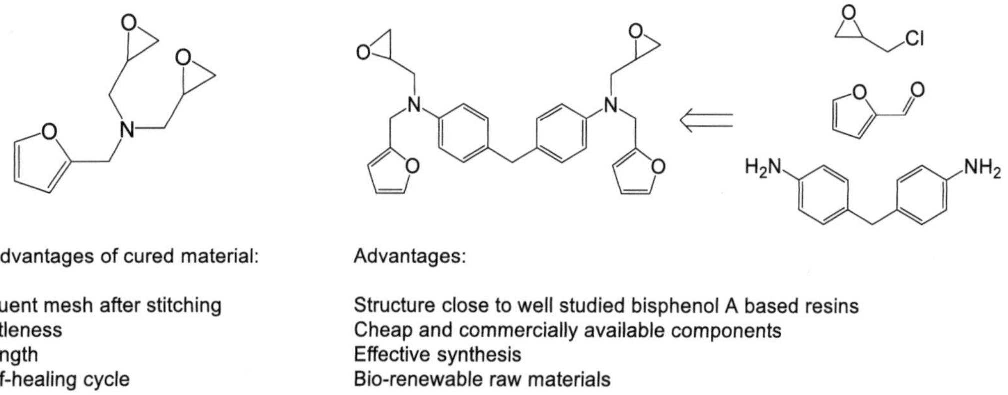

Figure 1. Comparison of two different approaches.

Simple strategy for the synthesis of the epoxy resin type 3b was proposed. For the synthesis of the desired compounds commercially available, inexpensive, industrially produced diamines (1a – c) such as 1,4-phenylenediamine, 4,4'-diaminodiphenylmethane, 4,4'-sulfonyldianiline and (±)-epichlorohydrin were used as starting compounds (Scheme 1).

A simple strategy for the synthesis of the epoxy resin type 3b was proposed. For the synthesis of the desired compounds commercially available, inexpensive, industrially produced diamines (1a – c) such as 1,4-phenylenediamine, 4,4'-diaminodiphenylmethane, 4,4'-sulfonyldianiline and (±)-epichlorohydrin were used as starting compounds (Scheme 1).

At the first stage, the formation of secondary amino group can be carried out via standard reduction amination procedure using $Na[BH(OAc)_3]$ as a reducing agent (Scheme 1, Method A, B). In order to increase the yield of the product 2b, an alternative two-step synthetic scheme was proposed (Scheme 1, Method B). Method B suggested isolation of the imine (2b') and its subsequent reduction with sodium borohydride to amine (2b). Thus, it was possible to increase the yield of compound 2b from 14 to 40%. Compounds (2a – 2c) were characterized by NMR (^1H, ^{13}C) and IR spectroscopy.

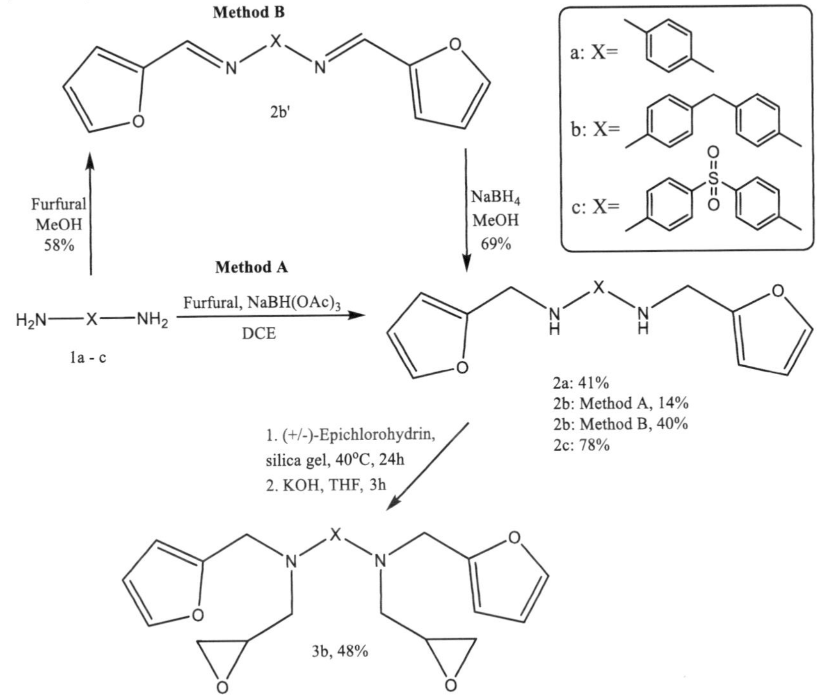

Scheme 1. Synthetic [pathway to polyfurans.

A series of epoxidation reactions were carried out under various conditions targeting 3b (p. 1–5, Table 1). The reaction conditions consisted of (±)-epichlorohydrin as a reagent and solvent, silica gel as a catalyst and KOH in THF as a base resulted in the highest yield of the product (p. 4, Table 1). This reaction requires a long time due to low reactivity of aromatic diamines. The reaction progress was monitored by IR spectroscopy. We used the characteristic absorption band at 3410 cm^{-1} in the spectrum of the compound 2b corresponding to the secondary amino groups as references. The disappearance of the band mentioned above indicated completion of the reaction. The compound 3c cannot be obtained under the same conditions, which is probably due to a more electron-deficient character of the secondary amino groups of compound 2c compared to 2b.

Table 1. Variation of reaction conditions

Substrate	Product	№	Conditions	%
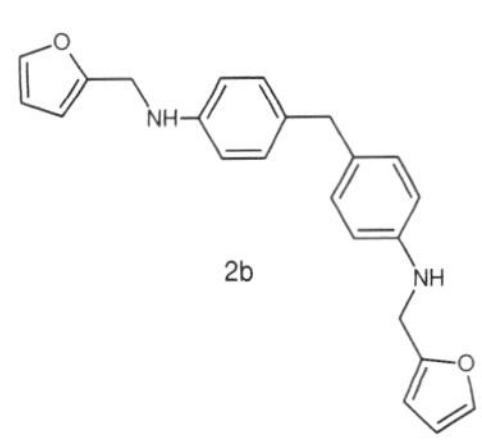	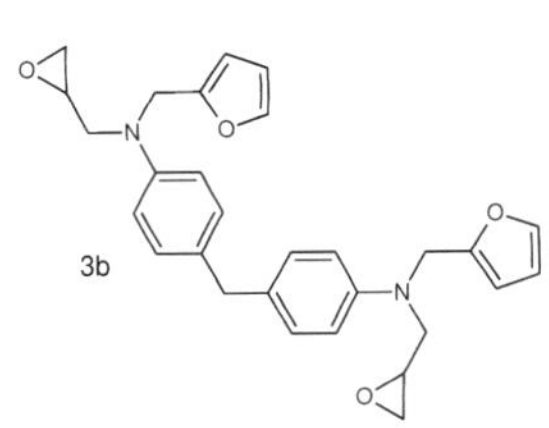	1.	NaOH, Na_2SO_4, $ZnCl_2$	-
		2.	KOH, K_2CO_3	-
		3.	CH_3COOH	-
		4.	KOH, K_2CO_3, $Et_4N^+Cl^-$, MW irradiation	-
		5.	1. Epichlorohydrin, silica gel, 40°C, 24 h. 2. KOH, THF, 3 h., r.t.	48

To demonstrate the ability of furan fragments to react reversibly with the maleimide derivatives via the Diels-Alder reaction we proposed the synthesis of a model compound — the product of the interaction of the difuran derivative 3b with N-phenylmaleimide (Scheme 2).

Adduct 3d was obtained from [4 + 2]-cycloaddition between compound 2b and N-phenylmaleimide after reflux for 24 h. Products of the Diels-Alder reaction are usually a mixture of diastereoisomers [1]: kinetic (endo) and thermodynamic (exo) reaction products. In the case of the compound 3d the formation of three isomers was possible (Figure 2): A (exo-exo), X (endo-endo) or Z (endo-exo). Only one set of signals was observed in the NMR spectra (^{1}H, ^{13}C) of the product 3d, corresponding to a symmetric product that is in agreement only with proposed structures A or X.

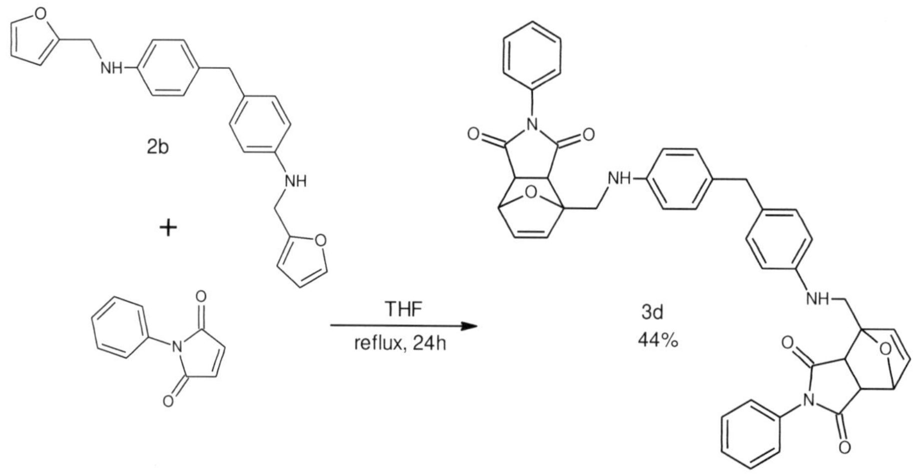

Scheme 2. Synthesis of the target compound (3d) by Diels–Alder [4+2]-cycloaddition.

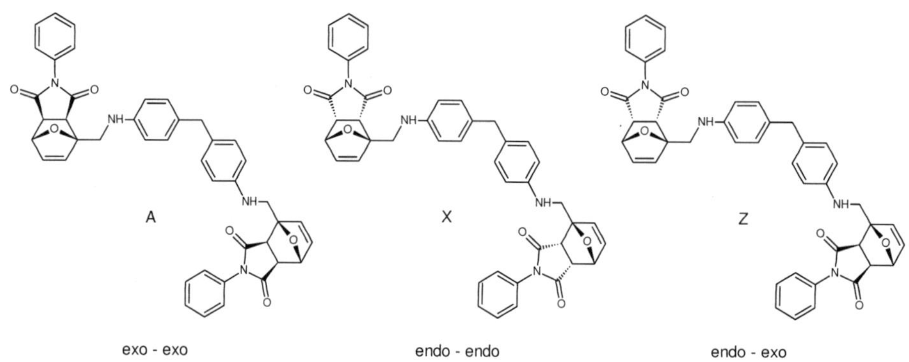

Figure 2. Possible isomeric products (3d) in the Diels-Alder reaction.

An NMR study was carried out using the HMQC, HMBC, NOESY, COSY, NOE methods (the spectra are shown below) for the complete assignment of signals of the obtained compound (3d). Using the spectra of two-dimensional correlation (HMQC, HMBC), as well as the ^{13}C DEPT spectrum, the assignment of ^{13}C and ^{1}H spectral signals is carried out unambiguously (Figure 3, 4).

It was previously shown [17], that the difference in chemical shifts and multiplicities of proton signals (H13 and H14, Figure 3) of the endo- and exo-isomers make it possible to determine the structure of the Diels-Alder reaction product unambiguously. Authors compared the vicinal constants of the spin-spin interaction $^3J_{HH}$, to determine the isomer. Constants 3JHH are dependent on several structural parameters, but most importantly on the H – C – C – H dihedral angle in the molecule. For protons 13 and 14 the constant should be approximately equal to 5.8 Hz in the case of the endo-isomer and close to zero for the exo one. According to the COZY spectrum, the vicinal constant of spin-spin interaction for H11 - H13 is zero, and the constant $^3J_{H12\text{-}H14}$ is less than 1 Hz. The combination of these factors leads to the fact that the signal at 5.23 ppm, corresponding to the bridge proton H14, degenerates into a singlet. Thus, the above facts and literature information [17-19] confirm our assumption about the preparation of the (exo-exo)-isomer.

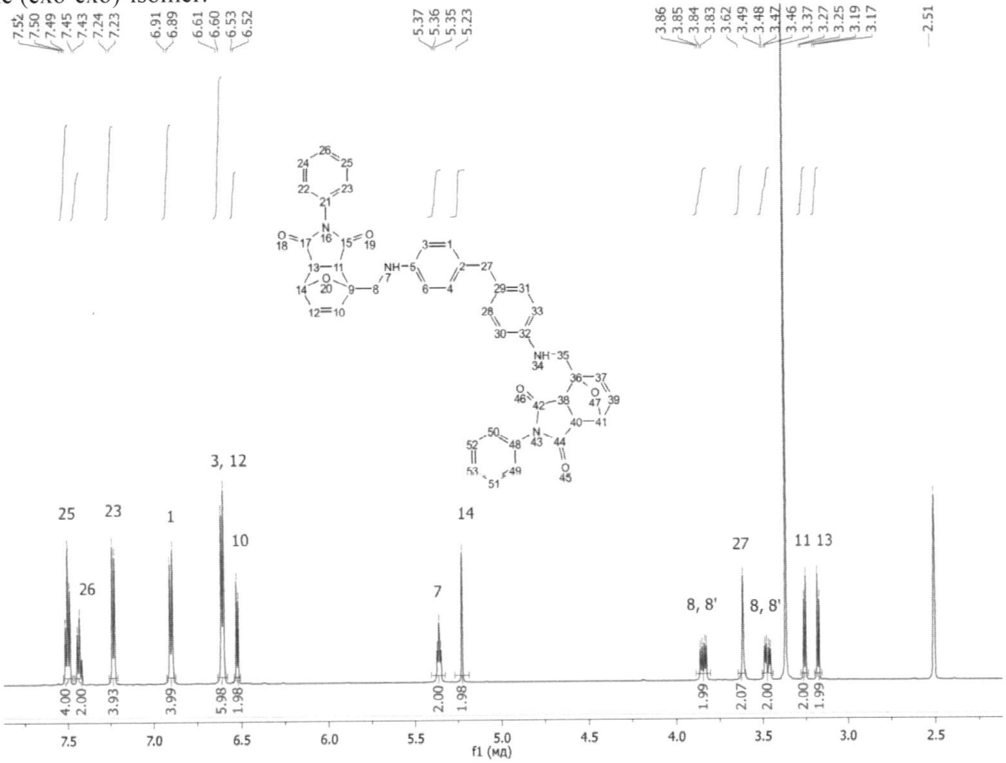

Figure 3. ^{1}H-NMR spectrum of compound 3d.

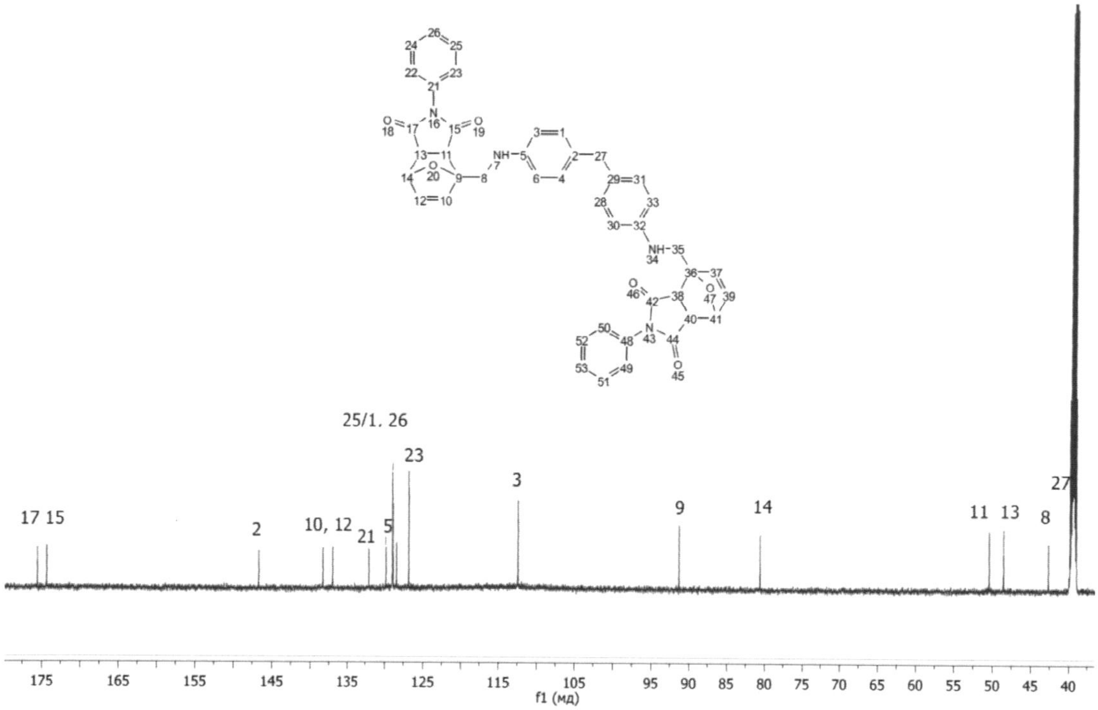

Figure 4. ^{13}C-NMR spectrum of compound 3d.

Reversibility of the Diels-Alder process was shown using DSC (Figure 5). It was shown that the Diels-Alder reverse reaction occurs at an unusually high temperature (173°C). This temperature was found to be significantly higher than values found in the literature for the furan-maleimide adduct (usually 110-130 °C). [20-22] We have suggested that such a high temperature of the reverse Diels-Alder reaction was achieved due to the formation of intramolecular hydrogen bonds between the protons of the NH groups and the oxygen atoms of the maleimide fragments. However, this phenomenon requires further studying.

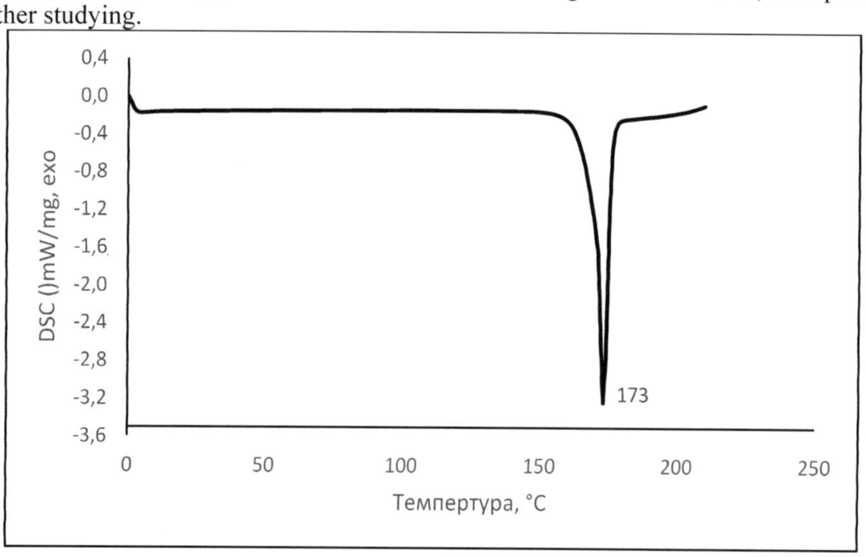

Figure 5. DSC curve of compound (3d). Heating from 0 to 210°C / 5K/min. T (rDA) = 173.20°C.

3. Experimental

Furfural (99.5%) was purchased from “Komponent-reaktiv” and purified by vacuum distillation. Sodium borohydride ((98%, Acros Organics), 1,4-phenylenediamine (98%, Sigma), 4,4'-diaminodiphenylmethane (97%, Aldrich), 4,4'-sulfonyldianiline (97%, Acros Organics), (±)-epichlorohydrin (99%, Fluka), aniline (99.8%, Acros Organics), maleic anhydride (99%, Aldrich) were purchased from Aldrich and used as received.

NMR spectra were recorded by a Bruker Avance 600 NMR Spectrometer (600.1 MHz), using residual proton signal of deuterated solvent as reference, chemical shifts were reported as parts per million downfield from tetramethylsilane (TMS). ATR-FTIR was performed on Nicolet iS10 spectrometer in the range of 4000 to 650 cm-1 on a germanium crystal. The thermal behavior was examined by DSC, with a NETZH DSC 204 F1 Phoenix calorimeter within a temperature range of 0 to 210 °C at heating/cooling rates of 5K min^{-1} in an argon atmosphere.

3.1. N,N'-difurfuryldiamines synthesis (2a-c)

3.1.1. N,N'-difurfuryl-p-phenylenediamine (2a).

To obtain sodium triacetoxyborohydride acetic acid (4.470 ml) was added dropwise to the solution of sodium borohydride (0.980 g, 25.79 mmol) in 1,2-dichloroethane (50 ml) maintaining the temperature between 0 and 5°C. Then the reaction mixture was stirred at room temperature for 3 h. Afterwards, 1,4-phenylenediamine (0.975 g, 9.03 mmol) (1a), furfural (1.490 ml, 18.05 mmol) and molecular sieves (4 Å) was added. The mixture was stirred at room temperature for 24 h. After the reaction completed (TLC monitoring), 50 ml of 0.1 N aq NaOH was added. The product was extracted with EtOAc (15 mL × 3), and the organic layer was dried (Na_2SO_4) and concentrated under reduced pressure. The product was purified by flash chromatography on silica gel eluting with Petroleum ether/EtOAc (1:1) to give an orange powder (0.998 g, 41 %). mp 74 – 76°C, (Lit. 75 – 77°C [1]). ^{1}H-NMR (DMSO-d6, δ, ppm., J, Hz): 4.11 (s., 4H), 5.24 (s., 2H), 6.24 (d., J = 3.1, 2H), 6.36 (dd., J1 = 1.8, J2 = 3.2, 2H), 6.48 (s., 4H), 7.54 (m., 2H). ^{13}C-NMR (DMSO-d6, δ, ppm.): 41.11, 106.62 110.33, 113.90, 139.99, 141.69, 154.06.

3.1.2. 4,4'-methanediylbis[N-(furan-2-ylmethyl) aniline] (2b)

Method A

To obtain sodium triacetoxyborohydride acetic acid (2.550 ml) was added dropwise to the solution of sodium borohydride (0.537 g, 14.13 mmol) in 1,2-dichloroethane (30 ml) maintaining the temperature between 0 and 5°C. Then the reaction mixture was stirred at room temperature for 3 h. Afterwards, 4,4'-diaminodiphenylmethane (0.979 g, 4.95 mmol) (1b), furfural (0.820 ml, 9.89 mmol) and molecular sieves (4 Å) was added. The mixture was stirred at room temperature for 24 h. After the reaction completed (TLC monitoring), 30 ml of 0.1 N aq NaOH was added. The product was extracted with EtOAc (= 15 mL × 3), and the organic layer was dried (Na2SO4) and concentrated under reduced pressure. The product was purified by flash chromatography on silica gel eluting with Petroleum ether/EtOAc (5:2) to give light yellow powder (0.250 g, 14 %). mp 52.4°C. 1H-NMR (DMSO-d6, δ, ppm., J, Hz): 3.58 (s., 2H), 4.18 (d., J = 5.9, 4H), 5.88 (t., J = 6.0, 2H), 6.25 (d., J = 2.9, 2H), 6.36 (m., 2H), 6.53 (d., J = 8.4, 4H), 6.86 (d., J = 8.4, 4H), 7.55 (s., 2H). 13C-NMR (DMSO-d6, δ, ppm): 39.64, 40.08, 106.79, 110.36, 112.36, 128.92, 129.75, 141.85, 146.34, 153.56.

Method B

Furfural (0.170 ml, 2.02 mmol) was added to the solution of 4,4'-diaminodiphenylmethane (0.200 g, 1.01 mmol) in MeOH (5 ml). The reaction was carried out with vigorous stirring at room temperature for 3 h. The obtained precipitate was collected by filtration, washed with MeOH and dried on air, giving a brown product (2b’) with yield 58% (0.207 g).

^{1}H-NMR (DMSO-d_6, δ, ppm., J, Hz): 3.96 (s., 2H), 6.69-6.71 (m., 2H), 7.12 (d., J = 2.7, 2H), 7.20 (d., J = 8.1, 4H), 7.27 (d., J = 8.2, 4H),7.93 (s., 2H), 8.42 (s., 2H). ^{13}C-NMR (DMSO-d_6, δ, ppm): 112.54, 116.79, 121.16, 129.54, 134.90, 139.38, 146.35, 147.83, 149.06, 152.02.
The product (2b') (1.000 g, 2.94 mmol) was dissolved in MeOH (50 ml). $NaBH_4$ (0.503 g, 13.24 mmol) was added at 0 °C. The reaction mixture was stirred at room temperature for 48 h. After the reaction completed (TLC monitoring), 30 ml of 0.1 N aq NaOH was added. The product was extracted with EtOA15 mL × 3), and the organic layer was dried (Na_2SO_4) and concentrated under reduced pressure. The product was purified by flash chromatography on silica gel eluting with Petroleum ether/EtOAc (5:2) to give light yellow powder (0.693 g, 69 %).

3.1.3. 4'-sulfonylbis[N-(furan-2-ylmethyl)aniline] (2c)
To obtain sodium triacetoxyborohydride acetic acid (4.470 ml) was added dropwise to the solution of sodium borohydride (0.98 g, 25.79 mmol) in 1,2-dichloroethane (50 ml) maintaining the temperature between 0 and 5 °C. Then the reaction mixture was stirred at room temperature for 3 h. Afterwards, 4,4'-sulfonyldianiline (2.239 g, 9.03 mmol) (1c), furfural (1.49 ml, 18.05 mmol) and molecular sieves (4 Å) was added. The mixture was stirred at room temperature for 72 h. After the reaction completed (TLC monitoring), 50 ml of 0.1 N aq NaOH was added. The product was extracted with EtOAc (15mL × 3), and the organic layer was dried (Na_2SO_4) and concentrated under reduced pressure. Product was obtained as light orange powder (2.847 g, 78 %). ^{1}H-NMR (DMSO-d_6, δ, ppm., J, Hz): 4.28 (d., J = 5.8, 4H), 6.30 (d., J = 3.2, 2H), 6.37 (t., J = 3.0, 2H), 6.69 (d., J = 8.8, 4H), 7.06 (t., J = 5.9, 2H), 7.51 (d., J = 8.8, 4H), 7.56 (d., J = 1.1, 2H). ^{13}C-NMR (DMSO-d_6, δ, ppm): 39.11, 107.34, 110.44, 111.56, 128.42, 128.58, 142.30, 151.65, 152.22.

3.2. N,N'-difurfurildiepoxydiamines synthesis (3a-c)

3.2.1. The product of the epoxidation of compound 2b (3b)
1.112 g (3.11 mmol) of 4,4'-methanediylbis[N-(furan-2-ylmethyl)aniline] (2b) was dissolved in (±)-epichlorohydrin (4 ml), silica gel and molecular sieves (4 Å) was added. After reaction mixture was stirred at 80°C for 10 h silica gel and molecular sieves was filtered and the excess of (±)-epichlorohydrin was removed under reduced pressure. To solution of obtained oil in THF (3 ml) was added 1N aq KOH (9.30 ml, 9.32 mmol) and tetramethylammonium chloride (0.010 g, 0.09 mmol), and then mixture was stirred at 50 °C for 5 h. Afterwards, the organic layer was dried (Na2SO4) and concentrated under reduced pressure; yield: 1.036 g (71%). 1H-NMR (DMSO-d6, δ, ppm., J, 600 MHz): 2.56 (dd., J1 = 2.6, J2 = 4.7, 2H), 2.71 (t., J = 4.2, 2H), 3.07 (m., 2H), 3.35 – 3.38 (dd., J1 = 4.7, J2 = 15.8, 2H, the overlapping signals of the compound 3b and water is detected), 3.59 – 3.62 (dd., J1 = 3.1, J2 = 15.8, 2H), 3.65 (s., 2H), 4.46-4.52 (m., 4H), 6.26 (d., J = 2.9, 2H), 6.35 (m., 2H), 6.76 (d., J = 8.5, 4H), 6.97 (d., J = 8.5, 4H), 7.55 (s., 2H). 13C-NMR (DMSO-d6, δ, ppm): 39.91, 45.63, 48.51, 50.83, 52.38, 107.53, 110.32, 113.16, 129.62, 130.90, 141.96, 146.79, 152.38.

3.3. Diels–Alder reaction

3.3.1. N-phenylmaleimide synthesis
A mixture of aniline (29.362 mL, 0,323 mol), maleic anhydride (32.855 g, 34.0 mmol) in dry acetone (200 mL) was stirred at room temperature for 1 h. The precipitated was filtered and washed with dry acetone (15 mL × 3), then dried in a desiccator overnight to give 3-phenylcarbamoyl-acrylic acid as pale yellow powder. 57.417 g, 93% yield. A mixture of 3-phenylcarbamoyl-acrylic acid (10.000 g, 52.36 mmol), acetic anhydride (25.700 mL, 272.62 mmol), sodium acetate (2.147 g, 26.18 mmol) was stirred at 50°C for 2 h, then the solution was diluted with H_2O (30 mL) and stirred at 70°C for 2 h. The precipitated was filtered, dried overnight in a desiccator to give 7.165 mg (80%) of final product. ^{1}H-NMR ($CDCl_3$, δ, ppm., J, Hz): 7.18 (s., 2H), 7.33-7.35 (m., 2H), 7.40-7.41 (m., 1H), 7.47-7.51 (m., 2H). ^{13}C-NMR ($CDCl_3$, δ, ppm): 126.81, 127.74, 128.88, 131.57, 134.65, 169.94 (C=O).

3.3.2. Product of Diels-Alder reaction (3d)
N-phenylmaleimide (0.193 g, 1.12 ммоль) was added to the solution of 4,4'-methanediylbis[N-(furan-2-ylmethyl) aniline] (2b) (0.200 г, 0.56 ммоль) in THF (5 ml). The reaction mixture was stirred at room temperature for 24 h, and then poured into ice. The reaction mixture was left for 30 minutes in an ice bath. The product was obtained by filtration, washed with THF and dried on air. Product of Diels-Alder reaction (3d) was obtained as white powder with yield 44% (0.172 g). ^{1}H-NMR (DMSO-d_6, δ, ppm., 600 MHz) δ_H(ppm): 3.18 (d., J = 6.5, 2H), 3.26 (d., J = 6.6, 2H), 3.46-3.49 (dd., J_1=5.5, J_2=14.0, 2H), 3.62 (s., 2H), 3.83-3.86 (dd., J_1=6.8, J_2=14.1, 2H), 5.23 (s., 2H), 5.36 (t., J=6.1, 2H, NH), 6.53 (d., J=5.5, 2H), 6.60 (m., 6H), 6.90 (d., J = 8.3, 4H), 7.24 (d., J = 7.7, 4H), 7.42 – 7.45 (t., J = 7.3, 2H), 7.49 – 7.52 (t., J = 7.7, 4H). ^{13}C-NMR (DMSO-d_6, δ, ppm): 39.42 (C_{27}. In ^{13}C-NMR spectrum peak of C_{27} hidden behind the solvent peak (DMSO-d_6). 2D HMQC experiment was conducted to determine the signal of C_{27}), 42.77, 48.64, 50.51, 80.72, 91.41, 112.50, 126.92, 128.50, 128.97, 129.04, 129.88, 132.13, 137.10, 138.38, 146.77, 174.39, 175.63. Found ($C_{43}H_{36}N_4O_6$): C, 73.21, H 5.17, N 8.02; Calculated: C 73.28, H 5.15, N 7.95%].

3.3.3. DSC-experiment.

For the experiment 12.8 mg of the compound 3d was used. Experimental conditions: heating from 0 to 210 °C / 5K/min. T (rDA) = 173.20°C.

Acknowledgments
This work was supported by Russian Foundation of Basic Research (RFBR) grants 18-29-18036 and 18-29-18037. NMR experiments were performed with the financial support from Ministry of Science and Higher Education of the Russian Federation using the equipment of Center for molecular composition studies of INEOS RAS.

References
[1] Mikhailin Yu A, Kerber M L, Gorbunova I Yu 2002 Binders for polymer composites *Int. Polym. Sci. Tech.* **29** 49-61
[2] Nelyub V A 2015 Determination of adhesion interaction between carbon fiber and epoxy binder *Polym. Sci., Ser.* D **8** 6-8
[3] Jones R M 2014 *Mechanics of composite materials*: CRC press
[4] Zhang W, Duchet J and Gérard J-F 2014 Self-healable interfaces based on thermo-reversible Diels–Alder reactions in carbon fiber reinforced composites *J. Col. Interf. Sci.* **430** 61-8
[5] Nelyub V A, Borodulin A S, Kobets L P and Malysheva G V 2016 Capillary hydrodynamics of oligomer binders *Polym. Sci., Ser.* D **9** 322-5
[6] Blaiszik B, Sottos N and White S 2008 Nanocapsules for self-healing materials *Compos. Sci. Tech.* **68** 978-86
[7] Tian Q, Yuan Y C, Rong M Z and Zhang M Q 2009 A thermally remendable epoxy resin *J. Mater. Chem.* **19** 1289-96
[8] Urdl K, Kandelbauer A, Kern W, Müller U, Thebault M and Zikulnig-Rusch E 2017 Self-healing of densely crosslinked thermoset polymers—a critical review *Prog. Org. Coat.* **104** 232-49
[9] Pramanik N B, Mondal P, Mukherjee R and Singha N K 2017 A new class of self-healable hydrophobic materials based on ABA triblock copolymer via RAFT polymerization and Diels-Alder "click chemistry" *Polymer* **119** 195-205
[10] Guo Y-k, Li H, Zhao P-x, Wang X-f, Astruc D and Shuai M-b 2017 Thermo-reversible MWCNTs/epoxy polymer for use in self-healing and recyclable epoxy adhesive *Chin. J. Polym. Sci.* **35** 728-38
[11] Mittal A, Black S K, Vinzant T B, O'Brien M, Tucker M P and Johnson D K 2017 Production of furfural from process-relevant biomass-derived pentoses in a biphasic reaction system *ACS Sus. Chem. Engin.* **5** 5694-701
[12] Bessonov I, Polezhaev A, Kuznetsova M, Nelub V, Buyanov I, Chudnov I and Borodulin A 2013

Rheological and thermal analysis of low-viscosity epoxy-furan composites *Polymer Science Series D* **6** 308-11

[13] Bessonov I, Kopitsyna M, Polezhaev A and Nelyub V 2016 A mechanistic study of the reaction between furfural-acetone resins and polyamines *Polym. Sci., Ser.* D **9** 17-21

[14] Ugryumov S 2009 Furane oligomers in plywood and wood-board production *Polym. Sci., Ser.* D **2** 106-8

[15] Ugryumov S 2017 A study of the viscosity of phenol–formaldehyde resin modified with furfural–acetone monomer FA *Polym. Sci., Ser.* D **10** 99-102

[16] Kopitsyna M, Bessonov I, Gusev S, Kireinov A, Kopitsyn D, Solodilov V and Kotomin S 2018 The Properties of Epoxy Binders Modified by Furan Resin and Polysulfone *Polym. Sci., Ser.* D **60** 299-306

[17] Froidevaux V, Borne M, Laborbe E, Auvergne R, Gandini A and Boutevin B 2015 Study of the Diels–Alder and retro-Diels–Alder reaction between furan derivatives and maleimide for the creation of new materials *Rsc Advances* **5** 37742-54

[18] Bai N, Saito K and Simon G P 2013 Synthesis of a diamine cross-linker containing Diels–Alder adducts to produce self-healing thermosetting epoxy polymer from a widely used epoxy monomer *Polym. Chem.* **4** 724-30

[19] Bai N, Simon G P and Saito K 2013 Investigation of the thermal self-healing mechanism in a cross-linked epoxy system *RSC Advances* **3** 20699-707

[20] Yuan Y, Yin T, Rong M and Zhang M 2008 Self healing in polymers and polymer composites. Concepts, realization and outlook: A review *Expr. Polym. Let.* **2** 238-50

[21] Chen X, Dam M A, Ono K, Mal A, Shen H, Nutt S R, Sheran K and Wudl F 2002 A thermally re-mendable cross-linked polymeric material *Science* **295** 1698-702

[22] Platonova E O, Vlasov E, Pavlov A A, Kireynov A, Nelyub V A and Polezhaev A V 2019 Self-healing polyurethane based on a difuranic monomer from biorenewable source **136** 47869

Research of the fatigue behavior of multilayer steel material based on steel 08x18Ni10 and 08x18 at high values of cycle stresses under conditions of pure bending

A I Plokhikh[1] and A A Minakov[1]

[1] Department of Materials Science, Bauman Moscow State Technical University, ul. Baumanskaya 2-ya, 5/1, Moscow, 105005, Russia

Abstract. In this paper, the authors studied the cycle durability of a multi-layered steel material based on stainless steels 08Cr18Ni10 and 08Cr18 at high values of stresses under the scheme of pure bending. The results showed that the cycle durability increases with increasing dispersion of the laminar structure. In addition, the authors studied the fractures after fatigue failure and analyzed their microstructure. The analysis showed that the nature of the fatigue failure of a multilayer steel material depends on the dispersion of the laminar structure.

Introduction

A new promising class of structural metallic materials are multilayer steel materials with ultrafine oriented multilayer (laminar) structure. Due to the high dispersion of individual layers (from micron to submicron thickness) and their large number (from several hundred to several thousand layers), such materials have a unique combination of physic and mechanical properties (especially strength and impact properties) [1-3].

One of the dangerous types of loading is cyclic loading. On the one hand, the fatigue characteristics of a material under cyclic loading will be large if it has greater strength [4]. On the other hand, it is known that the fatigue characteristics of a material strongly depend on its structural state. Subject to the Hall-Petch Act, $\sigma \sim h1/2$ (where σ – plastic strain strength, h – grain size), it can be said that the fatigue characteristics of the material will be large if it will have a greater dispersion of the structure. Thus, multilayer steel materials are very promising for their use in parts and structures operating under cyclic loading, for example, for helicopter torsions. Studies of such materials under cyclic loading with stresses less than their yield strength show that multilayer steel materials have high fatigue characteristics [5]. However, many parts and structures used in mechanical engineering work at high values of cyclic stresses, or they may experience short-term strong cyclic loading, i.e. overloads (for example, the same helicopter torsions). Therefore, the purpose of this work was to study the fatigue behaviour of multilayer steel materials in such conditions according to the scheme of pure flat bending.

Materials and methods for their preparation

In the production of multilayer steel material are widely used steel grades in mechanical engineering. Stainless steels of two grades were used for this study – 08Cr18Ni10 (austenitic class) and 08Cr18 (ferritic class). The production of multi-layer steel materials is carried out using the method of hot batch rolling according to the developed experimental technological route [2, 3].

In this technological route there are several repeating technological cycles (Fig. 1). In the first technological cycle, first, steel sheets with a thickness of 0.5 mm of each brand are taken and cut into dimensional cards with the surface treatment of these cards. After processing the surface of the card, they are assembled into a multi-layered package, which consists of 100 alternating steel sheets through one through 50 pieces of each brand. Then the multilayer package is evacuated and undergoes hot plastic deformation at a temperature of 1000 ° C. At the end of the first technological cycle, a steel laminated material with a thickness of 2 mm with a laminar structure is obtained, while the thickness of the individual layer is 20 μm and their number is 100. During the next technological cycle, the raw materials used are steel multilayer sheets obtained in the previous technological cycle, while each subsequent technological cycle consists of the same technological operations as the first cycle. In this study, two process cycles were used. At the end of the second cycle, the formed multilayer steel material with a thickness of 2 mm had a separate layer thickness of 1 μm, and the number of layers was about 1500.

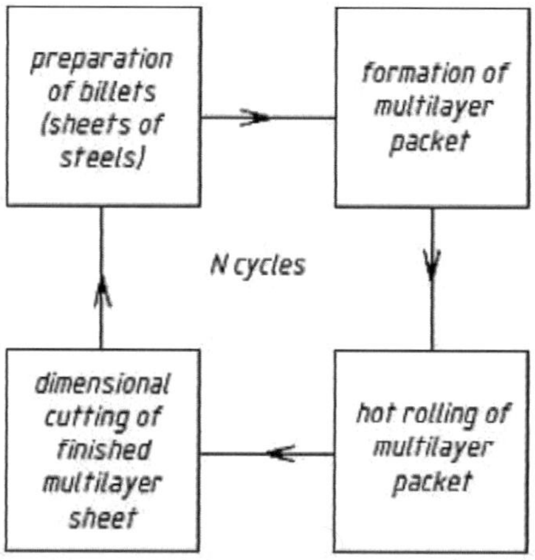

Fig 1. Diagram of the experimental technological route for production a multiplayer steel material

Experiment

After the first and second technological cycles, samples of type IV were prepared according to GOST 25.502-79. These samples were studied for cyclic durability according to the scheme of pure plane bending with symmetric alternating loading, implemented in Schenk-Erlinger machine (Fig. 2). For testing, samples (1) were placed in jaws (6), which were fastened with bolted joints on the drive (2) and measuring (3) levers. Samples are set so that their centre of inertia is exactly on the axis passing through the middle of the sample (0) of the drive lever (2). By moving the runner block of the drive motor armature (10), the static displacement of the sample is adjusted in the vertical direction until the sample's inertia centre coincides with axis passing through the middle of the sample (0). The load was applied on the drive lever (2), which is connected by a crank cod 7) with drive motor armature (9) and an eccentric mechanism (8). The bending moment acting during the test in the sample is transmitted to the measuring lever (2) and causes an alternating rotation around the ideal axis (0). Calibration of the alternating rotation is carried out using a micro indicator (5), the probe of which touches the measuring lever (3) attached to the testing machine (11) using a balance spring (4). Knowing the geometry of the tested samples and changing the position of the eccentric mechanism (8), the required voltage was set. Because specimens were tested at high stresses, it was set to 750 MPa, which is much more than the yield strength for a multilayer steel material 08Cr18Ni10+08Cr18 (655 MPa), but less tensile strength (800 MPa).

IOP Conf. Series: Materials Science and Engineering **683** (2019) 012003 doi:10.1088/1757-899X/683/1/012003

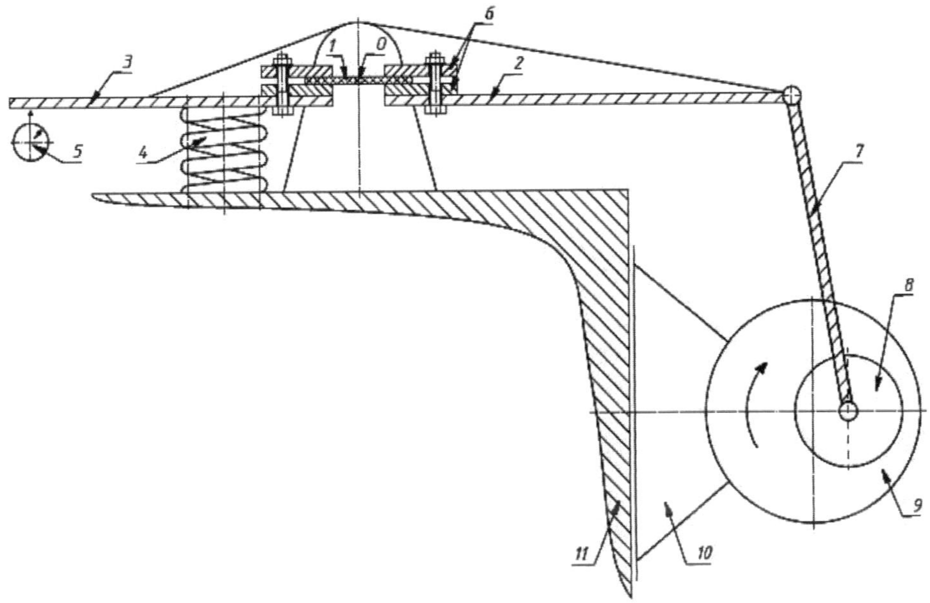

Fig 2. Schrenk-Erlinger testing machine diagram for the fatigue testing under cyclic loading according to the scheme of pure flat bending:
0 – axis passing through the middle of the sample; 1 – test sample; 2 – driving lever; 3 – measuring lever; 4 – balance spring; 5 – micro indicator; 6 – jaw; 7 – crank cod; 8 – eccentric mechanism; 9 – drive motor armature; 10 – runner block of the drive motor armature; 11 – testing machine.

VEGA TESCAN scanning electron microscope was used to study the fracture surface microstructure.

Results and discussion

The results of studies on cyclic durability showed that cyclic durability for a sample of a multilayer steel material 08Cr18Ni10 + 08Cr18 after the first technological cycle, having a number of layers equal to 100, amounted to 3,05·104 cycles to the destruction. For a sample of a multi-layered steel material 08Cr18Ni10 + 08Cr18 after the second technological cycle, having a number of layers of the order 1500 – 9,9·104 cycles to the destruction (fig. 3). Thus, the cyclic durability increased by a little more than 3 times with an increase in the number of layers in a sample of a multilayer steel material (respectively, with a decrease in the thickness of an individual layer from 20 μm to 1 μm).

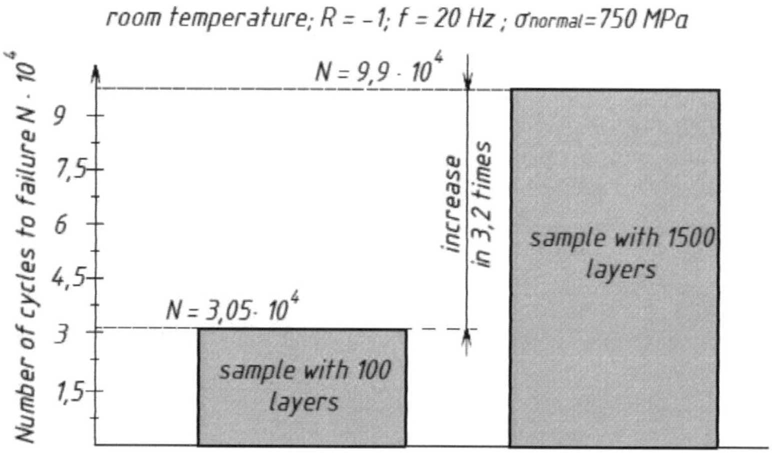

Fig 3. Test results on the cycle durability of steel laminate based on stainless steels 08Cr18Ni10 and 08Cr18 under the cycle loading according to the scheme of pure flat bending in the testing machine Achenk-Erlinger.

Studies of the microstructure of the surface of fatigue fractures showed that for a sample that had passed the first technological cycle, destruction occurred due to the initiation of fatigue cracks along the interface between the layers, and the propagation of fatigue cracks was carried out by opening the layers (Fig. 4 left). The development front of the main fatigue crack was perpendicular to the interface between the layers. For the sample that went through the second technological cycle, the destruction had a different character (Fig. 4, right). The development front of the main fatigue crack was parallel to the interface between the layers. This caused fatigue grooves to form along the layer along the border. In this case, a small part of the fatigue cracks originated along the interface between the layers and spread due to the opening of the layers.

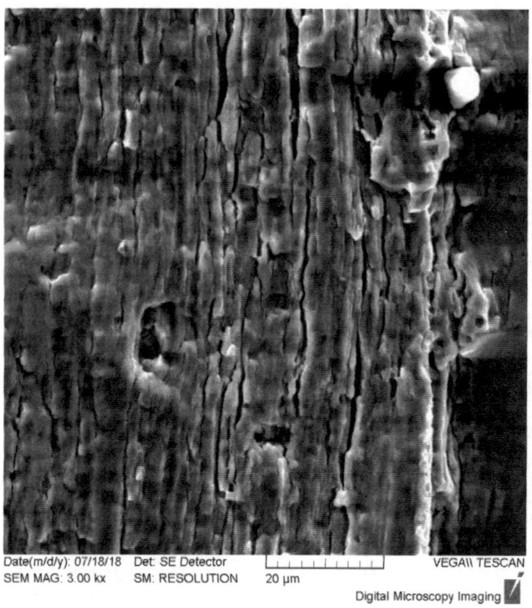

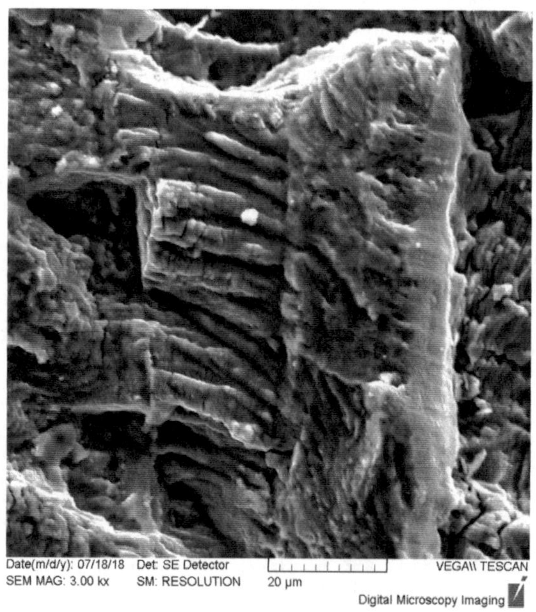

Fig 4. Microstructure of fatigue fractures of samples of the multilayer steel material 08Cr18Ni10+08Cr18: left – after the first technological cycle; right – after the second technological cycle

The research results showed that the fatigue behaviour of a multilayer steel material is most strongly influenced by the dispersion of its laminar structure. The increase in the dispersion of the laminar structure with an additional technological cycle made it possible to increase the cyclic durability by about three times. Analysis of fatigue fractures showed that the dispersion of the laminar structure also influenced the nucleation and propagation of fatigue cracks.

References

[1] Aksenova K V and etc. 2016 *Innovative metal materials* (Magnitogorsk: Publishing house Magnitogorsk State Technical University. G.L. Nosova) p 371

[2] Kolesnikov A G, Plokhikh A I, Komissarchuk Yu S and Mikhaltsevich I Yu 2010 *Metal Sci. Heat Treat.* **52** 273–78

[3] Tabatchikova T I, PlokhikhI A I, Yakovlev L and Klyueva S.Yu 2013 *Phys. Met. Metallogr.* **114** 580–92

[4] Hertzberg R V 1983 *Deformation and fracture mechanics of engineering materials, second edition* (UK: John Wiley, Chichester) p 715

[5] Schmiedt A, Luecker L, Kolesnikov A G, Plokhikh A I and Walther F 2017 *Materialpruefung.* **59** 123–29

IOP Conf. Series: Materials Science and Engineering **683** (2019) 012004 doi:10.1088/1757-899X/683/1/012004

Simulation of thermal conditions of anti-icing systems

Yu I Dimitrienko[1], V Yu Chibisov[1], M N Koryakov[1] and S V Lebedev[1]

[1]Bauman Moscow State Technical University, Moscow, Russia

E-mail: dimit.bmstu@gmail.com

Abstract The paper presents the results of modeling the conjugate processes of flow around the cold air flow, heat exchange between the flow and the air intake grating body (AIG) and heating of the structure by the anti-icing system (AIS). Modeling is carried out using a software package developed at the Computational Mathematics and Mathematical Physics Department of Bauman Moscow State technical University. A parametric numerical study of the thermal modes of the AIS at different flow rates and temperatures was carried out, as a result of which the power of the heating system was determined to maintain the temperature of the AIG body in a given mode. The developed method and the results of numerical simulation are used to create the AIS for marine vessels operating in the Arctic, in particular to protect against icing AIG marine ventilation systems for nuclear icebreakers series "Moscow".

Introduction

At present, the problem of the development of the Arctic territories of Russia and the development of the ocean shelf for the extraction of minerals and oil and gas products is of great importance. One of the major problems that designers of Arctic vessels have to overcome is the fight against icing of various critical parts of ship structures, including air intakes of ventilation and air conditioning systems of residential and industrial premises. The icing of the grates narrows the cross-section of the air intake ducts, reduces the amount of air entering the various ship systems and, in some cases, can disable them. For de-icing air-intake grids (AIG) develop special anti-icing system (AIS) heating [1-3]. The problems of modeling the ice accreation processes were considered in [1], but the problem of joint modeling of the processes of flow around the AIG with the AIS cold air flow and heat transfer processes in these systems were not considered. This work is devoted to the modeling of conjugate processes of gas dynamics and heat transfer in the structures of AIG and AIS.

General view and the device of the AIS heating of ship's AIG

Anti-icing heating system of ship's AIG was developed in the scientific and educational center "SIMPLEX" BMSTU together with Bi Pitron (St. Petersburg) "[4-6]. AIS AIG is designed to protect the air intake grilles of marine ventilation systems from icing when exposed to negative temperatures up to-60oC. A fragment of the design of the AIG with the developed AIS is shown in figure 1. AIG is a system of louvers interconnected by the supporting hollow struts, which feature an electric heating cable.

To calculate the heat exchange in the design of blinds with internal heating from electric heaters, it is necessary to calculate the convective heat flow withdrawn from the surface of the blinds when exposed to cold air flow, as well as the calculation of the coefficient of convective heat exchange between the surface of the housing and the cold air flow.

Figure 1-General view of the developed AIS module for heating ship's AIG

To model the processes of heat transfer to AIG, we consider the formulation of the conjugate problem of gas dynamics of the air flow around the structure and the thermal conductivity of the wall of the structure with the conditions of internal electric heating. This conjugate problem in the General formulation consists of a system of equations of dynamics of a linearly viscous heat-conducting gas (formed from the continuity equation, equations of motion, equations of energy and determining relations) describing the movement of cold air masses on the surface of the blinds AIG [7-9]:

$$\begin{cases} \dfrac{\partial \rho}{\partial t} + \nabla \cdot \rho \mathbf{v} = 0 \ , \\ \dfrac{\partial \rho \mathbf{v}}{\partial t} + \nabla \cdot \ \rho \mathbf{v} \otimes \mathbf{v} + p\mathbf{E} - \mathbf{T}_v \ = 0 \ , \\ \dfrac{\partial \rho E}{\partial t} + \nabla \cdot \ \ \rho E + p \ \ \mathbf{v} - \mathbf{T}_v \cdot \mathbf{v} + \mathbf{q} \ = 0, \qquad \mathbf{x} \in V_1 \end{cases} \tag{1}$$

$$p = R\rho\theta, \quad e = c_v\theta, \quad E = e + |\mathbf{v}|^2/2, \tag{2}$$

$$\mathbf{T}_v = \mu_1(\nabla \cdot \mathbf{v})\mathbf{E} + \mu_2(\nabla \otimes \mathbf{v} + \nabla \otimes \mathbf{v}^T),$$

$$\mathbf{q} = -\lambda \nabla \theta, \qquad \mathbf{x} \in V_1$$

Also the equations of thermal conductivity of the body AIG

$$\rho_s c_s \frac{\partial \theta}{\partial t} = \nabla \cdot \ \lambda_s \nabla \theta \ , \qquad \mathbf{x} \in V_2 \tag{3}$$

here marked with: ρ – gas density, t- is time, E is the density of full energy of the gas: $E = c_v\theta + |v|^2/2$, c_v — heat capacity of gas at constant volume, θ is the gas temperature, $|v|^2 = v^i v_i$ - is the square of the modulus of velocity, p – pressure, R –gas constant ($R = R/\mu$, μ –the molecular mass of gas R - universal gas constant), **E**- is the metric tensor, $\mathbf{T}_v$ - the tensor of viscous stresses in the gas, $\mathbf{q}$ - vector of heat flux,

$\mu_1, \mu -$ the coefficients of gas viscosity, $\lambda -$ gas thermal conductivity coefficient, ρ_s, c_s, λ_s - the density, specific heat and thermal conductivity coefficient of the material of the body AIG, ∇ - nabla –operator of Hamilton [10], V_1 - the conditional area of the air flow around AIG, V_2 - the area of design AIG.

Consider 4 cases of boundary conditions for the system of equations (1)...(3):

a) At the interface Σ_1 between the areas V_1 и V_2, which is a solid impermeable wall (surface of the body of the AIG), the conditions of adhesion, the balance of heat flow and the equality of temperatures of the air flow and the solid wall are set:

$$\mathbf{v} = 0, \quad \lambda \nabla \theta \big|_{\Sigma_{1-}} \cdot \mathbf{n} = \lambda_s \nabla \theta \big|_{\Sigma_{1+}} \cdot \mathbf{n} + \varepsilon_s \sigma \theta_w^4, \quad \theta \big|_{\Sigma_{1-}} = \theta \big|_{\Sigma_{1+}} \tag{4}$$

b) At the subsonic boundary Σ_2 f the air flow inlet to the region V_1 5 conditions are specified:

$$\rho = \rho_e, \ \mathbf{v} = \mathbf{v}_e, \ \theta = \theta_e \tag{5}$$

c) At the subsonic boundary Σ_3 air flow out of the area V_1 the following 5 conditions are specified:

$$\rho = \rho_e, \ \frac{\partial \mathbf{v}}{\partial n} = 0, \ \frac{\partial \theta}{\partial n} = 0 \tag{6}$$

where: $\frac{\partial \mathbf{v}}{\partial n} = \mathbf{n} \cdot \nabla \otimes \mathbf{v}$ - the normal derivative of the velocity vector.

d) On the internal (heated) boundary Σ_4 the solid wall is given the conditions of the heat inflow:

$$-\lambda_s \nabla \theta \big|_{\Sigma_4} \cdot \mathbf{n} = q_w \tag{7}$$

where q_w - the specified heat flux supplied to the electric heating, ε_s - integral coefficient of thermal radiation of a solid surface, σ - Stefan-Boltzmann coefficient.

The initial conditions for the system (1)–(3) have the form:

$$t = 0: \quad \rho\ 0, x^i\ = \rho^0, \quad \mathbf{v}\ 0, x^i\ = 0, \quad E\ 0, x^i\ = c_V \theta_0. \tag{8}$$

where ρ^0, θ^0 - set value.

A method for numerical solution of the conjugate problem of gas dynamics, heat exchange and heating of AIG

The numerical solution of the conjugate problem (1) - (8) was carried out in a model two-dimensional formulation-the normal cross-section of one support pillar of the AIG ASO was considered (Fig.2), in which the solution domains are both bounded by two concentric ellipses.

For the numerical solution of the conjugate problem, the following method was applied: an iterative cycle by "slow" time was introduced $\bar{t} = t / t_0$, , corresponding to the process of heat propagation in the wall of the structure of the AIG, where t_0 - characteristic heating time of the structure. Within this cycle, a "fast" time was introduced $\tau = t / t_y$, where t_y - - characteristic setting time of the gas flow. At each n-th step of the slow-time iteration $\bar{t}_{(n)}$ the calculation was carried out in 5 stages: 1) select the temperature $\theta_{w(n)}$ the surface of the structure at the n-m step, with this temperature by establishing the solution of the system of equations of gas dynamics (1), (2) with boundary conditions (5), (6), and instead of (4) set the conditions

$$\mathbf{v} = 0, \qquad \theta\big|_{\Sigma_{1-}} = \theta_{w(n)} \tag{9}$$

1) as a result V_1 gas, including temperature θ_e on the external conventional boundary of the wall viscous gas layer, as well as the heat flow from the air to the solid surface of the structure $q_- = -\lambda \nabla \theta\big|_{\Sigma_{1-}} \cdot \mathbf{n}$

2) heat transfer coefficient $\alpha = \dfrac{q_-}{\theta_e - \theta_{w(n)}}$, 3) the heat balance equation on the solid surface of the AIG structure (the second equation of the system (4)) was compiled, which was considered as a nonlinear algebraic equation for calculating the temperature $\theta_{w(n+1)}$ surfaces in the next iteration step:

$$\alpha(\theta_e - \theta_{w(n+1)}) = \alpha_s(\theta_{w(n+1)} - \theta_0) + \varepsilon_s \sigma \theta^4_{w(n+1)}, \tag{10}$$

here we have used Newton's equation for heat flow $q_- = \alpha(\theta_e - \theta_{w(n+1)})$ and $q_+ = \alpha_s(\theta_{w(n+1)} - \theta_0)$, where α_s - the heat transfer coefficient in a solid wall, the calculation method of which is described below; 4) the heat equation (3) with the boundary condition of the given temperature was solved $\theta\big|_{\Sigma_{1+}} = \theta_{w(n+1)}$. Then the transition to the next (n+1) step of the iterative cycle was carried out.

Optimization of the AIS design based on the results of numerical modeling

Were carried out several series of numerical simulations of the variant of warm-up designs shutters blinds AIG with AIS. The following parameters were varied: cold air temperature in the range $\theta_e = -10...-50^0 C$; flow rate of cold air in the range of 5 ... 20 m/s; the power of the electric heater W, falling on 1 element of the AIS lattice, in the range from 100 to 1000 watts.

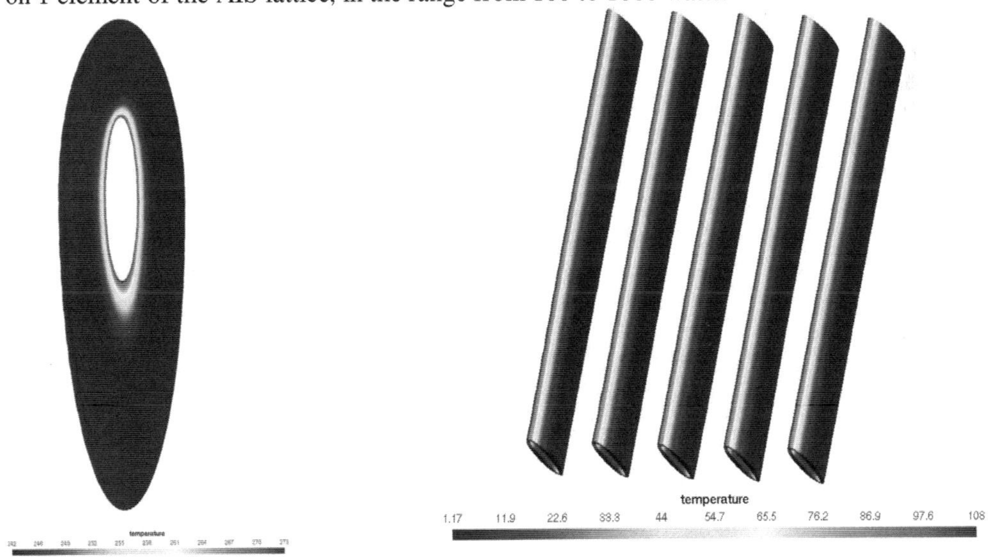

a) b)

Figure 2-the Calculated temperature field of the air flow flowing around the design of the blinds AIG at a temperature of cold air a) $\theta_e = -30^0 C$, air flow rate 20 m / s and heat flow density of the electric heating system q_W =210 W/m^2, b) $\theta_e = -50^0 C$, air flow rate 20 m / s and q_W =670 W/m^2

The heat flux density values were varied in the calculations $q_w = \gamma W / S$, where γ - thermal equivalent of electric energy transfer to thermal, S - the area of the heated surface of the shutters. Some results of numerical simulation are shown in Fig.2.

Mathematical modeling of thermal processes in the design of AIG allowed to find the optimal distance between the ribs. When increasing the distance greater than the optimum, the electric heating system does not provide the task - to maintain a positive temperature (+2 ° C) on the surface of the AIS.

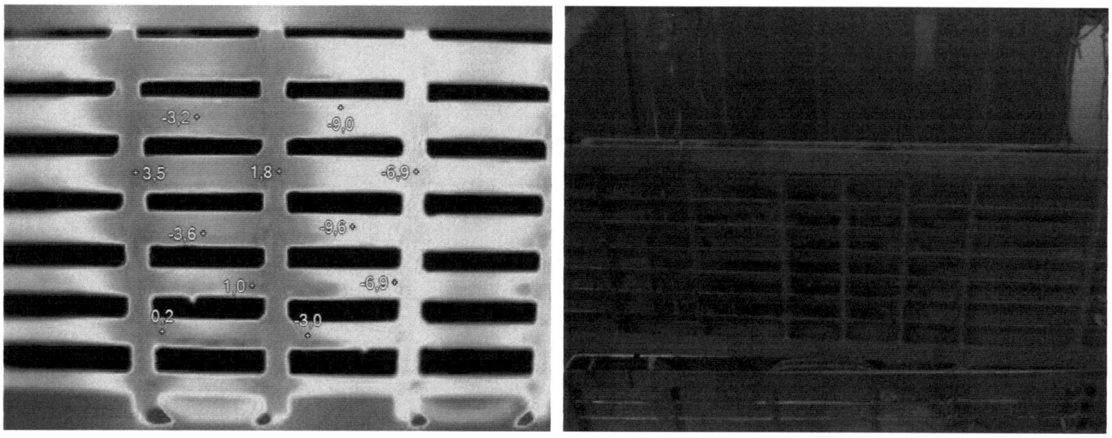

a) b)

Figure 3-a) the temperature field of AIG obtained by the infrared thermography,
b) the image of the AIG, obtained in visible light received by the camcorder.
Testing of AIG, ambient temperature -43 ° C

Some test results are shown in figure 3. It was found that this design of the air intake grating provides its protection from icing at ambient temperatures up to minus 40 ° C with the inclusion of one heating cable and up to minus 59.3 ° C with the inclusion of two heating cables.

The results of bench tests of the prototype of the AIG marine ventilation system confirm the effectiveness of the heating system when it operates at low ambient temperatures up to -59,3°C.

Conclusions

The work on the mathematical modeling of thermal modes of operation of air intake grids with electric heating system allowed to choose the optimal values of the design parameters of the AIS and to establish the most effective modes of electric heating of typical AIG ship ventilation systems at operating temperatures up to -60°C.

Acknowledgment

The work was carried out with support of state task of Ministry for Science and Education of the RF No. 9.3602.2017/ПЧ

References

[1] Lozowski E, Szilder K, and Makkonen L, 2000 Computer simulation of marine ice accretion. *Philosophical Trans of the Royal Society: Mathematical, Physical and Engineering Sciences*, Vol. 358, 2811–2845.

[2] Feher, L., and M. Thumm, Design of Avionic Microwave De-/Anti-Icing Systems//. in M.Willert-Porada(ed), Microwave devices. Springer 792 p., ISBN: 3540432523, pp. 695–702.- 2006.

[3] Petrenko, V.F., and C. Sullivan, 2003, Methods and Systems for Removing Ice from Surfaces. US Patent 6,653,598 B2, 25 November 2003.

[4] A G Kirchanov, Yu I Dimitrienko, V Yu Chibisov, I K Krasnov 2014 Method of control of icing of louvers of air intake grating. *RF patent* RUS 2563710 23.05.2014

[5] Kirchanov A G, Kokotkov V V, Dimitrienko Yu I, Ryizhkov A V, Gavrilov A Yu Sposob zashchity vozdukhozabornykh reshetok s zhalyuzi ot obledeneniya i ustroystvo dlya yego osushchestvleniya [Way of protection of air-intake lattices with blinds from frosting and the device for its implementation]. *RF Patent* 2563715. Publ. August 26, 2015
[6] Dimitrienko Yu I , Chibisov V Yu, Kirchanov A G, Voroshilov R Ya. 2015 Development and testing of anti-icing heating system for ship air-inlet grids *Engineering journal: science and innovation*, **12**(48), DOI: 10.18698/2308-6033-2015-12-1453
[7] Dimitrienko Yu. I., Koryakov M. N., Zakharov A. A., Syzdykov E. K. 2011 Development of the method of band-adaptive grids based on TVD schemes for solving gas dynamics problems. *Herald of the Bauman Moscow State Technical University Series Natural Sciences* **2** pp 87-97.
[8] Dimitrienko Yu. I., Kotenev V. P., Zakharov A. A. 2011 *The method of tape adaptive grids for numerical simulation in gas dynamics.* M.: Fizmatlit. 280 p.
[9] Dimitrienko Yu. I. 2011 Universal laws of mechanics and electrodynamics of continuous media. */Mechanics of continuous media: textbook.* Vol 2. M.: Baumann University press 559 p.
[10] Dimitrienko Yu.I. 2010 *Nonlinear Continuum Mechanics and Large Inelastic Deformations.* Springer. 722 p.

Modeling of the stability of thin composite plates based on an asymptotic theory

Yu I Dimitrienko[1], E A Gubareva[1] and A A Shurpo[1]

[1] Bauman Moscow State Technical University, Moscow, Russia

E-mail: dimit.bmstu@gmail.com

Abstract. The purpose of this work is the construction of the theory of stability of thin plates based on the asymptotic analysis of the general three-dimensional equations of the theory of elasticity and three dimensional stability theory, without introducing any assumptions regarding the nature of displacements and stresses. According to the proposed theory of stability calculation was made straight plate in uniaxial compression.

Introduction

The purpose of this operation is to construct stability theory on thin plates based common three-dimensional equations of the theory of elasticity and three dimensional stability theory, without introducing any assumptions regarding the nature of displacements and stresses, based only on analysis of basic equations asymptotic decompositions of small geometric parameter representing the ratio of the thickness of the plate to its length.

Problem statements for the base and varied states

According to the three-dimensional stability [1,2] theory is considered primary and varied plate state under the action of loads. The values for the main state of stress tensor $\boldsymbol{\sigma}^0$, tensor of small deformations $\boldsymbol{\varepsilon}^0$, vector movements $\mathbf{u}^0$ in base (steady) state have a three-dimensional object of the linear elasticity theory [19]:

$$\nabla\cdot\boldsymbol{\sigma}^0=0,\ \boldsymbol{\sigma}^0={}^4\mathbf{C}\cdot\cdot\boldsymbol{\varepsilon}^0,\ \boldsymbol{\varepsilon}^0=\frac{1}{2}(\nabla\otimes\mathbf{u}+\nabla\otimes\mathbf{u}^{\mathrm{T}}), \tag{1}$$

$$\mathbf{n}\cdot\boldsymbol{\sigma}^0\Big|_{\Sigma_\sigma}=\overset{0}{\mathbf{S}}{}^e,\quad \mathbf{n}\cdot\boldsymbol{\sigma}^0\Big|_{\Sigma_\pm}=-\tilde{p}_\pm\mathbf{n},\qquad \mathbf{u}\Big|_{\Sigma_u}=\mathbf{u}^e.$$

For varied state have the following problem of the three-dimensional stability theory [1,2]:

$$\nabla\cdot\boldsymbol{\sigma}-\boldsymbol{\sigma}^0\cdot\cdot(\mathbf{B}\cdot\mathrm{T})=0,$$

$$\boldsymbol{\sigma}={}^4\mathbf{C}\cdot\cdot\varepsilon(\mathbf{w}),\quad \varepsilon(\mathbf{w})=\frac{1}{2}(\nabla\otimes\mathbf{w}+\nabla\otimes\mathbf{w}^{\mathrm{T}}), \tag{2}$$

$$\mathbf{B}=\nabla\otimes\boldsymbol{\omega},\quad \boldsymbol{\omega}=\frac{1}{2}\mathrm{T}\cdot\cdot\boldsymbol{\Omega}(\mathbf{w}),\ \boldsymbol{\Omega}(\mathbf{w})=\frac{1}{2}(\nabla\otimes\mathbf{w}-\nabla\otimes\mathbf{w}^{\mathrm{T}}),$$

$$\mathbf{n}\cdot(\boldsymbol{\sigma}-\boldsymbol{\sigma}^0\cdot\mathrm{T}\cdot\boldsymbol{\omega})\big|_{\Sigma_\sigma}=0,\qquad \mathbf{n}\cdot(\boldsymbol{\sigma}-\boldsymbol{\sigma}^0\cdot\mathrm{T}\cdot\boldsymbol{\omega})\big|_{\Sigma_\pm}=0,\qquad \mathbf{w}\big|_{\Sigma_u}=0.$$

Here are the stress tensor, tensor of small deformations, vector displacements in varied state, normal vector, vector skew, symmetric rotation tensor, tensor Lewy- Chivites [3].
We will assume that the plate is thin and small parameter $\kappa = h/L << 1$ can be introduced, as the ratio of the total plate thickness h to the dimension L of the entire plate (e. g., to its maximum length). Also, let 's introduce non-dimensional global x_k and local ξ coordinates: $x_k = X^k / L$, $\xi = X^3/\kappa, k=1,2,3.$ Coordinate ξ varies across the thickness of the plate $-0.5<\xi_3<0.5$.
The basic premise is assumed corresponding to the real conditions of loading of the thin plates, that vector components of forces $\tilde{S}^0_j$ and pressures $\tilde{p}_\pm$ on the outer and inner surfaces of the plates are in the order of small $O(\kappa^3)$, i.e. $\tilde{p}_\pm = \kappa^3 p_\pm$, $\tilde{S}^0_j = \kappa^3 S^0_j$. In equations (1) and (2) assume moduli tensor $C_{ijkl}(\xi)$ components depending on the coordinates $\xi_3 = \xi$ as this tensor is different for different layers of the plate.

Asymptotic expansions
Solution of the problems (1), (2) for the main and varied states, following operations [4 -7], we will look for as asymptotic decompositions of the parameter κ

$$u_k = u_k^{(0)}(x_I) + \kappa u_k^{(1)}(x_I,\xi) + \kappa^2 u_k^{(2)}(x_I,\xi) + \kappa^3 u_k^{(3)}(x_I,\xi) + ... \quad (3)$$

$$w_k = w_k^{(0)}(x_I) + \kappa w_k^{(1)}(x_I,\xi) + \kappa^2 w_k^{(2)}(x_I,\xi) + \kappa^3 w_k^{(3)}(x_I,\xi) + ...\ k=1,2,3; \quad (4)$$

The Cauchy equation (3) decomposition will be substituted in the system (1), we obtain asymptotic decomposition for deformations:

$$\varepsilon^0_{ij} = \varepsilon^{0(0)}_{ij} + \kappa\varepsilon^{0(1)}_{ij} + \kappa^2\varepsilon^{0(2)}_{ij} + ... \quad (5)$$

Substituting the expression (5) in Hooke 's law in the system (1), we obtain the asymptotic voltage for decomposition:

$$\sigma^0_{ij} = \sigma^{0(0)}_{ij} + \kappa\sigma^{0(1)}_{ij} + \kappa^2\sigma^{0(2)}_{ij} + ..., \quad (6)$$

Substituting the decomposition, (3) (6) (5) Cauchy equation in relation with the balance of the system (1) and equating equations in terms κ^{-1} to zero and all other ratios between κ certain values, independent recurrent $h_i^{(0)}, h_i^{(1)}, h_i^{(2)}$ on ξ_l gives local tasks where a task sequence for the zero approximation is:

$$\sigma^{0(0)}_{i3/3} = 0,$$

$$\sigma^{0(0)}_{i3} = C_{i3KL}\varepsilon^{0(0)}_{KL} + C_{i3k3}\varepsilon^{0(0)}_{k3},\qquad \sigma^{0(0)}_{IJ} = C_{IJKL}\varepsilon^{0(0)}_{KL} + C_{IJk3}\varepsilon^{0(0)}_{k3}$$

$$\varepsilon^{0(0)}_{IJ} = \frac{1}{2}(u^{(0)}_{I,J} + u^{(0)}_{J,I}),\qquad \varepsilon^{0(0)}_{I3} = \frac{1}{2}(u^{(0)}_{3,I} + u^{(1)}_{I/3}),\qquad \varepsilon^{0(0)}_{33} = u^{(1)}_{3/3}, \quad (7)$$

$$\Sigma_{3\pm}: \sigma^{0(0)}_{i3} = 0;\qquad \Sigma_S: [\sigma^{0(0)}_{i3}] = 0,\quad [u_i^{(1)}] = 0,\quad \langle u_i^{(1)} \rangle = 0;$$

The solution of the problem of local approximation (7)-zero functions $u_j^{(1)}, \varepsilon_{kl}^{0(0)}, \sigma_{ij}^{0(0)}$ dependent on local coordinates ξ_l and input task means displacements $u_j^{(0)}(x_J)$. Similarly, there are solutions for the entire sequence of local tasks. The functions $h_i^{0(0)}, h_i^{0(1)}, h_i^{0(2)}$ find existence conditions of local decisions tasks [12]:

$$h_i^{0(0)} =< \sigma_{iJ,J}^{0(0)} > ,\ h_i^{0(1)} =< \sigma_{iJ,J}^{0(1)} > ,\ h_i^{0(2)} =< \sigma_{iJ,J}^{0(2)} > -\Delta p \delta_{i3},\ \Delta p = p_+ - p_- \quad (8)$$

The stability of the system in equation (2) can be written as follows:

$$\sigma_{ij,i} - \mathrm{T}^{jmk} B_{im} \sigma_{ij}^{0} = 0 \quad (9)$$

where the components $B_{\alpha\beta}$ of the gradient tensor from ω accompanying vector is :

$$B_{\alpha\beta} = \omega_{\alpha\beta} = \omega_{\beta,\alpha} = \frac{1}{\kappa} B_{\alpha\beta}^{(-1)} + B_{\alpha\beta}^{(0)} + \kappa B_{\alpha\beta}^{(1)} + \kappa^2 B_{\alpha\beta}^{(2)} + \kappa^3 B_{\alpha\beta}^{(3)} + .. \quad (10)$$

Substituting equations (6), (9), (10) in equation (2), collecting members at similar rates κ and equating them to constants $h_i^{(-1)} = 0, h_i^{(0)}, h_i^{(1)}$, we obtain local series resistance problems where local stability problem for approximation is zero :

$$\sigma_{3j/3}^{(0)} - \mathrm{T}^{jmk} B_{im}^{(-1)} \sigma_{ik}^{0(0)} = 0,$$

$$\sigma_{3k}^{(0)} = C_{3jKL}\varepsilon_{KL}^{(0)} + C_{3jk3}\varepsilon_{k3}^{(0)}, \qquad \sigma_{IJ}^{(0)} = C_{IJKL}\varepsilon_{KL}^{(0)} + C_{IJk3}\varepsilon_{k3}^{(0)},$$

$$\varepsilon_{IJ}^{(0)} = \frac{1}{2}(w_{I,J}^{(0)} + w_{J,I}^{(0)}), \qquad \varepsilon_{I3}^{(0)} = \frac{1}{2}(w_{3,I}^{(0)} + w_{I/3}^{(1)}), \qquad \varepsilon_{33}^{(0)} = w_{3/3}^{(1)},$$

$$\Sigma_{3\pm} : \sigma_{3j}^{(0)} = 0; \qquad \Sigma_S : [\sigma_{3k}^{(0)}] = 0, \quad [w_i^{(1)}] = 0, \quad < w_i^{(1)} > = 0;$$

The entire sequence of local stability problems similarly dismissed.

In view of the fact, that local stability problems are one-dimensional in a local variable ξ, their solution can be found analytically [15]:

$$u_I^{(1)} = -\xi u_{3,I}^{(0)} + 2S_{IKL}\varepsilon_{KL}^{0(0)},\ u_3^{(1)} = S_{KL}\varepsilon_{KL}^{0(0)},\ \sigma_{IJ}^{0(0)} = C_{IJKL}^{(0)}\varepsilon_{KL}^{0(0)},\ \sigma_{i3}^{0(0)} = 0$$

where the designated functions depend on ξ

$$S_{IKL} =< \int_{-\xi}^{\xi} C_{3j3I}^{-1} C_{3jKL} d\xi > - \int_{-\xi}^{\xi} C_{3j3I}^{-1} C_{3jKL} d\xi$$

$$S_{KL} =< \int_{-0.5}^{\xi} C_{33i3}^{-1} C_{i3KL} d\xi > - \int_{-0.5}^{\xi} C_{33i3}^{-1} C_{i3KL} d\xi$$

$$C_{IJKL}^{(0)} = C_{IJKL} - C_{IJk3} C_{k3i3}^{-1} C_{i3KL}$$

Substituting equation (8) into equation of equilibrium (1) multiplying functions $h_i^{0(0)}, h_i^{0(1)}, h_i^{0(2)}$ after administration at all $\xi\kappa$, and then integrating over the thickness, we obtain the following equation auxiliary :

$$\kappa(< \xi\sigma_{IJ,J}^{0(0)} > - < \sigma_{I3}^{0(1)} >) + \kappa^2(< \xi\sigma_{IJ,J}^{0(1)} > - < \sigma_{I3}^{0(2)} >) + ... = 0 \quad (11)$$

Let's introduce notation for forces T^0_{IJ} and moments M^0_{IJ} and shear forces Q^0_I in main and varied state:

$$T^0_{IJ} =< \sigma^{0(0)}_{IJ} > +\kappa < \sigma^{0(1)}_{IJ} > +\kappa^2 < \sigma^{0(2)}_{IJ} > \dots,$$

$$Q^0_I = \kappa < \sigma^{0(1)}_{I3} > +\kappa^2 < \sigma^{0(2)}_{I3} > + \dots$$

$$M^0_{IJ} = \kappa < \xi\sigma^{0(0)}_{IJ} > +\kappa^2 < \xi\sigma^{0(1)}_{IJ} > + \dots .$$

Then, the equation (11) can be written as a classical equilibrium equations plate where $\Delta\bar{p} = \kappa^2 \Delta p$

$$T^0_{IJ,J} = 0, \qquad Q^0_{J,J} = \Delta\bar{p}, \qquad M^0_{IJ,J} - Q^0_I = 0$$

The forces T_{IJ}, moments M_{IJ} ,shear forces Q_I in varied state have a similar decomposition while maintaining only main members. Thus, the resulting equations will assume a resistance:

$$\sum_{\gamma=1}^{2} T_{\alpha\gamma,\gamma} + (-1)^\alpha (< B^{(0)}_{13} \sigma^{0(0)}_{1\beta} > + < B^{(0)}_{23} \sigma^{0(0)}_{2\beta} >) = 0, \qquad \alpha, \beta = 1, 2, \qquad \alpha \neq \beta \ (12)$$

$$\sum_{\gamma=1}^{2} M_{\alpha\gamma,\gamma} - Q_\alpha + (-1)^\alpha (< \xi B^{(0)}_{13} \sigma^{0(0)}_{1\beta} > + < \xi B^{(0)}_{23} \sigma^{0(0)}_{2\beta} >) = 0 \qquad (13)$$

The averaged ratio for defining plates in main state are:

$$T^0_{IJ} = \bar{C}_{IJKL} \varepsilon^{0(0)}_{KL} + B_{IJKL} \eta^0_{KL} + K_{IJKLM} \varepsilon^{0(0)}_{KL,M},$$

$$M^0_{IJ} = B_{IJKL} \varepsilon^{0(0)}_{KL} + D_{IJKL} \eta^0_{KL} + \bar{K}_{IJKLM} \varepsilon^{0(0)}_{KL,M},$$

$$Q^0_I = K_{IJKL} \varepsilon^{0(0)}_{KL,J} + \kappa^2 < \sigma^{0(2)}_{I3} >$$

The averaged ratio defining plate to have the appearance of a state varied:

$$T_{IJ} = \bar{C}_{IJKL} \varepsilon^{(0)}_{KL} + \hat{B}_{IJKL} \eta_{KL} + \hat{K}_{IJKLM} \varepsilon^{(0)}_{KL,M} + \bar{V}_{IJKLM} B^{(0)}_{M3} \varepsilon^{0(0)}_{KL}$$

$$M_{IJ} = B_{IJKL} \varepsilon^{(0)}_{KL} + \hat{D}_{IJKL} \eta_{KL} + \hat{\bar{K}}_{IJKLM} \varepsilon^{(0)}_{KL,M} + \bar{\bar{V}}_{IJKLM} B^{(0)}_{M3} \varepsilon^{0(0)}_{KL}$$

where the plate tensors are averaged elastic constants

$$\bar{C}_{IJKL} =< C^{(0)}_{IJKL} >, \qquad B_{IJKL} = \kappa < \xi C^{(0)}_{IJKL} > \qquad D_{IJKL} = \kappa^2 < \xi^2 C^{(0)}_{IJKL} >$$

$$K_{IJKLM} = \kappa < \tilde{N}^{(0)}_{IJKLM} >, \ K_{IJKL} = -\kappa < \{C^{(0)}_{IJKL}\}_\xi >, \qquad \bar{K}_{IJKLM} = \kappa^2 < \xi \tilde{N}^{(0)}_{IJKLM} >.$$

Example for plate stability

Consider a classic problem of stability on the plate under the action of longitudinal compressive load $T^{(0)}_{11} \equiv -T^0 < 0$. The axis OX^1 is oriented in the direction of the longitudinal axis of the plate. In the plate in the main state, a state of uniaxial stretching (compression), which corresponds to the following solution

$$u^{(0)}_1 = \Pi_{1111} T^0_{11} X^1 + u_{10}, u^{(0)}_3 = 0, \ u^{(0)}_2 = 0, \ \eta_{KL} = -u^{(0)}_{3,KL} = 0$$

$$\sigma^{(0)}_{11} = T^0_{11}, \qquad \varepsilon^{(0)}_{11} = \Pi_{1111} \sigma^{(0)}_{11}, \qquad \varepsilon^{(0)}_{22} = \Pi_{1122} \sigma^{(0)}_{11}$$

where Π_{IJKL} components of the tensor compliances, back to C_{IJKL} the u_{10} constant of integration, determined from the boundary condition at the end of the plate.
The system of equations (12)-(13) stability theory has only two non-zero equations with two unknown functions $w_3^{(0)}$ and $w_1^{(1)}$

$$M_{11,1} - Q_1 = 0, \quad Q_{1,1} + F_{e3}^{(0)} = 0$$

Excluding from two equations, we obtain equation shear force Q_1

$$M_{11,11} + F_{e3}^{(0)} = 0 \qquad (14)$$

where

$$M_{11,11} = -D_{1111} w_{3,1111}^{(0)}, \; F_{e3}^{(0)} = -\sigma_{11}^{0(0)} \int_{-h/2}^{h/2} (w_{1/3,1}^{(1)}) dX^3 = -T_{11}^0 \int_{-h/2}^{h/2} (w_{1/3,1}^{(1)}) dX^3 \qquad (15)$$

Equation (15) will be substituted into equation (14) and equation gives overall shape plate stability theory

$$w_{3,1111}^{(0)} + k^2 w_{3,11}^{(0)} = 0 \qquad (16)$$

where

$$k^2 = \frac{T_{11}^0}{D_{1111}}$$

Equation (16) with the boundary conditions of the hinge fastening:

$$X^1 = 0: \; U_1^0 = 0, \; w_3^{(0)} = 0, \; M_{11} = 0;$$
$$X^1 = l: \; w_3^{(0)} = 0, \; M_{11} = 0$$

It has a minimum eigenvalue $k = \pi / l$ corresponds to the critical compressive load T^0 at which the loss of stability of the plate:

$$T_{\text{кр}}^0 = \frac{\pi^2 D_{11}}{l^2}$$

We classical Euler formula for the critical force.

Conclusions

Thus, herein were derived equations for stresses of deformation forces, shear moments and forces for the plate on the basis of three-dimensional equations of the mechanics of the media with the final deformations using the asymptotic theory and theory of Timoshenko shells and are generally local tasks to third order varied inclusive for main and local object plate condition and is approaching zero, manufactured by calculation of forward resistance plate proposed theory.

Acknowledgments

The work was carried out with support of state task of Ministry for Science and Education of the RF No. 9.3602.2017/ПЧ

References

[1] Dimitrienko Yu I 2014 Generalized three-dimensional stability theory. Part 2: small deformations *Herald of the BMSTU. Series: Natural sciences* **1**, pp. 17–26
[2] Dimitrienko Yu I 2013 Continuum Mechanics. *Fundamentals of solid mechanics* (Moscow) **4** 624
[3] Dimitrienko Yu I 2012 Continuum Mechanics. *Tensor analysis* (Moscow: BMSTU Publ.), **2** 463
[4] Dimitrienko Yu I 2014 Asymptotic theory of laminated thin plates *Herald of the BMSTU. Series: Natural sciences* **3** 86–100.

[5] Dimitrienko Yu I, Yakovlev D O 2013 Comparative analysis decisions asymptotic theory and laminated thin plates 3D elasticity theory *Engineering journal: science and innovation* **12**. Available at: http://engjournal.ru/catalog/mathmodel/technic/899.html
[6] Dimitrienko Yu I, Gubareva E A, Sborschikov C B 2013 Asymptotic theory design-orthotropic plates with two periodic structure *Mathematical modelling and numerical methods* **1** pp. 36 -57.
[7] Dimitrienko Yu I, Yakovlev D O 2014 *Composite Mechanics and Design* 2 **20** p 260–282.

IOP Conf. Series: Materials Science and Engineering **683** (2019) 012006 doi:10.1088/1757-899X/683/1/012006

Modeling of the effective universal constitutive relations for elastic laminated composites with finite strains

Yu I Dimitrienko[1], S B Karimov[1], D Yu Kolzhanova[1]

[1]Bauman Moscow State Technical University, Moscow, Russia

E-mail: dimit.bmstu@gmail.com

Abstract. The article considers the modeling results of layered composites with finite strains deformation according to the individual layers characteristics. The article proposes an asymptotic averaging method version for layered composites with finite deformations and periodic structure. The method allows calculating the effective deformation diagrams connecting the averaged Piola-Kirchhoff stress tensors components and the strain gradient.

Introduction

Composite materials, consisting of rubber-like or elastomeric matrices reinforced with fibers, dispersed particles or fabric fillers [1-3], are actively used in various industries. They are of considerable interest because they have successful combinations of properties, in particular, relatively high strength and a sufficiently large limiting deformation of destruction, due to the ability of rubbers to deform without failure in the region of large deformations. The calculation of the exact effective elastic characteristics of such composites is a rather complicated task due to the strong physical nonlinearity of the mechanical behavior of the phases of the composite, as well as the geometric nonlinearity of the problem [2, 4, 5]. The method of asymptotic homogenization is quite well developed; the numerical simulation of the micromechanics of composites has been successfully implemented, but mainly for linear problems [6, 7].

Asymptotic expansions for laminated elastic composites with finite strains

Consider a heterogeneous elastic solid medium V with finite deformations, which in the reference configuration $\overset{\circ}{K}$ has a periodic structure and for it can be selected a repeating element - a periodicity cell (PC) $\overset{\circ}{V}_{\xi}$, which consists of N-component, $\alpha=1,...,N$. The Euler coordinates of each material point in the reference and actual configurations are denoted as x^k and $\overset{\circ}{x}^k$, respectively, and Lagrangian coordinates are denoted as X^i. The latter are assumed to coincide with rectangular coordinates, i.e. $X^i=\overset{\circ}{x}^i$.

For a given inhomogeneous medium, we consider the problem of the nonlinear theory of elasticity in the Lagrangian description using the universal models A_n proposed by Yu.I. Dimitrienko, for medium with finite deformations [8,9]:

$$\overset{\circ}{\nabla}_i P^{ij} + \overset{\circ}{\rho} \overset{\circ}{f}^j = 0, \quad X^i \in \overset{\circ}{V},$$

$$P^{ij} = \overset{(n)}{\mathcal{F}}{}^{0ij}(F^{kl}, \overset{\circ}{X}{}^m), \quad X^i \in \overset{\circ}{V} \cup \Sigma, \tag{1}$$
$$F^k{}_l = \delta^k_l + \nabla_l u^k, \quad X^i \in V \cup \Sigma,$$
$$n_i[P^{ij}] = 0, \quad [u^i] = 0, \quad X^i \in \Sigma_{\alpha\beta},$$
$$n_i P^{ij} = t^j_e, \quad X^i \in \Sigma_1, \quad u^i = u^i_e, \quad X^i \in \Sigma_2.$$

Introduce a small parameter $\kappa = l/L << 1$, as the ratio of the characteristic size l of the PC to the characteristic size L of the entire composite (the sizes are defined for K) and also introduce local ξ^i Lagrangian coordinates in K which are related X^i to the following relations: $\xi^i = \bar{X}^i/\kappa$, $\bar{X}^i = X^i/L$.

The solution of problem (1) with respect to the displacement vector is sought as a quasiperiodic function of the coordinates and is written as asymptotic expansion in the parameter κ: $u^k(X^i, \xi^l) = u^k_{(0)}(X^i) + \kappa u^k_{(1)}(X^i, \xi^l) + \kappa^2 ...$

Find asymptotic expansions for the strain gradient, the defining relations and the Piola – Kirchhoff tensor, the equilibrium equations, and the boundary conditions:

$$F^k{}_l = F^k{}_l{}^{(0)}(X^i, \xi^l) + \kappa F^k{}_l{}^{(1)}(X^i, \xi^l) + \kappa^2 ...,$$
$$F^k{}_l{}^{(0)}(X^i, \xi^l) = \delta^k_l + u^{k(0)}{}_{,l} + u^{k(1)}{}_{/l}, \quad F^k{}_l{}^{(1)} = u^{k(1)}{}_{,l} + u^{k(2)}{}_{/l}$$
$$P^{ij} = P^{ij(0)}(X^k, \xi^l) + \kappa P^{ij(1)}(X^k, \xi^l) + \kappa^2 ...,$$
$$P^{ij(0)} = \overset{(n)}{\mathcal{F}}{}^{0ij}(F^k{}_l{}^{(0)}, \xi^m), \quad P^{ij(1)} = \frac{\partial}{\partial F^k{}_l{}^{(0)}} \overset{(n)}{\mathcal{F}}{}^{0ij}(F^k{}_l{}^{(0)}, \xi^m) F^k{}_l{}^{(1)}$$
$$\frac{1}{\kappa} P^{ij(0)}{}_{/i} + (P^{ij(0)}{}_{,i} + P^{ij(1)}{}_{/i} + \overset{0}{\rho} f^j) + \kappa(P^{ij(1)}{}_{,i} + P^{ij(2)}{}_{/i}) + \kappa^2 ... = 0, \tag{2}$$
$$n_i[P^{ij(0)}] + \kappa n_i[P^{ij(1)}] + \kappa^2 ... = 0, \quad [u^{k(0)}] + \kappa[u^{k(1)}] + \kappa^2 ... = 0, \quad X^i \in \Sigma_{\alpha\beta} \tag{3}$$
$$n_i P^{ij(0)} + \kappa n_i P^{ij(1)} + \kappa^2 ... = t^j_e, \quad X^i \in \Sigma_1, \quad u^{k(0)} + \kappa u^{k(1)} + \kappa^2 ... = u^i_e, \quad X^i \in \Sigma_2$$

Equating the terms in the expansions (2) - (3) with the same powers κ to zero a recurrent sequence L_κ of local problems of nonlinear elasticity is derived.

Local problem solution

The local problem for a layered composite in which the periodicity cell is a system of parallel layers orthogonal to the direction $O\xi$ takes the form:

$$P^{3j(0)}{}_{/3} = 0, \tag{4}$$
$$P^{ij(0)} = \overset{(n)}{\mathcal{F}}{}^{0ij}(F^k{}_l{}^{(0)}, \xi), \tag{5}$$
$$F^k{}_l{}^{(0)}(X^i, \xi) = \bar{F}^k_l + u^{k(1)}{}_{/3}\delta_{l3}, \tag{6}$$
$$[P^{3j(0)}] = 0, \quad [u^{k(1)}] = 0, \quad \xi = \xi_\alpha, \quad \alpha = 1...n-1, \tag{7}$$
$$\langle u^{k(1)} \rangle = 0,$$
$$[[u^{k(1)}]]_3 = 0. \tag{8}$$

The task is non-linear but one-dimensional - in it all functions depend only on ξ.
Integrating the equilibrium equations (4) can conclude that the stress $P^{3j(0)}$ are constant in the PC:

$$P^{3j(0)} = C^j = const\,, \tag{9}$$

From equation (6) it follows that of the 9 components of the deformation gradient $F^{k\,(0)}_{\,l}$ only 3 components depend on the coordinate ξ, and the remaining 6 coincide with the components of the averaged gradient:

$$F^{k\,(0)}_{\,3} = \bar{F}^k_3 + u^{k(1)}_{\ \ /3}, \quad F^{k\,(0)}_{\,L} = \bar{F}^k_L\,, \quad L = 1,2 \tag{10}$$

Substituting (9) into (5) obtain a system of three nonlinear algebraic equations which can be considered with respect to three components $F^{k\,(0)}_{\,3}$: $C^j = \overset{(n)}{\mathcal{F}}{}^{03j}(F^{k\,(0)}_{\,3}, \bar{F}^k_L, \xi)$. Write the formal solution of this system in the form $F^{k\,(0)}_{\,3} = \overset{(n)}{\mathcal{G}}{}^k(C^j, \bar{F}^k_L, \xi)$.
Using expression (10) obtain a system of three ordinary linear differential equations for displacements $u^{k(1)}$ $\bar{F}^k_3 + u^{k(1)}_{\ \ /3} = \overset{(n)}{\mathcal{G}}{}^k(C^j, \bar{F}^k_L, \xi)$ which is easily integrated

$$u^{k(1)} = \int_{-0.5}^{\xi} \overset{(n)}{\mathcal{G}}{}^k(C^j, \bar{F}^k_L, \xi^m)\,d\xi - \bar{F}^k_3\xi + B^k\,, \tag{11}$$

where B^k - are the integration constants.
Conditions (7) of the ideal contact for functions (9) and (10) are automatically fulfilled.
After substituting the expression (11) into the periodicity condition (8), we obtain an equation $< \overset{(n)}{\mathcal{G}}{}^k(C^j, \bar{F}^k_L, \xi^m) >= \bar{F}^k_3$ that can be considered as a nonlinear algebraic equation for the constants C^j. Write the formal solution of this equation: $C^j = \overset{(n)}{\mathcal{S}}{}^j(\bar{F}^k_3, \bar{F}^k_L)$ and find the ratio between the gradient $F^{k\,(0)}_{\,l}$ and the average gradient $\bar{F}^k_j$:

$$F^{k\,(0)}_{\,l} = \bar{F}^k_l + (\overset{(n)}{\mathcal{G}}{}^k(\overset{(n)}{\mathcal{S}}{}^j(\bar{F}^k_3, \bar{F}^k_l), \bar{F}^k_l, \xi) - \bar{F}^k_3)\ \delta_{l3} \tag{12}$$

Substituting expression (12) into (5), and integrating it over PC obtain the desired effective defining relations for the layered composite

$$\bar{P}^{ij} = \overset{(n)}{\mathcal{F}}{}^{0ij}(\bar{F}^k_m)\,, \tag{13}$$

where denotes the average stress and the averaged function of the defining relations

$$\bar{P}^{ij} =< P^{ij(0)} >, \quad \overset{(n)}{\mathcal{F}}{}^{0ij}(\bar{F}^k_m) =< \overset{(n)}{\mathcal{F}}{}^{0ij}(\bar{F}^k_l + (\overset{(n)}{\mathcal{G}}{}^k(\overset{(n)}{\mathcal{S}}{}^j(\bar{F}^k_3, \bar{F}^k_l), \bar{F}^k_l, \xi) - \bar{F}^k_3)\delta_{l3}, \xi) >$$

Universal models for effective constitutive relation of laminated composites

The tensor of defining relations for nonlinear-elastic components of a composite for models A_n of elastic media with finite deformations has a complex implicitly defined form and depends on the components of the strain gradient and Lagrangian coordinates (discontinuous). The tensor function (13) has an analytical expression for models A_V, A_I; for models A_{II}, A_{IV} was used the QUVF-system [11–13]

$$\overset{(n)}{\mathcal{F}}{}^{0ij}_{A} = \sum_{\gamma=1}^{3} \varphi_\gamma \overset{(n)}{E}{}^{0ijkl}_{A} \overset{(s)}{I}_{\gamma Akl}\,, \tag{14}$$

$$\overset{(s)}{I}_{1A} = E, \quad \overset{(s)}{I}_{2A} = EI_1\left(\overset{(n)}{C}\right) - \overset{(n)}{C}, \quad \overset{(s)}{I}_{3A} = \overset{(n)}{C}{}^2 - I_1\left(\overset{(n)}{C}\right)\overset{(n)}{C} + EI_2\left(\overset{(n)}{C}\right),$$

$$\varphi_1 = (l_1 + 2l_2) I_1\left(\overset{(n)}{C}\right), \quad \varphi_2 = -2l_2, \quad \varphi_3 = 0, \quad I^{(s)}_{1A} = \overset{(n)}{C}\cdot\cdot E, \quad I^{(s)}_{2A} = E\cdot\cdot \overset{(n)}{C}{}^2, \quad I^{(s)}_{3A} = \det \overset{(n)}{C},$$

$$\overset{(n)}{C}{}^{0ij}_{A} = \frac{1}{n-III}\sum_{\alpha=1}^{3} \lambda_\alpha^{(n-III)/2} Q^i_\alpha Q^j_\alpha - \frac{1}{n-III} g^{ij}, \quad \overset{(n)}{E}{}^{0ijkl}_{A} = \sum_{\alpha,\beta=1}^{3} E_{\alpha\beta} Q^i_\alpha Q^j_\beta Q^k_\beta Q^l_\alpha .$$

Computational results

The algorithm for the numerical finding of effective defining relations (13) [10] was implemented in the form of program code written in the C ++ programming language. The program produces a solution of a chain of systems of nonlinear algebraic equations by the method of multidimensional optimization.

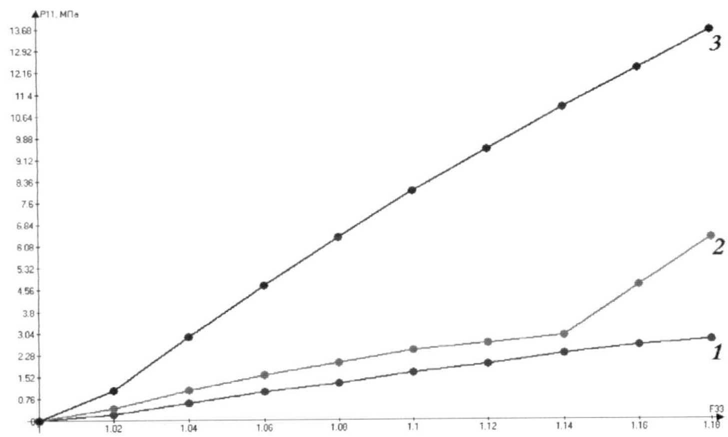

Figure 1 . Model A_I Deformation diagrams $\overline{P}_{11}\left(\overline{F}_{33}\right)$ with different relative thicknesses h_2: 1 - 0.0, 2 - 0.3, 3 - 0.5.

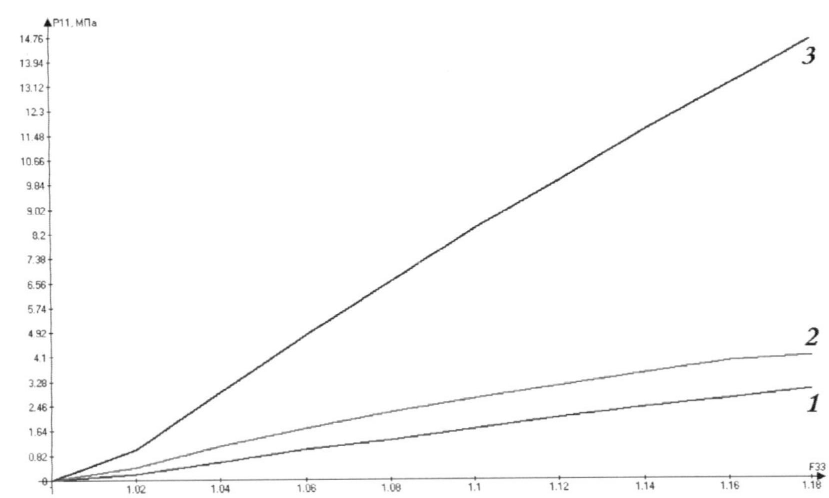

Figure 2. Model A_{II} Deformation diagrams $\overline{P}_{11}\left(\overline{F}_{33}\right)$ with different relative thicknesses h_2: 1 - 0.0, 2 - 0.3, 3 - 0.5.

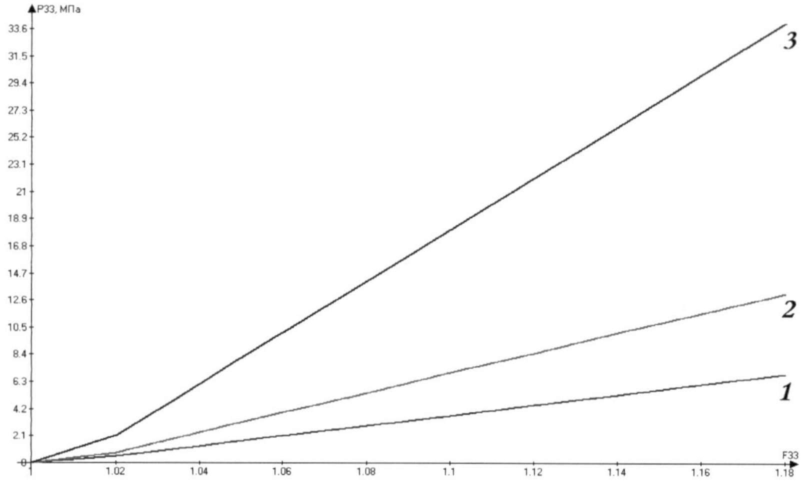

Figure 3. Model A_{IV} Deformation diagrams $\overline{P}_{33}\left(\overline{F}_{33}\right)$ with different relative thicknesses h_2 : 1 - 0.0, 2 - 0.3, 3 - 0.5.

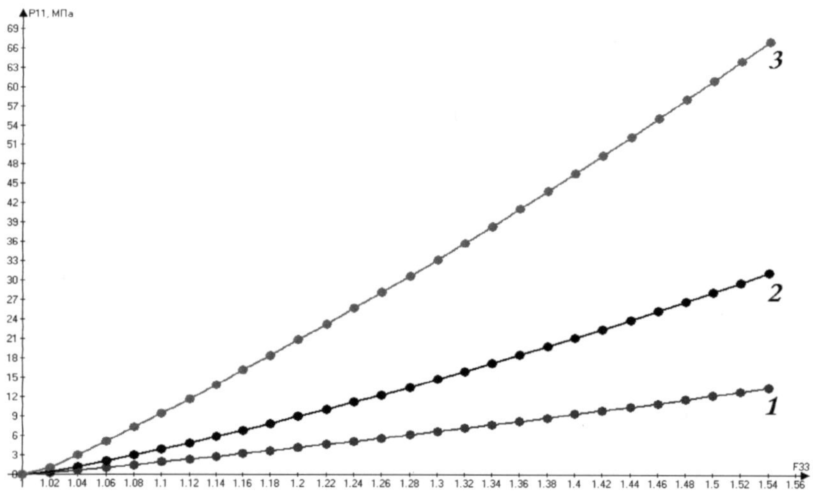

Figure 4. Model A_V Deformation diagrams $\overline{P}_{11}\left(\overline{F}_{33}\right)$ with different relative thicknesses h_2 : 1 - 0.0, 2 - 0.3, 3 - 0.5.

The figures show the graphs of the dependence of the Piola-Kirchhoff stress tensor on the components of the averaged strain gradient for various input data. The constructed functions represent a numerical experiment for separate layers of a composite with two different sets of elastic constants on the first and third layers: l1 = 100 MPa and l2 = 50 MPa, on the second layer: l1 = 20 MPa and l2 = 10 MPa. Let us set the following relations between the layers h1 = h3 and h2 = 1-2 h1.
Cases h1 = 0.5 and h1 = 0.0 correspond to homogeneous materials with the characteristics of the first or second layer, respectively.
For models A_I and A_V graphs the calculations are based on analytical formulas (14) and using the QUVF-system (solid continuous lines).

Conclusions

For nonlinear elastic composites with large deformations and a periodic structure has been developed a method of asymptotic averaging over a small geometric parameter using a universal representation of the governing relations for a complex of various medium models.

The local and averaged problem of nonlinear elasticity is formulated and an algorithm is developed for finding effective defining relations for the composite material as a whole.

The resulting algorithm allows us to calculate the effective strain diagrams of layered composites with finite deformations, connecting the components of the averaged Piola-Kirchhoff stress tensors and the strain gradient.

The numerical calculations performed for models of nonlinear elastic medium with finite deformations using various elastic constants and layer thicknesses confirm the feasibility and effectiveness of the proposed method.

Acknowledgments

The work was carried out with support of state task of Ministry for Science and Education of the RF No. 9.3602.2017/ПЧ

References

[1] Aboudi J 2004 Micromechanics-based thermoviscoelastic constitutive equations for rubber-like matrix composites at finite strains. *Int. J. Solids and Struct.* ed Elsevier **41** 5611–5629.

[2] Zhang B, Yu X, Gu B 2015Micromechanical modeling of large deformation in sepiolite reinforced rubber sealing composites under transverse tension. *Polym. Compos.*,. doi: 10.1002/pc.23596

[3] Dimitrienko Yu I, Dashtiyev I Z 2001 Models of viscoelastic behavior of elastomers at finite strains *Herald of the Bauman MSTU. Series: Natural sciences* **1** pp. 21–41.

[4] Ma C C M, Taib N H, Wua S H, Lina S H, Wua J F, Lina J M 1997 Creep behavior of carbon-fiber-reinforced polyetheretherketone (PEEK) [±45]4s laminated composites. *Composites Part B: Engineering* **28** 407–417.

[5] Takanoa N, Ohnishia Y, Zakoa M, Nishiyabub K 2000 The Formulation of Homogenization Method Applied to Large Deformation Problem for Composite Materials. *Int. J. of Solids and Structures* **37** 6517–6535.

[6] Dimitrienko Yu I, Yakovlev D O 2014 Asymptotic theory of Thermoelasticity of multilayer composite plates *Mekhanika kompositsionnykh materialov i konstruktsiy — Composite Mechanics and Design* **20** 260–282.

[7] Dimitrienko Yu I 2012 Asymptotic theory of thin multilayer plates *Vestnik MGTU im. N.E. Baumana. Seriya Estestvennye nauki —Herald of the Bauman MSTU. Series: Natural sciences* **3** pp. 86–100.

[8] Dimitrienko Yu I 2011 Universal laws of continuum mechanics and electrodynamics *Continuum Mechanics* (Moscow, BMSTU Publ.) **2** 560p.

[9] Dimitrienko Yu I 2013 Fundamentals of solid mechanics, *Continuum Mechanics* (Moscow: BMSTU Publ.) **4** 624 p.

[10] Dimitrienko Yu I, Gubareva E A, Kolzhanova D Yu 2015 Modeling of layered composites with finite strains by the method of asymptotic homogenization *Engineering Journal: Science and Innovation* **5**. DOI 10.18698/2308-6033-2015-5-1405

Computational modeling of the conjugated thermomechanical and aerogasdynamics processes for composite structures of high speed vehicles

Yu Dimitrienko[1], A Zakharov[1] and M Koryakov[1]

[1]Bauman Moscow State Technical University, Computational Mathematics and Mathematical Physics, Russian Federation

E-mail: dimit@bmstu.ru, azaharov@bmstu.ru, mkoryakov@bmstu.ru

Abstract. A conjugated problem statement for aerodynamics and thermomechanics of heat-shielding structures from thermally decomposing polymer composite materials is proposed. It is based on the iterative solution of the three types of detached problems: aerogasdynamics problem for viscous heat-conducting flows, internal heat and mass transfer and thermoelasticity of shell constructions. An example of the numerical solution of the conjugated problem is given. It is shown that due to the high temperatures of the aerodynamic heating of the structure made of a polymer composite material there can appear a polymer phase thermodecomposition and intensive internal gas generation into the structure of the material.

1. Introduction

The learning of hypersonic speeds is one of the most promising complex problems of the high-tech development. This problem can be identified such components as: the study of hypersonic aerodynamics of flights, study of the heat transfer on surfaces of aircraft constructions, thermophysics research of constructional materials, thermal strength study, analysis and development of constructional materials for hypersonic aircrafts, and the problems of hypersonic aeroelasticity, control and others. Significant amount of work [1, 5, 9, 10] described researches of hypersonic aerodynamics conditions, less studied problems of heat transfer [2, 6, 8] under hypersonic speeds. The more difficult problem is the high-temperature behavior of composite materials based on heat-resistant filaments and masters [3]. The complex conjugated problem of aerothermodynamics, heat transfer, thermal physics and thermal strength of hypersonic constructions is still practically uninvestigated, and there are relatively recent work that studied aeroelasticity of constructions under hypersonic speeds. However under the actual operating conditions of hypersonic vehicles the problems of aerothermodynamics, heat transfer and thermophysics of the constructions are conjugated through the boundary conditions on the surface of the constructions, so the parameters of the heat flux acting on the materials depend on the properties of these materials. By turn, thermal properties of the materials at high temperatures may depend on the mode of deformation of the constructions. So the significant level of thermal stresses in composite materials leads to the filament microcracking long before the full macrodestruction of constructions, thereby the gas permeability and thermal conductivity materials and temperature fields in the constructions are changed. Thus the development of the methods for solving the conjugated problem for the study of the real processes taking place in the constructions of hypersonic vehicles is necessary.

The general formulation of the conjugated problem of aerothermodynamics and thermomechanics consists of the three systems of equations

- the Navier-Stokes equations of an external gas flow;
- the internal heat and mass transfer equations;
- the equations of thermoelasticity of a shell.

2. Mathematical Formulation

2.1. *System of Gasdynamics Equations*

Consider the system of equations of a viscous heat-conducting gas (the Navier-Stokes equations) [1, 5, 7]:

$$\begin{aligned}
&\frac{\partial \rho}{\partial t} + \nabla \cdot \rho \mathbf{v} = 0, \\
&\frac{\partial \rho \mathbf{v}}{\partial t} + \nabla \cdot \left(\rho \mathbf{v} \otimes \mathbf{v} + p\mathbf{E} - \mathbf{T}_v \right) = 0, \\
&\frac{\partial \rho \varepsilon}{\partial t} + \nabla \cdot \left(\left(\rho\varepsilon + p \right) \mathbf{v} - \mathbf{T}_v \cdot \mathbf{v} + \mathbf{q} \right) = 0,
\end{aligned} \tag{1}$$

where ρ is the gas density, t is the time, v is the velocity vector, p is the pressure, E is the identity tensor, ε is the total energy per unit volume.

This system adds relations for the perfect gas, viscous stress tensor and heat flux vector

$$\begin{aligned}
&p = \rho \frac{R}{M} \theta, \quad e = c_v \theta, \quad E = e + |\mathbf{v}|^2 / 2, \\
&\mathbf{T}_v = \mu_1 (\nabla \cdot \mathbf{v}) \mathbf{E} + \mu_2 (\nabla \otimes \mathbf{v} + \nabla \otimes \mathbf{v}^T), \\
&\mathbf{q} = -\lambda \nabla \theta,
\end{aligned} \tag{2}$$

where R is the universal gas constant, M is the molecular weight of gas, θ is the gas temperature, cv is the specific heat at constant volume, μ1 and μ2 are the coefficients of viscosity, λ is the thermal conductivity of gas. The coefficients of viscosity and thermal conductivity are functions of temperature.

The boundary conditions on the solid surface, which is the interface of the gas and solid domains, are as follows

$$\begin{aligned}
&\mathbf{v} = \mathbf{0}, \quad \theta = \theta_w, \\
&-\lambda \nabla \theta \cdot \mathbf{n} = -\lambda_s \nabla \theta_s \cdot \mathbf{n} + \varepsilon_g \sigma \theta_{max}^4 - \varepsilon_s \sigma \theta_w^4,
\end{aligned} \tag{3}$$

where θw is the temperature of the solid surface, θmax is the maximum temperature into the boundary layer, $\nabla\theta_s$ is the temperature gradient on the solid wall from the construction, εg and εs are the emissivity of the heated gas and solid surface respectively and σ is the Stefan-Boltzmann constant.

2.2. *System of Equations of Internal Heat and Mass Transfer*

We consider a typical element which is made of a heat-resistant composite material consisting of a polymer master with heat-resistant filaments. There are physical and chemical processes of thermodecomposition in such composite master under high temperatures of aerodynamic heating. In these processes the gaseous products of thermal decomposition are generated, then they are accumulated in the pores of the material and filtered into the outer gas flow, as well as a new solid phase is formed. It is the phase of pyrolytic master which has significantly lower elastic-strength properties than the original polymer phase. The four-phase model to describe the internal heat and mass transfer and deformations of such composite is proposed in [3]. This model consists of:

- the equation of change of mass of polymer master phase:

$$\rho_b \frac{\partial \varphi_b}{\partial t} = -J; \tag{4}$$

- the equation of filtration of gaseous products of thermodestruction in pores of the composite material:

$$\frac{\partial \rho_g \varphi_g}{\partial t} + \nabla \cdot \rho_g \varphi_g \mathbf{v}_g = J\Gamma; \tag{5}$$

- the heat transfer equation in the thermodestruction composite:

$$\rho c \frac{\partial \theta}{\partial t} = -\nabla \cdot \mathbf{q} - c_g \nabla\theta \cdot \rho_g \varphi_g \mathbf{v}_g - J\Delta e^0; \tag{6}$$

where φb, φg are the volume concentrations of the phase of the initial polymer master and gas phase; ρb is the density of the phase of the initial polymer master which is assumed to be constant; ρg is the average pore density of the gas phase (variable); cg is the specific heat of the gas phase at constant volume, ρ and c are the density and specific heat of the composite as a whole, q is the heat flux vector, θ is the composite temperature for all phases in common; vg is the velocity vector of the gas phase; Δe0 is the specific heat of the master thermodecomposition; J is the mass velocity of the master thermodecomposition and Γ is the gas-producing factor of the master.

The equations (4)-(6) are added the relations between the heat flux vector q, the velocity vector of the gas in pore vg with the temperature gradient $\nabla\theta$ and the pressure gradient ∇p using the Fourier and Darcy laws, as well as the Arrhenius relation for the mass velocity of the master thermodecomposition J and Mendeleev-Clapeyron equation for the pore pressure of the gas phase p

$$\mathbf{q} = -\mathbf{\Lambda} \cdot \nabla\theta, \quad \rho_g \varphi_g \mathbf{v}_g = -\mathbf{K} \cdot \nabla p,$$
$$J = J_0 \exp\left(-\frac{E_A}{R\theta}\right), \quad p = \frac{\rho_g R\theta}{M_g}, \tag{7}$$

where J_0 is the pre-exponential factor, E_A is the activation energy of the thermodecomposition process, M_g is the molecular weight of the gas phase, **Λ** is the thermal conductivity tensor, **K** is the permeability tensor of the composite. They depend on the phase concentration.

The boundary conditions for the equations (4)-(6) on the heated surface of the construction are as follows

$$p = p_e, \quad \theta = \theta_e,$$

where p_e, θ_e are the pressure and temperature of the flow on the surface.
The boundary conditions of tightness and thermal insulation are specified on the rest of the composite surface

$$\mathbf{n} \cdot \nabla p = 0, \quad -\mathbf{n} \cdot \mathbf{\Lambda} \cdot \nabla\theta = 0.$$

2.3. *System of Equations of Thermoelasticity*

System of equations of thermoelasticity consists of [4]:

- the equilibrium equations of the shell;
- the kinematic relations.

3. Results and Discussion

A numerical solution of the conjugated problem was calculated for a hypersonic flow (M = 6) around the nose of a vehicle model flying at the altitude of 15 km.
Fig. 1 shows the distribution of the temperature of the flow near the body. The temperature reaches 1.600°K at the stagnation point on the nose of the vehicle and decreases monotonically as the distance from the stagnation point, but it remains rather high value. At the maximum distance the temperature

is about 800°K for the edge and lower generatrix and 1.000°K for the upper generatrix of the vehicle which has a large value of the cone angle.

The results of numerical calculations of the fields of internal heat and mass transfer in the shell element of the hypersonic aircraft are shown in Fig. 7. The curves show the distributions along the shell thickness (the dimensionless coordinate x = 0.5 + q3/h is introduced) at a control point on the bottom surface at the distance of 30r from the stagnation point, where r is the blunted body radius at the stagnation point. Different colors correspond to the 7 different time points t1,…,t7 of aircraft flight.

Fig. 3 shows maximum of the transversal stresses σ33 for the shell at time point t7.

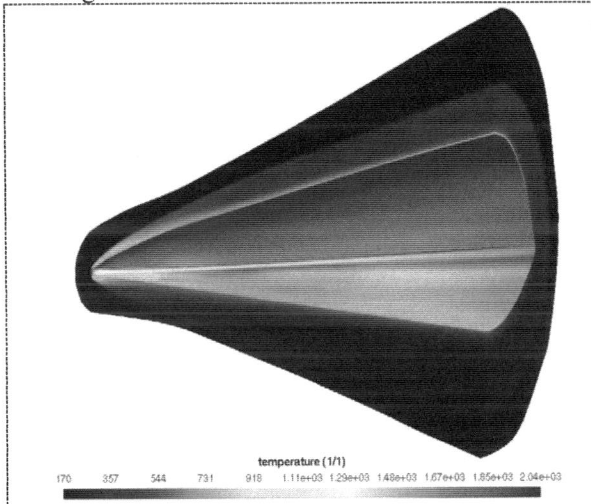

Figure 1. Temperature of the flow, θ (°K).

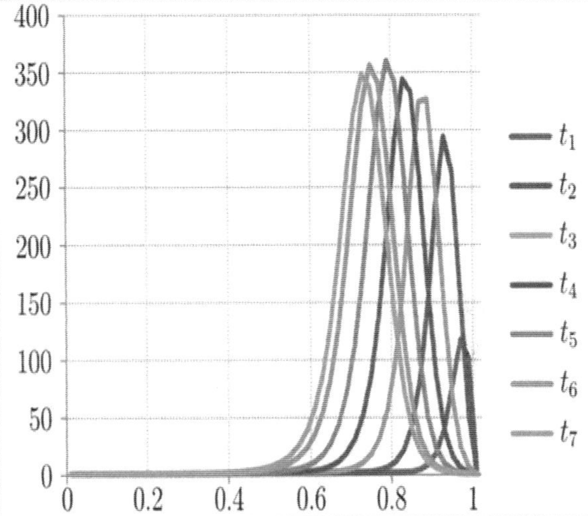

Figure 2. Pore pressure p, atm.

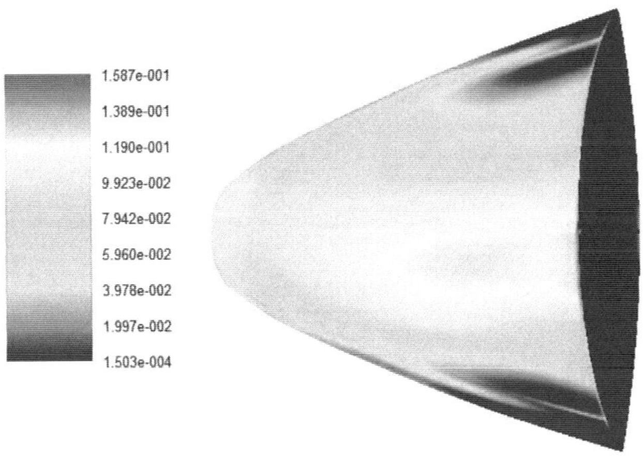

Figure 3. Maximum transversal stress, σ_{33} (GPa).

Acknowledgments
The work was carried out with support of state task of Ministry for Science and Education of the RF No. 9.3602.2017/ПЧ

References

[1] Anderson J D 2006 *Hypersonic and High-Temperature Gas Dynamics*. 2nd ed. American Institute of Aeronautics and Astronautics (Reston, Virginia)
[2] Shumaev V V and Kuzenov V V 2017 Development of the numerical model for evaluating the temperature field and thermal stresses in structural elements of aircrafts *J. Phys:Conf. Ser.* IOP Publishing Ltd 1 **891**
[3] Dimitrienko Yu I 1999 *Termomechanics of composites under high temperatures* ed Kluwer Academic Publishers (Dordrecht/Boston/London)
[4] Dimitrienko Y I, Minin V V and Syzdykov E K 2011 Modeling of Thermomechanical Processes in Composite Shells in Local Radiation Heating, *Composites: Mechanics, Computations, Applications* **2** 147-169
[5] Kuzenov V V 2018 et al Calculating processes of laminar and turbulent heat transfer around the elements of the aircraft *J. Phys.:Conf. Ser.* **980**
[6] Kandinsky R O, Prosuntsov P V 2015 Modelling combined heat exchange in the leading edge of perspective aircraft wing, *MATEC Web of Conferences* **23** 01019
[7] Grishin Y A, Bakulin V N 2015 New calculation schemes based on the large-particle method for modeling gas-dynamic problems, *Doklady Physics* **60** 555-558
[8] Leont'ev A I, Kuzma-Kichta Y A, Popov I A 2017 Heat and mass transfer and hydrodynamics in swirling flows (review), *Thermal Engineering* **64** pp. 111-126
[9] Gorskii V V, V A Sysenko, Dekermendzhi K Y, Accuracy of calculating the momentum-loss thickness in a laminar boundary layer on the surface of a hemisphere in a supersonic air flow Journal of Engineering Physics and Thermophysics 3 **90** pp. 535-540, 2017.
[10] Isaev S A, Schelchkov A V, Leontiev A I 2016 Baranov P A, Gulcova M E, Numerical simulation of the turbulent air flow in the narrow channel with a heated wall and a spherical dimple placed on it for vortex heat transfer enhancement depending on the dimple depth *International Journal of Heat and Mass Transfer* **94** pp. 426-448

Modeling of Non-Newtonian resin flows in Composite Microstructures

Yu I Dimitrienko[1] and Sh Li[1]

[1]Computational Mathematics and Mathematical Physics Department, Bauman Moscow State Technical University, Moscow, 105005, Russia

Abstract. The multi-scale modeling of the flow of the non-Newtonian viscosity resin in the composite structure is based on the asymptotic homogenization method, and a so-called local problem is obtained. The finite element method was used to calculate the local problem, and the distribution of single hole velocity, pressure and non-Newtonian viscosity was obtained.

1. Introduction

Plastic is a very widely used polymer composite applications, the main component is a resin. In the plastic manufacturing process, a resin-based polymer solution is usually injected into a specific mold. Especially for composites with resin as a reinforcing material, the flow of non-Newtonian resin in the composite microstructure is very important [1]. Therefore, in order to optimize these processes, numerical modeling of the flow of non-Newtonian resins in composite microstructures has become very important [2].

2. Main assumptions of the model

2.1. Geometric model

Consider a three-dimensional orthogonal composite structure [3]. (Fig. 1). It is assumed that the composite structure is characterized by periodicity. Wherein red indicates a solid material, and transparent color indicates a polymer reinforcing material (resin polymer solution).

2.2. The model assumptions

Take the following assumptions regarding the properties of the phases: (1) fluids are isotropic non-Newtonian viscous incompressible media; (2) the porous skeleton is assumed to be non-deformable, i.e. his movement is not considered; (3) fluid movement is isothermal; (4) mass forces are absent.

We introduce the following notation: l_0-characteristic size of periodicity cell (PC) V_ξ; x_0-characteristic size of the global entire porous medium V; $\kappa = l_0/x_0 << 1$-small parameter; $\vec{\xi} = \vec{x}/\kappa$-dimensionless local coordinates varying within a periodicity cell (PC) V_ξ.

3. Principal equations

The motion of non-Newtonian resins in the composite occupying the region V , within the framework of the assumptions made, is described by the system of the incompressible Navier-Stokes equations, which, together with the adhesion conditions on a solid surface, have the following form [4]: .

$$\nabla \cdot \vec{v} = 0 \tag{1}$$

$$\rho\left(\frac{\partial \vec{v}}{\partial t} + \vec{v} \cdot \nabla \otimes \vec{v}\right) = -\nabla p + \nabla \cdot \mu\left(\nabla \otimes \vec{v} + \nabla \otimes \vec{v}^T\right) \tag{2}$$

$$\vec{v}\big|_{\Sigma} = 0 \tag{3}$$

$$p\big|_{t=t_0} = p_0 \tag{4}$$

where ρ -density; $\vec{v}$ -velocity of fluid; p -pressure; μ -non-newtonian viscosity and $\otimes$ -tensor product. The strain rate tensor is defined as follow:

$$D = \frac{1}{2}\left(\nabla \otimes \vec{v} + \nabla \otimes \vec{v}^T\right)$$

The AI and AV models of Voigt isotropic viscous media [4] are considered, and non-Newtonian viscosity satisfies the following defining relation [1]:

$$\frac{\mu - \mu_\infty}{\mu_0 - \mu_\infty} = \left(1 + \lambda^2 I_2^2\right)^{\frac{n-1}{2}} \tag{5}$$

which depends on the second invariant of the strain rate tensor [5]:

$$I_2(D) = \sqrt{2D \cdot\cdot D}$$

Then we used a dimensionless form, where:

$$\bar{\rho} = \frac{\rho}{\rho_0},\ \bar{v} = \frac{\vec{v}}{v_0},\ \bar{p} = \frac{p}{p_0},\ \bar{t} = \frac{t}{t_0},\ \bar{x} = \frac{\vec{x}}{x_0},\ t_0 = \frac{x_0}{v_0}.$$

Here $\bar{p}$, $\bar{\rho}$, $\bar{v}$ are the dimensionless pressure, density and velocity of fluid, respectively; p_0, ρ_0, v_0 are their typical magnitudes. t_0 is the dimensionless time.

The equations for the flow of an incompressible fluid can be written in a dimensionless form as follows (the symbol is omitted below):

$$\nabla \cdot \vec{v} = 0 \tag{6}$$

$$\frac{\partial\vec{v}}{\partial t}+\vec{v}\cdot\nabla\otimes\vec{v}=-Eu\nabla p+\frac{1}{\mathrm{Re}}\nabla\cdot\mu\left(\nabla\otimes\vec{v}+\nabla\otimes\vec{v}^{T}\right) \qquad (7)$$

$$\mu=s+(1-s)\left(1+Cu^{2}I_{2}^{2}\right)^{\frac{n-1}{2}} \qquad (8)$$

$$\vec{v}\big|_{\Sigma}=0\ ,\ p\big|_{t=t_0}=p_0 \qquad (9)$$

where $Eu=\dfrac{p_0}{\rho_0 v_0^2}$ - Euler number, $\mathrm{Re}=\dfrac{\rho_0 v_0 x_0}{\mu_0}$ - Reynolds number, $Cu=\dfrac{\lambda v_0}{x_0}$ - Carreau number.

4. Asymptotic averaging

In the framework of the method of asymptotic averaging[6], the differentiation of quasi-periodic functions is performed in accordance with the rule of differentiation of a complex function:

$$\nabla f\rightarrow\nabla_{x}f+\frac{1}{\kappa}\nabla_{\xi}f \qquad (10)$$

where ∇_x, ∇_ξ -Hamilton operators with respect to $\vec{x}$ and $\vec{\xi}$ coordinates, respectively.

We introduce the operator of the mean value $\langle\cdot\rangle$ over the PC V_ξ for quasiperiodic in local $\vec{\xi}\in V_\xi$ coordinates of the functions f :

$$\langle f\rangle=\frac{1}{\varphi_p}\int_{V_{\xi p}} f\,dV \qquad (11)$$

where $\varphi_p=\int_{V_{\xi p}} dV$ is the volume fraction of the liquid in the PC (porosity), and $|V_\xi|$ is the volume of the PC V_ξ.

However, due to periodicity of the structure, the solution may be represented in the form of asymptotic expansion in terms of parameter κ [6,7]:

$$\vec{v}=\vec{v}^{(0)}(\vec{x},\vec{\xi})+\kappa\vec{v}^{(1)}(\vec{x},\vec{\xi})+\kappa^{2}\vec{v}^{(2)}(\vec{x},\vec{\xi})+\cdots \qquad (12)$$

$$p=p^{(0)}(\vec{x})+\kappa p^{(1)}(\vec{x},\vec{\xi})+\kappa^{2}p^{(2)}(\vec{x},\vec{\xi})+\cdots \qquad (13)$$

$$\mu=\mu^{(0)}(\vec{x},\vec{\xi})+\kappa\mu^{(1)}(\vec{x},\vec{\xi})+\kappa^{2}\mu^{(2)}(\vec{x},\vec{\xi})+\cdots \qquad (14)$$

Having substituted expansions (12)-(14) into Equations (1)-(5), then having collected in them terms at the same powers of κ and putting terms at the lowest powers of κ equal to zero, we obtain the local problems of the zero level 'over the periodicity cell':

$$\nabla_{\xi} \cdot \vec{v}^{(0)} = 0 \tag{15}$$

$$-\nabla_{\xi} p^{(1)} + \eta_0 \nabla_{\xi} \cdot \left(2\mu^{(0)} D^{(0)}\right) = \nabla_x p^{(0)} \tag{16}$$

$$D^{(0)} = \frac{1}{2}\left(\nabla_{\xi} \otimes \vec{v}^{(0)} + \nabla_{\xi} \otimes \vec{v}^{(0)T}\right) \tag{17}$$

$$\mu^{(0)} = s + (1-s)\left(1 + Cu^{0^2} Y^{(0)}\right)^{\frac{n-1}{2}}, \; Y^{(0)} = 2D^{(0)} \cdot\cdot D^{(0)} \tag{18}$$

$$\vec{v}^{(0)}\Big|_{\Sigma_{\xi sg}} = 0, \; \left\langle p^{(1)} \right\rangle = 0, \; [[\vec{v}^{(0)}]] = 0, \; [[p^{(1)}]] = 0 \tag{19}$$

where unknown $\vec{v}^{(0)}$ and $p^{(1)}$, the symbol $[[\cdot]]$ denotes the periodicity conditions. $\nabla_x p^{(0)}$ is considered as "input data". And $\mathrm{Re} = \kappa^0 \mathrm{Re}_0$, $Eu = \kappa^2 Eu_0$ and $Cu = \kappa^0 Cu_0$. Then $\eta_0 = \dfrac{1}{\mathrm{Re}^0\, Eu^0}$.

5. Results of Computational modeling

Next, the finite element method will be used to solve the local problems (16)-(19). We only consider finite elements that meet LBB conditions [8]. Here, the chemical industry raw material benzene is taken as an example for numerical study, so $\eta_0 = 0.0652$.

Tab. 1. Results for non-Newtonian fluids $n = 0.25$ at $\nabla_x p^{(0)} = (1,0,0)^T$.

Parameter	Value	Parameter	Value
Minimum pressure P	-0.30925494	Maximum pressure P	0.30260100
Velocity V_1	0.12414077	Velocity V_2	$9.06195241 \times 10^{-5}$
Velocity V_3	$4.74660504 \times 10^{-6}$	Viscosity $\left\langle \mu^{(0)} \right\rangle$	0.69590102
Porosity	0.62303864	Permeability	0.12414077

IOP Conf. Series: Materials Science and Engineering **683** (2019) 012008 doi:10.1088/1757-899X/683/1/012008

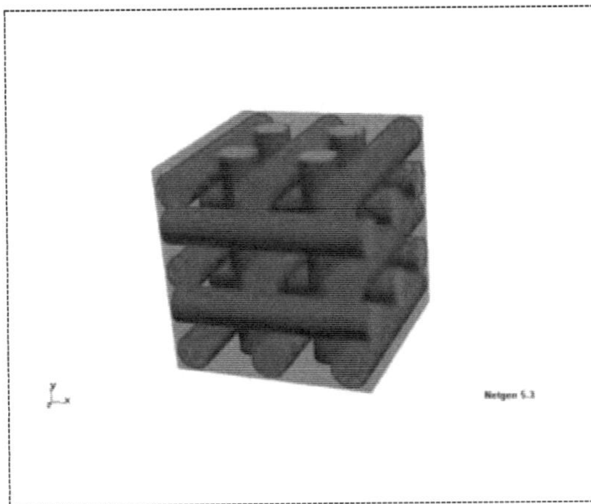

Fig. 1. V -periodic porous structure;.

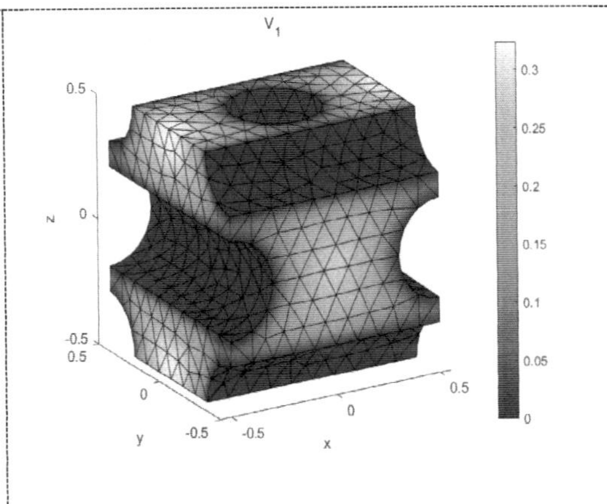

Fig. 2. Distribution of the velocity V_1.

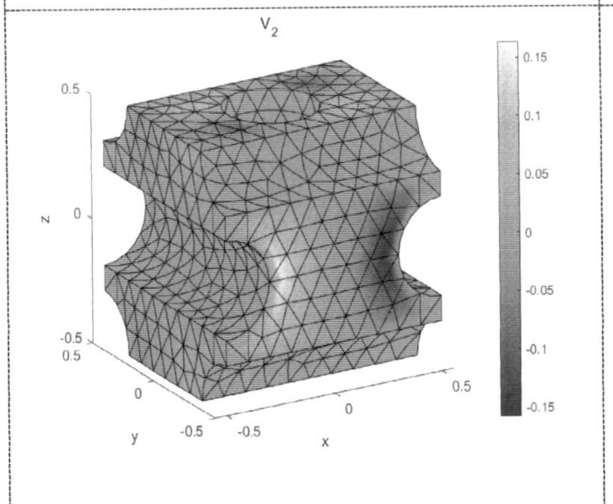

Fig. 3. Distribution of the velocity V_2.

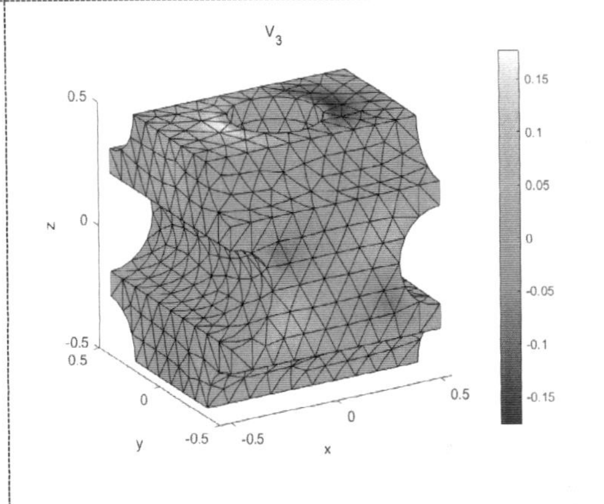

Fig. 4. Distribution of the velocity V_3.

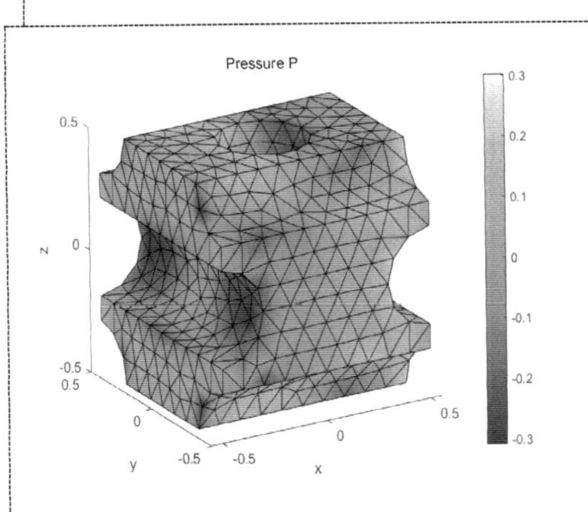

Fig. 5. Distribution of the pressure P.

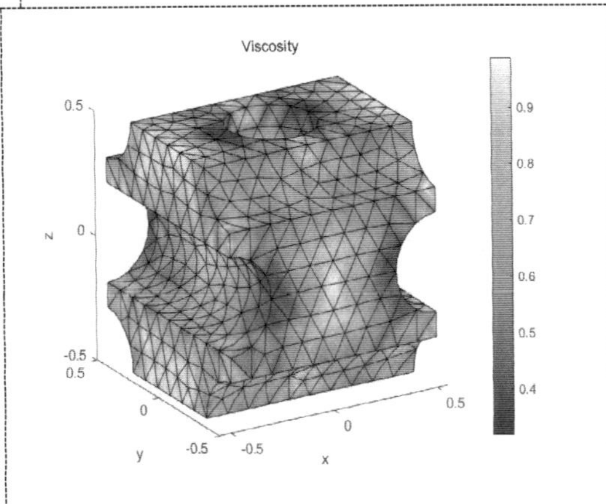

Fig. 6. Distribution of the viscosity μ.

6.Conclusions

Multi-scale modeling of non-Newtonian viscosity resin flow in a composite structure using an asymptotic homogenization method was developed. And the finite element method is used to calculate the local problem for the non-Newtonian viscosity resin flow in a composite microstructure.

Acknowledgments

The research was performed at Bauman Moscow State Technical University with the financial support of the Ministry of Education and Science of the Russian Federation under the Federal Target Program "Research and development on priority directions of scientific and technological complex of Russia for 2014-2020". Agreement # 14.577.21.0247 (unique identifier RFMEFI57717X0247).

References

[1] Bird R.B, Armstrong R C and Hassager O 1987 *Dynamics of polymeric liquids. Vol. 1: Fluid mechanics*. ed John Wiley 649

[2] Idris Z, Orgéas L, Geindreau C 2004et al. Microstructural effects on the flow law of power- law fluids through fibrous media. *Modelling and Simulation in Materials Science and Engineering* **12** pp. 995.

[3] Dimitrienko Yu I and Bogdanov I O 2018 Two-scale modeling of spatial flows of gas and weakly compressible liquid in porous composite structures. *IOP Journal of Physics: Conference Series* **1141** 012099

[4] Dimitrienko Yu I 2010 *Nonlinear continuum mechanics and large inelastic deformations*. Springer 747

[5] Dimitrienko Yu I 2002 *Tensor analysis and nonlinear tensor functions*. Springer 662

[6] Dimitrienko Yu I 2015 *Thermomechanics of composites under high temperatures*. Springer 434

[7] Dimitrienko Yu I, Dimitrienko I D 2013 Simulation of local transfer in periodic porous media. *European Journal of Mechanics/B-Fluids* **37** pp. 174-179.

[8] Zienkiewicz O C , Taylor R L., Zhu J Z 2005 *The Finite Element Method: Its Basis and Fundamentals: Its Basis and Fundamentals* Elsevier 733

IOP Conf. Series: Materials Science and Engineering **683** (2019) 012010 doi:10.1088/1757-899X/683/1/012010

Modeling of thermal stresses in inorganic matrix composite plates based on the asymptotic theory

Yu I Dimitrienko[1], S V Sborschikov[1], E S Egoleva[1], D O Yakovlev[1]

[1] Bauman Moscow State Technical University, Moscow, Russia

E-mail: dimit.bmstu@gmail.com

Abstract. A mathematical model has been developed for calculating a temperature non-stationary field and thermal stresses in thin multilayer plates of fabric composite materials on an inorganic matrix based on an alum-chromophosphate binder. The model allows to take into account changes in the microstructure in the matrix and composite fibers during high-temperature heating, taking into account the physical and chemical transformations occurring in them, as well as to calculate thermal stresses in thin plates from the tissue inorganic composite taking into account the kinetics of changes in the characteristics of composites during heating. Methods of multiscale homogenization and finite element modeling of microstresses in periodic cells were used to calculate the characteristics of the inorganic composite, and the method of asymptotic averaging was used to calculate the stresses in the plates. Some results of numerical simulation of the change in stresses in the composite plate during heating are given

1. Introduction

Composite materials based on inorganic binders (phosphate, alumino-phosphate, chromo-phosphate, magnesium-phosphate, alum-chromophosphate (ACP)) are a promising class of structural materials for the creation of heat-loaded structures [1-3]. consisting of oxide ceramic-fibers, and ACP with the addition of dispersed ceramic fillers. At high temperatures up to 1500 ° C, complex phase transformations occur in the matrix and fibers of such composite materials, leading to non-monotonous irreversible changes in all the thermal mechanical properties of the material. A multiscale model of the internal structure of the alum-chromophosphate composite was proposed in [3,4]. The purpose of this paper is to construct a mathematical model for calculating thermal stresses in thin composite plates based on transformations based on a multiscale structural model.

2. The mathematical formulation of the problem of calculating the thermal stresses in a composite plate under unsteady heating

Consider for a multilayer plate of inorganic composite a 3-dimensional problem of the linear theory of thermoelasticity, which is written in a dimensionless form as follows [5]

$$\nabla_j \sigma_{ij} = 0; \qquad \frac{C}{\kappa^2}\partial_t \theta = -\nabla_i q_i;$$
$$\sigma_{ij} = C_{ijkl}(\varepsilon_{kl} - \varepsilon^T_{kl}); \qquad q_i = -\lambda_{ij} g_j; \qquad (1)$$

$$\varepsilon_{ij} = \frac{1}{2}\left(\nabla_j u_i + \nabla_i u_j\right); \qquad g_j = \nabla_j \theta;$$
$$\Sigma_{3\pm} : \sigma_{i3} = -\kappa^3 p_\pm \delta_{i3}, \qquad q_3 = \pm q_{e\pm};$$
$$\Sigma_T : u_i = u_{ei}, \qquad q_I n_I = 0;$$

In system (1), the following are indicated: $\nabla_j = \partial / \partial x_j$ - the differentiation operator by dimensionless Cartesian coordinates, $x_k = \tilde{x}_k / L$, L - plate length, h - its thickness, $\kappa = h/L$, $p_\pm$-, - pressure, $q_{0\pm}$ - heat flux on the outer and inner surfaces of the plate $\Sigma_{3\pm}$ (their equation has the form $x_3 = \pm\kappa/2$), u_{ei} - given components of the displacement vector on the ends of the plate Σ_T, the ends are assumed to be heat insulated. The indices: *i, j, k, l = 1,2,3*, and large: *I, J, K* ... - take the values 1,2. Equations (1) also denote: ε_{ij} - the components of the small strain tensor, $\varepsilon_{kl}^T = \alpha_{kl}\Delta\theta$ - the components of the thermal strain tensor, which are functions of the temperature difference $\Delta\theta = \theta - \theta_0$, where θ_0 is the initial temperature, α_{kl} are the components of the thermal expansion tensor. The components of the tensor of the modulus of elasticity C_{ijkl}, thermal conductivity λ_{ij}, thermal deformation ε_{kl}^T, as well as the mass heat capacity $C = \rho c / Fo_0$ are variables, they depend on the variable phase composition of the plate material.

The following assumptions are made in system (1): 1) the parameter κ is small $\kappa \ll 1$, 2) the pressure on the outer and inner surfaces of the plate has the third order of smallness $O(\kappa^3)$ (i.e. $\sigma_{33} = -\kappa^3 p_\pm$), 2) the heating time is not too large, in the sense that the Fourier criterion $Fo = \lambda_0 t_0 / \rho_0 c_0 L^2$ of the process heating has one order of smallness with κ^2 (i.e., - $Fo = \kappa^2 Fo_0$, where the Fo_0 number is of the order of 1: $Fo_0 = O(1)$); 3) pressure $p_\pm$ and heat flux $q_{0\pm}$ - vary little over distances of the order of h.

3. Multi-level model of inorganic composite material at high temperatures

Composite materials based on ACP according to the general methodology for modeling composites [6] will be considered as a multi-level structure consisting of 4 structural levels. The first level of this structure is formed by the periodicity cells N 1 (PC1), which consist of 2 elements: oxide fibers and the matrix ACP, φ_f, φ_m — their volume concentrations, $\varphi_f + \varphi_m = 1$. In oxide fibers, phase transformations occur during heating: amorphous ("a") and crystalline ("l"), φ_a, φ_l - their concentration, related by the ratio $\varphi_a + \varphi_l = 1$.

At the 2nd level, the matrix is considered as ACP (polymer) with a filler in the form of ceramic particles based on chromium and aluminum oxides. At the 3rd level, ACP is considered as PC3, containing two phases, each of which corresponds to the transformation of the polymer along the aluminum and chromium chain, respectively, we denote the concentrations of the aluminum and chromium chains of phase transformations: $\varphi_{Al}, \varphi_{Cr}$, and $\varphi_{Al} + \varphi_{Cr} = 1$. For the 4th level PC, we introduce the following notation: φ_2 - volume concentration of polymer P-phase, φ_3 - metaphosphate

M-phase, φ_4 - orthophosphate O-phase, φ_5 - ceramic C-phase, φ_6 - concentration of gas g-phase. These phase concentrations are related by $\varphi_2+\varphi_3+\varphi_4+\varphi_5+\varphi_6=1$

The system of equations of continuity for the 6th environment has the following form:

$$\rho_2\frac{\partial\varphi_2}{\partial t}=-J_2,\qquad \rho_3\frac{\partial\varphi_3}{\partial t}=-J_2(1-\Gamma_2)-J_3,\qquad \rho_4\frac{\partial\varphi_4}{\partial t}=J_3-J_4,$$

$$\rho_5\frac{\partial\varphi_5}{\partial t}=J_4(1-\Gamma_5),\qquad \rho_3\frac{\partial\rho_6\varphi_6}{\partial t}+\nabla\varphi_6\rho_6\vec{V}_6=J_4\Gamma_5, \tag{2}$$

where are marked: $J_i=J_i^0\varphi_i\exp(-E_{Ai}/R\theta(t))$ - mass velocities of phase transformations, Γ_i - gasification coefficients, ρ_i - phase density, E_{Ai} - activation energy, J_i^0 - pre-exponential factors characterizing the rate of phase transformations, R - gas constant, $\vec{V}_6$ - filtration vector of gas phases in pores.

4. A model for the elastic properties of the matrix under high temperature

The modulus of elasticity of the inorganic matrix is calculated by the mixture formula

$$E_m=E_{m\_Al}\varphi_{Al}+E_{m\_Cr}\varphi_{Cr} \tag{3}$$

where E_{m_Al} and E_{m_Cr} - the elastic moduli of the aluminum-containing and chromium-containing parts of the matrix

$$E_{m\_Al}=\left(\frac{(1-S_C)}{E_1}+\frac{(S_C-S_O)}{E_2}+\frac{(S_O-S_M)}{E_3}+\frac{(S_M-S_P)}{E_4}+\frac{S_P}{E_5}\right)^{-1}$$

$$\varphi_5=1-S_C^3,\ \varphi_4=S_C^3-S_O^3,\ \varphi_3=S_O^3-S_M^3,\ \varphi_2=S_M^3-S_P^3,\ \varphi_6=S_P^3 \tag{4}$$

and E_i - the moduli of elasticity of the matrix phases i = 1 ... .6. For the modulus of elasticity of the C phase at high temperatures, we adopt the following model: $E_C=E_C\left(1-a_\theta\left(((\theta-\theta_0)/\theta_0)^n\right)\right)$, where n,a_θ are the constants.

5. Model changes in phase composition and elastic properties of oxide fibers under high temperatures

The change in the volume concentration of the "amorphous" phase of quartz fibers is described by the equations:

$$\rho_a\frac{\partial\varphi_a}{\partial t}=-J_a,\qquad J_a=J_a^0\varphi_a\exp\left(-\frac{E_{Aa}}{R\theta(t)}\right) \tag{5}$$

where: ρ_a is the density of the amorphous phase, E_{Aa} is the activation energy of the amorphous phase, J_a^0 is the preexponential factor.

The change in the elastic modulus E_f of oxide fibers during heating is determined by two factors:
1) a change in the elastic properties of the fiber in the amorphous state at relatively low temperatures;
2) physical and chemical processes of crystallization at high temperatures. The first factor causes a

reversible change in the elastic modulus E_f of the fibers, the second leads to irreversible changes after cooling.

$$E_f = \left(\frac{1-S_l}{E_1} + \frac{S_l}{E_2}\right)^{-1}, \quad E_1 = E_l, \quad E_2 = (1-S_l)E_l + S_l E_a, \quad \varphi_l = (1-S_l^3); \qquad \varphi_l = S_l^3 \tag{6}$$

here E_l, E_a are the elastic moduli of the fiber phases.

To calculate the effective elastic characteristics of PC on the 1st and 2nd levels, a finite-element method for solving problems on periodicity cells was used [7, 8].

6. Effective elastic characteristics of the textile composite

Modern reinforcing filaments consist of a large number of fibers, which are also interconnected by a matrix. To calculate the components of the tensors of the elastic moduli of the threads $C_{ijkl}^{(\alpha)}$, we use the model [9]. To determine the components of the tensor of effective elastic moduli of a composite C_{ijpq}, we use the solution of a series of local problems L_{pq} on a periodicity cell. These modules $C_{ijpq}(E_f, E_m)$ depend on the elastic moduli of the matrix and fibers, taking into account (3) and (6), this dependence is possible as a function of the phase concentrations $C_{ijpq}(\varphi_i, E_i)$.

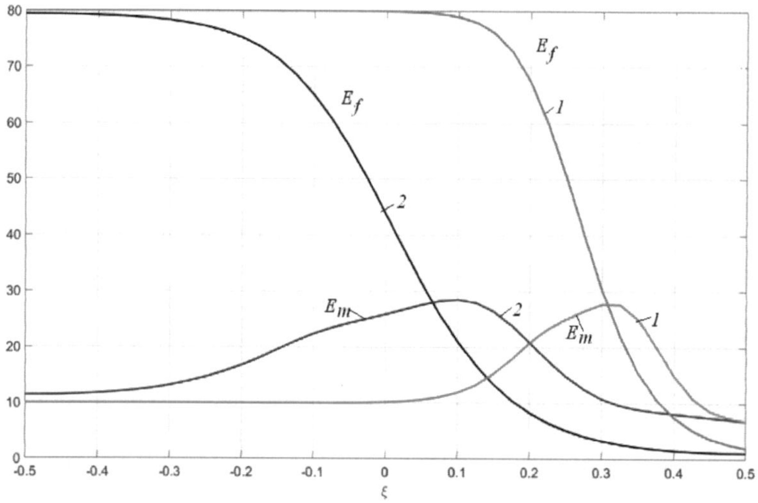

Figure 1. Distribution of the elastic modulus of the matrix and fiber (dimensionless values) across the plate thickness (in cross section x1 = 0.25) of the inorganic composite for two points in time t_2 and t_4 of non-stationary heating

7. Method for Solving Problem (1)

To solve problem (1), the method of asymptotic expansions was used, with which explicit analytical formulas were found for the stresses in an inorganic composite plate under bending pressure and uneven heating.

$$\sigma_{IJ} = \frac{C_{IJ11}^{(0)} \Delta\tilde{p}}{24\kappa^2 D_{11}} x(x-1),$$

$$\sigma_{I3} = \frac{\Delta\tilde{p}}{\kappa D_{11}}(x-1/2)\left\{\xi C^{(0)}_{I111}\right\}_{\xi}, \qquad \{f\}_{\xi} = \int_{-0.5}^{\xi} (<f>-f)d\xi \tag{7}$$

$$\sigma_{33} = -(\tilde{p}_{-} + \Delta\tilde{p}(\xi+0.5) - \frac{\Delta\tilde{p}}{D_{11}}\left\{\left\{\xi C^{(0)}_{1111}\right\}_{\xi}\right\}_{\xi}, \qquad <f> = \int_{-0.5}^{0.5} fd\xi$$

Here $D_{11} = <\xi^2 C^{(0)}_{1111}>$ is the flexural rigidity of the plate, and also denoted by: $C^{(0)}_{IJ11} = C_{IJ11} - C_{IJ33}C^{-1}_{3333}C_{3311}$, $\Delta\tilde{p} = \kappa^3\Delta p$, $\tilde{p}_{-} = \kappa^3 p_{-}$, $\Delta p = p_{+} - p_{-}$, a $\xi = x_3/\kappa$.

Figure 1 shows the graphs of the distribution of the dependence of the elastic modulus E_f, E_m of the fibers and the matrix in the composite across the thickness of the plate with uneven heating.
A feature of the elastic modulus of an ACP matrix is the increase in its values in the temperature range from 600 to 1100 ° C, which is associated with higher values of the elastic moduli of the meta- and orthophosphate phases. Figure 2 shows the distributions of the elastic moduli of a fabric composite based on glass fibers and AHFS over the plate thickness during heating. Figure 3 shows the distribution of flexural stress calculated from (7) for several moments of non-stationary heating.

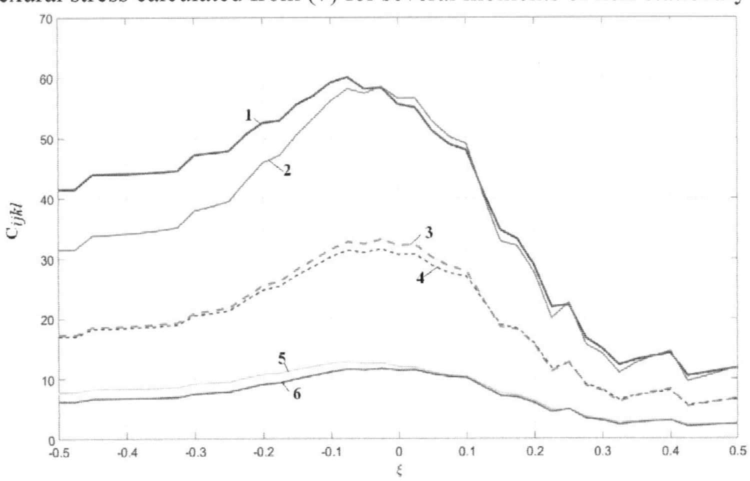

Figure 2. Distribution of the components Cijkl of the elastic modulus tensor (dimensionless values) over the plate thickness in the section $x_1 = 0.25$ for the time t_4 (1-C_{1111}, 2-C_{2222}, 3- C_{1122}, 4-C_{1133}, 5-C_{2323}, 6-C_{1212}).

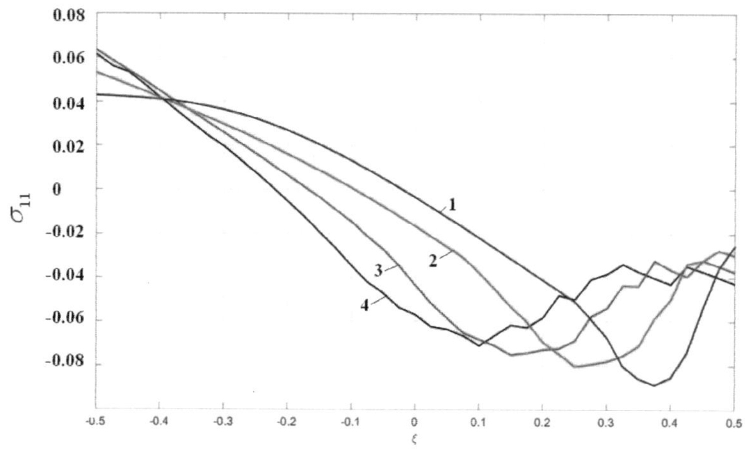

Figure 3 . Flexural stress distribution σ_{11} (dimensionless value) across the plate thickness in the cross section of the plate with coordinate $x_1 = 0.25$ for different moments of the dimensionless time t_i = 1,2,3 and 4.

8. Conclusions

The developed mathematical model makes it possible to predict the change in time of the stresses in a thin plate of inorganic composites during unsteady uneven heating to high temperatures, taking into account the physical and chemical transformations occurring in them. The results of numerical simulation have shown that due to the complex non-monotonic nature of the temperature dependence of the elastic properties of the inorganic matrix, the stresses in the plate with uneven heating also have a complex, non-monotonic change over time.

Acknowledgments

The work was carried out with support of state task of Ministry for Science and Education № 9.5387.2017/БЧ

References

[1] Abzgil'din F Yu, Tesvyatskiy S G 1980 *Asbo-phosphate materials* ed Naukova dumka (Kiev) 99
[2] Kopeykin V A 1974 *Technology and properties of phosphate materials* ed Stroyizdat (Moscow) 224
[3] Dimitrienko Yu I, Sborschikov S V, Egoleva E S, Matveeva A A 2013 Modeling of thermo-elastic properties of composites with aluminochromic phosphate *matrices Science and Education. Electronic scientific and technical Journal* **11** pp.497-518
[4] Dimitrienko Yu I, Sborshikov S V, Gubareva E A, Egoleva E S, Matveeva A A 2015 Multiscale Modeling of High Temperature elastic-strength properties of inorganic matrix Composites *Mathematical Modeling* **27** pp. 3-20.
[5] Dimitrienko Yu I, Yakovlev D O 2015 The Asymptotic Theory of Thermoelasticity of Multilayer Composite Plates. *Composites: Mechanics, Applications. An International Journal* **6** pp.13-51.
[6] Dimitrienko Yu I 2015 *Thermomechanics of Composites Structures under High Temperatures* ed Springer 367
[7] Dimitrienko Yu I, Gubareva E A , Sborshikov S V 2015 Numerical simulation of composite material thermal expansion by homogenization method *Engineering Journal: Science and Innovation* **12**
[8] Dimitrienko Yu I, Sborschikov S V, Belenovskaya Yu V, Aniskovich V A, Perevislov S N 2013

Modeling of Microdestruction and Strength of ceramic Composites on the base of reaction bonded SiC *Science and Education. Electronic scientific and technical Journal.* **11** pp.475-496.

[9] Dimitrienko Yu I, Yurin Yu V 2017 Finite Element Modeling of Damage and Durability of Composite Structures with Local Delaminations *Mathematical Modeling and Computational Methods* **3** pp 49–70.

Modelling of filtration of liquid binder in the composite textile structures under RTM processes

Yu I Dimitrienko[1] and I O Bogdanov[1]

[1]Department of Computational Mathematics and Mathematical Physics, Bauman Moscow State Technical University, 5 Baumanskaya 2-ya, Moscow, 105005, Russia

E-mail: dimit.bmstu@gmail.com, biofamily_7394@mail.ru

Abstract. The paper deals with three-scale modelling of the filtration of a weakly compressible liquid binder in the tissue material of the filler in the production of composite materials based on the RTM method. The technique is based on the application of the asymptotic homogenization method for the Navier-Stokes system of equations. Local filtration problems for the threads level and for the monofilaments level are obtained.

1. Introduction

One of the most effective methods for manufacturing products from composite materials is the technology of impregnating fabric reinforcing filler with a liquid binder in a mould (RTM, Resin Transfer Molding). The quality of composite structures obtained using this method largely depends on the parameters of the manufacturing technology. In this regard, it is extremely important to adequately model the flow of liquid binder in the porous space of a tissue filler, which has a complex microstructure of pores formed by interlacing the fibres and their bundles – threads.

It should be noted that in the overwhelming majority of publications, the study of the flow of liquids in porous structures is studied in the framework of the phenomenological theory of filtration, which is based on the Darcy law. In this case, the permeability coefficients of the porous medium are determined either experimentally or by using various empirical and approximate relations for describing local filtration processes. In this case, rather rough estimates of real processes occurring inside pores with complex geometry are obtained, which leads to large deviations in determining the permeability. Therefore, an important part of the filtration study is the analysis of local processes of spatial fluid flow in a single pore using direct analysis of the Navier-Stokes equations. At the same time, a direct solution of the Navier-Stokes equations is not possible due to the complexity of the computational domain and the need to use extremely fine grids. Therefore, in this case, the asymptotic averaging method is used.

Questions of the study of the movement of liquids and gases in porous structures, in particular, in relation to RTM technology, are considered in many papers, for example, in [1–4]. However, it should be noted that the study of the behaviour of mediums within separate pores with a complex geometric shape and a multiscale structure has not been studied enough

2. Geometric models of computational domains.

We assume that the porous medium (the fabric material of the filler) has the property of periodicity, i.e. it is possible to select geometrically repeating regions, called periodicity cells. In addition, we will assume that the porous medium has several spatial levels. The zero spatial level is, in fact, the porous medium itself, formed by interlacing the filaments (see Figure 1). The first spatial level is the level of periodicity cells V_ξ^1 (hereinafter referred to as PC1) formed by fibre bundles (see Figure 2, a). Finally, the second spatial level is the level of the periodicity cells V_ξ^2 (hereinafter referred to as PC2) formed by monofilaments (see Figure 2, b).

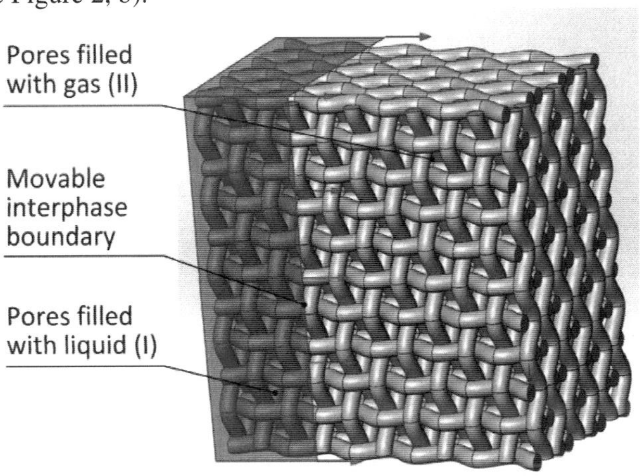

Figure 1. Model of fabric material formed by interlacing threads

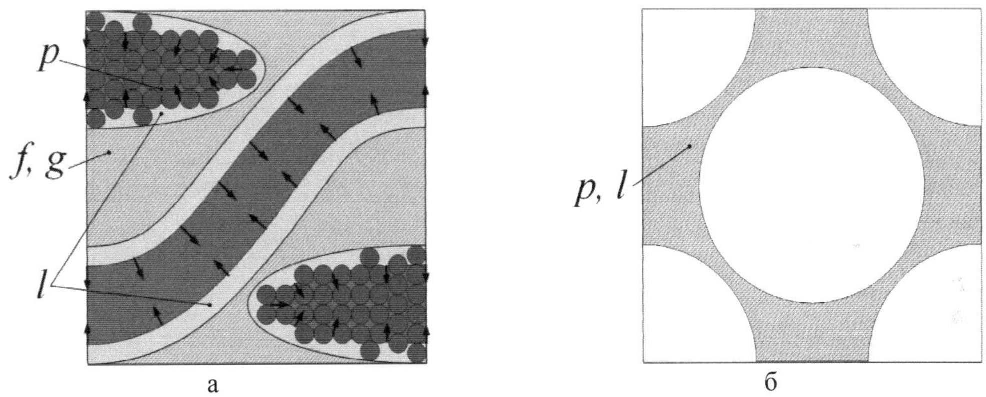

Figure 2. Patterns of periodicity cells:
a) the first spatial level (PC1); b) the second spatial level (PC2)

3. General formulation of the filtering problem.

The original exact equations of motion of a weakly compressible fluid have the form:

$$\frac{\partial \rho_i}{\partial t} + \nabla \cdot (\rho_i \mathbf{v}_i) = 0,$$
$$\kappa\left(\frac{\partial \rho_i \mathbf{v}_i}{\partial t} + \nabla \cdot (\rho_i \mathbf{v}_i \otimes \mathbf{v}_i)\right) = \nabla \cdot \mathbf{T}_i. \quad (1)$$

Here $i = l, f, g, p$, where l – fluid in threads from the area I (see Figure 1); f – fluid between threads from the area I; g – gas in area II; p – gas contained in threads from region I. κ – small task parameter.

The equation of state for the liquid phase is:

$$p_l = p_{0l} + K\left(\frac{\rho_l}{\rho_{0l}} - 1\right), \tag{2}$$

where ρ_{0l} – fluid density at the initial time, $\mathrm{kg/m^3}$; p_{0l} – initial pressure in the fluid, Pa; K – bulk compression module, Pa. For the case of gas, the Mendeleev-Clapeyron equation of state is used:

$$p_g = \rho_g R\theta_0, \tag{3}$$

where R – gas constant, $\mathrm{J/(kg\cdot K)}$; $\theta_0 = const$ – constant temperature, K.

We supplement the system with defining relations. For an isotropic linear-viscous fluid (gas), the stress tensor has the form:

$$\mathbf{T}_i = -p_i\mathbf{E} + \kappa\boldsymbol{\tau}_i(\mathbf{D}_i) \tag{4}$$

where p_i – pressure, Pa; $\boldsymbol{\tau}_i(\mathbf{D}_i) = \lambda_i(\nabla\cdot\mathbf{v}_i)\mathbf{E} + 2\mu_i\mathbf{D}_i$ – viscous stress tensor; λ_i, μ_i – viscosity coefficients, $\mathrm{Pa\cdot s}$; $\mathbf{D}_i = \frac{1}{2}\left(\nabla\otimes\mathbf{v}_i + \nabla\otimes\mathbf{v}_i^T\right)$ – strain rate tensor, $\mathrm{s^{-1}}$; $\mathbf{E}$ – metric tensor.

At the interface Σ_{fg} between liquid and gas, the following conditions are set:

$$\left(-p_f\mathbf{E} + \boldsymbol{\tau}_f\right)\cdot\mathbf{n} = \left(-p_g\mathbf{E} + \boldsymbol{\tau}_g\right)\cdot\mathbf{n},\ \mathbf{v}_f = \mathbf{v}_g. \tag{5}$$

Adhesion conditions are set at the interface between the liquid and the gas and the monofilament:

$$\mathbf{v}_i = \mathbf{0}. \tag{6}$$

The initial formulation (1) should also be supplemented with conditions on the external surfaces of the porous medium, as well as with initial conditions.

4. Statement of local problems for the first and second spatial levels.

Let $\hat{l}_1$ – the linear size of the periodicity cell V_ξ^1 of the medium and $\hat{l}_2$ – the linear size of the periodic cell V_ξ^2 of the medium. Let us introduce small parameters $\kappa_1 = \hat{l}_1/\hat{L} << 1$, $\kappa_2 = \hat{l}_2/\hat{l}_1 << 1$ and two types of dimensionless coordinates: global (slow) $\tilde{\mathbf{x}} = \mathbf{x}/\hat{L}$ (~ we omit) and local (fast) $\boldsymbol{\xi}_1 = \mathbf{x}/\kappa_1$ and $\boldsymbol{\xi}_2 = \boldsymbol{\xi}_1/\kappa_2$. Wherein $\kappa = \kappa_1\kappa_2$. Differentiation rules for new coordinates will take the form:

$$\nabla \to \nabla_x + \kappa_1^{-1}\nabla_{\xi_1},\ \nabla_{\xi_1} \to \nabla_{\xi_1} + \kappa_2^{-1}\nabla_{\xi_2} \tag{7}$$

Taking into account (7) for (1), we obtain:

$$\begin{gathered}\frac{\partial\rho_i}{\partial t} + \nabla_x\cdot(\rho_i\mathbf{v}_i) + \kappa_1^{-1}\nabla_{\xi_1}\cdot(\rho_i\mathbf{v}_i) = 0,\\ -\nabla_x p_i - \kappa_1^{-1}\nabla_{\xi_1}p_i + \kappa\left(\nabla_x\cdot\boldsymbol{\tau}_i + \kappa_1^{-1}\nabla_{\xi_1}\cdot\boldsymbol{\tau}_i\right) = 0,\end{gathered} \tag{8}$$

In accordance with the general concept of the asymptotic homogenization method [5–8], the solution of the problem (8) is sought in the form of asymptotic expansions in powers of a small parameter κ_1 in regions I and II:

IOP Conf. Series: Materials Science and Engineering **683** (2019) 012011 doi:10.1088/1757-899X/683/1/012011

$$\begin{aligned}
\rho_i(\mathbf{x},\xi_1,t) &= \rho_i^{(0)}(\mathbf{x},\xi_1,t)+\kappa_1\rho_i^{(1)}(\mathbf{x},\xi_1,t)+O(\kappa_1^2),\\
p_i(\mathbf{x},\xi_1,t) &= p_i^{(0)}(\mathbf{x},\xi_1,t)+\kappa_1 p_i^{(1)}(\mathbf{x},\xi_1,t)+O(\kappa_1^2),\\
\mathbf{v}_i(\mathbf{x},\xi_1,t) &= \mathbf{v}_i^{(0)}(\mathbf{x},\xi_1,t)+\kappa_1\mathbf{v}_i^{(1)}(\mathbf{x},\xi_1,t)+O(\kappa_1^2),\\
\boldsymbol{\tau}_i(\mathbf{x},\xi_1,t) &= \boldsymbol{\tau}_i^{(0)}(\mathbf{x},\xi_1,t)+\kappa_1\boldsymbol{\tau}_i^{(1)}(\mathbf{x},\xi_1,t)+O(\kappa_1^2)
\end{aligned} \tag{9}$$

After substituting (9) into (8) and grouping the terms with the same powers of the parameter κ_1, the desired local problem can be obtained for the first spatial level (PC1):

$$\begin{cases}
\nabla_{\xi_1}\cdot\mathbf{v}_i^{(0)}=0,\\
-\nabla_x p_i^{(0)}-\nabla_{\xi_1}p_i^{(1)}+\kappa_2\nabla_{\xi_1}\boldsymbol{\tau}_i^{(0)}=0,\ \xi_1\in V_\xi^1,\\
i=f,l,g,p.
\end{cases} \tag{10}$$

Further, taking into account the second expression in (7), we obtain from (10):

$$\begin{cases}
\nabla_{\xi_1}\cdot\mathbf{v}_i^{(0)}+\kappa_2^{-1}\nabla_{\xi_2}\cdot\mathbf{v}_i^{(0)}=0,\\
-\nabla_x p_i^{(0)}-\nabla_{\xi_1}p_i^{(1)}-\kappa_2^{-1}\nabla_{\xi_2}p_i^{(1)}+\kappa_2\left(\nabla_{\xi_1}\boldsymbol{\tau}_i^{(0)}+\kappa_2^{-1}\nabla_{\xi_2}\boldsymbol{\tau}_i^{(0)}\right)=0,\ \xi_2\in V_\xi^2,\\
i=l,p.
\end{cases} \tag{11}$$

The solution (11) will also be sought in the form of asymptotic expansions:

$$\begin{aligned}
\mathbf{v}_i^{(0)}(\mathbf{x},\xi_1,\xi_2) &= \mathbf{v}_i^{(0)(0)}(\mathbf{x},\xi_1,\xi_2)+\kappa_2\mathbf{v}_i^{(0)(1)}(\mathbf{x},\xi_1,\xi_2)+O(\kappa_2^2),\\
p_i^{(1)}(\mathbf{x},\xi_1,\xi_2) &= p_i^{(1)(0)}(\mathbf{x},\xi_1,\xi_2)+\kappa_2 p_i^{(1)(1)}(\mathbf{x},\xi_1,\xi_2)+O(\kappa_2^2).
\end{aligned} \tag{12}$$

Then, similarly can obtain the desired local problem second spatial level (PC2):

$$\begin{cases}
\nabla_{\xi_2}\cdot\mathbf{v}_i^{(0)(0)}=0,\\
-\nabla_x p_i^{(0)}-\nabla_{\xi_1}p_i^{(1)(0)}-\nabla_{\xi_2}p_i^{(1)(1)}+\nabla_{\xi_2}\boldsymbol{\tau}_i^{(0)(0)}=0,\ \xi_2\in V_\xi^2,\\
p_i^{(1)(1)}=A_i\rho_i^{(1)(1)},\\
i=l,p.
\end{cases} \tag{13}$$

Note that the following approximation for the continuity equation in the problem on the PC2 is:

$$\nabla_{\xi_1}\cdot\mathbf{v}_i^{(0)(0)}+\nabla_{\xi_2}\cdot\mathbf{v}_i^{(0)(1)}=0. \tag{14}$$

We will seek a solution of the problem (13) in the form:

$$\mathbf{v}_i^{(0)(0)}=-\mathbf{V}_i^{(0)(0)}(\xi_1,\xi_2)\cdot\left(\nabla_x p_i^{(0)}+\nabla_{\xi_1}p_i^{(1)(0)}\right). \tag{15}$$

We introduce the averaging operator over the region $V_i^2\subset V_\xi^2$ occupied by the i-th phase in the region V_ξ^2:

$$\langle\cdot\rangle_i^2=\frac{1}{|V_i^2|}\int_{V_i^2}(\cdot)\,dV,\ i=l,p. \tag{16}$$

Averaging using (16) equations (14) and (15), we obtain the filtration equations at the second spatial level:

$$\nabla_{\xi_1} \cdot \left\langle \mathbf{v}_i^{(0)(0)} \right\rangle_i^2 = 0,$$
$$\left\langle \mathbf{v}_i^{(0)(0)} \right\rangle_i^2 = -\mathbf{K}_{2i} \cdot \left(\nabla_x p_i^{(0)} + \nabla_{\xi_1} p_i^{(1)(0)} \right), \tag{17}$$
$$i = l, p.$$

To the problem (17) should add conditions on the interface Σ_{fl}:

$$\left\langle \mathbf{v}_l^{(0)(0)} \right\rangle_l^2 = \mathbf{v}_f^{(0)},\ p_l^{(1)} = p_f^{(1)}. \tag{18}$$

Note that the solution of the problem (17), in turn, can be represented in the form:

$$\left\langle \mathbf{v}_i^{(0)(0)} \right\rangle_i^2 = -\mathbf{V}_2^{(0)(1)}\left(\xi_1\right) \cdot \nabla_x p_i^{(0)},\ i = l, p. \tag{19}$$

and the solution of the problem (10) for f and g can be searched in the form:

$$\mathbf{v}_i^{(0)} = -\mathbf{V}_1^{(0)}\left(\xi_1\right) \cdot \nabla_x p_i^{(0)},\ i = f, g. \tag{20}$$

5. Results

In this work, we numerically simulated the flow of medium in porous composite structures using the finite element method [9–10]. Examples of the distribution of micro fields of pressure pulsations and components of the velocity vector, obtained by solving a local problem for a porous structure with a dimensionless filament radius of 0.125, are shown in Figures 3-4. The dimensionless permeability coefficients K_{11} and K_{33} for this structure turned out to be equal, respectively, $2{,}047 \cdot 10^{-3}$ and $1{,}385 \cdot 10^{-3}$ with porosity of 0.555.

Figure 3. Distribution of pressure $\tilde{P}^{(1)}$ in 1/8 of the tissue structure periodicity cell (PC1)

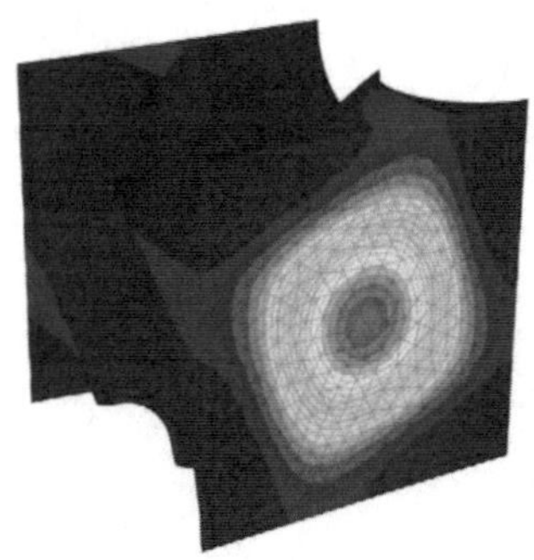

Figure 4. Distribution of the velocity component $\tilde{W}_1^{(1)}$ in 1/8 of the tissue structure periodicity cell (PC1)

6. Conclusions

The paper considers the method of three-scale modelling of filtration, based on the use of the asymptotic homogenization method. Statements of local filtration problems are obtained for two spatial levels: threads and monofilaments. A number of results are presented demonstrating the effectiveness of the considered algorithms.

Acknowledgments
The work was carried out with support of state task of Ministry for Science and Education of the RF № 9.5387.2017/БЧ

References

[1] Shargatov V A, Il'Ichev A T and Tsypkin G G 2015 Dynamics and stability of moving fronts of water evaporation in a porous medium *International J. of Heat and Mass Transfer* **83** 552-61

[2] Borodulin A S, Malysheva G V and Romanova I K 2015 Optimization of rheological properties of binders used in vacuum assisted resin transfer molding of fiberglass *Polymer Science, Series D. Glues and Sealing Materials* **8(4)** 300-3

[3] Golovatov D, Mikhaylov M and Bosov A 2016 Optimization of technological parameters of impregnation of load-bearing rod elements of reflector made of polymer composite materials by transfer molding method *Indian J. of Science and Technology* **9(46)** 11

[4] Loudad R, Saouab A, Beauchene P, Agogue R and Desjoyeaux B 2017 Numerical modeling of vacuum-assisted resin transfer molding using multilayer approach *J. of Composite Materials.* **51(24)** 3441-52.

[5] Hornung U 1997 *Homogenization and Porous Media* (New York: Springer-Verlag).

[6] Dimitrienko Yu I and Yakovlev D O 2014 Asymptotic theory of thermoelasticity of multilayer composite plates *Mechanics of Composite Materials and Structures.* **20(2)** 260-82.

[7] Dimitrienko Yu I and Dimitrienko I D 2013 Simulation of local transfer in periodic porous media *European J. of Mechanics – B-Fluids.* **1** 174-9.

[8] Dimitrienko Yu I and Bogdanov I O 2018 Two-scale modeling of spatial flows of gas and weakly compressible liquid in porous composite structures *J. of Physics: Conference Series.* **1141(1)** 012099

[9] Dimitrienko Yu I and Bogdanov I O 2016 Finite-element method for three-dimensional problems of elastic structures buckling theory *Herald of the Bauman Moscow State Technical University. Series Natural Sciences.* **6** 73-92.

[10] Zienkiewicz O C, Taylor R L and Zhu J Z 2013 *The Finite Element Method: Its Basis and Fundamentals: Seventh Edition* (Butterworth-Heinemann).

IOP Conf. Series: Materials Science and Engineering **683** (2019) 012012 doi:10.1088/1757-899X/683/1/012012

Distribution of electromagnetic fields in heterogeneous composite structures when exposed to direct current.

Yu I Dimitrienko[1], I K Krasnov1, K M Zubarev1

[1] Bauman Moscow State Technical University, Moscow, 105005, Russia

E-mail: dimit.bmstu@gmail.com, igorkrsnv@yandex.ru, zubarev.bmstu@mail.ru.

Abstract. In this paper, we study the electromagnetic fields in composite structures that occur when exposed to an external source of direct current, located on the surface of the composite. The article proposes an algorithm for solving the direct problem, based on the application of the finite element method, and a method for solving the inverse problem, which is described as an optimization problem. To search for the extremum of the target functional, an empirically developed method is used. The task is considered in a quasistatic approximation. The developed software was tested on a three-layer composite. The developed algorithms made it possible to solve a given problem with high accuracy.

Introduction

The solution of inverse problems and the study of electromagnetic fields are described in [1-14]. The purpose of the research in this paper is to determine the geometric and physical parameters of heterogeneous composite structures by comparing the measurement data of electromagnetic fields with the corresponding results of their numerical simulation. In the case of the application of electromagnetic fields, the problem is reduced to the determination of the coefficients of Maxwell's equations from some indirect data and refers to the classical inverse problems studied in numerous works, for example [8], [10], [13].

Now, the solution of inverse problems of restoring the parameters of multilayer structures is based on optimization principles. The essence of these methods is to minimize the objective functional of the difference between the measured values and the values obtained in the simulation in the presence of linear or non-linear constraints on the theoretical values of the parameters of the composite structure. Consequently, the search for a solution to the inverse problem is reduced to the multiple solution of the direct problem, which requires a lot of time. Also the complexity is added by a large number of variables and a multi-character character.

In this paper, we consider the problem of the distribution of the electric field to solve direct and inverse problems in composites characterized by the location of sources and measuring devices on the surface of the composite. Electrostatic or electromagnetic fields are investigated in various frequency ranges. The simulated fields are described by the Maxwell system of differential equations.

Solved problems for heterogeneous composite structures are considered in the 3-dimensional approximation. The purpose of research in this area is to determine the geometric characteristics and conductive properties of composites in their study.

The proposed computational algorithms are based on the approximation of multidimensional mixed boundary value problems by the finite element method (FEM) of high accuracy, on fast iterative processes of solving the generated systems of linear algebraic equations (SLAE) with sparse matrices of large order, and also on effective optimization methods. To solve the obtained SLAU, we used the biconjugate gradient method.

Mathematical formulation of a direct problem.

At points A and B, two electrodes are located on the surface of the composite. A current is supplied to these electrodes with opposite values I_k and $-I_k$, respectively. It is required to find the distribution of the electrostatic potential in a given region.

Electromagnetic fields described by the system of Maxwell's equations are investigated in various applications in papers [6], [7], [13], [14]. We study the fields constant in time. In the specified fields there are no volumetric sources, therefore, the electrostatic potential in the computational domain Ω is determined by the equation

$$L\varphi \equiv \nabla\sigma\,\nabla\sigma\varphi(\overset{\mathbf{1}}{x}) = 0, \quad \overset{\mathbf{1}}{x} \in \Omega \tag{1}$$

here the differential operator L is described in cylindrical coordinates for axisymmetric problems and in Cartesian coordinates for general three-dimensional statements. In the case of anisotropic media with different conductivities σ_x, σ_y, σ_z in the directions of the Cartesian axes x, y, z, we have

$$L\varphi = \frac{\partial}{\partial x}\sigma_x\frac{\partial\varphi}{\partial x} + \frac{\partial}{\partial y}\sigma_y\frac{\partial\varphi}{\partial y} + \frac{\partial}{\partial z}\sigma_z\frac{\partial\varphi}{\partial z} = 0 \tag{2}$$

At different parts of the outer boundary of the computational domain, for physical reasons, boundary conditions of the first or second kind are set:

$$\varphi|_{\bar{\Gamma}_i} = \varphi|_{\overset{+}{\Gamma}_i}, \quad \left.\frac{\partial\varphi}{\partial n}\right|_{\Gamma_{11}} = 0, \quad \sigma\left.\frac{\partial\varphi}{\partial n}\right|_A = I_k, \quad \sigma\left.\frac{\partial\varphi}{\partial n}\right|_B = -I_k, \quad \left.\frac{\partial\varphi}{\partial n}\right|_{\bar{\Gamma}_i} = \left.\frac{\partial\varphi}{\partial n}\right|_{\overset{+}{\Gamma}_i}. \tag{3}$$

We use the finite element method to solve this problem. We also use the Bubnov-Galerkin method. Then equation (2) can be rewritten as follows.

$$\int_\Omega [N]^T\left(\frac{\partial}{\partial x}\sigma_x\frac{\partial\varphi}{\partial x} + \frac{\partial}{\partial y}\sigma_y\frac{\partial\varphi}{\partial y} + \frac{\partial}{\partial z}\sigma_z\frac{\partial\varphi}{\partial z}\right)d\Omega = 0 \tag{4}$$

To begin with, we transform an equation containing only the first derivatives with respect to x, y and z. We do this for the x term using the first Green formula.

$$\frac{\partial}{\partial x}\left(\sigma_x[N]^T\frac{\partial\varphi}{\partial x}\right) = [N]^T\frac{\partial}{\partial x}\sigma_x\frac{\partial\varphi}{\partial x} + \frac{\partial[N]^T}{\partial x}\sigma_x\frac{\partial\varphi}{\partial x} \tag{5}$$

Using the expression, we can write

$$[N]^T\frac{\partial}{\partial x}\sigma_x\frac{\partial\varphi}{\partial x} = \frac{\partial}{\partial x}\left([N]^T\sigma_x\frac{\partial\varphi}{\partial x}\right) - \frac{\partial}{\partial x}[N]^T\sigma_x\frac{\partial\varphi}{\partial x} \tag{6}$$

Then the first term is transformed to the form

$$\int_\Omega [N]^T\frac{\partial}{\partial x}\sigma_x\frac{\partial\varphi}{\partial x}d\Omega = \int_\Omega \frac{\partial}{\partial x}\left([N]^T\sigma_x\frac{\partial\varphi}{\partial x}\right)d\Omega - \int_\Omega \frac{\partial[N]^T}{\partial x}\sigma_x\frac{\partial\varphi}{\partial x}d\Omega \tag{7}$$

Further, using the Ostragradsky-Gauss formula we get

$$\int_{\Omega} \frac{\partial}{\partial x}\left([N]^T \sigma_x \frac{\partial \varphi}{\partial x}\right) d\Omega = \int_{S} [N]^T \sigma_x \frac{\partial \varphi}{\partial x} l_x \, dS \tag{8}$$

Transforming in a similar way all the terms in (5) we get the following expression:

$$\int_{S} [N]^T \left(\sigma_x \frac{\partial \varphi}{\partial x} l_x + \sigma_y \frac{\partial \varphi}{\partial y} l_y + \sigma_z \frac{\partial \varphi}{\partial z} l_z\right) dS - \int_{\Omega} \frac{\partial [N]^T}{\partial x} \sigma_x \frac{\partial \varphi}{\partial x} + \frac{\partial [N]^T}{\partial y} \sigma_y \frac{\partial \varphi}{\partial y} + \frac{\partial [N]^T}{\partial z} \sigma_z \frac{\partial \varphi}{\partial z} d\Omega = 0$$

in which the integral over the surface is expressed in terms of the derivative along the normal $\partial \varphi / \partial n$

$$\int_{\Omega} \left(\frac{\partial [N]^T}{\partial x} \sigma_x \frac{\partial \varphi}{\partial x} + \frac{\partial [N]^T}{\partial y} \sigma_y \frac{\partial \varphi}{\partial y} + \frac{\partial [N]^T}{\partial z} \sigma_z \frac{\partial \varphi}{\partial z}\right) d\Omega - \int_{S} [N]^T \frac{\partial \varphi}{\partial n} dS = 0 \tag{9}$$

The first integral in the resulting expression contributes to the stiffness matrix, the second integral contributes to the column vector of the free terms. An unknown function φ is defined by the relation.

$$\varphi = [N][\Phi] \tag{10}$$

To solve the problem, we divide the original domain into tetrahedral finite elements.

Calculations of the direct problem

Expression (10) is a system of linear algebraic equations. To solve the resulting SLAE, we used several methods, namely the Seidel method, the Cholesky method, and the biconjugate gradient method.. As a result of the comparison, it was found that the results obtained using the method of biconjugate gradient method have a higher accuracy.

To test the developed algorithm, we chose a three-layer composite. The resistivity is respectively equal $\sigma_1 = 10$ Ом·м, $\sigma_2 = 0.1$ Ом·м, $\sigma_3 = 1000$ Ом·м. The thickness of the composite was 0.2 meters. The composite used for testing is shown in Fig. 1.

IOP Conf. Series: Materials Science and Engineering **683** (2019) 012012 doi:10.1088/1757-899X/683/1/012012

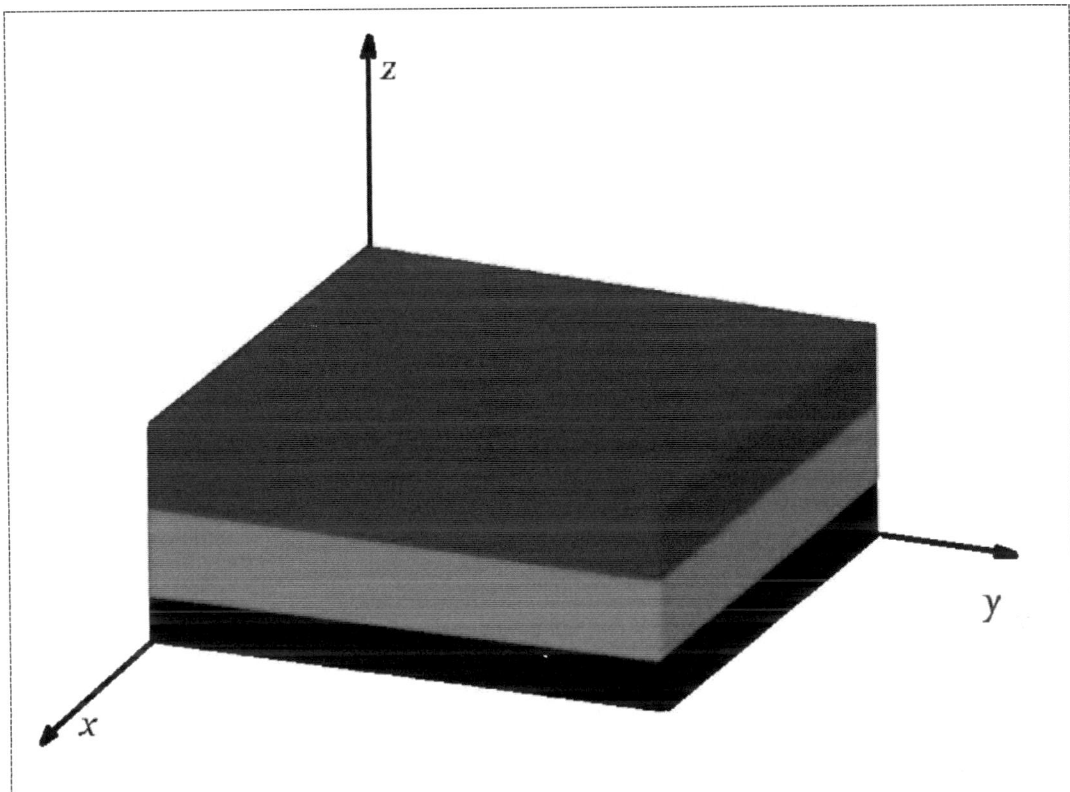

Figure 1. Heterogeneous composite

When solving the problem, the potential value was obtained which corresponds to the physical meaning. As an illustration of the solution, Fig. 2 shows the lines of the equal potential level. In practice, only the potential values on the surface of the composite can be measured. Clots on the surface of the composite correspond to the location of the electrodes.

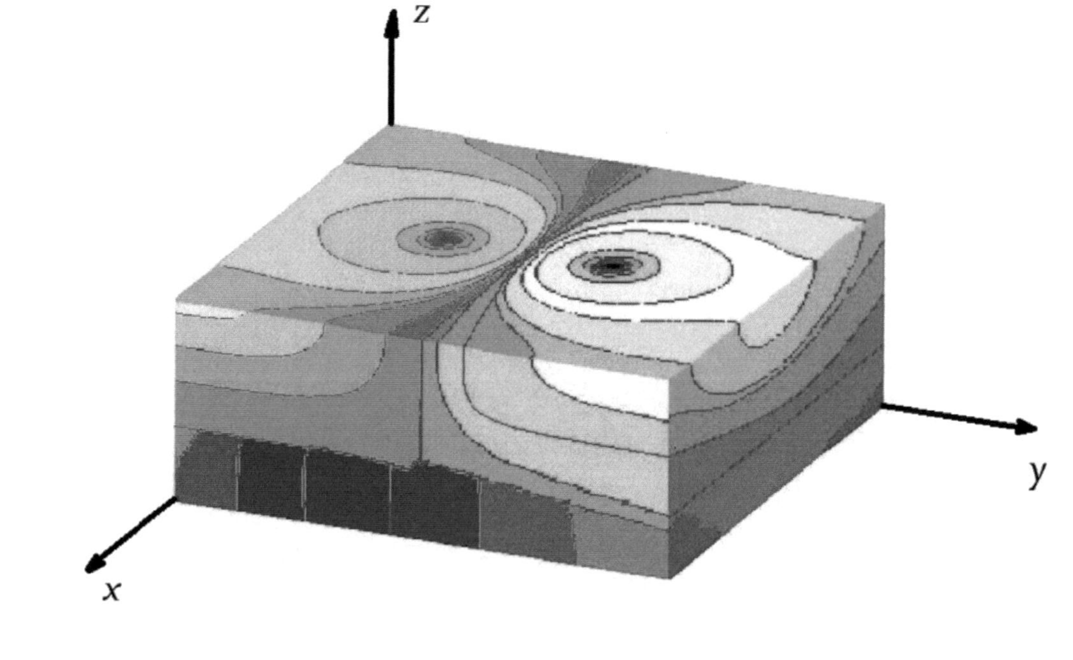

Figure 2. Lines of equal medium potential

Inverse task

The inverse problem can be formulated as an optimization problem with the objective functional

$$F(\sigma_1,...,\sigma_n) = \sqrt{\sum_{i=1}^{n} (\hat{\varphi}_Э - \varphi_T(\sigma_1,...,\sigma_n))^2} \to \min$$

Optimization methods for solving inverse problems are applied in various fields [1], [2], [3]. When solving the inverse problem, we used an algorithm developed experimentally. This solution method is described in [13]. The algorithm is as follows: for a start, suppose that the computational domain contains one layer with useful resistance σ, we obtain a solution for a given computational domain. Then the computational domain is randomly divided into two parts with resistivity σ_1, σ_2, and solve the inverse problem for the obtained computational domain. Then, in the same way, we continue to break the computational domain into new regions with specific resistivity σ_1, σ_2,...,σ_n. One of the possible schemes for the sequential separation of the computational domain is presented in Fig. 3.

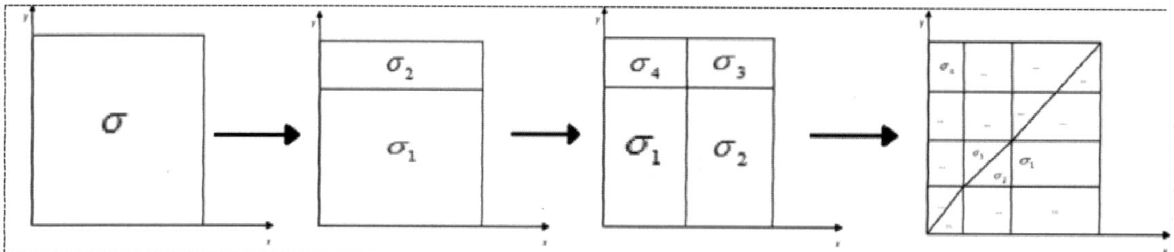

Figure 3. Successive division of the computational domain

IOP Conf. Series: Materials Science and Engineering **683** (2019) 012012 doi:10.1088/1757-899X/683/1/012012

The optimization problem was solved by the Hook-Jeeves method, as well as the brute force method. For the condition of stopping the optimization, we take the difference between the values of resistivity in the adjacent steps. To complete it is necessary that the deviation was less $\delta = 0.001$. The criterion for stopping the search for the solution of the inverse problem is the difference between the theoretical and experimental potential values at the nodal points. Provided that the difference is less $\varepsilon = 0.0001$, then the search for composite parameters is terminated. In the case when the minimum difference between the theoretical and numerical values does not change significantly over 10 steps, the search for a solution to the inverse problem stops and returns to the beginning. This algorithm is based on the analysis of existing optimization methods for solving inverse task [1], [8], [13]. The program implementation was carried out using existing methods [4], [5].

Calculations of the inverse problem

For the experimental data in solving the inverse problem, we take the values obtained in solving the direct problem for the composite selected above.

As an illustration of the results of solving the inverse problem, we present the composite model obtained by solving the inverse problem.

In fig. 4 depicts a composite with physical and geometric, obtained by solving the inverse problem. The error of the solution in all parameters does not exceed 5 percent; the most significant deviation in the analysis of the results obtained was found in the conductivity of the third layer.

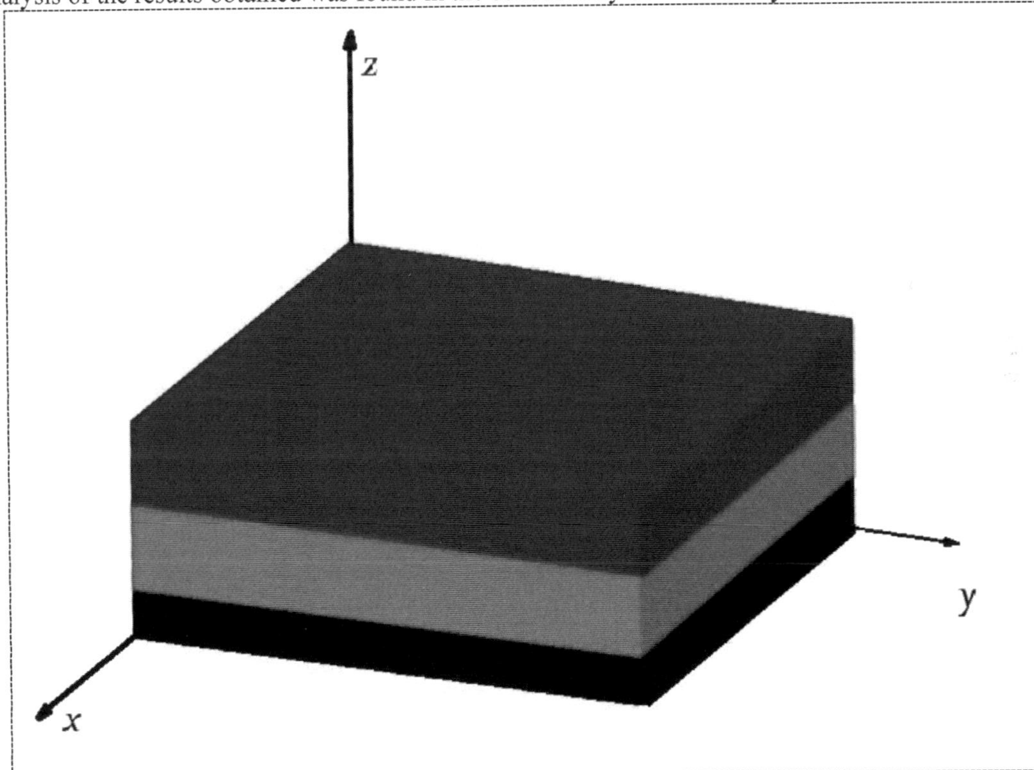

Figure 4. The composite model obtained by solving the inverse problem

Conclusions

The paper proposed a method for studying electromagnetic fields in heterogeneous composite structures, similar to geological exploration methods. To solve the direct and inverse problems, algorithms were

developed that allow one to calculate the potential values and the physical parameters of the composite with a different number of layers and in a wide range of thicknesses.
As a result of applying the new algorithm for solving the inverse problem, it was possible to significantly reduce the time spent on finding a solution, but even in this case, the task remains very resource-intensive.
It should be noted that the results of this work can also find application in the study of composites with a nonlinear structure.
It is assumed that the described algorithms can help in determining the characteristics of laying fabric composites, as well as, when finalizing, the developed methods can be used to study cracks in composites, taking the crack as a separate layer.

Acknowledgments
The work was carried out with support of state task of Ministry for Science and Education № 9.5387.2017/БЧ

References

[1] Zarubin V S, Sergeeva E S 2018 Application of Mathematical Modeling to Determine the Thermoelastic Characteristics of Nano-Reinforced Composites *Mathematical Models and Computer Simulations* **10** p. 288-298
[2] Sulimov V D, Shkapov P M, Sulimov A V 2016 Optimization of singular values of parameter dependent matrices using hybrid algorithms, *Herald of the Bauman Moscow State Technical University, Series Natural Sciences* **5** p. 46-66
[3] Fetisov D A 2016 On the construction of solutions of terminal problems for multidimensional affine systems in quasicanonical form *Differential Equations* **52**(12), p. 1638-1649
[4] Kozlov O S, Skvortsov L M 2016 MVTU software package in scientific research and applied developments *Mathematical Models and Computer Simulations* **8**(4), p. 358-368
[5] Agasiev T, Karpenko A 2017 The Program System for Automated Parameter Tuning of Optimization Algorithms *Procedia Computer Science* 103, p. 347-354
[6] Tikhomirov, V G, Gudkov A, Petrov V, Yankevich V, Evseenkov A 2017 Simulation of electric field distribution in GaN HEMTs for the onset of structure degradation *Proceedings of the 11th International Workshop on the Electromagnetic Compatibility of Integrated Circuits*, EMCCompo 7998094 p. 115-118
[7] Gorelik V S, Pyatyshev A Yu 2016 Electromagnetic waves in metamaterials placed in the magnetic field *Herald of the Bauman Moscow State Technical University, Series Natural Sciences* **1** p. 36-44
[8] Forsgren A , G i l l P.E., Wright M H 2002 Interior methods for nonlinear optimization *SIAM Review* **44** p. 525-597.
[9] Burger M, Muhlhuber W 2002 *Numerical approximation of an SQP-type method for parameter Identification*, SIAM J . Numer. Anal. **40** p. 1775-1797.
[10] Hoppe R H 2002 Shape and topology optimization in materials sciences, *Proceeding of ICCM* (Novosibirsk: NCC Publ.) p. 17-31
[11] P M van den Berg, A L van Broekhoven, A Abubakar. 1999 *Extended contrast source inversion,Inverse Problems* **15** p.1325–1344.
[12] Avdeev A V , Lavrentiev M M , Prijmenko V 1999 *Inverse problems and some applications (*Novosibirsk: ICM&MG Publ.)
[13] Dimitrienko Yu I, Krasnov I K, Zubarev K M 2018 *Mathematical model operation of an electric field in the nonuniform mediums at intubation by a direct current,* IOP Journal of Physics: Conference Series **1141**
[14] Dimitrienko Yu I, Gubareva E A , Markevich M N, Sborshikov S V 2016 Mathematical modeling of dielectric properties of nanostructural composite using asymptotic homogenzing method. *Vestnik MGTU im. N.E. Baumana. Seria Estestvennye nauki – Herald of the Bauman Moscow*

State Techical University. Series: Natural Sciences **1** p. 76–89.

Theory of composite cylindrical shells under quasistatic vibrations, based on an asymptotic analysis of the general viscoelasticity theory equations

Yu I Dimitrienko[1], E A Gubareva[1], A E Pichugina[1]

[1]Bauman Moscow State Technical University, Moscow, 105005, Russia

E-mail: dimit@bmstu.ru, eagubareva@mail.ru, a.e.pichugina@gmail.com

Abstract. A theory of thin multilayer anisotropic viscoelastic cylindrical shells with allowance for quasistatic oscillations, based on the application of asymptotic analysis of small geometric parameters to the general three-dimensional equations of the viscoelasticity theory for curvilinear coordinates, is proposed. Recurrent sequences of local problems of the viscoelasticity theory for shells under the quasistatic pressure oscillations are formulated and analytical solutions are obtained. It is shown that with this approach it is possible to obtain expressions for all 6 components of the stress tensor over the thickness of the shell. An example of the calculation of a cylindrical viscoelastic shell with axisymmetric bending with allowance for quasistatic oscillations is considered. The graphs of the dependences of the components of the stress tensor on the local coordinate are given for different angles of layers in the material. An algorithm is proposed for obtaining explicit analytical equations for calculating the distribution of the total stress tensor components over a cylindrical shell under the action of quasistatic oscillations.

1. Introduction

Polymeric composite materials have strongly pronounced viscoelastic properties, as compared with metals and alloys. This property provides a significantly higher energy dissipation during vibrations, making composite structures less noisy and more resistant to resonance phenomena. To solve engineering problems of viscoelastic composite structures, it is necessary to accurately calculate the components of the total stress tensor at any point in the structure, including interlaminar and trough – thickness stresses, due to low strength of composites in these directions. [1-6]. The classical theories of calculating thin composite shells are based on certain assumptions, the adequacy of which must be confirmed for each class of problems. A method for asymptotic averaging of the equations of the three-dimensional elasticity theory for thin multilayer plates and shells was developed in [7-15], which, without any hypotheses regarding the distribution of displacements or stresses over the thickness, allows us to obtain an explicit, mathematically valid expression for all 6 components of the stress tensor. In this paper we present the relations for the total stress tensor obtained with the help of the asymptotic theory for the case of thin viscoelastic cylindrical shells.

2. The equations of three-dimensional quasistatic oscillations of viscoelastic cylindrical composite shells

Let us consider the three-dimensional problem of quasistatic oscillations of viscoelastic media in the general form

$$\begin{cases} \nabla\cdot\sigma^*=0, \qquad в\ V \quad (1) \\ \sigma^*=C^*\cdot\cdot\varepsilon^*, \qquad в\ V\cup\Sigma \quad (2) \\ \varepsilon^*=\frac{1}{2}(\nabla\otimes u^*+\nabla\otimes u^{*T}), \qquad в\ V\cup\Sigma \quad (3) \\ n\cdot\sigma^*\big|_{\Sigma_\sigma}=t^*_{ne}, \qquad u^*\big|_{\Sigma_u}=u^*_{ue}. \quad (4) \end{cases},$$

where (1) are the equilibrium equations for quasistatic oscillations, (2) are constitutive relations, (3) are Cauchy relations, (4) are the boundary conditions, and $\sigma^*=\sigma'+i\sigma''$, $\varepsilon^*=\varepsilon'+i\varepsilon''$, $u^*=u'+iu''$ - are amplitudes of stress, strain and displacement oscillations, $C^*=C'+iC''$ - is complex elastic modulus tensor. We assume that the viscous properties of the composite are small compared with the elastic, then $|\sigma'|\ll|\sigma''|$, $|\varepsilon'|\ll|\varepsilon''|$, $|u'|\ll|u''|$, $|C'|\ll|C''|$.

Let us consider a thin multilayer viscoelastic shell for which the following relationship holds $æ=h/L<<1$: where h is the shell thickness, L is the shell length, æ is a small parameter, $\bar{R}=R/L$ is the dimensionless shell radius. We introduce $q_k=\{z,s,r\}$ - the cylindrical coordinates (s is the length of the arc in the circumferential direction), measured from the middle surface of the shell and $\bar{q}_k=q_k/L$ - the dimensionless coordinates, and the local coordinate $\xi=\bar{q}_3/æ$ along the thickness of the shell. The Lame parameters H_α for a cylindrical thin shell and their derivatives have the following form [15]: $H_1=1$; $H_2=1$, $H_3=1$, $\partial H_\alpha/\partial\xi=æH_{\alpha 3}$, $H_{13}=0$, $H_{23}=1/\bar{R}$, $\partial H_\alpha/\partial\bar{q}_\beta=0$.

Then the problem (1) - (4) for the case of a cylindrical shell takes the form [16]:

$$\sigma^*_{11,1}+\sigma^*_{12,2}+\frac{1}{\bar{R}}\sigma^*_{13}+\frac{1}{æ}\sigma^*_{13/3}=0, \tag{5}$$

$$\sigma^*_{12,1}+\sigma^*_{22,2}+\frac{2}{\bar{R}}\sigma^*_{23}+\frac{1}{æ}\sigma^*_{23/3}=0, \tag{6}$$

$$\frac{1}{æ}\sigma^*_{33/3}+\sigma^*_{13,1}+\sigma^*_{23,2}+\frac{1}{\bar{R}}(\sigma^*_{33}-\sigma^*_{22})=0; \tag{7}$$

$$\sigma^*_{IJ}=C^*_{IJKL}\varepsilon^*_{KL}+\tilde{C}^*_{IJk3}\varepsilon^*_{k3}, \quad \sigma^*_{i3}=C^*_{i3KL}\varepsilon^*_{KL}+\tilde{C}^*_{i3k3}\varepsilon^*_{k3}, \tag{8}$$

$$\varepsilon^*_{11}=u^*_{1,1},\ \varepsilon^*_{22}=u^*_{2,2}+\frac{u^*_3}{\bar{R}},\ \varepsilon^*_{12}=\frac{1}{2}(u^*_{1,2}+u^*_{2,1}), \tag{9}$$

$$\varepsilon^*_{33}=\frac{1}{æ}u^*_{3/3},\ 2\varepsilon^*_{13}=\frac{1}{æ}u^*_{1/3}+u^*_{3,1},\ 2\varepsilon^*_{23}=\frac{1}{æ}u^*_{2/3}-\frac{u^*_2}{\bar{R}}+u^*_{3,2}$$

$$\Sigma_{3\pm}:\sigma^*_{i3}=-\tilde{p}_\pm\delta_{i3} \qquad \Sigma_T:u^*_i=u^*_{ei} \qquad \Sigma_S:[\sigma^*_{i3}]=0 \qquad [u^*_i]=0 \tag{10}$$

where (5)-(7) are the equilibrium equations, (8) is the generalized Hooke's law, (9) the Cauchy relations, (10) are the boundary conditions, $\Sigma_{3\pm}$ - the outer and inner surfaces of the shell ($\xi=\pm0.5$), Σ_T - the end surface on which the displacement u^*_{ei} is specified, Σ_S - the interface between the layers of the shell, on which the conditions for the ideal contact of the layers are specified, $[u^*_i]$ - the jump in the functions, u^*_i - the dimensionless displacements, σ^*_{ij} - the dimensionless stresses, $C^*_{ijkl}(\xi)$ - the dimensionless elastic moduli, $\tilde{p}_\pm$ - dimensionless pressures related to the characteristic value of the modulus of elasticity E_0, ε^*_{ij} are small deformations. The shell layers are assumed to be elastic, with a cylindrical monoclinic symmetry of elastic properties [16].

3. Asymptotic solution

Let us consider the case when the pressure $\tilde{p}_\pm$ on the outer and inner surfaces of the shell is of the order of smallness $O(æ^3)$, $\tilde{p}_\pm = æ^3 p_\pm$., where $p_\pm$ is a finite value. This order of pressure, as a rule, corresponds to the real conditions of loading of thin shells. We will seek the solution of Problem (1)-(6) in the form of asymptotic expansions over small parameter $æ$ in the form of functions, depending on global and local coordinates:

$$u^*_k = u_k^{(0)}(\overline{q}_I) + \sum_{n=1}^{\infty} æ^n u_{ij}^{(n)} = u_k^{(0)}(\overline{q}_I) + æ u_k^{(1)}(\overline{q}_I,\xi) + æ^2 u_k^{(2)}(\overline{q}_I,\xi) + \dots \quad (11)$$

Using these asymptotic expansions, explicit expressions are obtained for the longitudinal σ^*_{IJ}, shear σ^*_{I3}, and transverse σ^*_{33} components of the stress tensor through deformations $\varepsilon^{(0)}_{KL}$ and curvatures η_{KL} of the zero approximation of the shell

$$\begin{cases} \sigma^*_{IJ} = \hat{C}^{(1)}_{IJKL}\varepsilon^{(0)}_{KL} + \hat{C}^{(0)}_{IJKL}\xi\eta_{KL} \\ -\sigma^*_{\alpha 3} = \hat{C}^{(2)}_{\alpha 3KL}\varepsilon^{(0)}_{KL} + \hat{R}^{(2)}_{\alpha 3KLJ}\varepsilon^{(0)}_{KL,J} + \hat{N}_{\alpha 3KL}\eta_{KL} + \hat{V}_{\alpha 3KLJ}\eta_{KL,J} \\ \sigma^*_{33} = \hat{C}^{(3)}_{33KL}\varepsilon^{(0)}_{KL} + \hat{R}^{(3)}_{33KLJ}\varepsilon^{(0)}_{KL,J} + \hat{E}^{(3)}_{33KLJM}\varepsilon^{(0)}_{KL,JM} + \hat{N}^{(3)}_{33KL}\eta_{KL} + \\ +\hat{V}^{(3)}_{33KLM}\eta_{KL,M} + \hat{W}^{(3)}_{KL}\eta^{(2)}_{KL} - (p_\pm + \Delta p(\xi + 0.5)) \end{cases} \quad (12)$$

In the formulas (12) we denote tensors $C^{(0)}_{IJKL}$ $\hat{C}^{(1)}_{IJKL}$, $\hat{C}^{(2)}_{\alpha 3KL}$ $\hat{R}^{(2)}_{\alpha 3KLJ}$ $\hat{R}^{(3)}_{33KLJ}$ $\hat{N}_{\alpha 3KL}$ $\hat{N}^{(3)}_{33KL}$ and others, depending only on the C^*_{IJKL}- modulus tensor of the elasticity layers of the shell, and the thicknesses of the layers C^*_{IJKL} ,and layer thicknesses.

The expressions for $\varepsilon^{(0)}_{KL}$ and η_{KL} in the cylindrical shell have the following form:

$$\varepsilon^{(0)}_{11} = u^{(0)}_{1,1}, \quad \varepsilon^{(0)}_{22} = u^{(0)}_{2,2} + u^{(0)}_3 / \overline{R}, \quad 2\varepsilon^{(0)}_{12} = u^{(0)}_{1,2} + u^{(0)}_{2,1},$$

$$\eta_{11} = -u^{(0)}_{3,11} \quad \eta_{22} = -(u^{(0)}_{2,2} + u^{(0)}_{3,22} / R^2) \quad -\eta_{12} = u^{(0)}_{3,12} / \overline{R} + u^{(0)}_{2,1}. \quad (13)$$

To find the displacements $u^{(0)}_k(\overline{q}_I)$, we obtain a system of averaged equations of the Kirchhoff-Love theory.

4. Example of calculation of stresses in a composite cylindrical shell

Consider the case of axisymmetric loading of a 4-layer composite cylindrical shell, under the influence of internal pressure and quasistatic vibrations . Each layer was a unidirectional material, the system of corners α_i of the reinforcement of the composite layers was chosen as follows

The figure 1 shows the distribution of the flexural stress σ_{11} along the thickness of the composite sheath. The figure 2-6 shows the distribution of the flexural stress σ_{22} along the thickness of the composite sheath at different angles α_i. Formulas (8) allow us to calculate the distributions of all 6 stresses in the shell.

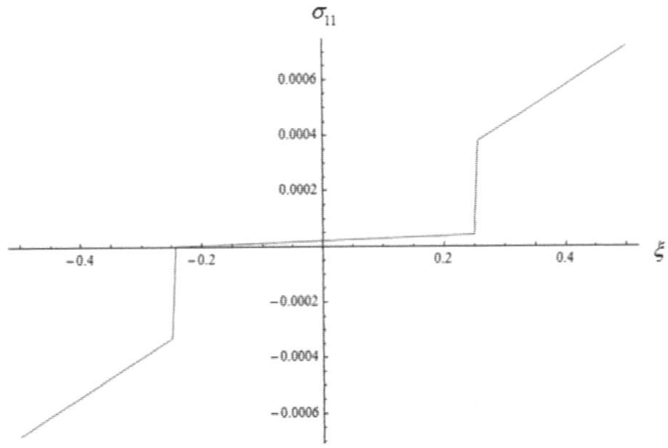

Figure 1. Distribution of stress σ_{11}, GPa across the thickness of the composite shell under axially symmetric bending.

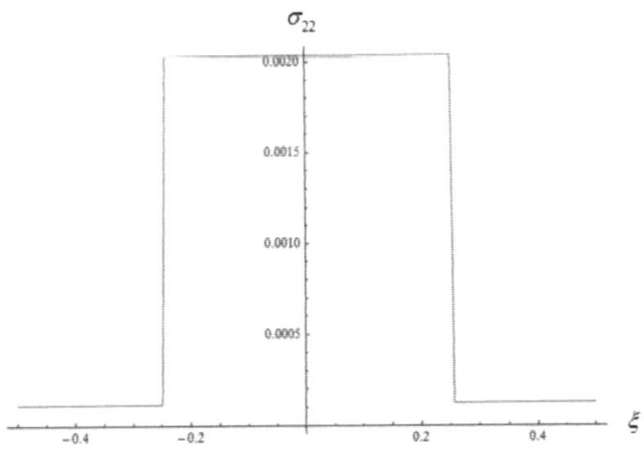

Figure 2. Distribution of stress σ_{22}, GPa across the thickness of the composite shell under axially symmetric bending: $\alpha_1 = 0^\circ$, $\alpha_2 = 90^\circ$, $\alpha_3 = -90^\circ$, $\alpha_4 = 0^\circ$

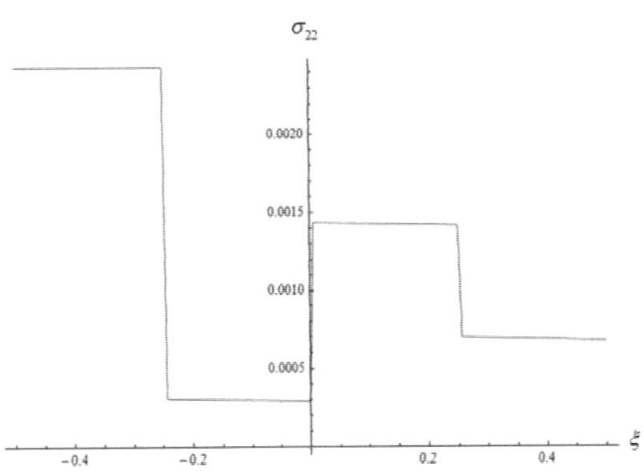

Figure 3. Distribution of stress σ_{22}, GPa across the thickness of the composite shell under axially symmetric bending: $\alpha_1 = 90^\circ$, $\alpha_2 = 30^\circ$, $\alpha_3 = 60^\circ$, $\alpha_4 = 45^\circ$

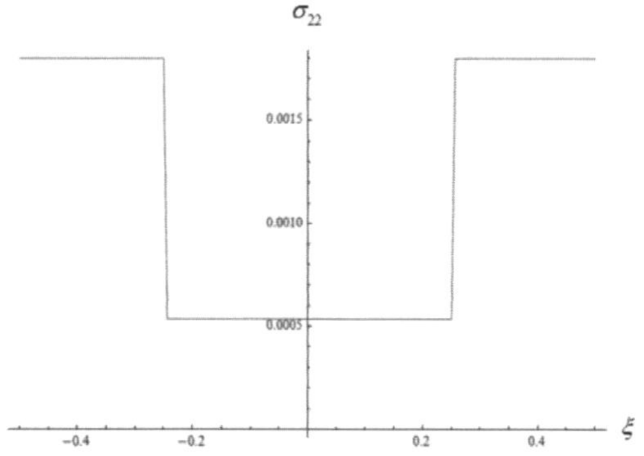

Figure 4. Distribution of stress σ_{22}, GPa across the thickness of the composite shell under axially symmetric bending: $\alpha_1 = 90^\circ$, $\alpha_2 = 45^\circ$, $\alpha_3 = 45^\circ$, $\alpha_4 = 90^\circ$

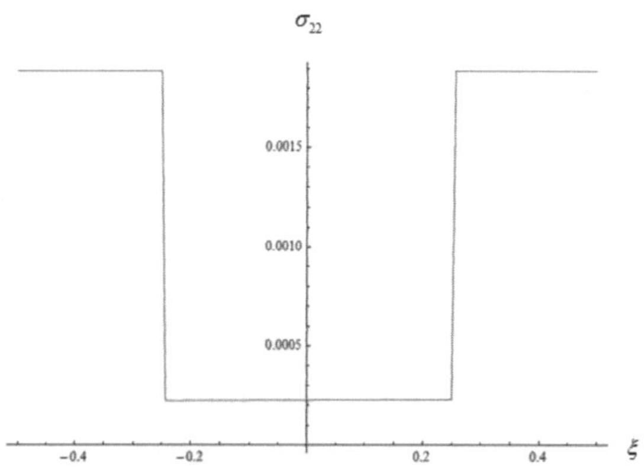

Figure 5. Distribution of stress σ_{22}, GPa across the thickness of the composite shell under axially symmetric bending: $\alpha_1 = 90^\circ$, $\alpha_2 = 30^\circ$, $\alpha_3 = 30^\circ$, $\alpha_4 = 90^\circ$

5. Conclusions

A theory of thin cylindrical composite visco-elastic shells is developed, based on the asymptotic analysis of three-dimensional equations of the theory of viscoelasticity without additional assumptions about the distribution of stresses and displacements along the thickness of the shell. Explicit, mathematically accurate analytical formulas are obtained for calculating the distribution of the components of all 6 components of the stress tensor over the thickness of a cylindrical shell. An example is given of calculating stresses in a cylindrical shell under axisymmetric bending of pressures

Acknowledgments

The work was carried out with support of state task of Ministry for Science and Education № 9.5387.2017/БЧ

References

[1] Vasilyev V V 1988 *Mechanics of constructions from composite materials* (Moscow: Mechanical engineering) 272 p
[2] Nikabadze M U, Ulukhanyan A R 2016 Analytical solutions in the theory of thin bodies *Advanced Structured Materials* **42** 319-61
[3] Van'ko V I 2016 On the load-carrying capacity of structural members *Journal of Applied*

Mechanics and Technical Physics **57(5)** 784-8

[4] Bakulin V N, Vinogradov Y I 2017 Analytical and asymptotic solution of boundary value problems in the mechanics of deformed shells under concentrated loading *Russian Aeronautics* **60(1)** 13-20

[5] Valishin A A 2017 Management and modeling of the influence of cracks on the temperature distribution in polymers and composites *Espacios* **38(48)** 38

[6] Ruslantsev A N, Portnova Ya M, Tairova L P, Dumansky A M 2016 Analysis of mechanical properties anisotropy of nanomodified carbon fibre-reinforced woven composites *IOP Conference Series: Materials Science and Engineering* **153(1)** 012003

[7] Dimitrienko Yu I, Yakovlev D O, Erasov V S, Fedonyuk N N, Sborschikov, Gubareva E A, Krylov V D, Grigoriev M M , Prozorovsky A A 2014 Development of a multilayer polymer composite with a discrete constructive-orthotropic filler *Composites and nanostructures* **6(1)** 32-48

[8] Grigolyuk E I, Kulikov G M 1988 Generalized model of mechanics of thin-walled constructions from composite materials *Mechanics of Composite Materials* **4** 698-704

[9] Levinski T, Telega J J 2000 *Plates, laminates and shells.Asymptotic analysis and homogenization.* (Singapore; London, World Sci. Publ.) 739 p

[10] Dimitrienko Yu I 2012 Asymptotic theory of multilayer thin plates. *Herald of the Bauman Moscow State Technical University Series Natural Sciences* **№3** 86-100

[11] Yu.I. Dimitrienko, D.O.Yakovlev 2015 The Asymptotic Theory of Thermoelasticity of Multilayer Composite Plates. *Composites: Mechanics,Applications. An International Journal*. **6.** 1 13-51.

[12] Yu. I. Dimitrienko, I. D. Dimitrienko and S.V. Sborschikov 2015 Multiscale Hierarchical Modeling of Fiber Reinforced Composites by Asymptotic Homogenization Method. *Applied Mathematical Sciences,* v.9 № 145 pp.7211-7220

[13] Dimitrienko Yu.I., Gubareva E.A., Markevich M.N., Sborschikov S.V. 2016 Mathematical modelling of dielectric properties of nanostructural composites using asymptotic homogenizing method *Herald of the Bauman Moscow State Technical University Series Natural Sciences*. **1** 76-89.

[14] Dimitrienko Yu I, Gubareva E A, Sborschikov S V 2014 Asymptotic theory of constructively orthotropic plates with a two-period structure *Matematicheskoe modelirovanie i chislennye metody – Mathematical Modeling and Computational Method* 1 pp. 36-57

[15] Dimitrienko Y I, Gubareva E A, Yurin Yu V 2016 Explicit formulas for the calculation of the c omplete tensor of the stresses in the monoclinic thin composite shells based on the a symptotic homogenization method *Inzhenernyy zhurnal: nauka i innovatsii — Engineering Journal:Science and Innovation* **12** (60) http://dx.doi.org/10.18698/2308-6033-2016-12-1557

[16] Dimitrienko Yu.I. 2010 *Nonlinear Continuum Mechanics and Large Inelastic Deformations.* Springer. 722 p.

IOP Conf. Series: Materials Science and Engineering **683** (2019) 012014 doi:10.1088/1757-899X/683/1/012014

Multi-parametric dynamic analysis of lightweight elastic laminates

J Kaplunov[1], D A Prikazchikov[1], L A Prikazchikova[1], A Nikonov[2], and T Savšek[3]

[1] School of Computing and Mathematics, Keele University, Staffordshire, ST5 5BG, UK
[2] Faculty of Industrial Engineering Novo mesto, Šegova ulica 112, 8000, Novo mesto, Slovenia
[3]TPV d.o.o., Kandijska cesta 60, 8000, Novo mesto, Slovenia
E-mail: j.kaplunov@keele.ac.uk, d.prikazchikov@keele.ac.uk

Abstract. Multi-parametric asymptotic analysis of dynamic phenomena in lightweight three-layered structures is performed. The presence of high contrast in densities of skin and core layers may lead to the small value of the lowest shear thickness resonance frequency, allowing a two-mode long wave low-frequency approximation of the exact dispersion relation. The range of uniform asymptotic validity of the two-mode approximation with respect to relative thickness and density are established. The theoretical predictions are illustrated numerically.

Introduction

Multi-layered structures with a strong vertical heterogeneity are of major importance for numerous high-tech domains in aerospace, automotive and civil engineering, see e.g. [1, 2] and references therein. Specific applications include, in particular, photovoltaic panels and laminated glass, see [3, 4, 5]. Another important example originating from automotive industry, is concerned with design of a new generation of environment-friendly lightweight vehicles, see [6, 7]. This inspires modelling of composite structural elements, containing lightweight components.

Multi-parametric asymptotic analysis of dispersion of elastic plates with high contrast in material and geometric parameters of the layers was studied in [8], see also a more recent contribution [9]. The key message of the cited papers is that due to contrast the lowest shear mode may be excited over the low frequency range, requiring substantial amendments to the established structural theories. In particular, instead of the classical Kirchhoff theory governing the fundamental vibration mode of a thin plate, a variety of two-mode approximations, strongly affected by the contrast parameters were discovered in [8]. The latter, in addition to the fundamental mode, also include the first low-frequency harmonic.

In this presentation we attempt to specify the approach in [8] for three-layered structures with a lightweight core component. In this case, we adapt a multi-parametric scheme assuming that the ratio of thickness of the skin and core component layers is expressed as a power of the main small parameter, defined as the relative core density.

The two-mode approximations of the full Rayleigh-Lamb-type dispersion relation, both uniformly and non-uniformly valid, are derived. The accuracy of the latter are tested by comparison with the exact solution. The obtained results address peculiarities of the vibration spectra characteristic of lightweight layered structures.

Statement of the problem

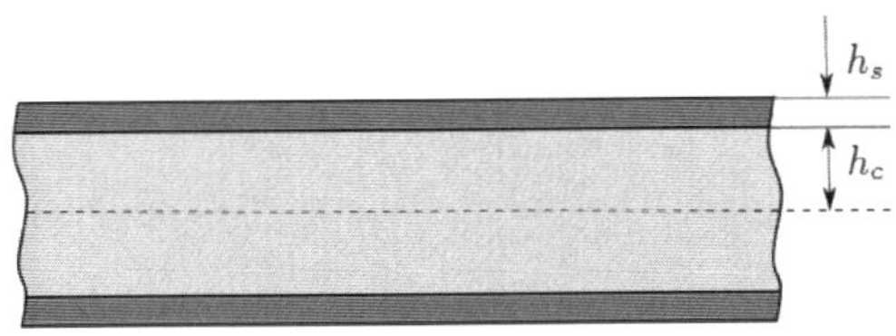

Figure 1. A lightweight sandwich plate.

Consider a sandwich symmetric plate with a light core, see Fig. 1, assuming that the ratio of densities of the core and skin layers

$$\rho = \frac{\rho_c}{\rho_s} \ll 1 \tag{1}$$

is a small parameter. In this case the inverse ratio of their thickness is expressed as

$$h = \frac{h_s}{h_c} \sim \rho^a, \tag{2}$$

where $0 \le a < 1$. In addition, for the sake of simplicity we assume for the ratio of shear moduli

$$\mu = \frac{\mu_c}{\mu_s} \sim \rho. \tag{3}$$

Then, the lowest shear resonant frequency

$$\Omega_{sh} \approx \left(\frac{\rho}{h}\right)^{1/2} \sim \rho^{(1-a)/2} \ll 1, \tag{4}$$

belongs to the low-frequency range, see also an earlier contribution [10] for a three-component rod.

The consideration below deals with a multi-parametric analysis (in ρ and a) of the full dispersion relation of the sandwich structure of interest at the long-wave low-frequency limit, when

$$K = kh_c \ll 1, \Omega = \omega h_c \sqrt{\frac{\rho_c}{\mu_c}} \ll 1, \tag{5}$$

where ω and k are angular frequency and wave number, respectively.

Asymptotic analysis

First, expand the dispersion relation in [8] over range (5) under the assumptions on the problem parameters adapted in the previous section. As a result, we arrive at the polynomial expression

$$\gamma_1\Omega^2 + \gamma_2 K^4 + \gamma_3 K^2\Omega^2 + \gamma_4 K^6 + \gamma_5\Omega^4 + \dots = 0, \tag{6}$$

where

$\gamma_1 = -\mu(h\mu_0 + \mu),$

$\gamma_2 = -\frac{4}{3}\mu(h(h^2 + 3h + 3)(\varkappa_s^2 - 1) + \mu(\varkappa_c^2 - 1)),$

$\gamma_3 = \frac{2}{3}(\mu\mu_0 h^3(2\varkappa_s^2 - 3) + 3h^2((\varkappa_s^2 - 1)(\mu^2 + 2\mu_0) - \mu_0\mu\varkappa_s^2) - 3\mu h(-2\varkappa_s^2 + \mu_0 + 2) + \mu^2(2\varkappa_c^2 - 3)),$

$\gamma_4 = \frac{4}{15}(\mu h(1 - \varkappa_s^2)(10h\varkappa_c^2(1 - \mu) + 5h^3\varkappa_s^2 + 10h^2 + 10\mu h + 5 + h^4) + 5h^4\varkappa_s^2(\varkappa_s^2 - 2) + \mu^2(1 - \varkappa_c^2) + 5h^4),$

$$\gamma_5 = \frac{1}{6}(\mu^2(3\mu_0(\varkappa_s^2+1)h^2+\varkappa_c^2+3)+\mu h\mu_0(h^2\mu_0(\varkappa_s^2+3)+3\varkappa_c^2+9)+6h^2\mu_0^2).$$

In the formulae above

$$\mu_0 = \frac{\mu}{\rho} \quad \text{and} \quad \varkappa_q = \frac{c_{2q}}{c_{1q}},$$

with

$$c_{1q}^2 = \frac{\lambda_q + 2\mu_q}{\rho_q}, \quad c_{2q}^2 = \frac{\mu_q}{\rho_q}, \quad q = c, s.$$

The leading order asymptotic approximation of (6) corresponding to the fundamental mode, can be written as

$$\rho^{1-a}G_1\Omega^2 + \rho^{1-a}G_2K^4 + G_3K^2\Omega^2 + \frac{1}{3}\rho^{2a}G_2K^6 = 0, \tag{7}$$

where $G_i \sim 1,\ i = 1,2,3$ are expressed as

$$G_1 = -h_0\mu_0^2, \qquad G_2 = -4h_0\mu_0(\varkappa_s^2 - 1), \qquad G_3 = 4h_0^2\mu_0(\varkappa_s^2 - 1),$$

with $h_0 = h/\rho^a$.
The shortened forms of (7) are

$$G_1\Omega^2 + G_2K^4 = 0, \quad \text{at} \quad \Omega \ll \rho^{1-a}, \qquad (0 \le a < 1), \tag{8}$$

$$\rho^{1-a}G_2K^2 + G_3\Omega^2 = 0, \quad \text{at} \quad \rho^{1-a} \ll \Omega \ll \rho^{1-2a}, \ \left(a \le \frac{1}{3}\right), \quad \text{or} \quad \rho^{1-a} \ll \Omega \ll \rho^{\frac{1-a}{2}}, \ \left(\frac{1}{3} < a < 1\right), \tag{9}$$

and

$$G_3\Omega^2 + \frac{1}{3}\rho^{2a}G_2K^4 = 0, \quad \text{at} \quad \rho^{1-2a} \ll \Omega \ll \rho^a, \left(a < \frac{1}{3}\right). \tag{10}$$

The leading order asymptotic behavior of the first harmonic with the cut-off frequency (4) takes the form

$$\rho^{1-a}G_1 + G_3K^2 + G_5\Omega^2 = 0, \quad \text{at} \ \Omega - \Omega_{sh} \ll 1 \ (0 \le a < 1), \tag{11}$$

where for $G_5 \sim 1$ we have $G_5 = h_0^2\mu_0^2$.

The derived formulae can be easily justified using the entries in Tables 1 and 2 below.

Table 1. Asymptotic behavior for the fundamental mode.

Order of γ_i	Terms	Fundamental mode

		$\Omega^2 \sim K^4$ $\Omega \ll \rho^{1-a}, 0 \le a < 1$	$\Omega^2 \sim \rho^{1-a} K^2$ $\rho^{1-a} \ll \Omega \ll \rho^{1-2a}, a \le \frac{1}{3}$ $\rho^{1-a} \ll \Omega \ll \rho^{\frac{1-a}{2}}, \frac{1}{3} < a < 1$	$\Omega^2 \sim \rho^{2a} K^2$ $\rho^{1-2a} \ll \Omega \ll \rho^{a}, a < \frac{1}{3}$
$\gamma_1 \sim \rho^{1-a}$	$\gamma_1 \Omega^2$	$\rho^{1-a} K^4$	$\rho^{2-2a} K^2$	$\rho^{1+a} K^4$
$\gamma_2 \sim \rho^{1-a}$	$\gamma_2 K^4$	$\rho^{1-a} K^4$	$\rho^{1-a} K^4$	$\rho^{1-a} K^4$
$\gamma_3 \sim 1$	$\gamma_3 K^2 \Omega^2$	K^6	$\rho^{1-a} K^4$	$\rho^{2a} K^4$
$\gamma_4 \sim \rho^{2a} + \rho^{1-a}$	$\gamma_4 K^6$	$(\rho^{2a} + \rho^{1-a}) K^6$	$(\rho^{2a} + \rho^{1-a}) K^6$	$\rho^{2a} K^6$
$\gamma_5 \sim 1$	$\gamma_5 \Omega^4$	K^8	$\rho^{2-2a} K^4$	$\rho^{4a} K^8$

Table 2. Asymptotic behavior for the lowest harmonic.

Order of γ_i	Terms	Harmonic $\Omega^2 - \Omega_{sh}^2 \sim K^2$ $\Omega - \Omega_{sh} \ll 1$
$\gamma_1 \sim \rho^{1-a}$	$\gamma_1 \Omega^2$	$\rho^{1-a}(K^2 + \rho^{1-a})$
$\gamma_2 \sim \rho^{1-a}$	$\gamma_2 K^4$	$\rho^{1-a} K^4$
$\gamma_3 \sim 1$	$\gamma_3 K^2 \Omega^2$	$K^2(K^2 + \rho^{1-a})$
$\gamma_4 \sim \rho^{2a} + \rho^{1-a}$	$\gamma_4 K^6$	$(\rho^{2a} + \rho^{1-a}) K^6$
$\gamma_5 \sim 1$	$\gamma_5 \Omega^4$	$(K^2 + \rho^{1-a})^2$

It is remarkable that as it follows from the Table 1 and also formulae (8)-(10), only at $0 \le a \le \frac{1}{3}$ the cut-off frequency belongs to the range of validity of the long-wave approximation (7) for the fundamental mode.

Thus, combining the expressions (7)-(11), we arrive at the uniformly valid two- mode approximation

$$\rho^{1-a} G_1 \Omega^2 + \rho^{1-a} G_2 K^4 + G_3 K^2 \Omega^2 + \frac{1}{3} \rho^{2a} G_2 K^6 + G_5 \Omega^4 = 0, \qquad 0 \le a \le \frac{1}{3}. \tag{12}$$

At the same time, at $\frac{1}{3} < a < 1$, the cut-off frequency (4) is outside the range of validity of (7), see Table 1 and formulae (8)-(10). Consequently, the corresponding two-mode expansion

$$\rho^{1-a} G_1 \Omega^2 + \rho^{1-a} G_2 K^4 + G_3 K^2 \Omega^2 + G_5 \Omega^4 = 0, \qquad \frac{1}{3} < a < 1, \tag{13}$$

is no more uniformly valid.

Numerical results

Let us illustrate the obtained results. Figs. 2 and 3 below are plotted for the values $\rho = 0.01, \mu = 0.01$, and $\varkappa_s = 0.59$.The two-mode asymptotic approximation (12) along with the exact dispersion curves for $a = \frac{1}{9}$, $h = 0.6$ and $a = \frac{1}{3}$, $h = 0.23$ are shown in Figs. 2(a) and 2(b), respectively. It is clear from Fig. 2(a) that the presented approximation is uniform, reaching its bound at $a = \frac{1}{3}$ in Fig. 2(b), see (12). Indeed, the vertical dotted lines, originating from the cutoff frequencies intersect the fundamental modes at $K < 1$.

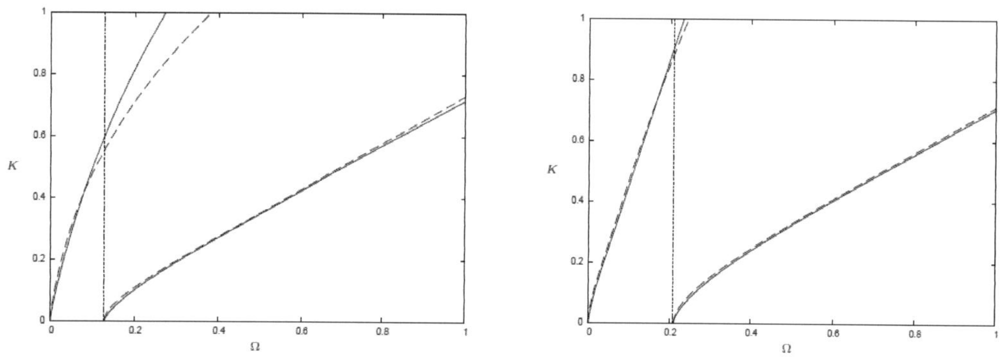

(a) (b)

Figure 2. Comparison of two-mode approximations (12) (dashed lines) and the exact dispersion curves (solid lines) for $\rho = 0.01, \mu = 0.01, \varkappa_s = 0.59$ with (a) $a = \frac{1}{9}$, $h = 0.6$; (b) $a = \frac{1}{3}$, $h = 0.23$.

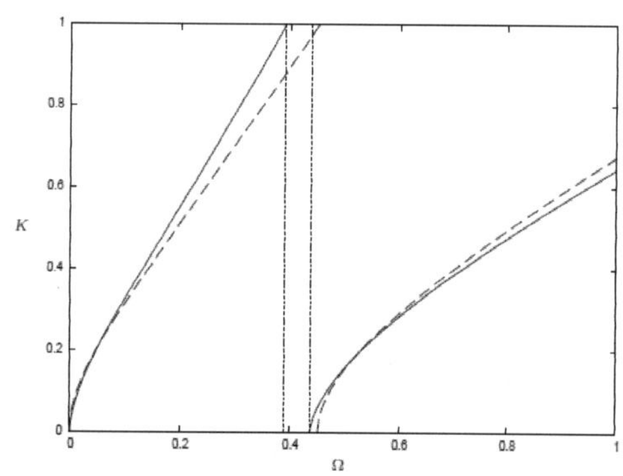

Figure 3. Comparison of two-mode composite approximation (13) (dashed lines) and the exact dispersion curves (solid lines) for $a = \frac{2}{3}$, $\rho = 0.01, \mu = 0.01$, $h = 0.05$, $\varkappa_s = 0.59$.

The composite expansion (13) versus the exact solution is shown in Fig. 3 for $a = \frac{2}{3}$ and $h = 0.05$. The non-uniform nature of (13), associated with a gap in frequency between the two vertical dotted lines can be clearly observed. Indeed, within this gap the long wave assumption ($K \ll 1$) is violated.

Concluding remarks

The developed approach is aimed at analysis and optimization of low-frequency vibration spectra of lightweight laminates. In particular, the consideration above revealed the range of parameters satisfying

$$\frac{h_s}{h_c} = \left(\frac{\rho_c}{\rho_s}\right)^a, 0 \le a \le \frac{1}{3}, \tag{14}$$

for which the asymptotic approximation (12) is uniform, whereas for the remaining region $\frac{1}{3} < a < 1$ the formula (13) provides only a composite approximation.

For plate bending, the shift of the first thickness shear resonance to the low frequency band results in its interaction with the fundamental mode. This makes the implementation of direct finite element computations more challenging, see e.g. [11], motivating a more elaborative theoretical insight.
The presented methodology is not restricted to the setup of a three-layered symmetric plate. It can be extended to other types of lightweight laminates. In addition, extra problem parameters, including a strong contrast in the wave speed for core and skin layers, may be taken into consideration. We also mention the generalizations to shell-type structures with thermo-viscoelastic core layers.
The derived two-mode expansions of the dispersion relation are an important preliminary step for establishing general models for lightweight multi-layered structures, including differential equations of motion along with properly justified boundary conditions.

Acknowledgments
Authors acknowledge the co-funding for this study from the Ministry of Education, Science and Sport of the Republic of Slovenia and European Regional Development Fund (European Commission): Program OP20.00362 EVA4GREEN - Environmentally Safe Green Mobility Car. LP acknowledges support by the Faculty Award by Keele University. The authors are grateful to M. Palsu for technical assistance.

References
[1] Davies J M (Ed.) 2001 *Lightweight sandwich construction* (Oxford: Blackwell).
[2] Yancey R N 2016 *Challenges, opportunities, and perspectives on lightweight composite structures: aerospace versus automotive. Lightweight Composite Structures in Transport* ed. J. Njuguna chapter 2 pp 35–52.
[3] Naumenko K and Eremeyev V A 2014 A layer-wise theory for laminated glass and photovoltaic panels. *Comp. Struct.* **112** 283–291.
[4] Viverge K, Boutin C and Sallet F 2016 Model of highly contrasted plates versus experiments on laminated glass. *Int. J. Solids Struct.* **102** 238–258.
[5] Aßmus M, Nordmann J, Naumenko K and Altenbach H 2017 A homogeneous substitute material for the core layer of photovoltaic composite structures. *Compos. Part B* **112** 353-–372.
[6] Lyu M Y and Choi T G 2015 Research trends in polymer materials for use in lightweight vehicles. *Int. J Prec. Eng. Manufact.* **16**(1) 213–220.
[7] Delogu M, Zanchi L, Dattilo C A and Pierini M 2017 Innovative composites and hybrid materials for electric vehicles lightweight design in a sustainability perspective. *Mater. Today Comm.* **13** 192–209.
[8] Kaplunov J, Prikazchikov D A and Prikazchikova L A 2017 Dispersion of elastic waves in a strongly inhomogeneous three-layered plate. *Int. J. Solids Struct.* **113** 169–179.
[9] Prikazchikova L A, Aydın Y E, Erbaş B, and Kaplunov J 2018 Asymptotic analysis of an anti-plane dynamic problem for a three-layered strongly inho- mogeneous laminate. *Math. Mech. Solids*, 1081286518790804.
[10] Kaplunov J, Prikazchikov D A and Sergushova O 2016 Multi-parametric analysis of the lowest natural frequencies of strongly inhomogeneous elastic rods. *J. Sound Vib.* **366** 264–276.
[11] Kaplunov J D, Nolde E V and Shorr B F 2005 A perturbation approach for evaluating natural frequencies of moderately thick elliptic plates. *J. Sound Vib.* **281** 905–919.

IOP Conf. Series: Materials Science and Engineering **683** (2019) 012015 doi:10.1088/1757-899X/683/1/012015

Inverse problems of restoring the geometric dimensions of a construction defects by thermal fields analysis

I.K. Krasnov[1] and A.A. Salnikova[1]

[1] Bauman Moscow State Technical University, Moscow, 105005, Russia

E-mail:igorkranv@yandex.ru, anna_salnikova@bmstu.ru

Abstract. The proposed work deals with the problem of finding the geometric dimensions of a cavity-type defect by measuring the thermal fields. As a structural element, a plate with a rectangular cavity is considered.

1. Introduction

The proposed work deals with the problem of finding the geometric dimensions of a cavity-type defect by measuring the thermal fields [1-22]. The method of solving this problem is based on the optimization approach. Optimization is carried out using methods of finding extremums without calculating the derivative, in this paper this is the Hook-Jeeves method. The results of solving this problem make it possible to improve the accuracy of determining the location of the defect and show the advisability of generalizing the problem under consideration to more complex structural elements. The results obtained in this paper can be used in problems of thermal non-destructive testing.

Thermal control is based on monitoring, measuring and analyzing changes in the temperature background on the surface and inside the monitored objects. Thermal control is applicable to the object under one main condition, namely, the presence in the controlled object of convection of heat flows. The process of recycling thermal energy to the environment, the allocation or absorption of heat within the object, movement through the entire volume of the object, invariably leads to a slow change in temperature relative to the surrounding space. The nature of the distribution of the temperature background along the surface of the monitored object is an important technical parameter in the thermal control method, since it contains data on the process of heat transfer, the constant and temporary modes of operation of the object, its external and internal arrangement, and the presence of hidden internal defects. The occurrence of heat fluxes in a controlled object is due to external influences and various internal causes.

The aim of the work is to increase the effectiveness of the method of nondestructive testing of aerospace products. One of the methods of nondestructive testing is thermal diagnostics, which makes it possible to draw conclusions about the internal structure of the objects under study from the measured thermal fields. Non-destructive testing methods are the main means of testing the reliability of assemblies and aggregates of aerospace and construction equipment.

Among the methods of non-destructive testing, the method of thermal diagnostics is one of the main methods for detecting a cavity-type defect, since it allows for diagnostics in a non-contact way.

2. Mathematical model

The model problem of thermal diagnostics in the following formulation is considered. An element of a product of aerospace engineering in the form of a plate with a cavity of rectangle type is considered.

IOP Conf. Series: Materials Science and Engineering **683** (2019) 012015 doi:10.1088/1757-899X/683/1/012015

The plate on one side is heated by a stationary flow, the two sides of the plate are thermally insulated, and the thermal imager measures the thermal flux of radiation from the fourth side. The boundary conditions [2-6] on each side are recorded.

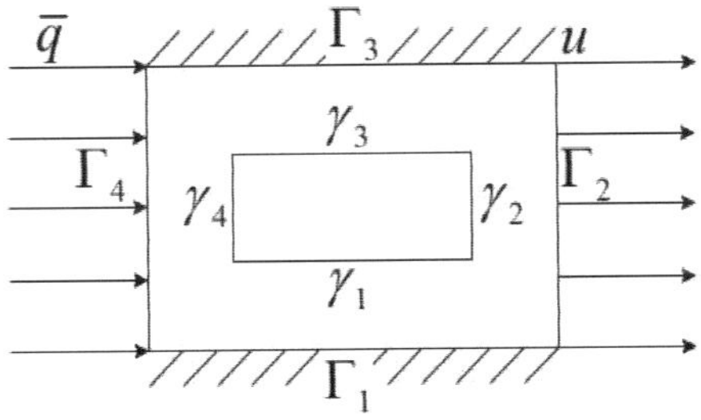

Figure 1. Type of plate boundary conditions

The problem under consideration reduces to a parabolic equation with nonlinear boundary conditions and taking into account reradiation. This problem was solved by the finite difference method together with the Fredholm integral equation of the second kind [1].

$$q(x) - k(x)\int_\gamma G(x,\xi)q(\xi)d\xi = q_0(x) \tag{1}$$

$$at\ \Gamma_1:\ K\frac{\partial u}{\partial n} = 0 \tag{2}$$

$$at\ \Gamma_2:\ K\frac{\partial T}{\partial n} + \alpha(T - T_c) + \sigma(T^4 - T_c^4) = 0 \tag{3}$$

$$at\ \Gamma_3:\ K\frac{\partial u}{\partial n} = 0 \tag{4}$$

$$at\ \Gamma_4:\ K\frac{\partial u}{\partial n} + q_0(x) = 0 \tag{5}$$

$$at\ \gamma_1,\gamma_2,\gamma_3,\gamma_4:\ K\frac{\partial u}{\partial n} + \alpha q(x) = 0 \tag{6}$$

The results of solving this problem make it possible to improve the accuracy of determining the location of the defect and show the advisability of generalizing the problem under consideration to more complex structural elements.

The obtained results showed the importance of accounting for reradiation in solving the problem of restoring geometric dimensions and location of defects in the plate.

3. Formulation of the inverse problem

The inverse problem is to restore the geometric dimensions of a plate with a cavity-type defect. This problem reduces to an optimization problem with four parameters (Figure 2).

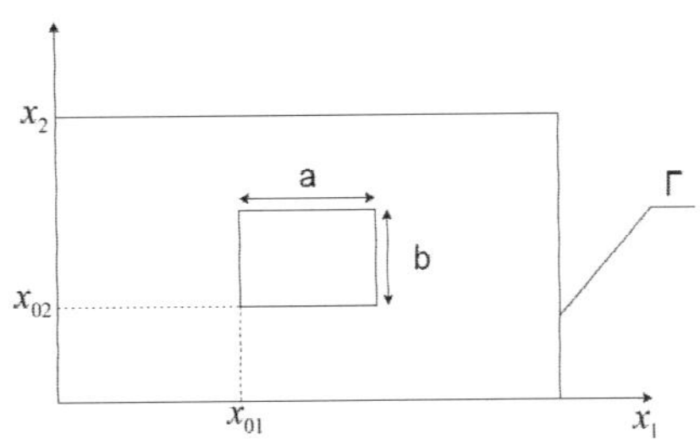

Figure 2. Parameters for restoring the geometric dimensions of the cavity

$$T(\Gamma) = A(x_{01,}x_{02,}a,b) \tag{7}$$

$$\left\| T(\Gamma) - T(x_{01}, x_{02}, a, b) \right\|^2 \underset{x_{01}, x_{02}, a, b}{\rightarrow} \min \tag{8}$$

4. The coordinated problem of heat exchange by radiation and thermal conductivity

Of greatest interest is the problem of taking into account reradiation for non-convex bodies, when it is necessary to take heat transfers into the body. Here, such a coordinated problem of taking into account heat transfer by radiation and thermal conductivity is considered under the conditions of the two-dimensional problem for the next model area (Figure 1)

For non-convex bodies or a system of bodies, it is necessary to take into account the heat fluxes that fall on the surface elements of the body from other parts of the surface. This leads to additional heat fluxes, an increase in surface temperature and, thus, an increase in the thermal radiation flux itself. In such conditions it is natural to talk about additional radiation, re-emission.

The task of calculating the processes of heat exchange must be considered in an agreed formulation. The intensity of thermal radiation depends on the surface temperature, which, in turn, depends on the heat fluxes on the surface, on the thermal radiation.

Of greatest interest is the problem of taking into account reradiation for non-convex bodies, when it is necessary to take heat transfers into the body. Here, such a coordinated problem of taking into account heat transfer by radiation and thermal conductivity is considered under the conditions of the two-dimensional problem for the next model area (Figure 1):

Let us consider the formulation of a stationary problem. The environment is assumed to be isotropic, so the thermal state is described by the equation:

$$\frac{\partial^2 u}{\partial x_1^2} + \frac{\partial^2 u}{\partial x_2^2} = 0 \tag{9}$$

We assume that the boundaries Γ_1, Γ_3 are thermally insulated.

We write the boundary conditions:

$$K\frac{\partial u}{\partial n} = 0,\ x \in \Gamma_1, \Gamma_3 \tag{10}$$

The boundary Γ_4 is supplied with heat flow, the boundary condition:

$$K\frac{\partial u}{\partial n} + q^*(x) = 0,\ x \in \Gamma_4 \tag{11}$$

On the remaining part of the boundary Γ_2 there is a heat exchange with the environment, the boundary condition:

$$K\frac{\partial u}{\partial n} + \alpha(u - u_c) + \sigma(u^4 - u_c^4) = 0,\ x \in \Gamma_2 \tag{12}$$

We rewrite the boundary condition in the following form:

$$K\frac{\partial u}{\partial n} + \alpha u + \sigma u^4 = \alpha u_c + \sigma u_c^4,\ x \in \Gamma_2 \tag{13}$$

$$K\frac{\partial u}{\partial n} + \sigma_1 u + \sigma_2 u^4 = \tilde{q},\ x \in \Gamma_2 \tag{14}$$

Where

$$\tilde{q} = \alpha u_c + \sigma u_c^4 \tag{15}$$

Where u - temperature of a solid at a point $(x_1, x_2) \in \Omega$,

u_c - ambient temperature,

coefficient $\sigma_1 > 0$ characterizes convective heat exchange with the environment, $\sigma_2 > 0$ - blackness.

The γ carries heat transfer by radiation, the boundary condition is written as::

$$K\frac{\partial u}{\partial n} + q_{рез}(x) = 0,\ x \in \gamma_1, \gamma_2, \gamma_3, \gamma_4, \tag{16}$$

Where K - coefficient of thermal conductivity,

$q_{рез}(x)$ - resultant radiation flux.

Let $q(x)$ be the radiation flux at the point x of the surface γ. It is defined as the sum of its own radiation $q_0(x)$ and part of the incident radiation, which is reflected. For $q(x)$ and part of the incident radiation, which is reflected:

$$q(x) - k(x)\int_\gamma G(x,\xi)q(\xi)d\xi = q_0(x) \tag{17}$$

- Fredholm integral equation of the second kind,

Where $0 < k(x) < 1$ - reflection coefficient.

The intrinsic radiation is due only to the body temperature, and in accordance with the Stefan-Boltzmann law, the right-hand side is determined by the expression:

$$q_0(x) = \sigma_2(x) g^4(x), \tag{18}$$

where $\sigma_2(x)$ - emissivity of the body.

$$q_{рез}(x) = q(x) - k(x)\int_\gamma G(x,\xi)q(\xi)d\xi \tag{19}$$

The kernel of the integral equation $G(x,\xi)$ is defined by the formula:

$$G(x,\xi) = \frac{1}{\pi r(x,\xi)}\cos\big(r(x,\xi), n(x)\big)\cos\big(r(\xi,x), n(\xi)\big) \tag{20}$$

Where $r(x,\xi)$ - distance between points x and ξ,

$\cos\big(n(\xi), r\big)$ - The cosine of the angle between the normal k and the segment joining x and ξ.

We note some properties of the integral equation (4), which is an integral Fredholm equation of the second kind. When $k(x) = const$ the kernel of the integral equation, as follows from (14), is symmetric. In addition, the kernel is nonnegative, i.e.

$$G(x,\xi) = G(\xi,x) \geq 0 \tag{21}$$

For positive $k(x) \neq const$ the integral equation (11) can easily be reduced to an integral equation with a symmetric kernel.

The asymmetry is caused by a factor $k(x)$. We divide equation (11) by $(k(x))^{\frac{1}{2}}$ and introduce a new unknown::

$$\hat{q}(x) = \frac{q(x)}{\left(k(x)\right)^{\frac{1}{2}}} \tag{22}$$

Equation (11) takes the form:

$$\hat{q}(x) - k(x)\int_{\gamma} \hat{G}(x,\xi)\hat{q}(\xi)d\xi = \hat{q}_0(x),\ x \in \gamma \tag{23}$$

Here the right-hand side:

$$\hat{q}_0(x) = \frac{q_0(x)}{\left(k(x)\right)^{\frac{1}{2}}} \tag{24}$$

And the kernel looks like this:

$$\hat{G}(x,\xi) = \left(k(x)k(\xi)\right)^{\frac{1}{2}} G(x,\xi) \tag{25}$$

Because of the symmetry of the kernel $G(x,\xi)$ the kernel $\hat{G}(x,\xi)$ of the integral equation is also symmetric.

5. Block diagram of the solution of the problem

We represent the algorithm of the Hook-Jeeves method in the form of a block diagram (Figure 3).

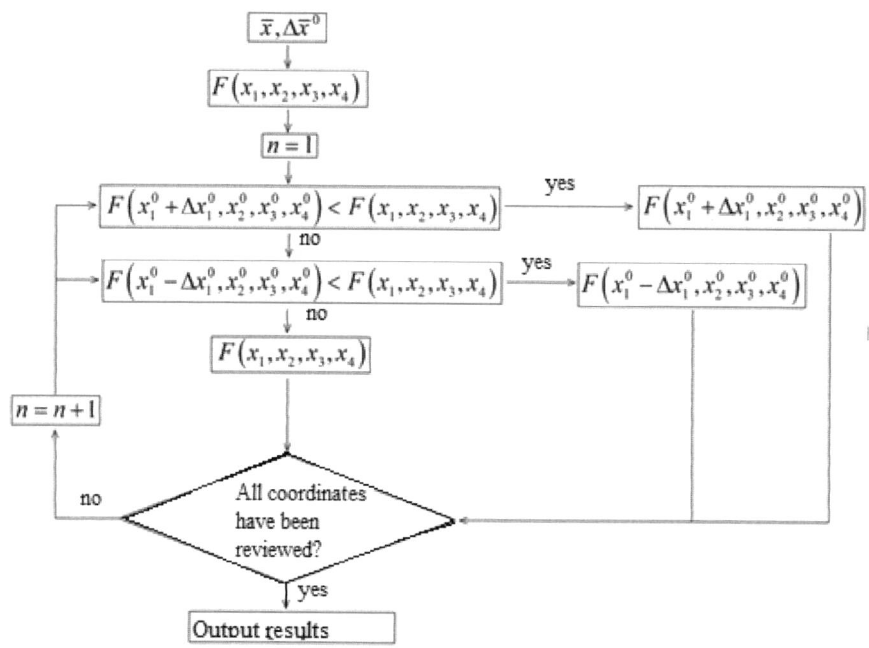

Figure 3. Block diagram of the Hook-Jeeves method for the inverse problem

x_1, x_2 - coordinates of the lower left corner of the cavity; x_3 - length of the upper and lower walls of the cavity; x_4 - length of the left and right walls of the cavity

The choice of the initial condition is as follows: the defect is located in the middle of the plate and its geometric dimensions are twice less than the specified dimensions of the plate.

6. The results of solving the inverse problem

The Hook-Jeeves method was programmed in C ++ in Visual Studio. The geometric dimensions of the cavity were restored. Figure 3 shows the results. Blue contour - the present location of the cavity

Figure 4. Results of solving the inverse problem

7. Conclusion

The results of solving this problem show the importance of taking into account reradiation and allow increasing the accuracy of defect location and show the expediency of generalizing the problem under consideration to more complex structural elements. The results obtained in this paper can be used in problems of thermal non-destructive testing.

Acknowledgments

The work was carried out with support of state task of Ministry for Science and Education № 9.5387.2017/БЧ

References

[1] Samarskij A A, Vabishhevich P N 2003 Vychislitel'naja teploperedacha. M.: Editorial URSS 784
[2] Kartashov Je M, Kudinov V A, Stefanjuk E V 2012 *Tehnicheskaja termodinamika i teploperedacha* ed M.: Jurajt, 576
[3] Karslou U, Eger D 1964 *Teploprovodnost' tverdyh tel*. M Nauka,. 488
[4] Galin N M , Kirillov P L 1987 *Teplomassoobmen* M.:Jenergoatomizdat 376
[5] Kartashov Je M 1989 *Analiticheskie metody v teploprovodnosti tverdyh tel* ed M.: Vyssh. shk. 550
[6] Vavilov V P 2009 *Infrakrasnaja termografija i teplovoj kontrol'* ed M.: ID Spektr 544
[7] Bloh A G, Zhuravlev Ju A and Ryzhkov L N 1991 *Teploobmen izlucheniem* ed M Jenergoatomizdat 432
[8] Bio M 1975 Variacionnye principy v teorii teploobmena ed M.: Jenergija 209
[9] Miheev M.A 1949 *Osnovy teploperedachi* M.-L. Gosjenergoizdat 397
[10] Gossorg Zh 1988 *Infrakrasnaja termografija. Osnovy. Tehnika. Primenenie.* M Mir 420
[11] Kljuev V. V. 1986 *Pribory dlja nerazrushajushhego kontrolja materialov i izdelij* M.: Mashinostroenie 488
[12] Drozdov V A, Suharev V I 1987 *Termografija v stroitel'stve* M Strojizdat 237
[13] Aliev R G 2005 *Uravnenija v chastnyh proizvodnyh* M Jekzamen 320
[14] Samarskij A A, Nikolaev E S 1978 *Metody reshenija setochnyh uravnenij* ed Moskva: Nauka 592
[15] Titov V A 1987 *Prjamye i obratnye zadachi himicheskoj termodinamiki* Novosibirsk: Nauka 144
[16] Alifanov O M 1988 *Obratnye zadachi teploobmena* M Mashinostroenie 280
[17] Dimitrienko Yu I 2012 Asymptotic theory of multilayer thin plates. *Herald of the Bauman Moscow State Technical University Series Natural Sciences* **3** 86-100

[18] Dimitrienko Yu I, Yakovlev D O 2015 The Asymptotic Theory of Thermoelasticity of Multilayer Composite Plates. *Composites: Mechanics,Applications. An International Journal.* **6** 13-51.

[19] Dimitrienko Yu I, Dimitrienko I D and Sborschikov S V 2015 Multiscale Hierarchical Modeling of Fiber Reinforced Composites by Asymptotic Homogenization Method. *Applied Mathematical Sciences* 145 **9** 7211-7220

[20] Dimitrienko Yu I, Gubareva E A, Markevich M N., Sborschikov S V 2016 Mathematical modelling of dielectric properties of nanostructural composites using asymptotic homogenizing method *Herald of the Bauman Moscow State Technical University Series Natural Sciences* **1** 76-89.

[21] Dimitrienko Yu I, Gubareva E A, Sborschikov S V 2014 Asymptotic theory of constructively orthotropic plates with a two-period structure *Matematicheskoe modelirovanie i chislennye metody – Mathematical Modeling and Computational Method* **1** 36-57

[22] Dimitrienko Yu I 2010 *Nonlinear Continuum Mechanics and Large Inelastic Deformations.* Springer 722

IOP Conf. Series: Materials Science and Engineering **683** (2019) 012016 doi:10.1088/1757-899X/683/1/012016

On the Mixed Steklov–Neumann and Steklov-type Biharmonic Problems in Unbounded Domains

Hovik Matevossian[1,2]

[1] Federal Research Center "Computer Science and Control", Russian Academy of Sciences, Vavilov str., 40, Moscow 119333, Russia

[2] Moscow Aviation Institute (National Research University), Volokolamskoe Shosse, 4, Moscow 125993, Russia

E-mail: hmatevossian@graduate.org

Abstract. We study the properties of generalized solutions in unbounded domains and the asymptotic behavior of solutions of elliptic boundary value problems at infinity. Moreover, we study the unique solvability of the mixed Steklov–Neumann and Steklov-type biharmonic problems in unbounded domains with a compact and non-compact boundaries (in particular, the exterior of a compact set, half-space, the domain with conical points) under the assumption that generalized solutions of these problems has a bounded Dirichlet (energy) integral with weight $|x|^a$. Depending on the value of the parameter a, we obtained uniqueness (non-uniqueness) theorems of these problems or present exact formulas for the dimension of the space of solutions.

Keywords: biharmonic operator; mixed Steklov–Neumann; Steklov-type problem; Dirichlet integral; weighted spaces

MSC: 35J35; 35J40; 31B30

1. Introduction

Let Ω be an unbounded domain in $\mathbb{R}^n$, $n \geq 2$, $x = (x_1, \ldots, x_n)$, and $|x| = \sqrt{x_1^2 + \cdots + x_n^2}$.

In the domain Ω we consider the following mixed problems for the biharmonic equation

$$\Delta^2 u = 0 \tag{1}$$

with the Steklov–Neumann boundary conditions

$$u\big|_{\Gamma_1} = \left(\Delta u + \tau \frac{\partial u}{\partial \nu}\right)\bigg|_{\Gamma_1} = 0, \quad \Delta u\big|_{\Gamma_2} = \frac{\partial \Delta u}{\partial \nu}\bigg|_{\Gamma_2} = 0, \tag{2}$$

and the Steklov-type boundary conditions

$$\frac{\partial u}{\partial \nu}\bigg|_{\partial\Omega} = \left(\frac{\partial \Delta u}{\partial \nu} + \tau\, u\right)\bigg|_{\partial\Omega} = 0, \tag{3}$$

where $\overline{\Gamma}_1 \cup \overline{\Gamma}_2 = \partial\Omega$, $\Gamma_1 \cap \Gamma_2 = \emptyset$, $\mathrm{mes}_{n-1}\Gamma_1 \neq 0$, $\nu = (\nu_1, \ldots, \nu_n)$ is the outer unit normal vector to $\partial\Omega$, $\tau \in C(\partial\Omega)$, $\tau \geq 0$, $\tau \not\equiv 0$, and $\tau > 0$ on a set of positive $(n-1)$ – dimensional measure on $\partial\Omega$, $\overline{\Omega} = \Omega \cup \partial\Omega$ is the closure of Ω.

Elliptic problems with parameters in the boundary conditions have been called Steklov or Steklov-type problems since their first appearance in [32]. For the biharmonic operator, these conditions were first considered in [2], [13] and [30], whose authors the isoperimetric properties of the first eigenvalue were studied.

Note that standard elliptic regularity results are available in [6]. The monograph covers higher order linear and nonlinear elliptic boundary value problems, mainly with the biharmonic or polyharmonic operator as leading principal part. The underlying models and, in particular, the role of different boundary conditions are explained in detail. As for linear problems, after a brief summary of the existence theory and L^p and Schauder estimates, the focus is on positivity. The required kernel estimates are also presented in detail.

In [5] and [6], the spectral and positivity preserving properties for the inverse of the biharmonic operator under Steklov and Navier boundary conditions are studied. These are connected with the first Steklov eigenvalue. It is shown that the positivity preserving property is quite sensitive to the parameter involved in the boundary condition. Moreover, positivity of the Steklov boundary value problem is linked with positivity under boundary conditions of Dirichlet and Navier type.

In [4], the boundary value problems for the biharmonic equation and the Stokes system are studied in a half space, and, using the Schwartz reflection principle in weighted L^q -space, the uniqueness of solutions of the Stokes system or the biharmonic equation is proved.

As is well known, if Ω is an unbounded domain, one should additionally characterize the behavior of the solution at infinity. As a rule, to this end, one usually poses either the condition that the Dirichlet (energy) integral is finite or a condition on the character of vanishing of the modulus of the solution as $|x| \to \infty$. Such conditions at infinity are natural and were studied by several authors (e.g., [10]- [12]).

In the present note, this condition is the boundedness of the weighted Dirichlet integral:

$$D_a(u,\Omega) \equiv \int_\Omega |x|^a \sum_{|\alpha|=2} |\partial^\alpha u|^2 \, dx < \infty, \quad a \in \mathbb{R}.$$

In various classes of unbounded domains with finite weighted Dirichlet (energy) integral, one of the author [14]- [27] studied uniqueness (non–uniqueness) problem and found the dimensions of the spaces of solutions of boundary value problems for the elasticity system and the biharmonic (polyharmonic) equation.

By developing an approach based on the use of Hardy type inequalities [8], [10]- [12], in the present note, we obtain a uniqueness (non–uniqueness) criterion for a solution of the mixed Dirichlet–Neumann and Dirichlet–Navier problems for the biharmonic equation.

Notation: $C_0^\infty(\Omega)$ is the space of infinitely differentiable functions in Ω with compact support in Ω.

We denote by $H^m(\Omega,\Gamma)$, $\Gamma \subset \overline{\Omega}$, the Sobolev space of functions in Ω obtained by the completion of $C^\infty(\overline{\Omega})$ vanishing in a neighborhood of Γ with respect to the norm

$$||u; H^m(\Omega,\Gamma)|| = \left(\int_\Omega \sum_{|\alpha|\le m} |\partial^\alpha u|^2 dx \right)^{1/2}, \quad m = 1, 2,$$

where $\partial^\alpha \equiv \partial^{|\alpha|}/\partial x_1^{\alpha_1} \dots \partial x_n^{\alpha_n}$, $\alpha = (\alpha_1, \dots, \alpha_n)$ is a multi-index, $\alpha_i \ge 0$ are integers, and $|\alpha| = \alpha_1 + \cdots + \alpha_n$; if $\Gamma = \emptyset$, we denote $H^m(\Omega,\Gamma)$ by $H^m(\Omega)$.

$\overset{\circ}{H}{}^m(\Omega)$ is the space obtained by the completion of $C_0^\infty(\Omega)$ with respect to the norm $||u; H^m(\Omega)||$;

$\overset{\circ}{H}{}^{m}_{loc}(\Omega)$ is the space obtained by the completion of $C_0^\infty(\Omega)$ with respect to the family of semi-norms

$$\|u; H^m(\Omega \cap B_0(R))\| = \left(\int\limits_{\Omega\cap B_0(R)} \sum_{|\alpha|\le m} |\partial^\alpha u|^2 \, dx \right)^{1/2}$$

for all open balls $B_0(R) := \{x : |x| < R\}$ in $\mathbb{R}^n$ for which $\Omega \cap B_0(R) \neq \emptyset$.

Let $\binom{n}{k}$ be the (n,k) - binomial coefficient, $\binom{n}{k}$=0 for $k > n$.

2. Mixed Steklov–Neumann Biharmonic Problem in the Exterior of a Compact Set

Let $\Omega = \mathbb{R}^n \setminus \overline{G}$ with the boundary $\partial\Omega \in C^2$, where G is a bounded simply connected domain (or a union of finitely many such domains) in $\mathbb{R}^n$, $0 \in G$.

Definition 2.1. *A solution of the homogenous biharmonic equation (1) in Ω is a function $u \in H^2_{loc}(\Omega)$ such that, for every function $\varphi \in C_0^\infty(\Omega)$, the following integral identity holds:*

$$\int_\Omega \Delta u \, \Delta\varphi \, dx = 0.$$

Lemma 2.2. *Let u be a solution of equation (1) in Ω such that $D_a(u,\Omega) < \infty$. Then*

$$u(x) = P(x) + \sum_{\beta_0<|\alpha|\le\beta} \partial^\alpha\Gamma(x) C_\alpha + u^\beta(x), \quad x \in \Omega, \tag{4}$$

where $P(x)$ is a polynomial, $\operatorname{ord} P(x) < m_0 = \max\{2, 2 - n/2 - a/2\}$, $\beta_0 = 2 - n/2 + a/2$, *$\Gamma(x)$ is the fundamental solution of equation (1),* $C_\alpha = \text{const}$, *$\beta \ge 0$ is an integer, and the function u^β satisfies the estimate:*

$$|\partial^\gamma u^\beta(x)| \le C_{\gamma\beta} |x|^{3-n-\beta-|\gamma|}, \; C_{\gamma\beta} = \text{const},$$

for every multi-index γ.

Remark 2.3. *As is known [31], the fundamental solution $\Gamma(x)$ of the biharmonic equation has the form*

$$\Gamma(x) = \begin{cases} C|x|^{4-n}, \; if\, 4-n<0\, or\, n\, is\, odd, \\ C|x|^{4-n}\ln|x|, \; if\, 4-n \ge 0\, and\, n\, is\, even. \end{cases}$$

Proof of Lemma 2.2 Consider the function $v(x) = \theta_N(x)u(x)$, where $\theta_N(x) = \theta(|x|/N), \theta \in C^\infty(\mathbb{R}^n)$, $0 \le \theta \le 1$, $\theta(s) = 0$ for $s \le 1$, $\theta(s) = 1$ for $s \ge 2$, while $N \gg 1$ and $G \subset \{x : |x| < N\}$. We extend v to $\mathbb{R}^n$ by setting $v = 0$ on $G = \mathbb{R}^n \setminus \overline{\Omega}$.

Then the function v belongs to $C^\infty(\mathbb{R}^n)$ and satisfies the equation

$$\Delta^2 v = f,$$

where $f \in C_0^\infty(\mathbb{R}^n)$ and $\operatorname{supp} f \subset \{x : |x| < 2N\}$. It is easy to see that $D_a(v, \mathbb{R}^n) < \infty$.

We can now use Theorem 1 of [9] since it is based on Lemma 2 of [9], which imposes no constraint on the *sign* of σ. Hence, the expansion

$$v(x) = P(x) + \sum_{\beta_0<|\alpha|\le\beta} \partial^\alpha\Gamma(x) C_\alpha + v^\beta(x),$$

holds for each a, where $P(x)$ is a polynomial of order $\operatorname{ord} P(x) < m_0 = \max\{2, 2 - n/2 - a/2\}$, $\beta_0 = 2 - n/2 + a/2$, $C_\alpha = \text{const}$ and

$$|\partial^\gamma v^\beta(x)| \le C_{\gamma\beta} |x|^{3-n-\beta-|\gamma|}, \quad C_{\gamma\beta} = \text{const}\,.$$

Therefore, by the definition of v, we obtain (4). The proof of Lemma 2.2 is complete.

Definition 2.4. *By a solution of the mixed Steklov-Neumann problem (1), (2), we mean a function* $u \in \overset{\circ}{H}{}^{2}_{loc}(\Omega) \cap \overset{\circ}{H}{}^{1}_{loc}(\Omega)$, $u = 0$ *on* Γ_1, *such that, for every function* $\varphi \in C_0^\infty(\mathbb{R}^n)$, $\varphi = 0$ *on* Γ_1, *the following integral identity holds:*

$$\int_\Omega \Delta u \, \Delta\varphi \, dx + \int_{\Gamma_1} \tau \frac{\partial u}{\partial \nu} \frac{\partial \varphi}{\partial \nu} \, ds = 0. \tag{5}$$

Theorem 2.5. *The mixed Steklov-Neumann problem (1),(2) with the condition* $D(u,\Omega) < \infty$ *has* $n+1$ *linearly independent solutions.*

Proof. For any nonzero vector A in $\mathbb{R}^n$, we construct a generalized solution u_A of the biharmonic equation (1) with the boundary conditions

$$u_A(x)\big|_{\Gamma_1} = (Ax)\big|_{\Gamma_1}, \quad \left(\Delta u_A + \tau \frac{\partial u_A(x)}{\partial \nu}\right)\Big|_{\Gamma_1} = \tau \frac{\partial (Ax)}{\partial \nu}\Big|_{\Gamma_1}, \quad \Delta u_A\big|_{\Gamma_2} = \frac{\partial \Delta u_A}{\partial \nu}\Big|_{\Gamma_2} = 0, \tag{6}$$

and the condition

$$\chi(u_A,\Omega) \equiv \begin{cases} \displaystyle\int_\Omega \left(\frac{|u_A(x)|^2}{|x|^4} + \frac{|\nabla u_A(x)|^2}{|x|^2} + |\nabla\nabla u_A(x)|^2 \right) dx < \infty \\ \qquad\qquad \text{for } n > 4, \\ \displaystyle\int_\Omega \left(\frac{|u_A(x)|^2}{||x|^2 \ln |x||^2} + \frac{|\nabla u_A(x)|^2}{||x| \ln |x||^2} + |\nabla\nabla u_A(x)|^2 \right) dx < \infty \\ \qquad\qquad \text{for } 2 \le n \le 4, \end{cases} \tag{7}$$

for $A, x \in \mathbb{R}^n$, where Ax denotes the standard scalar product of A and x.

Such a solution of problem (1), (6) can be constructed by the variational method [31], minimizing the functional

$$\Phi(v) = \frac{1}{2} \int_\Omega |\Delta v|^2 \, dx$$

in the class of admissible functions $\{v \colon v \in H^2(\Omega),\ v(x)\big|_{\Gamma_1} = (Ax)\big|_{\Gamma_1},\ \left(\Delta v + \tau \frac{\partial v(x)}{\partial \nu}\right)\big|_{\Gamma_1} = \tau \frac{\partial (Ax)}{\partial \nu},\ \Delta v\big|_{\Gamma_2} = \frac{\partial \Delta v}{\partial \nu}\big|_{\Gamma_2} = 0,\ v \text{ is compactly supported in } \overline{\Omega}\}$.

The validity of condition (7) as a consequence of the Hardy inequality follows from the results in [10]– [12].

Now, for any arbitrary number $e \neq 0$, we construct a generalized solution u_e of equation (1) with the boundary conditions

$$u_e\big|_{\Gamma_1} = e, \quad \left(\Delta u_e + \tau \frac{\partial u_e}{\partial \nu}\right)\Big|_{\Gamma_1} = 0, \quad \Delta u_e\big|_{\Gamma_2} = \frac{\partial \Delta u_e}{\partial \nu}\Big|_{\Gamma_2} = 0, \tag{8}$$

and the condition

$$\chi(u_e,\Omega) \equiv \begin{cases} \displaystyle\int_\Omega \left(\frac{|u_e(x)|^2}{|x|^4} + \frac{|\nabla u_e(x)|^2}{|x|^2} + |\nabla\nabla u_e(x)|^2 \right) dx < \infty \\ \qquad\qquad \text{for } n > 4, \\ \displaystyle\int_\Omega \left(\frac{|u_e(x)|^2}{||x|^2 \ln |x||^2} + \frac{|\nabla u_e(x)|^2}{||x| \ln |x||^2} + |\nabla\nabla u_e(x)|^2 \right) dx < \infty \\ \qquad\qquad \text{for } 2 \le n \le 4. \end{cases} \tag{9}$$

The solution of problem (1), (8) also is constructed by the variational method with the minimization of the corresponding functional in the class of admissible functions $\{v : v \in H^2(\Omega),\ v\big|_{\Gamma_1} = e,\ \left(\Delta v + \tau\frac{\partial v}{\partial \nu}\right)\big|_{\Gamma_1} = 0,\ \Delta v\big|_{\Gamma_2} = \frac{\partial \Delta v}{\partial \nu}\Big|_{\Gamma_2} = 0,\ v$ is compactly supported in $\overline{\Omega}\}$.

The condition (9) as a consequence of the Hardy inequality follows from the results in [10]–[12].

Consider the function $v(x) = (u_A(x) - Ax) - (u_e - e)$.

Obviously, v is a solution of problem (1), (2):

$$\Delta^2 v = 0, \quad x \in \Omega, \quad v\big|_{\Gamma_1} = \left(\Delta v + \tau\frac{\partial v}{\partial \nu}\right)\Big|_{\Gamma_1} = 0, \quad \Delta v\big|_{\Gamma_2} = \frac{\partial \Delta v}{\partial \nu}\Big|_{\Gamma_2} = 0.$$

One can easily see that $v \not\equiv 0$ and $D(v,\Omega) < \infty$.

To each nonzero vector $\mathbf{A} = (A_0, A_1, \ldots, A_n)$ in $\mathbb{R}^{n+1}$, there corresponds a nonzero solution $v_{\mathbf{A}} = (v_{A_0}, v_{A_1}, \ldots, v_{A_n})$ of problem (1), (2) with the condition $D(v_{\mathbf{A}},\Omega) < \infty$, and moreover,

$$v_{\mathbf{A}}(x) = u_A(x) - u_e - Ax + e.$$

Let $A_0, A_1, \ldots, A_n$ be a basis in $\mathbb{R}^{n+1}$. Let us prove that the corresponding solutions $v_{A_0}, v_{A_1}, \ldots, v_{A_n}$ are linearly independent. Let

$$\sum_{i=0}^{n} C_i v_{A_i} \equiv 0, \qquad C_i = \text{const}\,.$$

Set $W(x) \equiv \sum_{i=1}^{n} C_i A_i x - C_0 e$. We have

$$W(x) = \sum_{i=1}^{n} C_i u_{A_i}(x) - C_0 u_e,$$

$$\int_\Omega |x|^{-2} |\nabla W|^2\, dx < \infty, \quad n > 4,$$

$$\int_\Omega ||x| \ln |x||^{-2} |\nabla W|^2\, dx < \infty, \quad 2 \le n \le 4.$$

Let us show that

$$W(x) \equiv \sum_{i=1}^{n} C_i A_i x - C_0 e \equiv 0.$$

Let $T = \sum_{i=0}^{n} C_i A_i = (t_0, \ldots, t_n)$, where $A_0 = -e$. Then

$$\int_\Omega |x|^{-2} |\nabla W|^2\, dx = \int_\Omega |x|^{-2} (t_1^2 + \cdots + t_n^2)\, dx = \infty, \quad n > 4,$$

$$\int_\Omega ||x| \ln |x||^{-2} |\nabla W|^2\, dx = \int_\Omega ||x| \ln |x||^{-2} (t_1^2 + \cdots + t_n^2)\, dx = \infty, \quad 2 \le n \le 4,$$

if $T \neq 0$.

Consequently, $T = \sum_{i=0}^{n} C_i A_i = 0$, and since the vectors $A_0, A_1, \ldots, A_n$ are linearly independent, we obtain $C_i = 0,\ i = 0, 1, \ldots, n$.

Thus, the Steklov–Neumann problem (1), (2) with the condition $D(u,\Omega) < \infty$ has at least $n+1$ linearly independent solutions.

IOP Conf. Series: Materials Science and Engineering **683** (2019) 012016 doi:10.1088/1757-899X/683/1/012016

Let us prove that each solution u of problem (1), (2) with the condition $D(u,\Omega) < \infty$ can be represented as a linear combination of the functions $v_{A_0}, v_{A_1}, \dots, v_{A_n}$, i.e.

$$u = \sum_{i=0}^{n} C_i v_{A_i}, \qquad C_i = \text{const}\,.$$

Since $A_0, A_1, \dots, A_n$ is a basis in $\mathbb{R}^{n+1}$, it follows that there exists constants $C_0, C_1, \dots, C_n$ such that

$$A = \sum_{i=0}^{n} C_i A_i.$$

We set

$$u_0 \equiv u - \sum_{i=0}^{n} C_i v_{A_i}.$$

Obviously, the function u_0 is a solution of problem (1), (2), and $D(u_0,\Omega) < \infty$, $\chi(u_0,\Omega) < \infty$.

Let us show that $u_0 \equiv 0$, $x \in \Omega$. To this end, we substitute the function $\varphi(x) = u_0(x)\theta_N(x)$ into the integral identity (5) for the function u_0, where $\theta_N(x) = \theta(|x|/N)$, $\theta \in C^\infty(\mathbb{R})$, $0 \le \theta \le 1$, $\theta(s) = 0$ for $s \ge 2$ and $\theta(s) = 1$ for $s \le 1$; then we obtain

$$\int_\Omega (\Delta u_0)^2 \theta_N(x)\,dx + \int_{\Gamma_1} \tau \left|\frac{\partial u_0}{\partial \nu}\right|^2 \theta_N(x)\,ds = -J_1(u_0) - J_2(u_0), \tag{10}$$

where

$$J_1(u_0) = 2\int_\Omega \Delta u_0\, \nabla u_0\, \nabla\theta_N(x)\,dx, \qquad J_2(u_0) = \int_\Omega u_0\, \Delta u_0\, \Delta\theta_N(x)\,dx.$$

By applying the Cauchy–Schwarz inequality and by taking into account the conditions $D(u_0,\Omega) < \infty$ and $\chi(u_0,\Omega) < \infty$, one can easily show that $J_1(u_0) \to 0$ and $J_2(u_0) \to 0$ as $N \to \infty$. Consequently, by passing to the limit as $N \to \infty$ in (10), we obtain

$$\int_\Omega (\Delta u_0)^2 \theta_N(x)\,dx + \int_{\Gamma_1} \tau \left|\frac{\partial u_0}{\partial \nu}\right|^2 \theta_N(x)\,ds \to 0.$$

Using the integral identity

$$\int_\Omega (\Delta u_0)^2\,dx + \int_{\Gamma_1} \tau \left|\frac{\partial u_0}{\partial \nu}\right|^2 ds = 0,$$

we find that if u_0 is a solution of the homogeneous problem (1), (2), then $\Delta u_0 = 0$.

Therefore, we have

$$\Delta u_0 = 0, \quad x \in \Omega,$$

$$u_0\big|_{\Gamma_1} = \left(\Delta u_0 + \tau \frac{\partial u_0}{\partial \nu}\right)\bigg|_{\Gamma_1} = 0, \quad \Delta u_0\big|_{\Gamma_2} = \frac{\partial \Delta u_0}{\partial \nu}\bigg|_{\Gamma_2} = 0.$$

Hence, it follows [7, Ch.2] that $u_0 = 0$ in Ω. The relation

$$\int_{\partial\Omega} \tau \left|\frac{\partial u_0}{\partial \nu}\right|^2 ds = 0$$

implies that $u_0 \equiv 0$ on a set of a positive measure on $\partial\Omega$. The proof of the theorem is complete. □

Theorem 2.6. *The mixed Steklov–Neumann problem (1), (2) with the condition* $D_a(u,\Omega)<\infty$ *has:*

(i) the trivial solution for $n-2\le a<\infty$, $n>4$*;*

(ii) n *linearly independent solutions for* $n-4\le a<n-2$, $n>4$*;*

(iii) $n+1$ *linearly independent solutions for* $-n\le a<n-4$, $n>4$*;*

(iv) $k(r,n)$ *linearly independent solutions for* $-2r+2-n\le a<-2r+4-n$, $r>1$, $n>4$, *where*

$$k(r,n)=\binom{r+n}{n}-\binom{r+n-4}{n}.$$

The proof of Theorem 2.6 is based on Lemma 2.2 about the asymptotic expansion of the solution of the biharmonic equation and the Hardy type inequalities for unbounded domains [10]– [12]. In case *(iv)*, we need to determine the number of linearly independent solutions of the biharmonic equation (1), the degree of which not exceed the fixed number.

It is well know that the dimension of the space of all polynomials in $\mathbb{R}^n$ of degree $\le r$ is equal $\binom{r+n}{n}$ [28]. Then the dimension of the space of all biharmonic polynomials in $\mathbb{R}^n$ of degree $\le r$ is equal to

$$\binom{r+n}{n}-\binom{r+n-4}{n},$$

since the biharmonic equation is the vanishing of some polynomial of degree $r-4$ in $\mathbb{R}^n$. If we denote by $k(r,n)$ the number of linearly independent polynomial solutions of equation (1) whose degree do not exceed r and by $l(r,n)$ the number of linearly independent homogeneous polynomials of degree r, that are solutions of equation (1), then

$$k(r,n)=\sum_{s=0}^{r} l(s,n),$$

where

$$l(s,n)=\binom{s+n-1}{n-1}-\binom{s+n-5}{n-1}, \qquad s>0.$$

Further, we prove that the mixed Steklov–Neumann problem (1), (2) with the condition $D_a(u,\Omega)<\infty$ for $-2r+2-n\le a<-2r+4-n$ has equally $k(r,n)$ of linearly independent solutions.

3. Steklov-type Biharmonic Problem in a Half-Space

Let $x=(x',x_n)$, where $x'=(x_1,\dots,x_{n-1})\in\mathbb{R}^{n-1}$; then $\mathbb{R}^n_+=\{x=(x',x_n)\in\mathbb{R}^n : x'\in\mathbb{R}^{n-1}, x_n>1\}$ is an "open half–space" with the boundary $\partial\mathbb{R}^n_+ := \{x=(x',x_n)\in\mathbb{R}^n : x_n=1\}\equiv\mathbb{R}^{n-1}$. Let now $\Omega\equiv\mathbb{R}^n_+$ with the boundary $\partial\Omega\equiv\partial\mathbb{R}^n_+$.

Definition 3.1. *A function* u *is a solution of the Steklov-type biharmonic problem (1),(3) if* $u\in H^2_{loc}(\Omega)$, $\partial u/\partial\nu=0$ *on* $\partial\Omega$, *such that, for every function* $\varphi\in C_0^\infty(\mathbb{R}^n)$, $\partial\varphi/\partial\nu=0$ *on* $\partial\Omega$, *the following integral identity holds*

$$\int_\Omega \Delta u\,\Delta\varphi\,dx-\int_{\partial\Omega}\tau\,u\,\varphi\,ds=0.$$

Lemma 3.2. *Let* u *be a solution of equation (1) in* Ω *satisfying the boundary conditions (3) and the inequality* $|u(x)|\le C(1+|x|^k)$ *for all* $x\in\Omega$, *where* C *is a positive constant and* $k\ge 0$ *is an integer. Then* u *is a polynomial of degree at most* k, *i. e.,* $u(x)=P(x)$ *and* $\operatorname{ord}P(x)\le k$.

Proof. This lemma is an analogue of Liouville's theorem for systems of equations. It is valid for general Douglis–Nirenberg elliptic systems with constant coefficients. For $\Omega \equiv \mathbb{R}^n$ it was proved in [3] .

The proof for $\Omega \equiv \mathbb{R}^n_+$ is similar. Namely, in the theory of elliptic systems, the following Bernstein inequality is established:

$$\max_{|x|\le 1/2,\,|\alpha|=l} |\partial^\alpha u(x)| \le C_1 \max_{|x|\le 1} |u|, \quad l \ge 1, \tag{11}$$

where the constant C_1 is independent of u.

Let u be a solution of a homogeneous elliptic system with constant coefficients in the half-ball $|x| \le 1$, $x_n > 0$, and for $x_n = 0$ suppose that u satisfies the Shapiro–Lopatinskii zero boundary condition [1] (for more details, see [29]). Putting $x = \lambda y$ in (11) and using the homogeneity of the system and the boundary conditions, we get

$$\max_{|x|\le 1/2,\,|\alpha|=l} |\partial^\alpha u(x)| \le \lambda^{-1} \max_{|x|\le \lambda} |u|.$$

Hence, the hypotheses of Lemma 3.2 imply

$$\max_{|x|\le 1,\,|\alpha|=l} |\partial^\alpha u(x)| \le C_2\, \lambda^{k-l}.$$

Taking $l > k$ and letting λ tend to ∞, we see that $\partial^\alpha u(x) = 0$ for $|x| \le 1$. Hence, u is a polynomial of degree at most $l-1$ for $|x| \le 1$. Since the solution of an elliptic system is analytic, it follows that u is a polynomial in $\mathbb{R}^n_+$. The condition $|u(x)| \le C(1 + |x|^k)$ implies that the degree of this polynomial is at most k. The proof of Lemma 3.2 is complete. □

Theorem 3.3. *The Steklov–type biharmonic problem (1),(3) with the condition $D(u,\Omega) < \infty$ has n linearly independent solutions.*

Theorem 3.4. *The Steklov–type biharmonic problem (1),(3) with the condition $D_a(u,\Omega) < \infty$ has:*

(i) n linearly independent solution for $-n \le a < \infty$;

(ii) $k(r,n)$ linearly independent solutions for $-2r - n + 2 \le a < -2r - n + 4$, $r \ge 2$, where

$$k(r,n) = \binom{r+n}{n} - \binom{r+n-4}{n} - \binom{r+n-2}{n-1} - \binom{r+n-4}{n-1}.$$

The proof of Theorem 3.4 is based on Lemma 3.2, and in the case of *(ii)*, as in Theorem 2.6, it is necessary to determine the number of linearly independent solutions of a biharmonic equation, the degree of which do not exceed the fixed number. Further, the condition for the vanishing on the hyperplane $x_n = 1$ of the polynomial and its derivatives of x_n yields the equality to zero of certain polynomials of $(n-1)$ variables of degree $(r-j)$, $j = 1, 3$. Hence, the dimension of the space of solutions of the Steklov–type biharmonic problem is $k(r,n)$.

The results of the article are presented at the International Conference on International Conference on Composites "Advances in Composite Science and Technologies" (December 5–8, 2018, Moscow, Russia).

References

[1] Agmon S, Douglis A and Nirenberg L 1959, "Estimates near the boundary for solutions of elliptic partial differential equations satisfying general boundary conditions", *Commun. Pure Appl. Math.*, **12**, 623–727.

[2] Brock F 2001, "An isoperimetric inequality for eigenvalues of the Stekloff problem", *Z. Angew. Math. Mech. (ZAMM)*, **81**(1), 69–71.

[3] Douglis A and Nirenberg L 1955, "Interior estimates for elliptic systems of partial differential equations", *Commun. Pure Appl. Math.*, **8**, 503–538.
[4] Farwig R 1994, "A note on the reflection principle for the biharmonic equation and the Stokes system", *Acta Appl. Math.*, **34** 41–51.
[5] Gazzola F and Sweers G 2008, "On positivity for the biharmonic operator under Steklov boundary conditions", *Arch. Rational Mech. Anal.*, **188**(3), 399–427.
[6] Gazzola F, Grunau H.-Ch and Sweers G 2010, *Polyharmonic Boundary Value Problems: Positivity Preserving and Nonlinear Higher Order Elliptic Equations in Bounded Domains.* Lecture Notes Math. **1991**, Springer-Verlag.
[7] Gilbarg D and Trudinger N 1977, *Elliptic Partial Differential Equations of Second Order.* Berlin: Springer-Verlag.
[8] Egorov Yu V and Kondratiev V A 1996, *On Spectral Theory of Elliptic Operators.* Basel: Birkhauser.
[9] Kondratiev V A and Oleinik O A 1987, "On the behavior at infinity of solutions of elliptic systems with a finite energy integral", *Arch. Rational Mech. Anal.*, **99**(1), 75–99.
[10] Kondrat'ev V A and Oleinik O A 1988, "Boundary value problems for the system of elasticity theory in unbounded domains. Korn's inequalities", *Russ. Math. Surveys*, **43**(5), 65–119.
[11] Kondratiev V A and Oleinik O A 1990, "Hardy's and Korn's Inequality and their Application", *Rend. Mat. Appl.*, Serie VII. **10**(3), 641–666.
[12] Kon'kov A A 1995, "On the dimension of the solution space of elliptic systems in unbounded domains", *Russ. Acad. Sci. Sbornik Math.*, **80**(2), 411–434.
[13] Kuttler J R and Sigillito V G 1968, "Inequalities for membrane and Stekloff eigenvalues", *J. Math. Anal. Appl.*, **23**(1), 148–160.
[14] Matevosyan O A 1998, "On solutions of boundary value problems for a system in the theory of elasticity and for the biharmonic equation in a half-space", *Diff. Equations.*, **34**(6), 803–808.
[15] Matevosyan O A 2001, "The exterior Dirichlet problem for the biharmonic equation: Solutions with bounded Dirichlet integral", *Math. Notes*, **70**(3), 363–377.
[16] Matevossian O A 2001, "Solutions of exterior boundary value problems for the elasticity system in weighted spaces", *Sbornik Math.*, **192**(12), 1763–1798.
[17] Matevossian H A 2003, "On solutions of mixed boundary-value problems for the elasticity system in unbounded domains", *Izvestiya Math.*, **67**(5), 895–929.
[18] Matevosyan O A 2014, "On solutions of a boundary value problem for a polyharmonic equation in unbounded domains", *Russ. J. Math. Phys.*, **21**(1), 130–132.
[19] Matevossian H A 2015, "On solutions of the Dirichlet problem for the polyharmonic equation in unbounded domains", *P-Adic Numbers, Ultrametric Analysis, and Appl.*, **7**(1), 74–78.
[20] Matevosyan O A 2015, "Solution of a mixed boundary value problem for the biharmonic equation with finite weighted Dirichlet integral", *Diff. Equations*, **51**(4), 487–501.
[21] Matevossian O A 2015, "On solutions of the Neumann problem for the biharmonic equation in unbounded domains", *Math. Notes*, **98**, 990–994.
[22] Matevosyan O A 2016, "On solutions of the mixed Dirichlet–Navier problem for the polyharmonic equation in exterior domains", *Russ. J. Math. Phys.*, **23**(1), 135–138.
[23] Matevosyan O A 2016, "On solutions of one boundary value problem for the biharmonic equation", *Diff. Equations*, *52*(10), 1379–1383.
[24] Matevossian H A 2017, "On the biharmonic Steklov problem in weighted spaces", *Russ. J. Math. Phys.*, **24**(1), 134–138.
[25] Matevossian H A 2017, "On solutions of the mixed Dirichlet–Steklov problem for the biharmonic equation in exterior domains", *P-Adic Numbers, Ultrametric Analysis, and Appl.*, **9**(2), 151–157.
[26] Matevossian H A 2018, "On the Steklov–type biharmonic problem in unbounded domains", *Russ. J. Math. Phys.*, **25**:2, 271–276.
[27] Matevossian Hovik A 2019, "On the polyharmonic Neumann problem in weighted spaces", *Complex Variables & Elliptic Equations*, **64**:1, 1–7.
[28] Mikhlin S G 1977, *Linear Partial Differential Equations*, Vyssaya Shkola, Moscow (in Russian).
[29] Morrey C B 1966, *Multiple Integrals in the Calculus Variations*, Springer-Verlag, Berlin–New-York.
[30] Payne L E 1970, "Some isoperimetric inequalities for harmonic functions", *SIAM J. Math. Anal.*, **1**(3), 354–359.
[31] Sobolev S L 1988, *Some Applications of Functional Analysis in Mathematical Physics*, 3th ed., Nauka, Moscow; *Applications of Functional Analysis in Mathematical Physics.* AMS, Providence 1991.
[32] Stekloff W 1902, "Sur les problemes fondamentaux de la physique mathematique", *Ann. Sci. de l'E.N.S.*, 3^e serie, **19**, 191-259 et 455-490.

On the Modeling of Thin Bodies of Revolution

Nikabadze[1] M U, Bogatyrev[2] M A and Ulukhanyan[3] A R
[1,2] Lomonosov Moscow State University, Moscow, Russia
[1,3] Bauman Moscow State Technical University, Moscow, Russia

E-mail: nikabadze@mail.ru, bogatyrevmaxim@gmail.com, armine msu@mail.ru

Abstract. The new parametrization of the three-dimensional thin domain of an arbitrary body of rotation is considered, consisting in using several base surfaces in contrast to the classical approaches. The vector parametric equation is given. The geometric characteristics inherent to the new parameterization are determined. Expressions for the translation components of the unit tensor of the second rank and also relations connecting different bases and geometric characteristic generated by them are written. The determination of the moment of an arbitrary function is given. The motion equations and CR of the micropolar theory of elastic rotation bodies without a symmetry center of variable thickness are obtained under the new parametrization, from which equations are obtained for very thin and shallow rotation bodies, both variable and constant thickness. The equations of motion and CR of any approximation in the moments of unknown functions with respect to an arbitrary system of orthogonal polynomials and, in particular, for the system of Legendre polynomials are derived. The questions of setting initial-boundary problems are discussed.

1. Introduction

The use of composite materials of layered and fibrous structures in various branches of technology has led to the need to develop new methods for calculating and designing thin bodies. In connection with the use of thin bodies (one-, two-, three- and multilayer structures) there is a need to create new refined theories of thin bodies and improved methods for their calculation. One of the methods for constructing the theory of thin bodies is the analytical method. In particular, it is used the expansion in series of the system of orthogonal Legendre polynomials [1–20] (these decompositions are used equally to construct any theory of thin bodies [17]). In this work, we use the Legendre polynomial expansion for modeling micropolar bodies of revolution.

2. To new parametrization of the domain of a rotation body with one small size

We consider a new parametrization [21, 22] of the domain of a 3D Euclidean space that is bounded by two face surfaces $\overset{(-)}{S}$ and $\overset{(+)}{S}$ and a lateral surface Σ (Fig. 1). The face surface $\overset{(-)}{S}$ is called inner base surface, and the face surface $\overset{(+)}{S}$ is called outer base surface. The radius-vector of an arbitrary point of the domain of a thin body is introduced in the following form (Fig. 1)

$$\mathbf{r}(x',x^3)=\overset{(-)}{\mathbf{r}}(x')+x^3\mathbf{h}(x')=(1-x^3)\overset{(-)}{\mathbf{r}}(x')+x^3\overset{(+)}{\mathbf{r}}(x'),\;\; x'=(x^1,x^2),\;\; \forall x^3\in[0,1], \tag{1}$$

where relations $\overset{(-)}{\mathbf{r}}=\overset{(-)}{\mathbf{r}}(x')$, $\overset{(+)}{\mathbf{r}}=\overset{(+)}{\mathbf{r}}(x')$ are the vector parametric equations of the base surfaces $\overset{(-)}{S}$ and $\overset{(+)}{S}$, respectively, $x'=(x^1,x^2)$ is an arbitrary point on $\overset{(-)}{S}$, x^1 and x^2 is an curvilinear coordinates on inner base surface $\overset{(-)}{S}$. Vector $\mathbf{h}(x') = \overset{(+)}{\mathbf{r}}(x') - \overset{(-)}{\mathbf{r}}(x')$ performing topological

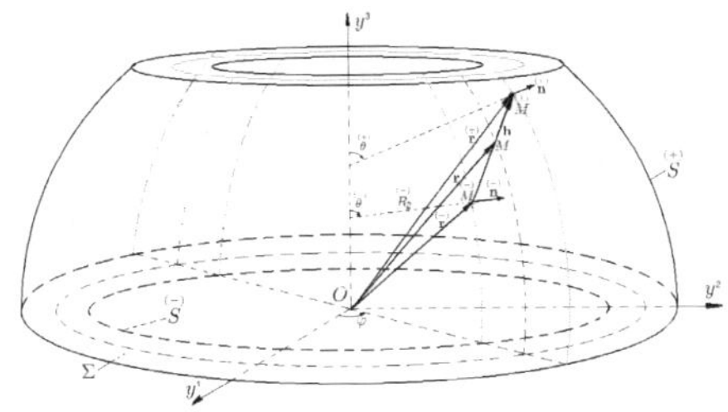

Figure 1. New parametrization of rotation body

mapping of the inner base surface $\overset{(-)}{S}$ on the outer surface $\overset{(+)}{S}$, in general is not perpendicular to the base surfaces. The relation (1) is exactly the vector parametric equation of the domain of a thin body. When presenting the material, the usual rules for tensor calculus are applied [21–23]. Uppercase and lowercase latin indices take values of $1, 2$ and $1, 2, 3$ respectively. In addition, short entries are used later on, e.g. $\overset{(*)}{M} \in \overset{(*)}{S}$, $* \in \{-, \emptyset, +\}$ or $\mathbf{r}_{\tilde{p}} = g^{\breve{q}}_{\tilde{p}}\mathbf{r}_{\breve{q}}$, $\sim, \smile \in \{-, \emptyset, +\}$ where $\emptyset$ denotes the empty set. The first entry means that if $* = -$ then $\overset{(-)}{M} \in \overset{(-)}{S}$; if $* = \emptyset$, then $M \in S$; if $* = +$, then $\overset{(+)}{M} \in \overset{(+)}{S}$. The second entry means that if $\sim = \emptyset$ and $\smile = -$, then $\mathbf{r}_{\underset{p}{+}} = g^{\bar{q}}_{\underset{p}{+}}\mathbf{r}_{\underset{q}{-}}$, etc. Going through all the values, one can get all the relations. Figure 1 shows the body of rotation with $x^1 = \theta$ (the angle between the rotation axis and the normal to $\overset{(-)}{S}$) and $x^2 = \varphi$ (azimuth angle). Further, we introduce the following notation for the derivatives of the $\mathbf{r}$ and $\overset{(*)}{\mathbf{r}}$ with respect to x^P at points $\overset{(*)}{M} \in \overset{(*)}{S}$, $* \in \{-, \emptyset, +\}$

$$\mathbf{r}_P \equiv \partial_P \mathbf{r} = \partial \mathbf{r}/\partial x^P, \quad \mathbf{r}_{\underset{P}{*}} = \partial \overset{(*)}{\mathbf{r}}/\partial x^P, \quad * \in \{-, +\}. \tag{2}$$

The pair of vectors $\mathbf{r}_{\underset{1}{*}}, \mathbf{r}_{\underset{2}{*}}$, $* \in \{-, \emptyset, +\}$, defined at points $\overset{(*)}{M} \in \overset{(*)}{S}$, $* \in \{-, \emptyset, +\}$ form two-dimensional covariant surface bases. It is well known that based on these bases [21–23], we can construct the corresponding contravariant bases $\overset{*}{\mathbf{r}}{}^1, \overset{*}{\mathbf{r}}{}^2$, $* \in \{-, \emptyset, +\}$. Taking into account the expression of the radius vector $\mathbf{r}$ (1) in the first relation of (2) and introducing denotation $\mathbf{h}_P \equiv \partial \mathbf{h}/\partial x^P \equiv \partial_P \mathbf{h}$ one can get

$$\mathbf{r}_P = \mathbf{r}_{\underset{P}{-}} + x^3\mathbf{h}_P = (1 - x^3)\mathbf{r}_{\underset{P}{-}} + x^3\mathbf{r}_{\underset{P}{+}}, \ \ \mathbf{r}_3 \equiv \partial_3 \mathbf{r} \equiv \partial \mathbf{r}/\partial x^3 = \mathbf{h}(x'), \ \ \forall x^3 \in [0, 1]. \tag{3}$$

Based on the second relation (3) we can assume

$$\mathbf{r}_{\underset{3}{-}} = \mathbf{r}_3 = \mathbf{r}_{\underset{3}{+}} \equiv \partial_3 \mathbf{r} = \mathbf{h}(x'), \quad \forall x^3 \in [0, 1]. \tag{4}$$

Relation (4) defines the spatial covariant bases $\mathbf{r}_{\underset{p}{*}}$, $* \in \{-, +\}$ at points $\overset{(*)}{M} \in \overset{(*)}{S}$, $* \in \{-, +\}$. Thus, the third basis vector of the spatial covariant bases at points $\overset{(*)}{M} \in \overset{(*)}{S}$, $* \in \{-, \emptyset, +\}$ is the same vector $\mathbf{h}(x')$. Given (4), relations (3) can be combined in the following form

$$\mathbf{r}_p(x', x^3) = \mathbf{r}_{\underset{p}{-}}(x') + x^3\mathbf{h}_p(x') = (1 - x^3)\mathbf{r}_{\underset{p}{-}}(x') + x^3\mathbf{r}_{\underset{p}{+}}(x'). \tag{5}$$

Three vectors $\mathbf{r}_{\underset{1}{\star}}, \mathbf{r}_{\underset{2}{\star}}, \mathbf{r}_{\underset{3}{\star}}$, $* \in \{-, \emptyset, +\}$ defined in the considered points $\overset{(*)}{M} \in \overset{(*)}{S}$, $* \in \{-, \emptyset, +\}$ form three-dimensional (spatial) covariant bases. It is well known that based on these bases [21–23], one can construct the corresponding contravariant bases $\mathbf{r}^{\overset{\star}{1}}, \mathbf{r}^{\overset{\star}{2}}, \mathbf{r}^{\overset{\star}{3}}$, $* \in \{-, \emptyset, +\}$. Further, the geometrical characteristics are written down as in [17, 21, 22] and in the important special case, when $\mathbf{h} \perp \overset{(-)}{S}$ $(\mathbf{h} = h\overset{(-)}{\mathbf{n}})$, for the rotation bodies we find

$$\begin{gathered}\overset{(-)}{\mathbf{r}}(\theta,\varphi) = a\mathbf{a} + f(a)\mathbf{k}_3, \quad \mathbf{a} = \cos\varphi\mathbf{k}_1 + \sin\varphi\mathbf{k}_2 \quad a = R_2\sin\theta, \\ \overset{(+)}{\mathbf{r}}(\theta,\varphi) = \overset{(-)}{\mathbf{r}}(\theta,\varphi) + \mathbf{h}, \quad \mathbf{r}(\theta,\varphi,x^3) = \overset{(-)}{\mathbf{r}}(\theta,\varphi) + x^3\mathbf{h}(\theta,\varphi),\end{gathered} \tag{6}$$

where to reduce the letter, instead of $\overset{(-)}{R}_2$ and $\overset{(-)}{\theta}$ the notations R_2 and θ are introduced, respectively. Further, introducing the notations $x^1 = \theta$ and $x^2 = \varphi$, it is easy to find the geometric characteristics (covariant and contravariant bases, Christoffel symbols, components of the unit tensor of the second rank, etc.) for the body of rotation under the new parameterization. To make shorter the work we will not go into detail, but we give some necessary relations without proof. Introducing $\mathbf{c} = \partial_\varphi\mathbf{a} = -\sin\varphi\mathbf{k}_1 + \cos\varphi\mathbf{k}_2$ $(\mathbf{a}\cdot\mathbf{c} = 0)$, according to (6) and easily proved formulas (the proof of the first formula can be found in [24])

$$\frac{da}{d\theta} = R_1\cos\theta, \ \partial_I f(a) = -R_1\cos\theta\delta_{1I}, \ \partial_I\mathbf{h} = g_{\underset{I}{+}}^{\bar{3}}\overset{(-)}{\mathbf{n}} - h\overset{(-)}{b}{}_{\underset{I}{-}}^{\bar{J}}\mathbf{r}_{\underset{J}{-}}, \quad h' = \frac{dh}{d\theta}, \ g_{\underset{I}{+}}^{\bar{3}} = h^{-1}h'\delta_{1I}$$

for the covariant and contravariant bases, and also for the orthonormal basis at the point $\overset{(-)}{M}$ we have expressions

$$\begin{gathered}\mathbf{r}_{\underset{1}{-}} = R_1\cos\theta\mathbf{a} - R_1\sin\theta\mathbf{k}_3, \quad \mathbf{r}_{\underset{2}{-}} = R_2\sin\theta\mathbf{c}, \quad \mathbf{r}_{\underset{3}{-}} = \mathbf{h} = h\overset{(-)}{\mathbf{n}} = h(\sin\theta\mathbf{a} + \cos\theta\mathbf{k}_3), \\ \mathbf{r}^{\bar{1}} = \frac{1}{R_1}(\cos\theta\mathbf{a} - \sin\theta\mathbf{k}_3), \ \mathbf{r}^{\bar{2}} = \frac{1}{R_2\sin\theta}\mathbf{c}, \ \mathbf{r}^{\bar{3}} = h^{-2}\mathbf{h} = h^{-1}\overset{(-)}{\mathbf{n}}, \ \sqrt{\overset{(-)}{g}} = hR_1R_2\sin\theta, \\ \mathbf{e}_{\underset{1}{-}} = \cos\theta\mathbf{a} - \sin\theta\mathbf{k}_3, \quad \mathbf{e}_{\underset{2}{-}} = \mathbf{c}, \quad \mathbf{e}_3 = \overset{(-)}{\mathbf{n}} = \sin\theta\mathbf{a} + \cos\theta\mathbf{k}_3, \quad \mathbf{e}_{\underset{p}{-}}\cdot\mathbf{e}_{\underset{q}{-}} = \delta_{\underset{pq}{--}},\end{gathered} \tag{7}$$

where $\delta_{\underset{pq}{--}}$ is the Kronecker delta, $\mathbf{e}_{\underset{1}{-}}, \mathbf{e}_{\underset{2}{-}}, \mathbf{e}_{\underset{3}{-}}$ is the orthonormal basis at the point $\overset{(-)}{M}$, $\overset{(-)}{b}{}_{\underset{I}{-}}^{\bar{J}}$ are the components of the second tensor of the inner base surface.

Taking into account $\mathbf{h}_1 = \partial_1\mathbf{h} = (h\cos\theta + h'\sin\theta)\mathbf{a} + (h'\cos\theta - h\sin\theta)\mathbf{k}_3$, $\mathbf{h}_2 = \partial_2\mathbf{h} = h\sin\theta\mathbf{c}$, similarly to (7) for the covariant bases at the points M and $\overset{(+)}{M}$, we obtain

$$\begin{gathered}\mathbf{r}_{\underset{1}{+}} = [(R_1 + h)\cos\theta + h'\sin\theta]\mathbf{a} + [h'\cos\theta - (R_1 + h)\sin\theta]\mathbf{k}_3, \\ \mathbf{r}_{\underset{2}{+}} = (R_2 + h)\sin\theta\mathbf{c}, \quad \mathbf{r}_{\underset{3}{+}} = \mathbf{r}_3 = h\overset{(-)}{\mathbf{n}} = h(\sin\theta\mathbf{a} + \cos\theta\mathbf{k}_3), \\ \mathbf{r}_1 = [(R_1 + x^3h)\cos\theta + x^3h'\sin\theta]\mathbf{a} + [x^3h'\cos\theta - (R_1 + x^3h)\sin\theta]\mathbf{k}_3, \\ \mathbf{r}_2 = (R_2 + x^3h)\sin\theta\mathbf{c}, \quad \sqrt{g} = h(R_1 + x^3h)(R_2 + x^3h), \quad \sqrt{\overset{(+)}{g}} = h(R_1 + h)(R_2 + h),\end{gathered} \tag{8}$$

Further according to (7) and (8) it is easy to find the components of the unit tensor of the second rank $g_{\underset{pq}{--}}$, $g_{\underset{pq}{-}}$, $g_p^{\bar{q}}$ and $g_{\underset{q}{-}}^{p}$. In fact, after simple calculations, for example, for the components

$g_{\bar{p}\bar{q}}$, $g_p^{\bar{q}}$ and $g_{\bar{q}}^{p}$ we find the next expressions:

$$\begin{aligned} & g_1^{\bar{1}} = 1 + hk_1x^3, \quad g_1^{\bar{2}} = 0, \quad g_1^{\bar{3}} = x^3 g_{\underset{1}{+}}^{\bar{1}}; && g_{\underset{1}{-}}^{1} = (1 + hk_1x^3)^{-1}, \quad g_{\underset{2}{-}}^{1} = 0, \quad g_{\underset{3}{-}}^{1} = 0, \\ & g_2^{\bar{1}} = 0, \quad g_2^{\bar{2}} = 1 + hk_2x^3, \quad g_2^{\bar{3}} = 0; && g_{\underset{1}{-}}^{2} = 0, \quad g_{\underset{2}{-}}^{2} = (1 + hk_2x^3)^{-1}, \quad g_{\underset{3}{-}}^{2} = 0, \\ & g_3^{\bar{1}} = 0, \quad g_3^{\bar{2}} = 0, \quad g_3^{\bar{3}} = 1; && g_{\underset{1}{-}}^{3} = \frac{-h^{-1}h'x^3}{1 + hk_1x^3}, \quad g_{\underset{2}{-}}^{3} = 0, \quad g_{\underset{3}{-}}^{3} = 1. \end{aligned} \tag{9}$$

Note that based on (8) it is easy to find the expressions for $g_{\underset{p}{+}}^{\bar{q}} = g_p^{\bar{q}}|_{x^3=1}$, $g_{\underset{q}{-}}^{\overset{+}{p}} = g_{\underset{q}{-}}^{p}|_{x^3=1}$, so we will not write them out. Further, it is easy to obtain the expressions for $\mathbf{r}^i$. In fact, we have

$$\mathbf{r}^I = \overset{(-)}{\vartheta}{}^{-1}\epsilon^{IJ}\epsilon_{LK}g_J^{\bar{K}}\mathbf{r}^{\bar{L}}, \quad \mathbf{r}^3 = \mathbf{r}^{\bar{3}} - g_J^{\bar{3}}g_{\underset{L}{-}}^{J}\mathbf{r}^{\bar{L}}, \quad \overset{(-)}{\vartheta} = (1 + hk_1x^3)(1 + hk_2x^3), \ k_I = R_I^{-1}. \tag{10}$$

Taking into account the appropriate relations (7) and (9), we obtain the expressions for $\mathbf{r}^3$ from the second formula (10), and from it we find the expression for $\mathbf{r}^{\overset{+}{3}}$. We will get

$$\begin{aligned} \mathbf{r}^3 &= \frac{1}{h(R_1 + hx^3)}\{[(R_1 + hx^3)\sin\theta - x^3h'cos\theta]\mathbf{a} + [(R_1 + hx^3)\cos\theta + x^3h'\sin\theta]\mathbf{k}_3\}, \\ \mathbf{r}^{\overset{+}{3}} &= \mathbf{r}^3|_{x^3=1} = \frac{1}{h(R_1 + h)}\{[(R_1 + h)\sin\theta - h'cos\theta]\mathbf{a} + [(R_1 + h)\cos\theta + h'\sin\theta]\mathbf{k}_3\}, \\ g^{33} &= \frac{1}{h^2(R_1 + hx^3)^2}[(R_1 + hx^3)^2 + (x^3h')^2], \quad g^{\overset{++}{33}} = \frac{1}{h^2(R_1 + h)^2}[(R_1 + h)^2 + (h')^2]. \end{aligned} \tag{11}$$

Here we have found expressions for g^{33}, $g^{\overset{++}{33}}$.

Let us find the expression for the unit normal to the surface S at the point M. Denoting it by $\mathbf{n}$, by the first formula (11) we get

$$\begin{aligned} \mathbf{n} = \mathbf{n} &= \frac{\mathbf{r}^3}{\sqrt{g^{33}}} = \frac{1}{\sqrt{(R_1 + hx^3)^2 + (x^3h')^2}}\{[(R_1 + hx^3)\sin\theta - x^3h'\cos\theta]\mathbf{a} + \\ &+[(R_1 + hx^3)\cos\theta + x^3h'\sin\theta]\mathbf{k}_3\}, \quad \overset{(+)}{\mathbf{n}} = \mathbf{n}|_{x^3=1}. \end{aligned} \tag{12}$$

Note that the relations (11) – (12) are used when writing boundary conditions.

3. Equations and constitutive relations of the theory of thin elastic bodies

Motion equations in terms of stress and moment stress tensors [17]

$$g_{\underset{M}{-}}^{P}N_P\mathbf{P}^{\bar{M}} + \partial_3\mathbf{P}^{\bar{3}} + \rho\mathbf{F} = \rho\partial_t^2\mathbf{u}, \quad g_{\underset{M}{-}}^{P}N_P\boldsymbol{\mu}^{\bar{M}} + \partial_3\boldsymbol{\mu}^{\bar{3}} + \underset{\sim}{\mathbf{C}}\overset{2}{\otimes}\underset{\sim}{\mathbf{P}} + \rho\mathbf{m} = \underset{\sim}{\mathbf{J}}\cdot\partial_t^2\boldsymbol{\varphi}; \tag{13}$$

$$N_P = \partial_P - g_P^{\bar{3}}\partial_3, \quad \partial_i = \partial/\partial x^i, \quad g_P^{\bar{3}} = x^3 g_{\underset{P}{+}}^{\bar{3}}, \qquad \mathbf{P}^{\bar{m}} = \mathbf{r}^m\cdot\underset{\sim}{\mathbf{P}}, \quad \boldsymbol{\mu}^{\bar{m}} = \mathbf{r}^m\cdot\underset{\sim}{\boldsymbol{\mu}},$$

$$g_{\underset{M}{-}}^{P} = \sum_{s=0}^{\infty}\underset{(s)}{A}{}_{\underset{M}{+}}^{\bar{P}}(x^3)^s, \quad \underset{(0)}{A}{}_{\underset{M}{+}}^{\bar{P}} = g_{\underset{M}{-}}^{\bar{P}}, \quad \underset{(1)}{A}{}_{\underset{M}{+}}^{\bar{P}} = g_{\underset{M}{-}}^{\bar{P}} - g_{\underset{M}{+}}^{\bar{P}}, \ \ldots, \ \underset{(n)}{A}{}_{\underset{M}{+}}^{\bar{P}} = (g_{\underset{N_1}{-}}^{\bar{P}} - g_{\underset{N_1}{+}}^{\bar{P}})\ldots(g_{\underset{M}{-}}^{\bar{N}_{n-1}} - g_{\underset{M}{+}}^{\bar{N}_{n-1}}).$$

Similarly to (13), from constitutive relations (CR) of micropolar linear elasticity theory one has [17]

$$
\begin{aligned}
&\underset{\sim}{\mathbf{P}} = \underset{\approx}{\mathbf{A}}^{\bar{M}\cdot}\cdot g_{\bar{M}}^{P} N_P \mathbf{u} + \underset{\approx}{\mathbf{A}}^{\bar{3}\cdot}\cdot\partial_3\mathbf{u} + \underset{\approx}{\mathbf{B}}^{\bar{M}\cdot}\cdot g_{\bar{M}}^{P} N_P \boldsymbol{\varphi} + \underset{\approx}{\mathbf{B}}^{\bar{3}\cdot}\cdot\partial_3\boldsymbol{\varphi} - \underset{\approx}{\mathbf{A}} \overset{2}{\otimes} \underset{\approx}{\mathbf{C}}\cdot\boldsymbol{\varphi},\\
&\underset{\sim}{\boldsymbol{\mu}} = \underset{\approx}{\mathbf{C}}^{\bar{M}\cdot}\cdot g_{\bar{M}}^{P} N_P \mathbf{u} + \underset{\approx}{\mathbf{C}}^{\bar{3}\cdot}\cdot\partial_3\mathbf{u} + \underset{\approx}{\mathbf{D}}^{\bar{M}\cdot}\cdot g_{\bar{M}}^{P} N_P \boldsymbol{\varphi} + \underset{\approx}{\mathbf{D}}^{\bar{3}\cdot}\cdot\partial_3\boldsymbol{\varphi} - \underset{\approx}{\mathbf{C}} \overset{2}{\otimes} \underset{\approx}{\mathbf{C}}\cdot\boldsymbol{\varphi};
\end{aligned}
\tag{14}
$$

$$
\underset{\approx}{\mathbf{A}}^{\bar{m}\cdot} = \mathbf{A}^{ij\bar{m}l}\mathbf{r}_i\mathbf{r}_j\mathbf{r}_l,\ \ \underset{\approx}{\mathbf{B}}^{\bar{m}\cdot} = \mathbf{B}^{ij\bar{m}l}\mathbf{r}_i\mathbf{r}_j\mathbf{r}_l,\ \ \underset{\approx}{\mathbf{C}}^{\bar{m}\cdot} = \mathbf{C}^{ij\bar{m}l}\mathbf{r}_i\mathbf{r}_j\mathbf{r}_l,\ \ \underset{\approx}{\mathbf{D}}^{\bar{m}\cdot} = \mathbf{D}^{ij\bar{m}l}\mathbf{r}_i\mathbf{r}_j\mathbf{r}_l.
$$

where $\underset{\sim}{\mathbf{P}}$ and $\underset{\sim}{\boldsymbol{\mu}}$ are the stress and moment stress tensors, $\underset{\approx}{\mathbf{C}}$ is the discriminant tensor [23], $\mathbf{u}$ and $\boldsymbol{\varphi}$ are the displacement and rotation vectors, ρ is the material density, $\mathbf{F}$ and $\mathbf{m}$ are the mass force and mass momentum densities, $\underset{\sim}{\mathbf{J}}$ is a tensor of inertia moment (inner property of the medium), ∇ is the Hamilton nabla-operator, $\partial_t = \partial/\partial_t$, t is a time, $\overset{2}{\otimes}$ is an inner 2-product [22, 23].

In this case, by virtue of the corresponding relations (9), we have relations

$$
N_1 = \nabla_1 - g_1^{\bar{3}}\partial_3,\quad N_2 = \nabla_2,\quad g_{\bar{M}}^{P} N_P \mathbf{P}^{\bar{M}} = g_{\bar{1}}^{1}(\nabla_1 - g_1^{\bar{3}}\partial_3)\mathbf{P}^{\bar{1}} + g_{\bar{2}}^{2}\nabla_2\mathbf{P}^{\bar{2}},
\tag{15}
$$

in virtue of which the equations (13) and CR (14) for rotation bodies of variable thickness are represented in the form

$$
\begin{aligned}
&g_{\bar{1}}^{1}(\nabla_1 - g_1^{\bar{3}}\partial_3)\mathbf{P}^{\bar{1}} + g_{\bar{2}}^{2}\nabla_2\mathbf{P}^{\bar{2}} + \partial_3\mathbf{P}^{\bar{3}} + \rho\mathbf{F} = \rho\partial_t^2\mathbf{u},\\
&g_{\bar{1}}^{1}(\nabla_1 - g_1^{\bar{3}}\partial_3)\boldsymbol{\mu}^{\bar{1}} + g_{\bar{2}}^{2}\nabla_2\boldsymbol{\mu}^{\bar{2}} + \partial_3\boldsymbol{\mu}^{\bar{3}} + \underset{\approx}{\mathbf{C}} \overset{2}{\otimes} \underset{\sim}{\mathbf{P}} + \rho\mathbf{m} = \underset{\sim}{\mathbf{J}}\cdot\partial_t^2\boldsymbol{\varphi};
\end{aligned}
\tag{16}
$$

$$
\begin{aligned}
\underset{\sim}{\mathbf{P}} =\ & [g_{\bar{1}}^{1}\underset{\approx}{\mathbf{A}}^{\bar{1}\cdot}(\nabla_1 - g_1^{\bar{3}}\partial_3)] + g_{\bar{2}}^{2}\underset{\approx}{\mathbf{A}}^{\bar{2}\cdot}\nabla_2]\cdot\mathbf{u} + \underset{\approx}{\mathbf{A}}^{\bar{3}\cdot}\cdot\partial_3\mathbf{u} +\\
&+ [g_{\bar{1}}^{1}\underset{\approx}{\mathbf{B}}^{\bar{1}\cdot}(\nabla_1 - g_1^{\bar{3}}\partial_3)] + g_{\bar{2}}^{2}\underset{\approx}{\mathbf{B}}^{\bar{2}\cdot}\nabla_2]\cdot\boldsymbol{\varphi} + \underset{\approx}{\mathbf{B}}^{\bar{3}\cdot}\cdot\partial_3\boldsymbol{\varphi} - \underset{\approx}{\mathbf{A}} \overset{2}{\otimes} \underset{\approx}{\mathbf{C}}\cdot\boldsymbol{\varphi},\\
\underset{\sim}{\boldsymbol{\mu}} =\ & [g_{\bar{1}}^{1}\underset{\approx}{\mathbf{C}}^{\bar{1}\cdot}(\nabla_1 - g_1^{\bar{3}}\partial_3)] + g_{\bar{2}}^{2}\underset{\approx}{\mathbf{C}}^{\bar{2}\cdot}\nabla_2]\cdot\mathbf{u} + \underset{\approx}{\mathbf{C}}^{\bar{3}\cdot}\cdot\partial_3\mathbf{u} +\\
&+ [g_{\bar{1}}^{1}\underset{\approx}{\mathbf{D}}^{\bar{1}\cdot}(\nabla_1 - g_1^{\bar{3}}\partial_3)] + g_{\bar{2}}^{2}\underset{\approx}{\mathbf{D}}^{\bar{2}\cdot}\nabla_2]\cdot\boldsymbol{\varphi} + \underset{\approx}{\mathbf{D}}^{\bar{3}\cdot}\cdot\partial_3\boldsymbol{\varphi} - \underset{\approx}{\mathbf{C}} \overset{2}{\otimes} \underset{\approx}{\mathbf{C}}\cdot\boldsymbol{\varphi}.
\end{aligned}
\tag{17}
$$

It should be noted that (16) and (17) represent the equations of motion and CR of the micropolar linear theory of elastic rotation bodies without a center of symmetry of variable thickness. For very thin and shallow rotation bodies $g_{\bar{1}}^{1} \approx 1$ and $g_{\bar{2}}^{2} \approx 1$, and $g_1^{\bar{3}} = 0$ for a constant thickness, therefore in these cases the equations of motion and the CR are simplified. It is not difficult to write them out, so we will not write them out to shorten the letter. However, we shall write out them below for very thin and shallow rotation bodies of constant thickness. Taking into account the above, we shall have

$$
\nabla_I\mathbf{P}^{\bar{I}} + \partial_3\mathbf{P}^{\bar{3}} + \rho\mathbf{F} = \rho\partial_t^2\mathbf{u},\quad \nabla_I\boldsymbol{\mu}^{\bar{I}}\partial_3\boldsymbol{\mu}^{\bar{3}} + \underset{\approx}{\mathbf{C}} \overset{2}{\otimes} \underset{\sim}{\mathbf{P}} + \rho\mathbf{m} = \underset{\sim}{\mathbf{J}}\cdot\partial_t^2\boldsymbol{\varphi};
\tag{18}
$$

$$
\begin{aligned}
&\underset{\sim}{\mathbf{P}} = \underset{\approx}{\mathbf{A}}^{\bar{I}\cdot}\cdot\partial_I\mathbf{u} + \underset{\approx}{\mathbf{A}}^{\bar{3}\cdot}\cdot\partial_3\mathbf{u} + \underset{\approx}{\mathbf{B}}^{\bar{I}\cdot}\cdot\partial_I\boldsymbol{\varphi} + \underset{\approx}{\mathbf{B}}^{\bar{3}\cdot}\cdot\partial_3\boldsymbol{\varphi} - \underset{\approx}{\mathbf{A}} \overset{2}{\otimes} \underset{\approx}{\mathbf{C}}\cdot\boldsymbol{\varphi},\\
&\underset{\sim}{\boldsymbol{\mu}} = \underset{\approx}{\mathbf{C}}^{\bar{I}\cdot}\cdot\partial_I\mathbf{u} + \underset{\approx}{\mathbf{C}}^{\bar{3}\cdot}\cdot\partial_3\mathbf{u} + \underset{\approx}{\mathbf{D}}^{\bar{I}\cdot}\cdot\partial_I\boldsymbol{\varphi} + \underset{\approx}{\mathbf{D}}^{\bar{3}\cdot}\cdot\partial_3\boldsymbol{\varphi} - \underset{\approx}{\mathbf{C}} \overset{2}{\otimes} \underset{\approx}{\mathbf{C}}\cdot\boldsymbol{\varphi}.
\end{aligned}
\tag{19}
$$

Applying the k-th order operator [17, 18] to (18) and (19), we obtain the equations of motion and the CR of the micropolar linear theory of very thin and shallow elastic bodies of constant thickness without a center of symmetry in moments of unknown functions with respect to any system of orthogonal polynomials

$$
\nabla_I \overset{(k)}{\mathbf{P}}{}^{\bar{I}} + \overset{(k)}{\mathbf{P}}{}^{\bar{3}\prime} + \rho\overset{(k)}{\mathbf{F}} = \rho\partial_t^2\overset{(k)}{\mathbf{u}},\quad \nabla_I\overset{(k)}{\boldsymbol{\mu}}{}^{\bar{I}} + \overset{(k)}{\boldsymbol{\mu}}{}^{\bar{3}\prime} + \underset{\approx}{\mathbf{C}} \overset{2}{\otimes} \overset{(k)}{\underset{\sim}{\mathbf{P}}} + \rho\overset{(k)}{\mathbf{m}} = \underset{\sim}{\mathbf{J}}\cdot\partial_t^2\overset{(k)}{\boldsymbol{\varphi}},\quad k \geq 0;
\tag{20}
$$

$$\begin{aligned}
&\overset{(k)}{\underset{\approx}{\mathbf{P}}}=\underset{\approx}{\mathbf{A}}^{\bar{I}\cdot}\cdot\partial_I\overset{(k)}{\mathbf{u}}+\underset{\approx}{\mathbf{A}}^{\bar{3}\cdot}\overset{(k)}{\mathbf{u}}{}'+\underset{\approx}{\mathbf{B}}^{\bar{I}\cdot}\cdot\partial_I\overset{(k)}{\boldsymbol{\varphi}}+\underset{\approx}{\mathbf{B}}^{\bar{3}\cdot}\overset{(k)}{\boldsymbol{\varphi}}{}'-\underset{\approx}{\mathbf{A}}\overset{2}{\otimes}\underset{\approx}{\mathbf{C}}\overset{(k)}{\boldsymbol{\varphi}},\\
&\underset{\approx}{\boldsymbol{\mu}}=\underset{\approx}{\mathbf{C}}^{\bar{I}\cdot}\cdot\partial_I\overset{(k)}{\mathbf{u}}+\underset{\approx}{\mathbf{C}}^{\bar{3}\cdot}\overset{(k)}{\mathbf{u}}{}'+\underset{\approx}{\mathbf{D}}^{\bar{I}\cdot}\cdot\partial_I\overset{(k)}{\boldsymbol{\varphi}}+\underset{\approx}{\mathbf{D}}^{\bar{3}\cdot}\overset{(k)}{\boldsymbol{\varphi}}{}'-\underset{\approx}{\mathbf{C}}\overset{2}{\otimes}\underset{\approx}{\mathbf{C}}\overset{(k)}{\boldsymbol{\varphi}}{}',\quad k\geq 0,
\end{aligned}\tag{21}$$

where the definition of the moment of the kth order $\overset{(k)}{F}$ and the expression for the operator "prime" $\overset{(k)}{F}{}'$ of any function F, for example, with respect to the system of Legendre polynomials are represented in the form

$$\overset{(k)}{M}(F)\equiv\overset{(k)}{F}(x')=(2k+1)\int_0^1 F(x',x^3)P_k(x^3)dx^3,\quad \overset{(k)}{F}{}'(x')=2(2k+1)\sum_{m=0}^{\infty}\overset{(k+2m+1)}{F}=$$
$$=(2k+1)\Big\{\big[\overset{(+)}{F}-(-1)^k\overset{(-)}{F}\big]-\sum_{p=0}^{k}[1-(-1)^{k+p}]\overset{(p)}{F}(x')\Big\},\ k\geq 0,\ \ \overset{(-)}{F}=F|_{x^3=0},\ \ \overset{(+)}{F}=F|_{x^3=1}.$$

Note that from (20) and (21) one can get the equations of motion and CR of any approximation. Of course, in order to correctly formulate initial boundary value problems for (20) and (21), depending on the type of initial boundary value problems, it is necessary to add the corresponding initial and boundary conditions in the moments that are obtained similarly to (20) and (21). We do not dwell on them in order to shorten the letter, but we refer the interested reader to [17, 25], in which the analogous questions for thin bodies of other configurations are described in detail.

4. Conclusion

Some questions of the new parametrization of the three-dimensional region of the body of revolution are considered. In particular, the vector parametric equation of the body of revolution is written out. Covariant and contravariant bases are constructed at the points of the inner and outer base surfaces, as well as at arbitrary points of the body. By virtue of these bases, expressions of some geometric characteristics necessary for the presentation of the material are found. The motion equations and CR of the micropolar theory of elastic rotation bodies without a symmetry center of variable thickness are obtained under the new parametrization, from which equations are obtained for very thin and shallow rotation bodies, both variable and constant thickness. Finally, equations of motion and CR of any approximation in the moments of unknown functions with respect to an arbitrary system of orthogonal polynomials and, in particular, for the system of Legendre polynomials are derived. The questions of setting initial-boundary problems are discussed.

Note also that of great interest are works on the thermomechanics of composite structures under high temperature [26–30], which can be used when considering problems of thermomechanics for thin structures.

Acknowledgement: The work was supported by the Russian Foundation for Basic Research (project no.18-29-10085-mk).

References

[1] *I. N. Vekua,* Shell theory, general methods of construction. Pitman Advanced Pub. Program, 1985.

[2] *A. E. Alekseev and B. D. Annin,* Equations of deformation of an elastic inhomogeneous laminated body of revolution // J. Appl. Mech. Techn. Phys. 2003. **44**, No. 3. 432-437.

[3] *Yu. M. Volchkov and L. A. Dergileva,* Reducing three-dimensional elasticity problems to two-dimensional problems by approximating stresses and displacements by Legendre polynomials // J. Appl. Mech. Techn. Phys. 2007. **48**, No. 3. 450–459.

[4] *G. V. Ivanov,* Theory of plates and shells. Novosibirsk State Univ., Novosibirsk, 1980 (in Russian).

[5] *J. Fellers and A. Soler,* Approximate solution of the finite cylinder problem using Legendre polynomials // AIAA J. 1970. **8**, No. 11.

[6] *Cicala Placido,* Sulla teoria elastica della parete sottile// Giornale del Genio Civile. 1959. **97**, No. 4, 6, 9

[7] *R. D. Mindlin and M. A. Medick,* Extensional vibrations of elastic plates // J. Appl. Mech. 1959. **26**, No. 4. 561–569.
[8] *T. V. Meunargiya,* Development of the method of I. N. Vekua for problems of the three-dimensional moment elasticity. Tbilisi State Univ., Tbilisi, 1987 (in Russian).
[9] *A. R. Ulukhanyan,* Representation of Solutions to Equations of Hyperbolic Type// Moscow university mechanics bulletin. 2010, Vol. 65, No. 2. 47–50. DOI: 10.3103/S0027133010020056
[10] *B. L. Pelekh, A. V. Maksimuk, and I. M. Korovaichuk,* Contact problems for laminated elements of constructions and bodies with coating Naukova Dumka, Kiev, 1988 (in Russian).
[11] *M. U. Nikabadze,* A variant of the theory of multilayer structures // Mech. Solids. 2001. No. 1. 143–158.
[12] *M. U. Nikabadze,* Mathematical modeling of multilayer thin body deformation//Journal of mathematical sciences. V. 187, No 3, 2012. P. 300-336.
[13] *M. U. Nikabadze and A. R. Ulukhanyan,* Statements of problems for a thin deformable three-dimensional body // Vestn. Mosk. Univ., Matem. Mekhan. 2005. No. 5. 43–49.
[14] *M. U. Nikabadze,* A variant of the system of equations of the theory of thin bodies // Vestn. Mosk. Univ., Matem. Mekhan. 2006. No. 1. 30–35.
[15] *M. U. Nikabadze,* Application of a system of Chebyshev polynomials to the theory of thin bodies // Vestn. Mosk. Univ., Matem. Mekhan. 2007. No. 5. 56–63.
[16] *M. U. Nikabadze,* Some issues concerning a version of the theory of thin solids based on expansions in a system of Chebyshev polynomials of the second kind // Mech. Solids. 2007. **42**. No. 3. 391-421.
[17] Nikabadze M U 2014 *Development of the Method of Orthogonal Polynomials in the Classical and Micropolar Mechanics of Elastic Thin Bodies* (Moscow University Press), 515 p. (in Russian). http://istina.msu.ru/media/publications/book/707/ea1/6738800/Monographiya.pdf
[18] Nikabadze M and Ulukhanyan A 2016, *Analytical Solutions in the Theory of Thin Bodies*, 319–361, in: Generalized Continua as Models for Classical and Advanced Materials, Advanced Structured Materials, Vol. 42, Eds.: H. Altenbach and S. Forest, Springer. DOI 10.1007/978 – 3 – 319 – 31721 – 2_15
[19] *M. Nikabadze, A. Ulukhanyan, T. Moseshvili, K. Tskhakaia, N. Mardaleishvili, Z. Arkania* 2019 On the Modeling of Five-Layer Thin Prismatic Bodies. *Mathematical and Computational Applications.* Vol. 24, 3. pp. 1–16 DOI: 10.3390/mca24030069
[20] *M. Nikabadze, A. Ulukhanyan, G. Sakhvadze* 2019 To mathematical modeling of deformation of micropolar thin bodies with two small sizes. *Journal of Physics: Conference Series.* Vol. 1205, 1. pp. 1–8 DOI: 10.1088/1742-6596/1205/1/012040
[21] Nikabadze M U 2015 On Several Issues of Tensor Calculus with Applications to Mechanics. *Contemporary Mathematics. Fundamental Directions*, **55** 3–194 (in Russian). http://istina.msu.ru/media/publications/book/e25/00c/10117043/M.U.Nikabadze.pdf
[22] Nikabadze M U 2017 Topics on tensor calculus with applications to mechanics, *J. Math. Sci.*, **225**(1) 1–194. DOI: 10.1007/s10958-017-3467-4
[23] Vekua I N 1978 *Fundamentals of tensor analysis and covariant theory,* (Nauka, Moscow) (in Russian)
[24] Chernina V S 1968 *Static of thin-walled shells of rotation* (Nauka, Moscow) 456 p.(in Russian).
[25] Vekua I N 1985 *Shell theory, general methods of construction.* (Pitman Advanced Pub. Program)
[26] Dimitrienko Y, Yakovlev D 2015 The asymptotic theory of thermoelasticity of multilayer composite plates *Composites-mechanics computations applications* **6** 1 pp. 13–51. DOI: 10.1615/CompMechComputApplIntJ.v6.i1.20
[27] Dimitrienko Y 2016 Thermomechanics of composite structures under high temperatures. *Solid Mechanics and its Applications* **224** DOI 10.1007/978-94-017-7494-9
[28] Dimitrienko Y 2016 Thermomechanics of composite structures under high temperatures. General equations of multiphase continuum mechanics for ablative composites *Solid Mechanics and its Applications* **224** pp. 31–59. DOI: 10.1007/978-94-017-7494-9-2
[29] Dimitrienko Y 2016 Thermomechanics of composite structures under high temperatures. Mathematical Model of Ablative Composites *Solid Mechanics and its Applications* **224** pp. 61–81. DOI: 10.1007/978-94-017-7494-9-3
[30] Dimitrienko Y, Dimitrienko I 2017 Modeling of thin composite laminates with general anisotropy under harmonic vibrations by the asymptotic homogenization method *Int. journal for multiscale computational engineering* **15** 3 pp. 219-237. DOI: 10.1615/IntJMultCompEng.2017019518

On modeling of three-layered thin bodies

Mikhail Nikabadze[1,2], **Armine Ulukhanyan**[2], **Andrey Khizhenkov**[2]

[1] Department of Mechanics and Mathematics, Lomonosov MSU
[2] Department of Basic Science, Bauman MSTU

E-mail: munikabadze@yandex.ru, armine_msu@mail.ru, a.hizhenkov@gmail.com

Abstract. Some questions about the new parameterization of three-dimensional thin body with one small size are considered. The vector parametric equations of the multilayered thin body are given. The geometric characteristics are considered. The representations of several differential operators, the equations of motion, the boundary conditions and the constitutive relations (CR) of classic and micropolar theories of elasticity are given. Some interlayer contact conditions are obtained. The definition of the k-order moment of any quantity with respect to the system of Legendre polynomials is given. The problems of the classic and micropolar theories of one-, two- and three-layered prismatic elastic thin bodies in moments of displacement and rotation vectors with respect to the system of Legendre polynomials are formulated. The problem of one-layered two-dimensional rectangular plate with two pinched edges under the distributed load are solved.

Keywords: *new parametrization, micropolar theory, thin body, theory of thin bodies, prismatic body, Legendre polynomials.*

1. Introduction

Many authors, for example, [1–10], as well as one of the authors of this work [11–19] used an analytical method by applying systems of orthogonal polynomials (Legendre and Chebyshev) to construct single-layer and multilayer theories of thin bodies. In addition, Nikabadze used this method to construct a theory of thin bodies with two small sizes [20, 21].

In this work, we construct the micropolar theory of multilayer thin bodies using the systems of orthogonal polynomials. Note that any problem of the thin body theory can be considered in a three-dimensional statement, which is more accurate than the two-dimensional one. However, it is not always possible to implement this approach in practice because of the high complexity of solving three-dimensional problems and a large variety of statements of problems being practically necessary. Due to the wide use of thin bodies (single-layer, two and more layer structures) in mechanical engineering, aircraft engineering, and rocket production, it becomes necessary to create new theories of the thin bodies within the framework of the classical as well as the micropolar theories and to improve the methods for their calculation. Therefore, the construction of complex theories of thin bodies and the development of effective methods for calculating them are important and relevant problems.

2. Parametrization of the multilayered thin body

We consider multilayered thin domain consists of at most countable set of layers. The layers are numbered in ascending order. There are two base regular surfaces on each layer. $\underset{\alpha}{\overset{(-)}{S}}$ and $\underset{\alpha}{\overset{(+)}{S}}$ is an inner and outer base surfaces of layer α. Thus surfaces $\underset{\alpha}{\overset{(-)}{S}}$ and $\underset{\alpha-1}{\overset{(+)}{S}}$ are the same.

For each layer α we introduce the new parametrization of thin body with one small size. Thus, for each layer α three vectors $(\mathbf{r}_{\underset{\alpha}{\tilde{}}1}, \mathbf{r}_{\underset{\alpha}{\tilde{}}2}, \mathbf{r}_{\underset{\alpha}{\tilde{}}3})$ defined in the considered points $\underset{\alpha}{\overset{(\sim)}{M}} \in \underset{\alpha}{\overset{(\sim)}{S}}$ form three-dimensional (spatial) covariant bases, where $\sim\ \in \{-, \emptyset, +\}$. It is well known that based on these bases [21–23], one can construct the corresponding contravariant bases $(\underset{\alpha}{\mathbf{r}}^{\tilde{1}}, \underset{\alpha}{\mathbf{r}}^{\tilde{2}}, \underset{\alpha}{\mathbf{r}}^{\tilde{3}})$. By the relation obtained for a new parametrization [21,23] we can obtain relation for each layer α by the next simple rule: if the quantity is belong to the layer α we put index α under this quantity. For example, the vector parametric equation of the layer α can be represented as follows

$$\underset{\alpha}{\mathbf{r}}(x^1, x^2, x^3) = \underset{\alpha}{\overset{(-)}{\mathbf{r}}} + x^3\underset{\alpha}{\mathbf{h}}(x^1, x^2) = (1 - x^3)\underset{\alpha}{\overset{(-)}{\mathbf{r}}}(x^1, x^2) + x^3\underset{\alpha}{\overset{(+)}{\mathbf{r}}}(x^1, x^2), \forall x^3 \in [0, 1].$$

The vector $\underset{\alpha}{\mathbf{h}}(x^1, x^2) = \underset{\alpha}{\overset{(+)}{\mathbf{r}}}(x^1, x^2) - \underset{\alpha}{\overset{(-)}{\mathbf{r}}}(x^1, x^2)$ performing topological mapping of the inner base $\underset{\alpha}{\overset{(-)}{S}}$ on the surface $\underset{\alpha}{\overset{(+)}{S}}$, in general is not perpendicular to the base surface. The end point of each vector $\underset{\alpha}{\mathbf{h}}(x^1, x^2)$ is the start point of the vector $\underset{\alpha+1}{\mathbf{h}}(x^1, x^2)$, $\forall\alpha$. So we have

$$\underset{\alpha+\delta}{\overset{(+)}{\mathbf{r}}} = \underset{\alpha}{\overset{(-)}{\mathbf{r}}} + \sum_{\nu=\alpha}^{\alpha+\delta} \underset{\nu}{\mathbf{h}} = \underset{\alpha}{\overset{(+)}{\mathbf{r}}} + \sum_{\nu=\alpha+1}^{\alpha+\delta} \underset{\nu}{\mathbf{h}}, \forall\alpha\delta.$$

Let the multilayer body consists of K layers. Thus we have following relations

$$\underset{K}{\overset{(+)}{\mathbf{r}}} = \underset{1}{\overset{(-)}{\mathbf{r}}} + \mathbf{h} = \underset{1}{\overset{(-)}{\mathbf{r}}} + \sum_{\nu=1}^{K}(\underset{\nu}{\overset{(+)}{\mathbf{r}}} - \underset{\nu}{\overset{(-)}{\mathbf{r}}}), \quad \mathbf{h} = \sum_{\nu=1}^{K}\underset{\nu}{\mathbf{h}} = \sum_{\nu=1}^{K}(\underset{\nu}{\overset{(+)}{\mathbf{r}}} - \underset{\nu}{\overset{(-)}{\mathbf{r}}}).$$

3. Basic relations of the theory of thin elastic bodies

As is known, motion equations of micropolar deformable rigid body are represented in the following form [24–26]

$$\nabla \cdot \underset{\sim}{\mathbf{P}} + \rho\mathbf{F} = \rho\partial_t^2\mathbf{u}, \quad \nabla \cdot \underset{\sim}{\boldsymbol{\mu}} + \underset{\approx}{\mathbf{C}} \overset{2}{\otimes} \underset{\sim}{\mathbf{P}} + \rho\mathbf{m} = \underset{\sim}{\mathbf{J}} \cdot \partial_t^2\boldsymbol{\varphi}, \tag{1}$$

where $\underset{\sim}{\mathbf{P}}$, $\underset{\sim}{\boldsymbol{\mu}}$ are stress and moment stress tensors, $\underset{\approx}{\mathbf{C}}$ – discriminant tensor, $\mathbf{u}$ – displacement vector, $\boldsymbol{\varphi}$ – rotation vector. $\underset{\sim}{\mathbf{J}}$ – moment of inertia (inner proerty of the medium), $\mathbf{F}$ – mass force density, $\mathbf{m}$ – mass momentum density, ρ – density of the medium, ∇ – Hamilton nabla operator, $\partial_t = \partial/\partial_t$, t – time, $\overset{2}{\otimes}$ – inner 2-product [21–23]. In the case of thin bodies theory with the new parametrization one can write these equations (1) in the following form [21]

$$\underset{M}{\overset{P}{g_{-}}} N_P\mathbf{P}^{\bar{M}} + \partial_3\mathbf{P}^{\bar{3}} + \rho\mathbf{F} = \rho\partial_t^2\mathbf{u}, \ \underset{M}{\overset{P}{g_{-}}} N_P\boldsymbol{\mu}^{\bar{M}} + \partial_3\boldsymbol{\mu}^{\bar{3}} + \underset{\approx}{\mathbf{C}} \overset{2}{\otimes} \underset{\sim}{\mathbf{P}} + \rho\mathbf{m} = \underset{\sim}{\mathbf{J}} \cdot \partial_t^2\boldsymbol{\varphi}, \tag{2}$$

where $\mathbf{P}^{\bar{m}} = \mathbf{r}^m \cdot \underset{\sim}{\mathbf{P}}$, $\boldsymbol{\mu}^{\bar{m}} = \mathbf{r}^m \cdot \underset{\sim}{\boldsymbol{\mu}}$, $N_P = \partial_P - g_P^{\bar{3}}\partial_3$. As is known [21], for thin bodies $\underset{M}{\overset{P}{g_{-}}} = \sum_{s=0}^{\infty} \underset{(s)M}{\overset{P}{A_{+}}}(x^3)^s$. We consider only the first series term (the 0th order approximation) $\underset{(0)M}{\overset{P}{g_{-}}} = \underset{(0)M}{\overset{\bar{P}}{A_{+}}} = \underset{M}{\overset{\bar{P}}{g_{-}}} = \delta_M^P$. Thus for each layer α from (2) for the 0th order approximation we have the equilibrium equations in the following form

$$\underset{\alpha}{N}_I \underset{\alpha}{\mathbf{P}}^{\bar{I}} + \partial_3 \underset{\alpha}{\mathbf{P}}^{\bar{3}} + \underset{\alpha}{\rho}\mathbf{F} = 0, \ \ \underset{\alpha}{N}_I \underset{\alpha}{\boldsymbol{\mu}}^{\bar{I}} + \partial_3 \underset{\alpha}{\boldsymbol{\mu}}^{\bar{3}} + \underset{\alpha}{\underset{\approx}{\mathbf{C}}} \overset{2}{\otimes} \underset{\widetilde{\alpha}}{\mathbf{P}} + \underset{\alpha}{\rho}\underset{\alpha}{\mathbf{m}} = 0. \tag{3}$$

CR of micropolar linear elasticity theory for materials without a center of symmetry have a look [21, 26]

$$\underset{\sim}{\mathbf{P}} = \underset{\approx}{\mathbf{A}} \overset{2}{\otimes} \underset{\sim}{\boldsymbol{\gamma}} + \underset{\approx}{\mathbf{B}} \overset{2}{\otimes} \underset{\sim}{\boldsymbol{\varkappa}}, \ \ \underset{\sim}{\boldsymbol{\mu}} = \underset{\approx}{\mathbf{C}} \overset{2}{\otimes} \underset{\sim}{\boldsymbol{\gamma}} + \underset{\approx}{\mathbf{D}} \overset{2}{\otimes} \underset{\sim}{\boldsymbol{\varkappa}}, \tag{4}$$

where $\underset{\sim}{\boldsymbol{\gamma}} = \nabla\mathbf{u} - \underset{\approx}{\mathbf{C}} \cdot \boldsymbol{\varphi}$ is the strain tensor, $\underset{\sim}{\boldsymbol{\varkappa}} = \nabla\boldsymbol{\varphi}$ is the tortion-bending tensor, $\underset{\approx}{\mathbf{A}}$, $\underset{\approx}{\mathbf{B}}$, $\underset{\approx}{\mathbf{C}}$, $\underset{\approx}{\mathbf{D}}$ – the fourth rank material tensors. If the material has a center of symmetry, then $\underset{\approx}{\mathbf{B}} = 0$, $\underset{\approx}{\mathbf{C}} = 0$. Using denotion $\underset{\sim}{\mathbf{A}}^{\cdot\cdot\bar{m}\bar{n}} = \underset{\approx}{\mathbf{A}} \overset{2}{\otimes} \mathbf{r}^{\bar{m}}\mathbf{r}^{\bar{n}}$ similarly to (3) from (4) for each layer α with a center of simmetry, one has

$$\underset{\widetilde{\alpha}}{\mathbf{P}} = \underset{\widetilde{\alpha}}{\mathbf{A}}^{\cdot\cdot\bar{M}\bar{l}} \underset{\alpha}{N}_M \underset{\alpha}{u}_{\bar{l}} + \underset{\widetilde{\alpha}}{\mathbf{A}}^{\cdot\cdot\bar{3}\bar{l}} \partial_3 \underset{\alpha}{u}_{\bar{l}} - \underset{\widetilde{\alpha}}{\mathbf{A}}^{\cdot\cdot\bar{p}\bar{q}} \underset{\alpha}{C}_{\bar{p}\bar{q}\bar{k}} g^{\bar{k}\bar{t}} \underset{\alpha}{\varphi}_{\bar{t}}, \ \ \underset{\widetilde{\alpha}}{\boldsymbol{\mu}} = \underset{\widetilde{\alpha}}{\mathbf{D}}^{\cdot\cdot\bar{M}\bar{l}} \underset{\alpha}{N}_M \underset{\alpha}{\varphi}_{\bar{l}} + \underset{\widetilde{\alpha}}{\mathbf{D}}^{\cdot\cdot\bar{3}\bar{l}} \partial_3 \underset{\alpha}{\varphi}_{\bar{l}}.$$

4. Boundary conditions and interlayer contact conditions

Let the multilayered body consists of K layers and $\overset{(-)}{\underset{\sim}{\mathbf{P}}} = \underset{1}{\underset{\sim}{\mathbf{P}}}|_{x^3=0}$, $\overset{(+)}{\underset{\sim}{\mathbf{P}}} = \underset{K}{\underset{\sim}{\mathbf{P}}}|_{x^3=1}$, $\overset{(-)}{\underset{\sim}{\boldsymbol{\mu}}} = \underset{1}{\underset{\sim}{\boldsymbol{\mu}}}|_{x^3=0}$, $\overset{(+)}{\underset{\sim}{\boldsymbol{\mu}}} = \underset{\widetilde{K}}{\boldsymbol{\mu}}|_{x^3=1}$. Than static boundary conditions on the inner and outer surfaces of a multilayered body are [21]

$$\underset{1}{\overset{(-)}{\mathbf{P}}}{}^{\bar{3}} = \sqrt{\underset{1}{g}^{\bar{3}\bar{3}}}\overset{(-)}{\mathbf{P}}, \ \ \underset{1}{\overset{(-)}{\boldsymbol{\mu}}}{}^{\bar{3}} = \sqrt{\underset{1}{g}^{\bar{3}\bar{3}}}\overset{(-)}{\boldsymbol{\mu}}, \ \ \underset{K}{\overset{(+)}{\mathbf{P}}}{}^{\bar{3}} - \underset{K\overset{+}{I}}{g}^{\bar{3}} \underset{K\bar{J}}{g}^{\overset{+}{I}} \underset{K}{\overset{(+)}{\mathbf{P}}}{}^{\bar{J}} = \sqrt{\underset{K}{g}^{\overset{+}{3}\overset{+}{3}}}\overset{(+)}{\mathbf{P}}, \ \ \underset{K}{\overset{(+)}{\boldsymbol{\mu}}}{}^{\bar{3}} - \underset{K\overset{+}{I}}{g}^{\bar{3}} \underset{K\bar{J}}{g}^{\overset{+}{I}} \underset{K}{\overset{(+)}{\boldsymbol{\mu}}}{}^{\bar{I}} = \sqrt{\underset{K}{g}^{\overset{+}{3}\overset{+}{3}}}\overset{(+)}{\boldsymbol{\mu}}.$$

Interlayer contact conditions for a full contact take the form

$$\underset{\alpha}{\overset{(+)}{\mathbf{P}}} = -\underset{\alpha+1}{\overset{(-)}{\mathbf{P}}}, \ \ \underset{\alpha}{\overset{(+)}{\boldsymbol{\mu}}} = -\underset{\alpha+1}{\overset{(-)}{\boldsymbol{\mu}}}, \ \ \underset{\alpha}{\overset{(+)}{\mathbf{u}}} = \underset{\alpha+1}{\overset{(-)}{\mathbf{u}}}, \ \ \underset{\alpha}{\overset{(+)}{\boldsymbol{\varphi}}} = \underset{\alpha+1}{\overset{(-)}{\boldsymbol{\varphi}}}, \ \ \alpha = 1, \dots, K-1.$$

5. Equations and CR in moments with respect to the Legendre polynomials

We expand all unknown functions in a series

$$P_{ij}(x^1, x^2, x^3) = \sum_{k=0}^{\infty} \overset{(k)}{P}_{ij}(x^1, x^2) P_k(x^3), \ \ \mu_{ij}(x^1, x^2, x^3) = \sum_{k=0}^{\infty} \overset{(k)}{\mu}_{ij}(x^1, x^2) P_k(x^3),$$

$$u_i(x^1, x^2, x^3) = \sum_{k=0}^{\infty} \overset{(k)}{u}_i(x^1, x^2) P_k(x^3), \ \ \varphi_i(x^1, x^2, x^3) = \sum_{k=0}^{\infty} \overset{(k)}{\varphi}_i(x^1, x^2) P_k(x^3),$$

where P_k are the Legendre polynomials with transformed argument orthogonal on segment [0,1]. Here, if we consider only the first $N + 1$ terms of these series we say that we deal with $(0, N)$ order approximation [21].

Using the motion equations in the form (3), the definition of the kth-order moment of tensor quantity with respect to the system of Legendre polynomials, recurrence relations for Legendre polynomials, one can obtain motion equations of the micropolar body in moments of unknown functions with respect to the system of Legendre polynomials. Similarly, one can obtain CR in moments with respect to the system of Legendre polynomials [21].

6. The problem of one-layered two-dimensional rectangular isotropic plate with two pinched edges

Formulation of the problem of the classic theory of elasticity for one-layered two-dimensional rectangular plate with two pinched edges (Figure 1.) of $(0, N)$ order approximation in moments of displacement and rotation vectors with respect to the Legendre polynomials consists of:

1) the motion equations of $(0, N)$ order approximation of the classic theory of elasticity

$$\partial_1 \overset{(k)}{P}_{I1} + \frac{2k+1}{h}[\overset{(+)}{P}_{I2} - (-1)^k \overset{(-)}{P}_{I2}] - \frac{2k+1}{h}\sum_{p=0}^{k}[1-(-1)^{k+p}]\overset{(p)}{P}_{I2} = 0, \ I=1,2, \ k=\overline{0,N}, \tag{5}$$

2) the system of CR of (0,N) order approximation of the classic theory of elasticity

$$\overset{(k)}{P}_{11} = (\lambda+2\mu)\partial_1 \overset{(k)}{u}_1 + \frac{2(2k+1)\lambda}{h}\sum_{p=0}^{N}\overset{(k+2p+1)}{u_2}, \quad \overset{(k)}{P}_{12} = \mu\partial_1 \overset{(k)}{u}_2 + \frac{2(2k+1)\mu}{h}\sum_{p=0}^{N}\overset{(k+2p+1)}{u_1},$$
$$\overset{(k)}{P}_{22} = \lambda\partial_1 \overset{(k)}{u}_1 + \frac{2(2k+1)(\lambda+2\mu)}{h}\sum_{p=0}^{N}\overset{(k+2p+1)}{u_2}, \quad k=\overline{0,N}, \tag{6}$$

3) the system of boundary conditions on the base surfaces $\overset{(-)}{S}$ and $\overset{(+)}{S}$

$$\overset{(-)}{P}_{12} = 0, \quad \overset{(-)}{P}_{22} = 0, \quad \overset{(+)}{P}_{12} = 0, \quad \overset{(+)}{P}_{22} = -q, \tag{7}$$

4) the systems of boundary conditions of $(0, N)$ order approximation on the side surface

$$\overset{(k)}{u}_I|_{x_1=0} = 0, \quad \overset{(k)}{u}_I|_{x_1=L} = 0, \quad k=\overline{0,N}. \tag{8}$$

7. The problem of three-layered two-dimensional rectangular isotropic plate with two pinched edges

Formulation of the problem of the classic theory of elasticity for three-layered two-dimensional rectangular plate with two pinched edges of $(0, N)$ order approximation in moments of displacement and rotation vectors with respect to the Legendre polynomials consists of:

1) the systems of motion equations of $(0, N)$ order approximation of the classic theory of elasticity for each layer

$$\partial_1 \underset{\alpha}{\overset{(k)}{P}}_{I1} + \frac{2k+1}{\underset{\alpha}{h}}(\underset{\alpha}{\overset{(+)}{P}}_{I2} - (-1)^k \underset{\alpha}{\overset{(-)}{P}}_{I2}) - \frac{2k+1}{\underset{\alpha}{h}}\sum_{p=0}^{k}(1-(-1)^{k+p})\underset{\alpha}{\overset{(p)}{P}}_{I2} = 0,$$
$$I=1,2, \quad \alpha=1,2,3, \quad k=\overline{0,N},$$

2) the systems of CR of $(0, N)$ order approximation of the classic theory for each layer

$$\underset{\alpha}{\overset{(k)}{P}}_{11} = (\underset{\alpha}{\lambda}+2\underset{\alpha}{\mu})\partial_1 \underset{\alpha}{\overset{(k)}{u}}_1 + \frac{2(2k+1)}{\underset{\alpha}{h}}\underset{\alpha}{\lambda}\sum_{p=0}^{N}\underset{\alpha}{\overset{(k+2p+1)}{u_2}}, \quad \underset{\alpha}{\overset{(k)}{P}}_{12} = \underset{\alpha}{\mu}\partial_1 \underset{\alpha}{\overset{(k)}{u}}_2 + \frac{2(2k+1)}{\underset{\alpha}{h}}\underset{\alpha}{\mu}\sum_{p=0}^{N}\underset{\alpha}{\overset{(k+2p+1)}{u_1}},$$
$$\underset{\alpha}{\overset{(k)}{P}}_{22} = \underset{\alpha}{\lambda}\partial_1 \underset{\alpha}{\overset{(k)}{u}}_1 + \frac{2(2k+1)}{\underset{\alpha}{h}}(\underset{\alpha}{\lambda}+2\underset{\alpha}{\mu})\sum_{p=0}^{N}\underset{\alpha}{\overset{(k+2p+1)}{u_2}}, \quad k=\overline{0,N},$$

3) the system of boundary conditions on the base surfaces $\overset{(-)}{S}$ and $\overset{(+)}{S}$

$$\underset{1}{\overset{(-)}{P}}_{12} = 0, \quad \underset{1}{\overset{(-)}{P}}_{22} = 0, \quad \underset{3}{\overset{(+)}{P}}_{12} = 0, \quad \underset{3}{\overset{(+)}{P}}_{22} = -q,$$

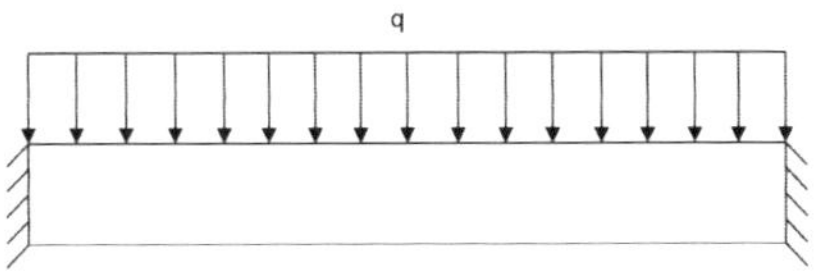

Figure 1. Rectangular plate with two pinched edges under the distributed load

4) the systems of boundary conditions of $(0, N)$ order approximation on the side surface

$$\underset{\alpha}{\overset{(k)}{u}}_{I}|_{x_1=0} = 0, \quad \underset{\alpha}{\overset{(k)}{u}}_{I}|_{x_1=L} = 0 \quad \alpha = 1, 2, 3, \quad k = \overline{0, N},$$

5) interlayer contact conditions in case of full contact

$$\underset{1}{\overset{(+)}{P}}_{I2} = \underset{2}{\overset{(-)}{P}}_{I2}, \quad \underset{2}{\overset{(+)}{P}}_{I2} = \underset{3}{\overset{(-)}{P}}_{I2}, \quad \sum_{p=0}^{N} \underset{1}{\overset{(p)}{u}}_{I} = \sum_{p=0}^{N} (-1)^p \underset{2}{\overset{(p)}{u}}_{I}, \quad \sum_{p=0}^{N} \underset{2}{\overset{(p)}{u}}_{I} = \sum_{p=0}^{N} (-1)^p \underset{3}{\overset{(p)}{u}}_{I}.$$

8. The problem of one-layered two-dimensional rectangular micropolar isotropic plate with two pinched edges

Formulation of the problem of the micropolar theory of elasticity for one-layered two-dimensional rectangular plate with two pinched edges (Figure 1.) of $(0, N)$ order approximation in moments of displacement and rotation vectors with respect to the Legendre polynomials consists of:

1) The motion equations of $(0, N)$ order approximation of the micropolar theory of elasticity

$$\begin{gathered}
\partial_1 \overset{(k)}{P}_{1I} + \frac{2k+1}{h}[\overset{(+)}{P}_{I2} - (-1)^k \overset{(-)}{P}_{I2}] - \frac{2k+1}{h} \sum_{p=0}^{k} [1 - (-1)^{k+p}] \overset{(p)}{P}_{I2} = 0, \;\; I = 1, 2, \\
\partial_1 \overset{(k)}{\mu}_{13} + \frac{2k+1}{h}[\overset{(+)}{\mu}_{12} - (-1)^k \overset{(-)}{\mu}_{12}] - \frac{2k+1}{h} \sum_{p=0}^{k} [1 - (-1)^{k+p}] \overset{(p)}{\mu}_{23} + \overset{(k)}{P}_{21} - \overset{(k)}{P}_{12} = 0, \\
k = \overline{0, N},
\end{gathered} \tag{9}$$

2) The system of CR of $(0, N)$ order approximation of the micropolar theory of elasticity

$$\begin{gathered}
\overset{(k)}{P}_{11} = (\lambda + 2\mu)\partial_1 \overset{(k)}{u}_1 + \frac{2(2k+1)\lambda}{h} \sum_{p=0}^{N} \overset{(k+2p+1)}{u_2}, \quad \overset{(k)}{P}_{22} = \lambda \partial_1 \overset{(k)}{u}_1 + \frac{2(2k+1)(\lambda + 2\mu)}{h} \sum_{p=0}^{N} \overset{(k+2p+1)}{u_2}, \\
\overset{(k)}{P}_{12} = (\mu + \alpha)\partial_1 \overset{(k)}{u}_2 + \frac{2(2k+1)(\mu - \alpha)}{h} \sum_{p=0}^{N} \overset{(k+2p+1)}{u_1} + 2\alpha \overset{(k)}{\varphi}, \\
\overset{(k)}{P}_{21} = (\mu - \alpha)\partial_1 \overset{(k)}{u}_2 + \frac{2(2k+1)(\mu + \alpha)}{h} \sum_{p=0}^{N} \overset{(k+2p+1)}{u_1} - 2\alpha \overset{(k)}{\varphi}, \\
\overset{(k)}{\mu}_{13} = (\delta + \beta)\partial_1 \overset{(k)}{\varphi}, \quad \overset{(k)}{\mu}_{23} = \frac{2(2k+1)(\delta + \beta)}{h} \sum_{p=0}^{N} \overset{(k+2p+1)}{\varphi}, \quad k = \overline{0, N},
\end{gathered} \tag{10}$$

3) The system of boundary conditions on the base surfaces $\overset{(+)}{S}$ and $\overset{(-)}{S}$

$$\overset{(-)}{P}_{21} = 0, \quad \overset{(-)}{P}_{22} = 0, \quad \overset{(-)}{\mu}_{23} = 0, \quad \overset{(+)}{P}_{21} = 0, \quad \overset{(+)}{P}_{22} = -q, \quad \overset{(+)}{\mu}_{23} = 0, \tag{11}$$

4) The systems of boundary conditions of $(0, N)$ order approximation on the side surface

$$\overset{(k)}{u}_{I}|_{x_1=0} = 0, \quad \overset{(k)}{u}_{I}|_{x_1=L} = 0, \quad \overset{(k)}{\varphi}|_{x_1=0} = 0, \quad \overset{(k)}{\varphi}|_{x_1=L} = 0, \quad k = \overline{0, N}. \tag{12}$$

Table 1

method	FEM	Polynomials								
		(0,0) approx	(0,1) approx	(0,2) approx	(0,3) approx	(0,4) approx	(0,5) approx	(0,6) approx	(0,7) approx	(0,8) approx
deflection, [mm]	-2,57	-0,77	-2,16	-2,44	-2,55	-2,56	-2,57	-2,57	-2,57	-2,57

Table 2

method	(0,0) approx	(0,1) approx	(0,2) approx	(0,3) approx	(0,4) approx	(0,5) approx
Classic theory, deflection, [mm]	-0,01	-4,87	-8,74	-8,74	-8,75	-8,75
Micropolar theory, deflection, [mm]	-0,01	-4,74	-8,50	-8,50	-8,50	-8,51

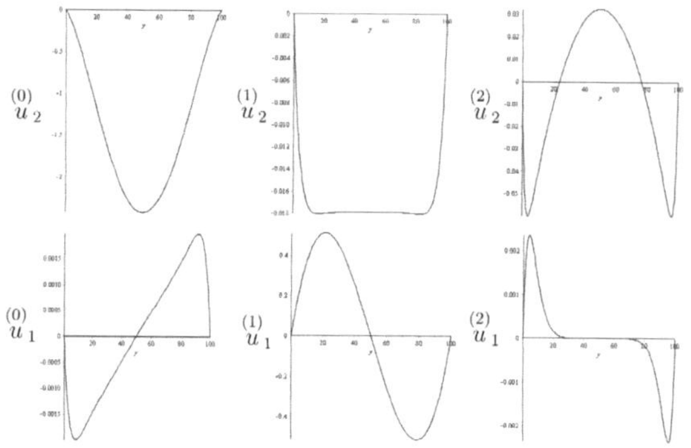

Figure 2. Numerical solution of the problem for (0,2) order approximation.

9. Numerical solution of the problem of one-layered two-dimensional rectangular isotropic plate with two pinched edges

Solving the problem (5) – (8) for classic elastic steel plate (Young's modulus $E = 21000\ kg/mm^2$, Poisson's ratio $\nu = 0.3$) at thickness $h = 10mm$, length $L = 100mm$, pressure $q = 100\ kg/mm$ we obtained the solutions for $(0,0)$th – $(0,8)$th approximations. The problem also was solved by finite element method. The maximum values of the deflection of a plate are presented in the Table 1. Solution of the problem for $(0,2)$th approximation is presented at the Figure 2.

For the problem (9) – (12) for micropolar isotropic plate ($E = 300MPa$, $\nu = 0.4$, $l_b = 0.33mm$, $N^2 = 0.02$) [27] at thickness $h = 1mm$, length $L = 100mm$, pressure $q = -0.01\ kg/cm$ we obtained solution for $(0,0)$th – $(0,5)$th approximations. Also we obtained solution for analogous isotropic classic plate ($E = 300MPa$, $\nu = 0.4$). The maximum values of the deflection of a plate are in the Table 2.

Note also that of great interest are works on the thermomechanics of composite structures under high temperature [28–32], which can be used when considering problems of thermomechanics for thin structures using the method described above.

Acknowledgement: this work was supported by the Russian Foundation for Basic Research (project no. 18–29–10085–mk).

References

[1] *I. N. Vekua,* Shell theory, general methods of construction. Pitman Advanced Pub. Program, 1985.
[2] *A. E. Alekseev and B. D. Annin,* Equations of deformation of an elastic inhomogeneous laminated body of revolution // J. Appl. Mech. Techn. Phys. 2003. **44**, No. 3. 432-437.
[3] *Yu. M. Volchkov and L. A. Dergileva,* Reducing three-dimensional elasticity problems to two-dimensional problems by approximating stresses and displacements by Legendre polynomials // J. Appl. Mech. Techn. Phys. 2007. **48**, No. 3. 450–459.
[4] *G. V. Ivanov,* Theory of plates and shells. Novosibirsk State Univ., Novosibirsk, 1980 (in Russian).
[5] *J. Fellers and A. Soler,* Approximate solution of the finite cylinder problem using Legendre polynomials // AIAA J. 1970. **8**, No. 11.
[6] *Cicala Placido,* Sulla teoria elastica della parete sottile// Giornale del Genio Civile. 1959. **97**, No. 4, 6, 9
[7] *R. D. Mindlin and M. A. Medick,* Extensional vibrations of elastic plates // J. Appl. Mech. 1959. **26**, No. 4. 561–569.

[8] *T. V. Meunargiya,* Development of the method of I. N. Vekua for problems of the three-dimensional moment elasticity. Tbilisi State Univ., Tbilisi, 1987 (in Russian).
[9] *A. R. Ulukhanyan,* Representation of Solutions to Equations of Hyperbolic Type// Moscow university mechanics bulletin. 2010, Vol. 65, No. 2. 47–50. DOI: 10.3103/S0027133010020056
[10] *B. L. Pelekh, A. V. Maksimuk, and I. M. Korovaichuk,* Contact problems for laminated elements of constructions and bodies with coating Naukova Dumka, Kiev, 1988 (in Russian).
[11] *M. U. Nikabadze,* A variant of the theory of multilayer structures // Mech. Solids. 2001. No. 1. 143–158.
[12] *M. U. Nikabadze,* Mathematical modeling of multilayer thin body deformation//Journal of mathematical sciences. V. 187, No 3, 2012. P. 300-336.
[13] *M. U. Nikabadze and A. R. Ulukhanyan,* Statements of problems for a thin deformable three-dimensional body // Vestn. Mosk. Univ., Matem. Mekhan. 2005. No. 5. 43–49.
[14] *M. U. Nikabadze,* A variant of the system of equations of the theory of thin bodies // Vestn. Mosk. Univ., Matem. Mekhan. 2006. No. 1. 30–35.
[15] *M. U. Nikabadze,* Application of a system of Chebyshev polynomials to the theory of thin bodies // Vestn. Mosk. Univ., Matem. Mekhan. 2007. No. 5. 56–63.
[16] *M. U. Nikabadze,* Some issues concerning a version of the theory of thin solids based on expansions in a system of Chebyshev polynomials of the second kind // Mech. Solids. 2007. **42**. No. 3. 391-421.
[17] *M. Nikabadze, A. Ulukhanyan* 2016 Analytical Solutions in the Theory of Thin Bodies. *Advanced Structured Materials, Generalized Continua as Models for Classical and Advanced Materials, H. Altenbach and S. Forest* . Vol. 42. pp. 319–361 DOI: 10.1007/978-3-319-31721-2$_1$5
[18] *M. Nikabadze, A. Ulukhanyan, T. Moseshvili, K. Tskhakaia, N. Mardaleishvili, Z. Arkania* 2019 On the Modeling of Five-Layer Thin Prismatic Bodies. *Mathematical and Computational Applications.* Vol. 24, 3. pp. 1–16 DOI: 10.3390/mca24030069
[19] *M. Nikabadze, A. Ulukhanyan, G. Sakhvadze* 2019 To mathematical modeling of deformation of micropolar thin bodies with two small sizes. *Journal of Physics: Conference Series.* Vol. 1205, 1. pp. 1–8 DOI: 10.1088/1742-6596/1205/1/012040
[20] *Kantor M M, Nikabadze M U, Ulukhanian A R* 2013 Equations of Motion and Boundary Conditions of Physical Meaning of Micropolar Theory of Thin Bodies with Two Small Cuts. *Mechanics of Solids.* Vol. 48, 3. pp. 317–328 DOI: 10.3103/S0025654413030084
[21] Nikabadze M U 2014 *Development of the Method of Orthogonal Polynomials in the Classical and Micropolar Mechanics of Elastic Thin Bodies* (Moscow University Press), 515 p. (in Russian). http://istina.msu.ru/media/publications/book/707/ea1/6738800/Monographiya.pdf
[22] Vekua I N 1978 *Fundamentals of tensor analysis and covariant theory,* (Nauka, Moscow) (in Russian)
[23] Nikabadze M U 2017 Topics on tensor calculus with applications to mechanics. *Journal of Mathematical Sciences.* Vol. 225, 1. 194 p. DOI: 10.1007/s10958-017-3467-4
[24] Nowacki W 1975 *Theory of elasticity* (Moscow, Mir) (in Russian).
[25] Kupradze V D, Gegelia T G, Basheleishvili M O and Burchuladze T V, 1976 *Three-dimensional problems of the mathematical theory of elasticity and thermoelasticity.* (Nauka, Moscow) (in Russian).
[26] Eringen A C 1999 *Microcontinuum field theories. 1. Foundation and solids.* (N.Y.: Springer- Verlag).
[27] Lakes R S 1995 Experimental methods for study of Cosserat elastic solids and other generalized continua *Continuum models for materials with micro-structure,* (ed. H. Muhlhaus, J. Wiley, N. Y. Ch. 1) p. 1-22.
[28] Dimitrienko Y, Yakovlev D 2015 The asymptotic theory of thermoelasticity of multilayer composite plates *Composites-mechanics computations applications* **6** 1 pp. 13–51. DOI: 10.1615/CompMechComputApplIntJ.v6.i1.20
[29] Dimitrienko Y 2016 Thermomechanics of composite structures under high temperatures. *Solid Mechanics and its Applications* **224** DOI 10.1007/978-94-017-7494-9
[30] Dimitrienko Y 2016 Thermomechanics of composite structures under high temperatures. General equations of multiphase continuum mechanics for ablative composites *Solid Mechanics and its Applications* **224** pp. 31–59. DOI: 10.1007/978-94-017-7494-9-2
[31] Dimitrienko Y 2016 Thermomechanics of composite structures under high temperatures. Mathematical Model of Ablative Composites *Solid Mechanics and its Applications* **224** pp. 61–81. DOI: 10.1007/978-94-017-7494-9-3
[32] Dimitrienko Y, Dimitrienko I 2017 Modeling of thin composite laminates with general anisotropy under harmonic vibrations by the asymptotic homogenization method *Int. journal for multiscale computational engineering* **15** 3 pp. 219-237. DOI: 10.1615/IntJMultCompEng.2017019518

Homogenization and multicontinuum models for high contrast composites

G P Panasenko[1]

[1] Institute Camille Jordan UMR CNRS 5208 and SFR MODMAD FED 4169, University of Lyon, 42023, Saint-Etienne, France

Abstract. We consider the high contrast periodic composite materials having high ratio $\omega >>1$ of physical constants of the components such as the heat conductivity or Young's moduli. The ratio of the characteristic microscopic size and the characteristic macroscopic size is a small parameter ε. If the component having high constants has the shape of small isolated particles then the homogenized (macroscopic, effective) model demonstrates a loss of the effective wave velocity. If the topology of the component having high constants corresponds to several (at least two) connected periodic sets disconnected one from other and if $\varepsilon 2\omega >> 1$ or $\varepsilon 2\omega = const$ then the macroscopic model is described by a multicontinuum, i.e. several continua in each point.

1.Introduction

The high contrast composites have components with high ratio of mechanical properties (for example, Young moduli or conductivities). Such materials have atypical properties which cannot be obtained in natural materials. Namely, they may have a considerable loss of the wave velocity compared to the wave velocity in the components, in some cases they are described by multicontinuum models (several continuums coexisting in each point of the space) or by frequency dependent models presenting the micro-resonance effects.

2.Loss of the effective velocity in composites with dispersed isolated heavy stiff inclusions.

Consider first the non-stationary elasticity equations in a composite material having the following periodic structure. Let ε be the ratio of the microscale size and the macroscale size. A periodic cell having the edge ε contains one stiff and heavy particle occupying a ball inside the cell, while the filament is soft and light. The ratio of the Young moduli of components as well as the ratio of densities is a large parameter $\omega >> 1$. Namely, let the density of inclusions be $\omega\rho_i$ and the density of the filament be $\rho_{m,}$ let the Young's modulus of inclusions be ωE_i let the Young's modulus of the filament be $E_{m.}$ Introduce the Poisson's ratios of the inclusions ν_i and of the filament $\nu_{m.}$ Define the density ρ (y) and the elasticity moduli a_{rj}^{kl} (y) as a function of a fast variable $y=x/\varepsilon$ equal to $\omega\rho_i$ and to

$0.5\ \omega E_i\ (1+ \nu_i)^{-1}(\ 2\ \nu_i\ (1\text{-}2\ \nu_i)^{-1}\delta_{rk}\delta_{jl}+\delta_{rj}\delta_{kl}+\delta_{rl}\delta_{kj})$ if x belongs to inclusions and respectively ρ_m and $0.5\ E_m\,(1+ \nu_m)^{-1}(\ 2\ \nu_m\,(1\text{-}2\ \nu_m)^{-1}\delta_{rk}\delta_{jl}+\delta_{rj}\delta_{kl}+\delta_{rl}\delta_{kj})$ if x belongs to the filament.

The wave propagation in such elastic medium is described by equations

$\rho(x/\varepsilon)u^k_{tt} - (a_{rj}^{kl}\,(x/\varepsilon)\ u_{xj}^{\ l})_{xr} = 0$ (summing up over i,j,r,l).

The homogenization of this model by the two stage homogenization technique [1] gives the effective density of order of $\omega >> 1$ while the stiffness is of order O(1). It means that the effective wave velocity is of order $(1/\omega)^{1/2} << 1$. It means that if there are some dissipative factors in the soft component then they will act for a longer time while the wave slows down. Such dissipative factors may be, for example, the friction between the parts of the soft material, modeled by the Robin type boundary conditions at the boundaries of periodic holes in the soft phase or a viscoelastic behavior of the soft phase, with short or long memory.

3.When the macroscopic model of a high contrast composite is a multicontinuum?

Another effect of the high contrast properties of the components is the multicontinuum models. Such models at macroscopic level are asymptotically derived from the microscopic description of a highly heterogeneous medium. To our knowledge the first asymptotic derivations of such models appear in [12],[14], [18]-[21] and later such models were obtained by means of two-scale convergence [1]-[3]. In particular, the double porosity model, (see [5] (introduced), [2], [8] (justified)) is one of examples, for wave propagation see [13]. In [23] we derive a multicontinuum wave propagation model in a laminated beam with contrasting stiffness and density of the layers using technique of multicomponent homogenization introduced in [19], the 3D theory of wave propagation in such media was developed in [20]. Consider an ε-periodic structure consisting of three periodic sets B_1, B_2, B_3 of thin cylinders parallel to the Cartesian axes x_1, x_2, x_3 respectively. The cylinders of the sets B_1, B_2, B_3 have no common points. Let the conductivity K be equal to positive constants ωK_1, ωK_2, ωK_3 in B_1, B_2, B_3 respectively and $K_0>0$ out of the union of B_1, B_2 and B_3. Let the density have a similar structure: $\omega\rho_1$, $\omega\rho_2$, $\omega\rho_3$ in B_1, B_2, B_3 and $\rho_0>0$ out of the union of B_1, B_2 and B_3. Consider the wave equation in such a structure. Then the asymptotic behavior of the solution depends on the parameter $\varepsilon^2\omega$. If this parameter is small (<<1) then the structure can be approximated by a classical homogenized wave equation [20] with the effective conductivities $\omega K_1\theta_1$, $\omega K_2\theta_2$, $\omega K_3\theta_3$ in three coordinate directions; here θ_i are the volume ratio of the set B_i; the effective density is $\omega\rho_1\theta_1+\omega\rho_2\theta_2+\omega\rho_3\theta_3$. If $\varepsilon^2\omega>>1$ then the effective macroscopic model is a multicontinuum: each set of highly conductive cylinders "lives" independently of other sets, so that in every point x there co-exist three one-dimensional continua described by three independent equations

$\omega\rho_i\theta_i u^i_{tt} - \omega K_i\theta_i u^i_{x_ix_i}=0$, i=1,2,3.

Finally, if $\varepsilon^2\omega=\eta$ is constant of order 1then these equations become weakly coupled:

$\omega\rho_i\theta_i u^i_{tt} - \omega K_i\theta_i u^i_{x_ix_i}+\omega\eta^{-1}D_k u^k=0$, i=1,2,3 (summing up over k),

where the coefficients D_i are defined by some special cell problems (see [20]). Such multicontinua demonstrate special atypical properties, in particular, they may have many different wave velocities (as many as there are sets of cylinders).

A similar effect takes place for composites having several disconnected three-dimensionally spaced stiff components or for a laminated rod with stiff and heavy layers alternating with the soft light layers [23]. In any case the multicontinuum effects take place for special topologies of the components, and they are absent in the case of isolated particles described in the first part of the paper.

Finally, some special effects are related to the frequency dependent materials. The spectral analysis of the high contrast composites was provided in the classical paper [25]. Later the theory was developed in papers [10], [11]. The sufficient conditions for the spectral gaps were pointed out.

4. Conclusion

The high contrast composites have atypical properties. Choosing an appropriate topology of the components and ratio of physical constants of components we can get special effects in the macroscopic behavior of the material. In particular, sufficient conditions for multicontinua at the macroscopic level is the presence of at least two unconnected (or weakly connected) highly conductive (very stiff) periodic sets crossing the whole material and the relation $\varepsilon^2\omega>>1$ or $\varepsilon^2\omega = \text{const}$.

Acknowledgments
The present work is supported by LABEX MILYON (ANR-10-LABX-0070) of the University of Lyon, within the program "Investissements d'Avenir" (ANR-11-IDEX-0007) operated by the French National Research Agency (ANR).

References
[1] Allaire G 1992 Homogenization and two-scale convergence. *SIAM J. Math. Anal.* **23** 1482 1518
[2] Arbogast T, Douglas J, Hornung U 1990 Derivation of the double porosity model of single phaseflow via homogenization theory. *SIAM J. Math. Anal.* 21 823 - 836
[3] Auriault J L 1991 Heterogeneous medium. Is an equivalent macroscopic description possible? *International Journal of Engineering Science* **29(7)** 785-795
[4] Bakhvalov N, Panasenko G 1989 *Homogenisation: Averaging Processes in Periodic Media* ed Kluwer (Dordrecht/Boston/London) 366 pp.
[5] Barenblatt G I, Zheltov I P, Kochina I N 1960 Basic concepts in the theory of seepage of homogeneous liquids in fissured rocks, *Prikl. Math. Mekh.* **24** 852-864
[6] Bellieud M, Gruais I 2005 Homogenization of an elastic material reinforced by very stiff or heavy fibers. Non local effects. Memory effects. *J. Math. Pures Appl.* **84** 55 - 96
[7] Bellieud M 2010 Torsion effects in elastic composites with high contrast. *SIAM J. Math. Anal.* **41** 2514 - 2553
[8] Bourgeat A, Luckhaus S, Mikelic A 1996 Convergence of the homogenization process for a double- porosity model of immicible two-phase flow, *SIAM J. Math. Anal.* **27** 1520-1543
[9] Boutin C, Royer P, Auriault J L 1998 Acoustic absorption of porous surfacing with dual porosity *International Journal of Solids and Structures* **35 (34)** 4709-4737
[10] Cherednichenko K D, Smyshlyaev V P, Zhikov V V 2006 Non-local homogenized limits for composite media with highly anisotropic periodic fibres. *Proceedings of the Royal Society of Edinburgh, section A- Mathematics* **136** 87-114
[11] Cherednichenko K, Cherdantsev M 2012 Two-scale Γ-convergence of integral functionals and its application to homogenisation of nonlinear high-contrast periodic composites. *Arch. Ration. Mech. Anal.* **204** 445 - 478
[12] Fenchenko V N, Khruslov E Ya 1981 The asymptotic of the solution of differential equations with a strongly oscillating coefficiejnt matrix not satisfying the condition of uniform limitation, *Dopovid. Akad.Nauk Ukr. RSR* (Ukrain) A **4** 23-27
[13] Figotin A, Kuchment P 1998 Spectral properties of classical waves in high-contrast periodic media, *SIAM J. Appl. Math.* **58** 683 - 702
[14] Khruslov E Ya 1981 On the convergence of the solutions of the second boundary value problems in weakly connected domains, *In Theory of Operators in Functional Spaces and its applications* ed Naukova Dumka (Kiev) 129-174
[15] Kosarev A Yu 1982 The asymptotics of the averaged characteristics of periodic elastic media with strongly variable properties, *USSR Doklady* **267** 38-42.
[16] Khruslov E A 1990 An averaged model of a strongly inhomogeneous medium with memory, *Russian Math. Surveys,* **45(1)** 211-212
[17] Panasenko G P 1983 Homogenization of fields in composite materials with high modulus reinforcement. *Vestnik Moscow University* **15(2)** 20-27
[18] Panasenko G P 1984 On the scale effect in spatially reinforced composites. *Proc. of the 2-nd USSR Conf. on Strength, Rigidity and Technology of Composite Materials.* (Yerevan, Yerevan Univ) **3** 22-24
[19] Panasenko G P 1988 Averaging of processes in strongly inhomogeneous media, *Doklady Akademii Nauk SSSR* **298** 76-79
[20] Panasenko G P 1990 Multicomponent homogenization of processes in strongly non-homogeneous structures, *Mathematics USSR Sbornik* **181** 134-142

[21] Panasenko G P 1990 Numerical-asymptotic multicomponent averaging method for equations with contrasting coefficients, *USSR Computational Mathematics and Mathematical Physics* **30** 134-142

[22] Panasenko G P 1995 Multicomponent homogenization of the vibration problem for incompressible media with heavy and rigid inclusions, *C.R. Acad. Sci. Paris,* **321** I 1109-1114

[23] Panasenko G 2018 Multicontinuum wave propagation in a laminated beam with contrasting stiffness and density of layers, *Journal of Mathematical Science* 232 503-515 https://doi.org/10.1007/s10958-018-3889-7

[24] Zhikov V V, Kozlov S M, and Oleinik O A 1994 *Homogenization of differential operators and integral functionals* ed Springer-Verlag (Berlin)

[25] Zhikov V V 2004 On spectrum gaps of some divergent elliptic operators with periodic coefficients *Algebra Analysis* **16** 34

Research of temperature and stress-strain state of mirror space antenna reflector with fins based on nonbraided polyester material

Y A Azhevsky[1], **D Novikov**[1]

[1] Dynamics and flight control of motion of rockets and spacecraft, Bauman Moscow State Technical University, 2-nd Baumanskaya, 5, b.1, 105005, Moscow, Russia

E-mail: e-mail: azhevsky.yaroslav@mail.ru, novikov.andrey.sm13@gmail.com

Abstract. A number of designs of reflectors of a mirror space antenna in the form of a thin-walled shell made of high-modulus carbon fiber reinforced with a developed finning system in the form of three-layer panels with a filler made of nonwoven polyester material is considered. Within the framework of the finite element modeling, a comparative analysis of serviceability under operating conditions (radiative heat transfer in open space conditions) has been carried out. Patterns of temperature distribution and thermal displacements along the reflector surface have been obtained. According to the results of the research, MSA reflector's design layout with the best indicators has been chosen.

Key words: mirror space antenna reflector, sandwich panels, non-braided polyester material, carbon fiber.

1. Introduction

Mirror space antennas were widely in demand at the beginning of the era of modern satellite communications. A lot of research has been conducted within the framework of the design of reflector structures used mainly for space and ground applications. Reflectors of mirror space antennas (MSAR) feature a relatively simple design and high reliability, while having a wide bandwidth of radio waves and a high directivity of action. These properties allow the use of mirror antennas as part of various-purpose spacecraft (SC) to provide high-speed radio channels [1-4].

The demand for high-speed communications in the last decade has grown significantly, which in turn leads to the development of new high-frequency bands with a reliable communication channel capable of transmitting data with short delays. A promising solution to this problem is the use of the V-band of the frequency spectrum (40-75 GHz frequency band). This frequency band will reduce the weight and size of satellite antennas, increasing energy efficiency and data rate. But one of the main restrictions for the mass production of MSAR structures operating in the V-band as the main means of communication for spacecraft is their high cost [5–8].

The high frequency of the radio communication range tightens the requirements for the quality of space satellite antenna designs. The permissible deviation of the MSAR surface from the calculated value should not exceed $\Delta = \Lambda/50$ (Λ – working length of the radio wave of the satellite antenna). Thus, for

correct operation of satellite communications, it is necessary to maintain the dimensionality of the reflecting surface of the MSAR at a level of 0.08 mm to 0.15 mm, depending on the selected operating frequency [9,10].
When in orbit around the Earth, a spacecraft is exposed to direct and reflected solar radiation, Earth's own radiation, which, coupled with periodic visits to the shadow part of the orbit, can lead to a temperature drop on the meter surface of the MSAR of 100 to 180 degrees. Such operating conditions can adversely affect the dimensional stability of the reflector design, which in turn negatively affects the quality of the transmitted (received) signal [11-16].
Severe requirements for the performance of satellite antennas cause greater attention to be paid to the use of modern materials. Composite materials, namely, carbon fiber-reinforced plastics (CFRP) were of great importance for the improvement of the MSAR designs. This material has high specific mechanical characteristics, low linear thermal expansion coefficients, and a relatively high thermal conductivity, which allows not only to reduce the antenna weight, but also to increase the stability of its shape and size [17, 18].

2. MSAR's design layout.

Since the beginning of the use of composite materials as basic for space satellite antennas, the design layout has found wide application in the form of a three-layer panel with thin-walled casings and a core of honeycomb filler (Fig. 1). As a rule, CFRP panels are used as skins, and aluminum foil, carbon or organoplastic serves as a material for honeycomb filler. Such an approach makes it possible to obtain a rigid, dimensionally stable structure with a linear density of 3 kg/m2 [19,20].

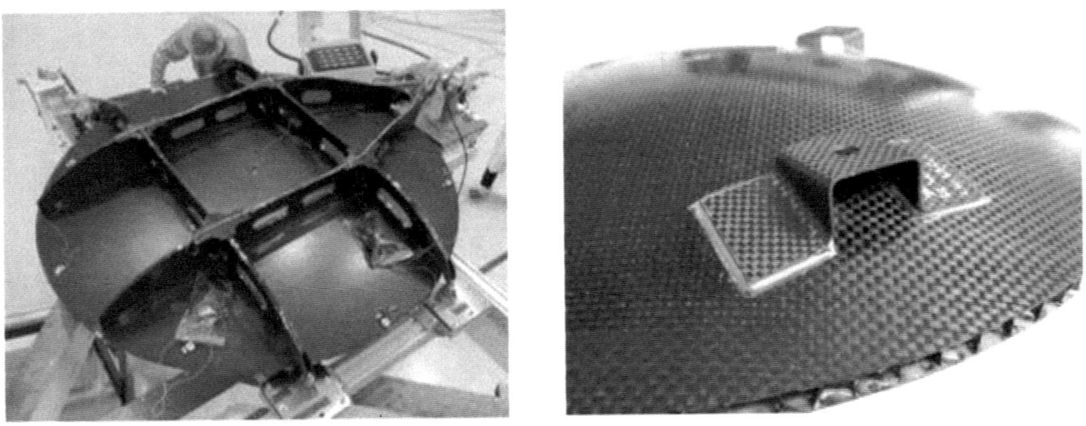

a b

Figure 1. MSAR with two layer-type design layouts:
(a) - reflector of a mirror space antenna manufactured by High Performance Space company;
(b) - High Performance Space's reflector with carbon fiber honeycomb filler

Another structural solution for the MSAR is the design layout in the form of a thin-walled composite shell with developed thin-walled spatial fins (Fig. 2). Structures of this type are made of carbon fiber. The fins enable to develop a rigid structure with high dimensional stability and running density at a level of 2.5 kg/m2 [21.22].

a b

Figure 2. MSAR with fin reinforcement design layout:
(a) - an experimental reflector by Bauman MSTU; (b) - reflector of the space satellite antenna "Artemis"

3. Development of MSAR design layout

The geometry of the location of the power set of the reflector of a mirror space antenna may vary depending on the conditions of fixing the satellite antenna and its deployment mechanism, with the design having to comply with the requirements for shape stability and size imposed for operating modes in orbital flight conditions. Given the domestic and foreign experience, a number of design layouts for reflectors has been designed with a variation in the arrangement of the power fins, implying attachment to the spacecraft through the central zone (Fig. 3).

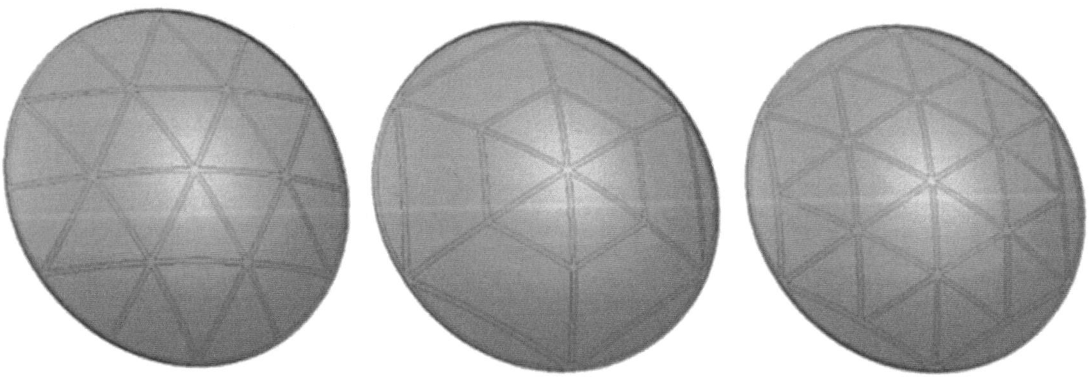

a b c

Figure 3. FSAR with three-layer fin reinforcement:
(a) Mark 1 layout; (b) Mark 2 layout; (c) Mark 3 layout

The above designs feature an original finning layout. The reflector itself is made as a thin-walled carbon fiber shell, and the fins are a localized three-layer carbon fiber panel with a filler of non-woven polyester material. The stiffening fins of the samples under investigation are 15 mm wide and 2.2 mm thick. The reflector shell is 0.6 mm thick.

4. Thermal and thermomechanical analysis of the MSAR layouts under investigation

The most important stage in the design of the MSAR is to recreate the operating conditions by simulating temperature and stress-strain states of the structure, typical for radiative heat transfer in open space.

Mathematical modeling of the heat exchange process was carried out through software package Siemens PLM NX using Space System Thermal solver. The results of the temperature state modeling for the finite element model were used as input data for determining the stress-strain state of the MSAR using the Siemens NX Nastran solver.

The following thermophysical and mechanical properties were used as initial characteristics for mathematical modeling of CFRP: thermal conductivity coefficient – 31 W/(m·K); specific heat – 1000 J/(kg·K); density – 1550 kg/m3; emissivity (degree of blackness) –0.85; absorption capacity – 0.735; coefficient of linear thermal expansion – 5.27 10–7 K–1; Young's modulus – 140 GPa; Poisson's ratio – 0.3. The studied options of reflectors were assumed to work as part of a satellite antenna in the geostationary orbit of the Earth [19,20].

Comparison of design options was made for a time point of 21,600 seconds of work in the geostationary orbit, starting with full shading by the Earth. This moment of time corresponded to the rotation of the spacecraft by 150° relative to the Earth – Sun axis, when the temperature difference on the MSAR surface reaches maximum values.

According to the results of the conducted finite element modeling, the following results on thermal and thermomechanical analysis were obtained (Fig. 4, 5, 6).

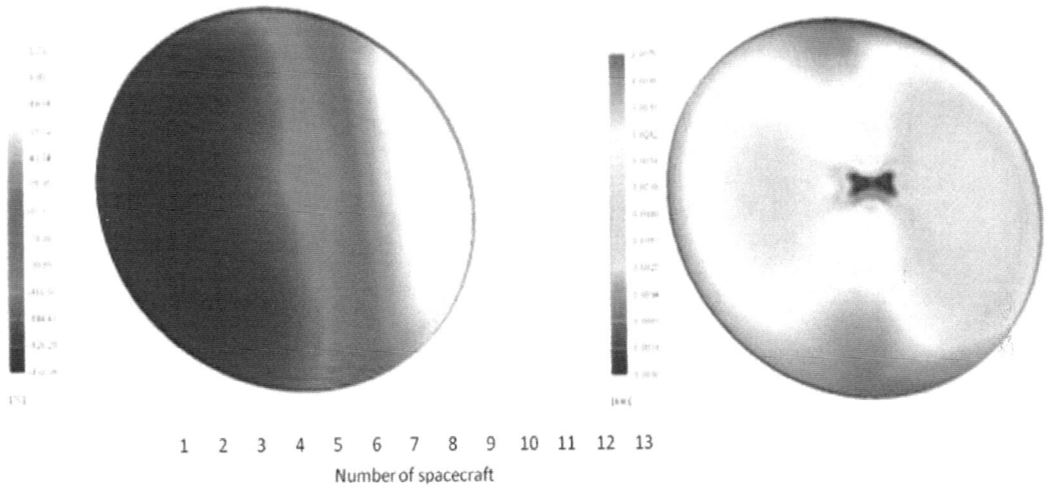

Figure 4. Temperature distribution (left) and thermal displacements (right) for the Mark 1 finning layout

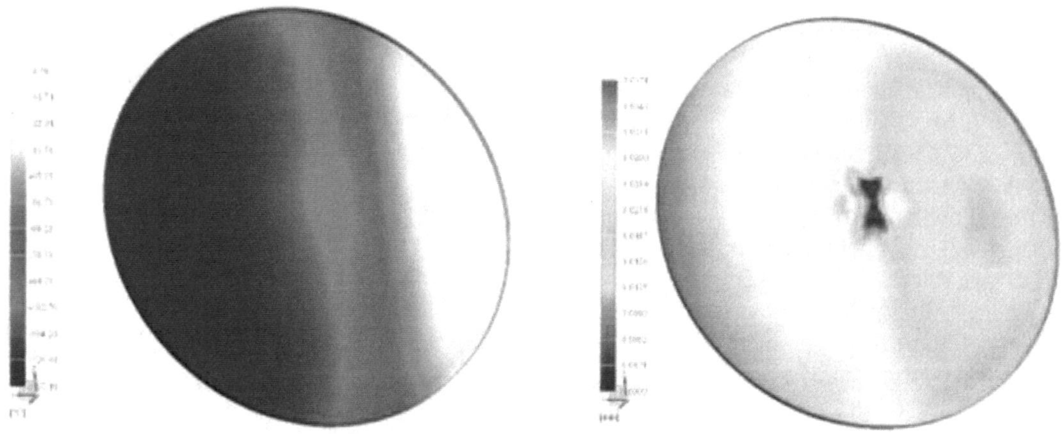

Figure 5. Temperature distribution (left) and thermal displacements (right) for the Mark 2 finning layout

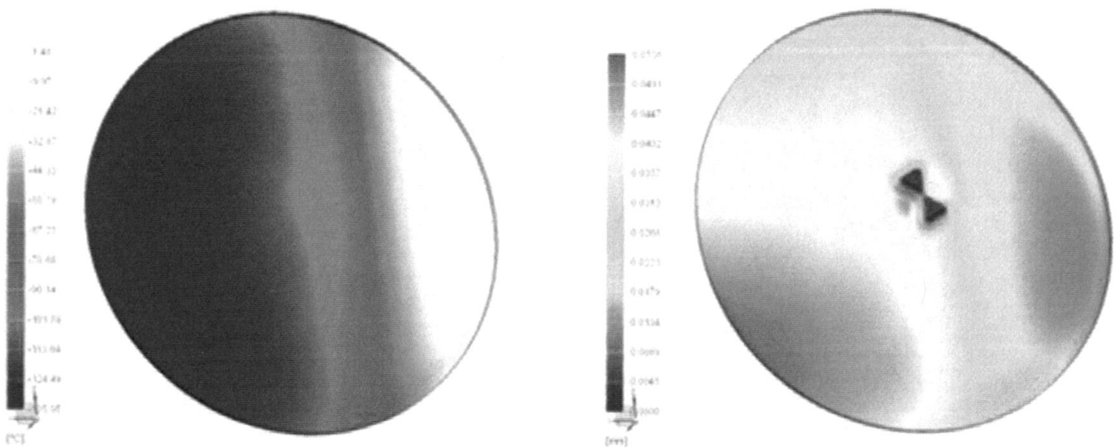

Figure 6. Temperature distribution (left) and thermal displacements (right) for the Mark 3 finning layout

As shown by the results of the finite element analysis, the smallest displacement values correspond to the Mark 1 power layout and amount to 0.037 mm, which satisfies the requirements for the dimensional stability of the structure of the space satellite antenna reflector in orbital flight conditions. The mass of the antenna reflector is 1.662 kg, the surface density is 1.354 kg/m2.
The finite element modeling shows that the MSAR design that has the "thin-walled shell with a three-layer rib reinforcement" layout can meet the requirements for the stability of shapes and sizes and is promising for further in-depth studies.

References

[1] Gao S, Rahmat-Samii Y, Hodges R, 2018 Advanced antennas for small satellites. *Proceedings of the IEEE.* DOI: 10.1109/JPROC.2018.2804664.

[2] Rao S, Shafai L and Sharma S 2013 Applications of Reflectors. *Handbook of Reflector Antennas and Feed Systems.* Volume 3.

[3] Sudhakar K, Rao S 2015 Advanced Antenna Technologies for Satellite Communications Payloads. *IEEE Transactions on Antennas and Propagation*. DOI: 10.1109/TAP.2015.2391283.
[4] Rao S 2003 Parametric design and analysis of multiple-beam reflector antennas for satellite communications. *IEEEAntennas & Propagation Magazine,* Volume 45, pp. 26-34.
[5] Keen K M 1976 Surface efficiency measurements on a high-modulus carbon fibre composite reflector antenna at L- and S-band frequencies. *Electronics Letters,* Volume 12, No. 7, 160–161.
[6] Nicholson, K J, Rowe W S T, Callus P J and Ghorbani K 2011 Split-ring resonator loading for the slotted waveguide antenna stiffened structure. *IEEE Antennas and Wireless Propagation Letters*. Volume 10, 1524–1527.
[7] Galehdar A, et al. 2012 Capacitively fed cavity-backed slot antenna in carbon-fiber composite panels. *IEEE Antennas and Wireless Propagation Letters*. Volume 11, 1028–1031.
[8] Galehdar A, et al. 2011 The effect of ply orientation on the performance of antennas in or on carbon fiber composites. *Progress In Electromagnetics Research*. Volume 116, 123–136.
[9] Reznik S V, Prosuntsov P V, Novikov A D 2018 Prospects for increasing dimensional stability and weighting efficiency of reflectors of mirror space antennas made of composite materials. *News of higher educational institutions*. DOI 10.18698/0536-1044-2018-1-71-83.
[10] Bitkina E V, Denisov A V and Bitkin V E 2012 Constructive-technological methods for creating dimensionally stable space composite structures of integral type. *Proceedings of the Samara Scientific Center of the Russian Academy of Sciences*. Volume 14. no. 4-2. p. 555—560.
[11] Manac'h L, Castel X and Himdi M 2012 Performance of a Lozenge monopole antenna made of pure composite laminate. *Progress In Electromagnetics Research Letters*. Volume 35, 115–123.
[12] Artner G. and Langwieser R 2016 Performance of an automotive antenna module on a carbonfiber composite car roof. *European Conference on Antennas and Propagation (EuCAP)*. Davos, Switzerland.
[13] De Assis R and Bianchi I 2012 Analysis of microstrip antennas on carbon fiber composite material. *Journal of Microwaves, Optoelectronics and Electromagnetic Applications*. Volume 11. No. 1, 154–161.
[14] Artner G, Gentner P K, Nicolics J, and Mecklenbrauker C F 2017 Carbon fiber reinforced polymer with shredded fibers: Quasi-isotropic material properties and antenna performance. *International Journal of Antennas and Propagation*. Volume 2017, Article ID 6152651.
[15] Mehdipour A et al. 2013 Mechanically reconfigurable antennas using an anisotropic carbon- fibre composite ground. *IET Microwaves, Antennas & Propagation*. Volume 7. No. 13, 1055–1063.
[16] Ka-Band &Q/V-Band Antennas. http://www.hps-gmbh.com/wp-content/uploads/2017/05/HPS_Ka-Band-QV-Band-Antennas-Flyer.pdf (assessed 20 October 2018).
[17] Nelyub V A 2015 Determination of adhesion interaction between carbon fiber and epoxy binder. *Polymer Science - Series D*. 8 (1), pp. 6-8.
[18] Nelyub V A, Borodulin A S, Kobets L P and Malysheva G V 2016 A study of structure formation in a binder depending on the surface microrelief of carbon fiber. *Polymer Science - Series D*. 9(3), pp.286-289.
[19] Composite honeycomb structure C-Core. URL:https://spacesolutions.esa.int/technology-transfer/esa-broker ermany/technology/composite-honeycomb-structure-ccore (assessed 20 October 2018).
[20] The Artemis Program. URL:https://www.esa.int/esapub/bulletin/bullet91/b91dick.htm (assessed 20 October 2018).
[21] Kim H C and See S K 1990 Electrical properties of unidirectional carbon-epoxy composites in wide frequency band. *Journal of Physics D: Applied Physics*. Volume 23. 916–921.
[22] Artner G, Gentner P K,Nicolics J, and Mecklenbrauker C F 2017 Carbon fiber reinforced polymer with shredded fibers: Quasi-isotropic material properties and antenna performance. *International Journal of Antennas and Propagation*. Volume 2017. Article ID 6152651.

IOP Conf. Series: Materials Science and Engineering **683** (2019) 012022 doi:10.1088/1757-899X/683/1/012022

Selection of composite material composition for non-evaporable getters of new generation

M V Aleksandrova[1], Y V Nikolyukin[2] and Y A Kurganova[1]

[1] Bauman Moscow State Technical University, 2-nd Baumanskaya, 5, b.1, 105005, Moscow, Russia
[2] IP "Nikolyukin Y.", Ryazan, Russia

E-mail: kurganova_ya@mail.ru

Abstract. The article is devoted to the issue of synthesizing materials of non-evaporable getters of the new generation with enhanced operational properties. The promising compositions of getter materials are considered, their basic properties are analyzed, and parameters that have a significant effect on their size are selected. According to the results of the literature analysis and patent search, a strategy was developed, on the basis of which three experimental compositions of the material for the non-evaporable getter were determined on the basis of the Ti-Zr system, a visualization-calculation model of the structure was constructed. Based on the traditional technology, the technology for the final product was developed. The results of studies of experimental samples of the material selected composition are presented and the level of acquired operational properties is predicted.

Keywords: getter, sorption, activation temperature, titanium, zirconium, powder metallurgy, mechanical activation.

1. Introduction

Objects that can adsorb and chemically bind all gases, except inert gases, are called getters. They are used in various fields of technology for pumping and maintaining a high vacuum. Performance of the relevant devices depend on the performance of getters [1]. Monitoring the market for such products has demonstrated a monopolization of the domestic manufacturer, which indicates the relevance of creating a competitive product of decent quality. Activity to all gases except inert, present in the atmosphere of an electrovacuum device, regardless of the nature of these gases (oxidizing or reducing), is assessed by sorption versatility, while the volumes and rates of gas evolution are important parameters in the design of the getter [2]. Traditionally, metals that have the ability to chemically react with non-inert gases are considered as getter materials. Inert in air material should be easily activated by heating under pumping during the manufacture of the device, with low vapor pressure at operating temperatures [3].

Gas absorbers are divided into unsprayed and evaporating [4]. The first ones are selected as the subject of study of this work. When activated, they remain in a solid state, have a longer lifespan, can be easily removed or replaced with new ones, which are significant advantages over evaporating ones.

It is known that increasing the maximum sorption capacity plays a key role in creating the most efficient material of the non-evaporable getter, since it is responsible for its performance, i.e. contributes to an increase for gas absorbed under these conditions for a certain operating period [5]. The sorption rate parameter is interrelated with the sorption capacity, it is similar to the so-called "pumping rate": at higher rates of gas sorption by the getter material, the evacuation process is accelerated, which indicates an increase in the productivity of both the product and the entire vacuum system as a whole. Lowering the temperature of material activation allows using the getter at less high temperatures, which expands the range of operating temperatures [6, 7].

Particularly widespread are the so-called porous getters based on transition metals of groups IV and V of the Periodic System of Chemical Elements: titanium, zirconium, vanadium. High sorption properties are provided by an increased specific surface, which contributes to a more intensive chemical reaction between the getter metal and the gas of the medium.

The choice of composition of the getter alloy is based on a number of conditions, the main of which are the final performance characteristics and total porosity. It is obvious that a highly porous structure can be rationally obtained by powder metallurgy methods. Materials, the total value of porosity which ranges from 30 to 55% received distribution in the technique. Alloys of this type are highly porous [8].

Currently, there are a number of successful combinations of such alloys gas getters: Zr-Al [9], Zr-V-Fe [10], Zr-Mn-Fe, and compositions Zr-V-Ca, Ti-Cr-Ca [11].

The effective performance of the composition is provided by the rational choice of the elemental composition [12]. The first component must include at least one element from the group of transition metals Ti, Zr, V, ensuring the sorption properties of the finished product. The second component is chosen from the group of Al, Mn, Fe, Ni, providing a level of mechanical properties, as well as "solidity" of the alloy. According to studies, the preferred mass ratio between the components of the alloy powder is from 5: 1 to 1: 2. The degree of chemical heterogeneity is determined based on the fact that the arithmetic average of the ratios of the concentrations of each of the elements of the first and second components in a randomly selected several pairs of points should not exceed 30 [5].

The purpose of this study is to obtain a prototype of an unsprayed getter of a new composition with a low activation temperature and enhanced performance properties.

2. Materials and research methods

Based on the analysis of technical literature and responses of the effective use of getters, Ti, Zr were selected as metals to create a prototype of an unsprayed getter, as elements providing enhanced sorption properties of the final product, and Al as a binder. The choice of a composition of two transition metals, rather than the use of only one of them, is due to the need to lower the activation temperature of the getter. It is known that gas absorbers made by sintering a powder of a single transition metal are high-temperature, their activation temperature exceeds 800°C. It is possible to lower the activation temperature using a combination of materials [1].

Summarizing the above recommendations, alloy compositions with a Ti: Zr ratio of 7:3, 5:5 and 3:7 by volume, respectively, were selected as prototypes. To improve the accuracy of calculation and estimate the required amount of aluminum binder, while maintaining the alloy's porosity at the required level, the package of titanium and zirconium powder on the 40 × 40 μm area was designed in the COMPAS-3D program (Figure 1).

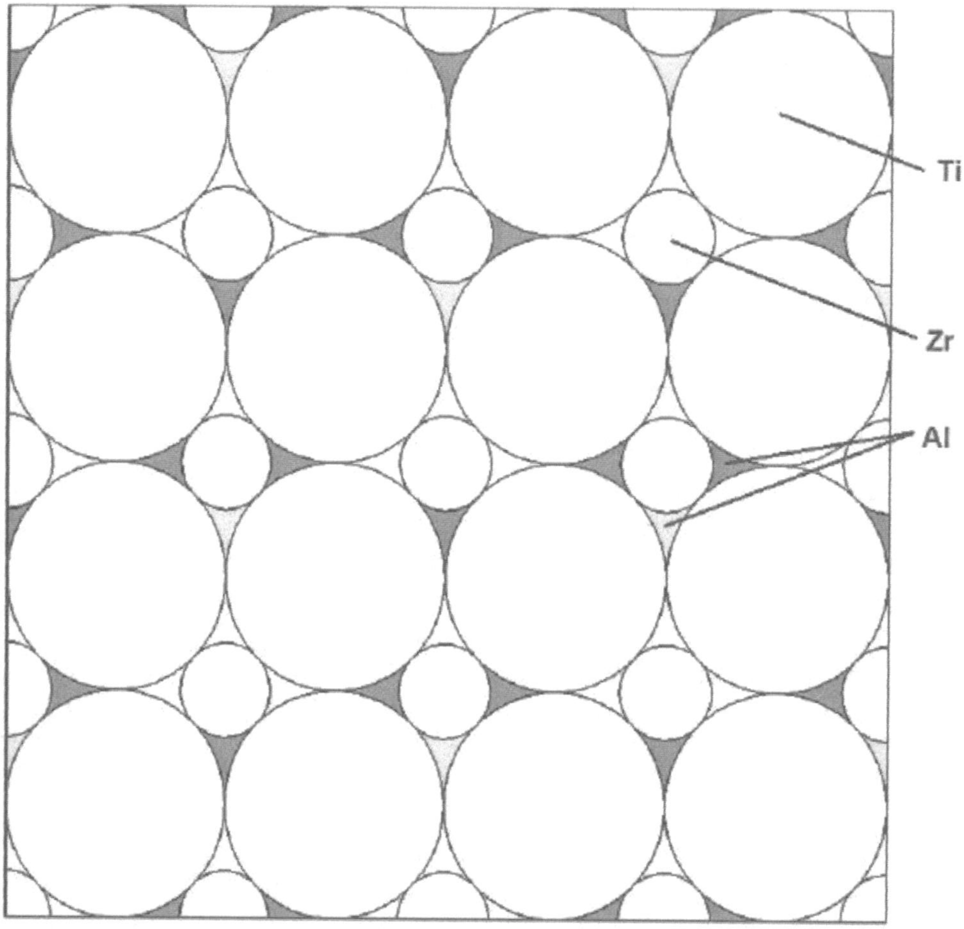

Figure 1. Estimated packing scheme of an experimental alloy Ti-Zr-Al (large grains - titanium Ø 10 microns, small grains - zirconium Ø 4 microns, color sections - aluminum bundle), area size - 40 × 40 microns.

Geometrical calculations have established that the aluminum content at the 5% level is sufficient for bonding powders of the selected chemical and fractional composition of the getter material alloy, provided that aluminum powder with a size of about 100 nm is used.
Taking into account the presented data, three model compositions of the alloys of the unsprayed getter prototypes were selected, which are presented in Table 1.

Table 1. Chemical composition of alloys for the manufacture of prototypes unsprayed getter with high sorption properties

Alloy Number	Ti content, vol%	Zr content, vol%	Al content, vol%
1	47.5	47.5	5
2	66.5	28.5	5
3	28.5	66.5	5

To obtain a getter, we considered a standard technology for obtaining powder non-evaporable gas getters (Table 2).

Table 2. Process map of the production of non-evaporable getters

Number	Process operation	Equipment	Mode		Notes
			t, °C	τ, h	
1	Powder preparation	Scales	20	0.5	Percentage of components is determined by the particle size
2	Powder mixing	Drum type ball mills	20	36	In polyvinyl alcohol
3	Drying	Furnace	Up to 60	-	The end of the drying is determined by the absence of sticking of the mixture on the cold spatula
4	Screening	Sieve	20	-	-
5	Molding	Auto press	20	-	-
6	Assembly of the charge	Special trays	20	0.5	-
7	Sintering	Furnace	According to the mode	-	Cooling with furnace to room temperature
8	Control	-	20	2	As per TU 1190
9	Mechanical restoration	CNC machine	20	-	Turning and grinding

Given the stated objectives, the technology was upgraded with the addition of a low-temperature component to the technological chain of the process of mechanical activation (Fig. 2).

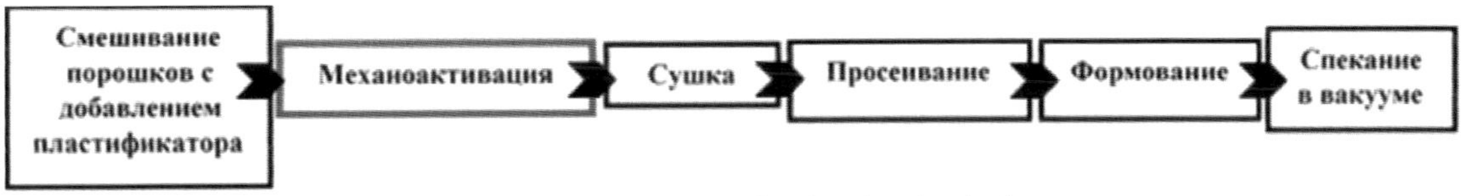

Figure 2. Scheme of the modernized technological process.

The task of increasing the sorption capacity and reducing the activation temperature is realized through the mechanical activation of the combined base: titanium powder of micron size and zirconium powder of submicron size. In our case, more plastic zirconium in the process of mechanical activation, in contrast to the more durable titanium powder particles when exposed to the shear forces of the interaction of powder particles, goes into a more dispersed state, significantly increasing its surface activity, as well as the sorption capacity and activity at lower temperatures.

As a result of the subsequent sintering, the obtained structure of the porous composite skeleton makes it possible to obtain the effect of increasing the sorption capacity of the getter with a decrease in its activation temperature.

As a result of testing modernized technology on selected compositions (Table 1), samples were obtained in the form of sintered discs with a diameter of 11– 13 mm and a height of 2–3 mm (Fig. 3).

IOP Conf. Series: Materials Science and Engineering **683** (2019) 012022 doi:10.1088/1757-899X/683/1/012022

Figure 3. Photos of prototypes.

Comparative analysis of the structure of the sintered samples of the second alloy, obtained using scanning electron microscopy in the laboratory of thin physical methods for studying the structure of materials of the MSTU named N.E. Bauman at the scanning electron microscope TESCAN VEGA.

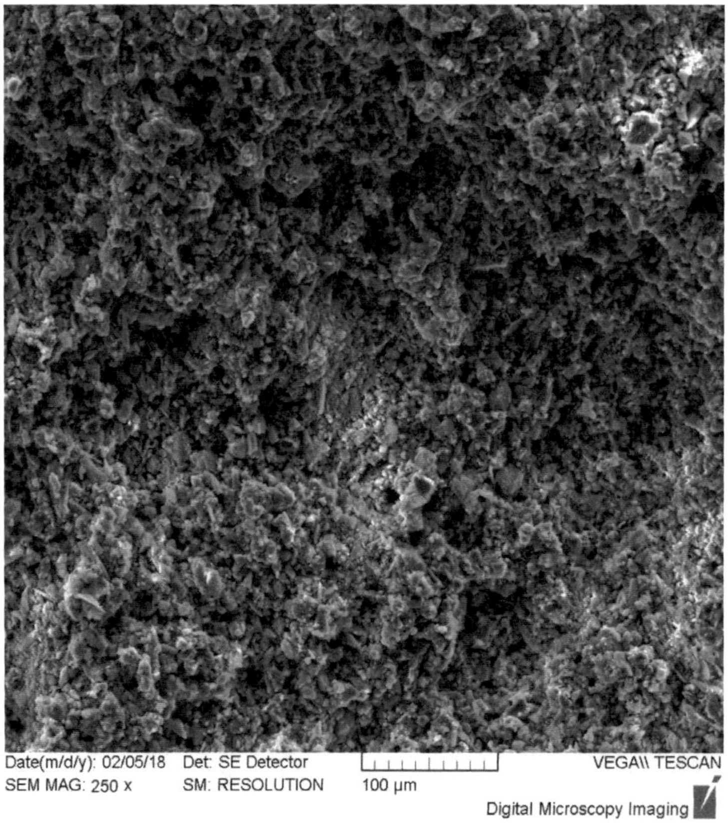

a)

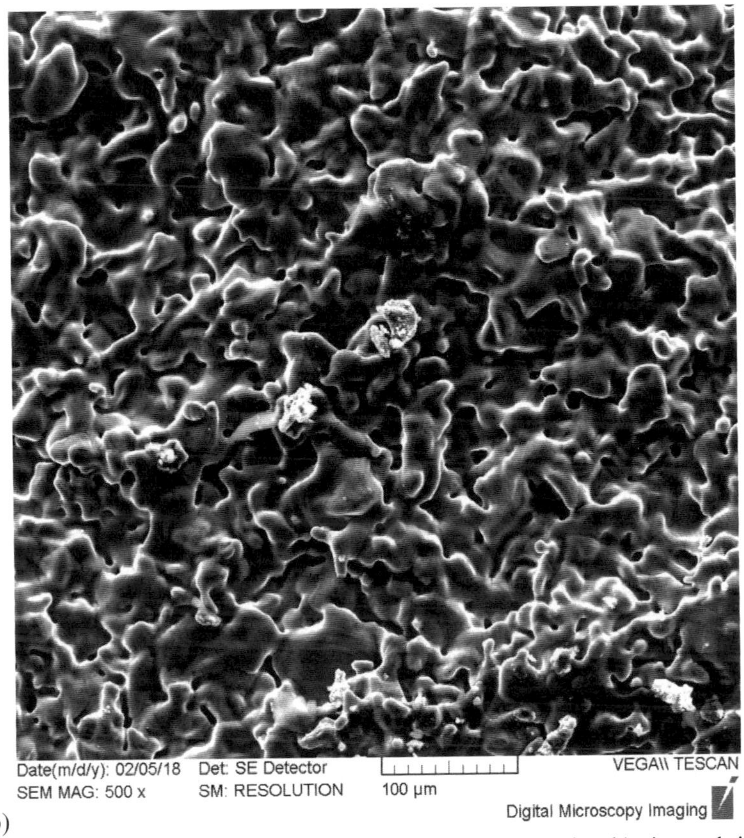

b)

Figure 4. Photographs of the structure of a) sintered titanium powder, b) sintered titanium powder with an aluminum binder (developed composition), obtained using scanning electron microscopy at magnification × 500.

It is seen that the structure of the sample with the addition of aluminum is more ordered, the grains are arranged integrally and monolithically, periodically interconnected.

In order to predict the performance properties of getters from the material, the proposed composition carried out an assessment of the porosity and density of the experimental samples obtained.

The porosity and density of the experimental samples were carried out according to the generally accepted method using a hydrostatic weighing method [13].

The measurement results are presented in Table 3.

Table 3. Average density and porosity of experimental samples

Density, g/cm3	Open porosity, %	Closed porosity, %	Total porosity, %
1.221	54.4	8.1	62.5

On the basis of the obtained data, the level of acquired operational properties is predicted, which is presented in Table 4.

Table 4. Projected performance characteristics of new generation gas absorbers

Characteristic	Properties of the existing analog	Predicted properties of the new material
Activation temperature, ° C	350 – 400	300 – 350
Specific maximum sorption capacity for hydrogen, m3 * Pa / g	7.1 ± 0,2	7.5 ± 0,2
Hydrogen sorption rate, l / s * g	6.6	6.9

The calculations demonstrate that the selected composition and the modernized manufacturing technology of the new non-evaporable getter are capable of meeting the requirements, with a certain increase in the operational properties.

3. Conclusions

Thus, as a result of the research work, a model composition of a composite material for a non-evaporable getter was chosen: with a ratio of Ti powder with a size of 10 μm to a Zr powder with a size of 4 μm 7: 3 by volume correspondence. Calculations have found that 5% of aluminum powder with a size of about 100 nm is sufficient for adhesion of powders of the selected chemical and fractional composition of the alloy. Modernization of the traditional technology by introducing the operation of mechanical activation ensures the achievement of a given porosity. Evaluation of the latter allowed to establish the feasibility and effectiveness of the technological changes made. Gas absorbers made from the selected material have higher performance characteristics than the existing compositions today.

References

[1] Aleksandrova M V 2018 The choice of the composition of the material of the non-evaporable getter with enhanced sorption properties. *All-Russian Scientific and Technical Conf. "Student Scientific Spring: Engineering technology.* conference materials URL: http://studvesna.ru/?go=articles&id=2167 .

[2] Bychkova M S 2018 Getter pumping of electrovacuum devices. *All-Russian Scientific and Technical Conference "Student Scientific Spring: Engineering technology.* conference materials: http://studvesna.ru/?go=articles&id=1268 .

[3] Glebov G D 1965 Gas absorption by active metals. *M.: Energy.* 184 p.

[4] Korzh I A 2016 The design and manufacturing technology of evacuated devices with unsprayed gas getters. *Radio communication technology.* Is.3. p.121-128.

[5] Reutova N P, Manegin S Yu, Pavlov V N 1989 Creation and study of metallic powder materials with an activation temperature below 550°C. *TsNIICHM named after I.P. Bardin. Moscow.* Final report on the Д-713-88.

[6] Avakumov E G 1980 Mechanical methods of activation of chemical processes. *Novosibirsk: Science.* 297 p.

[7] Kachenyuk M N and Smetkin A A 2014 The evolution of the structure of composite particles during the mechanical activation of powder mixtures based on titanium, silicon carbide and carbon. *Modern problems of science and education.* Is.6. URL: https://science-education.ru/ru/article/view?id=15969 .

[8] Reutov N P, Manegin S Yu, Akimenko V B 1998 The method of obtaining non-evaporated getter and getter, obtained by this method. *The patent №2118831 of the Russian Federation.* NTO "Technovak.

[9] Hendrik Johannes, Reinierus Perdijk, Johann Diedrich Fast, Jan Josephus Bernardus Fransen 1958 Method of producing a non-vaporizing getter. *Patent №2.855.368 USA.* US Patent Office.

[10] Claudio Boffito, Aldo Barosi, Alessandro Figini 1982 Non-evaporable ternary gettering alloy and method of use for the sorption of water, water vapor and other gases. *Patent №4.312.669 USA.*

[11] Claudio Boffito, Massimo Bolognesi 1993 Recovery of tritium deuterium from their oxides and intermetallic compound useful therein. *Patent №4.312.669 USA.*

[12] Kurganova Yu A and Kolmakov A G 2015 Structural metal matrix composite materials: a textbook. *M.: Publishing House of MSTU named after N. E. Bauman.* 141 p.

[13] GOST 18898-89. Powder products. Methods for determining the density, oil content and porosity. Eff. 1989-12-20 M.: Publishing house of standards, 1989.— 12 p.

[14] Adducts to produce self-healing thermosetting epoxy polymer from a widely used epoxy monomer. *Polymer Chemistry.* **4** 724-30.

[15] Bai N, Simon G P and Saito K 2013 Investigation of the thermal self-healing mechanism in a cross-linked epoxy system. *RSC Advances.* **3** 20699-707.

[16] Yuan Y, Yin T, Rong M and Zhang M 2008 Self healing in polymers and polymer composites. Concepts, realization and outlook: A review. *Express Polymer Letters.* **2** 238-50.

[17] Chen X, Dam M A, Ono K, Mal A, Shen H, Nutt S R, Sheran K and Wudl F 2002 A thermally re-mendable cross-linked polymeric material. *Science.* **295** 1698-702.

[18] Homer A and Winans C F 1939 Preparation of furfurylamines. *US Patent, 2175585 A.*

IOP Conf. Series: Materials Science and Engineering **683** (2019) 012023 doi:10.1088/1757-899X/683/1/012023

C/C composites developed from phthalonitrile based composites

V V Aleshkevich[1,2], A V Babkin[1,2] and V V Avdeev[1,2]

[1] Faculty of Chemistry, Lomonosov Moscow State University, Leninskie gory, 1-3, 119991 Moscow, Russia

[2] Institute of New Carbon Materials and Technologies, Leninskie gory, 1-11, 119991 Moscow, Russia

E-mail: vladislav.aleshkevich@gmail.com

Abstract. Carbonization of carbon fabric reinforced plastics (CFRPs) with phthalonitrile matrices at 1000°C was investigated. Four different carbonization modes were tested. It was explored, that a mode with step heat rate is most beneficial. Two different phthalonitrile resins were investigated. Elemental analysis of final carbon-carbon composite (C/C) matrix was carried out and the results showed up to 88.5 mass% carbon presence. Final C/C density reached 1.5991 after only one cycle of impregnation-carbonization. Final porosity of samples reached 9.2%. These results allow to consider phthalonitriles as a prospective carbon precursor for C/Cs manufacturing. Mechanical testing of the C/C samples was performed (Compressive strength P = 72.2MPa, E = 11.8 GPa). Tribological data was received (coefficient of friction = 0.31, wear rate = 30 μ/min). Thermal analysis was carried out (χ = 72.94 W/m*K, α = 85.8 cm2/s).

Keywords: Carbon-carbon composites, phthalonitriles, carbonization.

1. Introduction

Carbon-carbon composites (C/Cs) are the class of composite materials where both matrix and reinforcement are different forms of carbon. C/Cs not only retain mechanical strength at high temperatures, but even increase it [1]. One of the key features of C/C composites providing the most widespread application of these materials is the best tribological characteristics at high temperatures. The use of the brake discs made of C/C composites in heavy vehicles and aircrafts allows to significantly reduce a braking distance [2].

One of the most effective ways to obtain C/Cs is a carbonization of carbon-fiber reinforced plastics. Phenolic resins became common polymer precursors to form carbon matrix for C/C composites due to its low price. Despite high carbon content phenolic resins characterized by moderate char yield, namely 55-67% [3]. High porosity and low mechanical properties after carbonization require several cycles of impregnation and carbonization to obtain material with suitable characteristics. Thus, a process of brake disc manufacturing takes about 6-8 month from CFRP to C/C composite with target characteristics.

Phthalonitrile matrices demonstrate high char yields (70-80%) [4,5]. The development of low-melting resin formulations [6] allows to use cost-effective CFRP processing by vacuum infusion [7,8] or RTM [9]. Accordingly, phthalonitrile resins are prospective precursors for carbon matrices in C/C composites.

2. Materials and Methods

All the manipulations with oxidation and moisture sensitive compounds were carried out under inert atmosphere using the standard Schlenk technique. All the solvents were purchased from Alfa Aesar and purified according to the standard procedures [10] and stored under Ar. Dimethylacetamide (DMAc) was distilled over CaH2. Pyridine was distilled over CaH2 and stored under Ar. Resorcinol, 4-nitrophtalonitrile, 1,3-bis(4-aminophenoxy) benzene (APB) (Figure 1, 1), propargyl chloride, potassium bromide and potassium carbonate were obtained from Aldrich and used as received. 4-(3-(prop-2-ynyloxy) phenoxy) phthalonitrile (Figure 1, 2) was synthesized in accordance with the published procedures [11].

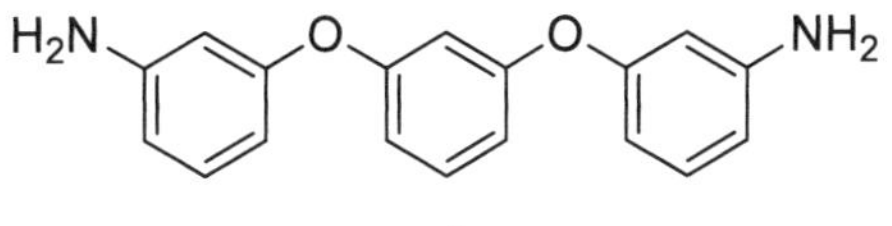

1

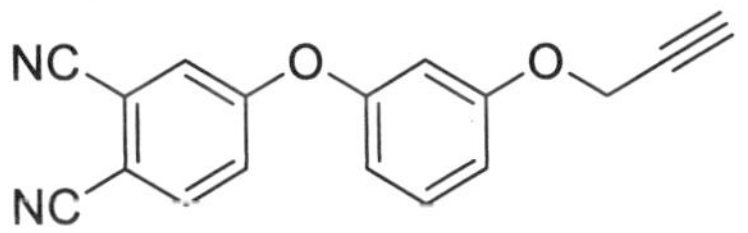

2

Figure 1. Structures of 1,3-bis(4-aminophenoxy) benzene (APB) and 4-(3-(prop-2-ynyloxy) phenoxy) phthalonitrile.

Carbon fabric 22502 from JSC INCMaT 2 × 2 twill wave from 3k HTA40 TohoTenax from Teijin fiber was used for CFRP preparation. For the matrix materials, phthalonitrile based resins obtained from ITECMA were used. Auxiliary materials for vacuum infusion were purchased from Airtech.

All NMR spectra were measured at 600 MHz (1H) and 151 MHz (13C) on a Bruker Avance III Ultrashield spectrometer. All NMR spectra were recorded in deuterated dimethylsulfoxide with reference to tetramethylsilane. Melt viscosity was measured with MCR 302 Rheometer with cone 10 at 200 r/min. Composite density was measured with a Vibra ht-220-CE according to hydrostatic weighing method in pure octane. STA data for samples mass loss was measured with a Netzsch STA 449 at heating rate of 10°C/min in a range of 40–980°C with Ar or air as s purge gases. Porosity analyses were performed on a Microactive Autopore v9600 (capillary nitrogen adsorption) and an an ASAP 2020 Plus (mercury porosimetry). SEM microphotographs were obtained using a Vega 3 (Tescan) microscope with secondary electron detector at operating voltage of 20kV. Interlaminar shear strength was measured with a Tinius Olsen 50ST according to GOST 32659-2014. Compression strength and elasticity modulus were measured with an Instron 5985 according to GOST 9550-81. Tribological characteristics were measured with a Bruker UMT-TRIBOLAB and CETR UMT2. Differential scanning calorimetry (DSC) was performed on a Netzsch DSC 449 at heating rate of 10°C/min and an Ar purge rate of 50 mL/ min. DSC was applied for determination of specific heat capacity of the resulting C/Cs according to ASTM E1269-11. Heat conductivity for resulted C/Cs was measured with a Netzsch LFA 457. Elemental analysis was measured with an automatic analyzer CE1106.

Vacuum infusion process was used to manufacture CFRP samples. Commercial phthalonitrile-based resin PN-3M (ITECMA, Russia) [12]and a formulation PPN, contained 1 and 2 in the ratio of 1:10, were used. The full process of infusion was described earlier[7]. Vacuum infusion process was performed at 150°C for PN-3M and at 120°C for PPN. Curing process was conducted at 180°C for 8h and post-curing was conducted at 375°C for both resins. Obtained composite materials as well as samples of the cured resins were carbonized. This process was carried out in a high temperature oven with different modes under nitrogen atmosphere. Nitrogen was supplied to the furnace using polyamide pipes going from cryogenic storage Dewar's tank filled with liquid N2. The samples were put into the fine powder coke to avoid probable oxidation. Tested modes of the carbonization process are shown in Table 1.

Table 1. Carbonization modes

Mode	Action	Duration, min	Ending temperature, °C	Heating speed, °C/min
1	Heating	100	1000	10
	Hold	60	1000	—
2	Heating	1960	1000	0,51
	Hold	60	1000	—
3	Heating	860	450	0,52
	Heating	2880	800	0,12
	Heating	400	1000	0,5
	Hold	60	1000	—
4	Heating	11760	1000	0,08
	Hold	60	1000	—

3. Results and Discussion

3.1. Carbonization modes specifying

TGA analysis of cured resins was performed to investigate the carbonization in more details (Figure 2). The highest mass loss rate was observed at temperatures from 450 to 800°C: 16.53% for PN-3M and 17.30% for PPN. The bulk of gaseous products is expected to be removed in this interval. TGA data are shown in Table 2.

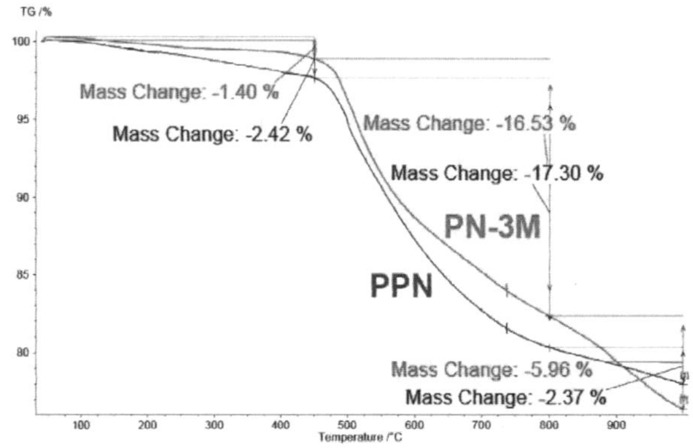

Figure 2. TGA data for PN-3M and PPN mixtures.

Table 2. TGA data for used post-cured resins

Name	T5%, °C	Mass residue, %
PN-3M	509	76.5
PPN	499	78.5

It is undesirable to use high heating speeds during carbonization, because bond cleavage, rearrangement and side products removal (e.g. H2) are taking place. It leads to shrinkage, cracking and mechanical stress gathering [13]. Low heating rates (below 10°C/min) are necessary to minimize this side processes. It was previously shown that the heating rate significantly affects the properties of the final C/Cs[14]. Cured thermoset samples of PN-3M and CFRPs with this matrix were used to investigate different carbonization modes (Figure 3). The objective of the first mode was to investigate an influence of high heating rate on carbonization of CFRPs with phthalonitrile matrices. The fourth mode was knowingly slow. The heating rate of the second mode was medium. The third mode considered the TGA data. It had a slowing step in a temperature interval from 450 to 800°C.

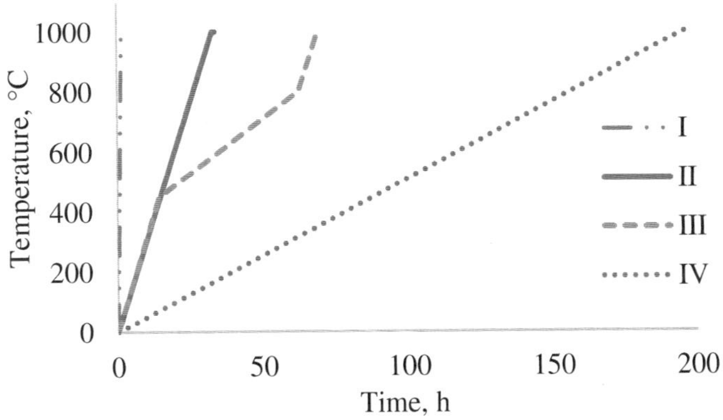

Figure 3. Carbonization modes.

Table 3. Carbonization data for C/Cs, made from CFRPs with PN-3M matrix

	Name	Mass loss, %	SEM pore area percentage, %	Interlaminar shear strength, MPa
Cured thermoset	PN 3M, I	25.98	—	—
	PN 3M, II	26.61	—	—
	PN 3M, III	26.76	—	—
	PN 3M, IV	26.25	—	—
CFRP	PN 3M, I	11.76	1.4	4.3±0.3
	PN 3M, II	14.69	4.3	5.5±0.3
	PN 3M, III	11.88	1.0	6.5±0.6
	PN 3M, IV	13.32	1.9	5.0±0.4

SEM analysis showed no pores or cracks in the original CFRP (Figure 4). The porous structure was obtained during the carbonization process. According to the micro-photographs pore size and pore distribution of the obtained C/Cs by different carbonization modes vary a little. To quantify the mechanical properties of the samples of composites, interlayer shifts were measured (Table 3). Obtained C/Cs by the third carbonization mode demonstrated the best results. It confirmed the importance of decreasing heat speed in the interval of the most mass loss. For further study all samples were carbonized in the third mode.

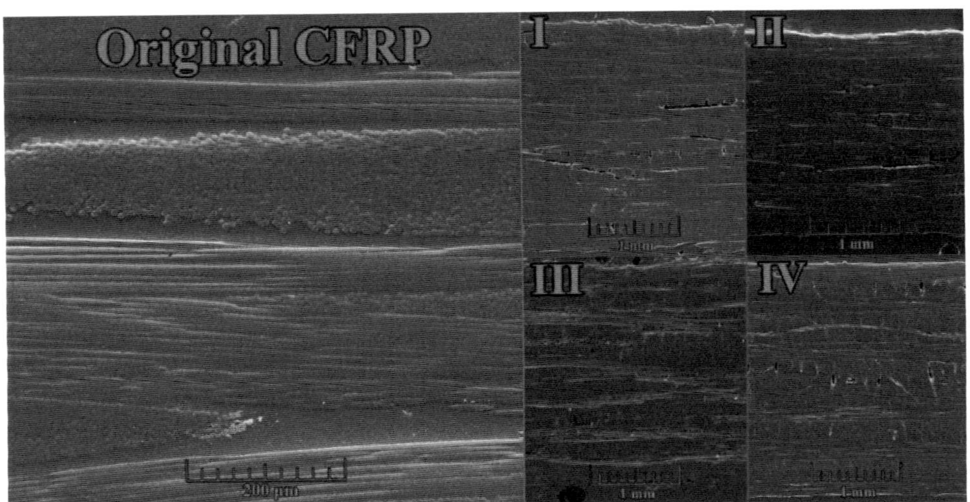

Figure 4. SEM microphotographs of original CFRP and C/C composites, made by different modes.

To explore the scope of phthalonitrile resins, which could be applied as precursors for carbon matrices formation, another phthalonitrile resin composition designated as PPN was tested (PPN, C/C). The formulation consisted of 4-(3-(prop-2-yn-1-yloxy)phenoxy)phthalonitrile containing two thermosetting groups which were described previously [11] (Figure 1, 2) and were already applied for CFRP fabrication [15], so the samples were obtained according to the reported procedure. Carbon content of PPN resin is higher than for PN-3M resin. It should allow to obtain denser samples of final CFRP. However, the porosity and mechanical properties were worse than for PN-3M, C/C (Table 4). It is also proved by comparing the density values.

Table 4. Comparing carbonization data from PN-3M, C/Cs and PPN, C/C

Resin	Mass loss, %	ISS, MPa	Density, g/cm3
PPN, C/C	29.51	4.6±0.1	1.5392
PN 3M, C/C	36.55	6.5±0.6	1.5991

To define the composition of C/Cs the elemental analysis was performed (Table 5). Despite worse mechanical properties and lower density of PN-3M- C/C, the elemental analysis data shows the higher carbon content for PPN.

Table 5. Elemental analysis data for obtained C/Cs

Sample	w(C), %	w(H), %	w(N), %
PPN, C/C	88.54	1.12	3.08
PN-3M, C/C	86.31	0.93	3.84

Capillary adsorption of nitrogen (micropores) and mercury porosimetry (meso- and macropores) methods were used to estimate a final porosity of C/Cs, derived earlier in this work (Table 6). Pore area (PA) for macropores is higher for PN-3M, C/C, but average pore diameter (APD) for macropores is higher for PPN, C/C. Micropores differ a little for PPN, C/C and PN-3M, C/C. General porosity is higher for PPN, C/C. Thus, pores in PN-3M, C/C are evenly distributed in volume, that can be useful for secondary impregnation-carbonization cycles.

Table 6. Porosity data for PN-3M, C/C and PPN, C/C

	N2		Hg		
CFRP matrix	PA, m2/g	APD, nm	PA, m2/g	APD, nm	Porosity, %
PPN, C/C	44.7	2.06	1.34	274	14.1
PN-3M, C/C	39.8	2.07	2.57	89.2	9.2

The compression tests on PN-3M, C/C were carried out (Table 7). The compressive strength (CS) and compression modulus (CM) were compared with a commercial C/Cs, derived by carbonization CFRPs with phenolic resin matrices.

Table 7. Compression data for PN-3M, C/C

	PN-3M, C/C	Literature data
CS, MPa	72,2 (noncarbonized — 403,9)	38 [16]
CM, GPa	11,8 (noncarbonized — 59,21)	11 [16]

One of the main applications of C/Cs is manufacturing of brake disks for heavy vehicles. In that case, tribological tests on PN-3M, C/C were carried out (Table 8). Obtained data shows the competitiveness of C/Cs, obtained in this research, with C/Cs, described in literature [17].

Table 8. Tribological data for PN-3M, C/C

	PN-3M, C/C	Literature
Coefficient of friction at 200 °C	0.31	0.26
Wear rate, μ/min	30	28

During friction process kinetic energy transforms into heat. Thus, the ability of brakes to reject heat is very important. To estimate it, thermal tests were conducted on PN-3M, C/C: LFA method (thermal diffusivity), DSC method (specific heat capacity) and hydrostatic weighing method (density) (Table 9). The resulting values thermal conductivity were compared with phenolic-based C/Cs with PAN fiber, described in literature [16].

Table 9. Thermal characteristics of PN-3M, C/C

$\chi=\alpha*Cp*\rho$	
Thermal diffusivity (α), cm2/s	85.8
Specific heat capacity (C_p), J/g*K	0.514
Density (ρ), g/cm3	1.6540
Thermal conductivity (χ), W/m*K	72.94
Literature data of thermal conductivity, W/m*K	45 [16]

4. Conclusions

C/C composites obtained from CFRPs with phthalonitrile matrices were investigated. The optimal carbonization conditions were selected according to interlaminar shear strength and mass loss data. The values of final porosity after one cycle are reaching 9.2%. The values of final density after one cycle is reaching 1.5991. Thus, utilization phthalonitrile resins for manufacturing C/Cs allow decreasing

Acknowledgments

The work was supported by Government of the Russian Federation Ministry of Education and Science. Contract No.14.607.21.0204 (RFMEFI60718X0204).

References

[1] Manocha L M 2003 High performance carbon-carbon composites. *Sadhana.* 28 349–358. doi:10.1007/BF02717143.

[2] Mohanty R, Mohanty R M 2013 Climate Based Performance of Carbon-Carbon Disc Brake for High Speed Aircraft Braking System. *Def. Sci. J.* **63** 531–538. doi:10.14429/dsj.63.3932.

[3] Mitchell S J, Pickering R S, Thomas C R 1970 The effect of phenol/formaldehyde ratio on the char yield from phenolic resins. *J. Appl. Polym. Sci.* **14** 175–183. doi:10.1002/app.1970.070140117.

[4] Babkin A V, Zodbinov E B, Bulgakov B A, Kepman A V, Avdeev V V 2015 Low-melting siloxane-bridged phthalonitriles for heat-resistant matrices. *Eur. Polym. J.* **66** 452–457. doi:10.1016/J.EURPOLYMJ.2015.03.015.

[5] Bulgakov B A, Babkin A V, Dzhevakov P B, Bogolyubov A A, Sulimov A V, Kepman A V, Kolyagin Y G, Guseva D V, Rudyak V Y and Chertovich A V 2016 Low-melting phthalonitrile thermosetting monomers with siloxane- and phosphate bridges. *Eur. Polym. J.* doi:10.1016/j.eurpolymj.2016.09.013.

[6] Babkin A V, Sulimov A V, Bulgakov B A, Kepman A V 2017 Phosphorus-containing phthalonitrile resin for vimp and rtm processing. *SAMPE Conf. Proc., Society for the Advancement of Material and Process Engineering, Seattle, WA*. pp. 713–724.

[7] Bulgakov B, Sulimov A, Babkin A, Timoshkin I, Solopchenko A, Kepman A, Avdeev V 2017 Phthalonitrile-carbon fiber composites produced by vacuum infusion process. *J. Compos. Mater. Tech. Phys.* **51** 4157–4164. doi:10.1177/0021998317699452.

[8] Bulgakov B A, Belsky K S, Nechausov S S, Afanaseva E S, Babkin A V, Kepman A V and Avdeev V V 2018 Carbon fabric reinforced propargyl ether/phthalonitrile composites produced by vacuum infusion. *Mendeleev Commun.* **28**. doi:10.1016/j.mencom.2018.01.014.

[9] Bulgakov B A, Sulimov A V, Babkin A V, Afanasiev D V, Solopchenko A V, Afanaseva E S, Kepman A V and Avdeev V V 2017 Flame-retardant carbon fiber reinforced phthalonitrile

composite for high-temperature applications obtained by resin transfer molding. *Mendeleev Commun.* **3** 257–259. doi:10.1016/j.mencom.2017.05.013.

[10] Tietze L F, Eicher T, Diederichsen U, Speicher A, Schützenmeister N Based on (expression) : Tietze, Reactions and syntheses : in the organic chemistry laboratory, n.d.

[11] Bulgakov B A, Sulimov A V, Babkin A V, Kepman A V, Malakho A P, Avdeev A V 2017 Dual-curing thermosetting monomer containing both propargyl ether and phthalonitrile groups. *J. Appl. Polym. Sci.* **133** 44786. doi:10.1002/app.44786.

[12] High temperature resin PN-3M, (n.d.).

[13] Luo R, Huai X, Qu J, Ding H, Xu S 2003 Effect of heat treatment on the tribological behavior of 2D carbon/carbon composites. *Carbon N. Y.* **41** 2693–2701. doi:10.1016/S0008-6223(03)00291-4.

[14] Sastri S B, Keller T M 1999 Phthalonitrile polymers: Cure behavior and properties. *J. Polym. Sci. Part A Polym. Chem.* **37** 2105–2111. doi:10.1002/(SICI)1099-0518(19990701)37:13<2105::AID-POLA25>3.0.CO;2-A.

[15] Bulgakov B A, Belsky K S, Nechausov S S, Afanaseva E S, Babkin A V, Kepman A V, V and Avdeev A V 2018 Carbon fabric reinforced propargyl ether/phthalonitrile composites produced by vacuum infusion. *Mendeleev Commun.* **28** 44–46. doi:10.1016/J.MENCOM.2018.01.014.

[16] Manocha L M 2013 Carbon-Carbon Composites. *Second Edi, Elsevier.* doi:10.1016/B978-0-12-385469-8.00011-3.

[17] Deng H, Li K, Li H, Wang P, Xie J, Zhang L 2010 Effect of brake pressure and brake speed on the tribological properties of carbon/carbon composites with different pyrocarbon textures. *Wear.* **270** 95–103. doi:10.1016/J.WEAR.2010.09.010.

[18] Adducts to produce self-healing thermosetting epoxy polymer from a widely used epoxy monomer. *Polymer Chemistry.* **4** 724-30.

[19] Bai N, Simon G P and Saito K 2013 Investigation of the thermal self-healing mechanism in a cross-linked epoxy system. *RSC Advances.* **3** 20699-707.

[20] Yuan Y, Yin T, Rong M and Zhang M 2008 Self healing in polymers and polymer composites. Concepts, realization and outlook: A review. *Express Polymer Letters.* **2** 238-50.

[21] Chen X, Dam M A, Ono K, Mal A, Shen H, Nutt S R, Sheran K and Wudl F 2002 A thermally re-mendable cross-linked polymeric material. *Science.* **295** 1698-702.

[22] Homer A and Winans C F 1939 Preparation of furfurylamines. *US Patent 2175585 A.*

IOP Conf. Series: Materials Science and Engineering **683** (2019) 012024 doi:10.1088/1757-899X/683/1/012024

Requirements for technical support of measurements of local elastic characteristics of CCCM by indentation method

Y D Andreeva[1] **and I V Magnitskiy**[1]

[1] JSC "Kompozit", Pionerskaya Str., 4, Korolev, Moscow region, Russia, 141070

Abstract. The paper is devoted to the study of the applicability of the indentation method for determining local properties of the matrix of carbon-carbon composite materials. It is shown that from the point of view of preserving the integrity of the matrix, it is advisable to use an indenter with the maximum possible diameter of spherical blunting. The characteristic values of contact stiffness, which have a significant value, are determined, which imposes restrictions on the components of the power circuit of the indentation device. The laws of changes in the level of stresses in the sample under various conditions, including the friction coefficients in the contact, are obtained. Estimates of the influence of the structural heterogeneity of the material on the measurement results are given, and it has been established that with a rational organization of the experiment, they are negligible.

1. Introduction

Currently, aviation, high-speed transport, energy widely use spatially reinforced carbon-carbon composite materials (CCCM) [1].

In this context, the task of designing new CCCMs with the required properties is topical. In view of the high cost of such materials and the considerable duration of the cycle of their production [2], it is advisable to preliminarily model their properties, primarily physical and mechanical, which requires knowledge of the characteristics of their components. In spatially reinforced CCCM, such components are carbon rods and the carbon matrix.

Now to determine the properties of the components of composite materials scientists mainly use indirect methods, both purely theoretical [3] and computational-experimental, for example, identification [4]. The results obtained in this way usually agree well with the experimental values [5], but their reliability in each specific case depends on the adequacy of the physical and mathematical models used to the phenomena observed in practice.

Experimental measurement of the elastic characteristics of the CCCM components is a complex task, since the rods and the matrix in them constitute a single material, and it is difficult to isolate a separate component and test it. Therefore, the development of methods that would allow to obtain the characteristics of the components in the composite.

An analysis of the literature has shown that methods of dynamic and kinetic indentation are often used to solve this problem [6–12]. Most common is the kinetic indentation, a process controlled by a special test setup, in which there is a continuous penetration of an indenter into the test sample under the action of a smoothly increasing load with its subsequent removal and recording the dependence of the tip movement on the load [13]. A method for measuring the set of physicomechanical characteristics of materials during indentation was proposed in [14] and currently, it is very popular and is widely used to determine the characteristics of hardness, stiffness and crack resistance in isotropic materials [15-17].

However, in the considered sources there are no comprehensive studies of the applicability of indentation methods for determining the local properties of the CCCM, including elastic ones. As a rule, the authors of papers associated with the corresponding measurements [6, 8] do not provide a justification for the correctness of the transfer of methods originally developed for metals and alloys to a heterogeneous material with fragile components.

In this paper, we will consider the issues of influence on the observed results of determining the local elastic modulus during indentation of the matrix of the CCCM of structure heterogeneity of the latter, including deviations related to the accuracy of the indenter positioning and characteristics of material sampling.

A model CCCM with a 4DL reinforcement pattern (hexagonal transversely- isotropic according to [18]) and a characteristic rod cross-section size of 0.7 mm was selected as a research object.

2 Task setting

The task was carried out in two stages. Ansys software package was used for calculations.

At the first stage, an indenter suitable for measuring the matrix of the CCCM was chosen its stressed state. Selection criteria in order of priority were considered minimum equivalent voltages and maximum displacements with the same efforts. Due to the fact that the pointed indenters near the top create in brittle materials, which include the CCC matrix, endless stresses [19], only spherically blunt indenters were considered.

To calculate the equivalent stresses, there were considered strength criteria of Mohr-Coulomb, Pisarenko- Lebedev and Botkin-Mirolyubov [20]. For our task, it was enough to make only a comparative assessment of the resulting stress values in the carbon matrix, so we established only general trends. For a more complete picture and taking into account the most unfavorable cases during the experiment, the coefficient of friction between the indenter and the sample and the Poisson's ratio of the matrix were varied.

The problem was solved in a two-dimensional axisymmetric formulation. A model of an indenter and a matrix was built, the hemispherical volume radii of which were 1:10 (figure 1), which ensured the absence of the influence of boundary conditions on the results. Geometric boundary conditions were set for rigidly fixing the matrix on the lower surface of the hemisphere and moving the back side of the indenter. The finite element grid had a thickening at the contact boundary. Verification of the constructed model by analytical calculation according to [23] showed satisfactory agreement.

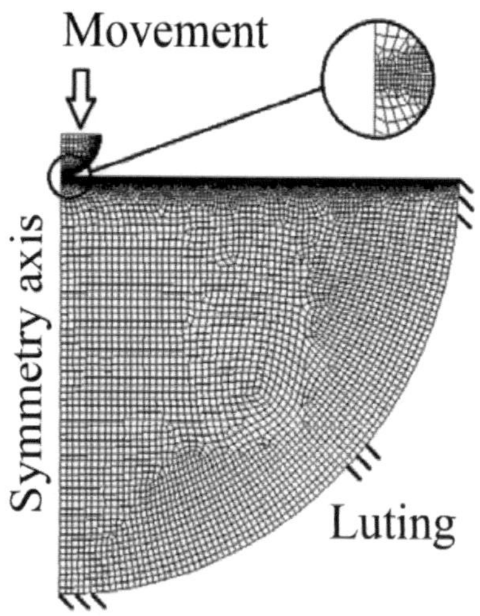

Figure 1. Calculated model of indenter and matrix in axisymmetric formulation.

The indenter was considered to consist of tungsten carbide with an elastic modulus of 656 GPa, and a Poisson ratio of 0.31 [21]. The properties of the CCCM components here and hereafter correspond to [22].

For a reasonable choice of the size of the indenter, we considered a number of spherical indenters, the sizes of which ranged from 0.10 mm to 0.30 mm in diameter with a step of 0.02 mm.
The choice of values was determined as the dimensions of the structural cell of the material under study, and the prevalence of the frame size indentors. At the second stage, a number of tasks related to the presence of the structure of the CCCM were solved.
The first of these was to determine the influence of the direction of indentation on the results of changes. For this purpose, the penetration of the indenter into the matrix on the structural cell of the CCCM in the material orthotropy planes of the material was simulated (Figure 2). Indentation was performed in the region as far as possible from the rods. A finite element grid was formed with condensation in the contact area. This choice of grid parameters ensured solution convergence.

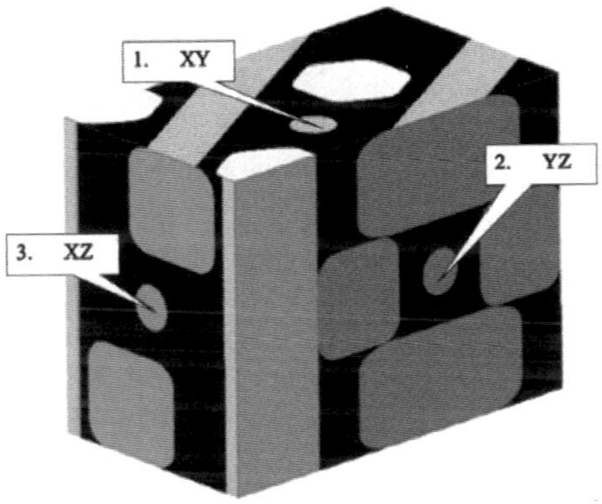

Figure 2. The considered structural cell of the material with the 4DL reinforcement pattern and the indentation plane of the 4DL structure matrix material.

Geometric boundary conditions were set on the cell. The face opposite to the loaded one was completely fixed. On the rear face of the indenter, displacements normal to this face were set, up to 10 μm. On the surface of contact of the indenter with the material, the contact problem was solved without friction.
The second task was to analyze the influence of the characteristics of rods on the results of measuring the parameters of the matrix. For its solution, the technical elastic constants of the rods were varied, and the Poisson ratios were considered constant, while the elastic and shear moduli were reduced by 2 and 5 times, and also increased by 2 times relative to the initial values. The finite element cell model was as described above.
Hereinafter, when assessing the influence of certain factors on the results of indentation, we used the value of the relative change in force equal to

$$b = \left|1 - \frac{P}{P_{ref}}\right|, \quad (1)$$

where P is the indentation power obtained for a given calculation point, and Pref is the reference value of the force calculated for indentation in the corresponding plane of the material with the initial values of the characteristics of the components for the current depth of the indenter.
The influence of a factor was considered significant if for it the value of the value b exceeded the criterion value b0. The latter was calculated as the maximum value of b obtained by varying the Poisson's ratio of the matrix in the range [0; 0.45], corresponding to the expected values of this quantity for the carbon material. This choice of the value b0 is due to the fact that the unavoidable error of measurement of the modulus of elasticity of the matrix by the indentation method is the unaccounted influence of its second elastic characteristic, the measurement of which is impossible in the same experiment. If the influence of the factor on the measurement results is less than this value, it is unobservable and, therefore, may not be taken into account. Such a choice of the value of b_0, despite

the obvious physical sense, is nevertheless conditional and intended only for a rough assessment of the materiality of the influence. The results obtained in this way are qualitative.

The last task was to assess the influence of the position of the CCCM volume boundary on the experimental results. Since the boundary of the real sample will differ from the boundary of the simulated cell, calculations were made for volumes of material in which the plane of indentation was shifted relative to the initial one in the direction of motion of the indenter by the value ½ and ¾ rod diameter. The size of the cell, the orientation of the planes, the boundary conditions and settings of the finite element grid remained the same.

3 Results

The results of calculations of the first stage showed that the values of equivalent stresses obtained by the three criteria considered coincide with high accuracy. Further general data are given.

Figure 3 shows the graphs of the dependences of the maximum values of equivalent stresses on the depth of penetration (loading diagrams) by indenters with a diameter of 0.1, 0.2 and 0.3 mm. It can be seen from them that with increasing diameter, the stress level decreases, which makes it possible to consider large indenters to be preferable. The equivalent stresses reach maximum values depending on the penetration depth either at a shallow depth under the tip of the indenter or on the surface near the area where the contact status of the indenter changes with the sample. These two dangerous areas can be considered equiprobable.

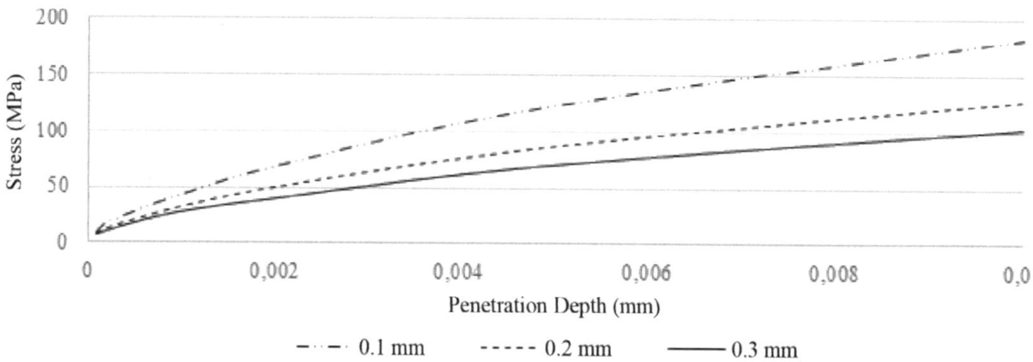

Figure 3. Graphs of dependencies of equivalent stresses on the penetration depth.

The results of calculating the maximum values of equivalent stresses for different values of the friction coefficient in the contact zone show that with increasing friction coefficient from 0 to 0.2, the stresses decrease, after which it will increase further to remain almost unchanged. The general view of the dependencies does not change. Based on this, a model without friction was used for further calculations due to its better convergence.

From similar results for different values of the Poisson's ratio of the matrix, it is clear that the nature of the dependences remains unchanged throughout the entire range of this magnitude considered, and the absolute values of equivalent voltages decrease monotonically. Since our task is to establish general laws, the Poisson coefficient value of 0.25 is considered further.

An important characteristic of the indenter, which determines the requirements for the means of measuring deformations and the power circuit of the device, is the stiffness of the contact area. According to the calculated load diagrams, this parameter can be estimated as a derivative of the current effort by the penetration depth. Table 1 shows values of this parameter for indenters of various diameters with a penetration depth of 10 μm.

The increase in diameter leads to an increase in stiffness, and, consequently, a decrease in the values of displacements under the same load, which complicates the process of measuring them. However, indenters of small diameters create large equivalent stresses in the material, which is dangerous from the point of view of its failure. In our case, it is more important to exclude possible damage to the

material, since this leads to the incorrectness of the physical model of the experiment and the inability to correctly interpret its results, therefore the indenter of the maximum allowable diameter is preferable. This leads to the need to use devices in power circuits for indentation of the most rigid elements.
The results of calculating the stiffness for different values of the Poisson's ratios provide the opportunity to estimate the value of b0 as of 0.16, with more significant deviations observed at large values of Poisson's ratio.

Table 1. Dependence of contact stiffness on sphere diameter.

Diameter of the indenter, mm	0.10	0.12	0.14	0.16	0.18	0.20	0.22	0.24	0.26	0.28	0.30
Rigidity, N/mm	89.3	97.4	104.9	111.8	118.3	124.4	130.3	135.8	141.2	146.3	151.3

Figure 4 shows the calculated loading diagrams of a CCCM cell with a 4-DL reinforcement pattern obtained for different planes by varying the stiffnesses of the rods included in its composition. Some non-monotonicity of the obtained curves is related to the error of calculations at small depths of penetration. The influence of the plane in which the indentation is carried out does not exceed 2%, which is significantly lower than the value of b0.
As can be seen from the figure, the effect of the elastic characteristics of the rods is also small, reaching 9% with a decrease in their moduli of elasticity and shear by 5 times. From these data, we can conclude that it is possible in principle to measure the local characteristics of a material using the indentation method, since the dependence of the obtained force values, and, therefore, the elastic moduli, on the material structure is negligible.

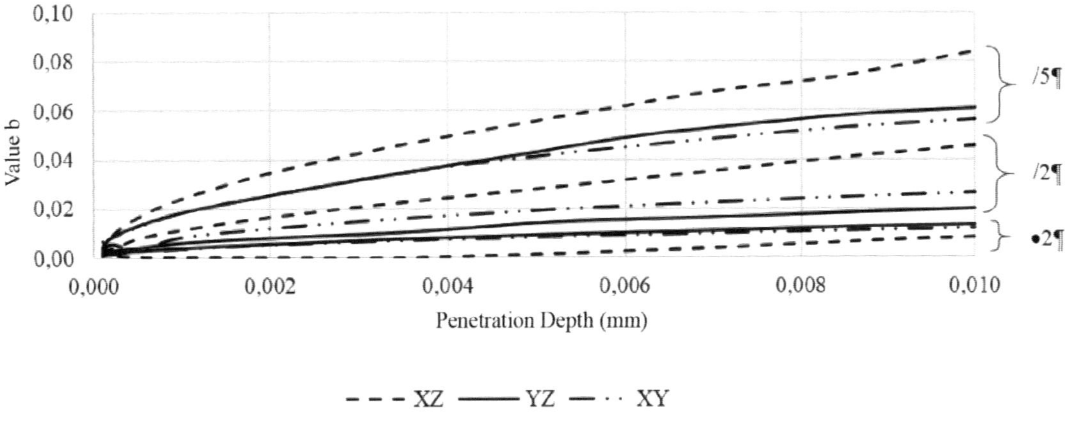

Figure 4. Load diagrams for the XZ, XY, YZ planes with variation of the stiffness of the rods.
The obtained estimates of the influence of the position of indentation surfaces relative to the structure of the CCCM on the force value with a penetration depth of 10 μm show that even when the sample surface is in the immediate vicinity of the rod, the value of the force deviates from the base by no more than 3%, which does not exceed the value b0. Thus, it is possible not to impose any requirements on the samples in this respect.
Therefore, with a rational choice of the size of the spherical indenter, the influence of factors related to the heterogeneity of the CCCM on the observed values of the elastic properties of the matrix does not

exceed the error associated with the uncertainty of the Poisson ratio and, therefore, it is permissible not to take it into account when conducting measurements.

4 Conclusion

This paper enables us to draw the following conclusions.

From the point of view of the strength of the matrix of CCCM, it is preferable to use the indenter of the largest possible diameter, however, the stiffness of the contact area when increasing the size of the indenter grows. Since the no-failure requirement is a priority for the material, the diameter of the indenter is determined by the characteristic dimensions of the structural heterogeneity of the material based on the possibility of falling into a specific component of the composite. The stiffness value of the contact area is very large, which imposes restrictions on the corresponding characteristics of the components of the instrument's power circuit for indentation, including the force sensor.

The dependence of the data obtained during the indentation of the data matrix on the heterogeneity of the local elastic properties of the CCCM is small. There is practically no influence of both the values of elastic constant rods and the position of the sample boundary relative to the structure of the material. This makes it possible not to take into account the presence of several components in the material with a rational choice of the indentation region and guaranteed hitting of the tip of the indenter to the region of exit to the matrix surface.

The results of this paper will be used to create a device for determining the local elastic properties of the CCCM matrix.

Reference

[1] Schurik A.G. 2009 *Artificial carbon materials* (Perm: PSU Publishing)

[2] Vasiliev V V , Protasov V D and Bolotin V V 1990 *Composite materials: Directory* by ed. V V Vasilyev and Yu M. Tarnopolskiy (Moscow: Mashinostroyeniye) 512 p

[3] Sokolkin Yu V, Tashkinov A A and Votinov A M 1996 *Technology and design of carbon-carbon composites and structures (*Moscow: Science) 237 p

[4] Alfutov N A, Zinoviev P A and Tairova L P 1989 *Strength calculations* (Moscow: Mashinostroyeniye v. 30) ed. O.F. Korsun pp.16–31

[5] Zinoviev P A, Tairova L P and Lebedeva O V 2002 A coupled analysis of experimental and theoretical results on the deformation and failure of composite laminates under a state of plane stress *Composites Science and Technology* **62** pp 1711–23

[6] Gogolinsky K V, Shypsha V G, Syasko V A and others. 2018 Application of the method of instrumental indentation for determining the elastic characteristics of reinforcing components of spatially reinforced carbon-carbon materials *Natural and technical sciences* **4** pp. 201–5

[7] Kossovich E L, Epstein S A, Skuratnik V L and Gavrilova D I 2016 The effect of different scales in assessing the mechanical properties of heterogeneous organic composites of natural origin by the method of indentation *Practical biomechanics. Materials reports of the All-Russian Conference of Young Scientists with international participation* (Izhevsk: Bukva) p. 83–5

[8] Kren A P., Matsulevich O V., Stepanova K A and Kinzhagulov Yu Yu 2017. The application of the dynamic indentation method to control the mechanical characteristics.of composite materials *Modern methods and devices for quality control and diagnostics of objects. Collection of articles of the 6th International Scientific and Technical Conference p. 265–9*

[9] Serizawa H, Kohyama A, Watanabe K et al 1996 Elastic FEM Analysis of Fiber Push-Out Test for C/C Composites *Materials Transactions* **37** pp 409–13

[10] Li X, Yang Q and Liu Z 2013 Interfacial mechanics of fiber push-out test: nano-indention technique and cohesive element modeling *13th International Conference on Fracture*

[11] Leatherbarrow A and Wu H 2012 Mechanical behaviour of the constituents inside carbon-fibre/carbon-silicon carbide composites characterised by nano-indentation *Journal of the European Ceramic Society* **32** pp 579–88

[12] Hardiman M, Vaughan T J and McCarthy C T 2017 A review of key developments and pertinent issues in nanoindentation testing of fibre reinforced plastic microstructures *Composite structures* **180** pp 782–98
[13] GOST R 8.748 - 2011 (ISO 14577-1: 2002) Metals and alloys. Measurement of hardness andother characteristics of materials with instrumental indentation.Part 1 Test method p. 28
[14] Oliver W C and Pharr G M 1992 An improved technique for determining hardness and elastic modulus using load and displacement sensing indentation experiments *Materials Research Society* **7** pp 1564–83
[15] Batanova O A and Matvienko Yu G 2015 Patterns of crack formation when flaking the sample edge and indentation of fragile materials *Physical mesomechanics* **18** (226)
[16] Volina A A, Kravchuk K S, Rusakov A A and Useinov A S 2014 Use of diamond tips with a small sharpening angle for the study of mechanical properties and surface relief. *News of higher educational institutions. Series: Chemistry and chemical technology* **57** p. 100-2
[17] Ignatovich S R, Zakiev I M and Zakiev V I 2008 Determination of the micromechanical characteristics of the surface of materials using a nano-indenter "MIKRON" *Bulletin of the Kharkov National Automobile and Road University* **42** p. 86–9
[18] Tarnopolsky Yu M, Zhigun I G and Polyakov V A 1987 *Spatially reinforced composite materials: Directory* (Moscow: Mashinostroyeniye) p. 224
[19] Johnson K. 1989 *Mechanics of Contact Interaction. Transl. from English* (Moscow: Mir) p. 510
[20] Filonenko-Borodich M M 1961 *Mechanical theory of strength* (Publishing House of Moscow University), ed. by Kapustin V.S. p. 89
[21] Babichev A P., Babushkina N A., Bratkovskiy A M and others 1991 *Physical values: Directory* (Moscow: Energoatomisdat) ed. Grigoriev I.S. and Meilikhova E.Z.p. 1234
[22] Vagin V P, Dvoretsky A E, Magnitskiy I V and others. 2015 Problems and tasks of non-destructive testing of parts, assemblies and assembly units made of carbon-carbon and ceramic-matrix composite materials of a new generation *Non-destructive control of composite materials: Collection of works of the 1st distance scientific and technical conference NKKM-2014 "Instruments and methods for non-destructive quality control of products and structures made of composite and heterogeneous materials"* (St. Petersburg: Sven 2015) p. 45–56
[23] Tymoshenko S P and Gudier J 1979 Theory of elasticity, transl. from English (Moscow: Science) p. 576

IOP Conf. Series: Materials Science and Engineering **683** (2019) 012025 doi:10.1088/1757-899X/683/1/012025

Modelling of composite energy absorbing structures for automotive load-bearing systems

G A Harutyunyan and A B Kartashov

1 Bauman Moscow State Technical University, 2-nd Baumanskaya, 5, b.1, 105005, Moscow, Russia

E-mail: georgy.arut@bmstu.ru, kartashov@bmstu.ru

Abstract. Due to the increasing demands for environmental friendliness and fuel economy of cars, the task of reducing their weight is becoming topical. One of the promising ways to solve this problem is the use of composite materials for the manufacture of automotive body. The body of the car is subjected not only to static and slowly changing dynamic loads, but also can be subjected to impact loading. Therefore, it is necessary to assess the effectiveness of the use of composite materials for the manufacture of elements responsible for energy absorption. The paper analyzes the features of energy absorption during the destruction of composite structures, describes the ongoing processes and gives recommendations on the design of composite front rails. To assess the effectiveness of the composite front rail, a simulation of process of its crushing is carried out. To confirm the performance of the design, a crash test of a car with a composite energy absorbing structure is simulated. The applicability and high efficiency of composite energy absorber are confirmed

1. Introduction

Currently, more and more of the world's largest automotive manufacturers are moving towards the use of polymeric fiber composite materials in the vehicle's load-bearing system. One of the applications of composite materials is the fabrication of energy-absorbing structures in the front and rear parts of passenger cars, designed to absorb energy and reduce accelerations in the event of a collision [1]. These structures are used in such cars as Mercedes-Benz SLR McLaren, Lexus LFA, Dallara Stradale Aston Martin Vanquish, and Lotus Elise. Carbon fiber structures find the widest application.

From the point of view of ensuring passive safety, the use of composite materials has significant differences from metals [2, 3]. Energy absorption in composite structures does not occur by deforming with the formation of plastic folds, but due to gradual stable destruction with the formation of a large number of fragments.

2. Mathematical model

This paper proposes a technique for modeling composite energy-absorbing structures for load-bearing systems of automobiles, and also substantiates the basic structural solutions for composite energy-absorbing elements.

A carbon-fiber energy-absorbing structure is being modeled for the vehicle's load-bearing system, as shown in Figure 1.

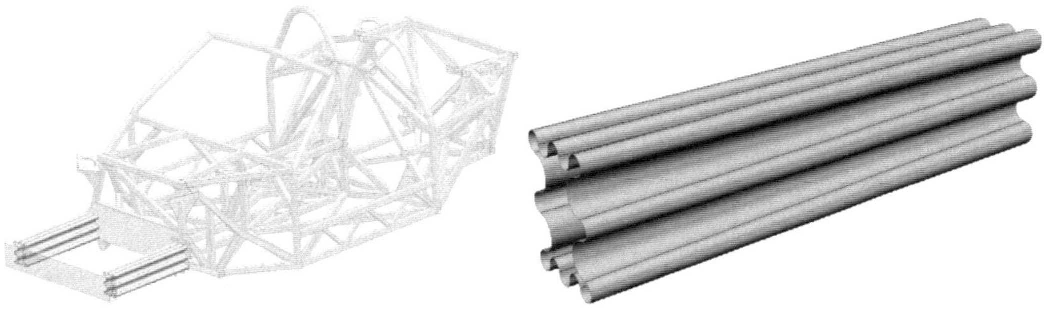

(a) (b)

Figure 1. Car's load-bearing system with energy absorbing elements.

The energy absorbing structure comprises two longitudinal composite frame rails and a cross member connecting them. Frame rails have a corrugated cross section, which contributes to increasing the rigidity of the walls and the uniform formation of fragments.

First, the frame rail is considered separately from the rest of the structure, and then the crash test of the frame rail as part of the vehicle is simulated.

Currently, shock loading is simulated using the finite element method. Since the task at hand includes the destruction of materials, complex contact loads, as well as high dynamic loads, a Radioss solver is used, which is included in Altair Hyperworks finite element analysis software package, and implements integration by an explicit method [4, 5, 6].

For the mathematical description of the work of composite materials, phenomenological failure criteria are widely used, which are formulated on the basis of processing a large amount of experimental data. Tsai-Wu failure criterion related to quadratic failure criteria is of wide use. The function of the failure criterion reads:

$$F(\sigma) = F_1\sigma_1 + F_2\sigma_2 + F_{11}\sigma_1^2 + F_{22}\sigma_2^2 + F_{44}\sigma_{12}^2 + 2F_{12}\sigma_1\sigma_2,$$

where σ_i – stresses in different directions,

F_i – stress-dependent failure criterion coefficients. The Tsai-Wu criterion enables to determine the current state of the material:

$F(\sigma)<1$ – zone of elastic strains;

$F(\sigma) = 1$ – beginning of the plastic zone;

$F(\sigma)>1$ – zone of plastic strain.

When calculating structures that react mainly static and slowly varying loads, performance is considered when the failure criterion is less than 1. A feature of the calculation of structures exposed to impact failure is the need to simulate the work of a composite material with values of the failure criterion in excess of 1 and in the presence of significant failures. For this, a modified form of failure criterion is used. In the software package for the description of composite materials, a model of material No. 25 "Orthotropic shell and solid material" was chosen, based on the material model "CRASURV" [7] and taking into account a large number of features of composite materials. The dependence of the failure criterion coefficients on the magnitude of the energy of plastic strain is taken into account, and the dependence of the yield strength on the energy of plastic strain is introduced.

The model takes into account the accumulation of failures and softening in composite materials, being considered separately for tension and compression. Load curves are shown in Figure 2.

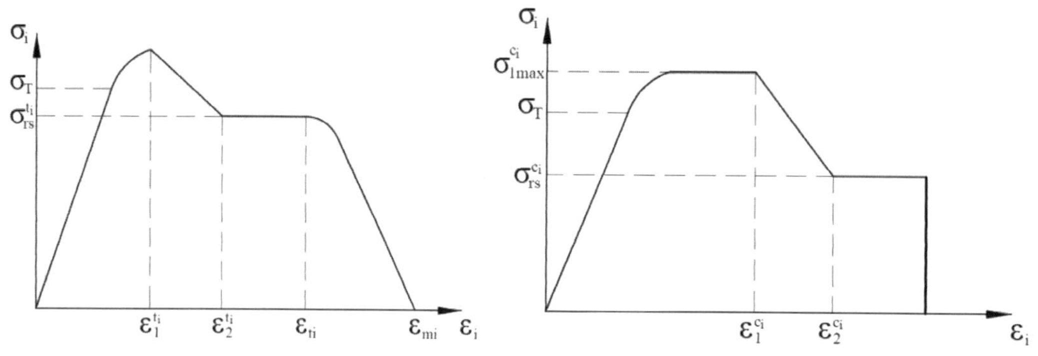

(a) (b)

Figure 2. Load curves (a) under tension, (b) under compression.

When in tension, the softening begins when the relative elongation reaches ε1ti and lasts until the relative elongation ε2ti, upon reaching which the stresses become equal to σrsti. Accumulation of failures begins with a relative elongation of εt1.

A similar process occurs during compression upon the achievement of relative deformations ε1ci and ε2ci.

The model provides two conditions for the failure of finite elements.

First, destructive failure in the process of stretching occurs when the relative strain exceeds εm. The value of the maximum relative strain is set separately for different directions of loading, and its value can be set so that the failure in this direction occurs earlier than softening or accumulation of failures.

The second criterion responsible for the failure and removal of the final element is the value of the energy of plastic strain. When the value of Wp is reached, the element fails. This is the only failure mechanism when subjected to compressive stresses. When subjected to tensile stresses, failure can occur due both to the achievement of maximum relative strain and to the achievement of the maximum energy of plastic strain.

The model of the material allows one set of parameters to take into account different forms of the loading curve, which are characteristic of composite materials when loaded in different directions. It can be described both as a brittle failure in the elastic zone, which is characteristic of stretching a unidirectional composite material along or across the fibers, or a plastic failure when sheared or compressed across the fibers, wherein the effects of hardening and softening appear. This feature allows to calculate the structures, the elements of which are in a complex stress state.

In the selected Radioss solver for four-node elements with reduced integration, QEPH (Quadrilateral ElastoPlastic Physical Hourglass Control) formulation is used [8].

To model a multilayer composite material, a special formulation for shell elements is used. In the Altair Hyperworks software package "Type 11" element is used for this purpose.

3. Results

The frame rail is considered separately to evaluate the efficiency of energy absorption and eliminate the influence of other structural elements. The design model and a fragment of the finite element model are shown in Figure 3.

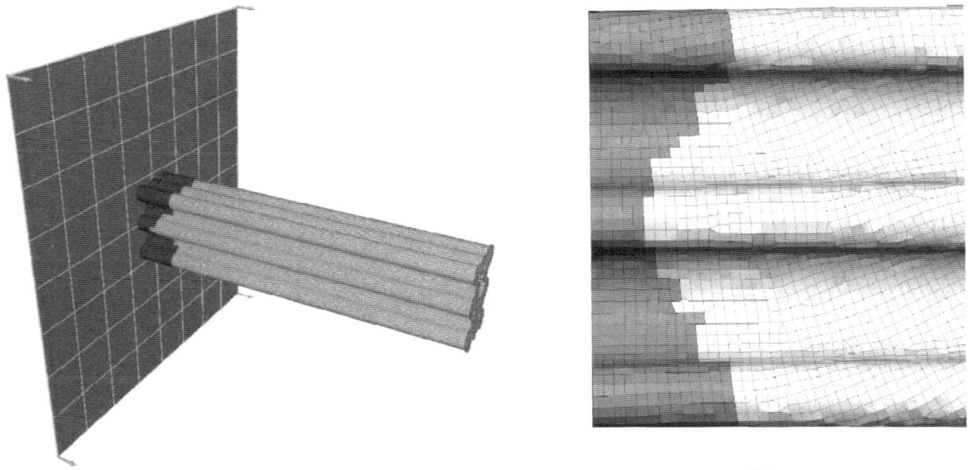

(a) (b)

Figure 3. Computational model and fragment of the finite element model.

The frame rail is fixed at the base, the impact is produced by an absolutely rigid wall weighing 500 kg at a speed of 56 km/h. The splitting into finite elements is irregular, which increases the stability of the results in modeling. As an initiator of the failure, one of the ends of the frame rail includes an area with a reduced number of layers of composite material. The wall thickness of the frame rail is 3 mm, with the frame rail modeled as 10 successively alternating annular and longitudinal layers.
The intermediate stages of failure resulting from the simulation are shown in Figure 4.

(a) 25 ms (b) 50 ms

Figure 4. Intermediate stages of failure of the frame rail.

To evaluate the structure, the criterion of specific energy absorption is used, which is calculated as the ratio of the absorbed energy to the mass of the destroyed or deformed part.
The specific energy absorption was 42.3 kJ/kg, that is more than two times higher than the values characteristic for metallic energy absorbing structures. Figure 5 shows a comparison of reactions on a rigid wall that occur in the process of energy absorption of the carbon fiber-reinforced plastic frame rail under consideration and a similar steel structure.

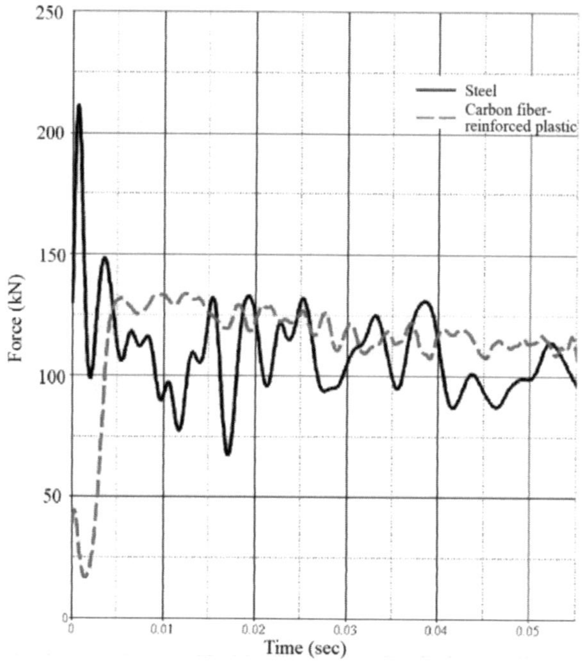

Figure 5. Comparison of reactions.

Both structures in the process of the failure create a reaction with a fairly constant average value, but the range of change in the magnitude of the reaction is much larger when the steel structure is deformed. This is due to the discreteness of the formation of plastic folds. Due to the uniform failure and the fact that the material entirely participates in the energy absorption, the carbon fiber-reinforced plastic frame rail creates a reaction of almost constant magnitude.
With the failure of the composite frame rail, a smooth increase in the reaction occurs without a pronounced peak at the beginning, which indicates the effectiveness of the applied initiator of the failure. The crash test is modeled on a vehicle with a composite energy absorbing structure to evaluate its performance and estimate the accelerations, allowing to draw conclusions about the safety of the vehicle. The design model is depicted in Figure 6.

Figure 6. Crash Test Design Model.

The boundary conditions correspond to the crash test under the UNECE rule No. 94 [9]. The impact is simulated with 40% overlap into a deformable barrier at a speed of 56 km/h. Vehicle weight is 1200 kg. A deformable obstacle is modeled by volume finite elements. For this purpose, the Altair Hyperworks software package uses material model No. 28 "Honeycomb orthotropic".
To assess the performance of passive safety, the CE model of the Hybrid III dummy is used. Seat belts are modeled to connect the dummy to the load-bearing system.
The stages of the collision process resulting from the simulation are shown in Figure 7.

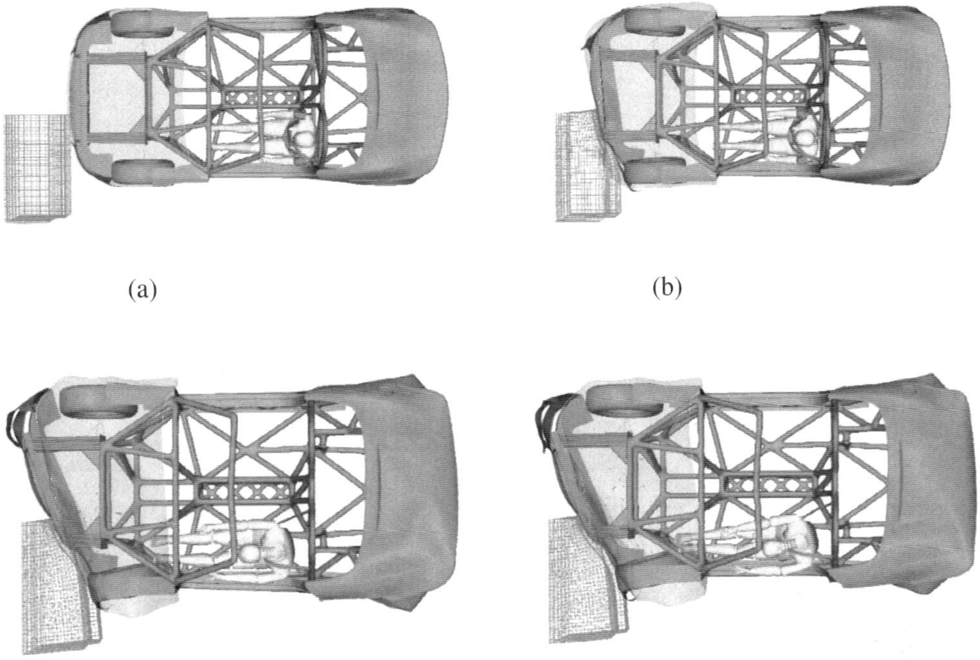

(a) (b) (c) (d)

Figure 7. Crash Test Stages.

The frame rail fails uniformly from the side of the obstacle. No undermining and cracking are observed where the EP passes to the main load-bearing system. The wheel did not have a significant influence on the process of energy absorption and the nature of the failure of the composite EP.
Based on the acceleration of the dummy's head, HIC (Head Injury Criterion) is evaluated, which is calculated by the formula:

$$HIC = \left[\frac{1}{t_2 - t_1} \int_{t_1}^{t_2} a dt \right]^{2,5} \cdot (t_2 - t_1),$$

where a is the acceleration of the dummy's head, t2 and t1 are the limits of the time interval, which are chosen so that the value of the criterion is the greatest. Figure 8 shows the acceleration of the dummy's head and the calculation of the HIC criterion.

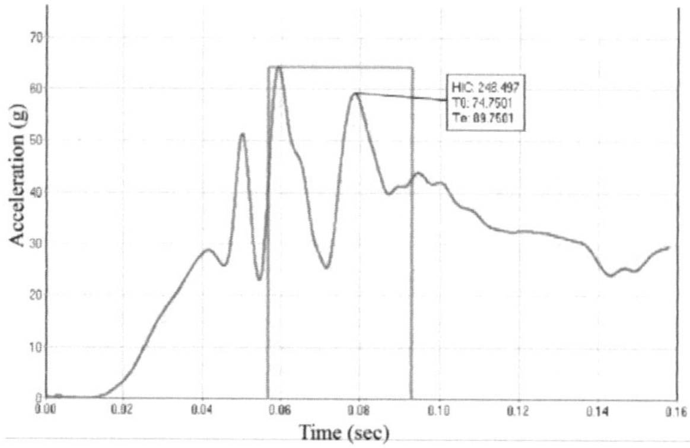

Figure 8. Acceleration of the dummy's head

The highest short-term acceleration value is 65g. The HIC criterion is 248.5 units within the 15 ms interval, which is significantly less than the permissible value of 1000 units.
Figure 9 shows the frame rail in the process of failure.

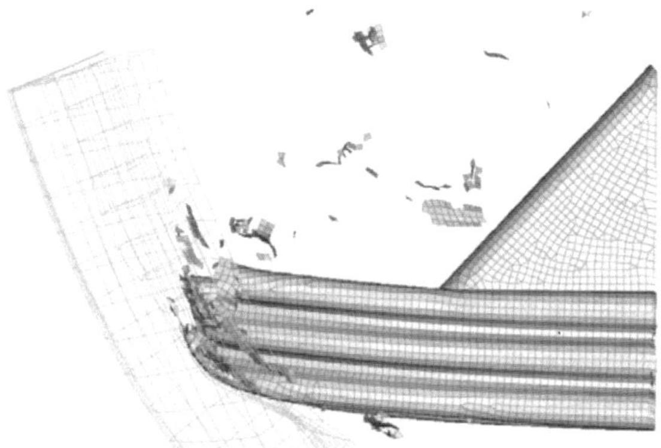

Figure 9. Frame rail in the process of failure.

4. Conclusion

The failure is observed by fragmentation with the separation of fragments. Due to the deformation of the barrier, the surface that contacts the frame rail is not perpendicular to the axis of the frame rail. Despite the presence of a force acting on the frame rail in the lateral direction, no catastrophic failure is observed. The front of the failure is where the frame rail contacts with an obstacle and oriented non-perpendicularly to the axis of the frame rail.
Thus, this paper defines approaches to modeling energy absorption in the process of the failure of composite structures. Significantly greater energy absorption efficiency of composite structures compared to that of the metal ones was demonstrated. The crash test was simulated on a vehicle fitted with a composite energy absorbing structure. The latter's performance and ability to ensure safety requirements were confirmed.

References

[1] Mikhailov M S, Yuryevich V, Aleksandrovich D 2017 Exploring fracture dynamics and passive safety of a superstructure made of composite materials. *International Journal of Applied Engineering Research* **10**, vol 12 2447-2456.

[2] Chermoshentseva A S, Pokrovskiy A M and Bokhoeva L A The behavior of delaminations in composite materials - Experimental results. *IOP Conference Series: Materials Science and Engineering*. 016 **1** vol 116

[3] Lurie S and Belov P 2014 Gradient effects in fracture mechanics for nano-structured materials. *Engineering Fracture Mechanics* Vol **130** 3-11.
[4] 2015 Practical aspects of finite element simulation. 3rd edition. Altair Engineering. 503 p.
[5] Buyanov I A, Rostovtsev M Y and Shelofast V V 2016 Validation of software for computer-aided engineering of composite constructions in relation to calculating the natural frequencies of asymmetrical laminated plates *Russian Aeronautics* **4** Vol 59 606-612.
[6] Dimitrienko Yu I, Sborshchikov S V and Sokolov A P 2013 Numerical simulation of microdestruction and strength characteristics of spatially reinforced composites. *Composites: Mechanics, Computations, Applications.* **4** Vol 4 345-364.
[7] Wang L H, Atluri S N 1994 An analysis of an explicit algorithm and the radial return algorithm, and a proposed modification, in finite plasticity. *Computational Mechanics.* Vol 13 **5** 380-389.
[8] UNECE Regulation No. 94 (document E/ECE/324/Rev.1/Add.93/Rev.1, E/ECE/TRANS/505). Uniform prescriptions concerning vehicle approval for driver and passenger protection in the event of a frontal collision. UN 2007 p86

IOP Conf. Series: Materials Science and Engineering **683** (2019) 012026 doi:10.1088/1757-899X/683/1/012026

Reduction of heat loss of pump-compressor pipes for oil production due to optimal thermal insulation based on basalt and glass fibers

M A Komkov[1], Y V Badanina[1] and A H Garashchenko[2]

[1] Bauman Moscow State Technical University, 2-nd Baumanskaya, 5, b.1, 105005, Moscow, Russia
[2] CENTRAL RESEARCH INSTITUTE FOR SPECIAL MACHINERY, Khotkovo, Moscow Region, Russia

E-mail: cm12@cm.bmstu.ru, m_komkov@list.ru, julia555-90@yandex.ru , a.n.gar@mail.ru

Abstract. The results of investigations of high-temperature and low-density heat insulation of pump-compressor pipes made of short basalt fibers and alumina bundles are presented. It is shown that in order to achieve a significant reduction in the thermal conductivity of the material from the initial staple fibers, it is necessary to grind them and clean them from impurities by liquid method. It is shown that the heat-insulating coating has restrictions on the thickness of the insulation wall and the size of the running mass of the pipe.

Keywords: heat-insulating coating, basalt super thin fibers, thermal conductivity coefficient, pump-compressor pipe.

1. Introduction

The efficiency of extraction of highly viscous and heavy oil reserves by industrially developed methods reaches an average of 55-75% of the initial geological reserves of oil in the subsoil. One of the effective directions in the development of highly viscous and difficult to extract oil reserves in the Russian Federation is more than 70% located at a depth of 2 ... 3 km [1, 2], is the use of steam-thermal effects on oil-bearing formations. Warming up of the formations is carried out by long-term injection of dry superheated steam with high initial parameters (temperature 420… 450 ° C under a pressure of 35 MPa or more [3]) to the wellhead with transition to an oil-bearing layer of steam with a temperature of 350… 400 ° C. Trouble-free operation of steam wells with respect to the extraction of heavy oil in the Far North is possible only without disrupting the geophysical balance of permafrost conditions [4, 5], which leads to the loss of an oil well. The solution of this problem is carried out on the way of creating heat-insulated tubing (tubing) with an initial steam temperature of 420 ° C and a pressure of 35 MPa [1-4]) so that in the near-surface zone of the well behind the casing the temperature of the cement stone will not exceed 60 °C.

The development of thermal insulation coatings on one side of the tubing is limited by the maximum diameter of thermal insulation, and on the other hand, by obtaining the minimum linear mass of the pipe. Section of thermally insulated tubing should freely enter the wellbore casing. The gap between the

regular part of the tubing is not less than 20 mm, and in the places of power grips of pipes with a hydraulic and spider key - not less than 12 ... 15 mm. Currently, two-layer steel vacuum-thermally insulated tubing with multi-layer screen insulation of tubing or "thermal cases" [2], with a 40 mm and 62 mm through diameter, are used for injection into the reservoir of superheated steam with temperatures up to 325 ° C and pressure of 16 MPa for outer diameter of the casing pipe 168 mm. However, the capabilities of "thermal cases" are limited by low working pressure and relatively low temperature, as well as significant running weight, respectively, equal to 19.6 kg / m and 32.3 kg / m, which limits the conditions for their use in wells more than 1.5 km deep. Furthermore, the pipe design is distinguished by the complexity of the manufacturing process and, as a consequence, the high cost of the product.
The aim of the work is to create a new composite construction of multilayer tubing with heat insulating coating based on basalt and glass fibers for pumping superheated to a temperature of 420 degrees Celsius and more under a pressure of 35 MPa into an oil reservoir with a depth of 3000 meters.
Analysis of the literature shows that highly porous heat insulating coatings (HIC) based on basalt super thin fibers (BSTF) having a low thermal conductivity [6] compared with other types of fibers (Figure 1 (a)), with high porosity and low density of the material HIC promising for use in the design of tubing. As the initial fibers for the manufacture of the HIC of the tubing, the tubing was selected by BSTF of the company "MINOL" [7] with diameters of 0.5 ... 3.5 μm (average dБВ = 1.92 μm), the initial length of staple fibers of lБВ =50...60 mm and low content of non-fibrous inclusions (≈ 8%) in canvas.
For thermal research, a thermal insulation material (TIM) was developed on the basis of short basalt fibers (lБВ= 0.5 ... 1.5 mm) obtained by grinding and cleaning the source fibers from impurities in the chamber of a liquid disintegrator (Figure 1 (b)), and 5-6% bundles of Al2O3 based on aluminum sulphate solution [8]. The designed TIM during heating in the temperature range of 60 ... 400 ° C (Figure 1 (c)) has a low thermal conductivity coefficient of an average of 0.042 W / (m K). At the same time, the density of such a TIM with a porosity of 94% is 156 kg / m3, and the compressive strength at 20% elastic deformation is 0.043 MPa [9, 10].
To determine the required thickness of the thermal insulation coating of the tubing, it is necessary to find the averaged thermal conductivity coefficient of the TIM depending on its density or porosity of the fibrous structure. We assume that the conductive heat flux from the hot tubing wall to the colder outer protective fiberglass shell will be transmitted due to the thermal conductivity of dry air and the material of the basalt fiber frame.

a b c

Figure 1. The dependence of thermal conductivity on temperature (a) for air (1), BSTF with a diameter of 1.92 μm (3) and felt from BSTF (2); initial basalt fibers and disintegrator chamber (b); the microstructure of the TIM of short basalt fibers with a bunch of Al2O3 (c).

Analysis of the experimental data obtained for the thermal conductivity coefficients of basalt fibers with a diameter of 1.5 ... 3.0 μm on temperature (Figure 1 (a)) shows that they can be averaged and approximated by a linear dependence of the form:

$$\lambda_{БВ}(T) = \lambda_{0Б}\left[1 + a\left(T - T_0\right)\right] \quad (1)$$

where T - the current temperature; $\lambda_{0Б}$ - coefficient of thermal conductivity of the material of thin basalt fibers at a temperature T_0 = 20°C, $\lambda_{0Б}$ = 0.032 W / (m · K); a = 1 / (170 K) - approximation coefficient. For still air, experimental values (Figure 1 (a)) can also be expressed by a linear relationship:

$$\lambda_{возд}(T) = \lambda_{0В}\left[1 + b\left(T - T_0\right)\right] \quad (2)$$

where $\lambda_{0В}$ - coefficient of thermal conductivity of air at a temperature T_0 = 20°C, $\lambda_{0В}$ = 0.0257 W / (m · K); b = 1 / (380 K) - coefficient of approximation. Without taking into account the heat losses due to radiation and the quiet state of the air in the pores of the heat-insulating material, the coefficient of conductive heat conduction of basalt HIC can be recorded on the basis of the mixture rule:

$$\lambda_{тип}^{ср} = (1-n)\cdot\lambda_{БВ}^{ср} + n\cdot\lambda_{возд}^{ср} = 0{,}0425 \text{ W / (m · K)}, \quad (3)$$

where n - the porosity of the fibrous insulation: $n = 1 - \rho_{тип}/\rho_{БВ}$; $\rho_{тип}$ - density of the heat insulation material; $\rho_{БВ}$ - the density of basalt fibers, $\rho_{БВ}$ = 2750 kg/m3. Average values of coefficients $\lambda_{возд}^{ср}$ = 0.0413 W / (m · K) and $\lambda_{БВ}^{ср}$ = 0.0587 W / (m · K) for the temperature range from 60 ° C to 400 ° C is determined on the basis of dependencies (1) and (2).

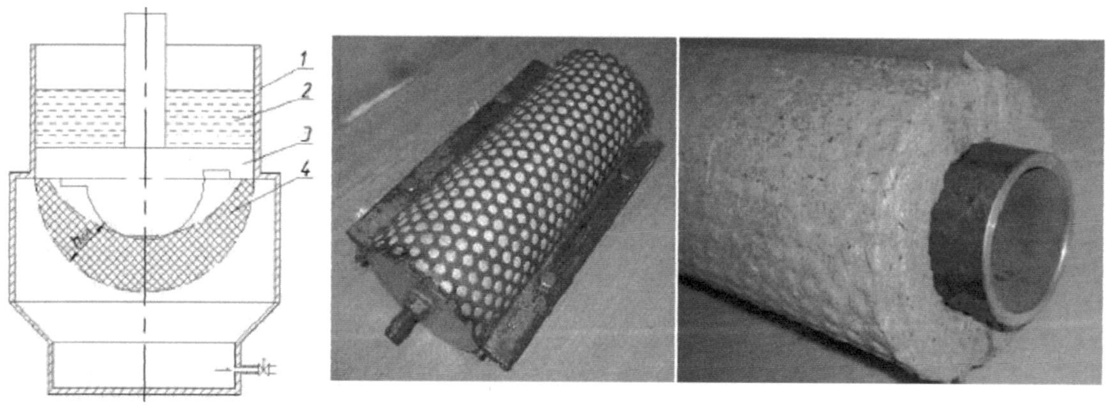

a b c

Figure 2. Scheme of forming cylindrical shells of short basalt fibers: 1 - chamber; 2 - pulp; 3 - punch; 4 - the initial layer of sediment fibers; $h_{сл}$ is the layer thickness (a); general view of the mold for drying and heat treatment (b); type of assembled insulating shells tubing ($d_{тр}$= 60 mm) from short basalt fibers with a bunch of Al_2O_3 (c).

Hydromassage formulation (Figure 2 (a)) in the manufacture of thermal insulation products: 0.95 kg of dry short basalt fiber is taken to contain 0.4 l of a 27% solution of aluminum sulphate Al2(SO4)3 18H2O and 0.07 l of a 25% aqueous solution of ammonia 6NH4OH. The formation of aluminum oxide occurs in the process of drying and heat treatment of the product (Figure 2 (b)) at a temperature of 500 ° C. In the process of filtration deposition, it is necessary to strictly control the acidity of the pulp, to hydrolyze and precipitate the fibers at pH = 7.6 ÷ 8.4. In this case, it is possible to guarantee the calculated amount

of alumina in the insulating material in the form of a bundle while maintaining a sufficiently high filtration rate.
Focusing on the characteristics of vacuum-thermally insulated tubing [3] (thermal case), for which the loss of heat flow along the length of the pipe is acceptable $q_l = 150$ W/m when taking into account (3), we define the required thickness of the thermal insulation coating of tubing from short basalt fibers $h_{тип}$ excluding radiant heat transfer in highly porous thermal insulation material:

$$h_{тип} = \frac{d_{тр}}{2}\left[\exp\left(2\pi \frac{\lambda^{ср}_{тип}\,\Delta T}{q_l}\right) - 1\right] = 24.94 \text{ mm}, \quad (4)$$

where $\Delta T = (T_1 - T_2)$ = 340°C – temperature difference on the outer and inner walls of the insulation. With a porosity of 94% and a density of basalt fibers $\rho_{БВ} = 2750$ kg/m3, the developed insulation material has a density $\rho_{ТИМ}$ = 165kg / m.
To determine the actual density of shell-type products obtained by filtration deposition of short fibers from the pulp (suspension) in the mold, the necessary measurements of the mass and geometrical parameters of the product were carried out. For the experiment, products were taken from 4 different batches. The geometrical parameters turned out to be almost equal and, with small errors, amounted to (Figure 2 (c)): $D_{тип}$ = 110±0.2 mm; dтр=60±0.2 mm; $l_{тип}$ = 255±0.1 mm; $h_{тип}$ = 25±0.1 mm. The average weight of the shells was equal to 0.1327 kg, the density of the material $\rho_{тип}$ =156±3.4 kg/m^3, and the porosity of fibrous the composition was 94.5%.
The total linear mass of the thermally insulated three-layer tubing with accounting (4) and the thickness of the fiberglass shell equal to 2 mm will be equal (6.8 + 1.47 + 1.44 = 9.7 kg / m). This value is 2 ... 3 times less than the linear mass of tubing tubing with vacuum insulation – "thermocases".
The assembly scheme of heat insulation from basalt shells (Figure 3 (a)) and a general view of the composite tubing (Figure 3 (b)), as well as the temperature on the surface of the GRP shell when tested on a thermal stand with maximum heating of the air flow to 600 o C, are shown in Figures 4 (a, b).

a

b

Figure 3. Assembling insulating shells with a metal part of the tubing (a); general view of two interconnecting thermally insulated tubing 2.5 m long with a wound glass-plastic shell (b).

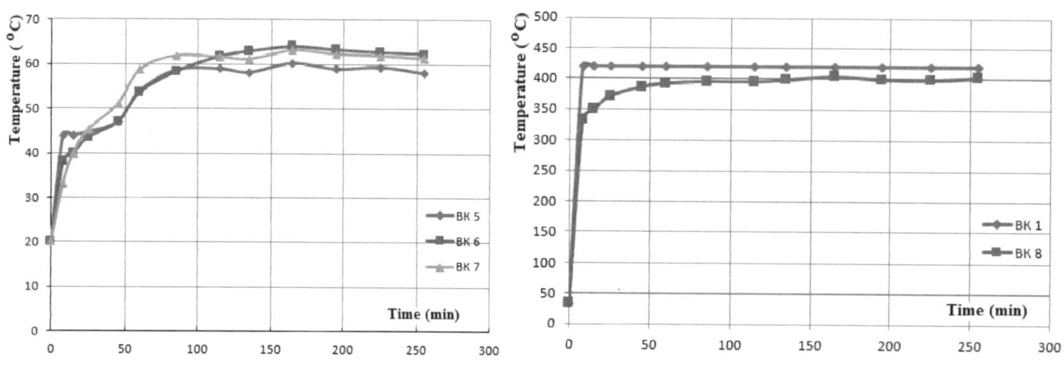

a b

Figure 4. Time variation of temperature sensor readings on the thermally insulated surface of the tubing (a): BK7 - at the inlet, BK6 - at the junction of pipes and BK5 - at the exit from the second pipe; when feeding tubing section with air heated to420 C from the fan heater (b): BK1 - at the inlet and BK8 - at the outlet of the pipe.

2. Conclusions

We can draw the following conclusions. The economic and technical effectiveness of multilayer composite insulated tubing of a new generation according to the results of bench tests of OOO "VNIITneft" is achieved by:

1. use of affordable, inexpensive and environmentally friendly short basalt super-thin fibers with a coefficient of thermal conductivity 1.5 ... 2.0 times smaller than that of quartz, glass or silica fibers;
2. the use of high-strength fiberglass protective-power shell made by the winding method, which provides high tensile strength (48 ton-force, 1.6 times more than the calculated one) and torsion (9540 N m, 3.82 more according to the project TOR) , with half the linear mass and cost of composite tubing compared to - "thermocases".

References

[1] Kalinin V 2012 Suite for oil kings. *Siberian oil.* № 4/91, p. 16-19.

[2] Knots B M 2007 Temperature conditions of working wells and thermal methods of oil production. *Moscow, Izhevsk, Izd-vo LKI.* 406 p.

[3] Moiseyev V A, Moiseyev A V, Komkov M A and Frolov V I 2012 High-temperature oil field steam line. *Exchange of intellectual property.* Volume XI, № 9, p. 57-60.

[4] [4] Komkov M A, Moiseev V A, Tarasov V A and Timofeev M P 2015 Reducing the negative impact on the biosphere in the extraction of heavy oil and environmentally friendly technology of steam injection supercritical parameters in oil reservoirs by creating new tubing with environmentally friendly heat-shielding coating. *Geophysical processes and the biosphere.* **14** No. 1. P. 70-79.

[5] Komkov M A, Moiseev V A, Tarasov V A and Timofeev M P 2015 Minimization of the Negative Influence on the Biosphere in Heavy Oil Extraction and Environmentally Clean Technology for the Injection of the Steam with Supercritical Parameters in Oil Strata on the Basis of New Environmentally Clean Tubing with Heat Resistant Coatings. *Izvestiya, Atmospheric and Oceanic Physics.***51** No. 8, pp. 819-825.

[6] Digris D D, Volynski A K and Kozlowski P P 1980 The basic technology of reception of basalt fibers and their properties. Basalt fiber composite materials and structures. *Kyiv, Naukova Dumka.* 81 p.

[7] Basalt super thin Minol fiber. URL: http://uteplitel-minol.ru/holst/ (accessed 05.09.2016).

[8] Badanina Yu V, Komkov M A Tarasov V A, Timofeev M P and Moiseyev A V 2015 Modeling and experimental determination of technological parameters of liquid-based thermal insulation of

pump-and-compressor pipes. *Science and education: scientific edition of the MSTU. N.Uh. Bauman.* No. 4. P. 13-28.

[9] Denisov N P and Badanina Yu V 2015 Research and preparation of hydro-mass for forming of heat-insulating coating of pump-compressor pipes from short basalt fibers. *Electronic journal of MSTU. N. Uh. Bauman. Youth scientific and technical Bulletin*. No. 4.

[10] Komkov M A, Badanina Yu V and Timofeev M P 2014 Development and research of heat-resistant coatings of pipelines from short basalt fibers. *Engineering journal: science and innovation*. №2(2).

Analysis of the information and diagnostic capability of processing technologies

A A Barzov[1], V A Belov[1] and A L Galinovski[1]

[1] Bauman Moscow State Technical University, 2-nd Baumanskaya, 5, b.1, 105005, Moscow, Russia

E-mail: vkoryanov@bmstu.ru

Abstract. The possibility of informational diagnostic application of various combinations of traditional processing technologies is considered. By means of expert analytical analysis the potential diagnostic performance of combinations of laser and ultrajet machining of materials is shown. Outlined prospects for the development of research results.

Keywords: diagnostics, quality, expert analysis, processing technology, ultra-jet-impact, laser, rational combination

1. Introduction

Dominant physical processes (ДФП) of form and structure-forming technologies such as fracture, friction, plastic deformation and phase transformations in the material being processed have, in addition to the direct functional purpose, significant information and diagnostic potential [1]. This circumstance is due to the similarity of the ДФП at the stage of manufacturing parts of the product and ДФП, which can lead to their failure during operation. Therefore, purposeful disclosure of the information potential of the ДФП processing technologies is a reserve in improving the quality of the manufacturing processes of various production facilities and ensuring their reliable operation. The essence of the control and diagnostic use of processing technologies is that the object of analysis (OA)

has a certain, usually local, physical and technological impact FF in the form of a functional-rational combination of ДФП, and the result of this influence RR has the necessary information about the material state parameters OA. Examples of such capplication of ДФП are well known, in particular, they include various methods for determining the hardness and microhardness of the surface layer of parts, evaluation of its wear resistance by frictional impact, etc.

2. Objects and research results

In recent years, other methods of mesodiagnostics have developed, which methodologically occupies an intermediate position in terms of the impact on the material under study between traditional invasive (destructive) tests and classical methods of non-destructive testing. In particular, in the works [2, 3] the information-physical capabilities of minimally invasive ultra-jet diagnostics (UJD) are analyzed in detail, which is based on the analysis of the results of local hydroerosion destruction of a high-speed jet of liquid (water) of the surface layer of OA material. However, the necessary scientific and applied research, aiming at the formation of a full-scale apparatus of information technologies of mesodiagnostics, based on the ДФП of various methods of processing materials, has not yet been carried out. This circumstance hinders the development of this potentially very promising direction of ensuring the quality of workmanship and the reliability of operation of products in the leading sectors of the industrial economy.

It should be emphasized that one of the options for the effective use of "processing" technologies in the development of methods and means of mesodiagnostics of the quality of materials and parts is to use a rational combination of ДФП, including physically different effects of FF on OA, mainly on its surface layer [4, 5].

The diagnostic result *R* obtained with such a consistently combined effect can significantly exceed the result of a single exposure in terms of informational significance.

Thus, using a rational combination of ДФП, it is possible to diagnose not only traditional indicators of the quality of a material, but also to evaluate its functional latent characteristics due to the manifestation of the effects of physico-technological inheritance [6, 7]. We formalize the above in the form of a visual linear performance model of hybrid diagnostic technologies [8, 9], based on a rational combination of the effect of ДФП on OA material.

Let the source material OA to be effected by the diagnostic impact F1, and then *F*2. In general, the result *R*12 can be represented as:

$$R12 = k_{\text{М}}F1 + k_{\text{Н}}F1 + k_{\text{М}}F2 \quad (1)$$

where: $k_{\text{М}}$ – is a parameter that comprehensively reflects the property and / or ability of an OA material to structurally respond to an applied impact; *k*n is the parameter that characterizes the hereditary change in the properties of the OA material, for example, an increase in the defectiveness of its surface layer at the site of the effect *F*1.

Then, the OA material in the zone of hybrid diagnosis is affected by *F*2, and then by *F*1. The formalized final result *R*21, by analogy with (1) we will present in the form:

$$R21 = k_{\text{М}}F2 + k_{\text{Н}}F2 + k_{\text{М}}F1 \quad (2)$$

For clarity, suppose that:

$$F2 > 1 \text{ and } R21 > R12 \quad (3)$$

Then subtracting from (2) the expression (1) in view of (3), we get

$$\Delta R = \Delta F \rightarrow k = \underline{\Delta R}\,\Delta F \quad (4)$$

where:

$$\Delta R = R21 - R12; \Delta F = F2 - F1 \quad (5)$$

Thus, a certain combination of ДФП allows you to get not only traditional information about the parameters of the state of the object of analysis from relationships like (1) and (2) - coefficient *k*m, but also to highlight the hereditary-latent characteristic, the functional quality of the material OA - the parameter kkn. In fact, it reflects the ability of a material to change its structural and phase state, for example, the degree of damage (defectiveness) under the influence of previous influences and thereby affect the performance of subsequent ones.

It is obvious that linear informational diagnostic models in the form of relations of the type (1) - (5) can be formulated for the most diverse combinations of structure and form-generating technologies with their characteristic DFT. However, the quantitative determination of the rationality of these combinations at this stage of the analysis of the information and technological capabilities of the

IOP Conf. Series: Materials Science and Engineering **683** (2019) 012027 doi:10.1088/1757-899X/683/1/012027

proposed device mezodiagnostics is almost impossible [10-13]. Therefore, to solve the problem of defining diagnostically rational combinations of "processing" technologies, we will use the method of expert-analytical analysis (EAA.) This method makes it possible to give subjective and quantitative assessments of functionally difficultly formalized categories, which provides flexibility and the necessary accuracy of EAA when considering the most diverse problems [14].

In this case, to use the EAA apparatus, it is necessary to formulate criteria for assessing the diagnostic quality of individual varieties of "processing" technologies, with subsequent processing of the results of expert questionnaires using the standard weighted-sum method:

$$U_j = \sum_{i=1}^{n} k_i u_{ji} \tag{6}$$

where: Uj – the final expert and analytical assessment of the considered quality parameter OA, in this case the diagnostic capabilities of the -th "processing" technology; ki - significance of the ith criterion of the quality assessed; uji - specific expert-criterial evaluation of the technology under consideration; $j = 1,2 \dots m$ - total number of compared combinations of technologies; $i = 1,2 \dots n$ - the number of criteria that in this case characterize the diagnostic capabilities of a particular processing technology.

As a result of the preliminary EAA, criteria for evaluating the expected diagnostic quality of technologies were formulated and their significance in points was determined (0 - min; 10- max points). The results are presented in Table 1 and do not need detailed comments.

Table 1. Significance of performance evaluation criteria and hybrid diagnostics of the quality of materials and structures

Criterion abbrv.	ИТ	ТР	ФВ	МП	ВМ	ЖЦ	ТО	КО	ЗО	СИ
Rate of Significance	9,2	9,0	8,9	4,8	3,1	6,8	7,5	6,8	9,6	4,1
Spread of opinions	±0,3	±0,2	±0,3	±0,6	±0,7	±1,1	±0,6	±0,5	±0,2	±0,8
Ratio of orientation	0,06	0,04	0,05	0,21	0,34	0,38	0,28	0,31	0,03	0,41

Table 1 adopts the following abbreviations for the significance criteria:

ИТ – information content of technology, i.e. the fundamental possibility of obtaining the necessary information about the object of analysis;

ТР – technical feasibility, as a set of available methods and means of implementing this type of hybrid diagnosis;

ФВ – functionality; consist in providing technological flexibility and variability of the implementation of the diagnostic procedure;

МП – the scale of application of the considered variant of hybrid diagnostics; is determined by the number and type of diagnosed materials used in various industries;

ВМ – the possibility of modernization is associated with the prospect of further improvement of the considered hybrid diagnostics;

ЖЦ – the life cycle of the analyzed hybrid diagnostic technology is determined by the expected duration of its effective use in practice;

ТО - the technical support of the considered hybrid diagnostics lies in the possibility and accessibility of the technological and instrumental support tools used for its realization;

Further, typical representatives of processing technologies were selected and EAA potential of their hybrid diagnostic quality was carried out using a weighted sum method, according to [15, 16]. It was taken into account that when assessing the relevance of a future study of a specific combination of processing methods for information technology purposes, the criteria for the cost and study of the analyzed version of the future hybrid diagnostic method in (6) are entered with a (-) sign. This is fully consistent with the logic of the problem being solved by the EAA by definition of rational, in this case, two homogeneous or different types of structure and form-building technologies, with its characteristic composition of the ДФП.

Table 2 contains the final results of this very large-scale work of several experts, specialists in the field of manufacturing manufacturing technologies and methods for controlling the quality of engineering products.

Table 2. Matrix of possible combinations of processing technologies in the synthesis of new hybrid diagnostics

ItemNo.	РДВ КДВ	ОР	ПД	ЭЭ	ЭХ	ХО	УО	УЗ	ЛО	ЭЛ	ПР
1	ОР	39	41	33	26	26	0	0	35	0	28
2	ПД	41	34	37	29	34	0	0	29	26	31
3	ЭЭ	37	39	69	32	27	34	36	29	27	27
4	ЭХ	26	29	44	35	0	41	0	34	0	0
5	ХО	24	34	41	26	0	26	0	31	0	0
6	УО	47	51	52	39	38	74	41	69	33	29
7	УЗ	27	21	41	28	31	41	38	0,27	0,26	0,32
8	ЛО	22	39	36	34	27	27	29	0,36	0	0
9	ЭЛ	21	22	31	27	0	26	0	0	0	0
10	ПР	0	0	0	0	0	0	0	0	0	0

The following abbreviations are used in Table 2:

РДВ - working defect-forming effect; КДВ - control and diagnostic effects; ОР - machining; ПД - plastic deformation; ЭЭ - electroerosive processing of conductive materials; ЭХ - electrochemical machining of metals and alloys; ХО - chemical treatment of the type of chemical milling; УО - ultra jet hydrophysical treatment; УЗ - ultrasonic treatment; ЛО - laser treatment; ЭЛ electron beam processing; ПР - plasma cutting of materials. Score (0) means obvious extremely low technology rating ($R < 0{,}25$) of the considered combination РДВ and КДВ. Rating values obtained by ЭАА were normalized to the maximum value Ui at $uji = 10$ by the dependence (6). Therefore, the rating values of the elements of the synthesis matrix of hybrid technologies are presented in% of this maximum achievable value. Based on a review of the data in Table 2, the following features of the ЭАА performed can be distinguished.

1. The integral rating of the effective and universal application of processing technologies for the synthesis of hybrid diagnostic methods suggests that the potential of the greatest significance is in the ultra-jet (УО) and laser (ЛО) methods of form and structure formation.

In addition, electroerosion effects (ЭО) for conductive materials and ultrasonic effects (УЗ) for fragile materials such as ceramics and some types of composites, such as carbon-carbon compositions, are very effective.

2. It should be noted the practical futility of using a plasma jet (ПР) as a structural-functional element of the hybrid diagnostic effects.

Some positions in Table 2 are controversial, but in general they adequately reflect the picture of rating assessment of the potential of the diagnostic use of processing technologies.

According to the results of an expert-analytical study, it is possible to draw certain conclusions and formulate preliminary generalizations.

1. According to the ЭАА presented in Table 2, the possible combinations of processing technologies have the greatest informational and diagnostic potential of combinations of ДФП, which approach as closely as possible the real thermal-force, including pulsed or continuous-cyclic, operational effects on the material of the structure. These combinations include: laser ultrajet diagnostics ($R86$) electroerosion and ultrasound diagnostics ($R37$).

The last type of hybrid diagnostics is very promising for assessing the quality of conductive fragile materials such as carbon-carbon compositions.

2. Hybrid ultra-jet diagnostics of structural materials operating in non-thermally stressed conditions may be of considerable scientific and practical interest. However, adding ultra-jet diagnostics with the help of thermal impact on the material [17–20], it is very likely that the potential of the effective-scale application of this information technology to ensure the quality of products from a wide variety of materials, primarily composite ones, will significantly increase.

3. The prospect of the formation of new, modified types of hybrid mesodiagnostics of the quality of materials and products from them consists in the additional introduction of special methods of damaging or strengthening effects on the material to the main ДФП. A typical example of the validity of this statement is the analysis of changes in the physicomechanical properties of polymer composites after thermal and / or radiation exposure, through the use of ultra-jet hydrophysical diagnostics. The possibility of express-diagnosing the effectiveness of various strengthening technologies should include the possibility of express determination of the optimal modes of chemical-thermal treatment by the method of laser-ultra-jet diagnostics at the preparatory stage of manufacturing products from the corresponding structural materials.

Conclusions

Thus, the analysis performed showed the potential information and physical performance of the minimally invasive application of known structure and form-generating technologies in solving diagnostic problems of ensuring the quality of typical and promising functionally different materials and products from them.

List of abbreviations

ДФП – dominant physical process;
ОА – object of analysis;
УСД – ultrajet diagnostics;
ЭАА – expert and analytical analysis;
КО – staffing; consists in the difficulty of training specialists for the implementation of the analyzed type of hybrid diagnostics;
ЗО – the cost of technology implementation; comprehensively reflects the totality of financial investments in the implementation of this type of hybrid diagnostics;
СИ – the degree of knowledge of the technology under consideration, the higher the value of this criterion, the less relevant its further scientific and applied research.

Acknowledgments

The studies were carried out as part of a grant from the President of the Russian Federation for state support of leading scientific schools (НШ-3778.2018.8) and a grant from the RFFI 18-29 18081.

References

[1] Emissionnaya tehnologicheskaya diagnostika [Emissive technological diagnostics]. *Moscow, Mashinostroenie Publ.* 384 p.
[2] Galinovski A L, Veltishchev V V, Belov V A, Mugla D R and Bolgarchuk I O 2018 Experimental studies on the choice of rational technological regimes and parameters of hydroabrasive cutting in conditions of equipment operation in the submerged position. *Reference book. Engineering journal with the application.* No.7 (256). pp. 3-9.
[3] Galinovski A L, Mulyar S G and Hafizov M V 2012 Primenenie gibridnoi diagnostiki dlya ocenki ekspluatacionnyh svoistv kompozicionnoi keramiki [Application of hybrid diagnostics for

evaluation of performance properties of composite ceramics]. *News of higher educational institutions Engineering Publ.* No.9 pp. 65-69. (in Russian).

[4] Abashin M I, Vinokurova E V, Galinovski A L and Korshunov S V 2014 Global trends in improving the system of training engineers. *Vestnik MGOU Publ.* No.2 pp. 57-65. (in Russian).

[5] Abashin M I, Galinovski A L, Zosimov M V and Moiseev V A 2016 Analysis of trends in the development of engineering education in countries with a high level of innovation and technological development. *Vestnik of Kostroma State University. Series: Pedagogy. Psycology. Sociokinetics. Publ.* No.3 pp. 22-28. (in Russian).

[6] Barzov A A, Galinovski A L and Abashin M I 2012 Faktornaya model' ul'trastruinoi gidroerozii [Factor model of ultra-jet hydroerosion]. *News of higher educational institutions. Engineering Publ.* No.10 pp. 63-68. (in Russian).

[7] Abashin M I, Barzov A A and Galinovski A L 2012 Analiz fiziko-tehnologicheskih osobennostei processa ul'trastruinoi diagnostiki [Analysis of physical and technological features of ultra-jet diagnostics process]. *Bulletin of Moscow state technical University. N. Uh. Bauman. Series: natural Sciences Publ.* No.6 pp. 7. (in Russian).

[8] Abashin M I, Galinovski A L, Zosimov M V and Moiseev V A 2016 Analiz tendencii razvitiya inzhenernogo obrazovaniya v stranah, imeyushih vysokii uroven' innovacionnogo i tehnologicheskogo razvitiya [Analysis of trends in the development of engineering education in countries with a high level of innovation and technological development]. *Vestnik of Kostroma State University. Series: Pedagogy. Psycology. Sociokinetics Publ.* No.3 pp. 22-28. (in Russian).

[9] Barzov A A, Galinovsky A L and Abashin M I 2012 Faktornaya model ultrastruynoy gidroerozii [Ultimate jet erosion factor model]. *Proceedings of higher educational institutions. Mashinostroyeniye Publ.* No.10, pp. 63-68. (in Russian).

[10] Abashin M I, Barzov A A and Galinovski A L 2012 Analysis of the physical and technological features of the process of ultra-jet diagnostics. *Bulletin of the Moscow State Technical University. N.E. Bauman. Series: Natural Sciences.* No.6 P. 7.

[11] Abashin M I, Barzov A A, Galinovski A L 2015 Tekhniko-ekonomicheskoye obosnovaniye vypolneniya kontrolno-diagnosticheskikh operatsiy. *Fundamentalniye I prikladniye problem tekhniki i tekhnologi.* No.1 (309) P.133-139. (in Russian).

[12] Galinovskiy A L 2008 Analiz innovatsionnogo potentsiala ultrastruinykh tekhnologiy v oboronnykh otrasliakh promyshlennosti. *Oboronnaya tekhnika.* No.6 P.54.

[13] Abashin M I, Barzov A A, Bochkarev S V, Galinovski A L and Maslov B G 2014 Primeneniye ultrastruynoi diagnostiki dlia otsenki kachestva svarnych shvov. *Svarochnoye proizvodstvo.* No.9 P.26-29. (in Russian).

[14] Tarasov V A and Galinovskiy A L 2013 Problemy i perspektivy razvitiya gidrostruinykh technologiy raketno-kosmicheskogo mashinostroyeniya. *Inginernyi zhurnal: nauka I innovazii.* No.3 (15). P.23. (in Russian).

[15] Li C, Galinovsky A L and Zarubina O V 2018 Development of manufacturing technology and ultra-fast diagnostics of bimetallic cutting tools for the utilization of aerospace engineering composite materials. *Mashinostroenie: innovative aspects of development materials of the I international scientific-practical conference.* P.84-5. (in Russian).

[16] Galinovsky A L 2014 Simulation of the process of diagnostics of nanostructured coatings with an adhesiometer and ultra-jet method. *Bulletin of the Rybinsk State Aviation Technological Academy. P.A. Solovyov.* P. 9-16. (in Russian).

[17] Barzov A A and Galinovsky A L 2009 Polyfunctional capabilities of ultra-jet technology for processing materials and liquids. *Vestnik UGATU. Mashinostroenie.* Volume 12, 4 (33) P.116-120. (in Russian).

[18] Barzov A A, Galinovsky A L, Salnikov S K and Elfimov V M 2010 Improving the performance of ultrajet processing by optimizing the kinematic factor. *Vestnik MGTU im. N.E. Bauman. Special edition.* P.217-29. (in Russian).

[19] Abashin M I, Galinovsky A L and Sudnik L V 2013 Determination of hardness and resistance to abrasion of steel samples by the ultra-jet method. *News of higher educational institutions. Mechanical Engineering.* No.9 P. 37-43. (in Russian).

[20] Galinovsky A L 2011 Study of the physical and technical features of the interaction of ultraformal and obstacles. *Fundamental and applied problems of engineering and technology.* No.4-2 (288) P. 131-5.

IOP Conf. Series: Materials Science and Engineering **683** (2019) 012028 doi:10.1088/1757-899X/683/1/012028

The information-physical mechanism of diagnostic of the functional coatings erosive wear by water jet

A A Barzov[1], A L Galinovsky[2] and A S Provatorov[2]

[1] Physics department, Lomonosov Moscow State University, Leninskie Gory, 1, p. 2, RU-119991, Moscow, Russia.
[2] Department of Special machinery, Bauman Moscow State Technical University, 2nd Bauman Str., 5, building 1., RU-105005, Moscow, Russia

Abstract. This paper discusses process of the functional coatings erosive wear by water jet with the purpose of coatings quality characteristics diagnosing. The probabilistic model of this process is considered. The main model feature is the surface and volume damages analysis after water jet influence. The experiments show the directly dependence between these coatings damages types. This fact highlights the water jet diagnostic efficiency.

Keywords: functional coating, water jet influence, erosive wear, probabilistic model.

1. Introduction
Successful development of engineering surfaces for critical products, mainly in the field of technological quality assurance of various functional coatings requires the creation of a physically adequate apparatus for their control and express diagnostics. Expert-analytical analysis showed [1–4] that ultra-jet hydrodiagnostics (UJD) has the greatest performance potential. Technologically, the UJD is based on an assessment of the degree of influence of the quality parameters of the surface layer of the object of study on the information-physical characteristics of its local hydroelectric destruction process under the action of a high-speed ultrajet fluid (water), with a power flux density of more than 0.1 MW/mm2.
This situation is explained by the fact that under ultrajet impact in the surface layer there is a stress-strain state (SSS) of the material, which is model-like for most practically important cases during its operation: friction, variable power loads, etc. Therefore, the UJD has certain informational advantages compared to the diagnostics of the quality of the surface layer through the use of an air-abrasive or hydro-abrasive jet, as the main tool of the physical-technological implementation, operating-simulated SSS.

2. Formulation of the problem
This problem consists of the analysis of the correlation of the two main components of the process of ultra-jet hydro erosion of the surface layer of the product being diagnosed:
- Variable wedging action of high ($P > 100$ MPa) water pressure (UJ) on the kinetics of nucleation, development and fusion of surface crack-like micro- and macro-defects between themselves and underlying similar discontinuities in the subsurface layer;
- Formation by the classical fatigue mechanism of critical values of crack-related damage to the material of the surface layers of the object of diagnosis in the context of intense variations of its SSS, characteristic of the hydrodynamic effects of high-speed UJ.

Note that the distinction between the significance of the surface-proppant and volume-fatigue mechanisms of ultrahydro-erosion material, schematically depicted in Fig. 1 for most operating ultrajetting technologies (UJT) is mainly of scientific and methodological value, in particular, in the UJ-cutting of sheet materials.

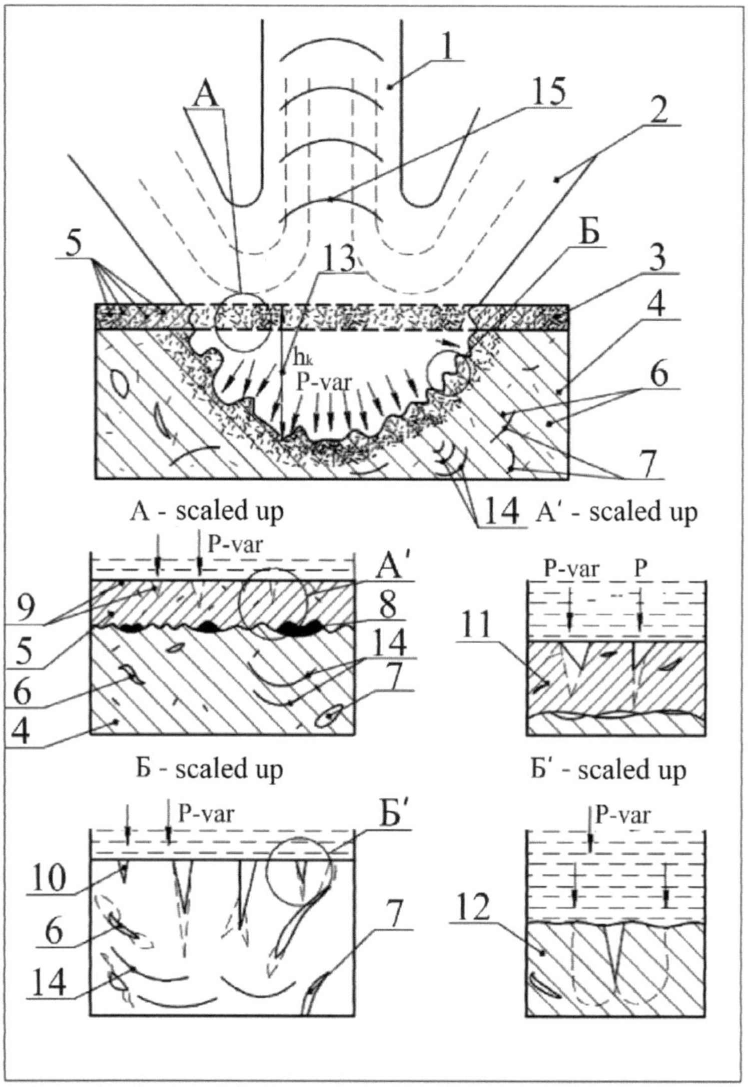

Figure 1. Hydrodynamic picture of the occurrence, development and merging of discontinuities in the surface layer of the object of analysis - the mechanism of its hydroelectric UHJ destruction.

In Figure 1:
1 - diagnostic ultra-hydrojet incident on the surface of the object of analysis (UHJ) with a speed Vc, which creates a hydrodynamic pressure P-var in the process of braking in the zone of the UJ-impact;
2 - hydraulic jets, reflected and having a speed *V*отр from the surface of the object of analysis, and hydro cavity formed under UHJ;
3 - functional coating applied to the surface of the object of analysis;
4 - source material of the object of analysis;
5 - micro nesting (micro and submicrocracks) in the structure of the functional coating 3;
6 - micro-continuity (micro- and submicrocracks) in the starting material of the object of analysis;

7 - technological potentially dangerous defects in the surface layer of the object of analysis;
8 - crack-like initial, nascent and developing defects at the interface between the surface of the object of analysis and its functional coating;
9 - existing, incipient and developing microcracks (defects, discontinuities) on the surface of a functional coating under the variable effect of hydraulic pressure P;
10 - existing, incipient and developing microcracks (defects, discontinuities) on the free surface of the hydrodynamic contact between the object of analysis and the inhibited UHJ;
11 - contours of the dispersible UHJ particles from the surface of the functional coating 3;
12 - the contours of the dispersible UHJ of the solid-phase particles from the surface layer of the object of analysis;
13 - microrelief *RK* and topography (depth h_k) of surface hydro cavity, formed in the process of diagnosing UHJ;
14 — elastic deformation waves — acoustic emission waves (AE) generated in the material of the object of analysis by the hydrodynamic interaction of the UHJ with its solid-state surface;
15 - AE waves propagating through the body of the UHJ from the zone of its shock-dynamic interaction with the surface of the object of analysis.

However, in the case of UJD of functional coatings, the role of each of these mechanisms is directly related to the informative performance of the entire procedure for express analysis of their operational quality. This is explained by the fact that the first "wedging mechanism" of the diagnostic factor of ultra-jet impact is closely related to the "micro and sub microficiency" of the functional coating itself. This component of the quality of the coating manifests itself in the form of the ability to withstand operational loads, leading to a violation of its continuity and an increase in damage, mainly during friction-force contact of structural elements in friction units, working surfaces of cutting tools, etc.

The second, no less significant mechanism of fatigue damage accumulation in the subsurface layer of the substrate material is physically responsible for the ultrahydro erosion local microfragmentation of the surface layer of the object of diagnostic exposure to UJ and, most importantly, damage to its interface with functional coating. It is the kinetics of this process that very often determines the technological quality of the functional coating, which consists in ensuring high values of the adhesive interaction between it and the underlying, protected layers of the main structural material of the product. Therefore, a detailed analysis of the coating peeling parameters during UJD is an information-significant sign of its operational and technological quality. This circumstance allows you to quickly work out the rational modes of formation of various modes of formation of various functional coatings without conducting time-consuming and expensive full-scale experimental studies, as well as carry out current, for example, selective or 100% control of the stability of the technology for their production. Moreover, with a 100% control, the UJ is applied to the contact-inoperative coverage area, for example, at a certain distance of their cutting edges of the forming tool.

Note that, if necessary, it is possible to quickly obtain information about the abrasive resistance of functional coatings by introducing highly hard particles into the UJ, for example, in the form of fine silicon carbide, aluminum oxide (corundum), etc.

Thus, the UJD by applying the fluid UJ (water) or high-speed jet of the corresponding abrasive-liquid suspension is an effective tool for the rapid analysis of the qualities of the most diverse functional coatings and methods of physical and technological hardening of the surface layer of product parts.

3. Analytical studies

To formalize the role of the surface wedging and volume fatigue mechanisms of erosive destruction of the surface layer of the coated material, we use the probabilistic approach, which is due to the specific variety of interaction processes of the characteristic physical and technological factors of the UJD operation. Moreover, in a deterministic formulation, the solution of this problem, taking into account the scale factor of strength of solids, can in principle be found using the well-known surface fracture modeling apparatus proposed, for example, in [5–6].

Taking into account the parallel processes of the formation of surface and subsurface crack-like damage to the material area being analyzed, the probabilistic distribution of microdefects, reflecting the large-scale microdestruction, based on the dependencies obtained in [7–8], we can write the total initial probability ratio in the form:

$$P_c(t) = P_S + P_V - P_S P_V \quad (1)$$

where: P_c is the total probability of destruction and separation at the current time of analysis t of the considered fragment (microparticle) of the surface material under hydrodynamic effects of UJ; P_S *and* P_V – respectively, the probability of the processes of surface and (S) bulk (V) local fragmentation (microfracture, dispersion), material in the impact zone of UJ. Physically P_c means the ratio of dispersed particles destroyed at the moment of time t to the maximum possible amount.
Using, after appropriate functional and physical adaptation, the results of the theoretical analysis performed in [9] P_S *and* P_V can be defined as:

$$P_S(t) = 1 - \exp[-\int_0^t c_S k_S(\tau) d\tau] \quad (2)$$
$$P_V(t) = 1 - \exp[-\int_0^{Ot} c_V k_V(\tau) d\tau] \quad (3)$$

where: $c_S(\tau)$ and $c_V(\tau)$ – accordingly, the concentration of microdefects, for example, subcritical cracks on the surface of the applied coating (S) and in the subsurface layer (V), as shown in Fig. 1; k_S and k_V – are respectively variable parameters, functionally determining the effects of SSS on the formation kinetics of c_S and c_V.
In fact, (1) determines the total probability of surface and subsurface hydro erosion fragmentation of the surface layer of the object of analysis, and (2) and (3) describe the kinetics of their change during the UJD, and in latent form these relations are functions of the coordinates included in the structure of c_S and c_V.
Considering in the first approximation that c_S and c_V are characterized by some averaged quasi-constant values, and the intensity of the local erosional fragmentation of the material over time is much less than the frequency of the variable-wave change of its SSS under the hydrodynamic action of the UJ ratio (1) with (2) and (3) for small c_S and c_V will look like:

$$P_c \sim k_S c_S + k_V c_V \quad (4)$$

Thus, (4) is a functional interpretation of the independence of the change in the crack-like defects of the surface of the functional coating and the subsurface layer of the substrate material under the hydrodynamic effects of the UJ in the process of implementing the UJD operation. Physically (4) reflects the principle of linear summation of damage, known in the mechanics of fracture as the Bailey principle. Moreover, due to the smallness of the critical values $c_S(l)$ *and* $c_V(l)$, where l is the length of the crack-like defect close to the critical value, the law of rare events is valid - Poisson's law, functionally similar to the exponential term in (2) or (3) describing the law of sudden failures, which governs the kinetics of the processes of variable-nonstationary erosion fragmentation of the surface layer of a material under shock-wave hydrodynamic effects.
It should be emphasized that from the scientific and applied point of view, the linear relationship of type (4) allows direct experimental verification by varying the values of S and cV with all other conditions being equal, UJ-interactions.

4. Experimental results

We illustrate the above with the results of direct experiments on the UJD of functional coatings, the main methodological feature of which was the functionally independent change of cS and cV by special physical and technological methods.

At the first stage of the experiments, five samples of steel X18H10T were subjected to fatigue loading with a different number of cycles of force action on them. This led to varying degrees of damage to their material, the appearance and development of crack-like submicro- and microdefects on the surface and in the thickness of the samples. Then, the surface of the samples was subjected to scribing UJ-effect on the FLOW standard hydraulic unit at a working pressure of P = 420 MPa. As a result, hydro cavities were formed on the surface of the tested samples, the depth of which hk completely correlated with the normalized maximum value with the number of cycles of their previous fatigue loading:

$$h_k = k_э\bar{N} \sim 0{,}5\bar{N} \quad (5)$$

where:
$h_k = h_{ki} / h_{max}$ – normalized to the maximum value depth of the hydro cavity formed by the UJ on the surface of the i –th sample (i = 1.2 ... 5); h_{max}~3,50 mm – the maximum value of the hydro cavity on the 5th sample, which turned out to be in this case the maximum number of loading cycles on a specially created test- process unit ($N_{max} = 50 \cdot 10^3$cycles); $k_э$ –experimental coefficient of proportionality between the degree of damage (N) and its ability to resist; UJ - destruction, which is characterized by the value of h_k.
In this case, $k_э$~0.5. The obtained experimental results are presented in Fig. 2

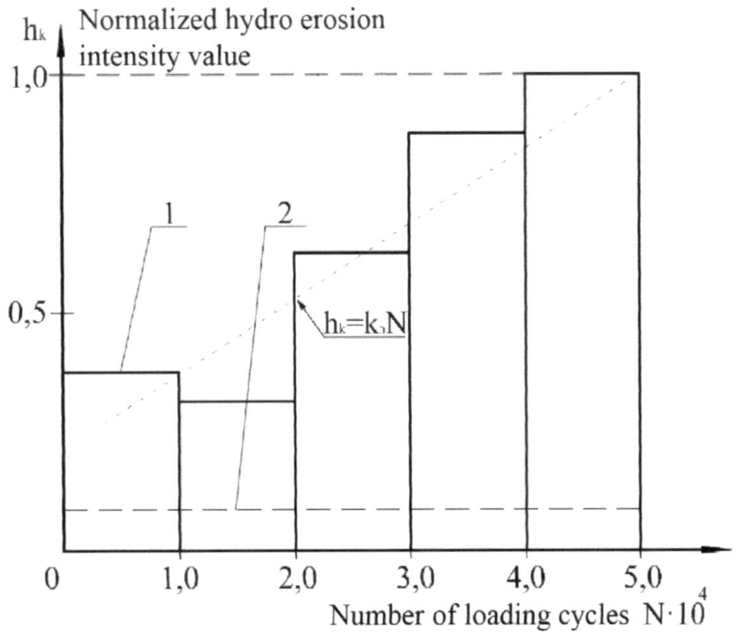

Figure 2. Experimental determination of the shielding effect of the coating on the surface of the UJS-hydroerosion.

In Figure 2:
1 - intensity of locally-surface hydroerosion UJ-destruction of samples subjected to various levels of fatigue-force loading;
2 - intensity of hydroerosion of samples, after applying a functional coating of TiN on their surface.
Note that the linear functional dependence of the type (5) has a completely independent information and diagnostic value, which was described in detail in [10], dedicated to the physical and technological capabilities of the UJD. Moreover, some decrease in hk when i = 2, i.e. the second test specimen showed

a slight increase in hydroerosion resistance, apparently due to the presence of a plastic hardening factor for the test material at N~104 loading cycles.

Thus, at the initial stage of the experiments, the variability of the parameters c_s and c_V was realized, due to the different degrees of fatigue damage of the diagnosed samples. In this case, this circumstance is described by the functional dependence of the type (5) reflecting the degree of hydro erosion, UJ impact on the material with different levels of the initial crack-like defects.

Then, a functional protective TiN coating was applied to the surface of all samples on a standard ion-plasma unit. This coating, according to the above description of 2 types of hydro erosion fracture mechanism, would have to screen the role of the surface - "wedging" effect of water pressure in the braking zone of the UJD. Direct experiments on single-time UJ-scribing of the surface of all samples with very different degrees of initial damage showed an almost identical, insignificant degree of their hydro-erosion destruction, mainly due to local-plastic displacement of material from the UJ-impact zone (*hk* <0.1). This experimental result makes it possible to recognize the methodically valid physically conditional gradation of the UJ-hydro-erosion mechanism of the material into two, basically, interrelated processes, due to differences in the kinetics of c_S and c_V, formation described by the probabilistic relations (1) - (3) and their actually determined type of lianerization (4).

It must be emphasized that in addition to the scientific value, the results presented in Fig. 2 may have serious technological application. Indeed, the targeted implementation of the shielding effect of the negative impact on the fatigue failure mechanism of the structural material of its surface defects by periodically applying a protective functional coating can be considered as a promising component of the renovation technology apparatus. This technology, based on the screening of the negative impact on the resource of fatigue failure of a construction of cracks emerging on its surface — initiating accumulations of dislocations, slip lines, etc., of micro- and sub-microdefects, is designed, along with thermo-gas-static and electropulse "healing" of structural micro-inhomogeneities, to reduce operational damage to the object of analysis. Moreover, only the exposure of the test sample in the working chamber of the ion-plasma unit, without coating, does not have practically significant effectiveness of shielding the effect of surface and bulk microdefects on the kinetics of the subsequent UJ-destruction of the test surface.

The final series of experiments consisted in artificially increasing the value *cs*, ceteris paribus, by scribing a diamond indenter with a different degree of graininess of the surface of samples with a TiN coating. The results of the subsequent UJ- impact presented in Fig. 3 showed:

- Close correlation between *cs* and the intensity of the hydroerosion process, which confirms the validity of the hypothesis about the physicotechnological dualism of the UJ-surface destruction mechanism;

– equidistant-delayed nature of dependencies h_k=f(t) due to the lack of influence of the functional coating after its dispersion in the zone of UJ-exposure;

erosion-high significance of the functional coating at the initial stage of the UJ-impact, even in the conditions of its preliminary discrete damage (see items 2 and 3 in Fig. 3).

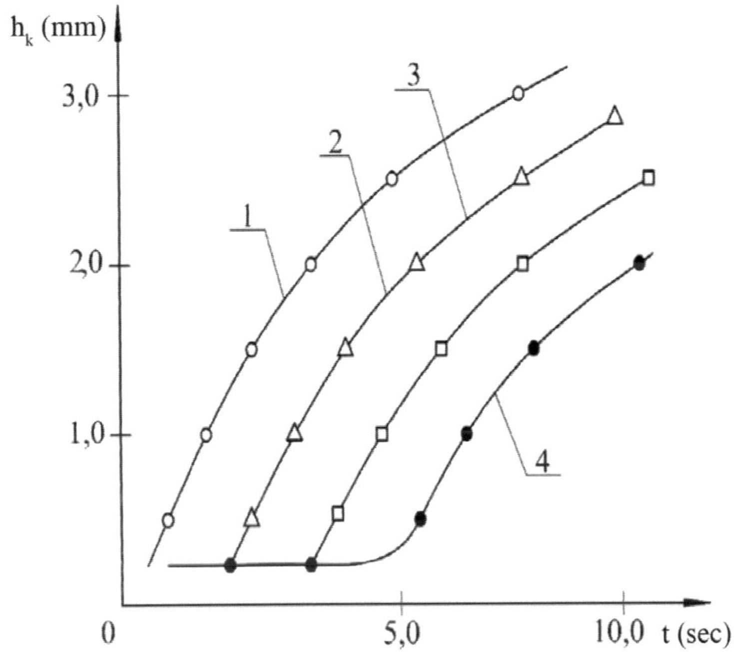

Figure 3. Experimental analysis of the effect of variations in the degree of surface damage of a functional coating on the intensity of hydro-erosional UJ-destruction.

In Figure 3 it is accepted:
1 - the original sample without coating;
2 - coating with frequent deep scratches;
3 - coating with rare shallow scratches;
4 - intact functional coating.
Note that by means of a specially implemented coating mode with low adhesion to the substrate, the principal possibility of ultra-jet express control was established using UJ of this most important physical and technological parameter determining the performance quality of virtually all types of functional protective coatings. In this case, there was a film detachment of the functional coating under the action of an UJ, practically without finely dispersed hydrodynamic dispersion of the material of the surface layer.
Summarizing the above, it is possible to draw certain conclusions and outline the prospects for the further development of research results.
1.Ultra-jet diagnostics allows to solve not only practical tasks of express control of the quality of functional coatings, but also allows more detailed study of the specifics of the mechanisms of surface hydroerosion destruction in physically extreme conditions of the physic-energy UJ- effects on the material of a solid-state object of study.
2.Experimental testing and methodical specification of algorithms for modeling the process of UJ-diagnosing the quality of functional coatings will allow to obtain numerical values of the parameters necessary for reliable calculations of the resource and operational characteristics of various products and, above all, their functional contact working surfaces.
3.The development of a physical and mathematical analysis of the problem of UJ-interactions, in particular, full-scale implementation of the procedure of simulation-probabilistic modeling and / or building finite-element models of UJ-hydroerosion, supplemented by the results of field experiments, will create an effective system for predicting and verifying the potential impact of promising functional

coatings and technologies for their production, including those with gradient coatings of nano-compositional structure.

5. Conclusions

Taking into account the scientific and applied value and the variability of the technological capabilities of ultra-jet hydrophysical diagnostics and quality control of various functional coatings, their technical totality can be classified as a minimally invasive express method for obtaining objective information about the state parameters of the surface layer of the object of analysis, complementing the modern apparatus and surface engineering tools.

Acknowledgments

The reported study was funded by RFBR, project number 19-38-90228.

References

[1] Abashin M I, Galinovsky A L, Moiseev V A, Osipkov A S, Provatorov A S 2014 On the development of a scientific and methodological base for obtaining and diagnosing nanostructured coatings. *Nanoengineering.* No.5 (35). p. 23-27.

[2] Barzov A A, Galinovsky A L, Abashin M I 2012 Factor model of ultra jet hydro erosion. *Engineering.* No.10. p. 63-68.

[3] Moiseev V A, Galinovsky A L, Abashin M I, Khakhalin A V 2014 Comparative analysis of methods for assessing the quality of nanostructured coatings. *Fundamental and applied problems of engineering and technology.* No.5. p. 134-141.

[4] Abashin M I, Barzov A A, Galinovsky A L, Shuteev V A 2011 Ultra-jet express diagnostics of materials and engineering products. *Scientific and technical statements of the St. Petersburg State Polytechnic University.* No.123. p. 141-147.

[5] Abashin M I, Galinovsky A L, Bochkarev S V, Tsaplin A I, Provatorov A S, Khafizov M V 2015 Ultra-jet simulation for coating quality control. *Physical mesomechanics.* **18**. No.1. p. 84-89.

[6] Abashin M I, Galinovsky A L and Khafizov M V 2014 Finite element modeling of thin film coating hydro erosion and experimental verification of its results. *Electrical Engineering. Energy. Engineering. Col. of sc. proc. I Intern. scientific conf. of young scientists.* p. 189-192.

[7] Abashin M I, Galinovsky A L, Barzov A A, Mazaeva I V, Sysoev N N and Sysoev P N 2015 Ultra jet hydrodynamics. *M.: Moscow State University named after M.V. Lomonosov.* 308p.

[8] Rabotnov Yu N 1959 On the mechanism of long-term destruction. *Questions of strength of materials and structures. M. : Publishing House of the Academy of Sciences of the USSR.* p. 5-7.

[9] Barzov A A, Galinovsky A L, Puzakov V S, Troshchy O A 2006 Probabilistic modeling in innovative technologies. *M.: Publishing house HT.* 100 p.

[10] Barzov A A, Galinovsky A L, Abashin M I 2012 Analysis of the physico-technological features of the process of ultra-jet diagnostics. *Engineering Journal: science and innovation.* No.11 (11). p. 14.

IOP Conf. Series: Materials Science and Engineering **683** (2019) 012029 doi:10.1088/1757-899X/683/1/012029

Thermophysical processes models in composite workpieces processed by microwave radiation

E S Belenkov[1], P V Prosuntsov[1] and S V Reznik[1]

[1] Bauman Moscow State Technical University, 2-nd Baumanskaya, 5, b.1, 105005, Moscow, Russia

E-mail: pavel.prosuntsov@mail.ru

Abstract. The solution that improves energy effectiveness and polymer composite materials parts quality by means of microwave heating with polymer matrix curing is proposed. Thermal state analysis of workpiece for various layout configurations of microwave heating equipment was conducted. It was found that the most uniform heating is obtained inside the rectangular chamber while workpiece is rotating around the transverse axis.

1. Introduction

Polymer composite materials (PCM) are widely used to achieve high mass efficiency of the aerospace structures. Curing of polymer matrix by means of autoclave and vacuum furnaces is an important stage of manufacturing of such structures. The quality of finished PCM products directly depends on thermal field uniformity in a workpiece during heating and annealing. Recently significant benefits of microwave radiation for PCM curing have been educed. These benefits include reduction of production time, saving energy costs and products quality enhancement [1–8] for components of transformable spacecraft [9-19]. To choose optimal heat treatment regime, the effects of PCM characteristics and heat removal conditions on thermal field of a workpiece need to be defined. Also the dependence of power and space distribution of microwave radiation with heating dynamic and temperature pattern uniformity is required.

2. Modeling Processes for Representative Volume Element

The curing process of the polymer matrix should be considered at two structural levels: macrolevel – operating area of the equipment with magnetrons and microlevel – representative volume element (RVE) of PCM. As RVE the model of fiber surrounded by matrix was used for unidirectional carbon-glass-aramid reinforced plastics. It was stated that a plane electromagnetic wave falls on one of the parallelepiped side face and passes through the RVE without reflection and dissipation. Energy absorption of electromagnetic radiation in dielectric materials is accompanied with heat generation. The intensity of heat generation depends on the power and frequency of radiation. At the same time, electric currents are induced in conductive materials. Electromagnetic wave propagation in RSV of PCM is described by Maxwell equations. On fiber-matrix interface (note, that fiber is made of dielectric material) the following boundary conditions were formulated:

$$H_{\tau 1} = H_{\tau 2};$$

IOP Conf. Series: Materials Science and Engineering **683** (2019) 012029 doi:10.1088/1757-899X/683/1/012029

$$E_{\tau 1} = E_{\tau 2};$$

$$D_{\tau 1} = \frac{\varepsilon_1}{\varepsilon_2} \cdot D_{\tau 2};$$

$$B_{\tau 1} = \frac{\mu_1}{\mu_2} \cdot B_{\tau 2},$$

where ε_1 and ε_2 – dielectric constants of fiber and matrix; μ_1 and μ_2 – magnetic constants of fiber and matrix; Hτ1 and Hτ2 – tangent vectors of magnetic field strength of fiber and matrix, A/m; Eτ1 and Eτ2 – tangent vectors of electric field strength of fiber and matrix, V/m; Dτ1 и Dτ2 – tangent vectors of electric induction of fiber and matrix, C /m2; Bτ1 and Bτ2 – tangent vectors of magnetic induction of fiber and matrix, T.

Electric field strength is close to zero for electroconductive carbon fiber. Hence, in this case, the boundary conditions on fiber-matrix interface are described as the following:

$$E_{\tau 1} = 0;$$

$$j_s = [H_{1\tau}, n_0],$$

where n0 – normal to conductive and dielectric medium; js – current density in conductor, A/m2.

The modeling was carried out in software package Ansys HFSS [20].

The analysis of the obtained results shows that in case of dielectric fiber the field distribution slightly changes in the fiber-matrix system (Figure 1). Modeling of electromagnetic processes in RVE of PCM with a conductive carbon fiber demonstrates that heat generation in the matrix is much less than in the fiber (Figure 2). Such significant difference can be explained by the occurrence of inducted currents in the carbon fiber. The conclusion is the following: in contrast to electromagnetic processes that take place in the PCM with dielectric fibers the major heat generation in PCM with conducting fibers occurs exactly in the fibers.

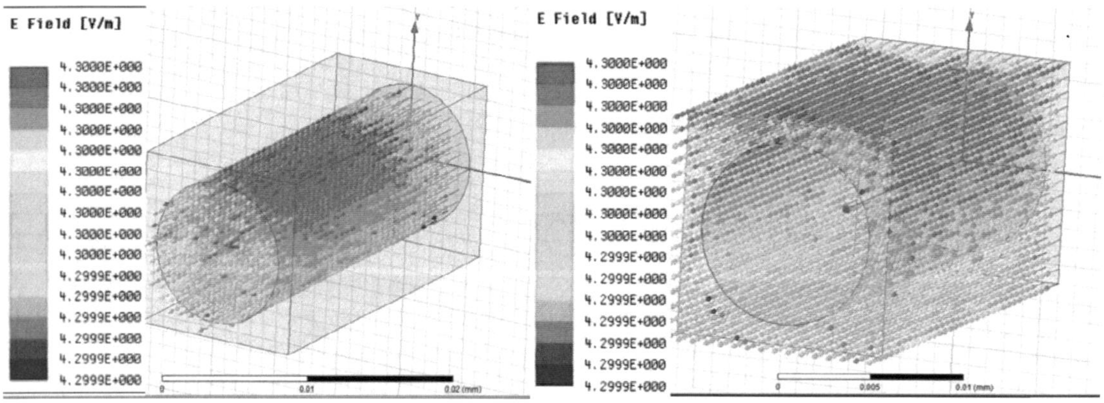

a) b)

Figure 1. The electric field distribution for glass fiber (a) and epoxy matrix for RVE of glass fiber reinforced plastic (b), V/m.

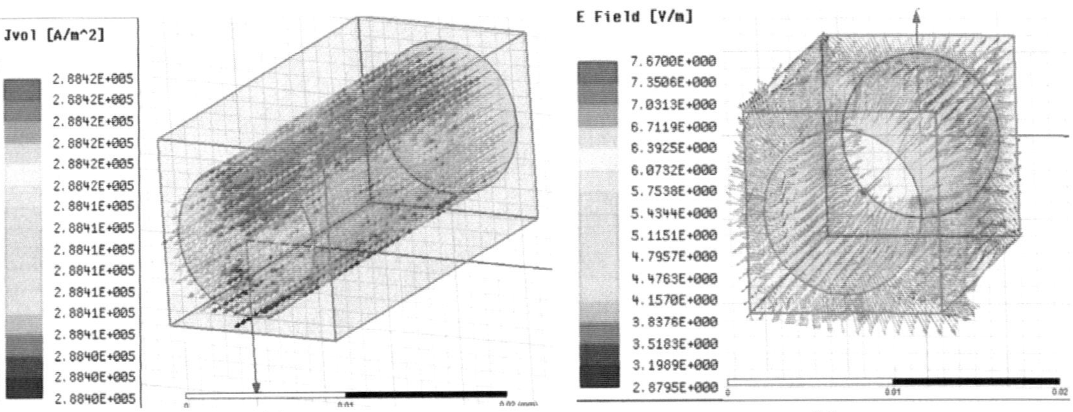

Figure 2. Electric current density distribution in carbon fiber (a), A/m2 and electric field intensity in epoxy matrix for RVE of carbon fiber reinforced plastic (b), V/m.

3. Homogenization Procedure of PCM Characteristics

Homogenization procedure was conducted to represent the PCM as a homogeneous structure material during thermal and electromagnetic processes modeling in microwave heating equipment. To find effective dielectric constant of fiber-matrix system the first order Rayleigh scatter was used [21]:

$$\varepsilon_{эфф} = \varepsilon_1 \cdot \sqrt{\frac{1-\Delta_{12}\cdot\nu}{1+\Delta_{12}\cdot\nu}};$$

$$\Delta_{12} = \frac{\varepsilon_1 - \varepsilon_2}{\varepsilon_1 + \varepsilon_2},$$

where ν – volume fraction of reinforcement in PCM. Mixture rule and condition of heat generation equality in heterogeneous and homogeneous material model were used to obtain other characteristics.

4. Modelling of Heating of PCM Workpiece

Hollow cylinder made of glass fiber reinforced plastic was used as a model for thermal and electromagnetic processes analysis in microwave radiation equipment. It was assumed that cylinder workpiece was placed in a chamber with one or more radiation sources and had the following sizes: outer radius – 90 mm, inner radius – 85 mm, length – 1200 mm. The workpiece material was considered as homogeneous and isotropic. Also heat energy on side faces of workpiece was dissipated by convective and radiation heat transfer with environment. Moreover, thermal and electromagnetic characteristics were assumed constant as well as ambient temperature during whole microwave radiation exposure period.

Heat transfer mathematical model for cylinder workpiece was formulated as following:

$$c\rho\cdot\frac{\partial T(r,z,\phi,\tau)}{\partial\tau} = \frac{\partial}{\partial z}(\lambda\frac{\partial T(r,z,\phi,\tau)}{\partial z}) + \frac{1}{r}\frac{\partial}{\partial r}(r\cdot\lambda\frac{\partial T(r,z,\phi,\tau)}{\partial r}) + \frac{1}{r^2}\frac{\partial}{\partial\phi}(\lambda\frac{\partial T(r,z,\phi,\tau)}{\partial\phi}) + q_{mw}$$

where c – specific heat, J/(kg·K); ρ – density, kg/m3; λ – thermal conductivity, W/(m·K); T – temperature, K; qmw – microwave absorbed energy, Вт/м3; τ – time, s; r,z,ϕ – point location in a model volume in a cylindrical coordinate system.

Initial and boundary conditions were defined as follows:

$$T(r,z,\phi,0)=T_f\,;$$
$$-\lambda\frac{\partial T\ (R_1,z,\phi,\tau)}{\partial r}=-\alpha_f\cdot(T\cdot(R_1,z,\phi,\tau)-T_f)-\varepsilon_w\cdot\sigma_0\cdot(T^4(R_1,z,\phi,\tau)-T_f^{\,4})\,;$$
$$-\lambda\frac{\partial T\ (R_2,z,\phi,\tau)}{\partial r}=-\alpha_f\cdot(T\cdot(R_2,z,\phi,\tau)-T_f)-\varepsilon_w\cdot\sigma_0\cdot(T^4(R_2,z,\phi,\tau)-T_f^{\,4}),$$

where ε_w – material emissivity factor; σ_0 – Stefan-Boltzmann constant, W/(m2·K4); Tf – ambient temperature, K; αf – heat transfer coefficient, W/(m2·K).
The modeling results obtained for a chamber equipped with a single 2.4 kW magnetron operating at a frequency 2.45 GHz indicate that electric field intensity distribution was extremely non-uniform (Figure 3). Consequently, the workpiece heating process is non-uniform as well. The difference in the electric field intensity in the order of 104 V/m for workpiece local regions can cause local temperature gradients up to dozens degrees.

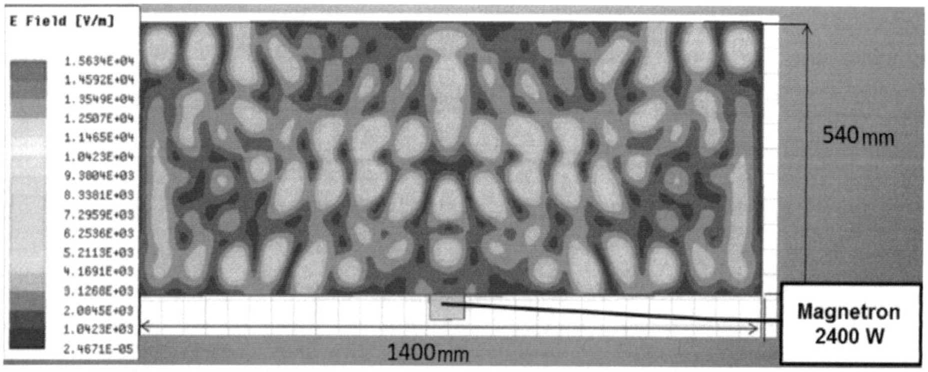

Figure 3. Electric field distribution in the operating area with single magnetron, V/m.

The chamber design with four microwave radiation power sources of 0.6 kW each was considered for reduction of non-uniform electric field intensity distribution. In this case, as follows from Figure 4, electric field intensity distribution becomes more uniform. Thus, temperature gradients inside the chamber decreases. Nevertheless, installation of additional radiation sources leads to complication of the equipment design and treatment cost increase. Therefore, other microwave radiation equipment versions were considered by varying operating area dimensions, magnetrons placement location and using different alternates of workpiece movement during processing. The preferred variant was the chamber with design that enables the workpiece rotation around the transverse axis during the microwave radiation process. Based on the obtained results it should be noted that more uniform electric field intensity distribution (Figure 5) and workpiece temperatures were acquired (Figure 6). Consequently, the workpiece rotation around the transverse axis can lead to a significant increase of temperature uniformity distribution and, accordingly, to increase the quality of the PCM product.

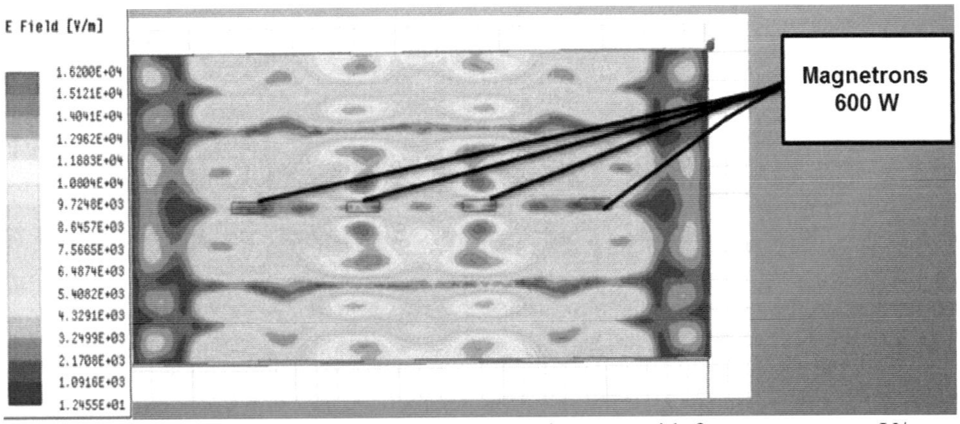

Figure 4. Electric field distribution in the operating area with four magnetrons, V/m.

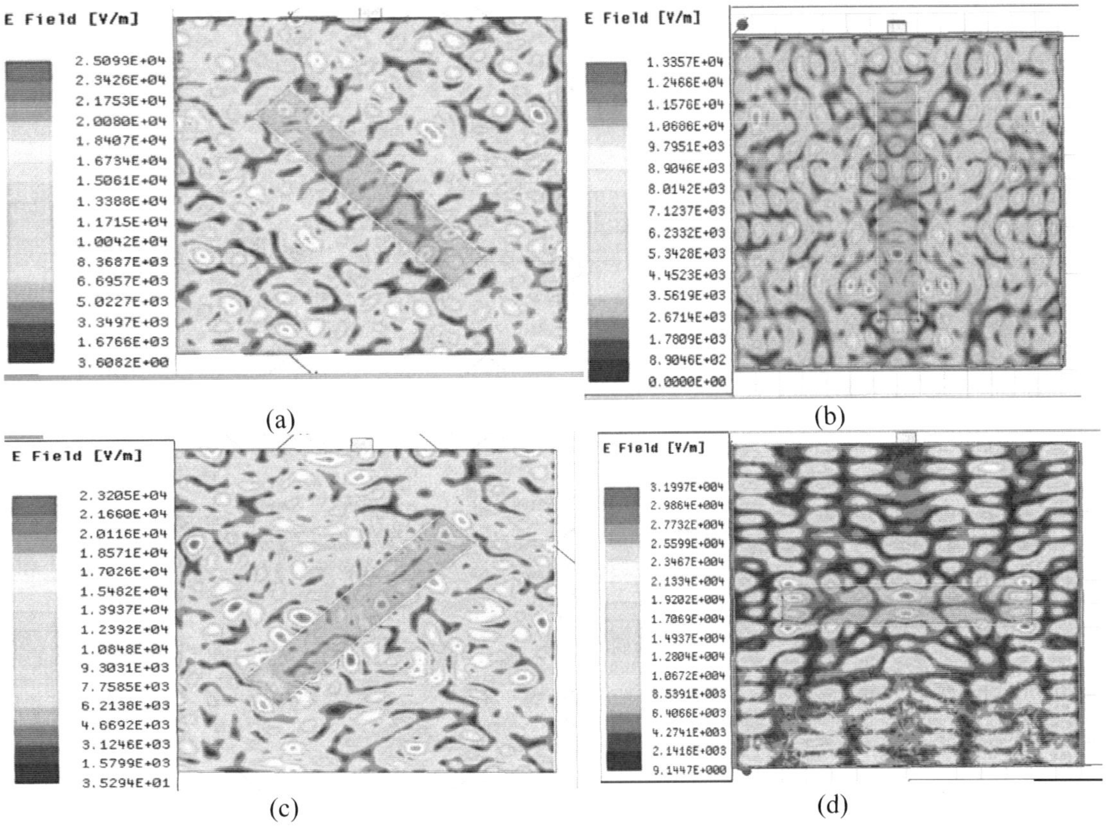

Figure 5. Distribution of the electric field strength when the workpiece rotates around the transverse axis in the microwave heating chamber, V/m: (a) angle of rotation 45 °C;(b) angle of rotation 90 °C;(c) angle of rotation 135 °C; (d) -rotation angle 180 °C.

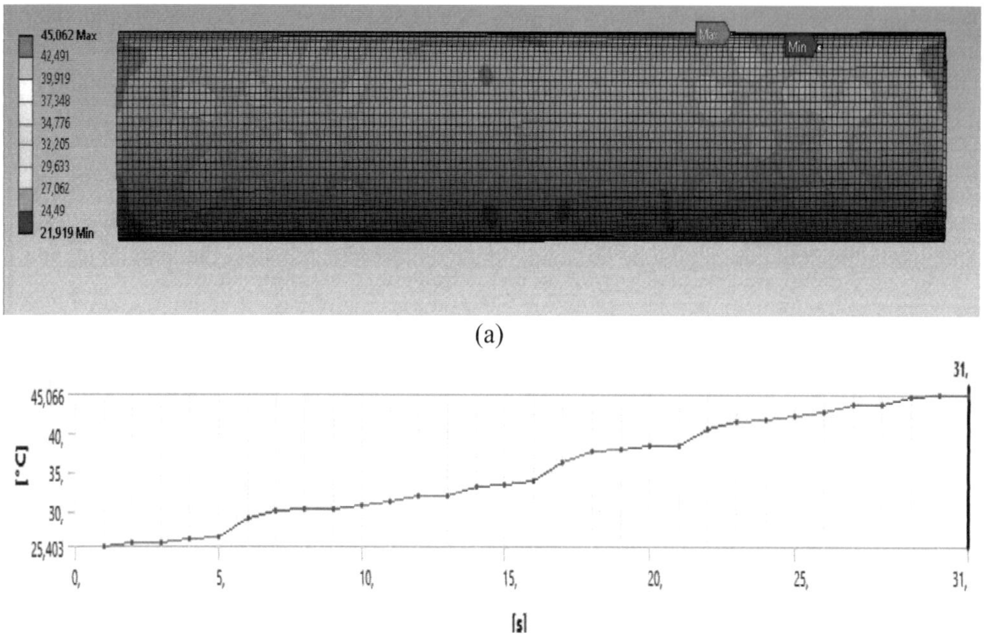

(a)

(b)

Figure 6. Temperature state of workpiece rotating around the transverse axis in microwave radiation installation, °C: (a) temperature field at the moment 30 s; (b) time dependence of the maximum workpiece temperature.

5. Conclusions

Physical and mathematical models of the electromagnetic radiation transfer and the heating of RSV of PCM were developed. These models take into account electromagnetic field distribution in each separate component during microwave radiation exposure and allow to make the analysis of thermal state of PCM RSV. The homogenization methods for RSV of PCM characteristics were proposed for material effective properties analysis and subsequent application on macrolevel model. Physical and mathematical models for electromagnetic radiation transfer and heating of a cylinder workpiece made of glass fiber reinforced plastic were developed. These models consider the non-uniform distribution of the electromagnetic field in the processed workpiece of PCM part during microwave radiation exposure. Modeling of the workpiece temperature state for various design-layout schemes of the microwave heating installation was carried out. Analysis of the results showed that the most uniform heating is achieved in a chamber equipped with moving elements, ensuring the rotation of the workpiece around its transverse axis.

References

[1] Museok K 2016 Microwave curing of carbon-epoxy composites: Process development and material evaluation. *London: Imperial College London, Department of Aeronautics.* p 150.

[2] Day R J and Samoladas E 1998 Micromechanics of model carbon epoxy composites cured using microwave heating. *Science and Engineering of Composite Materials.* 7 (1-2) 23–26.

[3] Navarrete A, Mato R B and Cocero M J 2012 A predictive approach in modeling and simulation of heat and mass transfer during microwave heating. *Application to SFME of essential oil of Lavandin Super Chemical Engineering Science.* **68** 192–201.

[4] Morozov G A et al 2011 Microwave treatment thermoreactive and thermoplastic polymers. *Physics of Waves Processes and Radio Engineering Systems (Fizika Volnovikh Processov i Radiotekhnicheskie Sistemi).* 3 114–121.

[5] Mikhailovskii K V and Reznik S V 2014 A prediction of the temperature regimes binder curing of details from polymer composite materials by microwave radiation. *Thermal Processes in Engineering (Teplovie Processi v Technike).* **6** 8 363–368.

[6] Reznik S V and Rumyantsev S A 2014 Mathematical modelling of temperature state of cylinder details from polymer composite materials under microwave heating. *Science and Education (Nauka i Obrazovanie).* 1 6–21.

[7] Guzeva T A 2013 Upgrade of the curing organic plastics technological regimes under microwave radiation. *Thesis on Cand. of Sci. (Moscow: Bauman MSTU).* p 163.

[8] Tsibizova T Y and Guzeva T A 2016 Automatic control systems for technological processes of curing products made of polymeric compositesю *Polymer Sci.* D **9** 1 22–26.

[9] Zimin V N, Koloskov I M, Meshkovsky V E and Usyukin V I 2001 Investigation of natural oscillations for self-deployable truss space antennae. *Trans. Mod. and Sim.* **30** 497–504.

[10] Gottero M, Sacchi E, Scialino G L, Reznik S V and Kalinin D Y 2005 The large deployable antenna (LDA) a review of thermal aspects. *SAE Technical Papers 2005. Proc. 35th Int. Conf. on Environmental Systems, ICES 2005 (11-14 July 2005, Rome, Italy).* DOI: 10.4271/2005-01-2985.

[11] Kotik A, Usyukin V, Vinogradov I and Arkhipov M 2017 Simulation of reflecting surface deviations of centimeter-band parabolic space radiotelescope (SRT) with the large-size mirror. *Proc. of SPIE - The Int. Soc. for Optical Engineering.* **10567,** 105671A 6 p DOI: 10.1117/12.2308118.

[12] Meguro A, Harada S and Watanabe M 2003 Key technologies for high-accuracy large mesh antenna reflectors. *Acta Astronautica.* **53** 11 pp. 899-908 DOI: 10.1016/S0094-5765(02)00211-4.

[13] Reznik S V, Timoshenko V P, Shulyakovsky A V and Denisov O V 2013 Thermal-vacuum tests of hollow composite rods intended for structures in space. *Polymer Science.* D **6** 3 pp. 242–245.

[14] Reznik S V, Prosuntsov P V, Mikhailovsky K V and Shafikova I R 2016 Material science problems of building space antennas with a transformable reflector 100 m in diameter. *IOP Conference Series: Materials Science and Engineering.* **153** 1. 012012 10 p. DOI: 10.1088/1757-899X/153/1/012001.

[15] Mironov R A, Reznik S V, Rukavishnikov R V, Shishulina V A and Zavaruev V A 2016 Optical characterisation of metallic meshes for space antennas transformable reflectors. *IOP Conference Series: Materials Science and Engineering.* **153** 1, 012013, DOI: 10.1088/1757-899X/153/1/012013.

[16] Sayapin S N 2017 Contactless online method for checking-up a net-shaped curtain's tension of a radio-reflecting surface of large umbrella reflector antenna. *J. of Machinery Manufacture and Reliability.* **46,** 6 pp. 519-526.

[17] Sayapin S N and Shkapov P M 2016 Kinematics of deployment of petal-type large space antenna reflectors with axisymmetric petal packaging. *J. of Machinery Manufacture and Reliability.* **45**, 5, 387-397.

[18] Golovatov D, Mikhaylov M and Bosov A 2016 Optimization of technological parameters of impregnation of load-bearing rod elements of reflector made of polymer composite materials by transfer molding method. *Indian J. of Science and Technology.* **9,** 46 No 107492 DOI: 10.17485/ijst/2016/v9i46/107492.

[19] Golovatov D, Mikhaylov M and Bosov A 2016 Computational studies of connections of spatial composite structures of space-rocket techniques. *Indian J. of Science and Technology.* **9** (46), No 107495 DOI: 10.17485/ijst/2016/v9i46/107495.

[20] Bankov S E, Kurushin A A Raschet antenn i SVCH struktur s pomoshchyu HFSS Ansoft. [Numerical analysis of antennas and microwave structures by HFSS Ansoft]. *Moscow, ZAO «NPP RODNIK» Publ.* p 256.

[21] Emetc Yu P Elektricheskie harakteristiki trekhkomponentnoj diehlektricheskoj sredy [Electrical characteristics of a three-component dielectric medium]. *Journal of Applied Mechanics and Technical Physics.* Volume 114, No 3(9) pp 1126–1136.

IOP Conf. Series: Materials Science and Engineering **683** (2019) 012030 doi:10.1088/1757-899X/683/1/012030

The method of reinforcement of holes of mechanical fasteners with composite liners

F A Nasonov[1,2], B B Morozov[2], S V Bukharov[1], AV Zinin[1], K D Kharchenko[1] and G S Piskunov[2]

[1] Moscow Aviation Institute, Moscow, Russia
[2] Sukhoi Design Bureau, Moscow, Russia

E-mail: nasonovf2006@mail.ru, mbbbox@gmail.com, bukharovsv@mail.ru, all-zin@yandex.ru

Abstract. Developed and experimentally justified constructive and technological method of increasing strength and reliability of point mechanical connection of composite panels by thermocompression installation along the contour of the hole of the intermediate element-less rigid composite liner. The effectiveness of the method is provided by reducing the stress concentration in the hole area and confirmed finite element analysis and results of mechanical tests of samples of connection.

1. Introduction

The efficiency of point mechanical joints of parts and assemblies made of polymer composite materials (PCM) is currently limited by a significant reduction in strength in the areas of joint and holes for fasteners due to increased stress concentration. It is noted that the destruction of composite panels with holes for fasteners occurs at loads of 65 ... 70 % of the limit for a solid panel, and after the installation of fasteners – at 40...50 % [1, 2]. Another negative factor that increases the stress concentration and does not allow to realize the advantages of composite structures is the local damage of the composite package when forming holes for fasteners with a mechanical tool [3].

2. Objects and methods of researching

Work was carried out on the production of experimental samples from the panels. Materials panels – epoxy carbon fiber brands VKU-30. The scheme of the reinforcement panel – [00/00/±450/900/±450/00/00]4S, standard regular zone structural element of the airframe of the aircraft. In this regard, it seems promising to develop methods to increase the bearing capacity of the point mechanical connection of composite elements by reducing the stress concentration on the basis of a complex of design and technological solutions [4, 5], primarily by reinforcing the mounting holes along the entire contour with the help of composite liners into which the connecting elements are installed (Fig. 1).

IOP Conf. Series: Materials Science and Engineering **683** (2019) 012030 doi:10.1088/1757-899X/683/1/012030

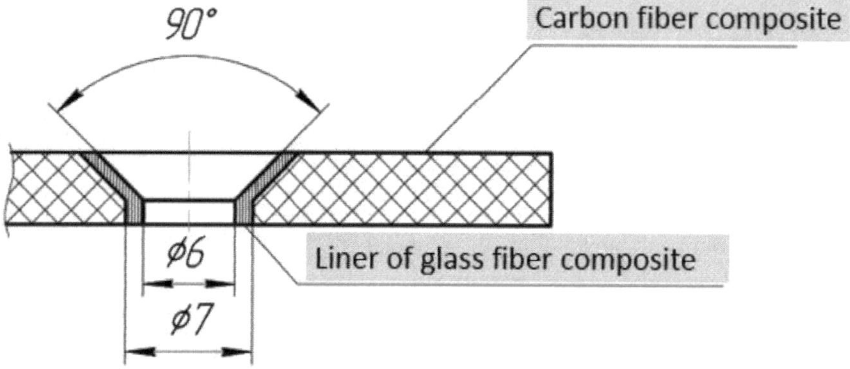

Figure 1. The glass fiber composite liner, istalled to the contour of the hole in carbon composite panel.

Liner material – epoxy fiberglass composite VPS-32K10. The scheme of reinforcement of fiberglass composite liners - circumferential direction ("spiral").
Two design variants of the point connection without reinforcement sleeves – in the form of a cylindrical hole and in the form of a conical countersink bolt with a countersunk head were considered.
The technology of unloading of the zone of point fastening of joints of composite structures by creating a combined connection with an intermediate element - a composite liner, installed by thermal expansion resin transfer molding (TERTM) method along the contour of the hole (Fig. 2).

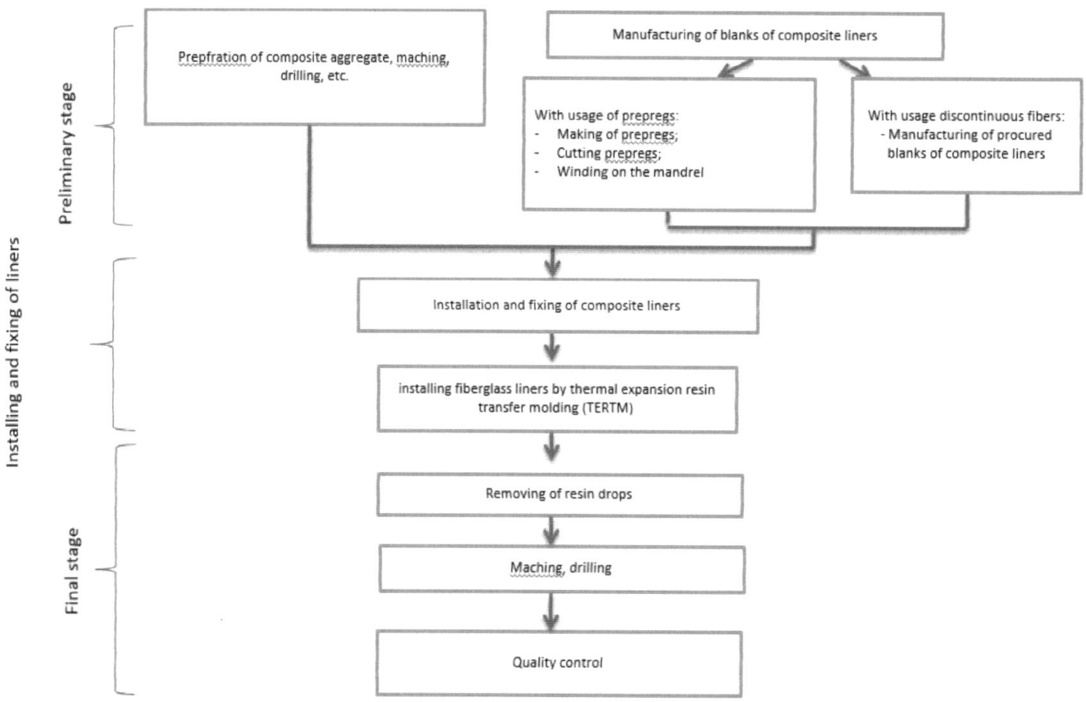

Figure 2. The schematic diagram of the process of installation of composite liners.

3. Results of researching and discussion

The liner made of a less stiff composite than combine the design and ensure the com-patibility of deformation of the entire mounting system, increase the bearing resistance of the connection and allow to reduce the level of stress concentration in the contact area.Intermediate liners made of less rigid composite than the connected structures ensure the compatibility of the deformation of the entire fastening system, increase the resistance to crumple of the joint and allow to lower the level of stress concentration in the contact zone.

To evaluate the effectiveness of the proposed technology by finite element modeling methods in the MSC environment Nastran comparison of stress-strain state in models of carbon-plastic panels containing a free hole for fasteners and reinforced with a fiberglass sleeve is performed (Fig. 3).

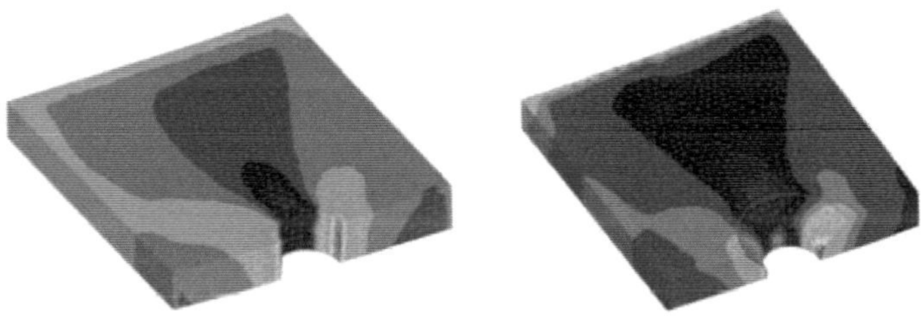

Figure 3. Distribution of equivalent stresses in samples without defects with cylindrical and countersunk holes.

Modeling of operational defects in the area of the hole according to the method of fictitious cracks and finite element calculation of the strength of composite panels with reinforced holes to notches (cracks) confirm these possibilities, because calculations found that for the plates with a cut reinforcement sleeve cylindrical hole in the presence of a crack on the circuit reduces the maximum stress level by 18.7 – 25 %, and for holes with countersink - 31.2 – 37.4 %, and for this type of connection with the sleeves the presence of cuts has virtually no effect on the level of stress [4 - 6].

4. Conclusions

Structural and technological method of increasing of load-carrying ability of point joints for fasteners with defects of production and operational origin by installing composite fiber-glass liners is promising due to its high reinforce and recovery effect for the connection of elements of high-loaded composite structures, in particular for aviation purposes, which are extremely important indicators of strength and reliability. The computer simulation and experimental study of strength properties of the models of composite compounds with different variants of the technological parameters of the installed bushings indicate the possibility of increasing of load-carrying ability of even relatively undamaged samples.

References

[1] Komarov G V 2006 Connection of parts from polymeric materials. *Profession, S.Petersburg, Russia*. 591 p.

[2] Zinin A V, Dedov M N and Abramova E A 2005 Fatigue and fracture of bolted joints of composite structures under low-cycle loading. *Proceedings of the IX international scientific conference dedicated to the 45th anniversary of the Siberian state aerospace University named after academician M. F. Reshetnev, Russia*. pp. 150 – 152.

[3] Zinin A V, Kaykov K V, Morozov L N and Pustokhailov A 2017 A Tomographic study of damage of composite structures in the zone of contact interaction of bolted connections. *Dynamic and technological problems of structural mechanics and continuum mechanics. Proceed-ings of the XXIII international Symposium. A. G. Gorshkov, Russia*. pp. 38-39.

IOP Conf. Series: Materials Science and Engineering **683** (2019) 012030 doi:10.1088/1757-899X/683/1/012030

[4] Nasonov F A 2016 Development of a method of repair of apertures under split connections in products from carbon fiber plastic by means of installation of a fiberglass liners. *New technologies, materials and the equipment of the Russian aerospace branch. All-Russian scientific-practical conference: in 2 volumes, Russia.* pp. 779-784.

[5] Vashukov Y A, Peresypkin V P 2005 Finite element modeling of the stress state in the reinforcement of openings in structural elements made of composite materials. *Proceedings of the Samara scientific center Russian Academy of Sciences, 7, No. 2 . In: Proceedings of the Samara scientific center Russian Academy of Sciences.* No. 2, pp. 36 -44.

[6] Kotov P I, Zinin, A V and Sukhov S V 2012 Practical fracture mechanics. *Moscow.: MATI.* **2**, (218).

Advanced repair methods using wear-resistant polymeric composite materials

V A Ivanov[1] and A B Tulinov[2]

[1] Bauman MSTU

[2] ZAO MMK "Mosintrast"

Abstract. The article is devoted to the technology of eliminating defects in equipment used in various industries and utilities. Possible defects are listed, the types of wear and the degree of their influence on the equipment performance are determined. To repair the defects described, it is proposed to use repair composite materials (RCM) made on a polymer base with various fillers, with the advantage given to composites with ceramic fillers as the most wear-resistant and most conducive to eliminating various types of wear in equipment and piping systems. To solve the problems posed, concrete composite materials are available that are available on the domestic market. It also emphasizes the cost-effectiveness of restoration work using RCM.

Introduction

Process and auxiliary equipment in various industries and public utilities in the process of operation is constantly subjected to various types of wear. The systems of water and heat supply of enterprises are subject to especially intensive wear. They are usually located in the field of activity of the Chief Technologist or Chief Mechanic. The main types of defects in water distribution systems are abrasive, corrosive and cavitation wear of equipment and devices such as pumping pumps, dampers, valves, fittings of various designs and sizes.

Traditional methods of eliminating such defects are welding, brazing and spraying. These recovery methods are expensive, require significant energy costs and are not always effective [5].

The need for technology repair technology and auxiliary equipment that does not require significant financial costs, labor and material resources, is constant. Such requirements are met by the restoration technology using wear-resistant repair composite materials (WRRCM).

Materials and methods

Composite materials with wear-resistant properties can be used when carrying out maintenance and preventive measures in such industries as light industry, utilities, energy, engineering, chemical and oil refining industry, auto and ship repair and others [9, 12].

Polymer-based wear-resistant composites have high physicomechanical and chemical properties, which make it possible to significantly reduce the cost and laboriousness of repairs and reduce the consumption of energy resources and materials for its implementation when restoring technological and auxiliary equipment of various industries, which is based on the following features of their application:

- when repairing using composites, there is no need to disassemble components and assemblies;
- the use of composite materials most often allows not only to exclude welding, surfacing or brazing, but also to carry out repair work on such products and components that cannot be restored using other known methods;
- the use of polymer composites makes it possible to carry out parts restoration, eliminating complex

IOP Conf. Series: Materials Science and Engineering **683** (2019) 012031 doi:10.1088/1757-899X/683/1/012031

technological processes of applying materials and subsequent processing [6].

At the same time, as practice shows, when carrying out restoration work on the proposed energy-saving technology using WRRCM the efficiency and quality of these works increase.

All these advantages confirm the feasibility of the wide use of repair composite materials by the services of the Chief Technologist and the Chief Mechanic at enterprises of various industries.

Process equipment operating in contact with liquids, over time, is subjected to various types of wear under the action of the dynamic pressure of the aquatic environment and heat carriers, the shock effect of abrasive particles (sand, etc.) contained in them, and corrosion and cavitation processes. This leads to the appearance of defects (Fig. 1) in the form of sinks, cracks, thinning of the walls of the devices and often their destruction and, as a result, depressurization and leakage of liquids [13,14].

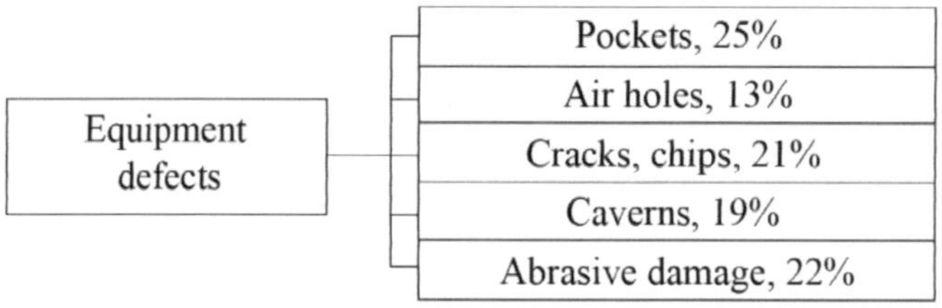

Fig. 1. Defects of equipment

As a result of wall thinning resulting from wear (Fig. 2), flow sections change, resulting in pressure losses in the systems.

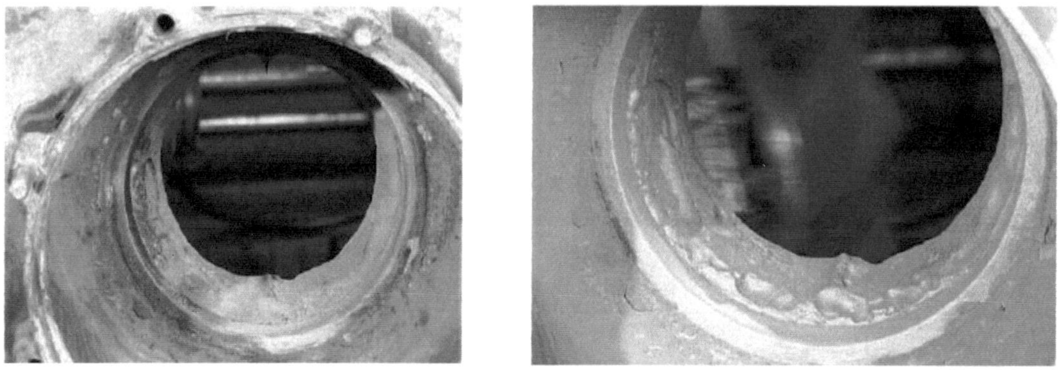

Fig. 2. Wear of the inner surface of the centrifugal pump insert

As a result of various types of wear - abrasive, corrosive, cavitation, the characteristics of the equipment in use change, which negatively affects its performance and often leads to failure. To eliminate the results of wear, proposals were made for the selection of appropriate composite materials and the technology for their use, ensuring the restoration of equipment in various industries.

The analysis showed that at present the main measures to restore worn-out equipment and devices are either its replacement, which is not economically feasible due to significant material and other costs, or repair work using energy-intensive technological processes, such as welding, surfacing, spraying.

Repair technologies using wear-resistant repair composite materials (WRRCM), working on the principle of cold welding (CW), eliminate the need for expensive tooling and equipment and are often an adequate substitute for welding, spraying, welding and soldering [11]. The rational use of their physical and chemical properties will significantly reduce the cost and the complexity of work by 50-60% and reduce by 50% the consumption of materials for repairs [7,8].

A significant number of repair composites have high adhesion to almost the whole variety of both metal and non-metal machine-building materials for various purposes, including functional and structural materials [1, 3].

The presence in the composition of the filler gives it new properties that reveal the technological advantages of cold welding, as compared to thermal (traditional: welding, soldering, brazing, etc.), and adhesive methods for the formation of joints of parts in restoration work [2].

Repair work with the use of polymeric materials in various industries can be divided into three main areas:

- the use of polymeric materials to eliminate defects and flaws in the main products of the enterprise, for example, injection molding;
- repair of the main and auxiliary process equipment of the enterprises themselves

(pumping, ventilation equipment, processing machines, pipelines, technological tanks);
repair of life support systems in urban utilities (including the application of wear-resistant coatings based on composite materials with ceramic fillers).

The technological process of applying repair composites for the restoration of parts, components and equipment in general differs from energy-intensive technical processes, which it can successfully replace in a number of specific cases [15].

Table 1 presents the technological route with a list of mandatory operations, each of which is distinguished by its specificity and the requirement of the mandatory fulfillment of certain conditions necessary for the qualitative performance of the technological process.

Table 1. Process map of repairing parts with metal-polymer materials

Item No.	Description **of actions**	**Process transitions**
1	Preparing the part	Correction of the form. Surface cleaning. Degreasing.
2	Preparation of composition	Manually. Mechanized way. With the help of tubes.
3	Applying the composition	Manually (brush, spatula). Mechanized way.
4	Layer formation for composition	Under pressure. W/o pressure Fixed and movable mandrels.
5	Curing of the composition	Cold Hot Mixed
6	Part processing	Deburring and sagging. Mechanical processing.

The conducted analysis of the possibilities of using WRRCM in various industries revealed a wide area of their use.

The works related to the elimination of the results of various types of wear are of great interest. In this case, as shown by the results of experiments, it is most appropriate to use the composite material "Chester-Ceramic", which was carried out in practice. The restoration method proposed in the article was used at Zhytomyr Cardboard Factory, LLC when carrying out repair measures to eliminate wear caused by the effects of corrosion on the seating surfaces of the press shaft of a paper-making machine (PM). Such a defect led to the beating of the shaft, since not only the inner surface of the shaft was worn out, but also the outer surface of the axle mounted inside the shaft, as well as the mounting holes were broken.

Figure 3 shows the restored trunnion with the applied WRRCM "Chester Keramik-T", and Figure 4 shows its mounting on the shaft.

Fig. 3. Trunnion with WRRCM applied

Fig. 4. Mounting trunnion on the shaft

The process map of the repair work is presented in table 2.

Table 2. Process map to restore the press shaft of PM

No	Operation name	Used tooling, tool, materials.
1	Detection, external inspection	Measuring tape, caliper
2	Cleaning the trunnion and shaft surfaces	Electric drill with bits, metal brushes, sandpaper
3	Degreasing.	Cleaner F-5 f. "Chester Molecular", rags
4	Application of the release agent on the inner surface of the shaft	Release agent f. "Chester Molecular", brush, hot air gun
5	Measurement of the inner and outer diameter of the shaft	Caliper
6	Preparation of repair mortar	Tinax plate, spatulas, material Chester Ceramic (2 components)
7	Application of composite material on the trunnion and shaft surfaces	Spatulas, composite material (mixed)
8	Assembly of trunnion with shaft	Lifting device, wrenches, torque wrench
9	Removal of excess material	Spatulas, rags
10	Visual inspection after assembly	

Analysis of the WRRCM available on the market revealed that the most preferred company supplying the range of repair composites to Russia is the Polish company Chester Molecular. The materials of this manufacturer are distinguished by stable high quality, as well as in their operational and physico-mechanical parameters are not inferior to the materials supplied to our country by such well-known companies like Durmetal (Switzerland), Belzona (USA), Diamant (Germany), Loctite (USA) and others, but significantly cheaper than these analogues, i. e. according to the most important price / quality ratio, they turn out to be the most acceptable.

Analysis of the characteristics of repair composite materials of various manufacturers showed that composites with ceramic fillers are most preferable to eliminate wear of metal surfaces in equipment. Such materials are composite materials of the company “Chester Molecular” (Poland), namely: “Metal Ceramic T” - a pasty consistency; “Metal Ceramic F, FHT, FSL” - these three types of materials have a liquid consistency. The main physicomechanical characteristics of these materials are presented in table 3.

Table 3. Physical and mechanical properties of materials Metal Ceramic

Material groups	Metal Ceramic			
Names of materials	Ceramic T	Ceramic F	Ceramic FHT	Ceramic FSL
Consistency	Paste	Liquid		
Specific composition weight, (g / cm3)	2.03	1.90	2.00	2.00
Colour	Grey	Grey, blue	Grey	Grey, blue
Mixing proportions (basis; activator) by weight by volume	2,5:1 2:1	9:1 packaging	9:1 packaging	9:1 packaging
Viability of the composition at 20 °C, min	20	15	20	50
Cure time before machining, h	3.5	3.5	4	7
Maximum temperature resistance, ° C	200	200	220	200
Operating temperature, °C	-50 +150	-50 +150	-50 +180	-50 +150
Tensile Strength, MPa in compression (according to ICO 604) in bending (according to ICO 178) in shear (according to ICO 4587)	 144 90 22.8	 120 110 22.9	 120 105 15.4	 120 110 24
Period of storage, month	36			
Packing, kg	1; 2; 5.	0.5; 1; 3.	0,5; 1.	0.5; 1; 3.

Chester-Ceramic materials are successfully used to restore ship propellers, rotor blades, to restore corroded surfaces. Therefore, they can be used to restore the equipment of water and heat supply systems, such as pump housings, heat exchangers, dampers, valves and other units.

Wear resistance should be attributed to the operational characteristics of composites. In order to determine it with the studied materials, a series of experiments was conducted. Samples for these tests are made of aluminum corners with a face size of 35x15 mm (Fig. 5-b). In the center of one of the sides of the corner, a groove was cut in the holder for the 77 MT-1 friction machine (Fig. 5-a) [10].

a) b)

Fig. 5. 77 MT-1 (a) friction machine and sample with applied material for testing

For the study of materials in friction with the reciprocating motion of a moving specimen (wear resistance test), a 77 MT-1 friction machine was used, shown in Fig. 5-a. The ultimate loads for a material or coating can be determined on a machine with a reciprocating motion 77 MT-1. The 77MT-1 friction machine consists of three units: electric motor, gearbox and friction machine itself [4]. The general view of the friction pair is shown in Fig. 6

Fig. 6. Friction pair

Results

The test results of the WRRCM samples were determined, taking into account the average mass loss gэ g, from the results of weighing before and after wear of all samples tested at a certain speed and duration of wear:

$$g_э = \frac{1}{n}\sum_{i=1}^{n} g_{эi} ,$$

where gэi is the mass loss for individual samples, g; n is the number of samples [4].

During the wear resistance tests, the following results were obtained:

IOP Conf. Series: Materials Science and Engineering **683** (2019) 012031 doi:10.1088/1757-899X/683/1/012031

Table 4. Sample test results report with applied materials of Chester Metal Ceramic F, FSL, FHT, T
The initial mass of the sample is m0 = 8.9 g.

Item No.	Sample mass in the test process, g				Time of the test phase (h)
	Ceramic F	Ceramic FSL	Ceramic FHT	Ceramic T	
1	8.8	8.8	8.8	8.8	1
2	8.7	8.7	8.8	8.8	2
3	8.6	8.7	8.8	8.7	3
4	8.6	8.6	8.7	8.7	4
5	8.6	8.6	8.6	8.7	5
6	8.5	8.5	8.6	8.6	6
7	8.5	8.5	8.6	8.5	7
8	8.5	8.5	8.5	8.5	8
9	8.5	8.4	8.5	8.5	9
10	8.4	8.3	8.4	8.4	10

The results of experiments for materials Ceramic F and Ceramic T are reflected in the graph (Figs. 7-8). The result graphs for the materials Ceramic FSL and Ceramic FHT are of a similar nature.

Change in mass of the sample during the test

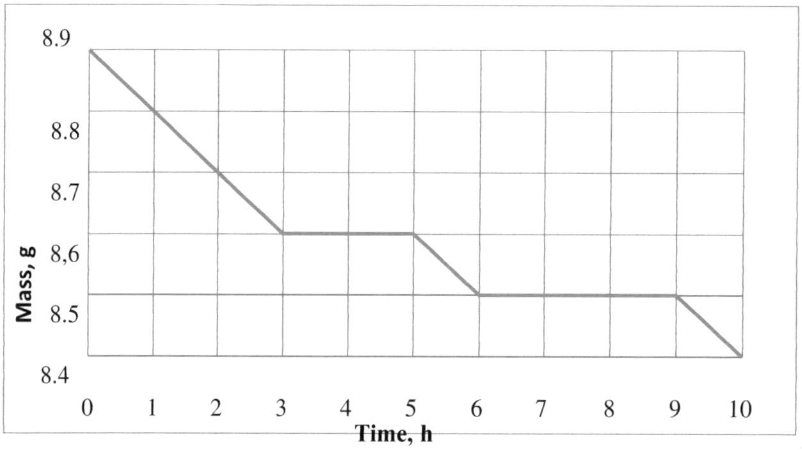

Fig.7. Weight loss of a sample with Ceramic F applied material within 10 hours

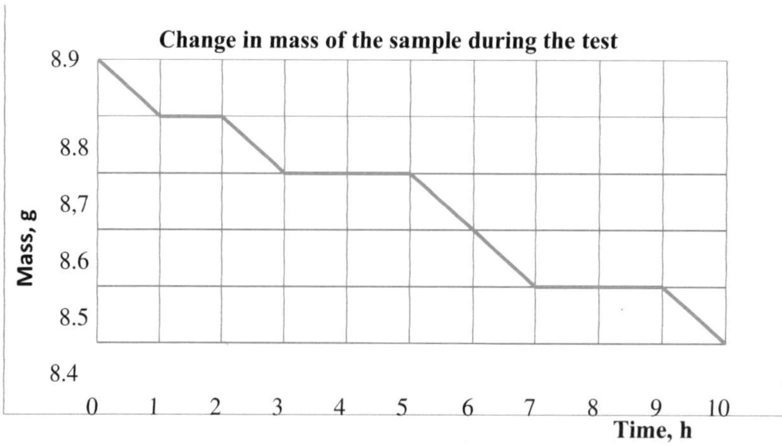

Fig.8. Weight loss of a sample with Ceramic T applied material within 10 hours

Thus, depending on the thickness of the applied layer of wear-resistant composite material, it is possible to predict the service life of reconditioned equipment about 8–12 months.

Conclusions

The economic effect from the use of repair composite materials in the repair of standard units and parts of technological and auxiliary equipment over the past 10 years amounted to 15 million rubles. Experimentally obtained high physicomechanical and operational indicators of the used repair compositions, as well as the economic efficiency of their implementation, confirm the wide range of applications and the rationality of this technology when carrying out repair and restoration work in various industries.

References

[1] Andrews E H, Kinloch A J 1974 Mechanics of elastomeric adhesion J Polymer Sci ,Polymer Symp **46** pp 1-14.

[2] Astrop, A. W., 1974 *Adhesives and steel replace solder and brass*. Mach. Prod. E., **124** (3192), 106— 112.

[3] Bascom W D, Patrick R L 1974 *The Surface Chemistry of Bonding Metals with Polymer Adhesives*. Adhesives Age.**17**.- **№10** 25-29.

[4] Garkunov D N 2001.Tribotechnika (wear and weariness) (Textbook. - 4th ed., Rev. and add. M. "Publishing House of the Moscow Agricultural Academy", 616 p., Ill. 280).

[5] Goncharov A B, Kulagin M V 2003 *Composite materials for repair work.*"Pulp.Paper. Cardboard", **9-10.**

[6] Ivanov V A , Tulinov A B 2018 *Repair of cylinder blocks of internal combustion engines using repair composite materials.* Industrial Service. no. 3 (68).

[7] Ischenko A A 2004 New polymeric materials in the practice of repair of industrial equipment. Bulletin of Engine Building **3** 130-132.

[8] Ischenko A A 2007 *Technological basis for the restoration of industrial equipment with modern polymeric materials.* Mariupol: PGGU 250.

[9] Goncharov A B, Tulinov A B , Ivanov V A 2016 *Restoration of cone crusher support bowl wear.* (Mining information and analytical bulletin (scientific and technical journal). № 9. p 22-28.

[10] Malysheva G N 2000 *Method of assessing the durability of adhesive joints.* Metal technology. **1** 10-16.

[11] Motovilin G V 1974 et al. *Restoration of automotive parts with polymeric materials.*Moscow, Transport, 180 .

[12] Nabatnikov Yu F, Goncharov A B, Tulinov A B, Ivanov V A 2017 *Progressive technological methods of cone crusher recovery. Mining Journal..* **6** 68-72.

[13] Tulinov A B, Ivanov V A 2013 *Possibilities of eliminating defects in heating equipment with composite materials.* Service in Russia and abroad. **1 (39)** 162-173.

[14] Tulinov A B, Ivanov V A 2013 *Restoration of surfaces of heat supply units with composite materials with ceramic fillers.* In the collection: Modern problems of tourism and service. Materials of the All-Russian Scientific Conference of graduate students and young scientists.. 166-169.

[15] Tulinov A B, Ivanov V A 2018 *The results of the study of physical and mechanical properties of anaerobic materials.* (Industrial Service.. **1 (66)** 13-18.

High density polyethylene modified with cyclohexyl dichloro-phosphonate

A Kh Shaov[1], A M Kharaev[1], T A Borukaev1, A S Borodulin[2], and A N Kalinnikov[2]

[1]Kabardino-Balkarian State University named after H.M. Berbekova, Nalchik

[2]Interdisciplinary Engineering Center "Composites of Russia" of Bauman Moscow State Technical University, FGAOU VO

Bauman Moscow State Technical University (National Research University)", Moscow.

E-mail: ah_shaov@mail.ru

Abstract. A chemical modification of high-density polyethylene with cyclohexylphosphonic acid dichloride was performed during high-temperature extrusion. The physicomechanical properties and thermal stability of the modified polyethylene were studied.

Keywords: high density polyethylene, chemical modification, cyclohexyl dichlorophosphate, mechanical properties, thermal stability

Introduction

Organic phosphorus compounds are widespread stabilizers and modifiers of polymeric materials of various structures and classes. The most studied among them are derivatives of trivalent phosphorus. As regards pentavalent phosphorus compounds, they are studied to a much lesser extent.
This paper is an integral part of the whole scientific field that we have been developing over the past 20 years or so [1-26].

Experimental

When extruding a mixture of polyethylene and an organophosphorus compound, phosphorylation (chemical modification) of the polymer occurs at a multiple bond, which is always formed during the heat treatment of polyethylene. This is confirmed by the fact that at low concentrations of cyclohexyl dichlorophosphonate a decrease in unsaturation in polyethylene is noticeable. A further increase in the concentration of dichlorohydride introduced into polyethylene leads to an increase in polymer unsaturation. We associate the latter circumstance with the subsequent process of dehydrochlorination after the attachment of the phosphoryl group to polyethylene.
One of the main performance characteristics of polymeric materials is impact strength; therefore, modified high-density polyethylene was subjected to a study on high-speed three-point bending by the method of Charpy.
The nature of the effect of the organophosphorus compound on the deformation and strength properties of HDPE was evaluated by the following characteristics: impact strength (Ap), modulus of elasticity under bending (E), maximum relative deformation under bending (εp), breaking stress at bending (σp), values calculated using the well-known calculation formulas [1]:

$$A_p = W_p / B(D-a); \qquad E = PL^3 / [4\delta B(D-a)^3];$$
$$\varepsilon_p = 6\delta(D-a)/L^2; \qquad \sigma_p = 3P_\delta L / [2B(D-a)^2],$$

where Ad is the impact strength; Wd is the sample failure energy; B is the width of the sample; D is the sample thickness; a is the size of the notch; P is the load on the initial linear part of the diagram; δ is the flexure of the sample in its middle; L is the distance between the pendulum supports (span); εd is the fracture strain; σd is the breaking stress; Pd is the load of failure.

The results of studies to determine the physicomechanical characteristics of phosphorylated compositions based on high density polyethylene are shown in Figures 1-2.

It is found that the relative elongation at failure of the samples mainly remains at the level of the original polyethylene, regardless of the degree of extrusion. It is interesting to note that the limiting characteristic of the polymer's workability — the limit of forced elasticity σB — is somewhat higher or remains at the level of the pure polymer.

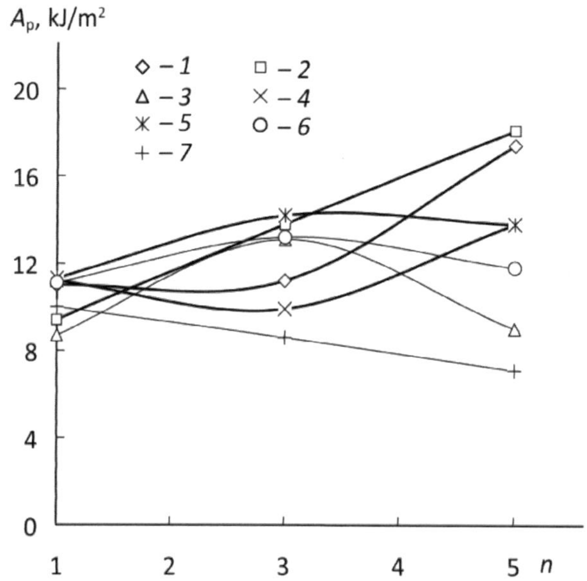

Fig. 1. Dependence of the impact toughness Ap on the extrusion ratio of n HDPE phosphorylated by cyclohexyl dichlorophosphonate. FOC content, wt%: 0 (1); 0,1 (2); 0,3 (3); 0,5 (4); 1 (5); 3 (6) and 5 (7).

The dependences of the toughness of phosphorylated HDPE from n shown in Figures 1-2 are significantly different from each other. From Figure 1 it can be seen that with respect to toughness when modifying polyethylene, the so-called "small additives effect" is not immediately observed when n = 1: impact strength remains at about the same level. Moreover, there is no significant increase in toughness for other compositions with cyclohexyldichlorophosphonate. The impact on the toughness of the multiplicity of processing is ambiguous: for the composition with 0.1% FOC, a linear increase in impact toughness is observed, in the initial HDPE, the value of toughness is stable to n = 3, and then at n = 5 its increase is observed. A weakly expressed extremum (maximum) is observed for compositions with 0.3–3% of an organophosphorus compound, and at its content of 5 %, a linear decrease in impact strength with increasing n occurs.

It should be noted that impact strength is an integral energy characteristic of the deformation-strength behavior of a material at high loading rates, and it largely depends on the contribution of plastic deformation mechanisms. Comparing the dependencies on n values of the elastic modulus (Fig. 2) and fracture stress, shown in Figure 1, one may notice a decrease in the modulus values for all compositions compared to the original HDPE and almost the same pattern for breaking strength. Bearing in mind the stability of the values of forced elasticity for almost all the studied samples, we can assume a complex effect on the deformation-strength properties of polyethylene with an organophosphorus compound as structural features and composition of FOC, which can interact with the polymer and chemically, and not least the process factor the multiplicity of processing. In this case, the observed decrease in toughness and modulus of elasticity can be attributed to the chemical interaction of FOC and the polymer matrix, leading to a decrease in the molecular mass distribution. Obviously, this effect of introducing cyclohexyl dichlorophosphonate into polyethylene prevails over plasticization, which could be attributed to a decrease in the elastic modulus.

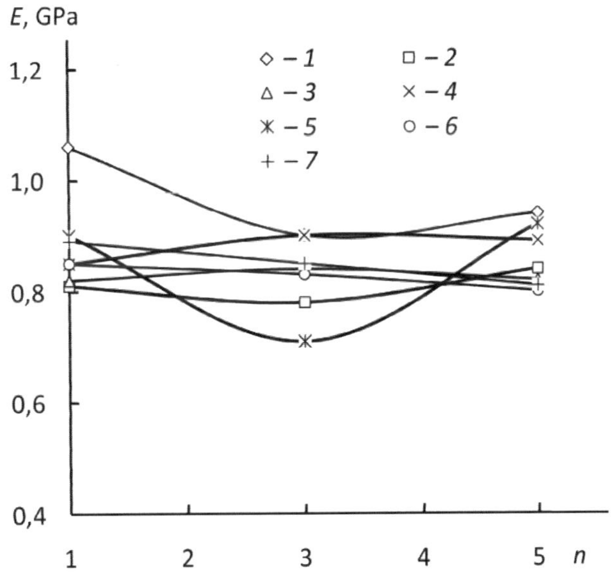

Fig. 2. Dependence of the modulus of elasticity in bending E from the rate of extrusion of n HDPE phosphorylated by cyclohexyl dichlorophosphonate. FOC content, wt%: 0 (1); 0,1 (2); 0,3 (3); 0,5 (4); 1 (5); 3 (6) and 5 (7).

During the processing of polyethylene unsaturated bonds in it can appear, which are the result and signs of the flow of destructive processes. The infrared spectrometry method allows detecting and quantifying three types of multiple groups: 1) vinylidene RRIC = CH2 in the region of 888 cm-1; 2) vinyl RCH = CH2 in the region of 909 cm-1; 3) trans-vinylene RCH = CHRI in the region of 965 cm-1.

Vinylidene groups are formed by β-cleavage of tertiary radicals, for example, RCH2C*(CH2RII)CH2R → RCH2C(=CH2)CH2RI + *RII. Vinyl groups can be formed during the decomposition of secondary radicals: RCH2C*HCH2R → RCH2CH=CH2 + *RI. Secondary radicals can also be sources of trans-vinylene groups according to the general scheme: RCHC*H=CHRI + *RII.

For the quantitative determination of the named groups with a multiple bond, the baseline method is used.

Another method for quantifying destructive processes in polyethylene can be the number of methyl groups per 100 carbon atoms (CH3/100C), also determined using the baseline method at the maxima of the absorption bands in the 1369 and 1378 cm-1 region on the IR spectra of film samples.

The content of methyl groups can serve as a characteristic of the molecular weight of polyethylene: the more methyl groups, the lower the molecular weight, i. e. during destructive processes, low molecular weight fractions are formed, and methyl groups can be not only terminal, but also lateral branches.

The experimental results obtained are shown in the form of graphical dependencies in Figures 3 through 9. It is established that the content of carbonyl groups in polyethylene after a single extrusion is 0.16%. The total number of double bonds per 100 carbon atoms is 0.280. Approximately 82% of multiple bonds are in the vinyl group. The value of CH3/100C in this case is 0.255.

Interesting results were caused by the phosphorylation of polyethylene by cyclohexyl dichlorophosphonate: with an increase in the concentration of the organophosphorus compound, the total content of double bonds noticeably increases, mainly due to the vinylidene and trans-vinylene groups. This fact indicates an increase in the splitting of secondary radicals and the β-splitting of tertiary radicals in the macrochain.

The number of methyl groups is predominantly reduced, i. e. in the decomposition of the macroradicals mentioned, along with groups with vinylidene and transvinylylene bonds, other radicals are formed, to which cyclohexyl phosphonic acid dichloride molecules join without the formation of terminal methyl groups.

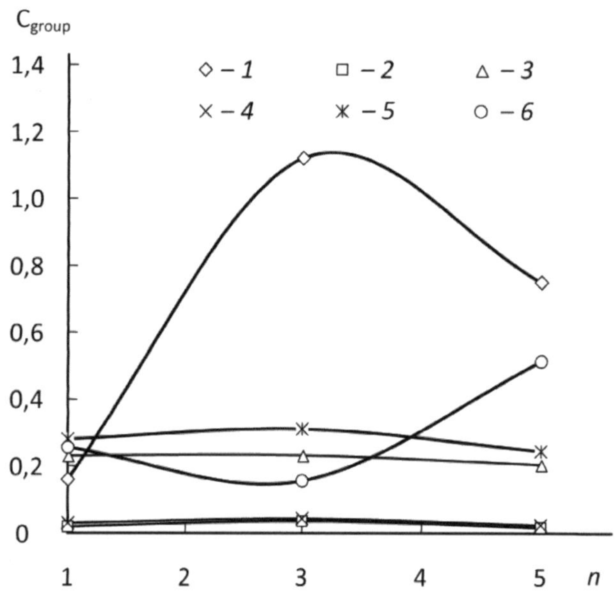

Fig. 3. Dependence of the concentration of different molecular groups on the extrusion ratio n for the initial HDPE: 1 – CO; 2 – (RRIC=CH2)/1000C; 3 – (RCH=CH2)/1000C; 4 – (RCH=CHRI)/1000C; 5 – (∑C=C)/1000C; 6 – CH3/100C

From a practical point of view, it is important to consider the possibilities of regulating the defects in the structure of polyethylene, having as a measure of defects the concentration in the polymer (or a composition based on it) of the indicated groups. It is interesting to trace the influence of two factors — the introduction of a modifying additive and the "technological" factor — the multiplicity of processing by extrusion.

Figures 3 through 8 show that the results of the simultaneous effect of these factors are ambiguous, but the possibility of controlling the structural properties of the polymer using additives and stabilizers and modifiers, as well as the necessary processing ratio, is traced. For a specific consideration of this issue, it is important to establish a correlation between the concentration of these groups and the level and level of operational characteristics of the modified material.

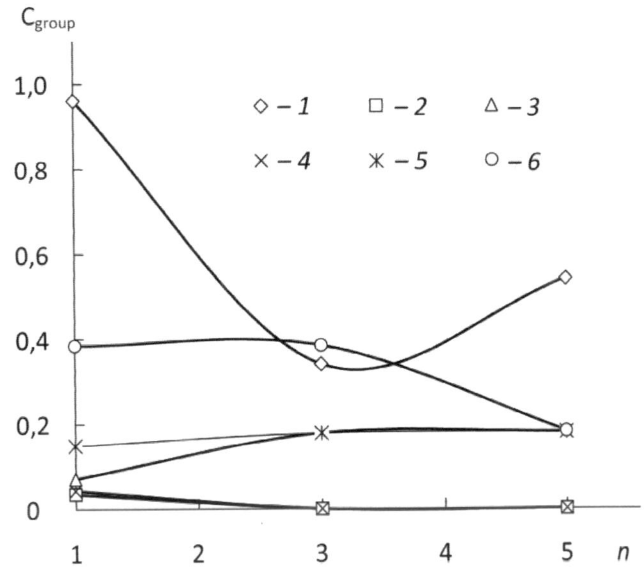

Fig. 4. Dependence of the concentration of different molecular groups on the extrusion rate n for samples of HDPE + FOC with a content of 0.1% cyclohexyldichlorophosphonate: 1 – CO; 2 – (RR^{I}C=CH2)/1000C; 3 – (RCH=CH2)/1000C; 4 – (RCH=CHR^{I})/1000C; 5 – (∑C=C)/1000C; 6 – CH3/100C

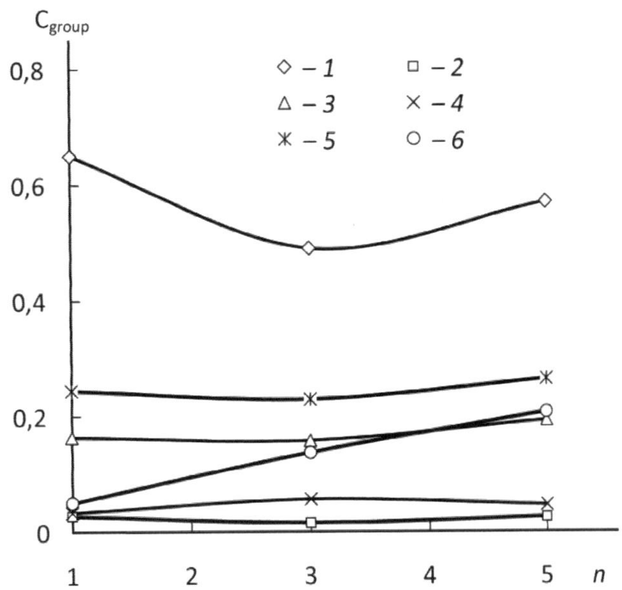

Fig. 5. Dependence of the concentration of different molecular groups on the extrusion rate n for samples of HDPE + FOC with a content of 0.3 % cyclohexyldichlorophosphonate: 1 – CO; 2 – (RR^{I}C=CH2)/1000C; 3 – (RCH=CH2)/1000C; 4 – (RCH=CHR^{I})/1000C; 5 – (∑C=C)/1000C; 6 – CH3/100C

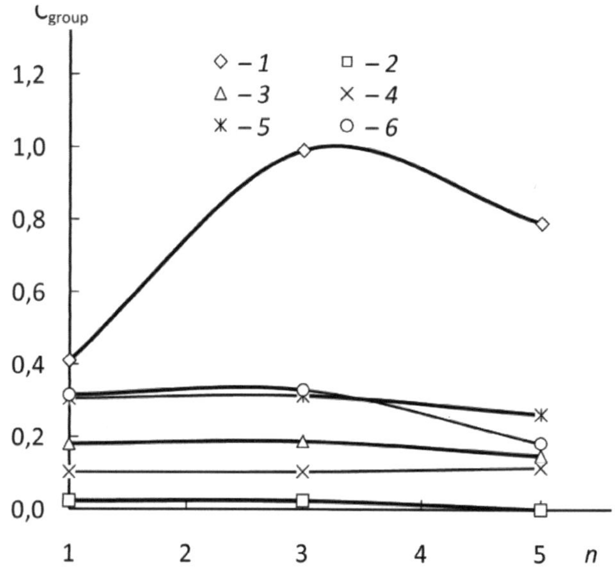

Fig. 6. Dependence of the concentration of different molecular groups on the extrusion rate n for samples of HDPE + FOC with a content of 0.5 % cyclohexyldichlorophosphonate: 1 – CO; 2 – ($RR^{I}C{=}CH_2$)/1000C; 3 – ($RCH{=}CH_2$)/1000C; 4 – ($RCH{=}CHR^{I}$)/1000C; 5 – (∑C=C)/1000C; 6 – CH_3/100C

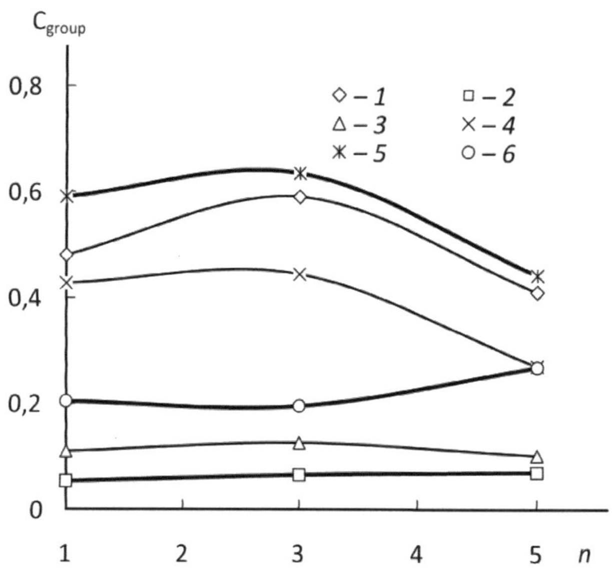

Fig. 7. Dependence of the concentration of different molecular groups on the extrusion rate n for samples of HDPE + FOC with a content of 3 % cyclohexyldichlorophosphonate: 1 – CO; 2 – ($RR^{I}C{=}CH_2$)/1000C; 3 – ($RCH{=}CH_2$)/1000C; 4 – ($RCH{=}CHR^{I}$)/1000C; 5 – (∑C=C)/1000C; 6 – CH_3/100C

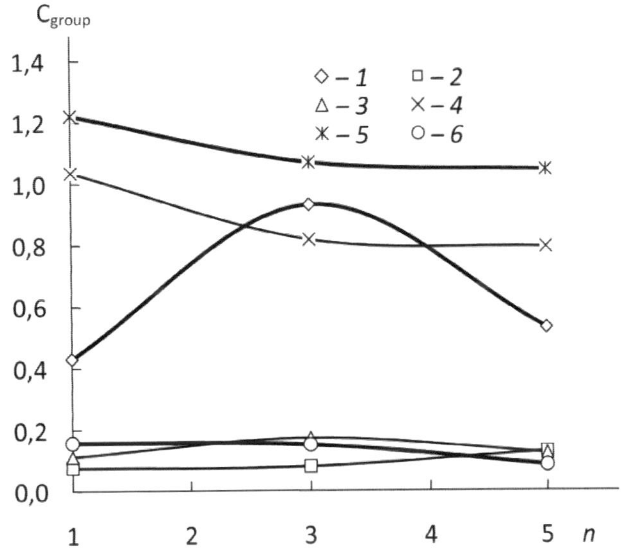

Fig.8. Dependence of the concentration of different molecular groups on the extrusion rate n for samples of HDPE + FOC with a content of 5 % cyclohexyldichlorophosphonate: 1 – CO; 2 – ($RR^{I}C{=}CH2$)/1000C; 3 – (RCH=CH2)/1000C; 4 – ($RCH{=}CHR^{I}$)/1000C; 5 – (∑C=C)/1000C; 6 – CH3/100C

Results and Discussion

1. The use of cyclohexyl dichlorophosphonate as a modifying and stabilizing additive to high-density polyethylene is fully justified, since the technology of the process is not complicated.
2. We characterized the interaction of the organophosphorus compound and the initial polymer matrix as plasticization, bearing in mind the effect of the phosphorus derivative on the efficiency of intermolecular, interchain interaction.
3. The factor influencing the behavior of the samples of compositions under high-speed loading is the multiplicity of extrusion, i.e. this procedure shows the technological "stock" of samples of different composition.
4. The possibility of subsequent heat treatment (extrusion) allows you to achieve a combination of properties - toughness, stiffness, strength.

This article has been prepared within the framework of Agreement No. 14.577.21.0241 dd September 26, 2017.
"Development of a new high-performance technology for obtaining heat-resistant dielectric superstructure polymers of long service life", unique identifier of the work (project) RFMEFI57717X0241, funded by the Ministry of Education and Science of the Russian Federation.

References

[1] Shaov A Kh 1999 *Modification of aromatic polyesters and polyolefins with organic derivatives of pentavalent phosphorus* Dissertation of a doctor of chemical sciences KBSU Nalchik 314
[2] Shaov A Kh, Dolbin I V, Bedanokov A Yu 1999 *Effect of cyclohexylphosphonic acid and its salts on the dielectric and physico-chemical characteristics of high-density polyethylene//* Bulletin of KBSU Series of Chemical Science Iss. 3 42-44
[3] Shaov A Kh, Sarieva Z I, Gurdaliev Kh Kh 2002 *The Research of the Character of the Influence of Phosphinates and Phosphinoxyde on Mechanical Properties of Polyethylene High Density* Nova Sci Publ Aging of polymers, polymer blends and polymer composites. New York 2 167-170

[4] Shaov A Kh, Kodzokova E Kh 2004 *Organic derivatives of pentavalent phosphorus as stabilizers and modifiers of polymeric materials (Review)* Part 1 Plast masses **12** 21-34
[5] Shaov A Kh, Kodzokova E Kh 2005 *Organic derivatives of pentavalent phosphorus as stabilizers and modifiers of polymeric materials (Review)* Part 2 Plast masses **3** 33-41
[6] Shaov A Kh, Gurdaliev Kh Kh , Kharaev A M 2006 *Polymeric nanocomposites, stabilized or ganic de- rivative of five-valent phosphorus* Polymers, polymer blends, polymer composites and filled polymers Synthesis, properties, application. Nova science publishers, Inc New York 121-134
[7] Ozden S, Borukaev T A, Shaov A Kh 2008 *Inhibition of Thermal Oxidation Destruction of PBT/HDPE Blends by Organic Compounds of Five-valent Phosphorus* J. Polymer Processing Society **23** Is.3 339-344
[8] Shaov A Kh, Teuvazhukova A N and Akezheva A A 2008 *The polymeric compositions stabilized nanodimensions phosphor organically by compounds* In: Modern Tendencies in organic and bioorganic chemistry Nova Science Publishers Inc 113-115
[9] Shaov A Kh,. Shetov R A, Tegaev R I 2010 *Research of physic mechanical and morphological proper- ties structures on the basis of polyethylene of high density and cyclohexyl phosphonic acid* Material science synthesis, properties, applicators (Polymer yearbook. **24**) R.A.Pehrick and G.E. Zaikov editors Nova Science Publishers Inc New York 263-272
[10] Shaov A Kh, Shetov R A, Tegaev R I, Balkizova L H, Beslaneeva A N 2011 *Investigation of physi- cochemical and morphological properties of structures based on HDPE and cyclohexyl phosphonic acid* J. of the Balkan Tribological Association **17**(2) 230-239
[11] Shaov A Kh, Beslaneeva A N, Karmokov A M, Shetov R A, Marshenkulov M A 2013 *Study of the physicomechanical properties of high density phosphorylated polyethylene* News of the Kabardino-Balkarian State University. Nalchik: Kabardino-Balkarian State University V. III, **5** 41-46
[12] Shaov A Kh, Beslaneeva A N, Balkizova L Kh, Shetov R A, Marshenkulov M A 2013 *Thermal stability study of phosphorylated HDPE by the IRS method* News of the Kabardino-Balka rian State University Nalchik: Kabardino-Balkarian State University V III **5** 114-119.
[13] Shaov A Kh, Beslaneeva A N, Mikitaev A K, Shetov R A, Balkizova L H 2014 *Study of the phys icomechanical properties of compositions based on high density polyethylene and organophos phorus compounds* Materials Science **9** 14-20
[14] Shaov A Kh, Beslaneyeva A N, Mikitaev A K, Shetov R A and Balkizova L Kh 2015 *Study of Physical and Mechanical Properties of Compositions Based on High Density Polyethylene and Organo- phosphorus Compounds* Inorganic Materials: Applied Research **6**(1) 67-72
[15] Beslaneeva A N, Shaov A Kh, Shetov R A 2015 *Study of the influence of calcium cyclohex ylphosphonate on the HDPE melt flow rate* News of the Kabardino-Balkarian State Univer sity Nalchik: Kabardino-Balkarian State University. **V**(6) 39-42
[16] Shaov A Kh., Borukaev T A, Kharaev A M, Shustov G B, Shetov R A, Khakyasheva E V 2016 *Photo-stabilization of high-density polyethylene with cyclohexylphosphonic acid and its po tassium salts* Proceedings of the Kabardino-Balkarian State University. Nalchik: Kabardino-Balkarian State University V VI **3** 58-63
[17] Beslaneeva A N, Shaov A Kh, Borukaev T A, Shetov R A and Binogerova F A 2016 *Rheological Properties of Compositions Based on High-Density Polyethylene and Calcium Cyclohexyl Phosphonate* J of Characterization and Development of Novel Materials New York **8**(4) 281-286
[18] Shaov A Kh., Borukaev T A, Kharaev A M, Mikitaev A K, Beslaneeva A N 2017 *Study of pho toaging of high density polyethylene containing cyclohexylphosphonic acid and its nickel salts.* Materials Science. **8** 15-19.
[19] Shaov A Kh ,Borukayev T A, Kharayev A M, Shetov R A, Mashukov N I, Shustov G B, Bazheva R Ch, Inarkieva Z I 2017 *Stabilization of high density polyethylene with calcium cyclo hexylphosphonate* /News of KBSU V VII **2** 70-74

[20] Shaov A Kh, Borukayev T A, Kharayev A M, Mikitayev A K and Beslaneyeva A N 2018 *Inves tiga- tion of Photodegradation of High-Density Polyethylene Containing Cyclohexyl Phos phonic Acid and Its Nickel Salts* Inorganic Materials Applied Research **9**(1) 125-129
[21] Nelyub V A 2016 *A study of the microstructure of dressed glass fibers* Polymer Science - Series D **9** (1) 96-100
[22] Nelyub V A 2015 *Determination of adhesion interaction between carbon fiber and epoxy binder* Polymer Science - Series D **8** (1) 6-8
[23] Nelyub V A, Borodulin A S, Kobets L P, Malysheva G V 2016 *Capillary hydrodynamic of oligo mer binder* Polymer Science – Series D **9**(3) 322-325
[24] Nelyub V A, Borodulin A S, Kobets L P, Malysheva G V 2016 *A study of structure formation in a binder depending on the surface microrelief of carbon fiber* // Polymer Science – Series D **9**(3) 286-289
[25] Borodulin, A S, Kalinnikov A N, Bazheva R C at all 2018 *Receipt and investigation of perfor mance characteristics of super constructions polyesters* International Journal of Me chanical Engineering and Technology **9** (13) 1117-1127
[26] Borodulin A S, Kalinnikov A N, Bazheva R C at all 2018 *Synthesis and properties of aromatic poly- ethersulfones* International Journal of Mechanical Engineering and Technology (IJMET) 9 Issue 13 1109-1116

Use of composite materials to increase the effectiveness of variable-geometry aircraft

P V Kruglov[1] and V I Kolpakov[1]

[1] Bauman Moscow State Technical University, 2 nd Baumanskaya st., 5, bld. 1, Moscow, Russia

E-mail: kpv17@bmstu.ru , kolpakov54@mail.ru

Abstract. The results of mathematical modeling of the functioning of the charges using composite materials forming high-speed aircraft of variable geometry are presented. The directional formation of a folded fin in the tail part of the element, providing the aerodynamic stability of the element on the trajectory, is considered as a way to increase the efficiency. The effect of charge parameters on the effectiveness of the action is shown. At the same time, steel claddings of variable thickness and claddings of polymeric material for the formation of a folded fin were considered. The effectiveness of the charges was determined by the shape of the element and its speed. It is found that with an increase in the width of the petal cladding the speed of the element decreases, a linear relationship is presented between the relative width of the cladding and the relative speed. It is revealed that the elongation of the element when varying the width of the cladding varies nonlinearly and there is a range of sizes of the cladding at which the elongation is minimal.

Keywords. Numerical simulation, high-speed aircraft with variable geometry, composite materials, dynamic deformation processes,metal cladding

Introduction

The advanced missile technology uses explosive formed projectile (EFP) [1-3]. With the explosion of such a charge, a high-speed elongated element is formed, changing its geometry during the flight [4-7]. This type of charges is also used in engineering charges for the destruction of concrete and brick buildings, while located at a certain distance from the charge. High-speed elongated elements are formed by high-speed deformation of metal claddings, made, as a rule, of ductile metals, by means of an explosive charge explosive (EX) [8-14]. Aerodynamic stability of high-speed elongated elements is ensured at the location of the center of mass of the element in front of its pressure center. This requirement is usually met by expanding the tail section of an element formed by the explosion, called the fin [15-18]. An additional stabilizing element are folds in the tail part of the element - stabilizers. The directional formation of stabilizers and the analysis of the conditions of their formation is of great scientific and practical importance, since it will allow the creation of high-speed elongated elements that are stable in flight. Abroad, similar elements are created using the claddings on the cladding [15-19]. The effectiveness of the action of projectile-forming charges is increased not only from the standpoint of changing the design parameters, but also by improving the manufacturing technology of cumulative cladding [20, 21], which is the most important part of the charge, as well as charge assembly technologies [22-24]. To study high-speed explosive processes, in order to reduce the cost of experimental research, methods of mathematical modeling of the functioning of the EFP are actively used [7-9, 25-28]. This paper presents the results of a study performed using numerical methods of continuum mechanics, which determines the influence of the structural parameters of the PFC on the shape and stability of high-speed elongated elements obtained from claddings based on the use of composite materials.

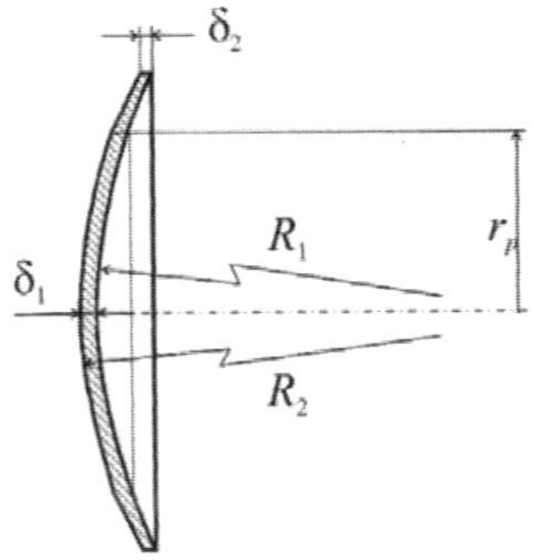

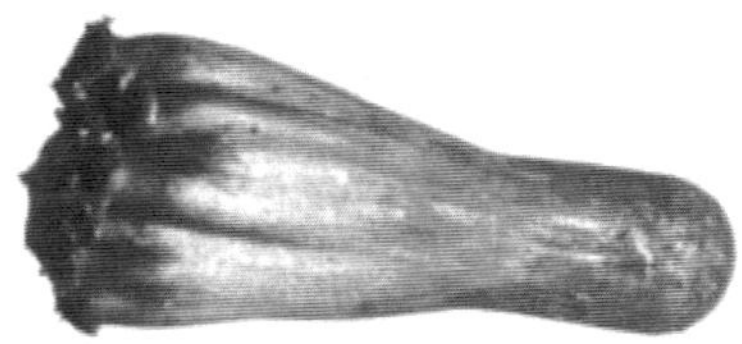

Fig. 1. Spheroconic cumulative cladding **Fig. 2.** The shape of the elongated high-speed element [1]:

EFP , as a rule, consists of a metal cumulative cladding (Fig. 1), explosive charge, body and detonator. The advantages of the elongated form of the element are its greater penetrating ability compared to compact elements of the same mass (Fig. 2). The effect is achieved due to greater length and less loss of speed when moving over long distances. Another important characteristic of a high-speed element is its fullness (Fig. 3) [8]. It is obvious that the higher the fullness of the element, the greater its penetrating effect. For the formation of stabilizers, it is proposed to use a cladding composition of a metal segment cladding and a polymer multilobe cladding installed on the outer surface of the cladding, from the side of the explosive charge.

To determine the parameters of the EFP that provide high-speed elongated elements of greater elongation (ie, greater efficiency) with developed stabilizers, a computational experiment was conducted based on the numerical solution of problems of continuum mechanics.

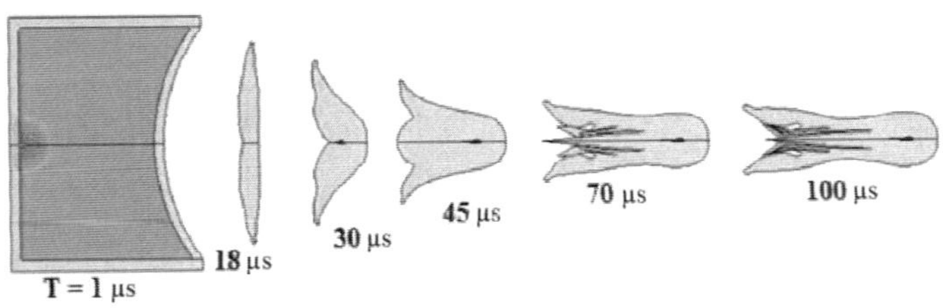

Fig. 3. The process of formation of high-speed elongated elements by the EFP explosion

Calculations were carried out for a model EFP with the following parameters: the diameter of the explosive charge (or the internal diameter of the shell) dз= 62 mm, height H = 50 mm, thickness of the shell δк = 3 mm. The cladding diameter is d=dз=62 mm, the cladding deflection is h = 10.45 mm. There werer used claddings of variable spheroconic thickness (Fig. 1) with such dimensions [8, 9]: with thickness in the central and peripheral parts was δ1 = 2.5 mm and δ2 =2.25 mm, respectively, the radius of transition of the central spherical surface in into the peripheral conical rp = 26 mm. The thickness of the cladding was δн = 1.0 mm, and the width was varied and was s = 2 ... 14 mm.

The following parameters were used in materials models: cladding - steel grades 08кп, 11кп (density ρ0 = 7.85 g / cm3, bulk compression modulus K0 = 175 GPa, dynamic yield strength Y = 0.6 ... 0.9 GPa, shear modulus G = 80 GPa, spall strength σ * = 1.65 GPa , the relative narrowing of the material $\psi = 0.45 \ldots 0.75$); housing - steel 45X; EX - TG40 (density ρBB = 1.68 g / cm3, detonation velocity D_{BB} = 7.85 km/s, heat of explosive transformation Q_{BB} =4.61 MJ / kg); cladding - plexiglass (density ρ0 = 1.15 g/cm3). In the charge, single-point initiation was applied at a distance of 49 mm from the cladding face.

As a result of the charge functioning, the pad degrades, forms an uneven pressure field on the surface of the cladding and stabilizers are created in high-speed element in those places of the cladding that are located opposite the cladding sections (Fig. 4).

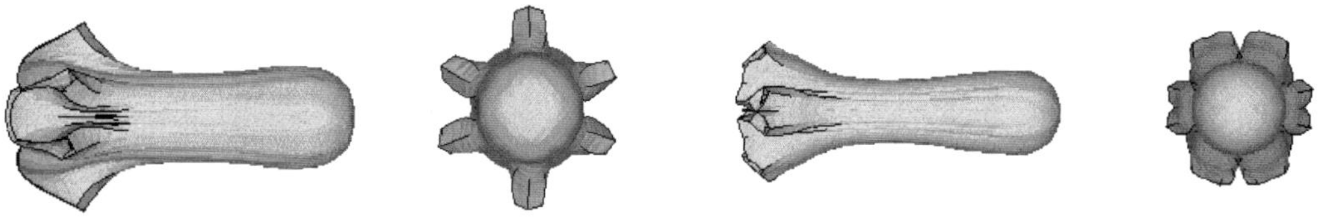

(a) (b) (c) (d)

Fig. 4. Options for the shape of high-speed elongated elements when using claddings of various widths

On the basis of mathematical modeling, high-speed elements with a velocity Vel ≈ 1.9 km / s (Fig. 4, a, b) and elongation l*=lэл/dэл ≈ 4 (ratio of the length of the element lel to its diameter del) were obtained (Fig. 4 , c, d) Figure 5a shows the dependence of the relative velocity of the element Vel * = Vel / Velmax (Velmax = 1.9 km / s) on the relative width of the cladding section s * = 2s / d. It is shown that with an increase in the width of the cladding section, the speed of the elongated element decreases by 1.5%, and the height of the stabilizers increases, which has a positive effect on the stability of the element movement. The elongation of the element also decreases with an increase in the section width, but after a certain width value, the elongation begins to grow again (Fig. 5, b).

The regression dependence between the relative width of the cladding section s * and the relative velocity of the element Vel * is described linearly

$$Vel^* = -0.041s^* + 1.0026. \quad (1)$$

The relationship between the relative width of the petal cladding s * and the elongation of the element l * is described in quadratic form

$$l^* = 11.49s^{*2} - 7.1306s^* + 2.9557. \quad (2)$$

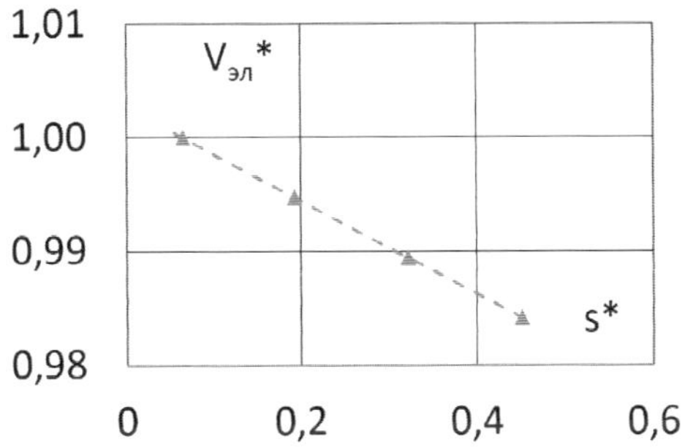

(a)

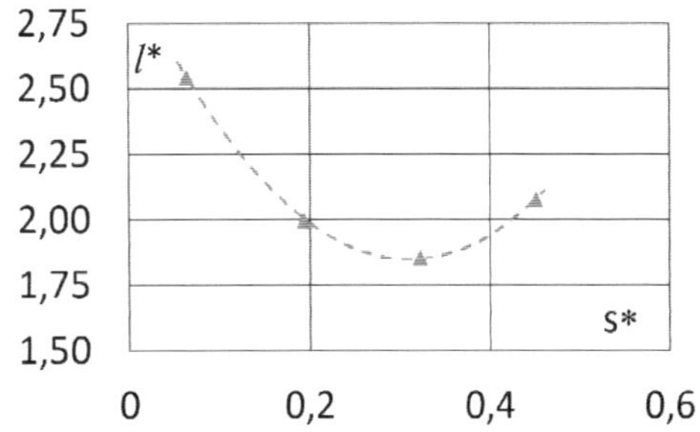

(b)

Fig. 5. The dependence of the velocity of the element (a) and elongation (b) on the relative width of the cladding section

Thus, the controversial effect of the polymer cladding with straight lobes on the efficiency of high-speed aircraft of variable geometry has been revealed. In particular, increasing the width of the cladding section leads to a decrease in speed, but at the same time to an increase in the height of the stabilizer fold. In turn, a decrease in speed leads to a decrease in the depth of the obstacle overcome by such an apparatus, and an increase in the height of the stabilizer increases the accuracy of hitting the barrier from a long distance. Therefore, further research in this area should be continued in the direction of studying the influence of other parameters of the cladding (geometrical, physicomechanical) in order to increase the efficiency of the charges that form the aircraft of variable geometry.

REFERENCES

[1] Ammunition V 1, Selivanov V V 2016 ed Moscow Publishing House of Bauman Moscow State Technical University 506
[2] Kolpakov V I 2012 *Mathematical modeling of the functioning of explosive devices* Science and Education **2** URL: http://technomag.edu.ru/doc/334177.html (date of application 10.12.2016)
[3] Kolpakov V I, Baskakov V D, Shikunov N V 2010 *Mathematical modeling of the operation of shell-forming charges with regard to technological asymmetries Defense technology* **1-2** 82–89
[4] Potapov I V, Kolpakov V I, Shikunov N V, Yakimovich G A 2010 *Mathematical modeling of the explosive formation of steel damaging elements Shock Waves in Condensed Matter* Int Conference Saint-Petersburg – Novgorod (Russia) 184–188
[5] Kolpakov V I 2012 *Features of the deformation and destruction of elongated damaging elements under explosive loading of cumulative claddings* Kolpakov V I XIII Khariton's thematic scientific readings: Proceedings of the Intern. conf. Sarov 532–536
[6] Kolpakov V I 2015 *Analysis of design features of charges with low segment cumulative claddings of tantalum alloys* Kolpakov V I Col. of abstr. of report from Intern. conf. «XVII Khariton's thematic scientific readings» Sarov: FGUP RFYATS-VNIIEF 288–290
[7] Kolpakov V I 2011 *Mathematical modeling of the process of explosive formation of an elongated striking element from high-density material* Kolpakov V I, Goryunov V V Defense technology **2-3** 30-34
[8] Kruglov P V, Kolpakov V I 2017 *Patterns of explosive formation of elongated high-speed elements of steel segment claddings*. Engineering Journal: science and innovations 1-19 URL: http://dx.doi.org/10.18698/2308-6033-2017-12-1714
[9] Kruglov P V, Kolpakov V I 2018 *Analysis of the effect of the profile thickness variation of metal segment claddings on the shape of high-speed elongated elements*. Engineering Journal: science and innovationsp 1-10 URL: http://dx.doi.org/10.18698/2308-6033-2018-7-1782.
[10] 2002 *Physics of the explosion*. **2** Ed. L.P. Orlenko. Moscow: Fizmatlit, **2** 656
[11] Held M 2002 *The shaped charge potential*. 20 Int. Symp. of Ballistics 81-90
[12] Sokolov S S, Sadovoy A A, Chaika T I 2004 *Mathematical modeling of the formation of high-speed compact metal elements*. VANT, ser. Mathematical modeling of physical processes 54–61.

[13] Lee V, Wang S, Lee V, Chen K 2016 The study of the formation of a sustainable explosively formed penetrator with fins. Applied Mechanics and Technical Physics **57**(5) 151–157
[14] Liu J, Gu W, Lu M, Xu H, Wu S 2014 *Formation of explosively formed penetrator with fins and its flight characteristics.* Defense Technology, **10** 119-123 DOI:10.1016/j.dt.2014.05.002.
[15] Mikhalev A N, Podlaskin A B 2009 *Investigations of promising long-range explosively formed penetrators* News of RARAN **1**(59) 3–9
[16] Mikhalev A N 2004 *Projectiles formed by the explosion: aerodynamic properties and methods of formation* (review according to foreign studies) Prepr. FTI named after A.F. Ioffe, St.Petersburg 1775 38
[17] Mikhalev A N, Podlaskin A B 2011 *Studies of flow and aerodynamics of explosively formed penetrators on a ballistic test bench.* Fundamentals of ballistic design:Proc. of All-Russia Scientific Tech. Conf. SPb **1** 100–108
[18] Bender D, Chhouk B, Fong R, Ng W, Rice B, Volkmann E 2001 *Explosively formed enetrators (EFP) with canted fins.* 19 Int. Symp. of Ballistics 755–762
[19] Aubry J and other. Core-forming explosive charge. US patent 4922825, 08.05.1990.
[20] Kruglov P V, Bolotina I A 2017 *Manufacturing technology of disk blanks of variable thickness for compact aircraft.* Engineering Journal: science and innovation 1-11 http://dx.doi.org/10.18968/2308-6033-2017-9-1674.
[21] Kruglov P V, Bolotina I A 2018 *Improving the manufacturing technology of disk blanks for metal cladding of compact aircraft*
[22] http://dx.doi.org/10.18698/2308-6033-2018-8-1787
[23] Kruglov P V, Tarasov V A 2012 *Method of generating design solutions for assembling products using oriented hypergraphs* E-Journal. Science and education **1**
[24] URL: http://technomag.bmstu.ru/doc/260312.html
[25] Kruglov P V, Tarasov V A, Bolotina I A 2012 *The method of forming a set of permissible options for assembling products based on the use of oriented hypergraphs* E-Journal. Science and education **2** URL: http://technomag.bmstu.ru/doc/339658.html
[26] Kruglov P V, Bolotina I A 2016 Application of oriented constraints hypergraphs in the design of manufacturing technology of high-precision structures Engineering Journal:*science and innovation Electronic scientific and technical publication* **5** http://dx.doi.org/10.18698/2308-6033-2016-05-1494.
[27] Hussein G, Hamid A, Hetherington D, Malik A K, Sanaullah K 2013 *Analytical study of the effectiveness of projectiles formed by the explosion.* Applied mechanics and technical physics, **54**(1) 13–24
[28] Pappu M, Murr L E 2002 *Hydrocode and microstructural analysis of explosively formed penetrators. J.of material science* **37** 233–248.
[29] Andreev S G, Boyko M M, Klimenko V Yu 2013 *The possibility of computer simulation of the operation of projectile-forming charges with aluminized explosives in the development of prototype ammunition.* Bulletin of Moscow State Technical University named after N.E. Bauman.Ser. Mashinostroyeniye **2** 108-126.
[30] Hussain G, Hameed A, Hetherington J G, Barton P C, Malik A Q 2013 *Hydrocode Simulation with Modified Johnson-Cook Model and Experimental Analysis of Explosively Formed Projectiles.* J. *of Energetic Materials* **31**143–155 DOI: 10.1080/07370652.2011.606453.

IOP Conf. Series: Materials Science and Engineering **683** (2019) 012034 doi:10.1088/1757-899X/683/1/012034

Evaluation of the effectiveness of the introduction of a discrete filler in the aluminum matrix melt

Yu A Kurganova[1], Yijin Chen1 and S P Shcherbakov[1]

[1] Faculty of "Engineering Technology", Bauman Moscow State Technical University, 5/1, 2nd Baumanskaya Street, Moscow, Russia, 105005

E-mail: kurganova_ya@mail.ru, chenyijin@yandex.ru

Abstract. Experimental samples of metallic composite materials filled with discrete Al2O3 fibers with a diameter of 10–20 nm were obtained by liquid-phase mechanical mixing. A technology has been developed for the effective introduction of a discrete filler into an aluminum melt, which ensures the absorption of the filler by the matrix. Device for the introduction of the filler was designed and manufactured. Recommendations were given on the choice of the composition and technological regimes for the introduction of a light discrete component of a composite material for sufficient assimilation by its matrix and ensuring the homogeneity of the structure. The structure and properties of the obtained samples of the new material have been studied. The changes made by the introduction of the selected nanodiscrete filler into the aluminum melt and the changes made by additional technological methods are analyzed. The effect of introducing discrete Al2O3 fibers into an aluminum melt was demonstrated.

1. Introduction

Aluminum composite materials (ACM) have found application in a number of industrial areas due to the unique combination of properties: low density and high specific strength [1-7]. Development of new functional and structural ACM, occupies a significant place in modern works of researchers around the world [1,3-5,7,8].

Particular attention is paid to the production of aluminum matrix composite alloys with nanoscale reinforcing phases [9-16]. Reducing the size of the filler to the nanoscale increases the length of its borders in a unit volume of the matrix melt, increasing the number of crystallization centers during cooling, providing the desired fine-grained structure. The nanofiller has a high sedimentation stability in the melt, and being in a suspended state, it blocks the diffusion of atoms to nascent and growing crystals. After the transition to the solid state, the nanofiller works as a barrier when dislocations move along the matrix, making it more difficult for them to move, the smaller the distance between the particles. The indicated effects lead to an increase in the mechanical properties with insignificant volumes of injected fillers [17, 18].

The main technological problems of the introduction of discrete fillers in aluminum are to overcome the surface tension forces and ensure the uniformity of their distribution. Analysis of the patent and specialized technical literature has shown that an effective way to increase wettability is doping a matrix alloy or applying special coatings on the reinforcing phase [17-21]. For example, in [21], when magnesium is added to an aluminum alloy, its segregation occurs at the interface, and thus wetting and adhesion are improved. In [19], a significant improvement in wetting in the Al – Al2O3 system was observed when layers of copper were sprayed onto a polycrystalline Al2O3 substrate.

Some technological approaches adapted to the use of a nanofiller are described in the literature, for example, the method of powder metallurgy, the method of mechanical doping, the method of spark plasma sintering, the liquid-phase method of mechanical mixing [1,3,5,7,9-18]. The simplest and cheapest way to obtain ACM is considered to be liquid-phase mechanical mixing, when the filler is introduced into the matrix heated to the molten state.

In the experimental part of this work, it was proposed to use copper as an effective component, which, on the one hand, can act as an alloying element, the effect of which on aluminum is well known, and on the other, an element providing transport functions when introduced.

2. Experimental part

2.1 Materials

Technical aluminum АД0 (1050 ISO) was chosen as a matrix. The chemical composition is presented in table 1.

Table 1. Chemical composition of АД0 according to GOST 4784-97, wt%

Al	Si	Fe	Cu	Mn	Zn	Ti	Mg
>99.50	<0.25	< 0.4	< 0.05	<0.02	<0.07	<0.05	<0.05

Discrete nanofibers of Al2O3 trademark "Nafen" were used as filler. The properties declared by the manufacturer are shown in table 2.

Table 2. Properties of Al_2O_3 nanofiber

Fiber	Phase	Diameter (nm)	Density (g/cm$^{3)}$	Thermal stability	Tensile strength (GPa)	Tensile modulus (GPa)
Al_2O_3	γ	10-20	3.98	up to 1200 °C	12	400

Experimental samples of the composite material were fabricated using a liquid-phase method of mechanical mixing. To solve the problem of the wettability of Al2O3 aluminum fibers, micron powders ПМС-1 copper according to GOST 4960-2009 with a diameter of about 10–20 μm were used, which in turn improved the absorption and distribution of Al2O3 nanofibers in the matrix. Transporting copper powder is pre-rubbed with filler, forming a certain conglomerate. The scheme for obtaining a conglomerate is shown in Figure 1.

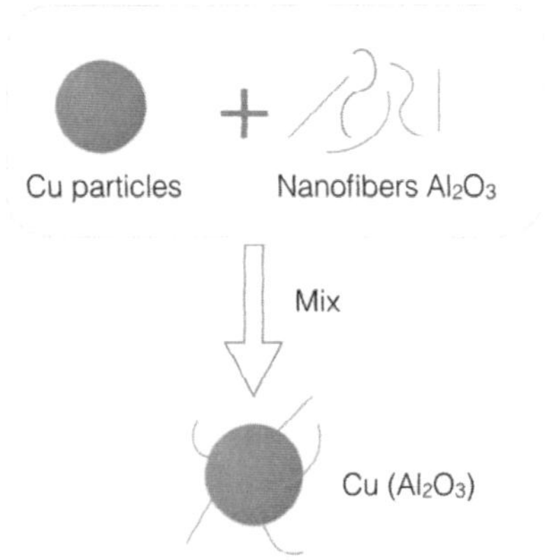

Figure 1. Scheme for obtaining a conglomerate

For comparative analysis, four compositions were melted under identical conditions: Table 3.

Table 3. Composition of the experimental samples

Sample no.	Composition of the samples (wt.%)	
	Matrix	Filler
1	АД0	-
2	АД0	1.14%Al_2O_3
3	АД0	1.198%Cu
4	АД0	1.26%(Al_2O_3+Cu)

3. Examination of samples of selected compositions

3.1 Structural Analysis

For comparative analysis under identical conditions, according to the plan of the experiment, four compositions were melted: matrix (АД0), matrix with the addition of Al2O3 nanofibers (АД0 + 1.14%), matrix with transport powder (АД0+ 1.198% Cu) and matrix with Al2O3 on a copper transport powder (АД0+ 1.26% (Cu + Al2O3)). The microstructures of the experimental samples after etching are shown in Figure 2.

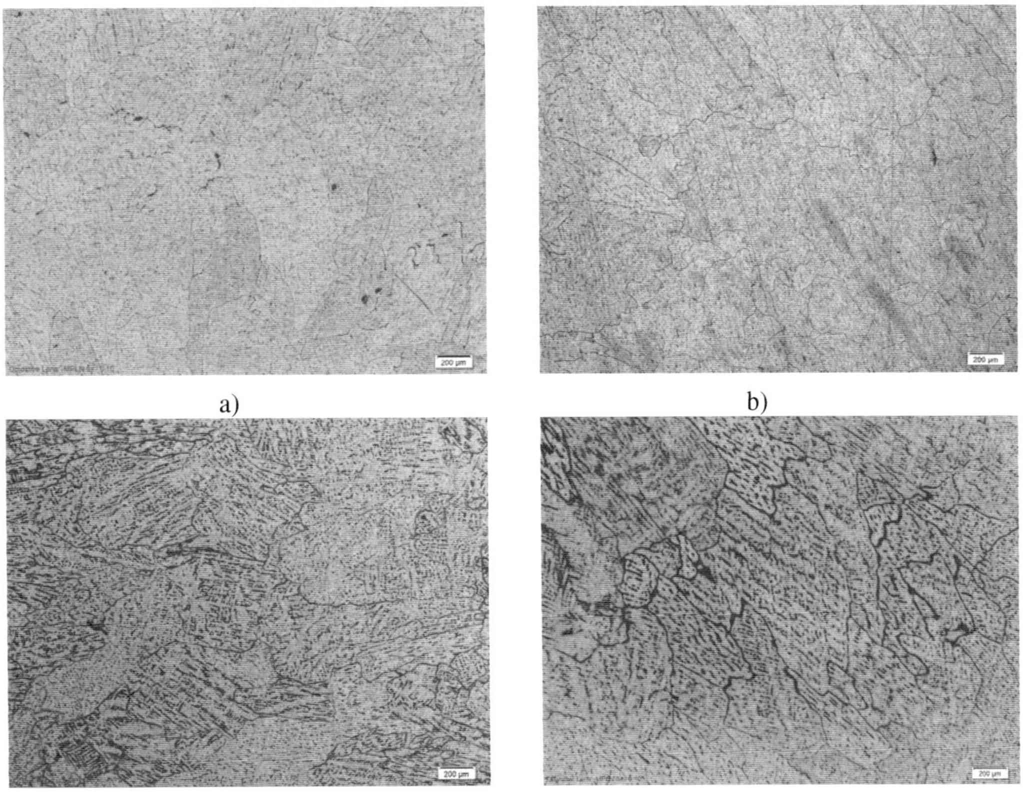

a) b) c) d)

Figure 2. The microstructure of experimental samples of АД0(a), АД0+1.14%Al_2O_3(b), АД0+1.198%Cu(c), АД0+1.26%(Al_2O_3+Cupowders)(d) after etching, Olympus GX51 metallographic microscope, zoom x50

It is seen that after the introduction of Al2O3 nanofiber, the structure is decreasing. When copper is added, decreasing is more efficient.

3.2 Hardness

Comparative measurements of the hardness of experimental samples according to the Brinell method (HBW) (GOST 9012-59) carried out on a Struers automatic hardness tester with a load of 62.5 kg, an indenter diameter of 5 mm and microhardness (HV0.1) on a DURASCAN 70 microhardness meter with a load of 100 g are presented in Table 4.

Table 4. Hardness and microhardness of experimental samples.

Samples	Brinell Hardness (HBW)	Microhardness ($HV_{0,1}$)
1. АД0	19.07±1.36	26.11±2.91
2. АД0+1.14% Al_2O_3	16.58±2.66	24.44±3.70
3. АД0+1.198%Cu	22.42±1.64	29.94 ±2.26
4. АД0+1.26%(Al_2O_3+ Cu)	26.87±1.54	31.86±3.16

The analysis of the results demonstrates an increase in the hardness of samples with copper powders, which is evidently due to the formation of the hardening phase CuAl2. The introduction of discrete Al2O3 fibers without copper carrier is not significant, but it reduces hardness. Compared to the hardness of the sample АД0 and the sample АД0+1.198% Cu, the hardness value of the samples АД0+1.26%(Cu+Al2O3) increases by 1.41 times and 1.20 times, respectively.

The microhardness value of the sample АД0+1.26%(Al2O3+Cu) increases 1.22 times compared with the microhardness of АД0. The decrease in microhardness of АД0+1.14%Al2O3 can be explained by the insufficient wettability of Al2O3 nanofibers with aluminum.

According to the results of microhardness measurements, probability distributions of microhardness of experimental samples were constructed (Figure 3). The accuracy of the repetition of the character of the distribution of the microhardness probability at 100 points over an area of 1 cm2 was demonstrated by sufficient structural homogeneity of the experimental samples. After the introduction of fillers, the symmetry of the probability distribution curves decreases. The sample АД0+1.26% (Al2O3+Cu powders) has a higher average value, but the asymmetry of the curve increases. This is due to the fact that after the introduction of fillers uniformity decreases.

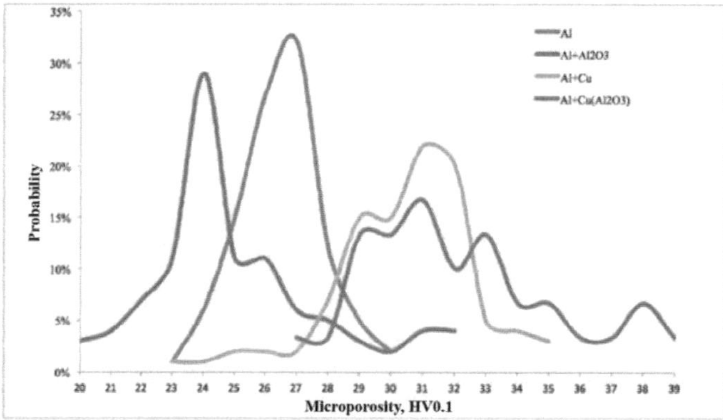

Figure 3. Microhardness probability distribution curves

4. Conclusions

Thus, in the course of the work done, a method for the effective introduction of discrete fibers of aluminum oxide with a diameter of 10–20 nm into the aluminum melt was proposed and implemented. The general patterns of changes in properties were found. Due to the use of transporting copper powder, it was possible not only to introduce nanofibers into aluminum melt, but also to increase the hardness of the material. Patent RF No. 179266 has been received for development.

References

[1] Kurganova Yu A and Kolmakov A G 2015 *Structural metal matrix composite materials* (Moscow: Publishing House of MSTU named after N. E. Bauman) p 141 (in Russian)

[2] Kurganova Yu A, Chernyshova T A, Kobeleva L I and Kurganov S V 2011 Service properties of aluminum-matrix precipitation-hardened composite materials and the prospects of their use on the modern structural material market *Russian Metallurgy (Metally)* **7** 663–666

[3] Chawla N and Chawla K K 2013 *Metal matrix composites* (New York: Springer) p 370
[4] Yijin Chen 2017 Metal matrix composite materials. Prospects for effective application and production (review) *Metal technology* 10 25–30 (in Russian)
[5] Tjong S C 2013 Recent progress in the development and properties of novel metal matrix nanocomposites reinforced with carbon nanotubes and graphene nanosheets *Mater. Sci. Eng. R. Rep.* **74** 281-350
[6] Miracle D B 2005 Metal matrix composites – From science to technological significance *Compos. Sci. Technol.* **65** 2526-40
[7] Moghadam A D, Omrani E, Menezes P L and Rohatgi P K 2015 Mechanical and tribological properties of self-lubricating metal matrix nanocomposites reinforced by carbon nanotubes (CNTs) and grapheme – A review *Compos. B Eng.* **77** 402-420
[8] Kablov E.N. 2012 Strategic directions for the development of materials and technologies for their processing for the period up to 2030 *Aviation materials and technologies* S 7–17 (in Russian)
[9] Mazaherya A, Abdizadeha H and Baharvandi H R 2009 Development of high-performance A356/nano-Al_2O_3 composites *Mater. Sci. Eng. A Struct. Mater.* **518** 61-64
[10] Sajjadi S A, Ezatpour H R and Parizi M T 2012 Comparison of microstructure and mechanical properties of A356 aluminum alloy/Al_2O_3 composites fabricated by stir and compo-casting processes *Mater. Des.* **34** 106-111
[11] Akbari M K, Baharvandi H R and Mirzaee O 2013 Nano-sized aluminum oxide reinforced commercial casting A356 alloy matrix: Evaluation of hardness, wear resistance and compressive strength focusing on particle distribution in aluminum matrix *Compos. B Eng.* **52** 262-268
[12] Akbari M K, Baharvandi H R and Mirzaee O 2013 Fabrication of nano-sized Al_2O_3 reinforced casting aluminum composite focusing on preparation process of reinforcement powders and evaluation of its properties *Compos. B Eng.* **52** 426-432
[13] Akbari M K, Mirzaee O and Baharvandi H R 2013 Fabrication and study on mechanical properties and fracture behavior of nanometric Al_2O_3 particle-reinforced A356 composites focusing on the parameters of vortex method *Mater. Des.* **46** 199-205
[14] Tahamtan S, Halvaee A, Emamy M and Zabihi M S 2013 Fabrication of Al/A206–Al_2O_3 nano/micro composite by combining ball milling and stir casting technology *Mater. Des.* **49** 347-359
[15] Wang J, Li Z, Fan G, Pang H, Chen Z and Zhang D 2012 Reinforcement with graphene nanosheets in aluminum matrix composites *Scr. Mater.* **66** 594-597.
[16] Kollo L, Leparoux M, Bradbury C R, Jaggi C, Carreno-Morelli E and Rodriguez-Arbaizar M 2010 Investigation of planetary milling for nano-silicon carbide reinforced aluminium metal matrix composites *J. Alloys Comp.* **489** 394-400
[17] Chen Y, Qin H, Kurganova Yu A and Gaaze V K 2018 A method for introduction of Al_2O_3 nanofiber into aluminum alloy *IOP Conf. Ser. Mater. Sci. Eng.* **347** 012050
[18] Kurganova Yu A and Scherbakov S P 2017 Influence of a discrete additive of aluminum oxide on structure and properties of aluminium alloy *Journal of mining institute* **228** 717–721
[19] Leon C A, Mendoza-Suarez G and Drew R A L 2006 Wettability and spreading kinetics of molten aluminum on copper-coated ceramics *J. Mater. Sci.* **41** 5081-87
[20] Serpova V.M., Shavnev A.A., Grishina O.I., Krasnov E.I., and Solyaev Yu.O. 2014 Wettability and interfacial interaction in MCM on an aluminum matrix reinforced with Al_2O_3 *Materials science* **12** 29–35 (in Russian)
[21] Jonas T R, Cornie J A и Russel K C. 1995 lInfiltration and wetting of alumina participate preforms by aluminum and aluminum-magnesium alloys *Metall. Mater. Trans. A* **26** 1491–97

Electric treatment of alumina composite materials

G N Minenko[1] and S L Timchenko[2]

[1]Russia, Moscow region Resource Center for Industrial Machinery

[2]Russia, Moscow, MSTU named after N.E. Bauman, Fundamental Science Faculty, Department of Physics

E-mail: minen29ko2009@yandex.ru, svtimchenko@yandex.ru

Abstract. A method of processing alumina composite materials using electric current is presented. Aluminum matrix composites based on aluminum alloy AK12 were obtained in a pass-through vessel with the addition of composite particles to the stream of molten material without the influence of an electric current and with the influence of a current. As a composite additive, graphite enriched with crystalline grade TG-1 and particles of titanium carbide (TiC) were used, the amount of the components being injected was 15 volume percent.

Measurements of the mechanical properties of the original AK12 alloy and composite material based on the AK12 alloy with the addition of composite particles carried out without an electric field and with exposure showed that the greatest increase in strength properties of alloys treated with an electric field was recorded for an aluminum matrix composite material with TiC particles, which amounted to 20% for ultimate strength and 24.5% for hardness. An increase in ductility was found in a composite material with a TG -1 composite additive, for it an increase in the relative elongation was 15.6%, while the electric current treatment of the composite material with a filler in the form of TiC did not lead to a change in the relative elongation of the sample material. In general, the increase in the strength properties of an aluminum matrix composite material with titanium carbide particles (TiC), after it was treated with an electric current, turned out to be slightly larger than that of an aluminum matrix composite material with particles of TG-1 enriched crystalline graphite. The physical mechanism of the influence of electric current on the properties of ALCM is given.

Keywords: *electric current, aluminum matrix composites, mechanical properties, dispersed hardening*

Introduction

The development of modern technology requires a qualitative improvement in the technical characteristics of machines and mechanisms, which can be ensured only under the condition of creating advanced manufacturing techniques for fundamentally new structural materials and their integrated use. At the present stage of development of engineering and technology, the improvement of methods for producing metal matrix, dispersion-strengthened composite materials (CM), as materials with reliable performance properties, is becoming important [1,2].

One of the promising directions in the technology of creating composite materials with a metal matrix is the use of the effects of external fields of various physical nature — ultrasonic, electromagnetic, and also elements of plasma technologies [1–8]. As it was shown in [9-16], the use of electromagnetic effects at the stage of modifying aluminum alloys, cast irons, steels, as well as at the stage of its crystallization during casting and thixolite allows reducing the grain size of the α phase and preventing the development of dendritic structure, increasing density and strength alloy, reduce porosity, increase its hardness. Also, the impact of the electric field is used in the preparation of aluminum matrix composite materials [6-8].

Experiments and results

The purpose of this work was to determine the degree of influence of electrophysical treatment of a composite material based on AK 12 alloy on its mechanical properties. The processing of the composite material based on the alloy AK12 with the electric current in the pass-through capacitance was carried out according to the electrical circuit shown in Figure 1.

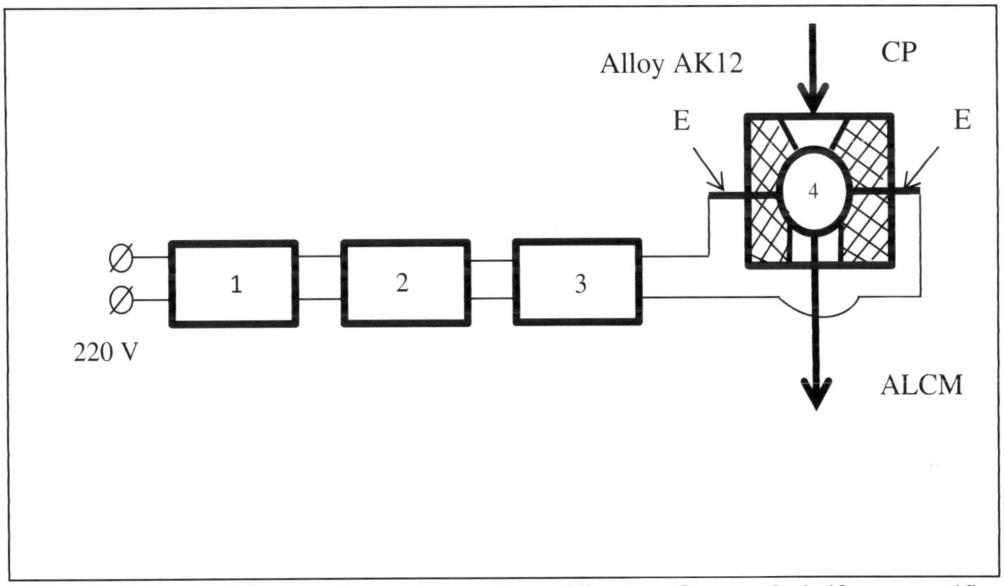

Figure 1. Electrical circuit of the experimental setup: 1 - welding transformer, 2 - half-wave rectifier, 3 - LC-filter with measuring instruments, 4 - through-passage reservoir with liquid alloy.

An aluminum matrix composite material (ALCM) (Figure 1) based on an aluminum alloy AK12 was obtained by adding particles of a composite (CP) to the stream of a melt. As the composite, graphite enriched with crystalline grades TG-1 and particles of titanium carbide (TiC) were chosen. The input components were 15 percent by volume. Two series of experiments were done: obtaining CM without electric field and with electric field.

Voltage was applied to tungsten electrodes (E, Figure 1) with a diameter of 6 mm, which are located in the internal cavity of the passage capacitance 2. An induction-capacitive LC-filter 3 with measuring instruments, a half-wave rectifier 2 made on a semiconductor power diode of type D151-160, and a welding transformer 1 of an alternating current brand TDM-161U2 were connected to the electrodes. This scheme made it possible to obtain direct electrical contact with a liquid alloy flowing through a through-passage reservoir (Figure 1, position 4). The values of inductance and electric capacitance of which the rectifier LC filter consisted were selected in such a way that the value of the ripple factor was 1.5. The used value of the pulsation coefficient was chosen to optimize the processing of the liquid alloy, taking into account the phenomenon of polarization of the electrodes. The specific power of the electric current was calculated by the power supplied to the electrodes, referred to the mass flow rate of the liquid composite (kg / s), poured from the pass-through capacity into the molds. The value of specific power for a given passage capacity was 650 ... 663 J / kg.

Measurements were made of the mechanical properties of the original AK12 alloy and ALKM on the basis of the AK12 alloy with the addition of composite particles, carried out without an electric field and with an impact, the results of which are presented in Table 1.

It was experimentally proved that the greatest increase in strength properties of untreated alloys (table 1) was recorded in ALCM with TiC particles, however, the relative elongation for CM with TG-1 was greater in value than CM with TiC. The impact of the electric field led (Table 1) to an increase in the strength properties and relative elongation of CM with TG-1, but the values of the strength characteristics of CM with TG-1 were less than those of CM with TiC.

Table 1. Data on the effects of treatment with an electric field on the mechanical properties of CM.

	Original raw alloys		
Mechanical properties of alloys	AK 12	AK12 + TG-1	AK12 + TiC
σ_B, MPa	149	161	182
δ, %	5.1	9,6	8,2
HB, MPa	5.2	4,9	5,3
	Electric field treated alloys		
σ_B, MPa	153	192	219
δ, %	5.9	11.1	8.1
HB, MPa	5.4	5.8	6.6

These changes in the mechanical properties indicate the influence of the electric field of the current on the process of forming the interface between the metal matrix and the dispersed inclusion. Hardening the interface structure causes an increase in strength of the CM structure.

The effect of the electric field of the current on the metal melt and its components is ambiguous. A number of physical phenomena occur that affect the intensity of the interaction between the composite filler and the liquid matrix. In our opinion, the action of the electric field of the current on the formation of the structure of the metal-matrix composite material can be represented by the following steps:

- set the stationary mode of flow of current in the melt containing the composite filler;
- particles of the dispersed material (DMP) under the influence of an electric field interact with the melt and the interface begins to form;
- as a result of the flow of electrical contact, surface phenomena, the charge of an electrical double layer that occurs at the interface of a metal melt with particles of a composite filler introduced into it increases, and qualitative changes of the interfaces occur.

To solve the problem of the further development of promising methods for obtaining CM, in which electrophysical processing is applied, it is necessary to develop and use a physical model of the influence of the electric field on the formation of the CM structure. Formalization of the physical model will allow optimizing the modes of electrophysical processing of CM, taking into account the used alloy and composite filler.

The physical model, which can explain the appearance of the effect of an electric field, should, in our opinion, take into account the peculiarities of physical phenomena associated with mass transfer in a metal melt and the occurrence of polarization electric charges on the surface of composite particles. As metallographic studies have shown, a physical model can be applied to describe the processes occurring in CM when it is processed by an electric field, which is schematically shown in Figure 2.

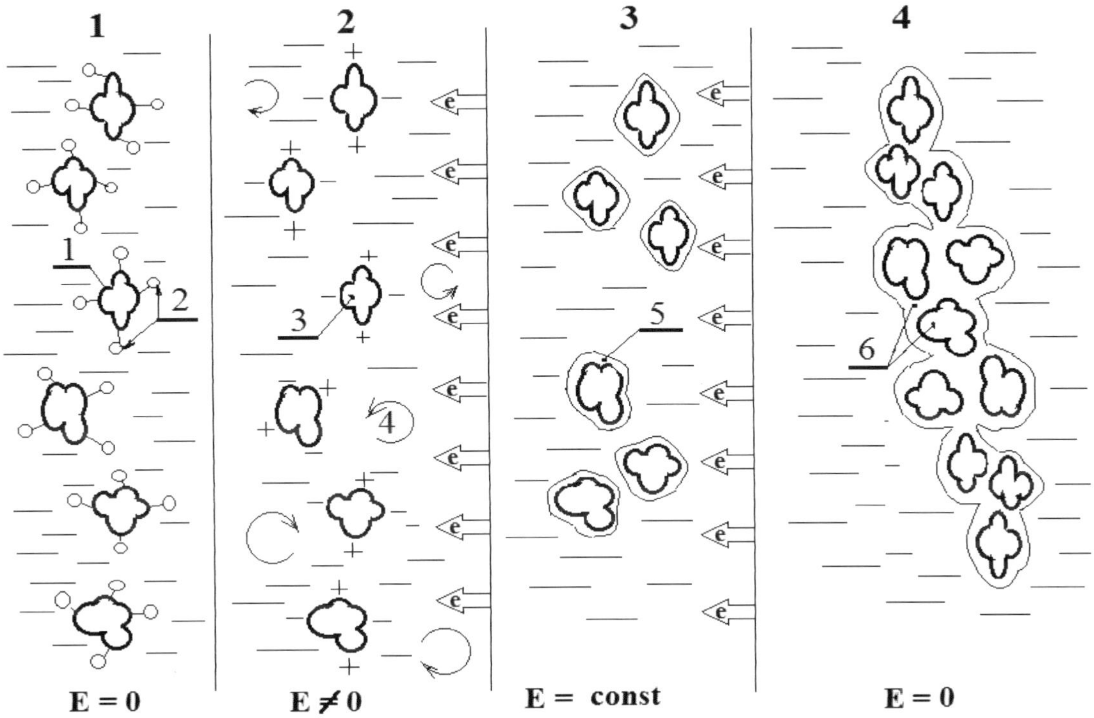

Figure 2. Formation of interphase boundaries.

At the first stage (Figure 2, E = 0, position 1), when the electric field strength E is zero, the reinforcing filler 1 in the form of composite particles is distributed over the volume of the melt. There is a capture by particles of a filler of atoms of chemical elements from a metal melt. The force field of the solid surface of the particles arising from contact with the liquid alloy is weakened at the protruding ends of the composite particles [1, 2, 6–8, 17–20] and, therefore, the interfacial energy changes and the wettability of the solid phase of CM by the liquid phase of the matrix melt. As a result, particles of CM are captured by atoms of chemical elements 2 from a metal melt.

At the second stage (Figure 2, E ≠ 0, position 2), the electric field strength E is not zero, polarization charges appear on the surface of the composite particles and energy fluctuation W occurs. The probability of this fluctuation of energy W according to Einstein's formula is equal to [17]:

$$W = \exp(-A/kT), \qquad (1)$$

where A is the electrical energy of the composite particle, k is the Boltzmann constant; T is the temperature of the metal melt.

In this case [17] work A is:

$$A = \frac{q_o \varphi}{2}, \qquad (2)$$

where q_o is the initial polarization electric charge (Кл), φ is the potential (B).

Then the volume density of the electric charge Q will be equal to

IOP Conf. Series: Materials Science and Engineering **683** (2019) 012035 doi:10.1088/1757-899X/683/1/012035

$$Q = \frac{q_o}{V}, \qquad (3)$$

where V is the particle volume of the composite.

Under the influence of the electron-ion flow in the melt created by an external electric field, on the surface of the particles of the composite 3 (Figure 2, E $\neq$ 0, position 2), an additional electric charge Δq occurs. This leads to an increase in the bulk density of the electric charge Q of particles CM:

$$\frac{dQ}{dt} = \frac{\mathbf{q}_o + \Delta\mathbf{q}}{Vdt}, \qquad (4)$$

where Δq is the increment of the particle charge of the composite.

Discussion

Also as a result of the flow of electric current, Joule heat is generated in the matrix melt, which leads to the formation of convective currents (Figure 2, E = 0, position 4), which intensify the movement of CM particles in the melt. These movements of composite particles contribute to a more uniform distribution in the volume of the matrix melt of reinforcing filler.

At the third stage (Figure 2, E = const, position 3), additional layers of metal ions 5 are formed on the surface of the charged particles of the composite under the influence of electrons and ions accelerated by the electric field, moving along the molten metal and causing an electric current. In this case, as is well known, the Joule heat value is proportional to the square of the current that occurs in the metal melt during processing.

When the voltage is turned off (Figure 2, E = 0, position 4), that is, when the passing capacity with the melt is disconnected from the source, the system relaxes to a new equilibrium state. At this final stage, there is a merging of charged particles of CM with each other due to the interaction forces obtained due to the charges formed by ions on their surface. Due to this, there are spatial formations of filler particles 6, which serve as a "skeleton" for the formed structure of CM. This hardening of the structure causes the growth of mechanical and operational properties, treated with an electric field CM, which leads to the appearance of the effect of external influence (EE) forming the structure of CM.

The result of these effects is the improvement of the conditions for the formation of interfacial boundaries, which determines the conditions for the process of adhesion of the particles of the composite with aluminum alloy and contributes to the hardening of the structure of alumomatric CM.

The above results on the effect of electric current on the properties of alumina composite materials are consistent with the results of research, which used the influence of electromagnetic fields, in particular, electric current on the properties of aluminum alloys, when the electromagnetic effect was in the processes of alloy modification and crystallization [9-16].

Conclusion

Thus, the electrophysical effect leads to an increase in mechanical properties due to the influence of the electric field on changes in the structure of the boundary layer formed between the inclusion of the composite and the metal matrix. The hardening of the CM structure is possibly due to physico-chemical phenomena occurring at the interface under the influence of an electric field. Using this method to obtain alumomatric CM will give an economic effect due to the hardening of the structure of the CM and increase the operational reliability of machine parts made of CM.

References

[1] Batyshev A I, Batyshev K A 2016 *Production of castings from composite materials with a metal matrix* Foundry **4** 17-23
[2] Krushenko G G 2011 The role of nanopowder particles in the formation of the structure of aluminum alloys *Metallurgy of mechanical engineering* **1** 20-24
[3] Yang Y, Lan J, Li X 2004 Study on bulk aluminum matrix nano-composite fabricated by ultrasonic dispersion of nano-sized SiC particles in molten aluminum alloy *Material Science and Engineering* **A380** 378-383
[4] Casati R, Vedani M 2014 *Metall Matri Composites Reinforced by Nano-Partikles-A* Review Metals. **4** 65-83
[5] Sanaty-Zadeh A 2012 Comparison between current models for the strength of particulate-reinforced metal matrix nanocomposites with emphasis on consideration of Hall-Petch effect *Material Science and Engineering* **A531** 112-118
[6] Minenko, G N 2018 A method of processing aluminum-matrix composite materials by an electric field. *Metallurgy of mechanical engineering* **4** 42-44
[7] Minenko G N 2018 The Mechanism of the Effect of an Electric Field on Composite Aluminum Matrix Composites. *"Casting of Ukraine"* **5** 2-3
[8] Minenko G N 2018 *Method of electrophysical processing of metal-matrix composite material. "Casting of Ukraine"* **7** 2-3
[9] Korobova N V, Aksenenko A Yu, Tarasov F Ye, Friesen V E, Luzgin V I and Fatkullin S M 2013 *On the electromagnetic effect on the aluminum melt when it is modified* Metallurgy of mechanical engineering. **1** 8–11
[10] Korobova N V, Aksenenko A Yu, Luzgin V I, Frizen V Ye, Shevchenko S Yu 2015 *The evolution of the microstructure of cast aluminum alloys under the hydrodynamic influence of electromagnetic fields on the melt* Procurement Production **7** 43-48
[11] Deev V B, Selyanin I F, Nokhrina O I, Bashmakova N V 2008 *The Influence of Temperature Processing and Electric Current on the Properties of Aluminum Alloys* Procurement Production in Mechanical Engineering **4** 50-53
[12] Kolchugina I Yu, Selyanin I F 2009 *The influence of external influences on the microstructure of the crystallizing alloy* Foundry **8** 13-15
[13] Deev V B, Selyanin I F, Ri Hosen, Tsorina S A and Ponomarev K V Effective 2012 *Technologies for processing of melts in the production of cast aluminum* alloys Liteischik Rossii **10** 19–21
[14] Timchenko S L 2011 *Study of alloy crystallization under the action of electric current* // Melts **4** 53-61
[15] Timchenko S L, Kobeleva L I 2013 Features of the dendrite structure of silumin AK12 during crystallization under the action of current Physics and Chemistry of Materials Processing.**4** 78-83.
[16] Timchenko S L, Zadorozhnyj N A 2016 *Features of the formation of the structure of the alloy AK 12 during its crystallization under the action of current* Foundry **8** 11-14
[17] Darken L S, Gurri R S 1962 *Physical Chemistry of Metals* M., Metallurgizdat 582
[18] Missol V 1978 *Surface energy phase separation.* M.: "Metallurgy" 176
[19] Damascin B B, Petriy O A 1983 *Introduction to electrochemical kinetics* – M.: "High school", 400
[20] Rusanov A I, Levichev S A, Zharov V T 1981 *The Superficial Separation of Substances: Theory and Methods*. - L.: Chemistry, 184

Stress-strain state of the interfacial layer in a visco-composite composite with longitudinal shear

A A Dudchenko[1,*], S A Lurie2[,3], S V Makovskij[1] and K K Shramko[1]

[1] Moscow Aviation Institute, Moscow, Russia

[2] Institute of Applied Mechanics of Russian Academy of Science, Moscow, Russia

[3] Lomonosov Moscow State University, Russia

[*] Corresponding author at Moscow Aviation Institute, Moscow, Russia

E-mail: a_dudchenko@mail.ru (A A Dudchenko)

Annotation. In this paper, we propose a three-phase method for determining the effective properties of reinforced by whiskerized fibers composite materials. The model takes into account the presence of a whiskerized interfacial layer between the fiber and the matrix. The calculation of the stress-strain state of the interfacial layer loaded with a longitudinal shear using the Eshelby self-consistent field method is given.

Key Words: *Fiber-reinforced interfacial layer, fibrous composites, Eshelby self-consistent method, micromechanics*

1. Introduction

Composite materials are widely used in aircraft and rocket structures, and also find wide application in other areas of technology, including transport, medicine, and others. Composite structures with various combinations of components in a structure have many advantages compared to traditional materials based on metal alloys. It is known that the mechanical properties of composites are controlled by the contact conditions between the fiber and the matrix (i.e. interface characteristics) in the composite [1, 2]. The influence of the structure on the nature of the interaction of the phases in the region of the contact of the phases turns out to be particularly significant. Various techniques have been developed to improve the quality of the interface, and the most common are: modifying the fiber surface, improving chemical interactions, or adding a third phase (interfacial layer) between the fiber and the matrix [3-5]. The ideas underlying these methods are to improve the interfacial adhesion properties and to increase the surface area of the fiber for more efficient transfer of loads between the fibers and the matrix and to further improve the properties of the composites.

Currently, technologies for the production of modern fiber composites are being actively developed, in which special microstructures containing nanofibers (whiskers) - nanowires [4, 6, 7] are grown on the circular surface of carbon fibers to increase the shear properties of the composite. For composite materials with whiskerized fibers, a higher ultimate strength and shear stiffness is realized compared to standard composites that do not have an additional microstructure on the fiber surface. In addition, the presence of these microstructures increases the transversal strength and rigidity, as well as the damping characteristics and electrical conductivity of the composites [8, 9, 10].

The aim of the work is to substantiate and develop a model of functional fibrous non-uniform composite materials based on an analytical solution and able to reliably predict the elastic mechanical properties of composites on the properties and phase concentrations taking into account the characteristics of micro- or nanostructures (whiskers) grown on the surface of the base fiber.

2. Approximate structural model of the whiskerized interfacial layer. Effective properties

Fig. 1 shows a structural model of a whiskerized interfacial layer, which is built to find the effective properties of a whiskerized interfacial layer, taking into account various characteristics of whiskers (i.e. length, density, diameter, properties of whiskers) in this layer.

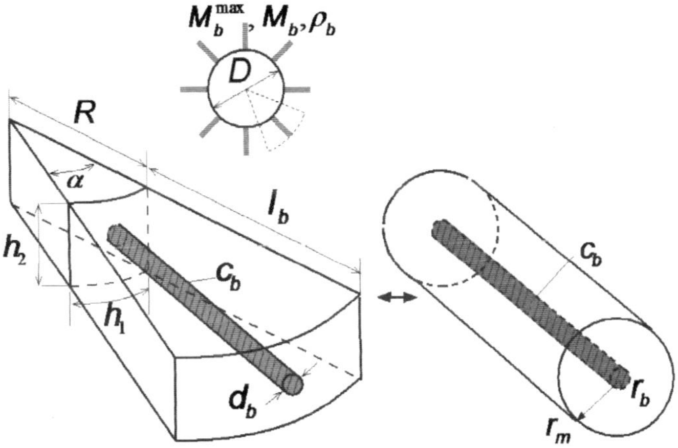

Fig. 1. An approximate structural model of a whiskerized interfacial layer.

Now we determine the maximum allowable number of whiskers $M_b^{\max}$, which can fit on the circumference of the fiber by the formula

$$M_b^{\max} = \pi D / d_b, \qquad (1)$$

where D is the diameter of the base fiber; d_b - diameter of whiskers. The angle between the whiskers α is given as

$$\alpha = 2\pi M_b, \qquad (2)$$

where M_b is the number of whiskers grown on the surface of the fiber. The distance h_1 between the whiskers is given by the formula

$$h_1 = \alpha R, \qquad (3)$$

where $R = D/2$ is the radius of the base fiber. The distance between the whiskers on the unit length of the fiber h_2 we write as:

$$h_2 = p h_1, \qquad (4)$$

where p is a constant that defines the relationship between h_1 and h_2. The volume of whiskers in the whiskerized interfacial layer V_b is determined by the formula

$$V_b = \pi M_b l_b d_b^2 / 4, \qquad (5)$$

where l_b is the length of the whiskers, and the volume of the whiskerized interfacial layer is

$$V_{bm} = \pi \left(r_2^2 - r_1^2 \right) h_2, \qquad (6)$$

where $r_1 = R$ and $r_2 = R + l_b$. So, using the relation (5) and (6), we determine the volume fraction of whiskers in this layer:

$$c_b = \frac{M_b d_b^2}{4(l_b + D)h_2}. \quad (7)$$

Since, for simplicity, the cell at the root of the whiskers is considered square, we have $p = 1$. Therefore, based on (2), (3) and (4), we can rewrite (7) in the form

$$c_b = \frac{M_b^2 d_b^2}{4\pi(l_b + D)D}. \quad (8)$$

The density of whiskers is related to the number of whiskers grown on a single fiber surface. However, for simplicity, we determine in the calculation that the density of whiskers is the ratio of the number of whiskers grown on the fiber to the maximum allowable number of whiskers that fit along the circumference of the fiber. We have

$$\rho_b = \left(M_b / M_b^{\max}\right)100\%. \quad (9)$$

After obtaining an estimate of the concentration of whiskers, we can use the well-known formulas of the three-phase method given in [11] for a two-phase composite to determine the effective properties of the whisker interfacial layer (see Fig. 2).

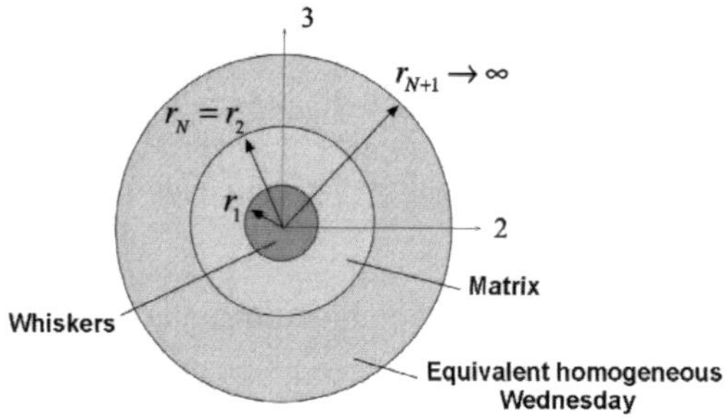

Fig. 2. Cell of the whiskerized interfacial layer using the three phase method.

Young's longitudinal modulus E_{11}^{eff}:

$$E_{11}^{eff} = c_b E_b + (1 - c_b)E_m + \frac{4c_b(1 - c_b)(v_b - v_m)^2 \mu_m}{(1 - c_b)\mu_m/(k_b + \mu_b/3) + c_b\mu_m/(k_m + \mu_m/3) + 1}, \quad (10)$$

Poisson's ratio (along the fibers) v_{12}^{eff}:

$$v_{12}^{eff} = c_b v_b + (1 - c_b)v_m + \frac{c_b(1 - c_b)(v_b - v_m)\left(\mu_m/(k_m + \mu_m/3) - \mu_m/(k_b + \mu_b/3)\right)}{(1 - c_b)\left(\mu_m/(k_b + \mu_b/3)\right) + c_b\left(\mu_m/(k_m + \mu_m/3)\right) + 1}, \quad (11)$$

Longitudinal shear modulus μ_{12}^{eff}:

$$\frac{\mu_{12}^{eff}}{\mu_m} = \frac{\mu_b(1 + c_b) + \mu_m(1 - c_b)}{\mu_b(1 - c_b) + \mu_m(1 + c_b)}, \quad (12)$$

Volumetric modulus of flat deformation K_{23}^{eff}:

$$K_{23}^{eff} = k_m + \frac{\mu_m}{3} + \frac{c_b}{1/\left(k_b - k_m + \frac{1}{3}(\mu_b - \mu_m)\right) + (1 - c_b)/(k_m + 4\mu_m/3)}, \quad (13)$$

Transverse shear modulus μ_{23}^{eff} :

$$A\left(\frac{\mu_{23}^{eff}}{\mu_m}\right)^2 + 2B\left(\frac{\mu_{23}^{eff}}{\mu_m}\right) + C = 0 \quad , \qquad (14)$$

where $\eta = 3 - 4v$ and

$$A = 3c_b\left(1-c_b\right)^2\left(\frac{\mu_b}{\mu_m}-1\right)\left(\frac{\mu_b}{\mu_m}-\eta_b\right)+$$
$$+\left[\frac{\mu_b}{\mu_m}\eta_m + \eta_b\eta_m - \left(\frac{\mu_b}{\mu_m}\eta_m - \eta_b\right)c_b^3\right]\left[c_b\eta_m\left(\frac{\mu_b}{\mu_m}-1\right)-\left(\frac{\mu_b}{\mu_m}\eta_m+1\right)\right],$$

$$B = -3c_b\left(1-c_b\right)^2\left(\frac{\mu_b}{\mu_m}-1\right)\left(\frac{\mu_b}{\mu_m}+\eta_b\right)+$$
$$+\frac{1}{2}\left[\eta_m\frac{\mu_b}{\mu_m}+\left(\frac{\mu_b}{\mu_m}-1\right)c_b+1\right]\left[\left(\eta_m-1\right)\left(\frac{\mu_b}{\mu_m}+\eta_b\right)-\right.$$
$$\left.-2\left(\frac{\mu_b}{\mu_m}\eta_m-\eta_b\right)c_b^3\right]+\frac{c_b}{2}\left(\eta_m+1\right)\left(\frac{\mu_b}{\mu_m}-1\right)\left[\frac{\mu_b}{\mu_m}+\eta_b+\left(\frac{\mu_b}{\mu_m}\eta_m-\eta_b\right)c_b^3\right],$$

$$C = 3c_b\left(1-c_b\right)^2\left(\frac{\mu_b}{\mu_m}-1\right)\left(\frac{\mu_b}{\mu_m}+\eta_b\right)+$$
$$+\left[\eta_m\frac{\mu_b}{\mu_m}+\left(\frac{\mu_b}{\mu_m}-1\right)c_b+1\right]\left[\frac{\mu_b}{\mu_m}+\eta_b+\left(\frac{\mu_b}{\mu_m}\eta_m-\eta_b\right)c_b^3\right].$$

2.1 The model of whiskerized fiber composites, based on the Eshelby self-consistency method (three-phase method).

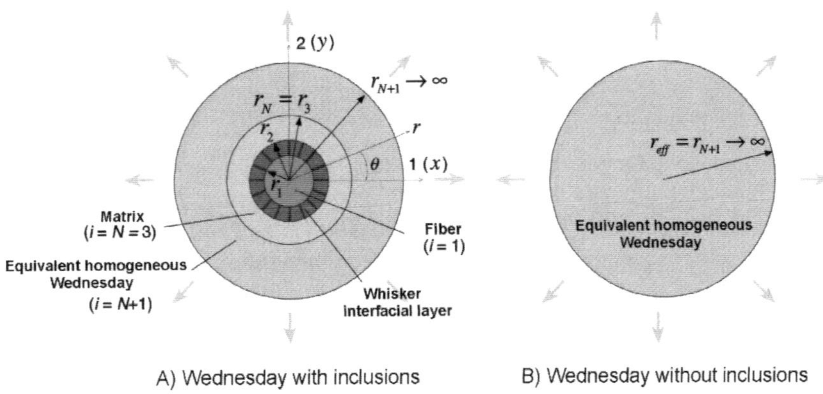

Fig. 3. (a) Cell of the whiskerized fiber composite; (b) cell of an effective composite (homogenized composite).

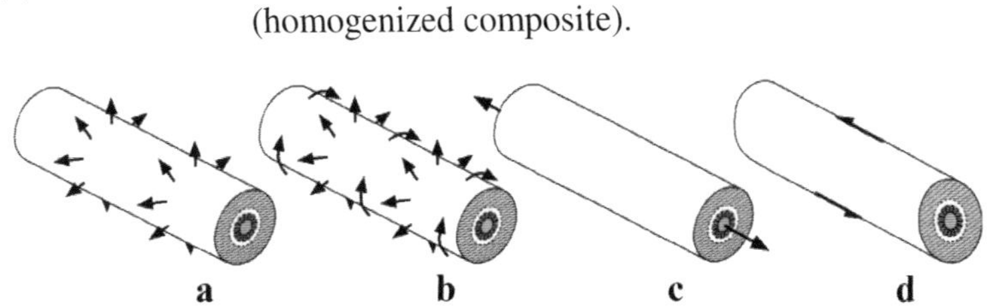

Fig. 4 The problem in the isotropy plane: (a) the problem of comprehensive tension (compression) across the fibers; (b) the problem of pure shear in the transversal plane and the problem in the perpendicular isotropy plane; (c) the problem of uniaxial tension (compression); (d) the problem of pure shear in the direction along the fibers.

Figure 3 shows the cells of the whiskerized fiber composite using the Eshelby self-consistent field method and its effective composite. The effective properties of such a composite were found when considering the difference in strain energy between media with and without inclusions. Christensen and Law showed that this method allows finding all the necessary effective constants, including the longitudinal and transverse Young modulus and shear modulus, the flat strain modulus, and Poisson's coefficients from solving problems in the isotropy plane and in the plane perpendicular to it (see. Fig. 4). In each case of loading, the general admissible displacement fields for the orthotropic phases of the composite are determined from the solution of the corresponding problem of the theory of elasticity. With the help of the Cauchy relations and the Hooke's equations, specific deformation and stress fields are found for all phases in the composite. Further, these fields of displacements and stresses are substituted into the boundary conditions at the boundaries of the phase contacts: the conditions of continuity of displacement and stress, the non-singularity conditions at the center of the fiber and the conditions at infinity (external boundary condition). Based on these conditions, a system of algebraic equations is formulated, from which, using the Eshelby integral formula

$$\int_{S_i} \left(\sigma_{ij}^{N+1} u_i^{eff} - \sigma_{ij}^{eff} u_i^{N+1} \right) n_j dS = 0 , \qquad (15)$$

all unknown constants are found and, therefore, all effective elastic moduli, which were also included in the number of unknown constants.

The tensors of elastic moduli ($C_{ijkl}^{(i)}$) phases in the composite that will be used for the proposed model are shown here. Based on Hooke's rule, the tensor of elastic moduli of the first phase ($i=1$) or base fiber transversely isotropic medium with axis of symmetry directed along its fiber axis has the form

$$\begin{pmatrix} \sigma_{rr}^{(1)} \\ \sigma_{\theta\theta}^{(1)} \\ \sigma_{zz}^{(1)} \\ \sigma_{\theta z}^{(1)} \\ \sigma_{rz}^{(1)} \\ \sigma_{r\theta}^{(1)} \end{pmatrix} = \begin{bmatrix} C_{11}^{(1)} & C_{12}^{(1)} & C_{13}^{(1)} & 0 & 0 & 0 \\ C_{12}^{(1)} & C_{11}^{(1)} & C_{13}^{(1)} & 0 & 0 & 0 \\ C_{13}^{(1)} & C_{13}^{(1)} & C_{33}^{(1)} & 0 & 0 & 0 \\ 0 & 0 & 0 & C_{44}^{(1)} & 0 & 0 \\ 0 & 0 & 0 & 0 & C_{44}^{(1)} & 0 \\ 0 & 0 & 0 & 0 & 0 & \frac{C_{11}^{(1)} - C_{12}^{(1)}}{2} \end{bmatrix} \begin{pmatrix} \varepsilon_{rr}^{(1)} \\ \varepsilon_{\theta\theta}^{(1)} \\ \varepsilon_{zz}^{(1)} \\ 2\varepsilon_{\theta z}^{(1)} \\ 2\varepsilon_{rz}^{(1)} \\ 2\varepsilon_{r\theta}^{(1)} \end{pmatrix} , \qquad (16)$$

where the indices of the components of the tensor of elastic moduli r, θ, z are replaced by indices $1, 2, 3$, respectively. For the second phase ($i=2$) or a whiskerized interfacial layer, the tensor of elastic moduli takes the form

$$\begin{pmatrix} \sigma_{rr}^{(2)} \\ \sigma_{\theta\theta}^{(2)} \\ \sigma_{zz}^{(2)} \\ \sigma_{\theta z}^{(2)} \\ \sigma_{rz}^{(2)} \\ \sigma_{r\theta}^{(2)} \end{pmatrix} = \begin{bmatrix} C_{11}^{(2)} & C_{12}^{(2)} & C_{12}^{(2)} & 0 & 0 & 0 \\ C_{12}^{(2)} & C_{22}^{(2)} & C_{23}^{(2)} & 0 & 0 & 0 \\ C_{12}^{(2)} & C_{23}^{(2)} & C_{22}^{(2)} & 0 & 0 & 0 \\ 0 & 0 & 0 & \frac{C_{22}^{(2)} - C_{23}^{(2)}}{2} & 0 & 0 \\ 0 & 0 & 0 & 0 & C_{66}^{(2)} & 0 \\ 0 & 0 & 0 & 0 & 0 & C_{66}^{(2)} \end{bmatrix} \begin{pmatrix} \varepsilon_{rr}^{(2)} \\ \varepsilon_{\theta\theta}^{(2)} \\ \varepsilon_{zz}^{(2)} \\ 2\varepsilon_{\theta z}^{(2)} \\ 2\varepsilon_{rz}^{(2)} \\ 2\varepsilon_{r\theta}^{(2)} \end{pmatrix} , \qquad (17)$$

It can be seen that the axis of symmetry of this transversely isotropic layer differs from the axis of the isotropy plane of the fiber. The obtained effective physical elastic moduli in (10-14) can be converted to elastic moduli $C_{ij}^{(2)}$ (17) using the following relations:

$$\begin{aligned} C_{11} &= E_{11} + 4v_{12}^2 K_{23}, \\ C_{12} &= 2K_{23}v_{12}, \\ C_{22} &= \mu_{23} + K_{23}, \\ C_{23} &= -\mu_{23} + K_{23}, \\ C_{66} &= \mu_{12} = \mu_{13}. \end{aligned} \quad , \qquad (18)$$

Further, the tensor of elastic moduli of an isotropic matrix or the third phase ($i = N = 3$) can be written as

$$\begin{pmatrix} \sigma_{rr}^{(3)} \\ \sigma_{\theta\theta}^{(3)} \\ \sigma_{zz}^{(3)} \\ \sigma_{\theta z}^{(3)} \\ \sigma_{rz}^{(3)} \\ \sigma_{r\theta}^{(3)} \end{pmatrix} = \begin{bmatrix} C_{11}^{(3)} & C_{12}^{(3)} & C_{12}^{(3)} & 0 & 0 & 0 \\ C_{12}^{(3)} & C_{11}^{(3)} & C_{12}^{(3)} & 0 & 0 & 0 \\ C_{12}^{(3)} & C_{12}^{(3)} & C_{11}^{(3)} & 0 & 0 & 0 \\ 0 & 0 & 0 & \frac{C_{11}^{(3)} - C_{12}^{(3)}}{2} & 0 & 0 \\ 0 & 0 & 0 & 0 & \frac{C_{11}^{(3)} - C_{12}^{(3)}}{2} & 0 \\ 0 & 0 & 0 & 0 & 0 & \frac{C_{11}^{(3)} - C_{12}^{(3)}}{2} \end{bmatrix} \begin{pmatrix} \varepsilon_{rr}^{(3)} \\ \varepsilon_{\theta\theta}^{(3)} \\ \varepsilon_{zz}^{(3)} \\ 2\varepsilon_{\theta z}^{(3)} \\ 2\varepsilon_{rz}^{(3)} \\ 2\varepsilon_{r\theta}^{(3)} \end{pmatrix} , \qquad (19)$$

The elastic moduli tensor for an equivalent homogeneous medium or external phase ($i = N + 1$) is assumed to be equal to the effective elastic modulus of the effective homogenized material and is the desired quantity. In addition, this medium has transversely isotropic properties along the axis of symmetry, directed along the axis of the base fiber, and therefore we have

$$\begin{pmatrix} \sigma_{rr}^{N+1} \\ \sigma_{\theta\theta}^{N+1} \\ \sigma_{zz}^{N+1} \\ \sigma_{\theta z}^{N+1} \\ \sigma_{rz}^{N+1} \\ \sigma_{r\theta}^{N+1} \end{pmatrix} = \begin{bmatrix} C_{11}^{N+1} & C_{12}^{N+1} & C_{13}^{N+1} & 0 & 0 & 0 \\ C_{12}^{N+1} & C_{11}^{N+1} & C_{13}^{N+1} & 0 & 0 & 0 \\ C_{13}^{N+1} & C_{13}^{N+1} & C_{33}^{N+1} & 0 & 0 & 0 \\ 0 & 0 & 0 & C_{44}^{N+1} & 0 & 0 \\ 0 & 0 & 0 & 0 & C_{44}^{N+1} & 0 \\ 0 & 0 & 0 & 0 & 0 & \frac{C_{11}^{N+1} - C_{12}^{N+1}}{2} \end{bmatrix} \begin{pmatrix} \varepsilon_{rr}^{N+1} \\ \varepsilon_{\theta\theta}^{N+1} \\ \varepsilon_{zz}^{N+1} \\ 2\varepsilon_{\theta z}^{N+1} \\ 2\varepsilon_{rz}^{N+1} \\ 2\varepsilon_{r\theta}^{N+1} \end{pmatrix} , \qquad (20)$$

Finally, the elastic modulus tensor for an effective composite (medium without inclusions or a homogenized composite that has equivalent transversely isotropic properties) has the same elastic moduli tensor of the equivalent homogeneous medium (i.e. $C_{ij}^{eff} = C_{ij}^{N+1}$), and we write this in the form:

$$\begin{pmatrix} \sigma_{rr}^{eff} \\ \sigma_{\theta\theta}^{eff} \\ \sigma_{zz}^{eff} \\ \sigma_{\theta z}^{eff} \\ \sigma_{rz}^{eff} \\ \sigma_{r\theta}^{eff} \end{pmatrix} = \begin{bmatrix} C_{11}^{eff} & C_{12}^{eff} & C_{13}^{eff} & 0 & 0 & 0 \\ C_{12}^{eff} & C_{11}^{eff} & C_{13}^{eff} & 0 & 0 & 0 \\ C_{13}^{eff} & C_{13}^{eff} & C_{33}^{eff} & 0 & 0 & 0 \\ 0 & 0 & 0 & C_{44}^{eff} & 0 & 0 \\ 0 & 0 & 0 & 0 & C_{44}^{eff} & 0 \\ 0 & 0 & 0 & 0 & 0 & \frac{C_{11}^{eff} - C_{12}^{eff}}{2} \end{bmatrix} \begin{pmatrix} \varepsilon_{rr}^{eff} \\ \varepsilon_{\theta\theta}^{eff} \\ \varepsilon_{zz}^{eff} \\ 2\varepsilon_{\theta z}^{eff} \\ 2\varepsilon_{rz}^{eff} \\ 2\varepsilon_{r\theta}^{eff} \end{pmatrix} , \qquad (21)$$

2.2 Definition of longitudinal shear modulus

2.2.1. Statement of the problem of pure shear in the direction along the fibers in cylindrical coordinates for an orthotropic multiphase medium.

It is assumed that the fiber axes are directed along the z axis (axis 3) in a cylindrical coordinate system. The effective longitudinal shear modulus is determined from the solution of the pure shear problem along the fibers (see Fig. 5). To begin with, it is necessary to find the general fields of displacement of orthotropic phases that satisfy the equilibrium equations of this problem. According to [12, 13, 14], the field of displacement along the fibers $u_z(r,\theta)$ is specified on the outer border of the cell. In this case, there are only phase movements along the fibers, which depend on the radius and angle, and the angular and radial movements are equal to zero:

$$u_z^{(i)}(r,\theta) \neq 0, \quad u_r^{(i)} = u_\theta^{(i)} = 0., \qquad (22)$$

where i is the phase of the composite.

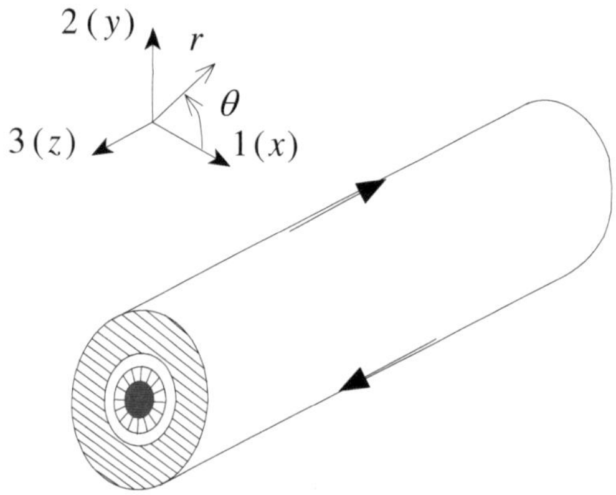

Fig. 5. Pure longitudinal shear problem.

For simplicity, we introduce the notation $u_z(r,\theta) = u_z$, and by the Cauchy deformation relations:

$$\varepsilon_{rr} = \frac{\partial u_r}{\partial r}, \quad \varepsilon_{\theta\theta} = \frac{1}{r}\left(\frac{\partial u_\theta}{\partial \theta} + u_r\right), \quad \varepsilon_{zz} = \frac{\partial u_z}{\partial z},$$

$$\varepsilon_{rz} = \frac{1}{2}\left(\frac{\partial u_z}{\partial r} + \frac{\partial u_\theta}{\partial z}\right), \quad \varepsilon_{\theta z} = \frac{1}{2}\left(\frac{\partial u_\theta}{\partial z} + \frac{1}{r}\frac{\partial u_z}{\partial \theta}\right), \quad \varepsilon_{r\theta} = \frac{1}{2}\left(\frac{1}{r}\frac{\partial u_r}{\partial \theta} + \frac{\partial u_\theta}{\partial r} - \frac{u_\theta}{r}\right), \qquad (23)$$

and (22) we get

$$\varepsilon_{zz} = \frac{\partial u_z}{\partial z}, \quad 2\varepsilon_{\theta z} = \frac{\partial u_\theta}{\partial z} + \frac{1}{r}\frac{\partial u_z}{\partial \theta} = \frac{1}{r}\frac{\partial u_z}{\partial \theta}, \quad 2\varepsilon_{rz} = \frac{\partial u_z}{\partial r} + \frac{\partial u_r}{\partial z} = \frac{\partial u_z}{\partial r},$$
$$\varepsilon_{rr} = \varepsilon_{\theta\theta} = \varepsilon_{r\theta} = 0. \qquad , \qquad (24)$$

With the help of Hooke's law

$$\begin{pmatrix} \sigma_{rr} \\ \sigma_{\theta\theta} \\ \sigma_{zz} \\ \sigma_{\theta z} \\ \sigma_{rz} \\ \sigma_{r\theta} \end{pmatrix} = \begin{bmatrix} C_{rr} & C_{r\theta} & C_{rz} & 0 & 0 & 0 \\ C_{r\theta} & C_{\theta\theta} & C_{\theta z} & 0 & 0 & 0 \\ C_{rz} & C_{\theta z} & C_{zz} & 0 & 0 & 0 \\ 0 & 0 & 0 & G_{\theta z} & 0 & 0 \\ 0 & 0 & 0 & 0 & G_{rz} & 0 \\ 0 & 0 & 0 & 0 & 0 & G_{r\theta} \end{bmatrix} \begin{pmatrix} \varepsilon_{rr} \\ \varepsilon_{\theta\theta} \\ \varepsilon_{zz} \\ 2\varepsilon_{\theta z} \\ 2\varepsilon_{rz} \\ 2\varepsilon_{r\theta} \end{pmatrix}, \qquad (25)$$

together with relations (24) we have

$$\sigma_{\theta z} = G_{\theta z} 2\varepsilon_{\theta z} = G_{\theta z}\left(\frac{1}{r}\frac{\partial u_z}{\partial \theta}\right),$$
$$\sigma_{rz} = G_{rz} 2\varepsilon_{rz} = G_{rz}\left(\frac{\partial u_z}{\partial r}\right). \qquad , \quad (26)$$

Substituting (26) into the equilibrium equation

$$\frac{\partial \sigma_{rr}^{(i)}}{\partial r} + \frac{1}{r}\frac{\partial \sigma_{r\theta}^{(i)}}{\partial \theta} + \frac{\sigma_{rr}^{(i)} - \sigma_{\theta\theta}^{(i)}}{r} + \frac{\partial \sigma_{rz}^{(i)}}{\partial z} = 0,$$
$$\frac{\partial \sigma_{r\theta}^{(i)}}{\partial r} + \frac{1}{r}\frac{\partial \sigma_{\theta\theta}^{(i)}}{\partial \theta} + \frac{2\sigma_{r\theta}^{(i)}}{r} + \frac{\partial \sigma_{\theta z}^{(i)}}{\partial z} = 0,$$
$$\frac{\partial \sigma_{rz}^{(i)}}{\partial r} + \frac{1}{r}\frac{\partial \sigma_{\theta z}^{(i)}}{\partial \theta} + \frac{\sigma_{rz}^{(i)}}{r} + \frac{\partial \sigma_{zz}^{(i)}}{\partial z} = 0, \qquad , \quad (27)$$

We get:

$$\frac{\partial}{\partial r}\left[G_{rz}\left(\frac{\partial u_z}{\partial r}\right)\right] + \frac{1}{r}\left[\frac{\partial}{\partial \theta}\left(G_{\theta z}\left(\frac{1}{r}\frac{\partial u_z}{\partial \theta}\right)\right)\right] + \frac{1}{r}\left[G_{rz}\left(\frac{\partial u_z}{\partial r}\right)\right] = 0$$
$$\Rightarrow G_{rz}\frac{\partial^2 u_z}{\partial r^2} + \frac{1}{r}G_{rz}\frac{\partial u_z}{\partial r} + \frac{1}{r^2}G_{\theta z}\frac{\partial^2 u_z}{\partial \theta^2} = 0$$
$$\Rightarrow r^2\frac{\partial^2 u_z}{\partial r^2} + r\frac{\partial u_z}{\partial r} + \lambda^2\frac{\partial^2 u_z}{\partial \theta^2} = 0$$
$$\Rightarrow r^2 u_z''(r) + r u_z'(r) + \lambda^2 u_z''(\theta) = 0. \qquad , \quad (28)$$

where $\lambda = \sqrt{G_{\theta z}/G_{rz}}$. Another form of differential equation (28) can be written as

$$r^2 R''(r)T(\theta) + rR'(r)T(\theta) + \lambda^2 R(r)T''(\theta) = 0$$
$$\Rightarrow \quad \frac{r^2 R''(r) + rR'(r)}{-R(r)} = \lambda^2\frac{T''(\theta)}{T(\theta)}, \qquad , \quad (29)$$

where the general solution of this equation has the form

$$u_z(r,\theta) = R(r)T(\theta) = u_z(r)u_z(\theta) \qquad , \quad (30)$$

Now we find the solution form for $u_z(r)$. Let the right side of equation (29) be given by some constant, i.e. $T''(\theta)/T(\theta) = K$, and then equation (29) has the form

$$\frac{r^2 R_z''(r) + r R_z'(r)}{-\lambda^2 R_z(r)} = K \qquad , \quad (31)$$

Let $K = -1$, and therefore, equation (31) takes the form

$$r^2 R_z''(r) + r R_z'(r) - \lambda^2 R_z(r) = r^2 u_z''(r) + r u_z'(r) - \lambda^2 u_z(r) = 0 \qquad , \quad (32)$$

By making in (32) a variable change

$$u_r = r^m, \quad u_r' = mr^{m-1}, \quad u_r'' = m(m-1)r^{m-2}, \qquad (33)$$

we arrive at a characteristic equation for m:

$$r^2\left[m(m-1)r^{m-2}\right] + r\left[mr^{m-1}\right] - \lambda^2 r^m = 0$$
$$\Rightarrow m^2 - \lambda^2 = 0$$
$$\Rightarrow m_{1,2} = +/-\lambda. \qquad , \quad (34)$$

Thus, the general solution of the differential equation (35) is

$$R(r) = u_r(r) = A_1 r^{\lambda} + A_2 r^{-\lambda}, \quad (35)$$

Next, look for the general form of solutions for $u_\theta(r)$. Similarly, let the left side of equation (32) be –1, and therefore we have

$$T''(\theta) + T(\theta) = 0, \quad (36)$$

and the general form of solving the differential equation (36) can be written as

$$T(\theta) = u_z(\theta) = B_1 \cos\theta + B_2 \sin\theta, \quad (37)$$

However, based on the state of cell deformation under the action of a pure longitudinal shear, the constant B_2 must be equal to zero, and therefore the solution (37) is reduced to

$$T(\theta) = u_z(\theta) = B_1 \cos\theta, \quad (38)$$

Thus, on the basis of (35) and (38) we have

$$u_z(r,\theta) = \left(A_1 r^{\lambda} + A_2 r^{-\lambda}\right) B_1 \cos\theta = \\ = \left(A_1 B_1 r^{\lambda} + A_2 B_1 r^{-\lambda}\right)\cos\theta., \quad (39)$$

So, the general allowable fields of displacement of an orthotropic multiphase composite are:

$$u_z^{(i)}(r,\theta) = \left(D_1^{(i)} r^{\lambda_{(i)}} + D_2^{(i)} r^{-\lambda_{(i)}}\right)\cos\theta, \quad (40)$$

where $D_1^{(i)}, D_2^{(i)}$ - unknown constants, which are determined from the boundary conditions; and $\lambda_{(i)} = \sqrt{G_{\theta z}^{(i)} / G_{rz}^{(i)}}$ - material phase constant.

Consequently, using the Cauchy relations, one can write the admissible deformation fields of orthotropic phases in the following form

$$2\varepsilon_{\theta z}^{(i)}(r,\theta) = \frac{1}{r}\frac{\partial u_r}{\partial z} = \frac{1}{r}\frac{\partial}{\partial z}\left[\left(D_1^{(i)} r^{\lambda_{(i)}} + D_2^{(i)} \frac{1}{r^{\lambda_{(i)}}}\right)\cos\theta\right] = \\ = \left(D_1^{(i)} r^{\lambda_{(i)}-1} + D_2^{(i)} r^{-\lambda_{(i)}-1}\right)(-\sin\theta), \quad (41)$$

$$2\varepsilon_{rz}^{(i)}(r,\theta) = \frac{\partial u_z}{\partial r} = \frac{\partial}{\partial r}\left[\left(D_1^{(i)} r^{\lambda_{(i)}} + D_2^{(i)} \frac{1}{r^{\lambda_{(i)}}}\right)\cos\theta\right] = \\ = \left(\lambda_{(i)} D_1^{(i)} r^{\lambda_{(i)}-1} + (-\lambda) D_2^{(i)} r^{-\lambda_{(i)}-1}\right)\cos\theta, \quad (42)$$

and using Hooke's law for (41) and (42), the admissible stress fields of orthotropic phases take the form

$$\sigma_{\theta z}^{(i)}(r,\theta) = G_{\theta z}^{(i)} 2\varepsilon_{z\theta}^{(i)} = -G_{\theta z}^{(i)}\left(D_1^{(i)} r^{\lambda_{(i)}-1} + D_2^{(i)} r^{-\lambda_{(i)}-1}\right)\sin\theta = \\ = -C_{44}^{(i)}\left(D_1^{(i)} r^{\lambda_{(i)}-1} + D_2^{(i)} r^{-\lambda_{(i)}-1}\right)\sin\theta, \quad (43)$$

$$\sigma_{rz}^{(i)}(r,\theta) = G_{rz}^{(i)} 2\varepsilon_{rz}^{(i)} = G_{rz}^{(i)}\left(\lambda_{(i)} D_1^{(i)} r^{\lambda_{(i)}-1} + (-\lambda_{(i)}) D_2^{(i)} r^{-\lambda_{(i)}-1}\right)\cos\theta = \\ = C_{55}^{(i)}\left(\lambda_{(i)} D_1^{(i)} r^{\lambda_{(i)}-1} + (-\lambda_{(i)}) D_2^{(i)} r^{-\lambda_{(i)}-1}\right)\cos\theta. \quad (44)$$

It can be seen that in equations (43) and (44), the tensors of elastic moduli G_{rz} and $G_{\theta z}$ are replaced by indices C_{55} and C_{44}, respectively.

In the case of isotropic or transversely isotropic media with the axis of the isotropy plane directed along the fibers, the material phase constant is equal to

$$\lambda_{(i)} = \sqrt{G_{\theta z}^{(i)} / G_{rz}^{(i)}} = \sqrt{C_{44}^{(i)} / C_{55}^{(i)}} = 1, \quad (45)$$

because in this case there is equality

$$G_{\theta z} = G_{rz} \text{ or } C_{44}^{(i)} = C_{55}^{(i)}, \quad (46)$$

In addition, the ratio

$$\mu_{23} = \mu_{32} = C_{44} = C_{55} = \mu_{13} = \mu_{31}, \quad (47)$$

is valid. Thus, taking into account (45), (46) and (47), we can rewrite (40-44) as follows:

$$u_r^{(i)}(r,\theta) = \left(D_1^{(i)} r + D_2^{(i)} r^{-1}\right)\cos\theta, \quad (48)$$

$$2\varepsilon_{\theta z}^{(i)}(r,\theta) = -\left(D_1^{(i)} + D_2^{(i)} r^{-2}\right)\sin\theta, \quad (49)$$

$$2\varepsilon_{rz}^{(i)}(r,\theta) = \left(D_1^{(i)} - D_2^{(i)} r^{-2}\right)\cos\theta, \quad (50)$$

$$\sigma_{\theta z}^{(i)}(r,\theta) = -\mu_{23}^{(i)}\left(D_1^{(i)} + D_2^{(i)} r^{-2}\right)\sin\theta, \quad (51)$$

$$\sigma_{rz}^{(i)}(r,\theta) = \mu_{23}^{(i)}\left(D_1^{(i)} - D_2^{(i)} r^{-2}\right)\cos\theta, \quad (52)$$

2.2.2 Method of obtaining an effective longitudinal shear modulus.

Consider again the pure shear problem for a layered structure and assume that a displacement field $u_z(r,\theta) = 2\varepsilon_0 r\cos\theta$ is specified in the direction along the cell at infinity. In that case only the allowable movement along the fibers occur, which have the following form

$$u_z^{(i)}(r,\theta) = \left(D_1^{(i)} r^{\lambda_{(i)}} + D_2^{(i)} r^{-\lambda_{(i)}}\right)\cos\theta, \quad (53)$$

Where $\lambda_{(i)} = \sqrt{C_{44}^{(i)}/C_{55}^{(i)}}$, (54)

for isotropic and transversely isotropic phases with an axis of symmetry directed along the fiber axis, $\lambda_{(i)} = 1$. From equations (43) and (44), the corresponding stress fields of each phase take the form

$$\sigma_{\theta z}^{(i)}(r,\theta) = -C_{44}^{(i)}\left(D_1^{(i)} r^{\lambda_{(i)}-1} + D_2^{(i)} r^{-\lambda_{(i)}-1}\right)\sin\theta,$$

$$\sigma_{rz}^{(i)}(r,\theta) = C_{55}^{(i)}\left(\lambda_{(i)} D_1^{(i)} r^{\lambda_{(i)}-1} - \lambda_{(i)} D_2^{(i)} r^{-\lambda_{(i)}-1}\right)\cos\theta. \quad (55)$$

Now we define the fields of displacements and stresses for the equivalent homogeneous medium, which correspond to media with transversely isotropic properties with an isotropy plane perpendicular to the fiber axis. From equations (48), (51) and (52), we have

$$u_z^{N+1}(r,\theta) = \left(D_1^{N+1} r + D_2^{N+1} r^{-1}\right)\cos\theta, \quad (56)$$

$$\sigma_{\theta z}^{N+1}(r,\theta) = -\mu_{23}^{N+1}\left(D_1^{N+1} + D_2^{N+1} r^{-2}\right)\sin\theta,$$

$$\sigma_{rz}^{N+1}(r,\theta) = \mu_{23}^{N+1}\left(D_1^{N+1} - D_2^{N+1} r^{-2}\right)\cos\theta. \quad (57)$$

Obviously, we have nine unknown constants consisting of eight unknown constants ($D_1^{(1)}, D_2^{(1)}, D_1^{(2)}, D_2^{(2)}, D_1^{(3)}, D_2^{(3)}, D_1^{N+1}$ and D_2^{N+1}) and one unknown module - the effective module of the longitudinal shear μ_{23}^{N+1}.

Next, to find all the constants, the condition of non-singularity in the center of the fiber ($r = 0$) is set:

$$D_2^{(1)} = 0, \quad (58)$$

continuity condition at the boundaries of the ideal contact of the phases:

$$u_z^{(i)}(r_{(i)},\theta) = u_z^{(i+1)}(r_{(i)},\theta), \quad \sigma_{rz}^{(i)}(r_{(i)},\theta) = \sigma_{rz}^{(i+1)}(r_{(i)},\theta), \quad (i = 1,2,\ldots N); \quad (59)$$

outer boundary condition at infinity ($r_{N+1} \to \infty$):

$$u_z^{N+1}(r_{N+1}) = 2\varepsilon_0 r_{N+1}\cos\theta \quad , \qquad (60)$$

where in the micromechanics of composites the value ε_0 is often taken as unity ($\varepsilon_0 = 1$); and the condition of the energy interaction over the surface at the contact ($r = r_N$):

$$\int_S \left(\sigma_{rz}^{N+1} u_z^{eff} - \sigma_{rz}^{eff} u_z^{N+1}\right)_{r=r_N} dS = 0 \quad , \qquad (61)$$

It is seen that the angular stress fields are not included in the Eshelby integral formula (60), because the angular displacements are equal to zero. First, we define the displacement and stress fields of the effective composite in the integrand (60). These fields have the same shape of the displacement and stress field of an equivalent homogeneous medium:

$$u_z^{eff}(r,\theta) = \left(D_1^{eff} r + D_2^{eff} r^{-1}\right)\cos\theta, \qquad (62)$$

$$\sigma_{\theta z}^{eff}(r,\theta) = -\mu_{23}^{eff}\left(D_1^{eff} + D_2^{eff} r^{-2}\right)\sin\theta,$$

$$\sigma_{rz}^{eff}(r,\theta) = \mu_{23}^{eff}\left(D_1^{eff} - D_2^{eff} r^{-2}\right)\cos\theta. \qquad (63)$$

Using equality (58), which is also applicable to an effective composite, where $D_2^{eff} = 0$, (62) and (63) lead to the following

$$u_z^{eff}(r,\theta) = D_1^{eff}\cos\theta \quad , \qquad (64)$$

$$\sigma_{\theta z}^{eff}(r,\theta) = -\mu_{23}^{eff} D_1^{eff}\sin\theta,$$

$$\sigma_{rz}^{eff}(r,\theta) = \mu_{23}^{eff} D_1^{eff}\cos\theta. \qquad (65)$$

It can be seen that on the basis of (59), which is also applicable for an effective composite together with (64), the unknown constant D_1^{eff} takes the form

$$D_1^{eff} = 2\varepsilon_0 \quad , \qquad (66)$$

Thus, based on (56), (57), (64), (65) and (66), formula (61) is reduced to

$$D_2^{N+1} = 0 \quad , \qquad (67)$$

where the cell length is considered equal to $2L$. Further, based on (56), (60) and (67), we can write the value D_1^{N+1} in the form

$$D_1^{N+1} = 2\varepsilon_0 \quad , \qquad (68)$$

Now we have five unknown constants - $D_1^{(1)}, D_1^{(2)}, D_2^{(2)}, D_1^{(3)}, D_2^{(3)}$ and one desired effective longitudinal shear modulus $\mu_{23}^{N+1} = \mu_{23}^{eff}$. Six unknown constants can be obtained using the conditions of continuity of displacement and stress (59) on the radii $r_1, r_2, r_3 = r_N$:

$$\begin{aligned}
&1)\quad D_1^{(1)} r_1^{\lambda_{(1)}} - D_1^{(2)} r_1^{\lambda_{(2)}} - D_2^{(2)} r_1^{-\lambda_{(2)}} = 0,\\
&2)\quad D_1^{(2)} r_2^{\lambda_{(2)}} + D_2^{(2)} r_2^{-\lambda_{(2)}} - D_1^{(3)} r_2^{\lambda_{(3)}} - D_2^{(3)} r_2^{-\lambda_{(3)}} = 0,\\
&3)\quad D_1^{(3)} r_3^{\lambda_{(3)}} + D_2^{(3)} r_3^{-\lambda_{(3)}} - 2\varepsilon_0 r_3 = 0,\\
&4)\quad C_{55}^{(1)}\lambda_{(1)} D_1^{(1)} r_1^{\lambda_{(1)}-1} - C_{55}^{(2)}\lambda_{(2)} D_1^{(2)} r_1^{\lambda_{(2)}-1} + C_{55}^{(2)}\lambda_{(2)} D_2^{(2)} r_1^{-\lambda_{(2)}-1} = 0,\\
&5)\quad C_{55}^{(2)}\lambda_{(2)} D_1^{(2)} r_2^{\lambda_{(2)}-1} - C_{55}^{(2)}\lambda_{(2)} D_2^{(2)} r_2^{-\lambda_{(2)}-1} - C_{55}^{(3)}\lambda_{(3)} D_1^{(3)} r_2^{\lambda_{(3)}-1} + C_{55}^{(3)}\lambda_{(3)} D_2^{(3)} r_2^{-\lambda_{(3)}-1} = 0,\\
&6)\quad C_{55}^{(3)}\lambda_{(3)} D_1^{(3)} r_3^{\lambda_{(3)}-1} - C_{55}^{(3)}\lambda_{(3)} D_2^{(3)} r_3^{-\lambda_{(3)}-1} - 2\varepsilon_0\mu_{23}^{N+1} = 0.
\end{aligned} \quad , \qquad (69)$$

It can be seen that the effective volumetric modulus of longitudinal shear is determined from the sixth equation (69)

$$\mu_{23}^{N+1} = \mu_{23}^{eff} = \frac{1}{2\varepsilon_0} C_{55}^{(3)} \left[\lambda_{(3)} D_1^{(3)} r^{\lambda_{(3)}-1} + (-\lambda_{(3)}) D_2^{(3)} r^{-\lambda_{(3)}-1} \right] \cos\theta. \quad (70)$$

Thus, if we generalize the result (70) for multiphase composites, then the effective volumetric modulus of longitudinal shear is

$$\mu_{23}^{eff} = \frac{1}{2\varepsilon_0} C_{55}^{(N)} \left[\lambda_{(N)} D_1^{(N)} r^{\lambda_{(N)}-1} + (-\lambda_{(N)}) D_2^{(N)} r^{-\lambda_{(N)}-1} \right] \cos\theta, \quad (71)$$

Where

$$\lambda_{(N)} = \sqrt{C_{44}^{(N)} / C_{55}^{(N)}}, \quad (72)$$

Table 1. Material parameters and configuration of CNT-whiskerized fiber composites

	PHASE		
	Base Fiber	**Whiskers**	**Matrix**
Materials	Carbon T-650	CNT	Epoxy
Dimensions			
- Diameter (μm)	5 μm	0,00051 – 0,00085	—
- Length (μm)	—	1 – 2	—
Properties			
- Young's longitudinal modulus, E_L (GPa)	241	1100	3
- Young's transverse modulus, E_T (GPa)	14,5	—	—
- Longitudinal shear modulus, μ_L (GPa)	22,8	—	—
- Transverse shear modulus, μ_L (GPa)	4,8	—	—
- Poisson's ratio, ν_{LT}	0,27	0,14	0,3

3. Results and Discussion

In order to determine the advantage of a composite with a whiskerized interfacial layer over the classical composite, it is necessary to determine the destructive stresses by the criterion of the matrix failure, the destructive deformations will be taken equal to 0.4%. The calculation is performed according to the two-phase method based on the Eshelby self-consistent field method.

The method of two phases:

Assume $\varepsilon_0 = Pt / \mu_{23}^{eff}$, where $Pt_1 = 6.35MPa$, $\mu_{23}^{eff} = 1.5911GPa$. When $\theta = \pi/2$ stress and strain are as follows:

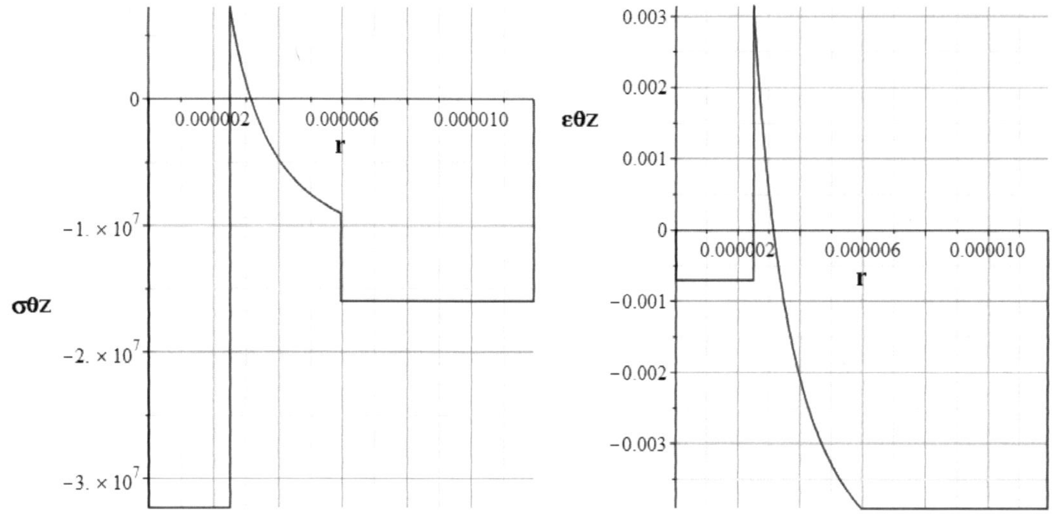

When $\theta = 0$:

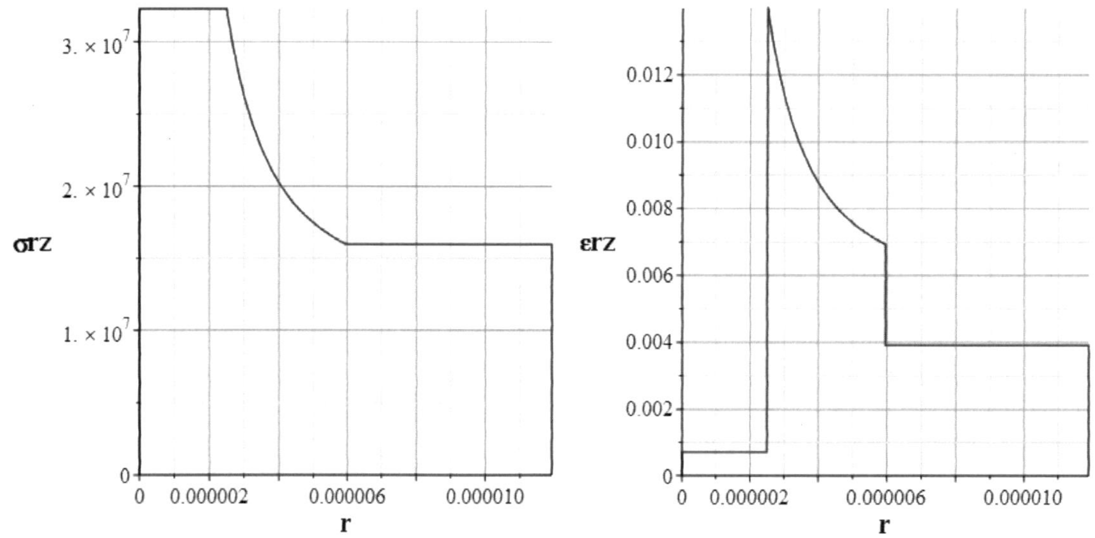

Fig. 6. Shear stresses and shear deformations using the two phase method.

Making the interphase layer matrix properties:

In the case when we accept the properties of the interfacial layer as equal to the properties of the matrix, the result is absolutely similar, which indicates the correctness of the model we have constructed.

Assume $\varepsilon_0 = Pt / \mu_{23}^{eff}$, where where $Pt_1 = 6.35MPa$, $\mu_{23}^{eff} = 1.5911GPa$. When $\theta = \pi / 2$ stress and strain are as follows:

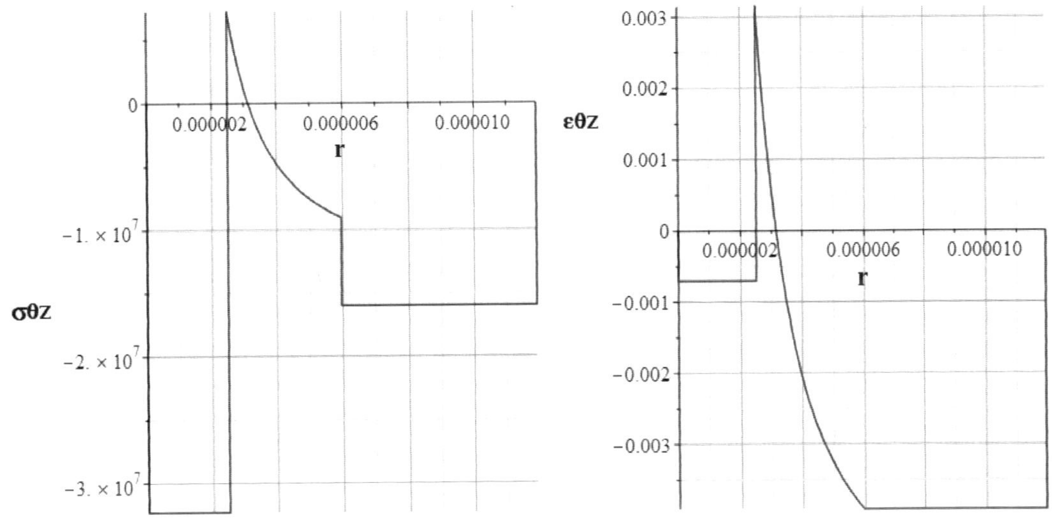

When $\theta = 0$:

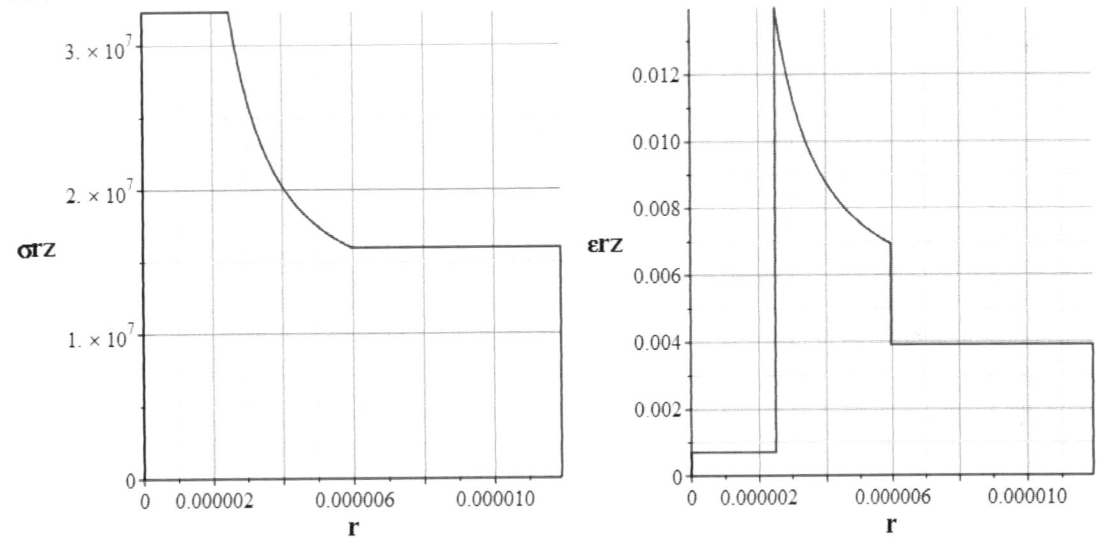

Fig. 7. Shear stresses and shear deformations with imparting matrix properties to the interfacial layer.

The method of three phases:

As can be seen from the results, the destructive load is $Pt_1 = 6.35MPa$. We take this value as a starting point for finding the destructive load for a three-phase composite with a whiskerized interfacial layer. At $Pt_1 = 6.35MPa$ the deformations in the matrix of the effective composite equal $\varepsilon = 0,4\%$. Due to the increased modulus of elasticity of the longitudinal shear of the composite with the interfacial layer compared to the classical, the next step we "load" the composite to the level of destructive deformations $\varepsilon = 0,4\%$.

Assume $\varepsilon_0 = Pt / \mu_{23}^{eff}$, where $Pt_2 = 8.13MPa$, $\mu_{23}^{eff} = 2.0358GPa$. At $\theta = \pi / 2$ stress and strain are as follows:

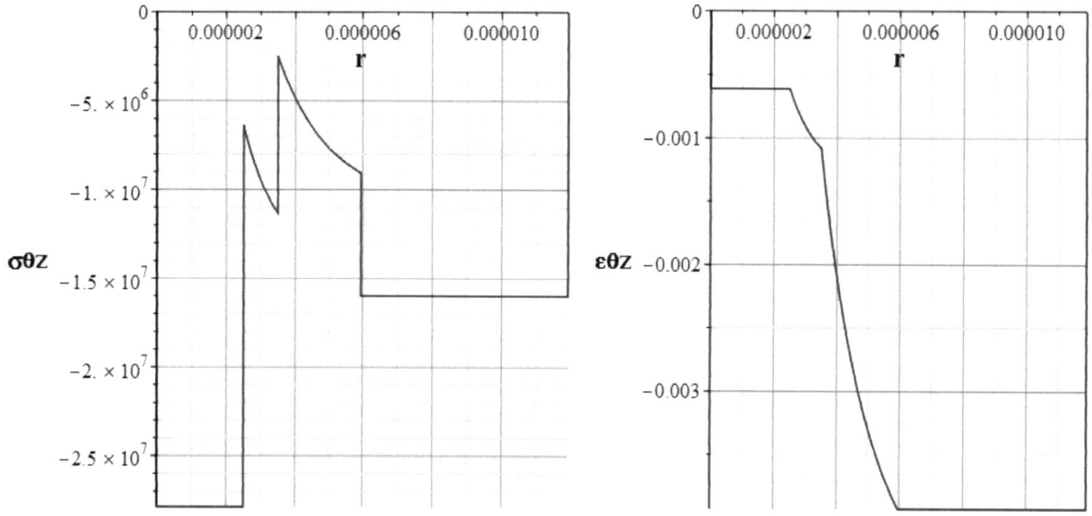

At $\theta = 0$:

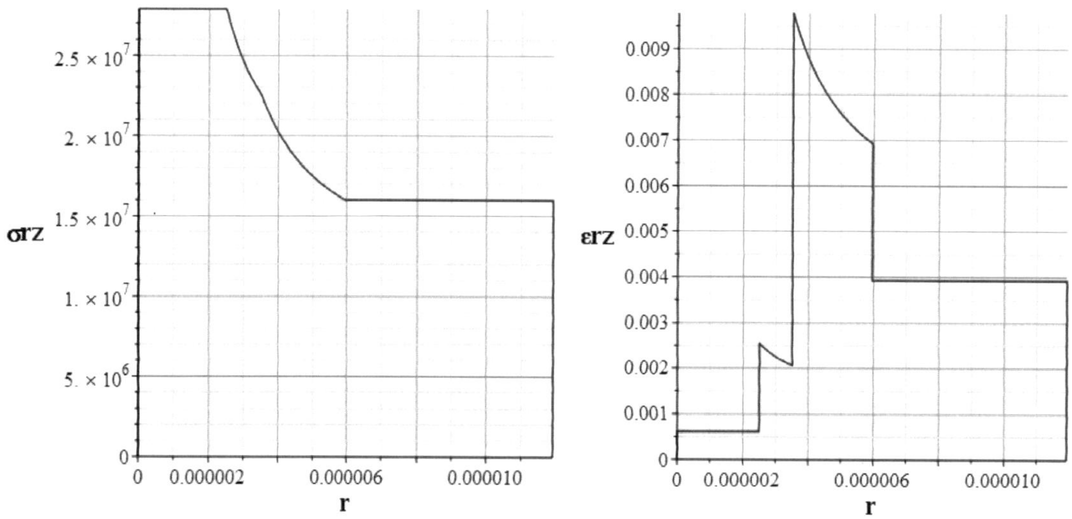

Fig. 8. Shear stresses and shear deformations with imparting matrix properties to the interfacial layer.

4. Conclusions

As can be seen, the destruction of the composite occurs under load $Pt_2 = 8.13MPa$, from which we conclude that the hardening coefficient of the composite by the presence of a whiskerized interfacial layer is $n_{упр} = \frac{Pt_2}{Pt_1} = \frac{8.13}{6.35} = 1.28$ times. In addition, there is a noticeable decrease in the level of stresses in the fiber, since part of the load is perceived by the interfacial layer, unloading the fiber, which has a positive effect on the strength characteristics of such a composite.

IOP Conf. Series: Materials Science and Engineering **683** (2019) 012036 doi:10.1088/1757-899X/683/1/012036

REFERENCES

[1] Vasilev V V, Tarnopolsky U M 1990 Composite materials. Handbook. *Moscow, Engineering*, 512 p.

[2] Lurie S A, Dudchenko A A 2010 Introduction to the Mechanics of Nanocomposites *MAI Moscow* 160

[3] Kim J K, Mai Y W 1998 Engineered interfaces in fiber reinforced composites *Elsevier Science Ltd* 18–33

[4] Lin Y, Ehlert G J, Sodano H A 2009 Increase interface strength in carbon fiber composites through a ZnO nanowire interphase *Advan. Funct Mater* 19:16 2654–2660

[5] Sager R J, Klein P J, Lagoudas D C , Zhang Q, Liu J , Dai L 2009 Effect of carbon nanotubes on the interfacial shear strength of T650 carbon fiber in an epoxy matrix. *Compos. Sci. Technol.* **69** 898–904.

[6] Wang Y, Tang Z, Liang X, Liz-Marzan L M, Kotov N A 2004 SiO_2-Coated CdTe nanowires: bristled nano centipedes. *Nano Letters* 4:2 225–231

[7] Galan U, Lin Y, Ehlert G J, Sodano H A 2011 Effect of ZnO nanowire morphology on the interfacial strength of nanowire coated fibers *Compos. Sci. Technol* **71** 946–954

[8] Garcia E J, Wardle B L, Hart A J, Yamamonj N 2008 Fabrication and multifunctional properties of a hybrid laminate with aligned carbon nanotubes grown in situ. *Compos. Sci. Technol.* 68:9 2034–2041

[9] Veedu V P, Cao A, Li X, Ma K, Soldano C, Kar S, Ajayan P M, Ghasemi-Nejhad M N 2006 Multifunctional composites using reinforced laminae with carbon nanotube forests. *Nature materials* **5** 457–462

[10] Alipour Skandani A, Masghouni N, Case S W, Leo D J, Al-Haik M 2012 Enhanced vibration damping of carbon fibers-ZnO nanorods hybrid composites. *Applied Physics Letters* 101:073111 1–4

[11] Christensen R M 1982 Introduction to the mechanics of composites *Mir, Moscow,* 334

[12] Tsukrov I, Drach B 2010 Elastic deformation of composite cylinders with cylindrically orthotropic layers. *Int. J. Eng. Sci.* **47** 25–33

[13] Chatzigeorgiou G, Siedel G D, Lagoudas D 2012 Effective mechanical of ‘‘fuzzy fiber’’ composites. *Composites B* **43** 2577–2593

[14] Lurie S, Minhat M 2014 Application of generalized self-consistent method to predict effective elastic properties of bristled fiber composites. Composites: Part B. **61** 26–40

IOP Conf. Series: Materials Science and Engineering **683** (2019) 012037 doi:10.1088/1757-899X/683/1/012037

Dual-curing propargyl/phthalonitrile monomers for composites by vacuum infusion process

S S Nechausov[1], A.P. Malakho[1], B A Bulgakov [1,2]

[1]Faculty of Chemistry, Lomonosov Moscow State University, Leninskie gory, 1-3, 119991 Moscow, Russia;

[2] Corresponding author Sergey Nechausov

Abstract. Four different bifunctional monomers were synthesized: two monomers containing N-propargylamine and phthalonitrile groups and two monomers containing O-propargyl ester and phthalonitrile groups. It was shown that only the latter one can be used to obtain composite materials by vacuum infusion (viscosity 40 mPa*s at 120 °C). The N-propargylamine containing monomers demonstrated high viscosity of the melts and degradation with allene evolution while curing. Carbon fiber reinforced plastics (CFRP) were successfully obtained from 4-[3-(prop-2-yn-1-yloxy)phenoxy]benzene-1,2-dicarbonitrile by a vacuum infusion molding process. Mechanical testing of the composites samples was performed. Influence of end-curing temperature on mechanical properties and morphology of composites was investigated. The curing temperature of 375 °C critically affects the properties of the composite due to the high shrinkage. Composites cured at 375 °C lost 18% of mechanical properties at 300 °C but cured at 330 only 10%. LOI for composites with a matrix of propargyl/phthalonitrile dual-curing polymers was equal to 75, which is higher than most materials obtained from thermosets.

Keywords: Carbon composites, vacuum infusion, phthalonitriles, propargyl resins, thermosetting copolymers.

Introduction

The use of composite materials in applications ranging from aerospace structures to automotive and marine applications instead of conventional materials is of great interest. Advanced composites show a wide range of desirable properties such as high stiffness and strength, dimensional stability, temperature resistance, and easy processing. One of the most efficient techniques in manufacturing of such composites are injection methods such as vacuum infusion (VIMP) and resin transfer molding (RTM). The main reasons why these methods have recently become so attractive in the fabrication technology are low cost of tooling and scalability to very large structures. VIMP and RTM also allows to minimize the void contents inside the composites what results in less defects content leading to the increase of mechanical properties of composites comparing to other molding techniques [1].

The development of bifunctional polymers with two types of thermosetting functional groups allows to vary the properties of polymer such as melting point, viscosity, curing conditions and customize them for a particular task [2]. It is known that phthalonitrile polymers have the best value of heat resistance among thermosetting polymers of high-temperature use . However, phthalonitrile monomers are often characterized by high melting and polymerization temperatures, which complicates the formation of composites. In addition, a high degree of crosslinking decreases the mechanical properties of phthalonitrile polymers. The introduction of additional functional groups in the polymer, such as propargyl, allows to eliminate the above-mentioned disadvantages of phthalonitrile polymers. Propargyl groups are of great interest as functional groups for modifying the polymers, primarily due to the final properties of the polymer obtained, its high mechanical characteristics and thermal stability. There are few reports about the possibility of copolymerization of propargyl and phtalonitrile groups [3,4]. It is assumed that intermediates formed during the Claisen rearrangement can react with phthalonitrile fragments through OH-groups. This copolymerization allows to decrease greatly the curing temperature of phatalonitrile without loss of properties.

Monomer Synthesis

All the manipulations with oxidation and moisture sensitive compounds were carried out under inert atmosphere using the standard Schlenk technique. All the solvents were purchased from Alfa Aesar and purified according to the standard procedures.

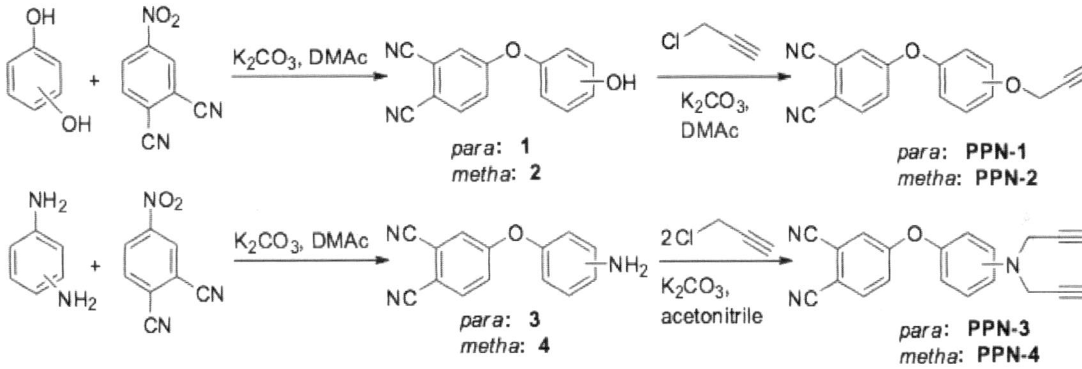

Figure 1. Scheme of synthesis propargyl/phthalonitrile bifunctional monomers PPN-1, PPN-2, PPN-3, PPN-4.

Monomers PPN-1 and PPN-2 was synthesized according to a two-step scheme (Figure 1). In the first step, precursor 1 or 2 reported earlier [5] was obtained by the reaction between hydroquinone or resorcinol and 4- nitrophthalonitrile. The second step was Williamson reaction between 1 or 2 and propargyl chloride in DMAc in the presence of K2CO3. The product was obtained with a total yield of 77% to 4-nitrophthalonitrile and characterized with 1H NMR spectroscopy and DSC.

Monomers PPN-3 and PPN-4 were synthesized in similar ways but for the alkylation raction acetonitrile was used instead of DMAc as solvent according to the method described in Ref. [6].

Analytical Methods.

NMR spectra were run on Bruker Avance 600 at 600 MHz for ^{1}H and 125 MHz for ^{13}C with DMSO-*d6* as solvent. Melt viscosity was measured with MCR 302 Rheometer with cone 10 at 200 r/min. Composite density was measured with Vibra ht-220-CE according to hydrostatic weighing method in pure octane. STA data for samples mass loss was measured with Netzsch STA 449 at a heating rate of 10°C/min in range 40–980°C and an Ar or air purge rate of 50 mL/min. SEM microphotographs were obtained using Vega 3 (Tescan) microscope with a lanthanum hexaboride cathode and secondary electron detector at operating voltage of 20kV. Differential scanning calorimetry (DSC) was performed on Netzsch DSC 449 at a heating rate of 10°C/min and an Ar purge rate of 50 mL/ min.

Mechanical Testing

Interlaminar and in plane shear strengths were measured with Tinius Olsen 50ST according to GOST 32659-2014. Compressive, tensile strengths and elasticity modulus were measured with Instron 5985 according to GOST 9550-81.

Vacuum infusion molding process (VIMP).
Carbon fabric 22502 from JSC INCMaT 2 × 2 twill wave from 3k HTA40 TohoTenax from Teijin fiber was used for CFRP preparation. Auxiliary materials for vacuum infusion were purchased from Airtech.
Vacuum infusion process was used to manufacture CFRP samples. Synthesised resin PPN-2 were used. The full process of infusion was described earlier [7]. Vacuum infusion process was performed at 120°C. Curing process was conducted at 180°C, 12h and post-curing was conducted at 375 °C (C-1) or 330 °C (C-2).

Results and Discussion
It's necessary that the resin has viscosity less than 800 mPa·s in order to be applied in vacuum infusion process. All four investigated resins satisfy this requirement at 130 °C beyond its melting point (Table 1). However the viscosity of PPN-3 resin is the highest of these four monomers. It was found that resins PPN-3 and PPN-4 are in accordance with Carnelley's rule which runs as follows: high molecular symmetry is associated with high melting point [8]. However for PPN-1 and PPN-2 this rule does not work. Perhaps this occurs due to the formation of intermolecular hydrogen bonds, as mentioned in works dedicated to phthalonitriles, in which uncharacteristic dependences of the melting point on the structure of molecules are also observed [9,10].

Table 1. Rheological and curing properties of investigated monomers.

Monomers	$T_{m.}$, °C	Viscosity at 130 °C, mPa·s	Curing parameters		
			T_{onset}, °C	T_{max}, °C	ΔH, J/g
PPN-1	98	39	244	282 303	877
PPN-2	116	47	242	283 301	800
PPN-3	133	567	243	288 298	1001
PPN-4	88	56	245	287 308	1094

IOP Conf. Series: Materials Science and Engineering **683** (2019) 012037 doi:10.1088/1757-899X/683/1/012037

The melting points were found by endothermal pick from DSC-curve (Figure 2).

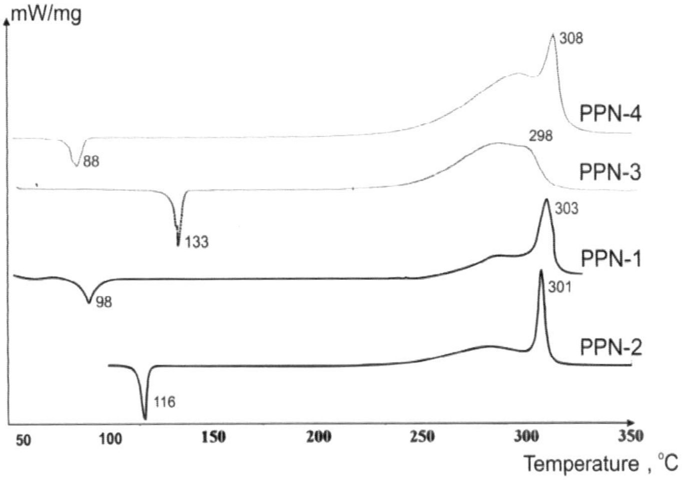

Figure 2. DSC-curves of investigated monomers.

For all monomers, a first exothermal broad peak at 285 °C is observed at DSC-curves, which relates to the copolymerization reaction described above [11]. The second one at 305 °C corresponds to the polymerization of phthalonitrile fragments. It was found that aminopropargyl resins PPN-3 and PPN-4 can't be cured without using any catalysts, because the degradation with allene emission occurs before reaching the temperature of curing (Figure 3). It's observed that the addition of trimerization and Claisen rearrangement of propargyl groups catalysts [12,13] decreases the allene emission but leads to rapid polymerization after the melting point (Figure 4). Therefore catalyst introduction is not suitable for obtaining composite by vacuum infusion. This process requires using resins which features low viscosity under melting point for the time of infusion. Monomer PPN-4 showed similar behavior as PPN-3.

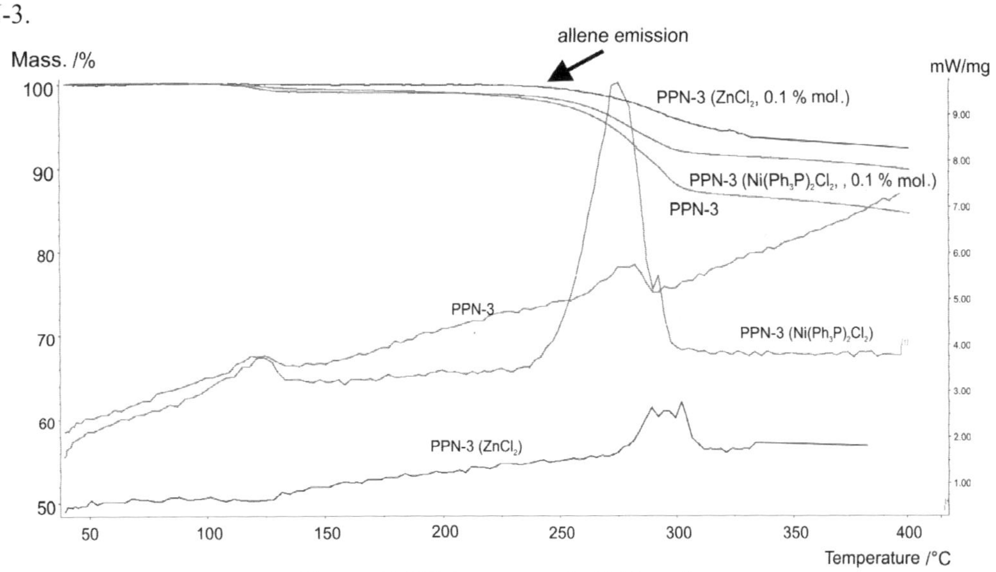

Figure 3. TGA-curves of PPN-3 resin.

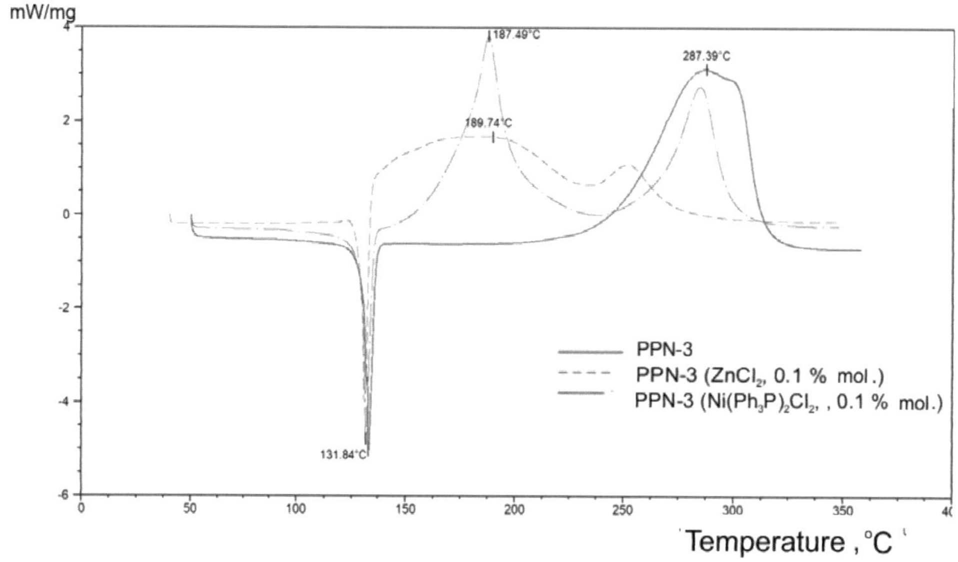

Figure 4. DSC-curves of aminopropargyl containing monomer **PPN-3**.

PPN-1 and PPN-2 resins were cured by following temperature program 180 °C – 8 h; 200 °C – 3 h; 220 °C – 3 h; 300 °C – 3 h; 330 °C - 3 h; 375 °C – 3 h. PPN-3 and PPN-4 resins were cured by the same program at another end-temperature 350 °C in the presence of ZnCl2, while the void formation during curing was practically absent. Thermal properties of all obtained resins are presented in table 2.

Table 2. Thermal properties of the thermal cured monomers (**PPN-3** and **PPN-4** cured in the presence of $ZnCl_2$)

monomers	$T_{5\%}$, °C	Y_c, 900°C, %	T_g, °C
PPN-1	465	70	380
PPN-2	451	72	432
PPN-3	442	68,1	376
PPN-4	421	63,7	352

Thus, the curing without decomposition of propargylated phthalonitrile amines PPN-3, PPN-4 is possible only in the presence of catalysts. Their thermal properties are superior to common propargyl resins [14], but they are inferior to polymers obtained from PPN-1 and PPN-2 propargyl monomers

Based on thermal properties, viscosity and cure data resin PPN-2 was chosen for composites manufacturing by VIMP. The CFRP were obtained from a mixture of PPN-2 and 10 % mass. 1,3-bis(4-aminophenoxy) benzene (APB). In our previous work the best properties were found for such formulation containing 10% APB [15]. Composites obtained by vacuum infusion process were post-cured at two different temperatures 375 °C (C-1), (typical temperature for phthalonitrile curing) and 330 °C (C-2). The mechanical properties of post-cured composites were tested; the results are presented in table 3. LOI for composites C-1 and C-2 was equal to 75, which is higher than most materials obtained from thermosets.

Table 3. Mechanical properties of composite post-cured at 375 °C and 330 °C.

Mechanical tests	**C-1** (375 °C post-cured)	**C-2**(330 °C post-cured)
σ_{11}^{+}, MPa/ E, GPa	548±14/67±1,9	742±15/62±0,9
σ_{11}^{-}, MPa/ E, GPa	325±22/63±3,6	485±39/59±8,4
τ_{13}, MPa, 25 °C	32±2,8	32±2,2
τ_{12}, MPa, 25 °C	57±3,6	72±1,5
τ_{13}, MPa, 300 °C	25±3,1	30±1,7
τ_{12}, MPa, 300 °C	47±2,7	59±3,0

The tensile strength of C-2 obtained at a maximum curing temperature 330 °C is about 50 % higher than C-1, obtained at a maximum curing temperature 375 °C. At the same time, for sample C-2, no microcracking are observed in the SEM images (Figure 5), this can be related to the decrease of shrinkage for composite cured at lower temperature [16]. Indeed the volume shrinkage of the PPN-2 resin during curing at 330 °C was found 0.63%, while curing at 375 °C - 5.6%.

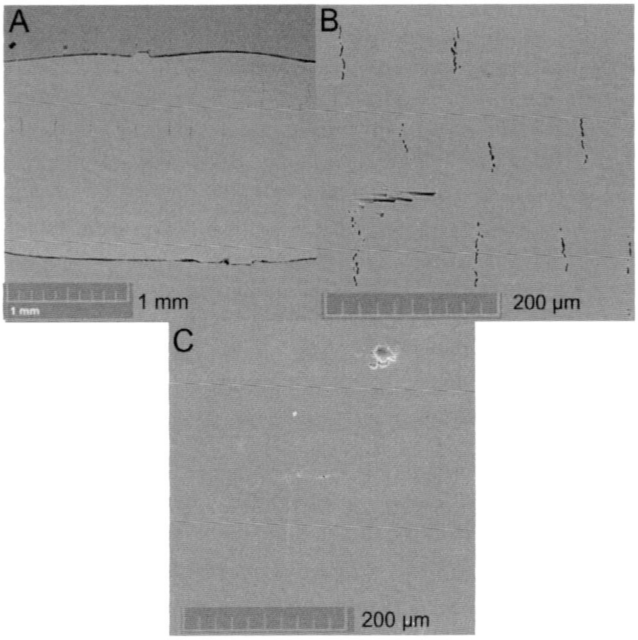

Figure 5. SEM microphotographs of C-1 and C-2 composites, made by different curing temperature.

Conclusions

CFRP were obtained from propargyl/phtalonitrile bifunctional monomer by vacuum infusion and the influence of post-cured temperature on mechanical properties and thermostability was investigated. It was found that the optimal post-cured temperature is 330 °C, this temperature provides no shrinkage while curing. The obtained composite retains good mechanical properties at 300 °C. It was established that aminopropargyl resins can't be cured without using any catalysts, curing resins were obtained using ZnCL2. Their thermal properities were investigated.

Acknowledgments

The work was supported by Government of the Russian Federation Ministry of Education and Science. Contract No.14.607.21.0204 (RFMEFI60718X0204)

References

[1] Goren A, Atas C 2008 *Manufacturing of polymer matrix composites using vacuum assisted resin infusion molding*, Reading. **34** 117–120. doi:10.1007/978-1-4939-2398-4_13.

[2] Mathias L J, Muthiah J 1994 *Synthesis and characterization of a new class of thermosetting resins:Allyl and propargyl substituted cyclopentadiene derivatives*, Polym. Compos.**15** 464–473. doi:10.1002/pc.750150611.

[3] Augustine D, Mathew D, Reghunadhan C P 2015 Nair, *One component propargyl phthalonitrile novolac: Synthesis andcharacterization*,Eur.Polym.J.**71**389–400doi:10.1016/j.eurpolymj.2015.08.013.

[4] Augustine D, Mathew D, Reghunadhan C P 2015 Nair, *Mechanistic and kinetic aspects of the curing of phthalonitrile monomers in the presence of propargyl groups*, Polym. United Kingdom **60** 308–317. doi:10.1016/j.polymer.2015.01.055.

[5] Liu X Q, Jikei M, Kakimoto M A 2001 *Synthesis and properties of AB-type semicrystalline polyimides prepared from polyamic acid ethyl ester precursors*, Macromolecules. **34** 3146–3154 doi:10.1021/ma001862n.

[6] Sakai N, Hori H, Ogiwara Y 2015 *Copper(II)-Catalyzed [4+1] annulation of propargylamines with N,O-acetals: Entry to the synthesis of polysubstituted pyrrole derivatives*, European J. Org. Chem. 1905–1909. doi:10.1002/ejoc.201500098.

[7] Bulgakov B, Sulimov A, Babkin A, Timoshkin I, Solopchenko A, Kepman A, Avdeev V 2017 *Phthalonitrile-carbon fiber composites produced by vacuum infusion process*, J. Compos. Mater doi:10.1177/0021998317699452.

[8] Carnelley T 1882 *XIII. Chemical symmetry, or the influence of atomic arrangement on the physical properties of compounds*,Philos.Mag.Ser.5.**13**112–130. doi:10.1080/14786448208627154.

[9] Zhou H, Badashah A, Luo Z, Liu F, Zhao T 2011 *Preparation and property comparison of ortho, meta, and para autocatalytic phthalonitrile compounds with amino group*, Polym. Adv. Technol. **22** 1459–1465. doi:10.1002/pat.2018.

[10] Chen X, Liu J, Xi Z, Shan S, Ding H, Qu X, Zhang Q 2017 *Synthesis and thermal properties of high temperature phthalonitrile resins cured with self-catalytic amino-contanining phthalonitrle compounds* **29** 1209–1221 doi:10.1177/0954008316673419.

[11] Bulgakov B A, Sulimov A V, Babkin A V,. Kepman A V, Malakho A P, Avdeev V V 2017 *Dual-curing thermosetting monomer containing both propargyl ether and phthalonitrile groups*, J. Appl. Polym. Sci. **134** 1–8 doi:10.1002/app.44786.

[12] Bulgakov B, Babkin A, Makarenko I, Tikhonov N, Kalugin D,. Kepman , Malakho A, Avdeev V 2015 *Ni(II) and Cu(II) based catalysts for propargylated novolac resins curing: Activity study and curing process simulation*, Eur. Polym. J. **73** 247–258 doi:10.1016/j.eurpolymj.2015.10.018.

[13] Jain S, Pandey N, Kishore D 2007 *Lewis acid catalyzed amino-Claisen rearrangement: A facile one pot synthesis of 2-allylarylamines from N-allylarylamines*, Indian J. Chem. - Sect. B Org. Med. Chem. **46** 529–531

[14] Nair C P R 2004 *Advances in addition-cure phenolic resins*, Prog. Polym. Sci. **29** 401–498. doi:10.1016/j.progpolymsci.2004.01.004.

[15] Bulgakov B A, Belsky K S, Nechausov S S, Afanaseva E S, Babkin A V, Kepman A V, Avdeev V V 2018 *Carbon fabric reinforced propargyl ether/phthalonitrile composites produced by vacuum infusion,* Mendeleev Commun. **28** doi:10.1016/j.mencom.2018.01.014.

[16] Nairn, John A 2000*"Matrix microcracking in composites." Polymer matrix composites* **2** 403-432.

IOP Conf. Series: Materials Science and Engineering **683** (2019) 012038 doi:10.1088/1757-899X/683/1/012038

Modern approaches to the manufacture of ceramic-matrix composite materials for long-term operation at temperatures above 1000°C. Review of world experience and capabilities of JSC "Kompozit"

P A Timofeev[1], I A Timofeev[1], E A Bogachev[1] and A N Timofeev[1]

[1] JSC "Kompozit", Pionerskaya Str., 4, Korolev, Moscow region, Russia, 141070
Abstract. The article describes modern approaches to the creation of ceramic-matrix composite materials for long-term operation in oxidizing environments. It provides information on advanced technologies for the formation of ceramic matrices and their recipes. The current problems and prospects for the production of such filaments in Russia are discussed.

Introduction

The development of gas turbine and rocket technology is inextricably linked with the development and implementation of new composite materials (CM). Due to high properties, such as strength and oxidation resistance, CM with a ceramic matrix — ceramic-matrix composite materials (CCM) are promising. At the same time, due to the specific operating conditions of gas turbine technology, the reinforcing filler of CCM must also be resistant to high temperatures, oxidative erosion flows and thermal cyclic loads. The main problem of the latest generation of domestic aircraft engines is the high percentage of air used for cooling the metal parts of the working area of the propulsion system. Considering the international experience of using CCM in GTE, it can be argued that the introduction of CCM into the design of domestic GTE will not only reduce air consumption for engine cooling, but also increase the operating temperature and reduce the mass of some nodes.

Objects and research results

Today, among high-temperature structural CMC, the leading positions are held by materials based on carbon reinforcing preforms. The JSC "Kompozit" collected the majority of world-known technologies for forming such preforms: surround braiding, contour braiding, robotic braiding, needle-punched preforms. These technologies allow you to create high-strength billet with a branched system of pores available for the formation in them of a different composition of matrices. High temperature oxidation resistant composite materials with ceramic matrix, usually carbide silicon, are called ceramic-matrix composite materials (CCM). In the interaction of silicon carbide at high temperatures, an oxide film of SiO2 forms with an oxidizing flow. It plays a key role in blocking access of oxygen-containing gases to the reinforcing filler. The effectiveness of the SiO2 film in terms of blocking the access of the oxidizing environment by it is limited by the processes of formation and evaporation. So, depending on the parameters of the working environment, materials like Cf/SiCm can effectively perceive temperatures from 1450°C to 1850°C for dozens of minutes.
There are several methods for formation of ceramic matrices. They can be divided into two groups according to the state of aggregation of the precursor entering the porous body - gas-phase and liquid-phase.

Gas-phase methods (chemical vapor infiltration - CVI) are based on the chemical process of pyrolysis of the gas precursor, and are most widely used in the world practice of making ceramic-matrices CM. Methyltrichlorosilane (MTS) or monomethylsilane (MMS) is commonly used as the gas phase [1, 2]. The technologies applying MTS and MMS have their advantages and disadvantages and are used in JSC "Kompozit" for the deposition of silicon carbide. MTS is significantly cheaper than MMS, provides a high growth rate of silicon carbide, less sensitive to the purity of a chemical reactor. However, the relatively high temperatures of the process and the presence of chlorine-containing compounds in the MTS significantly complicates the instrumentation and power consumption of the process and imposes increased requirements on personnel qualifications. Among the liquid-phase methods of obtaining the matrix, the methods LSI (Liquid Silicon Infiltration) [3, 4] are widely used - impregnation with a silicon-containing melt and PIP (Polymer Infiltration and Pyrolysis) [5] - impregnation with a silicone polymer followed by pyrolysis. Table 1 presents a comparison of these methods for the condition of the original carbon-carbon skeleton with a density of 1.2 g/cm3 and a CMC with a final density of ≈ 1.8 g/cm3.

Table 1. Comparison of various technologies for the formation of ceramic matrices CM.

	Vapor Deposition (CVI)	Melt impregnation (LSI)	Polymer impregnation (PIP)
Level of physicomechanical characteristics	High	Low	Average
Degradation of physical and mechanical properties of CM in the process of matrix formation	Missing	Possible depending on the structure of the workpiece.	Missing
Time to produce high density CMC (not less than 1.8 g/cm^3), h	600-800	8-40	150-300
Precursor cost (raw materials)	Low for MTS High for MMS	Low	Average
Cost of equipment	High for MTS Low for MMS	High	Low
Cost of the final product	High	Average	Average

A significant drawback of CVI and LSI technologies is the complexity of creating multicomponent matrices with their help. For example, the matrix composition Si-BC or Si-BCN by the specified methods is difficult to obtain. The CVI method allows creating multilayer matrices ([SiC/B4C]n or [SiC/HfC]n). Three or more component matrix using the LSI method can be obtained by adding submicron powders to the melt, such as boron powder in a silicon melt. The introduction of powder components beyond the solubility limit strongly changes the rheological properties of the melt and does not provide a high uniform distribution of all components in the bulk of the material. PIP methods stand out against the above mentioned. The composition of the formed matrix is determined, to a greater extent, by the composition of the starting polymer precursor. As a rule, polycarbosilanes or polysilazanes are used in the PIP method. At the stage of synthesis, various modifying additives can be added to the polymer structure, for example, boron, zirconium, titanium, which will become part of the composite matrix, which ensures their maximum uniform distribution in the ceramic matrix.

Multi-component incl. multilayer matrix based on silicon carbide, supplemented with nitrogen and boron compounds, is of great interest for aviation technology [6, 7]. The University of Bordeaux and the Snecma Corporation have developed and mastered the production of composite materials of the A410 and A500 grades with a multilayer matrix structure ([SiC/B4C]n (Figure 1). Materials have been developed to replace heat-resistant steels in the elements of adjustable nozzles [7].

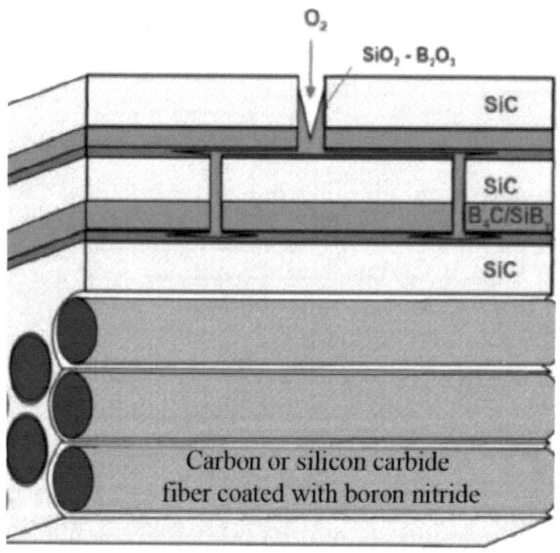

Figure 1. - Mechanism of protection against oxidation of the multilayer matrix CCM of A500 structure $C_f/[SiC\text{-}B_4C]_m$ [6].

Due to the relatively low temperatures of formation of the liquid glass phase of boron oxide, such a complex matrix provides excellent protection of carbon fibers from oxidation to temperatures of 1200°C in all operating modes of gas turbine technology for several hundred hours. This fully meets the requirements for existing and future GTE for US military aircrafts [7].

However, the maximum advantage and wider use of uncooled parts from CMC in GTE can be achieved at operating temperatures of up to 1500 (1600)°C and durations of several thousand for the military and several tens of thousands of hours for civilian GTE.

The solution to this problem is to replace the material of the frame - the transition from carbon to ceramic filler. Ceramic filaments were first developed in Japan in 1975 by Professor Yajima, and already in 1982, they became commercially available under the brand name Nicalon (Nippon Carbon, Japan) [8, 9]. These were the silicon carbide filaments of the so-called first generation with oxygen content of 12-15% (at.). Today, the first generation Nicalon filaments are still produced, however, the presence of oxygen in the structure leads to a relatively low temperature limit for the use of products made on their basis - no more than 1150°C. Further heating of this fiber leads to the formation of CO and SiO2, which catastrophically affects the strength of the fibers. Since the 1990s, Nippon Carbon has launched a second generation of carbide silicon fibers under the Hi-Nicalon brand, the oxygen content of which is less than 1% and the operating temperature reaches 1200 to 1250°C with a significant increase in operating time [10-11]. The best brands of SiC fibers today are third generation fibers - Hi-Nicalon Type S, Sylramic-iBN , Tyranno SA3. These are oxygen-free fibers of stoichiometric composition (C/Si at. ≈ 1: 1), and the temperature of their long-term operation can reach 1400°C [12].
In Russia, the issues of obtaining fibers of the composition of SiC and Si-C-N are carried out by JSC "GNIIHTEOS" and JSC "Kompozit". The properties of SiC fibers of JSC "GNIIHTEOS" are practically not inferior to the first generation of Japanese fibers. To date, the most pressing problems are the stabilization of the characteristics of a multi-filament yarn and the manufacture of various reinforcing structures from it (tape, 2D and 3D fabrics). The technology for producing SiC and SiCN fibers consists of the following main steps: pre-ceramic polymer manufacturing, polymer fiber molding, polymer fiber stitching, fiber ceramization. The presence of oxygen and its amount is determined mainly by the stages of crosslinking and ceramization, therefore today special attention is paid to these stages in order to obtain high-quality filaments. Work on the creation of SiCN and SiC fibers is also being carried out at JSC "Composite", where laboratory technologies for producing SiCN filaments (grade «KV-1») with an oxygen content of about 10 to 15% and oxygen-free (oxygen content less than 2%) SiC filaments (brand «KV-2»). The filaments KV-2 can withstand temperature changes up to 1400°C in an oxidizing environment without significant loss of strength characteristics and long-term temperatures up to 1800°C in an inert atmosphere without loss of geometric shape. Improving the strength, textile properties and increasing the volume of fiber production are the main tasks in these areas. According to our estimates, oxygen-free textile-processed multi-filament SiCN and SiC (third generation analog) filaments will appear in Russia in 2022. The presence of such filaments in combination with the extensive experience of JSC "Kompozit" in the formation of reinforcing preforms and ceramic matrices should ensure the creation of composite materials for long-term operation (thousands of hours) in the composition of promising aircraft engines.
In 2018, JSC "Kompozit", for the first time, based on Russian SiCO fibers manufactured by AO "GNIIHTEOS", manufactured samples of CMC SiCOf-BN/SiCm (Figure 2). Samples were made on the basis of unidirectional ribbons, the SiC matrix was formed using a combination of CVI and PIP methods. The obtained material samples had a density of ≈2.17 g/cm3, Vf ≈ 35 wt%, open porosity ≈10%.

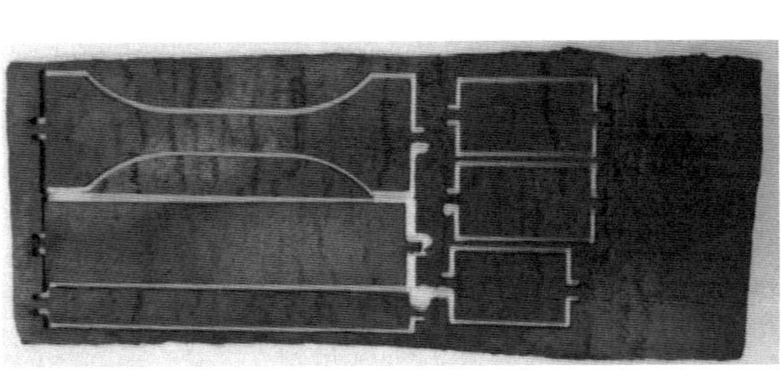

Figure 2. - Cutting of plates from CCM $SiCO_f$-BN/SiC_m into samples.

Conclusion

JSC "Kompozit" actively develops a new for Russia subclass of ceramic-matrix composite materials for gas turbine equipment. In 2020, it plans to form a complete technological readiness for the production of parts of the Cf/(SiBCN)m structure for gas turbine engines, the service interval of which is not more than 500 hours and the operating temperature does not exceed 1200°C. The expected performance limits are scheduled to be confirmed by bench tests as part of the GTE. In 2018, JSC "Kompozit" obtained laboratory samples of oxygen-free (oxygen less than 2%) silicon carbide fiber. The work will be continued in terms of reducing the oxygen content, optimizing the C/Si ratio, and increasing textual content with access to pilot production with a capacity of up to 1000 kg/year.

Reference

[1] Lacombe A, Bonnet C 1990 Ceramic Matrix Composites, key materials for future spaceplane techologies. *AIAA Second international aerospace planes Conference (*Orlando) p.p. 48-62.

[2] Timofeev A N, Bogachev E A, Gabov A V, Abyzov A M, Persin M I, Smirnov E P Method for producing a composite material, RF patent o.2130509, priority of 26.01.1998.

[3] Kostikov V I, Varenkov A N 2003 Ultrahigh-temperature composite materials, ed Intermet Engineering 560

[4] Ceramic matrix composites/ Edited by Krenkel W Wiley VCH Verlag GmbH & Co.2008.418 p.

[5] Timofeev P A 2017 Formation of matrix of composite materials from carbides, nitrides and silicon borides by the method of pyrolysis of polymer precursors: Diss.papers of Cand. of Tech.Sc Moscow: MSTU named after N.E. Bauman,. 126 p. (reg . ICD number AAAA-B18-418070290051-1).

[6] Dezellus O 2008 Materiaux composites matrice metalliqueset ceramiques, *Journal of phase equilibria and diffusion* **29** 155 –162.

[7] Christin F 2005 A global approach to fiber nD architectures and self-sealing matrices — from research to production / *Conference High Temperature Ceramic Matrix Composites* **5** 477-483

[8] Yajima S, Okamura K 1975 Hayashi J Structural Analysis in Continuous Silicon Carbide Fiber of High Tensile Strength, *Chemical Soc. of Japan* 1209-1212.

[9] Yajima S, Okamura K, Hayashi J 1975 Structural Analysis in Continuous Silicon Carbide Fiber of High Tensile Strength *Chemical Soc. of Japan* 1209-1212.

[10] Ishikawa T 1994 Recent Developments of the SiC Fiber Nicalon and its Composites, Including Properties of the SiC Fiber Hi-Nicalon for Ultra-High Temperature, *Composites Science and Tech.* **51** 135-144.

[11] Ichikawa H 2006 Advances in SiC fibers for high temperature applications, *Adv. In Science and Technology* **50** 17-23

[12] Bansal N P, Lamon J 2015 *Ceramic matrix composites: materials, modeling and technology* Ed Wiley-VCH Verlag GmbH & Co. 694

IOP Conf. Series: Materials Science and Engineering **683** (2019) 012039 doi:10.1088/1757-899X/683/1/012039

Influence of refractory nanosized particles of tungsten carbide and titanium on the structure and properties of the weld metal

A S Pankratov[1], N V Kobernik[2] and A V Samokhin[3]

[1] FSAU "NUTSSK at N. E. Bauman MSTU", 2nd Baumanskaya Str., 5, bldg. 1, Moscow, Russia, 105005
[2] FGBOU VO "N. E. Bauman MSTU" (National Research University), 2nd Baumanskaya Str., 5, bldg. 1, Moscow, Russia, 105005
[3] IMER RAS, Leninsky Prospect, 49, Moscow, Russia, 119334
Abstract. The paper considers the influence of nanosized particles of tungsten and titanium carbides on the structure and properties of the weld metal of welded joints made by automatic submerged-arc automatic arc welding. To add them into the weld pool, an alloying composition was used in the form of composite granules based on nickel powder (PNE-1 as per GOST 9722). The prospects of using nanosized tungsten carbide particles for modifying the weld metal with the aim to increase the values of impact strength were shown. However, titanium carbide in the same conditions leads to a decrease in the average value of impact strength, despite the modifying effect of its use.

1. Introduction
Submerged arc welding has found wide application in the welding industry, at the same time in order to increase productivity, two or more electronic wires can be used, at the torches of which independent arcs are burning (multi-arc submerged-arc welding). In particular, multi-arc processes have proven themselves in the production of main pipelines, and two-arc welding can be used to weld annular joints of pipelines during their installation [1, 2]. However, in this case, due to the greater heat input, a coarse structure can be formed of both the weld metal and the heat-affected zone, which leads to a decrease in the impact strength in these zones. Traditionally, to solve this problem, it is necessary to optimize the scheme of alloying of the weld metal through welding materials, as well as apply various process methods (for example, the use of additional filler wire). An alternative way of influencing the mechanical characteristics of the weld metal is its modification, which allows grinding the grain of the weld metal and, thus, improving the mechanical properties (in particular, increasing the impact strength values) [3,4,5].
An analytical review [3], conducted by the authors earlier, showed that recently there was an interest in using nanoscale particles for modifying the structure of the weld metal and built up metal of functional coatings. Therefore, the article is devoted to the study of the influence of nanoscale particles on the structure and properties of the weld metal during the implementation of two-arc welding under flux. For the addition of nanosized particles the filler material in the form of alloying composition was used, which was tested in [4]. In addition, the analogue of such a material in the form of a metal chemical additive (MHP) has a widely used analog in bridge construction [5–7].
The alloying composition represents composite granules obtained by mixing in a high-energy planetary mill powders of metals and their alloys and powders of nanoscale refractory compounds.

2. Research methodology

Experimental studies were carried out when welding plates 300x100x10 mm (LxWxT), made of St3sp as per GOST 380. The welded joint is of C-19 type according to GOST 8713 (with the remaining lining). The applied cutting allows to weld in one pass, the presence of blunting allows to control the dosage of the added alloying composition, and the steel lining will keep it from spilling out and does not require special measures to prevent burn-through and the formation of the back side of the seam.

The combination of electrode wire and welding flux is selected for reasons of making the welded joint in one pass, ensuring a chemical composition of the weld metal of 0.08% C; 1.7% Mn and 0.9% Si.

The alloying composition was prepared by mixing nickel powders (PNE-1 as per GOST 9722) with an average size of not less than 50 microns and nanoscale particles in a proportion of 30 wt% of nanoparticles and 70 wt% of nickel. Tungsten carbide (WC) and titanium carbide (TiC) particles were used as refractory particles. Nanoscale carbide particles were obtained by plasma-chemical synthesis in thermal plasma using electric arc plasma torch [8,9]. Powders consist of individual, non-porous particles of equiaxial form. The average particle size of titanium carbide is 50 nm, tungsten carbide is 70 nm. The powders are characterized by a particle size distribution close to log-normal [10] and single-phase.

The experiments were carried out according to the scheme of submerged arc welding with two wire electrodes (two-arc welding). The alloying composition is pre-laid into the groove before welding and then melted by an arc [11]. This scheme of the process allows avoiding the welding arc direct impact on particles and shields them at the expense of the liquid layer. In this case, the second arc creates an additional metal flow in welding pool, which intensively mixes it and allows you to get uniform distribution of particles by volume of the pool. To perform the experiments, the parameters of the mode were selected that ensure the formation of a weld in one pass that meets the requirements of GOST 8713 (Table 1).

Table 1. Mode parameters used in studies of two-arc submerged arc welding with the use of the alloying composition

Item Number	Parameter	Value range	
		first arc	second arc
1	Type of current	DC+[4]	AC[5]
2	Welding current per arc (A)	650 ± 20	650 ± 20
3	Arc voltage (V)	27 ± 1	32 ± 1
4	Welding speed (m/h)	45	45
5	Electrode stickout distance (mm)	30	40
6	Distance between electrodes (mm)	27	27
7	Electrode Tilt Angle	90°	60° (forward angle)
8	Balance[6] (%)	100	25

[4] DC+ - direct current of reverse polarity

[5] AC - alternating current

[6] Balance - the ratio of the arc burning time in reverse polarity to the duration of the arc burning period

The alloying composition was pre-laid in the groove while controlling the height of the filling. Two variants of the filling height were tested: 2 and 3 mm. As a result, joints characterized by the number of nanosized particles added in accordance with Table 2 were obtained. Also, to establish the degree of influence of nanoscale particles on the weld metal, experiments on welding with nickel powder filling (PNE-1 as per GOST 9722) were performed, as well as a variant without additional additives, the so-called "Basic".

After the welds were completed, samples were cut from them to prepare thin sections to study the structure of the weld metal. Also, samples of type X as per GOST 6996 were prepared for testing the impact strength of the weld metal at a temperature of "minus" 20°C. After testing, the fracture surface was subjected to fractographic analysis.

3. Results of research

The macrostructure of the weld metal has a pronounced column structure and is a ferritic-pearlitic mixture (Figure 1). Inside the columnar crystals, ferritic-pearlitic areas are observed, while at the periphery the ferritic component of different dispersion dominated. The use of nanoscale particles does not change the morphology of the structural components.

However, with the addition of nanoscale particles of tungsten carbide grain size decrease is observed: the average width of the primary crystals decreases by at least 2 times (Table 2). At the same time, the structure becomes more uniform (Figure 3), which should have a positive impact on the plastic characteristics. With the addition of titanium carbide, a decrease in the average width of the primary crystals is also observed, however, it does not exceed 30% (Table 2). Moreover, the addition of nanoscale particles of titanium carbide, making part of the alloying composition, leads to a variation of the widths of the primary crystals, which can adversely affect the stability of mechanical properties.

Analysis of the results obtained after tests for impact strength showed that the addition of a total of 1.10-1.56 wt% of nickel powder into the molten weld pool leads to a slight increase in the average impact strength of the weld metal: by 13 and 7%, respectively, compared to the base case (Figure 2, Table 2). At the same time, there is a decrease in the stability of its values. The insignificant effect of nickel on the impact strength of the weld metal is explained by the high manganese content in the weld metal, which is consistent with the paper [12].

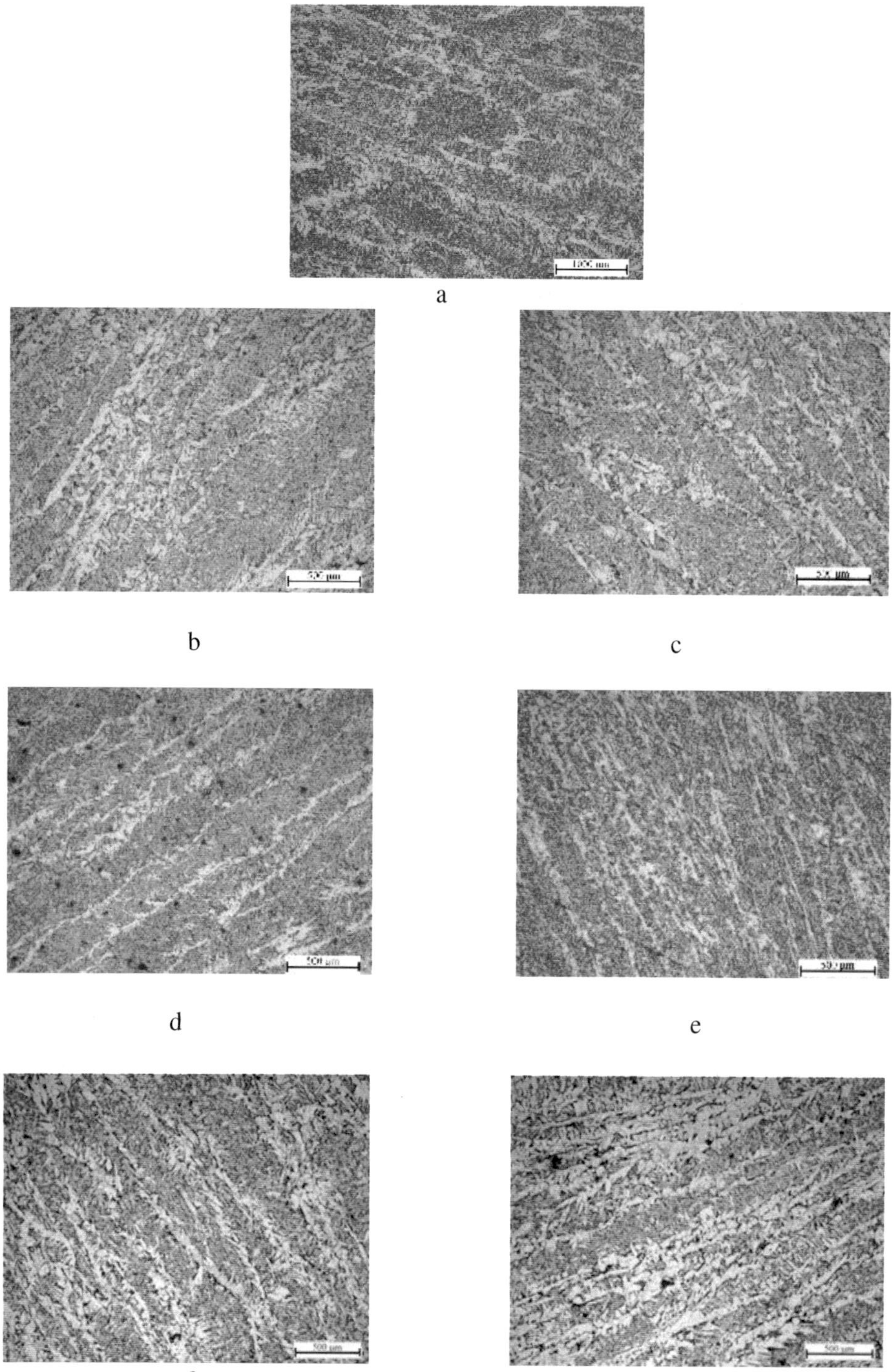

Figure 1. Structure of the weld metal of welded joints obtained without the use of the alloying compositions (a, x50), as well as with the use of the alloying compositions (x100) L / Ni (b, c), L / Ni-WC (d, e) and Л/Ni-TiC (f, g). b, d, f - *KK*kg = 0,03 vol%, в, д, ж - *KK*kg = 0.04 vol%

IOP Conf. Series: Materials Science and Engineering **683** (2019) 012039 doi:10.1088/1757-899X/683/1/012039

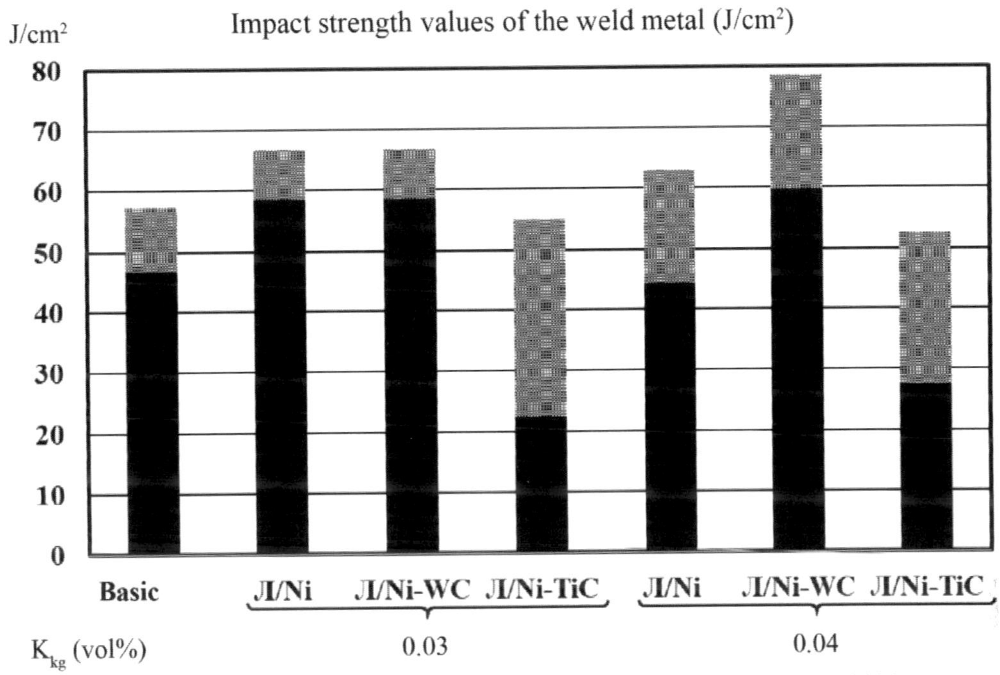

Figure 2. Results of impact strength tests at a temperature of "minus" 20°C:
■ – minimum impact strength value, J/cm2;
▩ – variation of impact strength values

The use of the alloying composition which included nanoscale particles of tungsten carbide leads to an increase in the average impact strength of the weld metal by 26% and 36% compared to the base case, and by 12% and 26% compared to the use of the nickel powder-based alloying composition, respectively (Figure 4, Table 2). It should be noted that with the addition of a smaller amount of the alloying composition with tungsten carbide, the spread of values decreases by 39% relative to the base case.

Thus, the addition of tungsten carbide through the alloying composition allows to modify the structure of the weld metal and increase the impact strength by 26-36% while reducing scatter values, which is explained by its influence on the structure.

The impact strength of the weld metal made using the L/Ni-TiC alloying composition turned out to be lower (by 12 ... 13% depending on the amount of the additive added) average values of the base case (Figure 4). At the same time, there is an increase in the spread of the impact strength. Also, nanoscale particles of titanium carbide level the positive effects of nickel on the impact strength of the weld metal.

The research has shown that the addition of nanosized particles of titanium carbide into the weld pool together with the alloying composition is not promising.

The fractographic analysis of fractures included the analysis of the ratio of the fragile and viscous component (Table 2) according to GOST 30456 and GOST R ISO 148. Analysis of the results indicates an increase in the proportion of the viscous component when using the L/Ni-WC alloying composition. On the contrary, the use of the L/Ni-TiC alloying composition leads to an increase in the fraction of the brittle component in the fracture.

4. Conclusions

1. The prospects of modifying the weld metal of welded joints by nanoscale tungsten carbide particles are shown.

2. The addition of nanoscale particles of tungsten carbide, being the part of the alloying composition, in the weld pool melt leads to an increase in the impact strength values of the weld metal. At the same time, this indicator stabilizes, which is associated with a decrease in the width of the primary crystal of the weld metal and an increase in the uniformity of dimensions. Thus, adding 0.29 wt% and 0.34 wt% of particles to the weld pool melt leads to an increase in the average impact strength of the weld metal at a test temperature of minus 20°C by 26% and 36%, respectively.
3. The addition of titanium carbide as part of the alloying composition into the weld pool melt leads to a decrease in the average impact strength values of the weld metal: by 12% when adding 0.05 wt% of the compound and by 13% when adding 0.16 wt% thereof. However, an increase in the impact strength values is observed. At the same time, the addition of titanium carbide contributes to the modification of the structure of the weld metal: the average width of the primary crystals is reduced by at least 30%.

Acknowledgments
The paper has been prepared thank to the financial support of applied researches by the Ministry of Education and Science of the Russian Federation (Agreement No. 14.578.21.0216 on the provision of a grant dated September 28, 2016, the unique identifier PNIER RFMEFI57816X0216).

References
[1] 2015 Welding in the construction and repair of steel vertical tanks. *РД-25.160.10-КТН-015-15 Main pipeline transportation of oil and oil products. Part 1. Moscow.* 150.
[2] Shakhmatov M V, Shakhmatov D M and Yakovlev D S 2015 Calculation of parameters of the mode for multi-arc welding of pipes under a layer of flux. *Welding and diagnostics.* **6** 14–16.
[3] Kobernik N V, Pankratov A S and Linnik A A 2015 Modern ideas on the modification of the weld metal with nanoscale particles (review). *Welding and diagnostics.* **5** 11–16.
[4] Kuznetsov V D, Smirnov I V and Shapovalov K P 2013 Effect of nano-additives on the structure and properties of the weld metal when welding high-strength low-alloy steels. *Advanced technologies and systems of mechanical engineering.* **2** (no. 46) 143–150.
[5] Guschin D A 2017 Improving the efficiency of the modification of the weld metal when welding low-alloyed steel under the flux with a metal-chemical additive. *Abstract Ph.D. Dissertation. Moscow, MSTU. N.E. Bauman.* 17.
[6] Boldyrev A M, Grebenchuk V G, Gushchin D A, Tkachev A G and Blinov S V 2013 Obtaining a metal-chemical additive with nanodispersed particles of titanium dioxide. *Nanotechnologies in construction.* **6** 53-66.
[7] Boldyrev A M, Grigorash V V, Guschin D A and Grebenchuk V G 2012 Study of the adhesion strength of particles in the modifying additive for welding bridge structures under flux. *Nanotechnology in construction.* **2**. 56-69.
[8] Samokhin A V, Alekseev N V, Kornev S A, Sinaiskii M A, Blagoveschenskiy Yu and Kolesnikov A V 2013 Tungsten carbide and vanadium carbide nanopowders synthesis in DC plasma reactor. *Plasma Chem. Plasma Process.* **33** (no. 3) 605–616.
[9] Samokhin A V, Kirpichev D E, Alekseev N V, Sinayskiy M A and Tsvetkov Yu V 2016 Synthesis of nitride, titanium carbide and carbonitride nanopowders in a plasma jet-confined reactor. *High energy chemistry.* **50** (no. 6) 491–497.
[10] Sinayskiy M A, Samokhin A V, Alekseev N V and Tsvetkov Yu V 2016 Enhanced characteristics of the dispersed composition for plasma chemical synthesis nanopowders. *Russian nanotechnologies.* **11** (no. 11-12) 110–115.
[11] Pankratov A S 2017 Modification of the weld metal by nanoscale particles of tungsten carbide and titanium nitride when welding under the flux of low-carbon low-alloy steels. *Thesis for the degree of a candidate of technical sciences (Moscow, MSTU named after N.E. Bauman).* 135.

[12] Leonov V V 2016 The design features of inflatable large-scale mirror concentrators for space high-temperature solar power plant. *Proceedings of the international astronautical congress*, IAC 67, Making Space Accessible and Affordable to All Countries.

[13] Zhang Z и Farrar A 1997 Influence of Mn and Ni on the Microstructure and Toughness of C- Mn-Ni Weld. *Metals Weld. J.* **5**. 183-196.

IOP Conf. Series: Materials Science and Engineering **683** (2019) 012040 doi:10.1088/1757-899X/683/1/012040

Silicon nitride-based ceramic composite materials for corrosion-resistant rolling bearings

S A Pakhomova[1] and A I Povalyayev[2]

[1] Bauman Moscow State Technical University, 2-nd Baumanskaya, 5, b.1, 105005, Moscow, Russia
[2] "Kompozit", Pioner st., 4, 141070 Korolev, Russia.

E-mail: mgtu2013@yandex.ru

Abstract. The technological process of obtaining material and manufacturing ceramic rolling elements of the bearing is considered. The composition of a silicon nitride-based ceramic composite material was investigated. The chemical inertness of ceramic materials determines the low adhesive wear in the mating parts of the supporting elements and makes it possible to use the rolling elements of the bearings in conditions of poor lubrication. Bearings with ceramic rolling elements show excellent running performance in emergency conditions and do not lead to sudden failures. The excellent corrosion resistance of ceramic materials allows them to be used in aggressive environments. The combination of such properties is accompanied by a decrease in temperature in the working area and allows the use of higher rotational speeds. The feasibility of using ceramics for the production of balls for bearings and the main advantages of using a ceramic material compared to metal products are shown. The process mode of hot pressing of blanks of ceramic rolling elements has been developed.

1. Introduction

Rolling bearings are classified into the following main types: ball, roller cylindrical, roller conic, double-row self-aligning, needle, thrust ball, thrust roller. Bearings are widely used in such mechanisms as electric motors, lifting and transporting and agricultural machines, aircraft, locomotives, wagons, machine tools, gear reducers and many machines and mechanisms.

Rolling bodies of bearings perceive compression loads, cyclic loads, and must comply with the conditions of operability, reliability and durability of the structure. The peculiarity of the bearings is that the materials from which they are made must have high strength, hardness, wear resistance, toughness. In critical structures most often bearings for special purposes are used, which should have improved properties. In some cases, steel, commonly used for the manufacture of bearings, do not meet the criteria of high heat resistance, heat resistance, corrosion resistance. Ceramic composite material meets these requirements [3, 4].

For harsh operating conditions of bearings, the most promising solution is the use of ceramic materials with high physicomechanical properties that are resistant to wear and change in geometric shape during operation. Ceramic materials are dominated by the following advantages over bearing steel: low density; high hardness; high modulus of elasticity; chemical inertness; the coefficient of dry friction in a pair of ceramics-steel is less than in a steel-steel pair; low thermal expansion coefficient, high heat capacity and heat resistance, as well as the fact that ceramics is a dielectric [5].

Due to the low density in combination with high physicomechanical and tribotechnical indices, the following operational properties provide the advantages of ceramic rolling bodies over steel rolling elements of bearings: reduction of centrifugal forces; reduce preload in the bearing; reduction of the radial clearance in the bearing; reduced vibration and noise; reduction of bearing operating temperatures. The chemical inertness of ceramic materials determines the low adhesive wear in the mating parts of the supporting elements. This property of ceramic materials makes it possible to use rolling elements of bearings in conditions of poor lubrication or even without the use of lubricants. Therefore, bearings with ceramic rolling elements show excellent running performance in emergency conditions and do not lead to sudden failures. The excellent corrosion resistance of ceramic materials allows them to be used in aggressive environments.

The low coefficient of friction in combination with favorable lubrication conditions allows the support elements to work under conditions of significant reduction of friction, which is accompanied by a decrease in temperature in the working area and allows the use of higher rotational speeds [6, 7].

Silicon nitride Si3N4 is the only compound of silicon and nitrogen and exists in two versions - α- and β-Si3N4. Silicon nitride is characterized mainly by the covalent type of chemical bond (approximately 70%). Here, in contrast to compounds with the ionic type of chemical bond, the processes of diffusion-viscous flow during the sintering of covalent compounds are shown to a minor extent. In connection with this circumstance, it is rather difficult to obtain high-density single-phase silicon nitride material using powder technology in practice.

2. Materials and methods of the experiment

The grinding of the charge components was carried out in the Pulverisette-5 planetary mill (Fritch Germany). The powders that were ground together were crushed and ground by carbide grinding balls with a diameter of 10 mm in the grinding cups of a planetary mill. At the same time, the impact energy is many times greater than the impact energy in conventional ball mills. Due to this, good efficiency and short grinding time are achieved. The rotation speed of the drum 400 rpm allows an order to reduce the grinding time by an order of magnitude compared to drum and vibratory mills.

Ceramic composite material consisted of a mixture of powders of silicon nitride, yttrium oxide and aluminum oxide in an environment of isopropyl alcohol with a dispersion of not more than 3 microns. Powder weight ratio: grinding bodies: fluid was 1:3:1. The chemical composition of the ceramic material for the manufacture of blanks of the tested rolling elements of the bearing is presented in Table 1.

Table 1. The chemical composition of the material based on silicon nitride and sintering additives

Composition number	Material composition		
	Silicon nitride, %	Aluminum oxide, %	Yttrium oxide, %
1	85.0	5.0	10.0
2	83.3	4.2	12.5
3	80.0	5.0	15.0

After grinding, the suspension was poured onto a baking sheet, placed in a heating chamber and dried at a temperature of (80 ± 5)°C for 5 h. The charge powder was separated from the grinding balls by sifting through a sieve No. 2, and then sieved additionally through a sieve No. 0063.

The compacts were made on a hydraulic press PYE 25X355 by compacting the batch powder in a 23 mm steel cylindrical single-press mold with a pressure of (30 ± 3) MPa in order to reduce shrinkage during sintering.

The resulting compacts were visually controlled for the absence of cracks, chips, and shells.

Making blanks of ceramic rolling bodies was carried out by the method of hot pressing. The resulting compacts were placed in a 12-nest graphite mold.

The working surfaces of the graphite mold were pre-treated with a suspension based on hexagonal boron nitride. Hot pressing was carried out on the press KCE HP W 200/250-2200-180.

The study of the fine structure was carried out on a scanning electron microscope

IOP Conf. Series: Materials Science and Engineering **683** (2019) 012040 doi:10.1088/1757-899X/683/1/012040

TESCAN version VEGA 3. The apparent density and porosity of the blanks was determined according to GOST 473.4 on the VIVAFLOW-50 device (Sartorius, Germany) by saturating ceramic samples with water followed by hydrostatic weighing. Hardness at a temperature of 20°C by Vickers was determined according to GOST 2990 by the static indentation of the diamond pyramid indenter into the sample surface. Using plasticine, the sample was fixed on the stage of the PMT-3 hardness tester (LOMO firm, Russia).
The strength under static bending at a temperature of 20°C was determined on a VTK-2 testing machine as per GOST 24409 on samples (width 10.0 mm, thickness 8.0 mm) made from a mixture. Distance between supports 100 mm. Each test was conducted at a constant loading rate until its failure. On a scale, the position of the arrow at which the specimen was destroyed was fixed.

3. Experimental results and discussion

Hot pressing of the ceramic composite material was carried out according to the regime shown in Figure 1. The hot-pressing mode of the blanks consisted of three main stages.

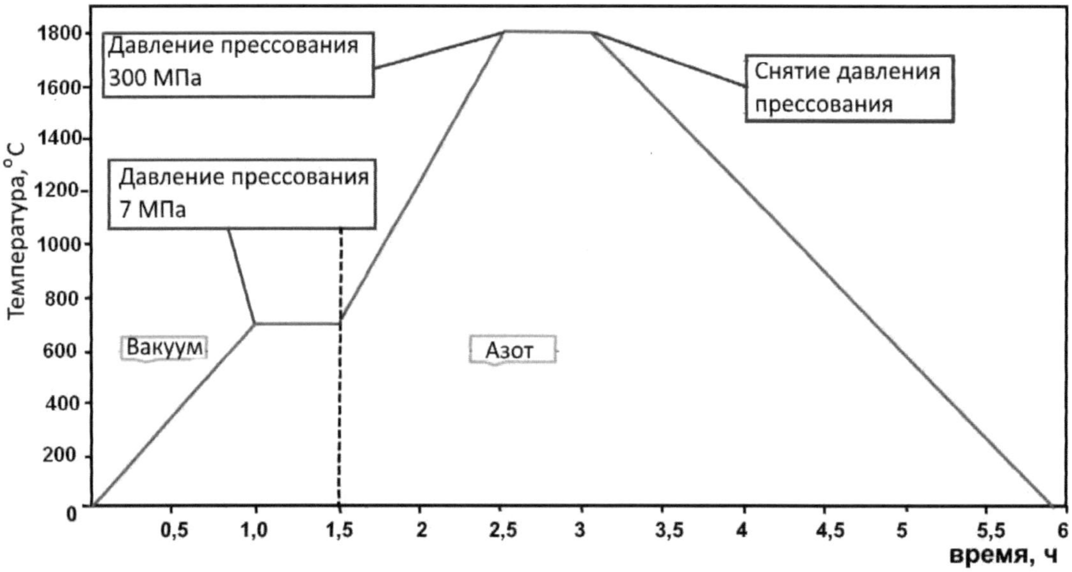

Figure 1. Technological scheme of hot pressing of blanks.

Technology of hot pressing included the following operations:
1. Raising the temperature to 700°C and holding for 20 minutes at this temperature. Heating and aging took place in a vacuum. At this stage, adsorbed gases, moisture and volatile impurities are removed.
2. At the end of curing at 700°C, nitrogen was fed into the press chamber and the pressing was loaded with a pressing pressure of 7 MPa. Then, the temperature was raised to (1750 ± 50) ° C with curing at a maximum temperature of 30 minutes. With the rise of temperature, at the same time gradually increase the pressure of pressing to 30 MPa. At this stage, the pressing shrinks, which is controlled by the movement of the plunger of the press and the readings of the press device, and the final sintering of the powder occurs. During the process of hot pressing the following parameters were controlled: vacuum discharge value and nitrogen pressure; temperature; shrinkage.
3. Cooling with a furnace. At a temperature of 1200 ° C, the pressure is removed from the mold. After cooling, the cylindrical blanks are removed from the mold, the surface of the blanks is subjected to sandblasting, after which the blanks are transferred to the control operation.
The technological stage of hot pressing is performed in order to give the product a predetermined density and the necessary physical and mechanical properties. The results of measurement of porosity, apparent

density, hardness and strength under static bending are shown in Table 1, and the microstructures of the studied samples are shown in Fig. 2.

Table 2. Test results of samples of blanks based on silicon nitride and sintering additives

Composition number	Properties of sample blanks of silicon nitride			
	Apparent density, g/cm3	Porosity, %	Hardness at a temperature of 20°C by Vickers, GPa	Static bending strength, MPa
1	3.26	0.06	15.70	629.00
2	3.23	0.08	16.10	637.00
3	3.25	0.05	15.30	644.00

Figure 2. Microstructure of ceramics based on silicon nitride and sintering additives: (a), (b), (c) - compositions 1, 2, 3 according to Table 2.

The microstructure of ceramics is characterized by a fine-grained matrix phase, represented by prismatic silicon β-nitride crystals with sizes of 0.5 to 1.0 μm and elongated prismatic crystals with widths of about 0.5 to 1.0 μm, length - about 5 to 8 microns.

The test results of samples of ceramics based on silicon nitride showed that the strength under static bending at a temperature of 20°C is (629 - 644) MPa, which significantly exceeds the strength values of steel balls [5]. The best results were shown by the material composition 2.

Silicon nitride has an extremely high chemical resistance. Silicon nitride is practically resistant to all acids, many molten metals, and water vapor. It is resistant to oxidation not only in air, but also in oxygen at moderately high temperatures [2]. The hardness and modulus of elasticity of silicon nitride is more than 1.5 times higher than similar characteristics of steel. This factor makes the bearing harder and increases its life when working in conditions of high pollution.

Rolling bodies of silicon nitride have a lower degree of thermal expansion than steel bodies of a similar size, which means that structures are less sensitive to temperature changes inside the bearing and are able to withstand large preload forces [8-10].

Thus, taking into account the physicomechanical and electrical insulating properties, it is advisable to use silicon nitride as a material for the manufacture of ceramic balls, components of rolling bearings.

4. Conclusions

1. The use of ceramics based on silicon nitride increases the strength of rolling elements of bearings. The selected chemical composition of ceramic material for the manufacture of blanks of ceramic rolling elements of the bearing, containing 83.3 wt. % silicon nitride,
4,2 wt. % aluminum oxide, 12.5 wt. % yttrium oxide.
2. The developed process of hot pressing to obtain the material of ceramic rolling elements of bearings, will ensure their strength under static bending

up to 644 MPa and hardness up to 16.1 gPa according to Vickers.
3. Ceramic composite material based on silicon nitride is a promising material for corrosion-resistant rolling bearings.

References

[1] Garshin A P, Goropyanov V M, Zaytsev G P and Semenov S S 2003 Keramika dlya mashinostroeniya [Ceramics for mechanical engineering]. *Moscow: Nauchtekhizdat publ.* 384.
[2] Guzman I Ya 2003 Khimicheskaya tekhnologiya keramiki [Chemical technology of ceramics]. *Moscow: Stroymaterialy publ.* 196.
[3] Salakhov A M and Salakhova R A 2012 Innovatsionnyye materialy: sovremennaya keramika [Innovative materials: modern ceramics]. *Kazan': Paradigma*. 383.
[4] Volovik A P, Kovarskiy M Ye and Zaytsev V A 2014 Vysokoskorostnyye podshipniki kacheniya generatora dlya gazoturbinnykh ustanovok [High-speed generator bearings for gas turbine installations]. *Voprosy elektromekhaniki. Trudy VNIIEM.* **139** p 3-6.
[5] Belova G S, Titova Yu V and Maydan D A 2018 Perspektivy ispol'zovaniya nitrida kremniya marki SVS-A3 dlya proizvodstva detaley podshipnikov [Prospects for the use of SVS-A3 silicon nitride for the production of bearing parts]. *Sovremennyye materialy, tekhnika i tekhnologii.* **2 (17)** pp 26-33.
[6] Serdobintsev Yu P, Kharkov M Yu and Nazzal 2014 A Review and analysis of uses of ceramic materials in various industrial fields. *Sovremennyye problemy nauki i obrazovaniya [Modern problems of science and education].* **2** p 231.
[7] Supancic P, Danzer R, Witschnig S, Polaczek E and Morrell R 2009 A new test to determine the tensile strength of brittle balls — The notched ball test. *Journal of the European Ceramic Society.* **29** pp 2447–59.
[8] Supancic P, Danzer R, Harrer W, Witschnig S and Schöppl O 2009 Strength Tests on Silicon Nitride Balls. *Key Engineering Materials.* **409** pp 193-200.
[9] Strobl S, Supancic P, Schöppl O and Danzer R 2014 A new strength test for ceramic cylinders – The Notched Roller Test. *Journal of the European Ceramic Society*. **34** pp 2575-84.
[10] Kurganova Yu A, Fetisov G P and Gavrilov G N 2015 Kompozitsionnyye materialy v aviatsii i ikh prognozirovaniye [Composite materials in aviation and their prediction]. *Tekhnologiya metallov [Metal technology].* **1** pp. 22-25.

IOP Conf. Series: Materials Science and Engineering **683** (2019) 012041 doi:10.1088/1757-899X/683/1/012041

Constitutive relations of the endochronic theory of thermoplasticity for high-temperature composites under plane stress state

B S Sarbayev

Bauman Moscow State Technical University, 2-nd Baumanskaya, 5, b.1, 105005, Moscow, Russia

E-mail: bssarbayev@mail.ru

Abstract. A variant of the constitutive relations of the endochronic theory of thermoplasticity for high-temperature composites is proposed. The relations for the plane stress state are considered. Examples of the use of the proposed relations in which the material behavior under complex loading is analyzed are given. The dependence of the elastic modulus of the material on the temperature of pre-heating is shown.

1. Introduction

The strength analysis of structural elements made with the use of high-temperature composite materials is one of the stages in the development of rocket-space and aviation equipment, engine-building, chemical industry and metallurgy. Problems such as the development of methods for calculating and design uncooled rocket motor nozzles, exit cone of jet engines, wing leading edges and nose cone, turbine blades, etc. can be solved with the use of adequate mathematical models of deformation of anisotropic structural materials, operating under intensive thermo-force loading.

A feature of modern high-temperature composite materials, for example, carbon-ceramic-matrix composites, is their irreversible deformation at both room and elevated temperatures. The deformation diagrams are non-linear, during unloading residual strains are formed [1,2]. To describe this effect, theories of thermoplasticity based on the traditional approach using the concepts of yield surface can be used. Such models have been developed in detail for ordinary isotropic structural materials [3,4]. In this models separation of the total strain tensor into elastic, thermal and plastic components are assumed. In this case, the elastic moduli, the coefficient of thermal expansion, and the yield strength depend on temperature. The deformation history of an isotropic body is described, as a rule, by the Odqvist parameter. In such theories, a refined concept of active and passive loading should be used. Algorithms for calculation of structural elements, founded on the concept of yield surface, should include a procedure for determination of the ultimate load at which the material is transferred from an elastic to a plastic state.

Here, for high-temperature composites, we propose a version of the relations of the endochronic theory of thermoplasticity, which describes the effects of nonlinear deformation. In this theory, the key role is played by the parameter of internal time, which phenomenologically describes the irreversible nature of the deformation [5]. Constitutive relations, including this time-like parameter, are written either as relations of hereditary type, or as differential non-linear relations. At the same time, it is not explicitly the concept of yield surface is used, and there is no separation of strains into elastic and plastic components. The theory leads to satisfactory results for materials having a heterogeneous structure at the microlevel, for example, for composites. As a rule, nonlinear deformation of such materials begins at relatively low stresses, which makes it difficult to experimentally study the yield surface in the deformation process.

It is well known that the endochronic theory of plasticity is effective technique for analysis of inelastic behavior of isotropic structural materials under various kind of loading conditions [6-8]. Excellent results in the description of the deformation of laminated composites, concrete under force loading have been obtained by use of this approach [9-11]. It is obvious that the development of an endochronic theory for the case of thermal and force loading of structural materials is a natural next step.

It should be noted that causes of irreversible strains may vary for structure materials. They may differ from the causes of plastic deformations for traditional isotropic structural materials, such as steels and alloys. In certain cases irreversible strains may be caused by damages of the microstructures (microcracks, fiber ruptures, delamination, etc.), and in other cases they may be caused by plastic microstrains of components of the composite materials. In this article, the deformation is studied by means of a phenomenological method. Processes, which take place at the microlevel and cause an irreversible deformation at the macrolevel, is not considered. Therefore, used term "plastic strain" is understood in the generalized sense as irreversible rate-independent strain.

2. Theoretical background

To describe the deformation effects of high-temperature composites under a thermal load, we propose a variant of the endochronic theory of thermoplasticity, which takes into account the history of loading. Constitutive relations shall be written in the following form

$$\sigma_{ij} = \int_0^{\xi} E_{ijkl}[z(\xi) - z(\xi')]\frac{d\varepsilon_{kl}}{d\xi'}d\xi' - \int_0^{\xi} B_{ij}[z(\xi) - z(\xi')]\frac{dT}{d\xi'}d\xi'. \qquad (1)$$

Here, σ_{ij} and ε_{ij} are the stress and small strain tensors, T is the temperature change, $E_{ijkl}[z(\xi)]$, $B_{ij}[z(\xi)]$ $[z(\xi)]$ - material functions, $z(\xi)$ - is the internal time for which equations $dz=d\xi/f(\xi)$ are valid, $f(\xi)$ is a material function, and $f(\xi)>0$, $f(0)=1$. The expression for the internal time measure ξ is represented as

$$(d\xi)^2 = r_{ijkl}d\varepsilon_{ij}d\varepsilon_{ij} + m^2(dT)^2, \qquad (2)$$

where r_{ijkl} is a positive definite symmetric tensor of rank IV of material parameters, $m > 0$ is a material parameter that describes irreversible deformation during temperature heating. In formula (2), the temperature change along with the strain tensor is considered as a loading parameter.

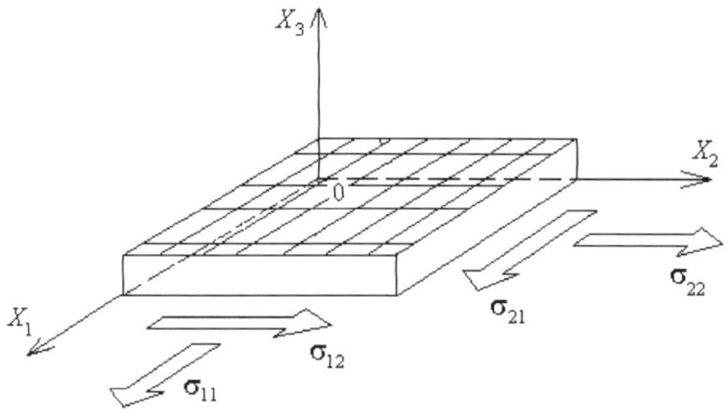

Figure 1. Plane stress state of an orthotropic body.

We assume that the composite at the macrolevel is an orthotropic body in a plane stress state (Figure 1). Subject to the use of exponential kernel, the constitutive relations (1) in the matrix form can be written as follows

$$\{\sigma\} = \int_0^{\xi} \exp\{-[\alpha_S][z(\xi) - z(\xi')]\}[D]\frac{d\{\varepsilon\}}{d\xi'}d\xi' - \int_0^{z} \exp\{-[\alpha_T][z(\xi) - z(\xi')]\}\{B\}\frac{dT}{d\xi'}d\xi'. \quad (3)$$

Here, $\{\sigma\}=(\sigma_{11},\sigma_{22},\sigma_{12})^T$ and $\{\varepsilon\}=(\varepsilon_{11},\varepsilon_{22},\gamma_{12})^T$ – are the stress and strain vectors, respectively, $[\alpha_s]=diag(\alpha_{s1},\alpha_{s2},\alpha_{s3})$, $[\alpha_T]=diag(\alpha_{t1},\alpha_{t2},0)$ – are the matrix of material parameters.
Stiffness matrix [D] is of the form

$$[D]=\begin{bmatrix} E_1' & E_2'\nu_{12} & 0 \\ E_1'\nu_{21} & E_2' & 0 \\ 0 & 0 & G_{12} \end{bmatrix}$$

where E_1, E_2, G_{12}, ν_{12}, ν_{21} – elastic characteristics of an orthotropic material, $E_{1,2}' = E_{1,2}/(1-\nu_{12}\nu_{21})$, $\{B\}=[D]\{\beta\}$, $\{\beta\}=(\beta_{11},\beta_{22},0)^T$ - vector of linear thermal expansion coefficients. The increment of the parameter ξ in accordance with the relation (2) is calculated by the formula

$$(d\xi)^2 = d\{\varepsilon\}^T[R]d\{\varepsilon\} + m^2(dT)^2 \quad (4)$$

Here, the positive definite matrix of material parameters [R] is represented as follows

$$[R]=\begin{bmatrix} r_{11} & r_{12} & 0 \\ r_{12} & r_{22} & 0 \\ 0 & 0 & r_{33} \end{bmatrix}$$

Material parameters are determined from the condition of the best agreement between experimental and theoretical results. For this, numerical procedures can be used, with the help of which the results of basic experiments are processed. These include force uniaxial loading along the OX1 and OX2 axes, under a pure shear in the OX1X2 plane, and also the heating of material specimen.
Further, for convenience of computation, we introduce the following form of the representation for constitutive relations (3)

$$\{\sigma\} = \{\sigma_s\} - \{\sigma_t\}, \quad (5)$$

where

$$\left.\begin{aligned} \{\sigma_s\} &= \int_0^{\xi} \exp\{-[\alpha_S][z(\xi) - z(\xi')]\}[D]\frac{d\{\varepsilon\}}{d\xi'}d\xi', \\ \{\sigma_t\} &= \int_0^{z} \exp\{-[\alpha_T][z(\xi) - z(\xi')]\}\{B\}\frac{dT}{d\xi'}d\xi'. \end{aligned}\right\} \quad (6)$$

Based on (5) and (6), the constitutive relations can be represented in the following differential form

$$\left.\begin{aligned} &\frac{d(\{\sigma\}+\{\sigma_t\})}{d\xi} + [\alpha_s]\frac{\{\sigma\}+\{\sigma_t\}}{f(\xi)} = [D]\frac{d\{\varepsilon\}}{d\xi}, \\ &\frac{d\{\sigma_t\}}{d\xi} + [\alpha_t]\frac{\{\sigma_t\}}{f(\xi)} = \{B\}\frac{dT}{d\xi}. \end{aligned}\right\} \quad (7)$$

Here, $\{\sigma_t\}=(\sigma_{11t},\sigma_{22t},0)^T$ is an auxiliary vector of thermal stresses. In coordinate form, the equations (7) are written as follows

$$\left.\begin{aligned}
&\frac{d(\sigma_{11}+\sigma_{11t})}{d\xi}+\alpha_{s1}\frac{\sigma_{11}+\sigma_{11t}}{f(\xi)}=E_1'\frac{d\varepsilon_{11}}{d\xi}+E_1'\nu_{21}\frac{d\varepsilon_{22}}{d\xi},\\
&\frac{d(\sigma_{22}+\sigma_{22t})}{d\xi}+\alpha_{s2}\frac{\sigma_{22}+\sigma_{22t}}{f(\xi)}=E_2'\nu_{12}\frac{d\varepsilon_{11}}{d\xi}+E_2'\frac{d\varepsilon_{22}}{d\xi},\\
&\frac{d\sigma_{12}}{d\xi}+\alpha_{s3}\frac{\sigma_{12}}{f(\xi)}=G_{12}\frac{d\gamma_{12}}{d\xi}.
\end{aligned}\right\}$$

For thermal stresses we have the following differential relations

$$\left.\begin{aligned}
&\frac{d\sigma_{11t}}{d\xi}+\alpha_{t1}\frac{\sigma_{11t}}{f(\xi)}=(E'\beta_{11}+E_1'\nu_{21}\beta_{22})\frac{dT}{d\xi},\\
&\frac{d\sigma_{22t}}{d\xi}+\alpha_{t2}\frac{\sigma_{22t}}{f(\xi)}=(E_2'\nu_{12}\beta_{11}+E_2'\beta_{22})\frac{dT}{d\xi}.
\end{aligned}\right\}$$

The increment of the parameter ξ describing the loading history is calculated by the formula

$$(d\xi)^2=r_{11}(d\varepsilon_{11})^2+r_{22}(d\varepsilon_{22})^2+2r_{12}d\varepsilon_{11}d\varepsilon_{22}+r_{33}(d\gamma_{12})^2+m^2(\mathrm{dT})^2.$$

It should be noted that for some materials, for example, for carbon-carbon composites based on woven filler, due to the smallness of the coefficients ν12 and ν21, the Poisson effect can be neglected. Then the constitutive relations can be simplified and written in this form.

$$\left.\begin{aligned}
&\frac{d(\sigma_{11}+\sigma_{11t})}{d\xi}+\alpha_{s1}\frac{\sigma_{11}+\sigma_{11t}}{f(\xi)}=E_1\frac{d\varepsilon_{11}}{d\xi},\\
&\frac{d(\sigma_{22}+\sigma_{22t})}{d\xi}+\alpha_{s2}\frac{\sigma_{22}+\sigma_{22t}}{f(\xi)}=E_2\frac{d\varepsilon_{22}}{d\xi},\\
&\frac{d\sigma_{12}}{d\xi}+\alpha_{s3}\frac{\sigma_{12}}{f(\xi)}=G_{12}\frac{d\gamma_{12}}{d\xi},\\
&\frac{d\sigma_{11t}}{d\xi}+\alpha_{t1}\frac{\sigma_{11t}}{f(\xi)}=E_1\beta_{11}\frac{dT}{d\xi},\\
&\frac{d\sigma_{22t}}{d\xi}+\alpha_{t2}\frac{\sigma_{22t}}{f(\xi)}=E_2\beta_{22}\frac{dT}{d\xi}.
\end{aligned}\right\} \tag{8}$$

Here for the parameter ξ we use a simpler expression

$$(d\xi)^2=r_{11}(d\varepsilon_{11})^2+r_{22}(d\varepsilon_{22})^2+r_{33}(d\gamma_{12})^2+m^2(\mathrm{dT})^2. \tag{9}$$

For some types of loading, the constitutive relations (7) allow us to obtain analytical expressions. In general, they should be integrated numerically.

Assuming σ12= σ22=0 and r11=1, β22=0 in equations (8) and (9), we obtain the relations for the case of uniaxial thermal force loading in the direction of the OX1 axis. Let us write them as follows

$$\left.\begin{aligned}
&\frac{d(\sigma_{11}+\sigma_{11t})}{d\xi}+\alpha_{s1}\frac{\sigma_{11}+\sigma_{11t}}{f(\xi)}=E_1\frac{d\varepsilon_{11}}{d\xi},\\
&\frac{d\sigma_{11t}}{d\xi}+\alpha_{t1}\frac{\sigma_{11t}}{f(\xi)}=E_1\beta_{11}\frac{dT}{d\xi},\\
&(d\xi)^2=(d\varepsilon_{11})^2+m^2(\mathrm{dT})^2.
\end{aligned}\right\} \tag{10}$$

Using the relations (10), we consider some deformation effects predicted by the proposed theory. In the calculations we will use the well-known expression $f(\xi)=1+b\xi$ which allows in some cases to obtain analytical solutions [5].

3. Deformation under complex loading

Let us consider the following variant of complex loading. Let at the first stage the material specimen is subjected to free heating, at which the force load is absent, i.e. σ11=0, dT≠0. In the initial state at ξ=0 we have σ11t=0, ε11=0, T=0. Next, we take into account that for heating dT>0, and when cooling take place - dT<0. Writing the third equation of system (10) in the form

$$\left(\frac{d\varepsilon_{11}}{d\xi}\right)^2+m^2\left(\frac{dT}{d\xi}\right)^2=1$$

and using the first two equalities of this system, we obtain the following differential equation

$$\frac{d\sigma_{11t}}{d\xi}=\frac{\pm\rho\sqrt{E_1^2(1+\rho^2)f^2(\xi)-(\alpha_{s1}-\alpha_{t1})^2\sigma_{11t}^2}-(\alpha_{s1}\rho^2+\alpha_{t1})\sigma_{11t}}{(1+\rho^2)f(\xi)}, \tag{11}$$

where ρ=β11/m. From the equation obtained, as well as from the first two equalities of system (10), the differential equations follow.

$$\frac{d\varepsilon_{11}}{d\xi}=\frac{\pm\rho\sqrt{E_1^2(1+\rho^2)f^2(\xi)-(\alpha_{s1}-\alpha_{t1})^2\sigma_{11t}^2+(\alpha_{s1}-\alpha_{t1})\sigma_{11t}}}{E_1(1+\rho^2)f(\xi)} \tag{12}$$

$$\frac{dT}{d\xi}=\frac{\pm\sqrt{E_1^2(1+\rho^2)f^2(\xi)-(\alpha_{s1}-\alpha_{t1})^2\sigma_{11t}^2}-\rho(\alpha_{s1}-\alpha_{t1})\sigma_{11t}}{\mathrm{m}\,E_1(1+\rho^2)f(\xi)}. \tag{13}$$

Equations (11) - (13) form a system of nonlinear differential ordinary equations, which can be integrated numerically for given initial conditions. When heating, one should take a plus sign, when cooling - minus.

Let us determine the deformation during the heating of the material specimen to a certain temperature To, with further cooling it to its initial state. Thus, within the interval $0\le\xi\le\xi_o$ there will be heating, and within $\xi_o\le\xi\le\xi_k$ - cooling. Value ξo corresponds to the time of the end of heating, and the value of ξк – the moment of return of the material specimen to its initial temperature state.

In the numerical example, let the specimen be heated to a temperature Tmax=2140°C, which corresponds to the value ξ0=0.005. For material parameters, let us take the following values: E1=20 GPa; αS1 =150; αt1 =30; b=5; β11=1.8·10-6 1/deg; m=0.24·10-6 1/deg. The system of equations integrates numerically via the Runge-Kutta method. Deformation diagram is shown in Figure 2, wherein $\bar{\varepsilon}_{11}=\varepsilon_{11}/\varepsilon_{max}$, $\bar{T}=(T-293^\circ\mathrm{K})/T_{max}$. εmax≈0.005 is obtained as a result of the calculation. As seen, at heating nonlinear deformation occurs while increasing tangential coefficient of temperature expansion $\beta_{11k}=d\varepsilon_{11}/dT$. Note that during cooling, the theory predicts formation of residual strains (see Figure 2). In this example, when cooled to the initial temperature, a residual strain of ε11r≈0.0017 is formed. This value corresponds to ξk≈0.0084.

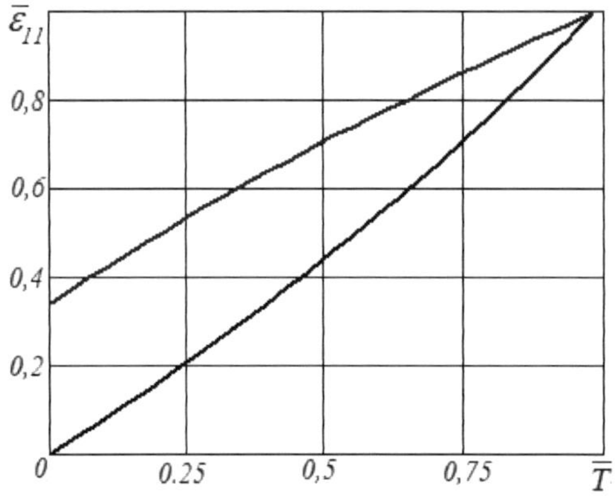

Figure 2. Strain change of the material specimen during heating and cooling

At the second stage, when ξ≥ξ0 the specimen is subjected to uniaxial tension at a constant temperature T=Tmax. In this case, we have $dT=0, d\varepsilon_{11}>0$. The system of equations (10) takes the following form

$$\left.\begin{aligned} &\frac{d\sigma_{11s}}{d\xi}+\alpha_{s1}\frac{\sigma_{11s}}{f(\xi)}=E_1, \\ &\frac{d\sigma_{11t}}{d\xi}+\alpha_{t1}\frac{\sigma_{11t}}{f(\xi)}=0, \\ &d\xi=d\varepsilon_{11}, \end{aligned}\right\}$$

where $\sigma_{11s}=\sigma_{11}+\sigma_{11t}$. This system of equations can be integrated analytically using the following initial conditions $\sigma_{11s}(\xi_o)=\sigma_{11t}(\xi_o)=\sigma^*_{11t}$. The value σ^*_{11t} was obtained in the numerical solution presented above. As a result, when $\xi\geq\xi_0$ we have the following expression

$$\sigma_{11}(\xi)=\frac{E_1(1+\beta\xi)}{\alpha_{1s}+\beta}\left[1-\left(\frac{1+\beta\xi_o}{1+\beta\xi}\right)^{\frac{\alpha_{1s}}{\beta}+1}\right]+\sigma^*_{11t}\left[\left(\frac{1+\beta\xi_o}{1+\beta\xi}\right)^{\frac{\alpha_{1s}}{\beta}}-\left(\frac{1+\beta\xi_o}{1+\beta\xi}\right)^{\frac{\alpha_{1t}}{\beta}}\right],$$

where $\xi=\xi_o+\varepsilon_{11}$.. By differentiating this expression with respect to ξ, we can obtain a formula for calculating the tangential stiffness of the material. At the initial moment of force loading at $\xi=\xi_0$ we will have

$$E_{1k}(\xi_o)=\frac{d\sigma_{11}}{d\varepsilon_{11}}=E_1-\sigma^*_{11t}\frac{\alpha_{1s}-\alpha_{1t}}{1+\beta\xi_o}. \tag{14}$$

Formula (14) shows that the tangent modulus at the initial moment of force loading decreases as a result of preheating. This deformation of the material is confirmed by experimental studies of composites [12]. In addition, formula (14) allows to reasonably choose the values of the material parameters α_{1s} and α_{1t}.

4. Conclusions

The proposed constitutive relations of the theory of thermoplasticity are the development of the endochronic approach in the theory of inelastic deformation of structural materials. In applications, it is convenient to write them in the form of differentially non-linear relations. In particular cases of loading, they can lead to relatively simple analytical relations. In the general case, they are integrated numerically. Constitutive relations describe the effects of nonlinear deformation of the material under thermal force loading, in particular, non-linear change of thermal deformations with free heating, the formation of residual strains during cooling, the dependence of the elastic modulus on the preheating temperature.

References

[1] Sarbayev B S, Baryshev A N 2017 *Vestnik MGTU im. N.E. Baumana. Seria «Mashinostroenie» (in Russian)* **4** 65.
[2] Bobrov A V, Sarbaev B S, Shirshov Yu Yu 2016 *J.of Mach.Manuf. And Reliab.* **45** 145.
[3] Kachanov L M 1969 Osnovy teorii plastichnosti (in Russian). *Moscow: Nauka.* 420.
[4] Birger I A, Demynushko I V 1968 Mechanika tverdogo tela. *Inzhenerniy Zhurnal (in Russian).* **6** 70.
[5] Valanis K C 1971 *Arch. of Mech.* **23** 517.
[6] Valanis K C, Lee C F 1984 Trans. ASME. *J. of Appl. Mech.* **51** 367.
[7] Wu H C, Yip M C 1981 Trans. ASME. *J. of Eng. Mater. And Technol.* **103** 212.
[8] Kadashevich U I and Pomitkin S P 2010 *Izvestia RAN. Mechanika tverdogo tela (in Russian).* **6** 123.
[9] Sarbayev B S 1995 *Comput. Mater. Science.* **4** 220.
[10] Pindera M -J, Herakovich C T 1983 Mechanics of Composite Materials. Recent Advances, ed. Z Hashin, C T Herakovich. *New York: Pergamon Press.* pp. 367-381.
[11] Bazant Z P, Bhat P D 1976 *J. of the Eng. Mech.Div.* **4** 701.
[12] Vasiliev V V, Morozov E V 2007 *Advanced Mechanics of Composite Materials (Amsterdam: Elsevier).* p 491.

IOP Conf. Series: Materials Science and Engineering **683** (2019) 012042 doi:10.1088/1757-899X/683/1/012042

Comparative analysis of averaging methods for obtaining composite material elastic characteristics

V S Zarubin[1], E S Sergeeva[1,2] and I V Magnitsky[2]

[1] Bauman Moscow State Technical University, 2-nd Baumanskaya, 5, b.1, 105005, Moscow, Russia

[2] JSC "Kompozit", Pionerskaya, 4, Korolev, Moscow Region, 141070, Russia

Abstract. Composites are widely used as engineering structural materials in different technical areas, given their higher mechanical properties combined with their smaller weight as compared with the "pure" materials. An important characteristic of a composite is a set of its elastic properties. To determine these properties, both experimental methods and mathematical simulation are used. Due to large material costs and time expenditures inherent in each experiment (fabrication of the material and the test sample, performance of the experiment and processing of the results), on the material design stage, as well as when determining the material application area, preference is given to the simulation methods.This paper is focused on the comparative analysis of the available mathematical simulation methods in terms of performance study of the unidirectional fiber-reinforced composite in the plane normal to the fiber orientation. For illustrative purposes, composites with two different kinds of matrix, polymeric and metal, are considered.

1. Introduction

Currently, composite materials are widely used in different technical areas. These materials are of particular importance in the aerospace and rocket applications where the structural items are exposed to concurrent extreme thermal and mechanical stresses. Alongside with the tough requirements to the thermal and thermomechanical characteristics of structural materials, in these industries, the structural weight is also an essential factor. With the use of the composite materials in different industries, it becomes possible to reduce significantly the structural weight, while retaining, or even improving, the thermomechanical properties of the structural item under development [1].

The most common types of the composite material matrix are polymeric and metal ones. For polymeric composites, various thermoplastics, thermosetting materials and elastomeric materials are most frequently used [2]. The typical metal composites are materials based on the aluminum, titanium, copper or nickel matrix [2].

When designing a composite-based structural item, among the most important characteristics to be considered are the elastic properties of the composite. Experimental investigation of the characteristics of nanocomposites, specifically, the components they are composed of, is often pretty resource-consuming, both in terms of the cost and from the viewpoint of the time required for experiment preparation, performance and result processing. Therefore, to investigate the properties of these materials, mathematical simulation methods are mainly used. These methods can be divided into three groups, namely, analytical methods, numerical methods without the use of the finite-element methods (FEM), and FEM-based methods.

Each group has its own advantages and shortcomings. For example, the analytical methods have been thoroughly studied, do not require too much computing capacity, and impose no requirements to the software (SW). However, the considerable disadvantage of these methods is the necessity of making numerous assumptions that are not always consistent with the reality. The numerical methods make it possible to perform computations in the most general formulation of the task, without making any simplifications or assumptions for the target of research and its properties. The main shortcoming of these methods are their high resource requirements. Hence it is reasonable to perform comparative analysis of the results obtained using different mathematical simulation methods.
In this paper, results obtained by using the best known analytical approaches used by Voigt-Reuss [3], Hashin-Shtrikman [3, 4], as well as the self-consistent method [5], the method outlined in [6], are compared with the results obtained by finite-element simulation using the ANSYS software.].

2. Formulation of the problem

The target of the research is an unidirectional fiber-reinforced periodic composite. The main difficulty when investigating the elastic behavior of such a composite is definition of the elastic properties in the plane normal to the reinforcement direction. This paper addresses a two-dimensional domain (the fiber-reinforced composite cross section normal to the reinforcement direction), with the periodic cell shown in Fig. 1.

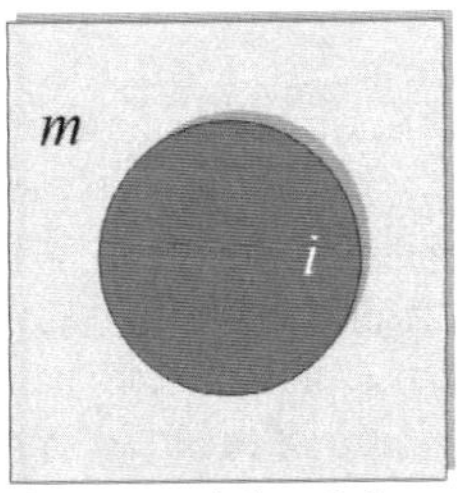

Figure 1. Periodic cell of the region. The materials of the matrix and the inclusion are assumed to be isotropic.

3. Description of the methods. Main relationships

One of the most common and simple approaches to describing the elastic behavior of a composite is the Voigt-Reuss approach that provides the upper and lower estimates for the possible values of the material elastic moduli.
The upper estimate can be obtained by averaging the stresses in the inhomogeneous linearly elastic isotropic medium, assuming its strain state homogeneity. This hypothesis was made by V. Voigt in 1910. The lower estimate can be determined by averaging the deformations in the inhomogeneous linearly elastic isotropic medium, assuming its stress homogeneity. This hypothesis was made by A. Reuss in 1929.
For bulk elasticity modulus K and shear modulus G of a two-component isotropic composite, the Voigt-Reuss model takes the following form:

$$K_+ = 1 - C_V + C_V \frac{K_i}{K_m}, G_+ = 1 - C_V + C_V \frac{G_i}{G_m},$$
$$K_- = \left(1 - C_V + C_V \frac{K_m}{K_i}\right)^{-1}, G_- = \left(1 - C_V + C_V \frac{G_m}{G_i}\right)^{-1}. \quad (1)$$

Here C_V – volume concentration of inclusions, K_+, K_-, G_+, G_- – upper and lower estimates of the bulk elasticity modulus and the shear modulus, correspondingly, of the composite according to the Voigt-Reuss method, K_i, K_m, G_i, G_m – bulk elasticity modulus of and shear modulus of the inclusion and the matrix, respectively.

In this paper, for the convenience of presentation and subsequent analysis, Young's modulus E of the composite, associated with bulk elasticity modulus K and shear modulus G via relation $E = \dfrac{9KG}{3K+G}$, was chosen.

Another approach to estimating the possible values of the material elasticity moduli is known as the Hashin-Shtrikman method [3, 4]. This approach based on the assumption of inhomogeneity of the stress and strain fields involves two functionals: the minimizable (Lagrange functional) and maximizable (the Castigliano functional) ones that coincide with each other over the true distributions of displacements and stresses. The estimates obtained by this method are the optimal ones that can be obtained without taking into account the geometry of the particles of the composite components. For elastic deformation, this approach also has an alternative name, namely, "dual variational formulation of the elasticity problem in an inhomogeneous solid".

In case of a two-component composite consisting of linearly elastic components, the Hashin-Strikman method relations have the following form:

$$\overset{\circ}{K}_{-} = K_m + \frac{C_V(K_i - K_m)}{1+(1-C_V)a_m(K_i - K_m)}, \quad \overset{\circ}{G}_{-} = G_m + \frac{C_V(G_i - G_m)}{1+(1-C_V)b_m(G_i - G_m)},$$
$$\overset{\circ}{K}_{+} = K_i + \frac{(1-C_V)(K_m - K_i)}{1+C_V a_i(K_m - K_i)}, \quad \overset{\circ}{G}_{+} = G_i + \frac{(1-C_V)(G_m - G_i)}{1+C_V b_i(G_m - G_i)}, \tag{2}$$
$$a_m = \frac{3}{3K_m + 4G_m}, \quad a_i = \frac{3}{3K_i + 4G_i}, \quad b_m = \frac{6(K_m + 2G_m)}{5G_m(3K_m + 4G_m)}, \quad b_i = \frac{6(K_i + 2G_i)}{5G_i(3K_i + 4G_i)}.$$

Here $\overset{\circ}{K}_{+}, \overset{\circ}{K}_{-}, \overset{\circ}{G}_{+}, \overset{\circ}{G}_{-}$ – upper and lower estimates of the bulk modulus of elasticity and the shear modulus of the composite, correspondingly, according to this approach.

One of the methods taking into account the particle geometry of the composite component is the self-consistency method. It involves consideration of the interaction between a single inhomogeneity element (an inclusion or a particle of the matrix) and the homogeneous isotropic medium featuring the design elastic moduli values. With subsequent averaging of the parameter distribution perturbations in these elements over the composite volume, it is possible to obtain the design relationships for the elastic characteristics of the composite. For the two-dimensional formulation of the problem, it is necessary to consider reinforcement based on infinite circular cylinders.

For the target material, the main relations of this method for modules K and G have the following form:

$$C_V u_{i1} + (1-C_V)u_{m1} = 0, \; C_V u_{i2} + (1-C_V)u_{m2} = 0. \tag{3}$$

Here

$$u_{i1} = \mathbf{u}_i \cdots \mathbf{V}, \; u_{i2} = \mathbf{u}_i \cdots \mathbf{D}, \; u_{m1} = \mathbf{u}_m \cdots \mathbf{V}, \; u_{m1} = \mathbf{u}_m \cdots \mathbf{D},$$
$$\mathbf{u}_i = (\mathbf{C}_i - \mathbf{C} + \mathbf{C} \cdot\cdot \mathbf{W}_i)^{-1} \cdot\cdot (\mathbf{C} - \mathbf{C}_i), \; \mathbf{u}_m = (\mathbf{C}_m - \mathbf{C} + \mathbf{C} \cdot\cdot \mathbf{W})^{-1} \cdot\cdot (\mathbf{C} - \mathbf{C}_m),$$
$$\mathbf{C} \cdot\cdot \mathbf{W} = (3K + 4G)\mathbf{V} + 5/3((3K+4G)/(K+2G)\mathbf{D}),$$

where $\mathbf{W}$ – isotropic Eshelby's tensor, $\mathbf{C} = 3K\mathbf{V} + 2G\mathbf{D}$, $\mathbf{C}_i = 3K_i\mathbf{V} + 2G_i\mathbf{D}$, $\mathbf{C}_m = 3K_m\mathbf{V} + 2G_m\mathbf{D}$ – tensors of the elasticity coefficients of the composite, inclusion and matrix, respectively, V and D – volumetric and deviator components, respectively, of the unit fourth-rank tensor. Tensor $\mathbf{W}_i$ corresponds to square matrix N of the sixth order that involves seven independent elements [5]:

$$N_{11} = N_{22} = QD_{11} + RD_1, \; N_{12} = N_{21} = QD_{11}/3 - RD_1, \; N_{33} = QD_{33} + RD_3, \; N_{31} = N_{32} = b^2 QD_{11} - RD_3,$$
$$N_{13} = N_{23} = QD_{13} - RD_1, \; N_{66} = QD_{11}/3 + RD_1, \; N_{44} = N_{55} = Q(1+b^2)N_{13}/2 + R(1-D_1)/2, \tag{4}$$

where

$$D_1 = \frac{b\left(b\left(1-b^2\right)^{1/2} - \text{arcch}b\right)}{2\left(1-b^2\right)^{3/2}}, D_3 = 1-2D_1, D_{13} = \left(D_1 - D_3\right)/\left(3\left(1-b^2\right)\right), D_{11} = \left(1-3D_{13}b^2\right)/4, D_{33} = 1/3-2D_{13}.$$

The other elements of this matrix are zero.

Here $Q = 3/2(1-\nu)$, $R = (1/2-\nu)/(1-\nu)$, $\nu = (3K/2-G)/(3K+G)$, and b is the ratio of the cylinder radius to its length for the case of circular cylinder-based reinforcement. To take into account the assumption about unlimited cylinder length, a support study was carried out to show that where $b \le 0.01$, any variations in the coefficients involved in the matrix N elements can be neglected (see Fig.2), therefore, value $b = 0.01$ can be deemed adequate for describing an infinitely long circular cylinder.

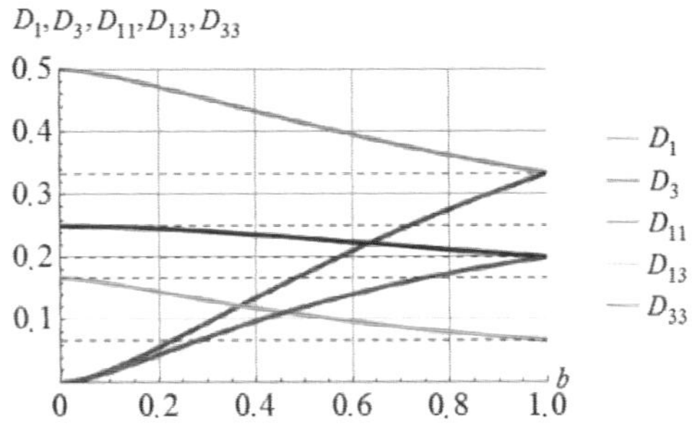

Figure 2. Dependence of auxiliary coefficients D_i involved in matrix N elements on b values.

For fibrous materials, there is an approach described by A.M. Skudra and F.Ya. Bulavs [6]. This method was developed for reinforced plastics. The composite elastic modulus in the direction transverse to the square-type fiber packing (which corresponds to the problem under consideration) can be determined as follows:

$$E = \left[(1-r_i)\frac{1}{1-v_m^2} + r_i I\right]E_m, \tag{5}$$

where

$$I = \frac{1}{a}\left[\frac{\pi}{2} - \frac{2a}{\sqrt{a^2-b^2}}\text{arctg}\sqrt{\frac{a-b}{a+b}}\right],\ a = \frac{1-v_m^2}{E_m},\ b = r_i\left[\frac{1-v_i^2}{E_i} - a\right],\ r_i = \sqrt{\frac{2\pi}{C_V}}.$$

Here E_m, E_i, n_m, n_i – Young's moduli and Poisson's coefficients of the matrix and inclusion, respectively.

For the numerical solution of the problem, the ANSYS finite element software was used. For the selected periodic cell of the composite, the relevant geometric boundary conditions were applied to realize the desired homogeneous uniaxial tension and shear according to the method described in [7]. To analyze the scale effect, simulation was performed for a single, four and nine cells of periodicity of the composite. Processing of the results involved building of a system including the stress and strain vectors, constructing of compliance matrix S for the composite, and determining of the Young modulus from relation $E = 1/S_{11}$.

4. Analysis of the results

As an example of reinforcement for a composite with a polymer and a metal matrix, elastic parameter values GPa and GPa, respectively, were chosen. As reinforcing inclusions for the matrices, those with elastic parameter values GPa and GPa, respectively, were chosen.

Figures 3 and 4 show the results obtained for the polymer and metal matrices.

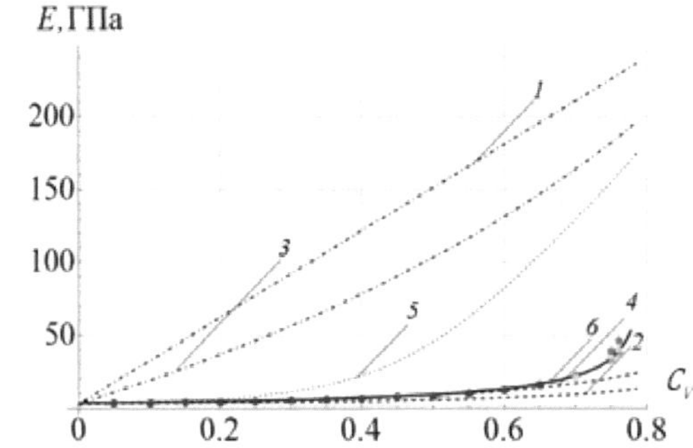

Figure 3. Composite with a polymer matrix.

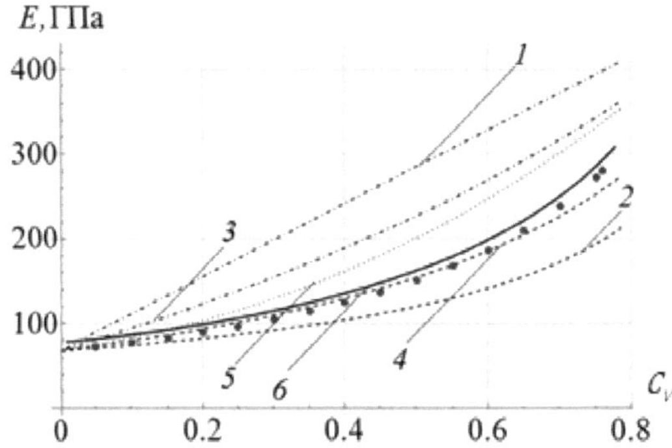

Figure 4. Composite with a metal matrix.

In Figures 3 and 4, 1 and 2 – upper and lower estimates obtained with the Voigt-Reuss method, 3 and 4 – upper and lower estimates obtained with the Hashin-Shtrikman method, 5 – estimate obtained with the self-consistency method, 6 – estimate obtained by the Skudra-Bulavs method. In Figure 3, the red dots represent the results obtained by finite element modeling using the ANSYS software for a single periodic cell, the blue dots – those obtained for four periodic cells, while the green ones – for nine periodic cells. In Figure 4, the blue dots represent the results of numerical simulation with four periodic cells.

As can be seen from Figures 3 and 4, with the Voigt-Reuss and Hashin-Shtrikman methods, it is possible to estimate the range of possible elasticity modulus values for the composite.

As a result of comparative analysis of the estimates obtained using analytical and numerical methods, it was found that with the volume fraction of inclusions from 0 to 15-20%, the self-consistency method, as well as the Skudra-Bulavs and Reuss methods, provide results that differ from each other by less than 10%.

It was also found that for the composites whose elastic moduli differ by several orders, the Skudra-Bulavs method show good agreement with the numerical simulation. For other ratios of the elasticity moduli of the composite components, the difference between the values obtained by these methods was more than 15%.

5. Conclusions

The existing methods of mathematical modeling, their application areas and limitations for their applicability were analyzed, as exemplified by an isotropic two-component material corresponding to the isotropy plane of an unidirectional fibrous composite reinforced with fibers of the same diameter. The most widely used analytical approaches to description of the composite elastic properties (the Voigt-Reuss method, the Hashin-Shtrikman model, the self-consistency method, and the Skudra-Bulavs method) were analyzed, and numerical modeling of the elastic behavior of such a material was carried out (using the ANSYS finite element software). Influence of different combinations of composite

components elasticity modulus values was investigated. For this purpose, two cases were considered, namely, where the Young's modulus of the inclusion exceeds the Young's modulus of the matrix by several orders (about 100 times) (which corresponds to reinforcement of a soft matrix, e.g. a polymer one, by hard inclusions), and where the difference in the composite components elasticity moduli does not exceed 10 times (which in turn corresponds to reinforcement of a metal matrix).
For numerical simulation, the optimal number of grid elements was chosen based on test analysis. Furthermore, influence of the scale factor was investigated, and the optimal number of periodic cells was determined.
The results presented in this paper illustrate the applicability limits of the most frequently used analytical methods in comparison with the numerical one, which is especially important when choosing an approach to solving a specific application task.

Acknowledgments
The work was performed within the framework of implementation of the basic part of the governmental task of the Ministry of Education and Science of the Russian Federation (Project 9.7784.2017/BP) and RFBR grant 18-38-00618.

References

[1] Bhiksha J 2014 Using Composite Materials in the Defense and Aerospace Industry (Transl. from English by V. Rentyuk). *Vestnik Elektroniki.* No. 1. p. 24 – 27.
[2] 1988 Handbook of Composites vol 1: Transl. from English by A.B. Geller et al., ed f J. Lyubin, B.E. Geller. *Moscow: Mashinostroyeniye.* 448.
[3] Shermergor T D 1977 Theory of Elasticity of Microheterogeneous Media. *Moscow: Nauka.* 400.
[4] Hashin Z, Shtrikman S 1962 On some variational principles in anisotropic and nonhomogeneous elasticity. *J. Mech. Phys. Solids.* Volume 10. № 4. 335 – 343 pp.
[5] Eshelby J 1963 Continuum Theory of Dislocations: Transl. from English. *Moscow: IL.* 248.
[6] Skudra A M, Bulavs F Ya 1978 Structural Theory of Reinforced Plastics. *Riga: Zinatne Publishers.* 192.
[7] Pobedriya B E 1984 Mechanics of composites materials. *Moscow: Moscow State University Publishers.* 336.

IOP Conf. Series: Materials Science and Engineering **683** (2019) 012043 doi:10.1088/1757-899X/683/1/012043

Ultra-jet diagnostics of functional-latent factors of composite materials

A A Barzov[1], A L Galinovski[1] and V A Belov[1]

[1] Bauman Moscow State Technical University, 2-nd Baumanskaya, 5, b.1, 105005, Moscow, Russia

Abstract. It is proposed to use a probabilistic modeling scheme to identify the role of latent factors influencing the functional parameters of products made from polymer composite materials. The principles in detection of latent factors are formulated, and the high efficiency of application of ultra-jet diagnostics technology for analyzing the significance of various structurally latent factors is shown.

Keywords: polymer composite material, ultra-jet diagnostics, latent influence factor, probabilistic model, identification principles.

1. Introduction

For the latent factor in the structure of an object of analysis (OA), we mean it's physically implicit property and / or the hidden possibility to have a significant impact on the functional quality, for example, the performance of the operation of this OA.

As a rule, latent influence factors (LIF) are characteristic of technically and / or technologically complex systems, which include almost all types of composite materials.

(CM), primarily based on the polymer matrix - PCM. Figure 1 shows the block diagram of the classification of dominant LIF types in relation to quality analysis of PCM. A distinctive feature of this representation of LIF is the potential possibility of their identification and analysis through the use of the apparatus and tools of ultra-jet hydrophysical diagnostics (UJD) [1, 2], functional features of which are discussed in detail in a number of studies, in particular, in works [3-5].

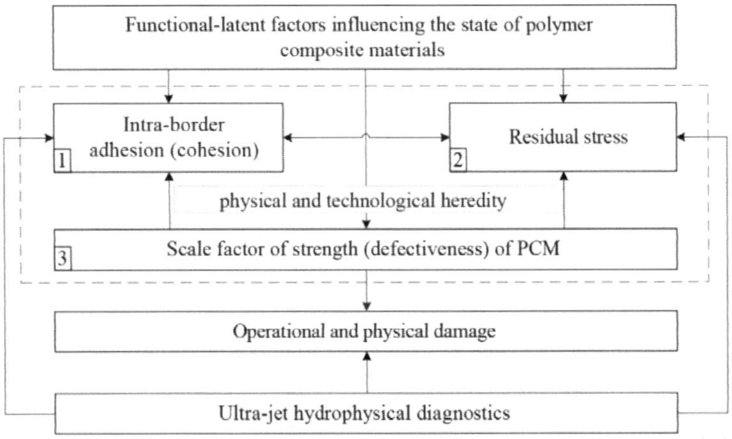

Figure 1. Phenomenological classification of LIF and the role of UJD in their information identification.

2. Systematization of influence factors

Indeed, the physical and technological basis of the UJD is controlled minimally invasive, including step-by-step, hydro erosion local destruction of the surface layers of OA, for example, products made of PCM. According to the results of the analysis of the informative features of this physico-diagnostic influence, by studying the characteristic features of hydroerosion products, the factors influencing the specificity of this process are identified, including those having a structurally latent nature. In [6–9], it was shown that the informative parameters of the UJD are unambiguously related to the degree of damage to the surface layer of OA in the broad sense of the term, summarizing the results of the effect on OA of several, including dissimilar factors. It is the presence of a significant number of physicochemical, constructive-technological and other factors influencing the quality of PCM products, as well as the difficulty in forming functional interrelationships between explicit and latent state parameters of this class of OA, it is quite reasonably to suggest the effectiveness of using the probability theory apparatus to determine the required regularities between their input and output characteristics. To illustrate this point, we transform the phenomenological classification of LIF presented in Figure 1 into a block diagram of the functional interconnectedness of LIF presented in Figure 2 and which is quite a realistic base case for building the corresponding probabilistic model. Let us comment in more detail on some of the features of the interaction between the LIF, which are shown in this structural diagram.

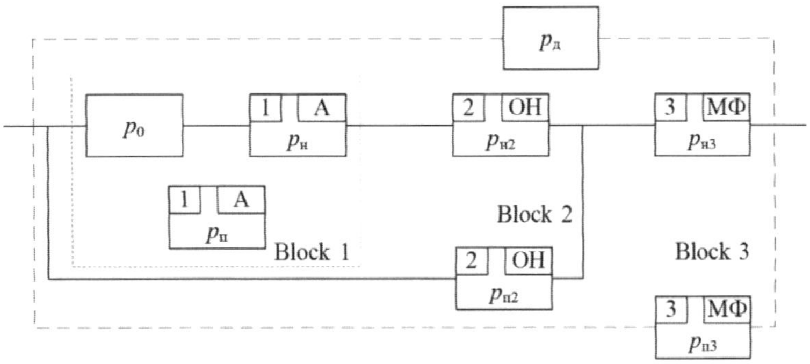

Figure 2. A schematized version of the interaction of various LIF to build a probabilistic model for their analysis.

The initial internal block 1 is an illustration of the latent influence of the adhesion-cohesive factor (ACF) in the structural-elementary PCM cell, which has an initial probability of reliable operation with the required quality level, and the element reflects the positive influence of good adhesion between the PCM components: a filler and the binding media on this characteristic of operational and technological quality. The second block, by analogy with the first, takes into account both the positive, and negative latent influence of the residual stress factor in the PCM of a particular product on its functional quality. The third block of the structural scheme reflects, respectively, the negatively positive role of the scale factor (SF) of PCM products in the form of elements. The structural element parallel to the LIF interaction scheme examined in the form shows the conditionally unambiguous positive role of the UJD in the information and methodological support of the procedure for determining the characteristic probability values of the above-mentioned latent elements and blocks. For example, for products, parts and other OA made from traditional PCM, the standard procedure for estimating the level and topography of the occurrence of residual stresses of the first kind by the classical Davidenkov method is practically impracticable. However, using a weakly invasive microfragmenting effect of ultra-jet (UJ) on the surface layer of OA, direct implementation of the engineering evaluation of the residual stress factor (RSF) using the Davidenkov method is possible. In addition, the dependence of the intensity of the UJ-hydroerosion process on the magnitude and sign of the residual stresses opens up a physically determined very potentially effective opportunity for the express analysis of information and diagnostic distinguishing features of the RSF.

3. Generalized probabilistic model

It should be emphasized that the effectiveness of the UJD for the identification and analysis of various LIF for typical and prospective PCM, depending on their physical and technological nature, will also be different. Nevertheless, by creating special conditions for conducting model comparative experiments, the necessary effectiveness of solving information technology tasks to identify the role of LIF is ensured [10-14]. Moreover, these experiments should be planned on the basis of the analysis of a specific, previously constructed model of interactions and functional features of one or several studied LIFs. For example, for an experimental study of the role of the latent adhesion factor, it is necessary, at a minimum, to change, in particular by screening, the level of adhesion between the fiber (filler) and the binder (matrix) with all other conditions of PCM formation being equal. Therefore, the success of the procedure for constructing a generalized model of accounting for the influence of LIF on the quality of PCM actually predetermines the purposeful effectiveness of planning and conducting experimental research, mainly of a comparative nature.

Based on this, we will form a probabilistic model of interaction between the three abovementioned LIFs: adhesive (ACF), residual tension (RSF) and scale factor (CF) strength (defectiveness) of the analyzed PCM, while not taking into account the roles of UJD in their identification. According to the block diagram presented in Figure 2, this probabilistic model will have the form:

$$P = 1 - \{[1 - (1 - p * p_{\text{н}2})(1 - p_{\text{п}2})] * p_{\text{н}3}\} * (1 - p_{\text{п}3}) \tag{1}$$

$$p = 1 - (1 - p_0 * p_{\text{н}1})(1 - p_{\text{п}1}) \tag{2}$$

where: P – total probabilistic assessment of the functional performance of PCM and / or products from it, taking into account the positively negative role of three types of LIF: ACF, RSF and CF.

Conventionally, in model (1) and its initial block, the role of the UJD is not taken into account in detail, which should actually be present in each of the analyzed blocks of the entire structural scheme. In particular, the role of UJD can be reflected in a generalized model of the form:

$$P_0 = 1 - (1 - P)(1 - p_{\text{д}}) \tag{3}$$

where: P – determined from (1) and (2), and $p_{\text{д}}$ – schematically shown in Figure 2.

We give a numerical example. Without a UJD, let the probabilistic assessment of the role of LIF, performed empirically, in particular according to the results of expert and analytical analysis to be $P = 0{,}75$. At the same time, the reliability of the UJD data factors is also equal to $p_д = 0{,}75$. Then the final assessment of the significance of LIF, according to (3) will be: $P_0 \sim 0{,}94$, i.e. the error in determining the role of LIF will not exceed 4%, which is quite acceptable in engineering calculations and the corresponding design and technological solutions.
Methodological interest is the analysis of any block from the standpoint of the neutral role of LIF in the formation of functional parameters of OA, in particular, products made of PCM. Using the example of block 1 (Figure 2), we determine the priority correlation between the positive and negative effects of the adhesion-cohesive factor in the PCM unit cell. For this, it is necessary to assume that the influence $p_{н1}$ and $p_{п1}$ is in fact of no effect due to their mutual functional annihilation and the final probabilistic assessment of the role of the analyzed LIF has not changed. Formally, this provision can be represented as:

$$p = 1 - [(1 - p * p_н)(1 - p_п)] \tag{4}$$

where, to simplify recording, it is accepted $p = p_0$; $p_н = p_{н1}$; $p_п = p_{п1}$ (see block 1 in Figure 2).
For the detailed solution (4), it is logical to assume that the probabilities of the positive $p_п$ and negative $p_н$ effects of the latent adhesion factor form a complete group of events and their total probability is equal to one: $p_п + p_н = 1{,}0$. Then, taking into account this circumstance, the decision (4) relatively to $p_п$ looks like:

$$p_п = \frac{2p-1}{p} \tag{5}$$

or for negative manifestations of LIF:

$$p_н = \frac{1-p}{p} \ . \tag{6}$$

And from (5) we can obtain an obvious restriction for p:

$$2p - 1 \geq 0 \ . \tag{7}$$

The last relation physically means that with a small level of influence of source-dominant factors, i.e. when $p < 0{,}5$, it is practically impossible to balance the negative and positive manifestations of LIF.
Thus, using the model of the system of interaction of an aggregate of LIFs by means of a targeted experimental and physical influence on it, the possibility is realized not only of assessing the role of individual factors in the overall picture of functional relationships, but also of their very likely synergistic manifestations. First of all, this provision refers to the structural-boundary effects of the physicochemical interactions of the matrix with the filler [15-20]. Moreover, the instrument and toolkit of the LIFs technology can become a tool for analyzing the surface structural manifestations of LIF, including synergistic ones.

4. Graphical illustration

As a typical example, Figure 3 provides a graphical illustration of an engineering methodology for rapid assessment of the level of residual stresses in the surface layer of OA, in particular products made of PCM. This technique organically combines the basic principles of UJD:

• the principle of the need for additional impact on the object of diagnosis in order to identify the analyzed LIF in accordance with its formalized description within a certain functional model;
• the principle of a relative comparison of the results of UJ-impact to OA, in particular the surface of the product from PCM, in "natural" conditions of operation and during model experiments;
• the principle of extrapolation of actually observed dependencies to the zone of determining the predicted value of LIF with hypothetically functional OA parameters.

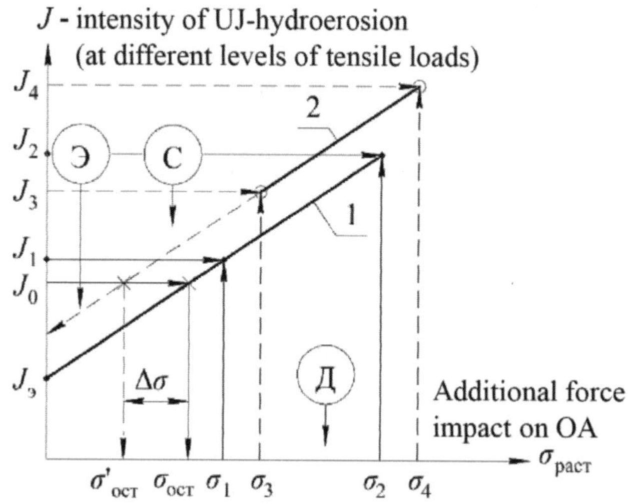

Figure 3. Graphic illustration of the algorithm for determining residual stresses as a characteristic LIF.

In the figure it is accepted:
1 – $J = f_1(\sigma)$ in the presence of reference value: $J_э = f(\sigma_{ост} = 0)$;
2 – reference value $J_э = f(\sigma_{ост} = 0)$ is absent and dependence $J = f_2(\sigma)$ is built with functional latent participation of $\sigma_{ост}$;
$J_1 J_2, J_3, J_4$ – according to the intensity of the UJ-hydro erosion for the reference sample (J_1, J_2) and the real OA (J_3, J_4)
$\sigma_1 \sigma_2$, σ_3, σ_4 – mechanical stresses OA from the action of additional power multi-level loads. Designations are similar for $J(\sigma)$;
$J_0 = f(\sigma_{ост})$ – method of determination $\sigma_{ост}$ using dependency 1; $J_0 = f'(\sigma'_{ост})$ – the same, but for dependence 2;
$\Delta\sigma$ – error in the determination $\sigma_{ост}$ in the absence of data on UJD of the reference sample.
Conventionally, we denote these principles by the terms: complementarity (Д), comparability (С) with the standard and extrapolation (Э). These terms methodologically illustrate the UJ-algorithm for determining the level of residual stresses $(\sigma_{ост})$, presented in Figure 4.

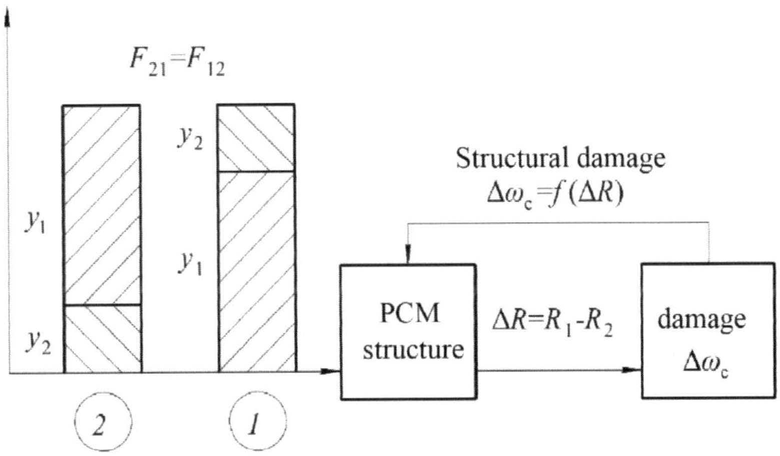

Figure 4. Graphic illustration of the method for express determination of structural damage ω_c of various PCM as a significant LIF.

In the figure it is accepted:

$y_1 + y_2 = y_2 + y_1$ equality of total UJD impact DCR F_{12} and F_{21};
R_1 & R_2 - respectively, the results of exposure F_{12} and F_{21};
ΔR & $\Delta\omega_c$ - respectively, the information-significant difference in the results of UJ-impact and structural damage of PCM.

A distinctive feature of this algorithm are two options for the definition: more accurate – in the presence of a reference OA sample with a conditionally zero level and only real OA, for which the assumption is a priori:

$$\sigma_{\text{ост}} < \sigma_{\text{раст}} \tag{8}$$

where $\sigma_{\text{раст}}$ – additional tensile stresses created in the conditions of a model experiment on the UJD of a sample of PCM or another material when it is significant multi-level tension by external force.

5. Conclusions

The above example of defining the role of LIF in the form of residual stresses in the structure of PCM clearly illustrates the information and diagnostic capabilities of the formulated principles of their functional identification. To identify other LIFs, such as the factor of physical and technological heredity and / or operational damage (defectiveness) of OA material, these principles should be supplemented with the following provisions:

♦the principle of non-commutativity of the results of consecutive-variable UJ impacts of the so-called K-principle of nonlinear diagnostics;

♦ the principle of performance of physical analogies of the additional impact - the principle - A, according to which the level of the most diverse influences on the PCM can be formalized along the abscissa axis: thermal, radiation, etc.

Schematically, Figure 4 illustrates the methodically feasible possibility of detecting LIF in the form of damage to the surface layer of a PCM product. The distinctive feature of this technique is comparing the results of 2 different-level effects in the analysis of which the functional potential of the K-principle is used to obtain comprehensive information about the damage (defectiveness) of an OA surface during its UJD.

Thus, the use of probabilistic modeling apparatus of the role of LIF in the formation of the output characteristics of products from PCM allows for quite effective information verification of the proposed models through the targeted application of UJD tools and the corresponding principles of identification analysis.

Acknowledgments
The studies were carried out as part of a grant from the President of the Russian Federation for state support of leading scientific schools (НШ-3778.2018.8) and a grant from the RFFI 18-29 18081.

References

[1] Barzov A A 2005 Emissionnaya tehnologicheskaya diagnostika [Emissive technological diagnostics]. *Moscow, Mashinostroenie Publ.* 384.

[2] Galinovski A L, Veltishchev V V, Belov V A, Mugla D R and Bolgarchuk I O 2018 Experimental studies on the choice of rational technological regimes and parameters of hydroabrasive cutting in conditions of equipment operation in the submerged position. *Reference book. Engineering journal with the application.* #7 (256). pp. 3-9.

[3] Galinovski A L, Mulyar S G and Hafizov M V 2012 Primenenie gibridnoi diagnostiki dlya ocenki ekspluatacionnyh svoistv kompozicionnoi keramiki [Application of hybrid diagnostics for evaluation of performance properties of composite ceramics]. *News of higher educational institutions Engineering Publ.* #9 pp. 65-69. (in Russian).

[4] Abashin M I, Vinokurova E V, Galinovski A L and Korshunov S V 2014 Mirovye tendencii sovershenstvovaniya sistemy podgotovki inzhenerov [Global trends in improving the system of training engineers]. *Vestnik MGOU Publ.* # 2 pp. 57-65. (in Russian).

[5] Abashin M I, Galinovski A L, Zosimov M V and Moiseev V A 2016 Analiz tendencii razvitiya inzhenernogo obrazovaniya v stranah, imeyushih vysokii uroven' innovacionnogo i tehnologicheskogo razvitiya [Analysis of trends in the development of engineering education in countries with a high level of innovation and technological development]. *Vestnik of Kostroma State University. Series: Pedagogy. Psycology. Sociokinetics. Publ.* #3 pp. 22-28. (in Russian).

[6] Barzov A A, Galinovski A L and Abashin M I 2012 Faktornaya model' ul'trastruinoi gidroerozii [Factor model of ultra-jet hydroerosion]. *News of higher educational institutions. Engineering Publ.* #10 pp. 63-68. (in Russian).

[7] Abashin M I, Barzov A A and Galinovski A L 2012 Analiz fiziko-tehnologicheskih osobennostei processa ul'trastruinoi diagnostiki [Analysis of physical and technological features of ultra-jet diagnostics process]. *Bulletin of Moscow state technical University. N. Uh. Bauman. Series: natural Sciences Publ.* #6 pp. 7. (in Russian).

[8] Abashin M I, Galinovski A L, Zosimov M V and Moiseev V A 2016 Analiz tendencii razvitiya inzhenernogo obrazovaniya v stranah, imeyushih vysokii uroven' innovacionnogo i tehnologicheskogo razvitiya [Analysis of trends in the development of engineering education in countries with a high level of innovation and technological development]. *Vestnik of Kostroma State University. Series: Pedagogy. Psycology. Sociokinetics Publ.* #3 pp. 22-28. (in Russian).

[9] Barzov A A, Galinovsky A L and Abashin M I 2012 Faktornaya model ultrastruynoy gidroerozii [Ultimate jet erosion factor model]. *Proceedings of higher educational institutions. Mashinostroyeniye Publ,* #10, pp. 63-68. (in Russian).

[10] Abashin M I, Barzov A A and Galinovski A L 2012 Analysis of the physical and technological features of the process of ultra-jet diagnostics. *Bulletin of the Moscow State Technical University. N.E. Bauman. Series: Natural Sciences.* #6 P. 7.

[11] Barzov A A, Eroeeva V V, Sysoev N N, Sysoev P N and Yablochnikov S L 2018 Analysis of the influence of structural-latent factors on the potential of their functional effectiveness. *Preprint Faculty of physics, Moscow state University. M. V. Lomonosov.* #5, 33p. (in Russian).

[12] Nelyub V A 2016 A study of the microstructure of dressed glass fibers. *Polymer Science – Series D.* # 9 (1), pp. 96-100.

[13] Nelyub V A 2015 Determination of adhesion interaction between carbon fiber and epoxy binder. *Polymer Science – Series D.* # 8 (1), pp. 6-8.
[14] Nelyub V A, Borodulin A S, Kobets L P and Malysheva G V 2016 A study of structure formation in a binder depending on the surface microrelief of carbon fiber. *Polymer Science – Series D.* # 9 (3), pp. 286-9.
[15] Li S, Galinovsky A L and Zarubina O V 2018 Development of manufacturing technology and ultra-fast diagnostics of bimetallic cutting tools for the utilization of aerospace engineering composite materials. *Mashinostroenie: innovative aspects of development materials of the Iinternational scientific-practical conference.* P.84-5. (in Russian).
[16] Galinovsky A L 2014 Simulation of the process of diagnostics of nanostructured coatings with an adhesiometer and ultra-jet method. *Bulletin of the Rybinsk State Aviation Technological Academy. P.A. Solovyov.* P. 9-16. (in Russian).
[17] Barzov A A and Galinovsky A L 2009 Polyfunctional capabilities of ultra-jet technology for processing materials and liquids. *Vestnik UGATU. Mashinostroenie.* Volume 12, 4 (33) P.116-120. (in Russian).
[18] Barzov A A, Galinovsky A L, Salnikov S K and Elfimov V M 2010 Improving the performance of ultrajet processing by optimizing the kinematic factor. *Vestnik MGTU im. N.E. Bauman. Special edition.* P.217-29. (in Russian).
[19] Abashin M I, Galinovsky A L and Sudnik L V 2013 Determination of hardness and resistance to abrasion of steel samples by the ultra-jet method. *News of higher educational institutions. Mechanical Engineering.* #9 P. 37-43. (in Russian).
[20] Galinovsky A L 2011 Study of the physical and technical features of the interaction of ultraformal and obstacles. *Fundamental and applied problems of engineering and technology.* #4-2 (288) P. 131-5.

Change in the structure of the aluminium alloy under the action of direct electric current

S L Timchenko[1] and N A Zadorozhnyj[2]

[1,2] Fundamental Science Faculty, Department of Physics, Bauman Moscow State Technical University, 2-nd Baumanskaya, 5, b.1, 105005, Moscow, Russia

E-mail: [1]svtimchenko@yandex.ru, [2]nikazador@mail.ru

Abstract. The capabilities of the method for producing cast aluminum alloys, including composite alloys, which include the transmission of direct electric current at the crystallization stage, are shown, and the results are studied for experiments on the crystallization of aluminum alloy (silumin: Si – 10.47 %; Cu – 1.75 %; Fe – <0.2 %; Cr – 0,.02%; Mg - <0.02; Mn - <0.02 %; Ni – 0.08 %) under the action of direct electric current. Comparative study of the structure and porosity of the aluminum alloy, the crystallization of which occurred under the influence of electric current and without current, showed that the electric current leads to a change in the structure of the alloy, reducing porosity, creating a finely differentiated eutectic with a decrease in the dendritic parameter. The mechanism of influence of electric current on the movement of inclusions in the melt is based on the diffusion process. Within the framework of considering the diffusion flux created by the electric field of the current, the speed of the inclusion of crystalline silicon in the melt was calculated in a linear approximation, which shows the possibility of applying this effect in obtaining aluminum matrix composite materials.

Keywords: crystallization, electric current, electrotransfer, silumins, aluminum matrix composite materials, eutectic, dendritic parameter, porosity, hardness.

1. Introduction

At present, it is relevant to improve existing and create new technologies for the production of products with more reliable performance properties. These properties are usually laid at the stage of choosing the material of the product, the material that will provide its necessary properties and manufacturing technology of a particular product. In order to control the physical properties of the material and the operational properties of the product at various stages of the technological process of obtaining the product, additional external effects are used: mechanical, ultrasonic, electromagnetic [1-12]. Also the technologies for producing products from cast aluminum alloys, composite materials use the effect of electric current at the stage of modification, crystallization [2-5, 12].

Experimental problems of the effect of electric current on the physical properties of metals were solved already in the 19th century. Namely, the first experimental studies of non-thermal influence of electrical current on the physical properties of metals were carried out as early as 1844 by the German scientist G. Wertheim [13]. Wertheim conducted two series of experiments. In the first series, he measured the elongation of wire samples of various metals with a constant external mechanical load under the conditions of transmission of electric current (j ~ 107 - 108 A/m2), and also measured the temperature of wires. In the second series of experiments, he measured the elongation of the samples in the process of their deformation when using additionally only thermal effects, allowing to create a fixed sample temperature similar to the temperature created by the flowing current in the first series of experiments. According to the results of measurements of the elongation of the samples, the elastic modulus of the studied material was indirectly determined. It was revealed that the values of the elastic modulus of the same material are different when it is deformed under the action of current and without passing the current. The presence of the difference in the values of the modulus of elasticity of the material served as evidence of the non-thermal effect of electric current on the magnitude of the elastic modulus of the metal.

Then, more than 100 years later, scientists returned to the question of the application of electric current in technological processes for the treatment of hard-to-deform materials with the aim to improve their properties, such as strength, hardness, ductility, as noted in the works of O.A. Troitsky, N. Conrad et al. [14-19] observed a decrease in the deforming force in the form of a sharp decrease in the resistance of metal crystals to deformation under impact of the current of high density in a pulsed mode, which was named the electroplastic effect. The effect manifested itself in the form of characteristic discharges of a deforming force in the diagrams of stretching or compressing zinc single crystals or deformation jumps, which is associated with the accumulation of a sufficient number of dislocations, the movement of which can be detected by a jump in deforming force and indicates a nonthermal, anisotropic effect of electric current. The presence of impurities and an increase in their concentration led to an increase in the jump of the additional mechanical stress under the action of current. This is due to the loss of stability of a larger number of dislocations under the action of current, that is, upon doping of crystals, the uniformity of shear formation and the decrease in the effective size of regions of incomplete shear are increased. Electric current causes the structure to transform during plastic deformation. Its effect on one of the recrystallization process parameters — the recrystallization temperature, as well as on the result of the recrystallization process — grain size and hardness — was determined.

Starting from the 70s of the 20th century to the present, scientists study and put into practice the effect of electric current on the processes of modification and crystallization in order to control the properties of the alloy [2-12]. It was found that the use of electromagnetic effects at the stage of phase transition during crystallization of castings from non-ferrous alloys or at the stage of their modification allows changing the structure of the material and the properties of the product [2, 3, 5]. This is due to the fact that the treatment of the melt with an electric current during the crystallization stage makes it possible to influence the rate of heat and mass transfer processes, the formation of phases and the structure of the material of the product, and to regulate the distribution of alloying components. The electric field causes directional diffusion flows of ions in metals and alloys, leading to the movement of inclusions, activating the process of dissolution of pores. Being sinks or places of origin of vacancies, the interphase boundary regions affect the movement of vacancies and determine the formation of alloy properties. The use of electric current at the stage of crystallization, alloying, modification, allows by creating additional heat fluxes to intensify the processes occurring at the interphase boundaries both in the production of melts and in the production of cast composite materials. For example, when combining solid and liquid phases, as well as various components that are in a state of melt.

2. Experiments and results

For this purpose, a study was carried out on the structure and porosity of the aluminium alloy, which crystallized under the influence of an electric current. For the experiment, a setup was used that allows for a method of producing shaped castings under conditions of direct current flow at the stage of filling the mold with the melt and subsequent solidification [6]. The electric current was passed directly through the melt in the process of filling the mold with it due to the sequential arrangement of the contacts in the mold. The crystallization of castings took place in sandy forms. As the material of the casting, an aluminum alloy was used the chemical composition of which was determined using spectral analysis on an LAES MATRIX atomic emission spectrometer with laser excitation. The chemical composition of the aluminum-based alloy was as follows: Si – 10.47 %; Cu – 1.75 %; Fe – <0.2 %; Cr – 0.02%; Mg - <0.02; Mn - <0.02 %; Ni – 0.08 % and, in accordance with the chemical composition, it can be attributed to silumin.

Silumins is usually characterized by low solubility of silicon at high and low temperatures of the alloy, and therefore it is not possible to harden the alloy by heat treatment. In this case, modifying is a reliable method for improving mechanical properties. Modification of the structure, in terms of the transmission of electric current through the melt, the imposition of electromagnetic fields, ultrasonic effects, causes a change in the crystal structure of castings, increases the density of the material, provides a fine-crystalline structure and a high level of mechanical properties of aluminum alloys [1-3, 11, 12].

It should be borne in mind that casting aluminum alloys have a high tendency to oxidation and saturation with hydrogen, which leads to such types of scrap castings as gas porosity, the presence of slag and oxide inclusions. The tendency of aluminum alloys to interact with gases is due to increased activity of aluminum. Upon receipt of castings by sand casting, gases, nitrogen, oxygen, carbon and its oxides condensed on the surface penetrate the casting material at high temperatures. Particularly affected gas porosity in this case is the surface layer of the casting. The formation of gas pores is also possible due to the presence of undissolved gas bubbles in the aluminum melt.

In this regard, it was necessary to check the effect of electric current on the porosity of the sample material. Porosity was calculated not directly, but indirectly, as the relative difference in the density of the material due to current processing during crystallization:

$$\frac{\Delta\rho}{\rho} = \frac{\rho(I \neq 0) - \rho(I = 0)}{\rho(I \neq 0)} \cdot 100\% \qquad (1)$$

During the crystallization of castings in the presence of current for a given alloy, the relative difference in the density of the material, in relation to the samples, crystallization of which occurred in the absence of the electric current (I = 0 A), was: $\Delta\rho/\rho \approx 3{,}80\%$ at a current of I = 66 A.

It can be assumed that a decrease in the porosity of the alloy during crystallization in the presence of electric current may be due to the activation of filtration processes under the action of electric current, which compensate for the insufficient supply of castings. On the other hand, the electric current passed through the melt during the crystallization stage activates the process of mass transfer in the melt due to the influence on the electron-ion subsystem of the alloy [20, 21]. An electric current, representing an orderly movement of electrons, which will influence the formation of the dendritic structure of the alloy. As a result of the impact of electric current on the alloy during the crystallization process, the size of the dendritic cell should change and the hardness will increase.

The study of the structure of the material of the samples was carried out with an optical microscope (Leika DM ILM). The structure parameters were calculated using an image analysis computer program (Qwin). The dendritic parameter and the area of the solid solution were chosen as parameters for the comparative analysis.

The microstructure of the aluminum alloy crystallization which took place without the action of the electric current I = 0 is shown in Figure 1 (a) and Figure 1 (b) shows the microstructure of the test silumin, crystallization is held by passing through the sample electric current of I = 66 A. The study showed that the transmission of electric current in the process of alloy crystallization leads to a decrease in the size of the dendritic parameter (DP) and to a change in the distribution of the eutectic in the alloy. The structure of the crystallized alloy without current (Figure 1 (a)) represents the characteristic dendritic cells of α-solid solution and eutectic. The structure of the alloy which crystallization occurred during of the DC electric current the flow through the melt (Figure 1 (b)) has changes in the size of the dendritic cells. So in the absence of a current (I = 0) the value of the dendritic alloy parameter is on average 32.0 μm, and during crystallization with current passing by a force I = 66 A the value of the dendritic parameter reaches 20 microns, the area of α- solid solution changes by about 2 times. Eutectic has a more dispersed structure with a characteristic change in the length of the boundaries between eutectic colonies. In some areas, one can observe a more subtle differentiated structure of the eutectic (Figure 1 (b)).

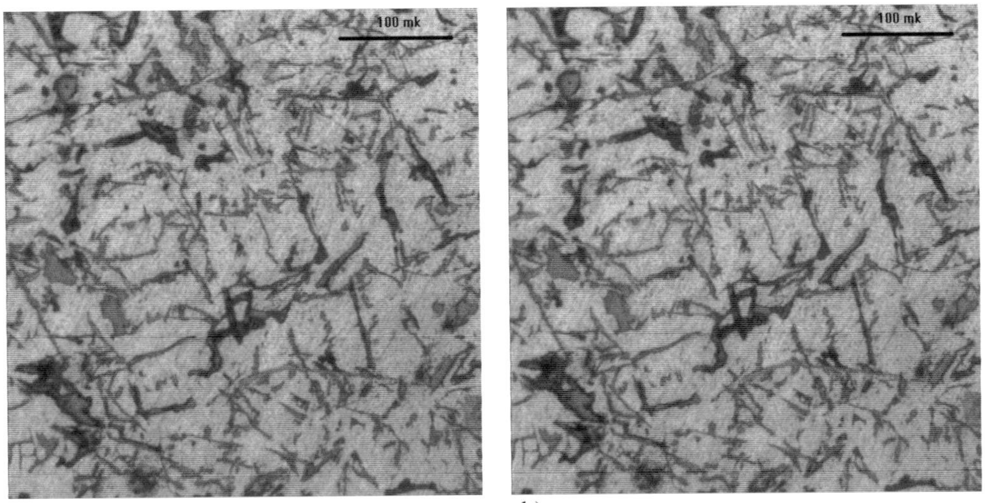

a) b)

Figure 1. (a) I = 0A, transverse section, x400, DP = 32 micron; (b) I = 66 A, transverse section, x400, DP = 20 microns.

Thus, there is a change in the configuration of the dendritic structure and its partial destruction. A fine-differentiated eutectic appears in the structure; no pronounced dendritic cells of α-solid solution are observed. Changes in the structure of the alloy due to the influence of the electric field of the current on the structure of the alloy through the electron and ion subsystems and less related to external alloy cooling conditions.

3. Discussion of the experimental results

An important causal process affecting the features of the crystallization of metal melts in the presence of electric current during crystallization is electrotransference [21], as the phenomenon of directional migration of ions of the melt under the action of electric current. The unusual manifestation of electrical transfer at the interface between solid and liquid phases, that is, during a phase transition is associated with an increase in the mobility of impurities in the liquid-phase region. In a liquid melt, the influence of current on the formation of the microstructure of the alloy will be carried out through the ion and electronic subsystems.

The influence of the electric field on diffusion jumps of crystal ions is associated with the forces that act on the diffusing ion. Since the probabilities of jumps in the direction of the force and in the opposite direction are different, a directed diffusion electrotransfer of metal ions appears. At least two forces act on the ion: the first from the electric field of the current $\boldsymbol{F_i}$, and the second from the side of the drifting electrons – $\boldsymbol{F}_{ie}$. In this case, the resultant force acting on ion determined as

$\boldsymbol{F} =_0 (z - z^*) \cdot e\boldsymbol{E}$ (z_0 – is the true ion charge, z^* is the effective charge) [21].

To substantiate the physical mechanism of the influence of electric current on the redistribution of the eutectic in the alloy under study, a diffusion model was chosen. Analysis of inclusions movement was performed in the studied alloy at its crystallization under the influence of a constant electric current of I = 66 A. For this purpose, a diffusion flux created only by the electric field of the current was considered. In the linear approximation [21], the speed of movement of the inclusion $\boldsymbol{\upsilon}$ will be proportional to the electric field strength $\boldsymbol{E}$:

$$\boldsymbol{\upsilon} = -2\chi_e \frac{De\,z}{f\,kT}\mathbf{E}, \qquad (2)$$

where $\chi_e = \dfrac{\lambda_e - \lambda_{oe}}{2\cdot\lambda_e + \lambda_{oe}}$ - coefficient determining the contribution of the electrical conductivity of the alloy, λ_e and λ_o

- the electrical conductivity of the matrix and the inclusions, which in cubic crystals are scalars; D is the diffusion coefficient; k is the Boltzmann constant; $e{\cdot}z$ – is the ion charge; f - correlation factor. The correlation factor takes into account that with a vacancy diffusion mechanism, the probability of an ion returning after a diffusion jump to its previous position exceeds the probability of its transition to a new position. In monatomic crystals, the f factor differs little from unity. For crystals with a face-centered cubic lattice, $f = 0.781$.

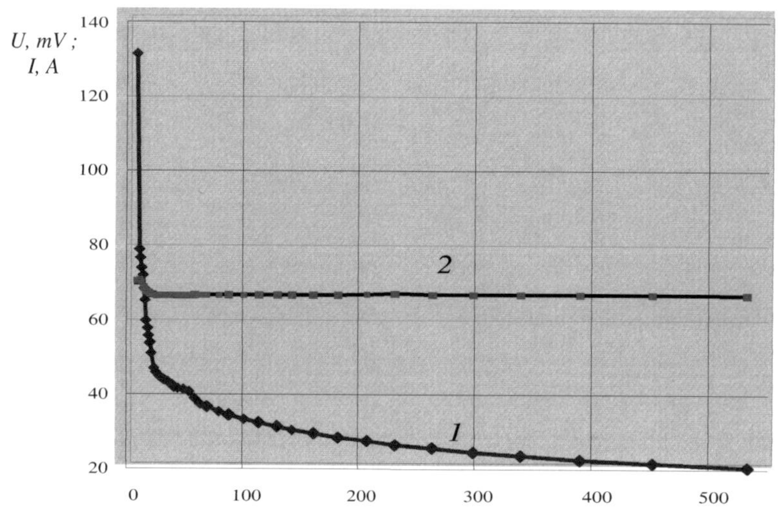

Figure. 2. The time dependences of the voltage drop U (t) (1) on the sample (alloy AK12) and current I (t) (2).

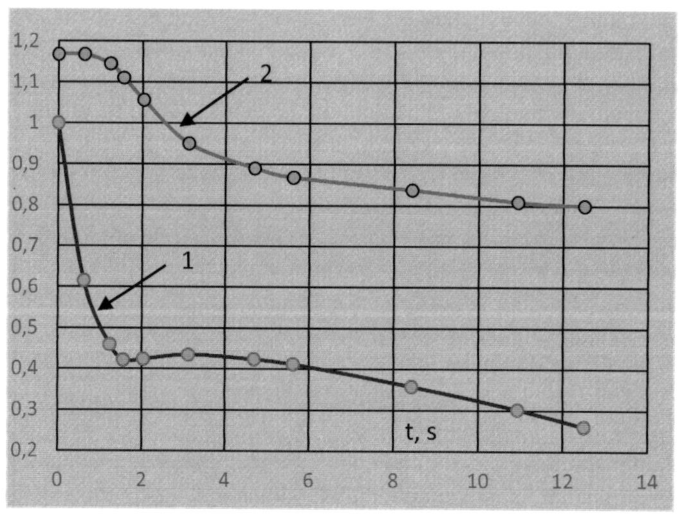

Figure. 3. The relative speed of movement of the inclusions v(t)/v(0) (curve 1), the ratio of the melt temperature to the eutectic temperature T/T_{evt} (curve 2).

The field strength in the sample was calculated from the experimental time dependences of the voltage drop and the current strength in it (Figure 2). Silicon was chosen as a transferring inclusions. When calculating it was taken into account that silicon with respect to aluminum is an non-conductive inclusion, since $\lambda_{oe} << \lambda_e$, and therefore the coefficient χ_e, characterizing the contribution of the electrical conductivity, was accepted as $\chi_e = ½$. In fig. 3 the curve 1 shows the relative speed of movement of inclusions $v(t)/v(0)$ in the alloy AK12, and curve 2 - the ratio of the melt temperature to eutectic temperature T/T_{evt}, T_{evt} = 850 K. Visible interval $1{,}0 \leq t \leq 4{,}5$ c, wherein the speed of movement is virtually constant inclusions. By increasing the range of the lifetime of a given state, the electric field can influence the movement of inclusions in the melt.

As a result of the movement of vacancies during the transmission of electric current in the melt, pore dissolution will be observed, which is proved by the results of these studies - the porosity of the aluminum alloy, which crystallized under the influence of a direct electric current, decreases by about 4%.

4. Conclusions

Thus, the transmission of electric current during the crystallization of the alloy leads to a decrease in the size of the dendritic parameter and the formation of a finely differentiated eutectic in the alloy. The detected changes in the structure of the alloy are due to the influence of the electric field of the current on the structure of the alloy through the electron and ion subsystems and are not related to the external conditions of the cooling of the alloy. The change in the kinetics of dendritic growth and the destruction of the dendritic structure should affect the filling rate of the interdendritic space, which will reduce the likelihood of shrinkage porosity. Considering that electric current is a source of electric, magnetic, mechanical and thermal energy, the use of this type of influence at the stage of phase transition will allow to influence the dynamics of the process of forming a crystalline structure, to control the formation of alloys properties, as well as cast aluminum composite materials.

References

[1] Golodnov A I, Sulitsin A V and Khazan G L 2009 The influence of vibration processing on the structure and copper properties. *Melts.* №5. - p. 38-44.

[2] Kolchugina I Yu, Selyanin I F 2009 The influence of external influences on the microstructure of the crystallizing alloy. *Foundry production.* №8, p. 13–15.

[3] Deev V B, Selyanin I F, Hosen Ri, Tsorina S A and Ponomarev K V 2012 Effective Technologies for processing of melts in the production of cast aluminum alloys. *Liteischik Rossii.* №10. Pp. 19–21.

[4] Aguilar von J,Grüneberg C, Nacke B, Langejürgen M, Sommer B, Niehues J and Haimerl M 2007 Erzeugungeines Gradientengefüges durch elektromagnetisch induzierte Seigerung von Primarsiliciumkristalen in AlSi-Legierungen. *Giesserei.* Volume 94. № 3. P. 26–31.

[5] Korobova N V, Aksenenko A Yu, Tarasov F Ye, Friesen V E, Luzgin V I and Fatkullin S M 2013 On the electromagnetic effect on the aluminum melt when it is modified. *Metallurgy of mechanical engineering.* № 1. Pp. 8–11.

[6] Zadorozhnyj N A, Makeyev V P, Rybkin V A, Sidorenkov V V and Timchenko S L 2004 Russian Patent 2220816C2 MPK7 B22 D 27/02 Method for controlling the process of forming shaped castings. *Iss.10.01.04.Bul. No. 1 - Discoveries, inventions, trademarks.* No. 1.

[7] Timchenko S L 2011 Study of alloy crystallization under the action of electric current. *Melts.* №4. - P.53-61.

[8] Timchenko S L, Kobeleva L I and Zadorozhnyj N A 2011 Effect of electric current on the structure and properties of aluminum alloy. *FIZHOM.* №6. - Pp.82-87.

[9] Timchenko S L, Kobeleva L I and Zadorozhnyj N A 2013 Effect of electric current on the porosity of aluminum alloys. *Melt.* №4. Pp. 58-68.

[10] Timchenko S L 2013 Effect of electric current on the dendritic structure of aluminum alloy. *Engineering magazine: science and innovation.* No. 8 (20). p. 47.

[11] Timchenko S L and Zadorozhnyj N.A 2016 Features of the formation of the structure of the alloy AK 12 during its crystallization under the action of current. *Foundry.* № 8. Pp. 11-14.

[12] Dorofeev A V, Kilin A B and Tertishnikov A S 2002 Processing of aluminum melts with electric current. *Liteischik Rossii.* №2. Pp. 19–21.

[13] Wertheim G Annalen der Physik und Chemie 1848. – B.11/11. - S/1–114.

[14] Spitsyn V I, Troitsky O A 1985 Electroplastic deformation of metals. *M.: Science.* 160 p.

[15] Baranov Yu V, Troitsky O A and Avraamov Yu S 2001 Physical bases of electropulse and electroplastic processing and new materials. *M.: MGIU.* 844 p.

[16] Gordeev V F, Pustogarov A V and Kucherov Ya R and others 1984 Effect of electric current on the orientation of the structure of metals during recrystallization. *Bul. SO AS USSR. Ser. Tech. Sciences.* №16, 3 - p.137-139.

[17] Vdovin E E, Kasumov A Yu 1988 Direct observation electrotransport dislocations in a metal. *FTT.* Volume 30, issue 1. p. 311-314.

[18] Conrad H, Karam N, Mannan S, Sprecher A F 1988 Effect of electric current pulses on the recrystallization kinetics of copper. *Scr. Met.* Volume 22. P.235–238.

[19] Sidorenkov V V, Timchenko S L 1992 Method of texturing metals. *AS. No. 1770399. B.I.* № 39.

[20] Geguzin Ya E 1962 Macroscopic defects in metals. *M.: Metallurgy.* 252 p.

[21] Fix V B 1969 Ion conductivity in metals and semiconductors. *M., "Science".* 296 p.

IOP Conf. Series: Materials Science and Engineering **683** (2019) 012045 doi:10.1088/1757-899X/683/1/012045

Application of a New Method of Non-Adhesion Defect Simulation in Multilayer Honeycomb Structures from Polymer Composite Materials for Various Nondestructive Testing Methods

D Rusakov[1] and V Chernushin[1]

[1] JSC ORPE Technologiya named after A.G. Romashin, 15 Kievskoye Shosse, 249031 Obninsk, Russia

Abstract. A new method of non-adhesion defect simulation in multilayer honeycomb structures from polymer composite materials is discussed. This simulation method is particularly relevant for honeycomb structures with large cells. An experimental model for the comparison of two simulation methods, a new one and flat-bottom hole simulation, was fabricated. Advantages and features of application of the new simulation method for the nondestructive testing methods such as impedance, free oscillation, and through-transmission ultrasonic and thermal methods are shown.

Key words: Nondestructive testing, polymer composite materials, defect simulation, impedance method, free oscillation method, through-transmission ultrasonic method, thermal method.

1. Introduction

Strict requirements for the quality of aerospace products particularly made of polymer composite materials (PCM) result in the need to work to the detection limit of nondestructive testing methods (NDT) [1-6, 14]. To assess detectability of the defect with minimal area it is necessary to simulate the defect with the same area in the reference block. Moreover, there should be a minimal difference between such a simulator and the real defect. For each NDT method a proper defect simulation method should be selected. Simulation which is not suitable for a special NDT method provides incorrect information on the method detection for the structure in question.

2. Rib-shorting simulation method

For the non-adhesion defect simulation in multilayer honeycomb structures from polymer composite materials, which is particularly relevant for honeycomb structures with the cell area of 0.8 cm2 and more, there is a new method [8, 9] (hereafter the rib-shorting simulation method), consisting in shorting of a certain number of honeycomb cell ribs by a height to 3 mm, and bonding of the honeycomb with skins, wherein the adhesive film in the shorting area is removed (Fig. 1).

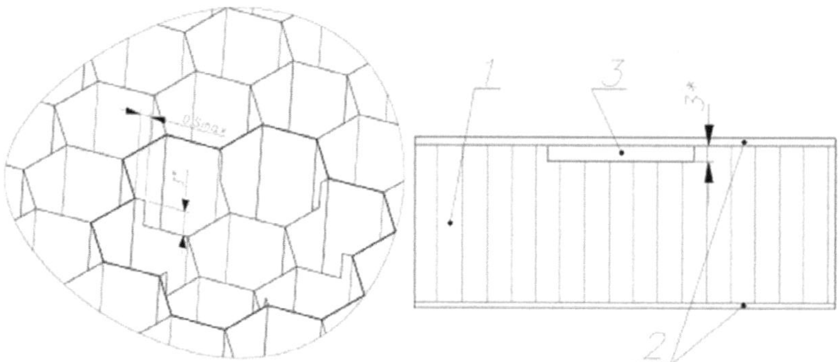

Figure 1. Example of the sandwich structure with PCM skins (2) and honeycomb core (1) with non-adhesion defect simulation (3) by rib-shorting (shorting of 12 honeycomb cell ribs).

Fig. 2 (the top view is schematically shown) provides certain variants of defects (1-4) with a various number of shorted ribs: 1, 3, 12, 24.

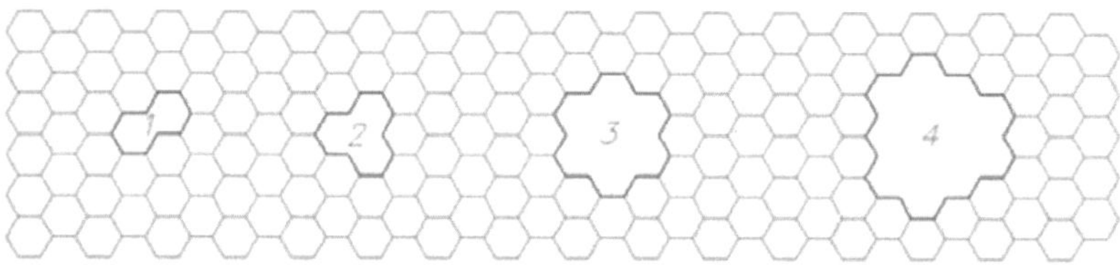

Figure 2. Examples of defect simulators by rib-shorting (shorting of a various number of honeycomb cell ribs).

The structure of the defect simulators fabricated by rib-shorting is most similar to that of the real defect. The ribs forming the outline of the defect simulator remain undamaged and provide sufficient structural stiffness in the defect-free area adjacent to the boundary of the defect simulator, which is critical for the low-frequency acoustic NDT methods: impedance and free oscillation ones [1, 2-5]. The area of the defect simulator formed has fixed precision and changes in steps (Fig. 2). Applicability of the rib-shorting method for honeycomb structures with large cell area (from 0.8 cm2) consists in the minimum error of the areas of the defect simulators formed since the area of one cell of the honeycomb becomes comparable to that of the defect simulator [9].

The rib-shorting method allows for the bilateral specimen testing. It enables to save material and fabricate smaller reference blocks for complex multilayer structures. For the through-transmission ultrasonic method this rib-shorting simulation method is the most preferable as compared with other common defect simulation means: films and embedded elements. The experiments also confirmed in the article [10] showed that что reliable identification of such defect simulators as inclusions (films, film bags) was not provided by the through-transmission ultrasonic method. It is attributed to an accidental adhesive inflow, and insufficient ultrasonic attenuation. However, simulation with embedded elements (wedges) does not permit to obtain a defect simulator with the standardized area.

3. Experimental results

The rib-shorting method can be applied for adjustment of flaw detectors in the thermal NDT method. In accordance with the new developed defect simulation technology a specimen was fabricated, in which defect simulators were additionally formed for comparison by milling the flat-bottom holes. As a result of active thermal control and using a method of thermogram array conversion described in [11-13] a thermogram shown in Fig. 3 was obtained.

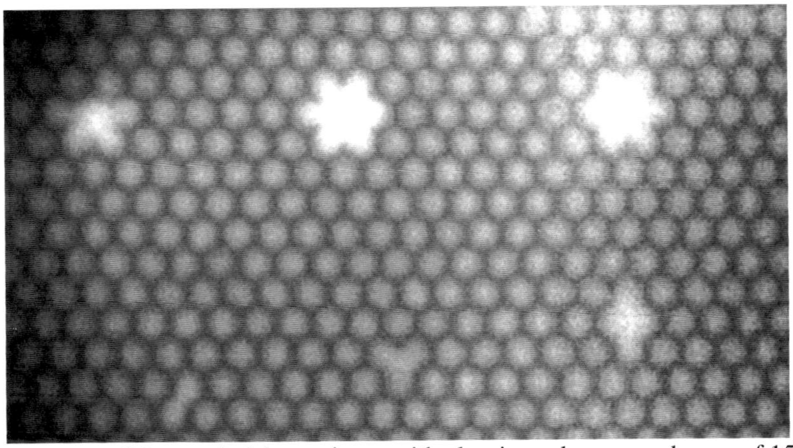

Figure 3. Thermogram of the sandwich specimen with aluminum honeycomb core of 15 mm high, the cell area ~ 0.8 cm^2 and PCM skins of 0.7 mm thick with defect simulators: flat-bottom holes (top) and rib-shorting – the cell ribs shortened by 3 mm (bottom).

Four defect simulators formed by milling the flat-bottom holes are much more clearly detected on the thermogram than the simulators formed by rib-shorting. It is attributed to the fact that during the milling the adhesive film is removed with partial skin damage, and by reason of removing the back wall of the sandwich structure, the front wall comes in direct contact with air, resulting in more intensive increase in the defective area surface temperature. As a result the temperature signal [14] from the defect simulators formed by milling the flat-bottom holes matches the defects such as skin delamination.

The defect simulators formed by rib-shorting indicate a temperature signal in the thermogram matching the signal from real defects such as non-adhesion. Besides, there is no false temperature signal amplification as is the case with flat-bottom hole simulation, which makes possible to avoid overestimation of the method detection.

None of the areas of the defects formed by the milling method after the determination from the thermogram did not match the area specified in the design and production of the specimen. Whereas the determined areas of the defects formed by rib-shorting differed from the design areas within statistical accuracy.

4. Conclusions

Thus, the new method of non-adhesion defect simulation (rib-shorting) for multilayer honeycomb structures from PCM which is particularly relevant for honeycomb structures with the cell area of 0.8 cm2 and more can be effectively applied for adjustment of flaw detectors for the following NDT methods:

а) Impedance and free oscillation method. Simulation by rib-shorting provides sufficient structural stiffness at the boundary of defect simulators. The structure of the defect simulator is the most similar to that of the real defect, and its area is strictly standardized but it changes in steps.

б) Through-transmission ultrasonic method. Simulation by rib-shorting allows for the bilateral testing, and does not have disadvantages of other simulation methods (embedded elements).

в) Thermal method. The temperature signal from the simulators formed by rib-shorting most closely matches the signals from the real defects such as non-adhesion.

For the multilayer honeycomb structures from PCM with the honeycomb core with small cell area (up to 0.8 cm2) it is possible to use a less precise conventional method of non-adhesion defect simulation by circular milling shorting of the honeycomb core until the skin is also bonded.

References

[1] Lange Iu V 1991 Acoustic low frequency methods and means of nondestructive testing of multilayer structures. *Moscow, Mashinostroenie Publ.*
[2] Cawley P and Adams R D 1988 The Mechanics of the Coin-Tap Method of Non-Destructive Testing. *J. of Sound and Vibration.* Volume 122, pp. 299-316.
[3] Wu H and Siegel M 2018 Correlation of Accelerometer and Microphone Data in the “Coin Tap Test”. *Robotics Institute, Carnegie Mellon University.* Date Views 03.10.2018 http://www.cs.cmu.edu/~whd/publications/imtc99.pdf.
[4] Murashov V V and Yakovleva S I 2017 Application of the acoustic free vibration method for control of the structures containing layers from non-metallic materials. *Kontrol. Diagnostika [Control. Diagnostics].* no. 10, pp. 28-35.
[5] Gryzagoridis J, Findeis D 2018 *Tap testing vs. Thermography.* Date Views 03.10.2018 http://www.ndt.net/article/ndtnet/2017/1_Gryzagoridis.pdf.
[6] Ibarra-Castanedo C, Bendada A and Maldague X 2007 Thermographic Image Processing for NDT. *IV Conferencia Panamericana de END. Buenos Aires.* pp. 1-12.
[7] Rusakov D Yu and Skomorokhov A O 2017 Automation of analysis of thermographic images in diagnostics of honeycomb core structure states. *Knowledge E Life Sciences.* pp. 350-356.
[8] Rusakov D Yu and Chernushin VA 2017 Method for simulating the defect of non-adhesive in multilayer structures. *Patent RF no.2667317.*
[9] Rusakov D Yu and Chernushin V A 2018 Simulation of defects by honeycomb understating in polymeric honeycomb structures. *Trudy MAI [Proc. of MAI].* no. 99. Date Views 03.10.2018 trudymai.ru/upload/iblock/a6b/Rusakov_CHernushin_rus.pdf.
[10] Trifonova S I, Generalov A S and Dalin M A 2016 Actual Technologies and instruments for polymer composite materials testing with through transmission ultrasonic method. *NDT World.* Volume 19, no.1, pp. 31-35.
[11] Rusakov D Yu and Skomorokhov A O 2017 Method for analyzing the results of active thermal control of products made of polymer composite materials. *Patent RF no. 2649247.*
[12] Rusakov D Yu and Skomorokhov A O 2017 Automation of analysis of thermographic images in diagnostics of honeycomb core structure states. *Collection of theses of the Future of Nuclear Power: XIII International conference. Obninsk.* pp. 195-197.
[13] Rusakov D Yu and Skomorokhov A O 2016 Analysis of results of thermal nondestructive control of products from polymeric composite materials. *Collection of theses of the XXI international scientific and technical conference "Structures and Technologies of Receiving Products from Nonmetallic Materials". Obninsk.* pp. 133-134.
[14] Vavilov V P 2006 Dynamic thermal tomography. *Zavodskaya laboratoriya. Diagnostika materialov [Plant laboratory. Diagnostics of materials].* no. 3, pp. 26-36.

Impact toughness of multilayer steel materials at low temperatures

D V Vlasova[1] and A I Plokhikh[1]

[1]Bauman Moscow State Technical University (National Research University); (Bauman MSTU), 5, 2nd Baumanskaya Str., bldg. 1, 105005, Moscow, Russian Federation

Abstract. The results of a number of studies show that there are ways to drastically change the temperature dependence of the impact strength of structural steels, which at first glance have a paradoxical character. It was established that when the test temperature decreases down to cryogenic temperatures, the work of failure of impact samples cut in certain directions of rolled products does not change, which indicates the absence of a formal viscous-brittle transition threshold and an increase in material reliability at low temperatures. The reason for this phenomenon is the layered structure formed during the hot batch rolling process, due to which complete failure of impact samples with a U-shaped concentrator does not occur.

1. Introduction

For most engineering steels, the transition from a viscous to a brittle state occurs at insignificant negative temperatures, which represents a significant danger in the operation of machinery and structures in conditions of low climatic temperatures. Given that cold resistance primarily depends on the viscosity of the material, the main methods aimed at its increase are associated with a decrease in the number of harmful impurities, the number and size of non-metallic inclusions, as well as additional doping. However, the results of a number of studies show that there are ways to drastically change the temperature dependence of the toughness of machine-building steels. The authors [1,2] found that with decreasing temperature the test work of failure of the impact sample increases, which indicates the absence of a formal threshold for viscous-brittle transition (VBT) and an increase in the reliability of the material at low temperatures. This becomes possible after a special type of thermomechanical processing in low alloyed steels due to the formation of a layered structure, which leads to an increase in the values of impact strength, and the formation of the fracture having a terraced structure.

It is also important that high values of toughness and strength at the same time can be achieved in multilayer materials [3-5]. This combination of characteristics makes it possible to use these materials in electronic devices, armor materials and various branches of engineering [6-8].

2. Materials and research methods

To study the temperature dependence of impact strength in the low and cryogenic temperatures the samples of multilayer materials consisting of 100 alternating steels with a thickness of 0.5 mm each were selected (50 layers of each grade): AISI 430 + AISI 304, W 108 + AISI 304, AISI 420 + AISI 304 [9]. Using the experimental technological route, including dimensional cutting of blanks from sheets, processing their surfaces, assembling sliced sheets into a package, evacuating the package and subsequent plastic deformation by hot rolling, the billets 10 mm thick were prepared. These compositions underwent two complete technological cycles. After the implementation of the first technological cycle, the resulting material consists of 100 layers, each layer of about 100 microns thick, and after the second - 1500 layers of approximately 5 microns thickness.

Studies were conducted on samples of toughness with U-shaped pin in the direction of hubs perpendicular and parallel to the rolling direction, in the temperature range from 20°C to -196°C.

3. Results and discussion

According to the test results, it was established that the investigated multilayer materials with 1500 layers have a significant margin of fracture toughness in the direction perpendicular to the rolling plane. Samples do not fail by a pile with a margin of work of 300 J and have a tear in the area of stress concentrator, and then subjected to bending deformation.

Such results can be explained by the fact that under dynamic loading of multilayer samples with a notch located in the direction of braking of the front of the crack, the interfaces between the layers serve as barriers to the propagation of the crack.

Splitting cracks arising in the way stop it, and for the further development of the fracture, the formation of microcracks on a new surface is necessary, which requires extra energy. The resistance to failure in this case increases, and the greater it is, the more layers in the structure of the material and the more often the effect of its separation manifests itself are.

These data are supported by the work of the authors of [10], who study the behavior of toughness in thirteen-layer Ti-6Al-4V material. The authors of [1-2,11] also believe that in crack limiter configurations attenuation and deflection of cracks are the dominant mechanisms and can lead to a noticeable increase in toughness. It should be noted that this effect of hardening by “peeling”, leading to an increase in impact toughness is observed not only in steels [1-2,12], but also in ceramics [13] and pyrolytic carbon [14].

When the temperature of the test is lowered in order to determine the threshold of VBT, the values of the reduced work in the direction perpendicular to the direction of hire are not reduced (Table 1), while the samples that have passed the second process cycle of production fail. Samples of compositions AISI 430 + AISI 304, W108 + AISI 304 have a formal threshold VBT is absent to the test temperature of -196°C. For samples of the composition AISI 420 + AISI 304 at this temperature, normal failure of the samples is observed, which for the studied materials can be considered the reaching the VBT threshold.

Table 1. Impact toughness values on compositions with 1500 layers.

Composition	Layer thickness μm	Concentrator, cut by relation to RD	KCU (Apr), J/cm^2		
			20°C	-70°C	-196°C
AISI 430+AISI 304		⊥	360 [a]	360 [a]	360 [a]
		//	--	80	70
W 108+AISI 304	5.0	⊥	340 [a]	325 [a]	360 [a]
		//	70	60	35
AISI 420+AIS I304		⊥	360 [a]	360 [a]	100
		//	70	60	50

[a] The sample does not completely fail.

The test results of the samples in the direction parallel to the direction of rolling are characterized by significantly lower values of impact strength, with the additional work of failure being expended on opening the interlayer boundaries with the formation of delamination cracks.

Similar results were obtained by the authors of [15], who established the advantage of the layered structure on steels of different phase composition at temperatures from 20°C to -196°C. They carried out experiments on composite five-layer samples of two stainless steels: ferritic steel 08Kh18T1 and austenitic steel 10Kh18AG19. In the experiments, the author found that the effect of increasing the toughness of the composite samples depends on the orientation of the sample on the test pile and is maximum if the direction of impact of the guillotine pendulum is perpendicular to the plane of the pairing plates.

Confirmation of the results obtained are also found in the works of the authors [1,2], presenting results showing the positive effect of a particular type of structure on the value of the impact strength of ferritic steels in the region of low climatic temperatures. This paper shows the results of the study of steel with different heat treatments:

QT - quenching and tempering at 500°C, TF - quenching at 500°C and deforming. In the experiments, a sample of TF steel showed an inverse temperature dependence of toughness at low temperatures (Fig. 1).

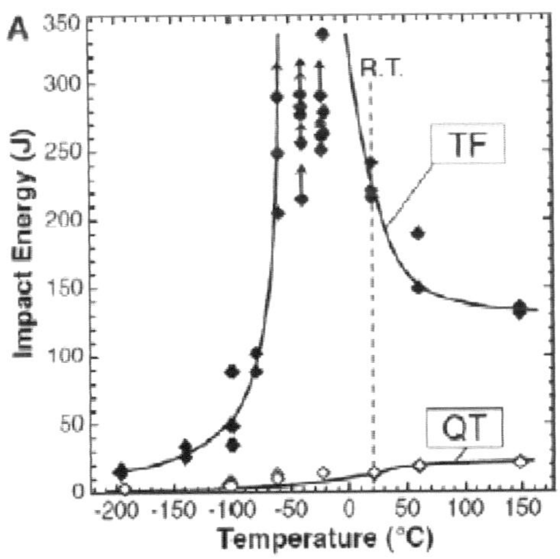

Figure 1. The dependence of the energy of failure of Charpy impact samples for two types of QT and TF treatment

There are other works in which data are given on the positive effects of a special kind of multilayer structures in steel brazed materials: an increase in the number of layers in the material leads to both increased toughness values and lower threshold of the viscous-brittle transition [16,17].

Thus, the results of a number of studies show that there are ways to drastically change the temperature dependence of the impact strength of structural steel, having at first glance a paradoxical character. It is shown that when lowering test temperature, the work of failure of the impact sample increases, which indicates the absence of a formal threshold for viscous-brittle transition and an increase in the reliability of the material at low temperatures. The formed layered structure is the main cause of such dramatic changes in the behavior of structural materials.

4. Conclusion

Complete failure of samples is absent in compositions AISI 430 + AISI 304, W108 + AISI 304 to temperatures equal to -196°C, and up to -70°C for samples of the composition AISI 420 + AISI 304, which indicates the absence of a formal threshold of VBT. However, the found effect is not a violation of the general theory, but serves as its confirmation.

Considering that with a decrease in temperature, a natural increase in the yield limit is observed, it becomes natural to increase work on the plastic strains of specimens that, due to the multi-layered structure, do not completely fail, but undergo bending deformation.

References

[1] Kimura Y, Inoue T, Yin F and Tsuzaki K 2008 Inverse Temperature Dependence of Toughness in an Ultrafine Grain-Structure Steel *Science* **320** 1057–60

[2] Inoue T, Yin F, Kimura Y and Tsuzaki K 2010 *Metall. Mater. Trans.* A **41** A 341–355

[3] Osman T M, Singh P M and Lewandowski J J 1994 Crack bridging in a laminated metal matrix composite *Scripta Metall. Mater.* **31** 607-612

[4] Pozuelo M, Carreño F, Cepeda-Jiménez C M and Ruano O A 2008 Effect of Hot Rolling on Bonding Characteristics and Impact Behavior of a Laminated Composite Material Based on UHCS-1.35 Pct C. *Metall. Mater. Trans.* A **39**

[5] Cepeda-Jiménez C M, Pozuelo M, García-Infanta J M, Ruano O A and Carreño F 2009 Interface Effects on the Fracture Mechanism of a High-Toughness Aluminum-Composite Laminate *Metall. Mater. Trans.* A **40** 69-79

[6] Li T, Grignon F, Benson D J, Vecchio K S, Olevsky E A, Jiang F, Rohatgi A, Schwarz R B and Meyers M A 2004 Modeling the elastic properties and damage evolution in Ti–Al_3Ti metal–intermetallic laminate (MIL) composites *Mater. Sci. Eng.* A **374** 10-26.

[7] Han J, Liao C, Jiang T, Spanheimer C, Haindl G, Fu G, Krishnakumar V, Zhao K, Klein A, and Jaegermann W 2011 An optimized multilayer structure of CdS layer for CdTe solar cells application. *Compd* **509** 5285-89

[8] Kang B C, Kim H Y, Kwon O Y and Hong S H 2007 Bilayer thickness effects on nanoindentation behavior of Ag/Ni multilayers. *Scripta Mater.* **57** 703-706

[9] Tabatchikova T I, Plokhikh A I, Yakovlev I L and Klyueva S Yu 2013 Structure and properties of a steel-based multilayer material produced by hot pack rolling *The Physics of Metals and Metallography* **114** Issue 7 580-92

[10] Cepeda-Jiménez C M, Carreño F, Ruano O A, Sarkeeva A A, Kruglov A A and Lutfullin R Ya 2013 Influence of interfacial defects on the impact toughness of solid state diffusion bonded
Ti-6Al-4V alloy based multilayer composites *Mater. Sci. Eng.* A **563** 28-35

[11] Kimura Y, Inoue T, Yin F and Tsuzaki K 2010 Delamination Toughening of Ultrafine Grain Structure Steels Processed through Tempforming at Elevated Temperatures. *ISIJ Int.* **50** 152-161

[12] Mintz B, Maina E and Morrison W B 2007 Origin of fissures on fracture surfaces of impact samples of HSLA steels with ferrite/pearlite microstructures *Materials Science and Technology* **23** 347–354

[13] Clegg W, Kendall K, Alford N, Birchall J and Button T 1990 A simple way to make tough ceramics. *Nature* **347** 455–457

[14] Sakai M, Bradt R C and Fischbach D B. 1986 Fracture toughness anisotropy of a pyrolytic carbon. *Journal of Materials Science* **21** 1491–1501

[15] Yakovleva I L, Tereshchenko N A, Mirzaev D A, Panov A V and Shaburov D V 2007 Impact toughness and plastic properties of composite layered samples as compared to monolithic ones *The Physics of Metals and Metallography* **104** Issue 2 203 – 211

[16] Embury J D, Pech N J, Wraith A E and Wright E S. 1967 The fracture of mild steel laminates *Transactions of the Metallurgical society of* AIME **239** 114-118

[17] Wright E S 1967 *Deformation and Fracture of plastically Anisotropic laminates*

IOP Conf. Series: Materials Science and Engineering **683** (2019) 012047 doi:10.1088/1757-899X/683/1/012047

Study of thermal expansion anomalies of multilayer steel materials at high temperatures

V E Kabantseva[1], M D Safonov[1] and A I Plokhikh[1]

[1]Bauman Moscow State Technical University (National Research University); (Bauman MSTU), 5, 2nd Baumanskaya Str., bldg. 1, 105005, Moscow, Russian Federation

Abstract. The paper presents the results of a study of the anisotropy of the thermal coefficient of linear expansion (CTE), which was found in a multilayer metallic material based on steels, made by hot batch rolling. It is established that the material consisting of alternating layers of steel, has abnormally low CTE values in the direction perpendicular to the rental plane. To explain this phenomenon, a hypothesis was proposed about the relationship between normal and tangential stresses arising at the interlayer boundaries of such material. The confirmation of the hypothesis of the leading role of reactive stresses in the manifestation of the specified invar effect arc thc results of metallographic analysis. It is also shown that the conduct of cyclic high-temperature heating provokes recrystallization at the interlayer boundaries caused by the difference in the thermal expansion coefficient values of the adjacent layers and the effect of high temperature.

1. Introduction

Currently, intensive research of materials scientists is aimed at creating methods to improve the performance characteristics of traditional materials by implementing a special type of structures in them. Such structures, possessing non-linear characteristics, are capable of adaptive response to external influences. Such effects are usually observed in composite materials and are the result of the interaction of local microscopic fields of elastic stresses, which is not found in materials with a homogeneous structure.

The study of the theoretical aspects of the creation of composite materials with anomalous characteristics was conducted by many researchers [1-9]. The detected effects in composites led to the intensification of work on the creation of materials with a negative Poisson ratio, as well as the search for materials with similar characteristics of the thermal coefficient of linear expansion (CTE). [10-13].

So, in particular, a mathematical model is known, according to which, CTE may have negative values in materials with a layered structure based on binary (iridium + invar) or triple (iridium + tungsten + invar) metal compositions. The main criterion when choosing materials for compositions is a significant difference in the value of CTE, which may differ by 4-5 times for selected materials [12,13].

However the technological aspects related to the compatibility of such materials, the interlayer diffusion of alloying elements, ways of creating an inextricable connection between the layers, due to which the manifestation of this effect would be possible, were not considered. Therefore, the idea of implementing the developed approaches in structural metallic materials having a layered structure, but obtained on the basis of available and technological materials, for example, on the basis of steel, is relevant.

2. Materials and methods

On the basis of the presented considerations, two model compositions composed of steels X5CrNI18+ 08kp and X5CrNI18 + U8 were investigated. It is known that the difference in the values of the thermal expansion coefficient and the modulus of normal elasticity is not as significant as that of the materials of the above mathematical model [12], and in the temperature range studied will differ by no more than 1.5 times [14].

The primary composite blanks consisted of 100 alternating sheets of specified steels with a thickness of 0.5 mm, i.e., 50 sheets of each grade, respectively. According to the previously developed experimental process map (Figure 1), blanks for a 100 mm wide strip and 10 mm thick were obtained [15]. Obtained after the first cycle strips 10 mm thick were rolled to a thickness of 2 mm, cleaned, assembled into a bag and, in accordance with the described process map, they went through the second treatment cycle by carrying out hot batch rolling to a thickness of 10 mm.

For dilatometric studies, samples were cut in three directions: along the rolling direction (RD), in the rolling plane (TD) and in the direction perpendicular to the rolling plane (ND).

The measurements were performed on a Dil-402C dilatometer manufactured by Netzsch (Germany) with a corundum holder and pusher in the temperature range 20 to 1200°C during heating and cooling of the furnace at a rate of 5 deg/min in an atmosphere of technically pure argon. The temperature was measured by a platinum – platinum – rhodium thermocouple (type S) located in the immediate vicinity of the sample, with an error not exceeding 5°C.

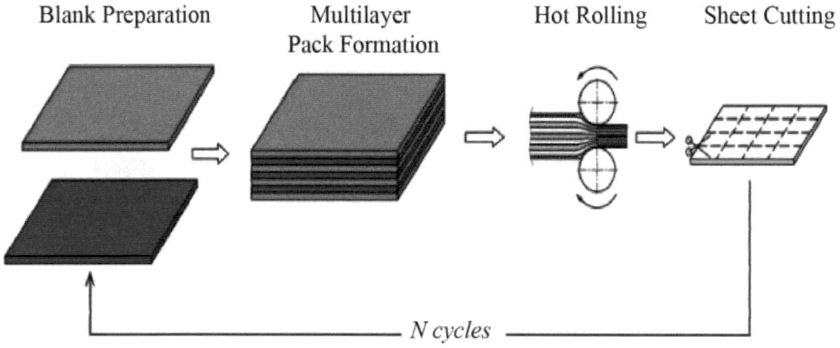

Figure 1. The flowchart of the experimental process route for obtaining multilayer metallic materials.

3. Results and discussion

The dilatometric study showed (Figure 2) that the test X5CrNI18 + U8 multilayer material, obtained after the first and second technological cycles, has an anomaly of thermal expansion ($\Delta L/L$) in the direction of ND. The transformation observed at a temperature of about 900°C in the direction of ND corresponds to the $\alpha \rightarrow \gamma$ transition in carbon steel, which is part of the composition under study. In Fig. 2, one can see that the anomaly in the direction of ND begins to manifest itself at temperatures of warm deformation, which coincides with the onset of the release of dislocations from the atmospheres of impurity atoms in low carbon steels.

It can be seen that after the completion of the second technological cycle, the detected invar anomaly is amplified and shifted to lower temperatures. To explain this effect, a hypothesis about the interaction of reactive tangential and normal stresses arising during heating has been put forward.

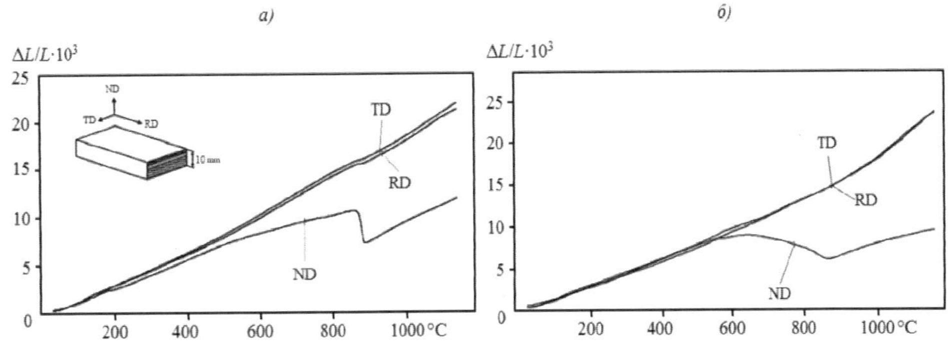

Figure 2. The dependence of the relative elongation of the multilayer material in the directions RD, TD and ND in samples with a thickness of 10 mm with the number of layers 100 (a) and 1500 (b).

In accordance with the hypothesis, a decrease in the thermal expansion coefficient values of the multilayer material is observed in the RD and TD rolling directions due to the mutual compensation of the thermal expansion of dissimilar layers (Figure 3, a). This effect is particularly pronounced in the range of the $\alpha \rightarrow \gamma$ transition of carbon steels. Thus, the decrease in the lattice parameters during the reverse transformation is almost completely compensated by the expansion of steel X5CrNI18, which is manifested in the absence of an anomaly in the dilatograms of the RD and TD directions. On the other hand, in the direction of ND, a cumulative enhancement of the effect of compression is observed, which is associated with the expansion of steel X5CrNI18, which is superimposed on the compressing layers undergoing $\alpha \rightarrow \gamma$ transformation (Figure 3, b). It is this compression that leads to the appearance of thermal expansion coefficient values in the direction of ND, corresponding to the values for classical invar alloys.

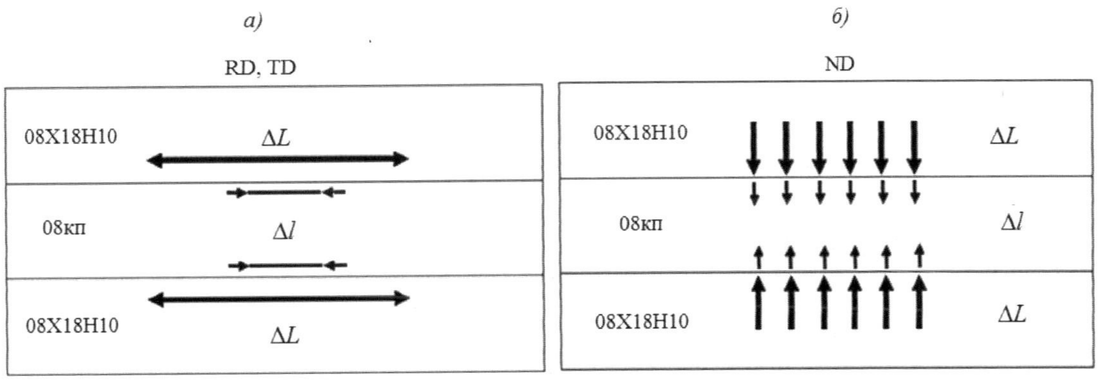

Figure 3. Diagram of the effect of tangent (a) and normal (b) stresses on the example of the composition X5CrNI18 + 08kp.

The analysis of the microstructure of 100 layered samples consisting of X5CrNI18 and U8 steels showed that after the completion of the first technological cycle, the modified structure consists of alternating layers of low carbon steel and X5CrNI18 layers of former steel (Figure 4, a). Characteristic of the X5CrNI18 steel layers is the presence of a large volume fraction of the highly dispersed carbide phase, which was formed as a result of the directional diffusion of carbon (Figure 4, b) [16, 17].

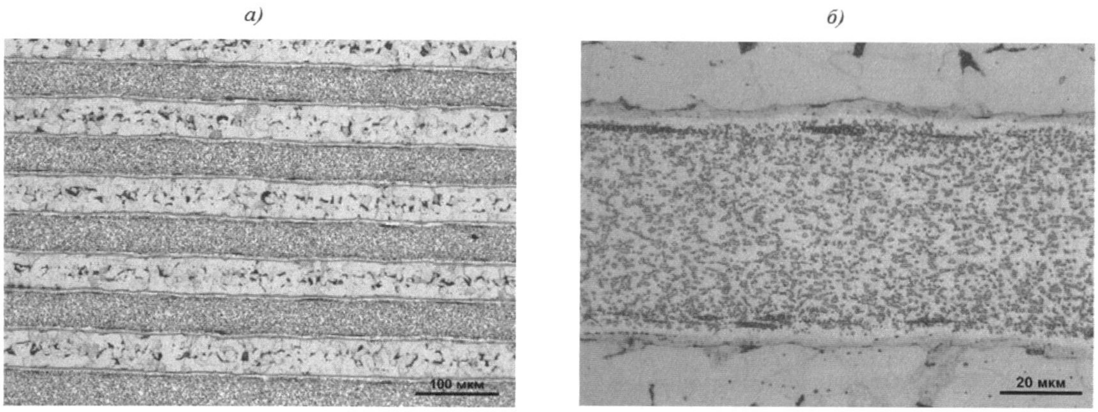

Figure 4. The microstructure of the composition X5CrNI18 + U8 (a) and the layer of steel X5CrNI18 (b).

To check this assumption, hundred-layer samples of composition X5CrNI18 + U8 were reheated to a temperature of 1200°C, which certainly exceeds the dissolution temperature of present chromium carbides. Figure 5a shows that the carbide phase completely dissolved in the layers of the multilayer material under study, which led to the active recrystallization with the formation of a coarse-grained structure.

At the same time, the layers that were previously U8 steel decreased in thickness — from an average value of 100 μm in the original hot rolled state to values of 85–90 μm (Figure 5, b).
Thus, the increase in the thickness of the layers that were previously X5CrNI18 steel occurred through the migration of interlayer boundaries in the direction of the layers that were previously U8 steel.

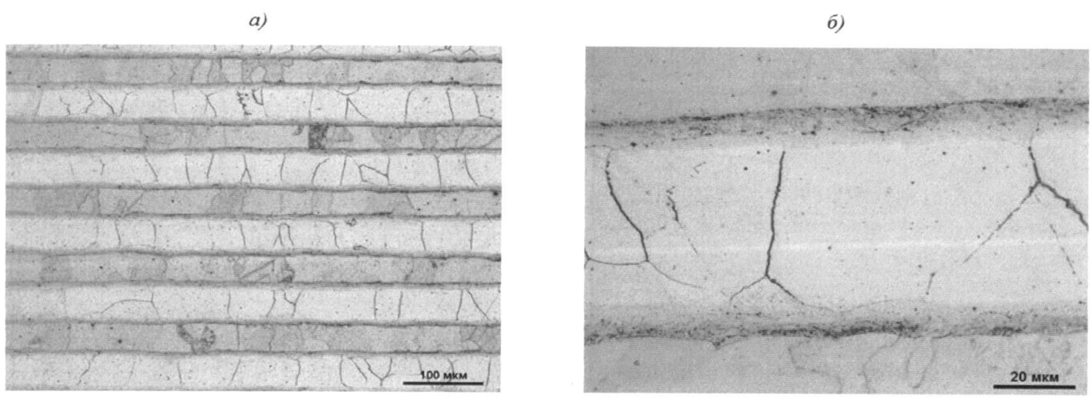

Figure 5. The microstructure of the composition X5CrNI18 + U8 after heating to a temperature of 1200°C.

Based on this, we can conclude that the driving force in this process is interlayer stresses arising from the difference in the CTE values of the mating layers, which for alloys with similar chemical composition at these temperatures differ by more than 1.5 times. The mechanism of relaxation of these stresses is dynamic recrystallization, which occurs at the boundary between the layers and is detected metallographically in the form of interlayers. In this case, the front of the moving boundary is directed towards the most plastic layer, which is caused by the lowest energy for the implementation of microplastic deformation, which precedes the onset of the recrystallization process.

To eliminate the influence of diffusion processes on the mechanisms of relaxation of interlayer stresses, similar studies were conducted on the composition X5CrNI18+08kp. It has been established

(Figure 6, b) that repeated single heating to a temperature of 1200°C also leads to the formation of an additional layer at the interface facing the 08kp steel. Threefold heating to 1200°C showed that the thickness of this layer increases, which indicates the development of recrystallization processes and confirms the suggestion of the leading role of interlayer stresses in the process of structure formation (Figure 6, c).

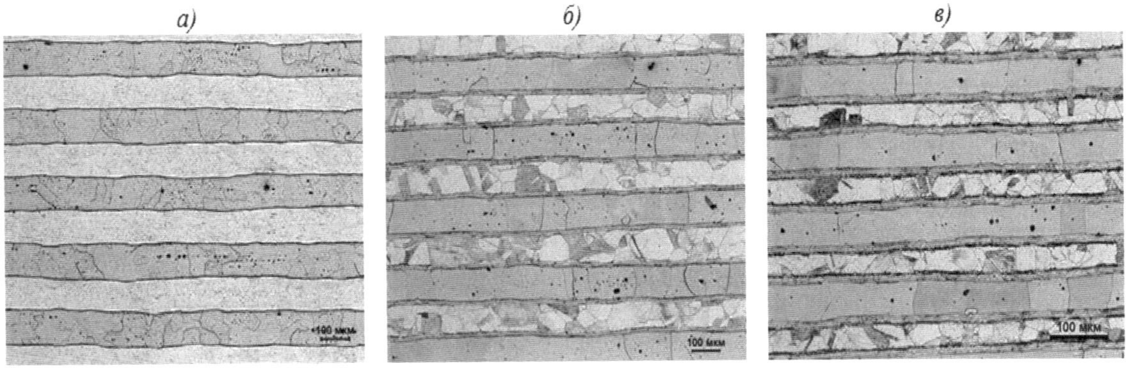

Figure 6. The microstructure of the composition X5CrNI18 + 08kp in the initial state (a) and after repeated once (b) and three times (c) heating to a temperature of 1200°C.

4. Conclusions

Thus as a result of the research carried out, the existence of a steel structural material, capable of demonstrating selective behavior in response to an external impact, which is manifested in a pronounced anisotropy of thermal expansion, was confirmed.

It has been established that with an increase in the number of layers in a multilayer material, the anomalous nature of thermal expansion in the direction perpendicular to the rolling plane increases and shifts to lower temperatures. It is shown that the reason for this anisotropy of properties is the stresses arising due to the difference in the values of the thermal expansion coefficient of the steels included in the composition of the original composition of the multilayer material. The mechanism of relaxation of these stresses is associated with dynamic recrystallization, which occurs at the boundary between the layers. The direction of migration of the interlayer boundary is directed towards the most plastic layer, which is caused by the lowest energy expenditures for the implementation of microplastic deformation, which precedes the onset of the recrystallization process.

References

[1] Khomenko O A 2007 The origin and characteristics of invar anomalies of physical properties. Fe – Ni alloys with FCC lattice *Physics of metals and metallography* **104** No. 2 p. 155-65

[2] Mary T A, Evans J S O, Vogt T & Sleight A W 1996 Negative thermal expansion from 0.3 K to 1050 K in ZrW_2O_8 *Science 272* 90–92

[3] Bodryakov V Yu and Povzner A A 2007 Invarion and insidious behavior of simple ferromagnets: thermodynamic modeling. *Journal of Technical Physics* **77** p. 65 - 71

[4] Nadutov V M, Vashchuk D L, Volosevich P Yu, Beloshenko V A, Spuskanyuk V Z 2012 Structure and properties of the Inv – FCC alloy Fe – 35% Ni after combined plastic deformation by hydroextrusion and drawing *High Pressure Physics and Technology* **22** No. 2 p. 125 - 37

[5] Mulyukov H Ya, Sharipov I Z, Bitkulov I H, Mulyukov R R 2002 Features of the temperature dependence of thermal expansion and saturation magnetization of the invar alloy Fe – 67.0%, Ni 32.5%, Co – 0.5% with nanocrystalline structure *Journal of Technical Physics* **72** is. 7 p. 75 – 8

[6] Wang Y, Gao J, Wu H, Yang S, Ding X, Wang D, Ren X, Song X 2014 Strain glass transition in a multifunctional b-type Ti alloy *Sci. Rep.* 4

[7] Kainuma R, Wang J, Omori T, Sutou Y, Ishida K 2002 Invar-type effect induced by cold-rolling

IOP Conf. Series: Materials Science and Engineering **683** (2019) 012047 doi:10.1088/1757-899X/683/1/012047

deformation in shape memory alloys. *Appl. Phys. Lett.* **80** 4348–50
[8] Monroe J, Gehring D, Karaman I, Arroyave R, Brown D and Clausen B 2016 Tailored thermal expansion alloys *Acta Mater* **102** 333 –341
[9] Bensoussan A, Lions J-L and Papanicolaou G C 1978 *Asymptotic Analysis for Periodic Structures,* North Holland Amsterdam
[10] E Sanchez-Palencia 1980 Nonhomogeneous Media and Vibration Theory, *Lecture Notes in Physics* **127**.
[11] Almgren R F 1985 An isotropic three-dimensional structure with Poisson's ratio. *J. of Elasticity* **15** pp. 427-430.
[12] Kolpakov A G and Rakin S I 1986 On the task of synthesis of a composite material of one-dimensional structure with given characteristics *Applied mechanics and technical physics* **no. 6** p. 143-150
[13] Kolpakov A G and Rakin S I 2004 Deformation characteristics of layered composites with nonlinear deformations *Applied mechanics and technical physics* №5 p. 157-166
[14] Plokhikh A I 2014 About the possibility of using multi-layer metallic materials for the manufacture of high-pressure cylinders *Bulletin of the PNIPU. Series Engineering, Materials Science.* **No. 4** p. 97 – 106
[15] Plokhikh A I, Vlasova D V, Khovova O M, and Polyanskiy V M 2011 Study of the effect of diffusion mobility of alloying elements on the stability of the structure of multilayer metallic materials, *Science and Education: electronic scientific and technical publication* No. 11
[16] Plokhikh A I and Putyrskiy S V 2014 Simulation of the process of plastic deformation of multilayer metallic materials, *News of Volgograd State Technical University. Series Problems of materials science, welding and strength in mechanical engineering* No.9 (136) p. 25-30
[17] Zubchenko A S, Koloskov M M, Kashirsky Yu V and others 2003 Steel and alloys grade guide, *Mechanical Engineering* p. 585 – 784

Study of the process of nitriding in multilayer materials based on steel

K B Polikevich[1] and A I Plokhikh[1]

[1]Bauman Moscow State Technical University named after N E Bauman (National Research University); (Bauman MSTU, 2nd Baumanskaya Str., bldg. 1, 105005, Moscow, Russian Federation

Abstract. The paper analyzes the processes of nitriding of multilayer metallic materials at various degrees of their structurization. The dependences of the influence of the thickness of the composite layer and the chemical composition of its materials on the parameters of the diffusion layer during chemical-thermal treatment are presented. It is shown that a decrease in the thickness of the laminar layer leads to an increase in the diffusion layer.

1. Introduction

Currently, a fairly large number of works are being carried out in the field of creating new materials that can improve the performance properties of machine parts. One of the types of such materials are materials with an oriented structure. Such structure can significantly increase the life of contractures and parts [1]. Multilayer metallic materials have a unique set of mechanical properties, however, additional hardening of contacting surfaces located in planes perpendicular to the section of the blanks may be promising [2]. Therefore, the actual task is the selection of raw materials for multilayer compositions, layer thicknesses, as well as modes of chemical-thermal treatment, which will be observed in the first place, the accelerated penetration of diffusing elements along the interlayer boundaries, deep into the saturable material, in accordance with the Fisher model.

2. Materials and research methods

To study the effect of chemical heat treatment by nitriding, the following compositions were select and prepared with alternating layers of steel: 1008+AISI304, 1008+AISI409, W108+AISI304, W108+AISI409. The compositions were obtained using hot batch rolling at a temperature of 1000 °C. After the first process cycle, which includes the initial assembly of a package of alternating layers of selected steels and rolling, the number of layers in the material is 100 pcs. After the second process cycle, in which the billet obtained in the previous step is cut to cards, from which a new package is assembled, the number of layers is already 1,500 pieces.

3. Results and discussion

In paper [3], the first experiments were carry out on the nitriding of a multilayer composition consisting of alternating layers of steel AISI304 and AISI409 in comparison with a control sample of hard-to-nitrided steel AISI304. At the same time, it was found that the depth of penetration of the diffusant, and accordingly, the depth of the diffusion layer increases with decreasing thickness of the layer of the multilayer material [4].

The results, which were obtain, can be explain by the fact that imperfections have a significant effect on the diffusion of alloying elements during the chemical-thermal treatment [5]. Thus, in metallic materials with a high density of dislocations, diffusion has higher kinetic parameters than in materials with a lower density of defects. The reason for the increase in the thickness of the nitrated layer in multilayer materials as compared with the control sample may be accelerated diffusion of nitrogen along the grain boundaries with subsequent penetration into the layer volumes. Due to the increased concentration of defects in the zone of contact of the layers (due to their crystallographic misorientation necessary to preserve the laminar structure), diffusion transfer occurs along the boundaries much faster than in the bulk of the layer. On this basis, it can be said that the idea of forming an oriented structure for subsequent chemical heat treatment can be quite effective, since it will allow increasing the penetration depth of the alloying elements and their speed.

Of interest is the analysis of the form of the meniscus of the nitrated layer, which depends on the thickness of the laminar layer, the nitriding regime, and the chemical composition of the layers in the composition.

On the example of composition AISI304+W108, it is seeing that with a decrease in the thickness of the laminar layer, the effect of “smoothing” the meniscus is observed, i.e. the vees become wider and shorter, and the angle of inclination to the border is larger (Figure 1, 2).

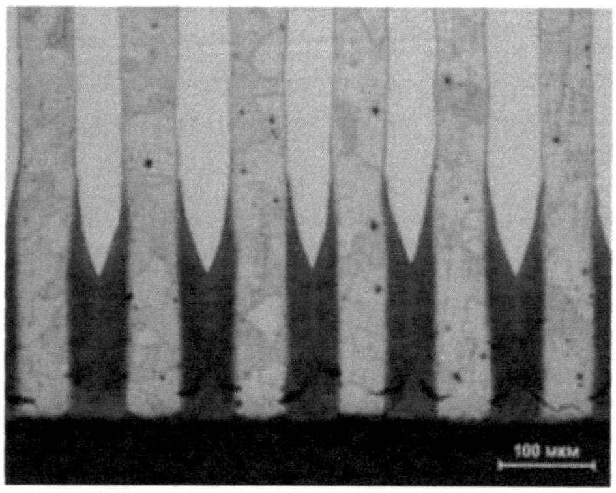

Figure 1. Structure of the nitrated layer in the multilayer material (composition AISI304+1008) after first process cycle

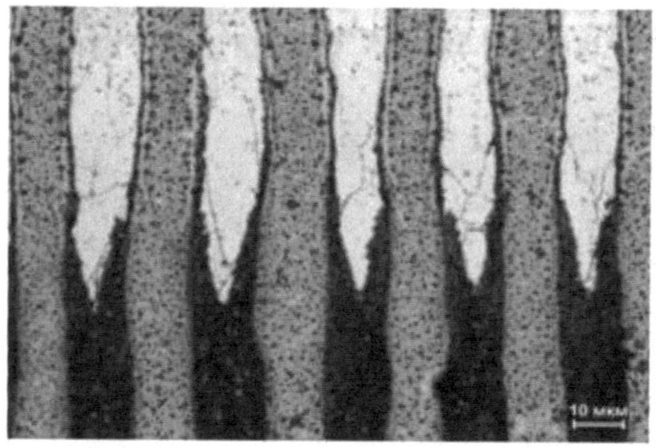

Figure 2. Structure of the nitrated layer in the multilayer material (composition AISI304+1008) after second process cycle

The chemical composition of the materials included in the composition has a significant effect on the shape of the meniscus and on the depth of the nitrated layer. Thus, the thickness of the laminar layer may not have the desired effect if hard-to-nitrate materials are present in the composition of the multilayer composite. This is observed in the composition AISI304 + W108 (Figure 3, 4). With a layer thickness of 10 μm with nitriding at a mode of $t = 540°C$, the exposure time $\tau = 40$ h, the depth of the diffusion layer is 10 μm, with a layer thickness of 2 μm, the depth of the nitrated layer also remains 10 μm.

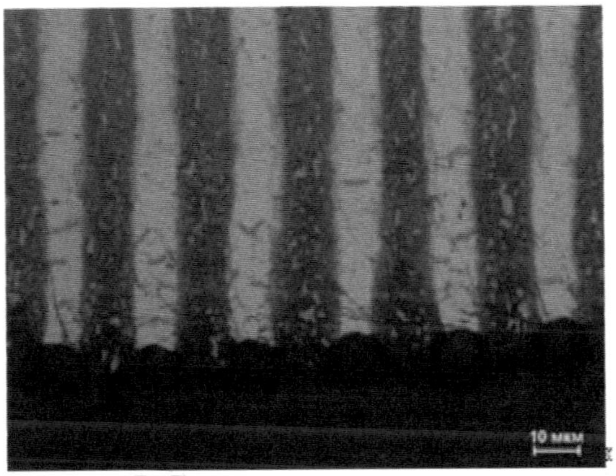

Figure 3. Structure of the nitrated layer in the multilayer material (composition ASI304+W108) after the first process cycle

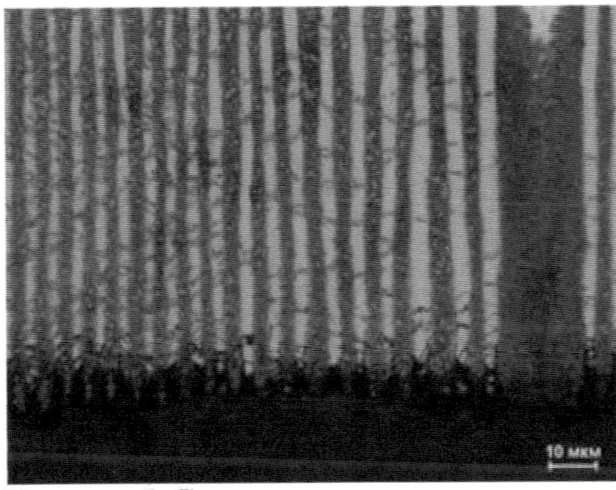

Figure 4. Structure of the nitrated layer in the multilayer material (composition AISI304+W108) after the first process cycle

Other results can be observed in the composition AISI304+1008. After the first process cycle, the layer thickness is 50 μm (Figure 5); during the second process cycle, the layer thickness is 10 μm (Figure 6). The depth of the diffusion layer in this case is 90 μm and 110 for the first and second process cycles, respectively.

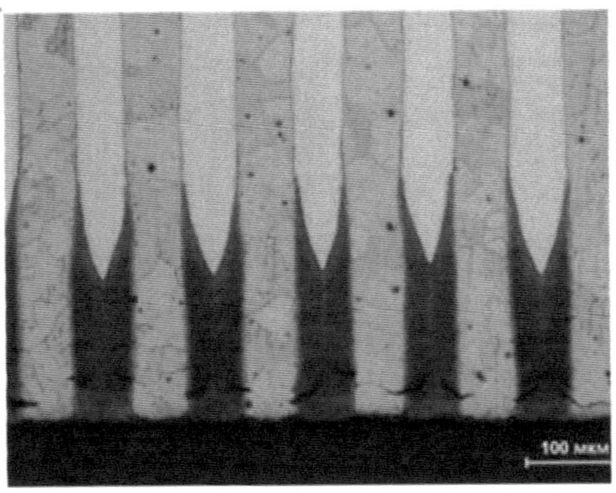

Figure 5. Structure of the nitrated layer in the multilayer material (composition 1008+AISI304) after the first process cycle

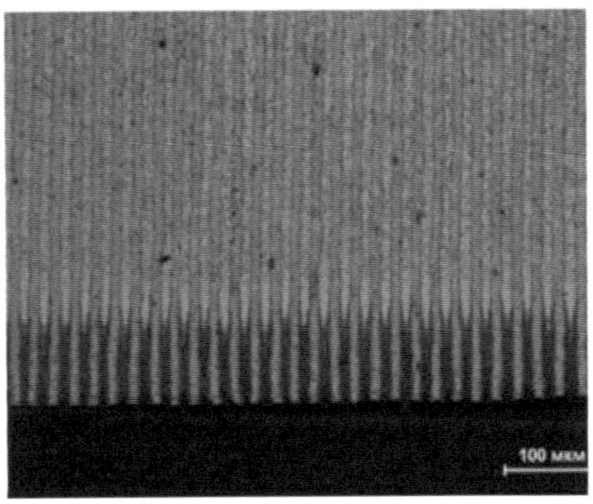

Figure 6. Structure of the nitrated layer in the multilayer material (composition 1008+AISI304) after the second process cycle

4. Conclusion

Thus, the study showed that the depth of penetration of the diffusant during nitriding depends on the thickness of the laminar material layers. It has been found that a decrease in the thickness of laminar layers from 100 to 20 μm leads to an increase in the diffusion layer of nitrogen by more than two times

References

[1] Kolesnikov A G, Plokhikh A I, Komisarchuk Yu S, Mikhaltsevich I Yu 2010 A study of special features of formation of submicro- and nanosize structure in multilayer materials by the method of hot rolling, *Metal Science and Heat Treatment*. **52**, N5–6, p. 273-278

[2] Plokhikh A I 2013 O vozmozhnosti primeneniya mnogoslojnykh metallicheskikh materialov dlya detalej mashin, uprochnyaemykh KHTO, *Izvestiya VolGTU* №6 13-17

[3] Petelin A L, Plokhikh A I 2013 Diffusion along grain boundaries in multilayer materials, *Steel in translation* №11 p.710-713

[4] Plokhikh A I, Safonov M D 2014 Issledovanie vliyaniya parametrov struktury mnogoslojnyx materialov na tolshhinu diffuzionnogo sloya pri azotirovanii, *Novye materialy i texnologii v mashinostroenii.* №20 p 55-58.

[5] Plokhikh A I, Vlasova D V, KHovova O M, Polyanskij V M 2011 *«Nauka i obrazovanie»* **№11** http://technomag.edu.ru/262116.html

Development of a new experimental method for the determination of inplane shear strength and elastic characteristics of polymer composite materials based on high-modulus carbon fiber-reinforced plastics having the ±45° plies

Nikolai Matyushevsky1, Alexei Popov[1]

[1]JSC ORPE Technologiya named after A.G. Romashin, 15 Kievskoye Shosse, 249031 Obninsk, Russia
Abstract. A new method for the determination of shear strength and modulus of high-modulus CFRP reinforced with ±45°plies was developed. The optimal geometrical parameters of the specimen to determine shear properties of high-modulus CFRP reinforced with the ±45° plies were determined by calculation and experimental methods. A comparative analysis of the results of the experimental determination of shear properties was performed.

Key words: Shear in plane of the sheet, polymer composite materials, high-modulus CFRP, stress-strain behavior, finite element method.

1. Introduction

Today there are a large number of standards for shear modulus and strength determination. Studies [1-7] were conducted on the comparison and analysis of the main methods of determining the shear properties of polymer composite materials (PCM). The diversity in standards and methods leads to substantial differences in the characteristics observed. Moreover, current test methods for laminated composites may not always ensure reliable determination of shear strength and elastic characteristics of advanced high-modulus carbon fiber-reinforced plastics (CFRP) [8].

One of the main objectives of developing the reinforced plastic shear test methods is to provide pure shear state in the specimens, and uniform distribution of shear stresses and strains on the whole working area of the specimen.

Ultimate shear strength is provided by the ±45° lamination pattern since the strength of CFRP with this lamination pattern depends not only on the matrix characteristics, but also on high fiber rigidity. When determining shear strength stress concentrators occur in the specimen from the contact of the specimen with the test fixture, and from the holes and cutouts. When testing high-modulus CFRP reinforced with the ±45° fibers, the impact of these stress concentrators increases. [9].

2. Rib-shorting simulation method

This work was aimed at the development of a method for determining the shear strength and modulus of high-modulus CFRP reinforced with the ±45° plies. A calculation analysis of stress-strain behavior (hereafter SSB) of specimens of various configurations with the ±45 reinforcement pattern was performed. SSB of the specimens was calculated by the finite-element method (hereafter FEM) in the NX/Nastran software application. The task was carried out using 3D Solid elements. The properties of CFRP reinforced with the ±45° plies were assigned to the elements.

The calculated SSB values were compared with the experimental data obtained during the shear test in plane of the sheet according to standards [10-14]. At the same time strain distribution in the specimens was determined both by strain gauging, and through an optical system for 3D strain measurement – VIC 3D. As a result of calculation and experimental research it was found that according to all the standards considered in the CFRP specimens reinforced with the ±45° plies there is no uniform tangential stress distribution in the specimen working area. Therefore, the results of strength and elastic properties determined according to the standards which are commonly used in practice can be insufficiently reliable.

To correctly determine shear strength and elastic characteristics one should choose such geometry with which there is pure shear state in the working area, and the impact of the stress concentrators on the characteristic determined would be minimal. The tests within the work showed that the highest and most stable values of shear strength and modulus were obtained when the tests were performed by the plate distortion in relation to the narrow strip (symmetrical shear) according to GOST R 50578-93, and this specimen was used as a basis for further research.

Calculation analysis of SSB of the specimens revealed that uniform shear strain distribution was mostly affected by the working area length-to-width ratio l/b, the presence of cutouts in the specimen working area, and the spherical radius R of the cutout. Figure 1 shows the working area length l as a function of its width b.

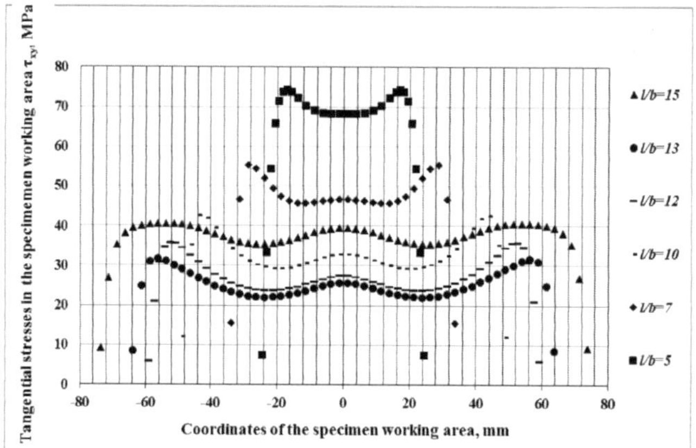

Figure 1 – Distribution of tangential stresses in the specimen working area

When the working area length-to-width ratio l/b > 10 there are no regions with uniform distribution of tangential stresses. The ratio l/b < 5 causes higher stress concentration in the specimen working area, which results in premature failure.

When analyzing the impact of the radii on the tangential stress distribution in the specimen working area it was determined that the spherical radius R > 10 mm increased heterogeneity. The spherical radius R<5 mm causes appearance of critical stress concentrators in the cutouts, resulting in the lowered values of the mechanical properties of the material. Based on the analysis performed the specimen geometry was determined to reliably determine shear properties of high-modulus CFRP reinforced with the ±45° plies.

3. Experimental results

A rectangular plate specimen of 112x73 mm with protective pads was accepted for experimental research. The working area consists of two strips of material with two pairs of U-cutouts on the opposite sides of these strips.

Figure 2 shows the diagrams of SSB of the selected specimen determined by FEM and an optical method using the Vic-3D system.

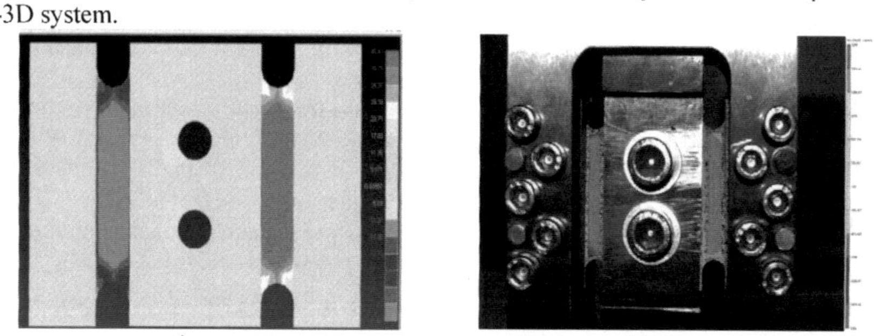

a) b)

a) Distribution of tangential stresses determined by FEM in NX/Nastran

b) Distribution of shear strains experimentally determined by an optical method

Figure 2 – Diagrams of SSB of the specimen obtained by calculation and experimental methods

It should be noted that in this specimen there is uniform tangential stress distribution nearly on the entire strip surface area.

The specimen shape developed enables to fix it in the test device without fixing holes in the gripping portion, which provides a minimal impact of the stress concentrators on the specimen, and a more uniform strain and stress distribution in the area of transition from the gripping portion to the test portion of the specimen.

After the calculation analysis the comparative mechanical tests for shear in plane of the sheet were conducted by the method developed and the most common standards [10-14]. The test specimens were fabricated by autoclaving from high-modulus CFRP VKU-25 (carbon filler HTS 40 and binder VSE1212) reinforced with the ±45° plies. The thickness of the test specimens was ≈2 mm.

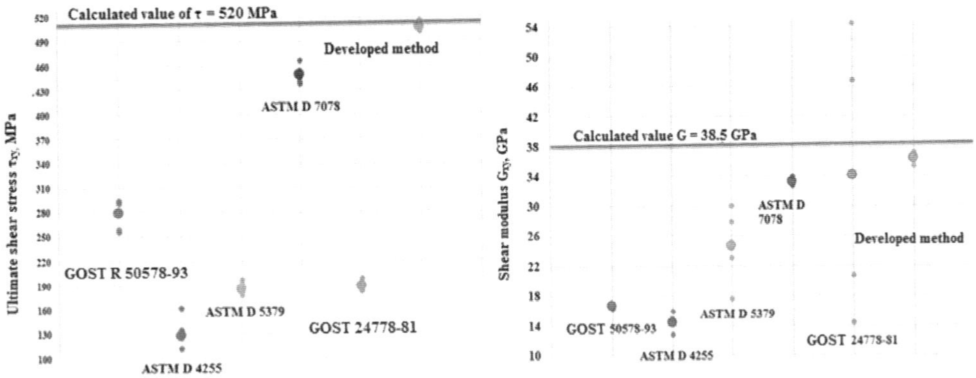

Figure 3 - Results of the tests for the shear in plane of the sheet of the high-modulus CFRP VKU-25 reinforced with the ±45° plies

The comparative analysis of the results of experimental determination of the advanced high-modulus CFRP shear properties showed that the method developed provides proper determination of the material shear properties and low coefficients of variation of shear modulus (CV=1.79%) and ultimate shear stress (CV=1.14%) values.

4. Conclusions

Thus, the method developed provides a reliable and precise strength and elastic modulus determination in shear in plane of the sheet through the uniform stress and strain distribution in the specimen working area. This is due to the specific geometrical shape of the specimen that is the presence of U-cutouts on the test specimen with the length and width of the specimen working area. This geometrical shape of the specimen is selected so that over the whole specimen working area that is between the U-cutouts the "pure" shear state is at most realized. The test method using this specimen was protected by the RF patent [15].

References

[1] Tarnapolsky Yu M and Kintsis T Ya *Metodi staticheskih ispitaniy armirovannih plastikov* Khimiya Publ (Moscow) 272

[2] Polilov A N 2016 *Experimental Mechanics of Composites* Bauman Moscow State Technical University Publ 121-131.

[3] Arnautov A and Bach T 1996 Determination of In-plane Shear Characteristics of Composites with the ±45 lamination. *Mechanics of Composite Materials* **32** 256-254.

[4] Adams D O, Moriarty J M, Gallegos A M, and Adams D F 2003 *Development and Evaluation of the V-Notched Rail Shear Test for Composite Laminates,* U.S. Department of Transportation Federal Aviation Administration. National Technical Information Service (NTIS) (Springfield, Virginia 22161) 90

[5] Gusenkov A P and Polilov A N 1992 *Modern Test Methods for Composite Materials*, IMASH RAN Publ (Moscow) 247

[6] Hodgkinson J M 2000 *Mechanical Testing of Advanced Fibre composites* ed Corporate Blvd pp. 100-122.

[7] Adams D F 2000 *Comprehensive Composite Materials* **5** pp. 113-148.

[8] Popov A G and Matyushevsky N V 2013 Abstracts of the XX Int. scientific and technical conf. *Structures and technologies of production from nonmetallic materials* (Obninsk) pp.146-148.

[9] Matyushevsky N V and Popov A G 2016 Proceedings of the 2nd International conference *Deformation and Failure of Composite Materials and Structures* Stolitsa Publ (Moscow) pp.219 – 220.

[10] GOST R 50578-93: Plate distortion shear test 11

[11] Standard Test Method for In-Plane Shear Properties of Polymer Matrix Composite Materials by the Rail Shear Method ASTM D4255/D4255M 15 A

[12] Standard Test Method for Shear Characteristics of Composite Materials by the V-Notched Beam Method ASTM D5379/D5379M 12

[13] Standard Test Method for Shear Properties of Composite Materials by V-Notched Rail Shear Method ASTM D7078/D7078M 12

[14] Methods of shear strength determination in plane of the sheet, 12 p GOST 24778-81

[15] Matyushevsky N V, Popov A G and Lisachenko N G Description of the RF patent no. 2617776 *Specimen for the determination of shear elastic modulus and ultimate strength of high-modulus CFRP in plane of the sheet* IPC G01N 1/28 (2006.01), G01N 3/24 (2006.01). Application no. 2016100524 of 11.02.2016 12

Surface express ultrajet diagnostics of space vehicle materials

A L Galinovsky[1], D R Mugla[1] and E S Golubev[1]

[1]FSEI HPE «Bauman Moscow State Technical University» (National Research University), Moscow, 105005, RF
Abstract. The article substantiates the possibility of using waterjet hydrophysical diagnostics for the express estimation of features of physical and mechanical properties of structural materials obtained by selective laser sintering. It is shown that, based on the results of ultrasonic diagnostics, functionally necessary information is obtained to improve the quality of implementation of all stages of products' life cycle in production of which it is promising to use additive technologies. On the example of the constructive element of the Russian Observatory " Millimetron "(Spektr-M project) it is shown that the approaches proposed by the author can be used in the process of technological preparation of rocket's production and space technology (RST), aircraft building, turbo-power engineering, etc.

Keywords: selective laser sintering, ultrajet diagnostics, anisotropy of material's properties, physical and mechanical properties

Introduction

Modern additive manufacturing of Rocket and space equipment products allows the manufacture of ready-for-use or requiring minor modifications of parts. Progressive company, leading participants in this market, emphasize that additive technologies can have broad prospects for implementation in various branches of technology. Currently, companies that produce only equipment for 3D printing, as a rule, also implement a full cycle of additive manufacturing of parts for the aerospace industry. This solves the problem of determining the characteristics of the materials used, carries out technological processes of finishing parts, including heat treatment, hot isostatic pressing, mechanical processing on CNC equipment, as well as quality control using modern diagnostic tools. Currently, economically capacious high-tech developments transferred to the relevant enterprises, primarily of the military-industrial profile.

In the Academician of RAS N.P. Aleshin works and scientists of "Bauman Moscow State Technical University", as well as the work of other authors [1-4] shows the need for adequate information and diagnostic support by non-destructive quality control methods for products made from materials obtained by additive technologies, in particular, selective laser fusion (SLF). However, the question of the operational determination of the geometrical microscale instability of the physic mechanical properties (PMP) of these materials remains open. Therefore, the task of rapid assessment of the anisotropy parameters of the PMP of structural materials obtained by SLS is relevant, since the quality of the working surfaces of products from them usually determines the operational and resource parameters of the structure as a whole, especially various types of detachable joints and friction mates of responsible use.

On the basics of the above, the research and scientific goal of the research formulated: to determine the functional and structural relationship between the instability of the physic mechanical properties of materials obtained by selective laser alloying and the specifics of their local fracture with the waterjet. To achieve this goal was necessary to solve the problem of methodological support of the waterjet technology, as well as to illustrate its functional and technological capabilities with specific examples.

Rationale for research performance

At the initial stage of research by using the methods of expert-analytical analysis proposed in [5, 6]. This method let to show the broad functionality of waterjet hydrophysical diagnostics. In the works [7, 8] was shown that by analyzing the products and results of local erosion-fatigue destruction by a high-speed waterjet of the surface layer of the diagnosed material it is possible to obtain objective information about the parameters of its state. Mainly about the PMP and operational life indicators, particular functional damage to the working surfaces of the analyzed structural element of the investigated product. In this regard, the possibility of a detailed analysis of the local micro geometric anisotropy of the surface properties of the products obtained by SLF opens up, since it's the instability of working surfaces formed by the PMP that is often the limiting factor of their operational and technological reliability.

In [7, 9], the general structure of the model of hydro erosion-fatigue fracture of the surface layer in the case of its PMP isotropy was proposed. Therefore, using the well-known algorithmic approach to the construction of this model, we introduce into it a significant physical and technological difference, due to the different-scale variation-cyclic instability of the PMP materials obtained by the SLF method. This difference consists in the frequency of changes in the parameters of technological microdefects of the structure of the material formed in the process of its fusion. Moreover, this periodicity is commensurate with the geometrical dimensions of the region of laser alloying of the initial dispersed powder supplied to the zone of structural and shaping structural material and at the same time the part itself at the SLF process operation as shows on the Figure 1.

Figure 1 - The nature of the change in the size of technological microdefects (micro- and submicrocracks) in the material obtained by the SLF method

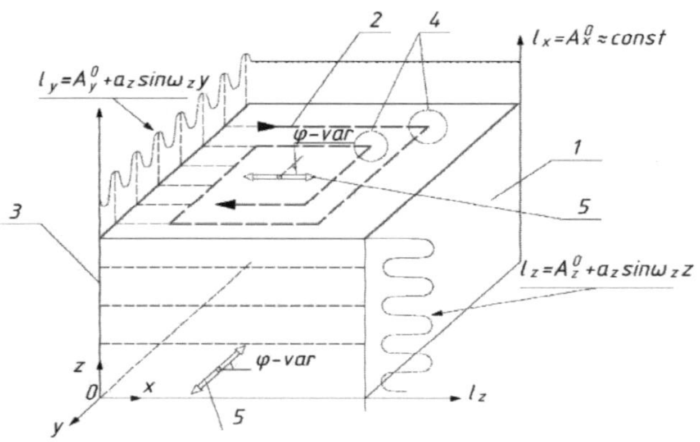

1. The analyzed sample obtained by the SLF technology;
2. The possible trajectory of the laser beam in the (xy) plane;
3. Laser-deposited layers of material in the axial direction (oz);
4. Zones of possible increased instability of the structure of the material, the occurrence of which is due to a change in the trajectory of the laser beam;
5. The trajectory of the scanning diagnostic effects of US on the surface of the sample.
6. lx, ly, lz - respectively, the expected change in the size of microdefects along the axes: ox, oy and oz (lz> ly> lx);
7. (--- →) - the direction of movement of the laser beam.
8. Moreover, in work [10, 11], the question of the quantitative coordinate instability of the PMP of images obtained by SLF, which as a whole determines their strength and, mainly, fatigue and surface-friction characteristics, was experimentally studied.

Methodical features of research

Using the phenomenology of the features of the distribution of initial technological defects, presented in Figure 1 and the algorithm of hydro erosion destruction, considered in [9] after performing the corresponding design procedures, we obtain the following results of the simulation of the UWJD process on the surface layer of a sample of material obtained by SLF, which are graphically shown in Figure 2. In Figure 2a shows a sectional view of the sample, with the profile of cavities formed as a result of diagnostic exposure, having a corresponding bottom surface micro relief. In Figure 2b is a diagram illustrating how the depths h of the cavity and the micro relief of its bottom R vary with the angle φ as a result of the diagnostic effect on the sample.

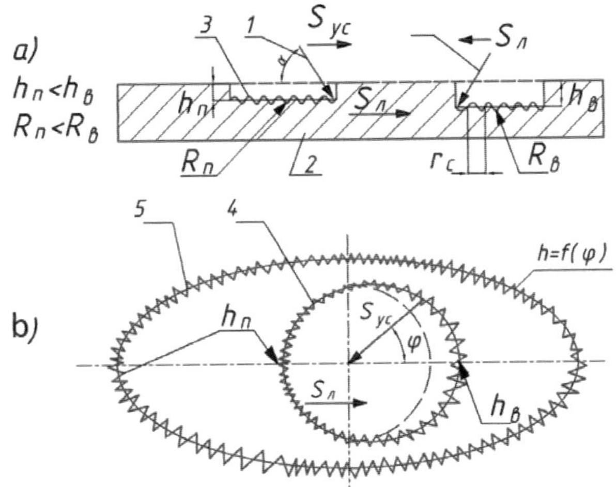

Figure 2 - A generalized diagram of changes in the depth h and the microrelief of the surface R of the bottom of the hydrocourt obtained under various conditions of the diagnostic effect of waterket on the material formed by SLF technology

In Figure 2a, it is accepted:
1 - control and diagnostic high-speed waterjet;
2 - the surface of the studied sample material;
3 - the microrelief of the surface of the bottom of the hydrocavity formed as a result of the WJ exposure to the material being diagnosed; ; $S_{\text{УС}}$ и $S_{\text{Л}}$- respectively, the direction of movement (feed rate) of the WJ and the laser beam ;
$h_{\text{п}}$ *and* $h_{\text{в}}$ are, respectively, the depth of the hydraulic tank when the directions coincide $S_{\text{УС}}$ и $S_{\text{Л}}$ (passing movement of WJ) and if the directions do not coincide - counter displacement of CSS. And $h_{\text{п}} > h_{\text{в}}$ and $R_{\text{п}} > R_{\text{в}}$.
In Figure 2b it is accepted:
4 - diagram of variation of the depth h of the cavity depending on the angle φ - the angular mismatch between; ; $S_{\text{УС}}$ и $S_{\text{Л}}$ (in the xoy plane)
5 - is a diagram of changes in the depth h of the cavity and profile (microrelief) of its bottom R depending on the angle φ in the planes yoz and xoz.
The averaged values of h and the variation of the R parameters — microgeometries of the cavern bottom profile with a diameter WJ (dc) substantially exceeding the size of the effective fusion zone (rc) largely depend on the planar factor of the formation of the material and the product as a whole, which is clearly shown in Figure 2b. This provision is fully correlated with the results of direct strength experiments

presented in [10-12], however, the efficiency of their receipt by the WJD is incommensurably higher and much less expensive.

Examples of the implementation of the capabilities of ultra-jet diagnostics

The characteristic result of determining the value of h and the variation of the parameters R - microgeometry of the profile of the bottom of the cavity are shown in Figure 3. Test experiments were performed on samples obtained from LPW 316 brand aluminum powder traditionally used in rocket and space technology [13–15]. To obtain the samples, the system RenAM 500M from Renishaw (United Kingdom) was used. To implement the WJD procedure, a waterjet cutting machine from Multicam WaterJet (USA) had used.

The technological parameters were as follows: the pressure of the multiplier was 350 MPa, the feed rates of the nozzle head, i.e. the rate of scribing of the USS of the surface being diagnosed is 2 mm / s, the distance from the cut of the focusing tube to the surface of the sample is 5 mm, the interaction of the water jet without abrasive passed at a right angle to the surface. An image of the profile of the hydrocave was taken with a Carl Zeiss LSM 700 laser scanning confocal microscope (Germany).

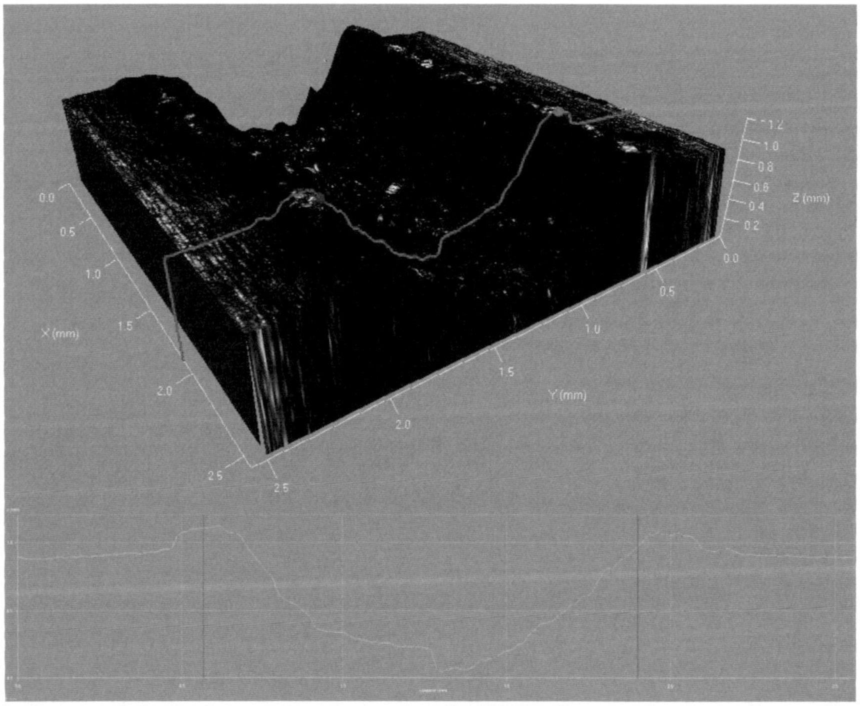

Figure 3 - Image of the hydrocoupler obtained at the WJD of a sample of aluminum alloy (1- the studied part of the specimen; 2- profile of the hydrocourt on its surface)

Note that the final diagrams of diagnostic results presented in Fig. 2b were obtained by summarizing the results of optical measurements of the geometric parameters of tracks left on the surface of the scanning ultra waterjet (UWJ) at different angles to the vector S_L - the direction of the laser fusion of the original powder for forming the part according to the SLF technology. Moreover, the angle α - formed by the UWJ and the diagnosed plane of the object under investigation plays a certain role in ensuring the effectiveness of the UWJD as a whole.

According to the results of the analysis performed, the following main conclusions some generalizations can be formulated:

1. The change in the microrelief of the surface of the cavity, its bottom part functionally and quantitatively copy the patterns of distribution of microdefects: lx, ly, lz. This circumstance is

the most important information and diagnostic parameter for the quality of additive technologies, in particular, SLF technology. The mechanism of this hereditary technological copying is described in detail in [14-15].

2. The quality of the SLF process itself can be judged by the difference in the depths of the erosive hydrocaverns formed on the surface of the sample during the "passing" and "counter" movement of the US with respect to the trajectory of the laser beam. This parameter is called the "kinematic factor" of diagnosis and applied to other conditions of its manifestation is also analyzed in [16-18].
3. In the zone of change of the path of movement of the beam - pos. 4 (see Figure 1), it is not yet possible to isolate clear patterns in the characteristics of hydro erosion destruction of the surface of the material obtained using the SLF technology.

Note that in the future, potentially very significant results on the UWJD of products manufactured using additive technologies, in particular, by using the SLF method, will be fully verified by their resource, for example, fatigue and / or friction tests, as well as after a certain period of operation of the products being analyzed. The WJ-diagnostic identification of the vector-kinematic nature of the layer-by-layer formation of operational and physical damage is highly likely, which is essential for reliable prediction of the residual life of products made from such materials, which are distinguished by special PMP, which are due to the technological specifics of their production. In this case, presented in Figure 1, the initial distributions of technological microdefects should be supplemented by the dependencies of their quantitative and dimensional changes during operation [19-20].

Conclusions

As a study result, the goal of the research had been resolved. The tasks on the scientific and methodological substantiation of the performance potential of ultra-jet hydro physical diagnostics of the parameters of anisotropic instability of PMP materials and parts realized by the SLS methods achieved. The physical and technological basis for achieving the goal and solving the tasks set was verified by field experiments on samples and models of real parts, the statement about the significant effect of the PMP spread of the surface layer of the material formed by the SLS method on the instrumentally diagnosed specificity of its local hydro erosion under the impact of the dynamic effect of the waterjet.
In conclusion, we note that the theoretical and experimental substantiation of the effectiveness use waterjet method in additive technologies not only complements the well-known and developing arsenal of tools for studying their functional capabilities, but is also a tool for obtaining new knowledge about the specificity of PMP materials derived from these advanced technologies, in particular local anisotropy, which must be taken into account at all stages of the life cycle of CT products.

Acknowledgments

The research was supported by the grant of the President of the Russian Federation for state support of leading scientific schools (NSH-3778.2018.8) and the grant of RFBR 18-29-18081.

References:

[1] Aleshin N P, Murashov V V, Evgenov A G, Grigoriev M. V, Simakov N A, Vasilenko S A and Krasnov I S 2016 The classification of flaws of metal materials synthsised by selective laser melting, and the capabilities of non-destructive testing methods for their detection. *Defektoskopia* **52** 48-55.
[2] Babentsova L P, Antsiferova I V 2017 Quality and ecology of the selective laser technology. *Master's journal* **1** 87-92
[3] Kosushkin P A 2017 Quality control in additive manufacturing. *Vector of high technology* **2** (31) 28-31
[4] Ryzhkov E V, Pavlov M D, Gusarov A V, Artyomenko Yu A, Vasiltsov V V 2011 Crack formation in ceramics under selective laser sintering. *Physics and Chemistry of Materials Treatment* 77-83.

[5] Bouncers R M, Jagiello O I, Korneev V M, Policarpo M P 2012 *Expert evaluation of quality control in engineering* (Moscow) 395

[6] Galinovsky A L, Samsonov K S, Sevryukova V A, Salakhutdinov R A 2017 Comparison of different methods of control and diagnostics of the quality of ceramics by the method of expert estimation. *Innovatics and expert examination: scientific works* **1** (19) 64-74.

[7] Abashin M I 2013 Accelerated determination of the quality parameters of the surface layer of the material of products by the results of the influence of supersonic liquid jet on it, MSTBU (Moscow) 17

[8] Abashin M I, Bochkarev S V, Tsaplin A I, Kobernik N V 2015 Ultra-jet diagnostics of the quality of weld. *Izvestiya vysshih uchebnyh zavedenij. Mashinostroenie* **12** (669) 52-61.

[9] Bochkarev S V, Tsaplin A I, Galinovskii A L, Abashin M I, Barzov A A 2017 Ultrajet diagnostics of the microstructure of a material under heat treatment. *Metal Science and Heat Treatment* **6** (744) 58-63.

[10] Kok Y 2018 Anisotropy and heterogeneity of microstructure and mechanical properties in metal additive manufacturing: A critical review, *Materials and Design* **139** 565–586

[11] Popovich A A, Sufiiarov V S, Borisov E V, Polozov I A, Masaylo D V, Grigoriev A V 2017 Anisotropy of mechanical properties of products manufactured using selective laser melting of powdered materials *Russian Journal of Non-Ferrous Metals* **58** 389-395

[12] Kok Y 2018 Anisotropy and heterogeneity of microstructure and mechanical properties in metal additive manufacturing: A critical review, *Materials & Design* **139** C 565–586.

[13] Sklyar M O, Turichin G A, Klimova O G, Zotov O G, Topalov I K 2016 Microstructure of 316L stainless steel components produced by direct laser deposition. *Steel in Translation* **46** 883-887

[14] Wang Z, Palmer T A, Beese A M 2016 Effect of processing parameters on microstructure and tensile properties of austenitic stainless steel 304L made by directed energy deposition additive manufacturing, *Acta Mater* **110** 226–235.

[15] Hussam E C, Bruno C, Branchu S, Xiaowei H, Hasco J Y, Guille R 2012 Direct Laser Fabrication process with coaxial powder projection of 316L steel. Geometrical characteristics and microstructure characterization of wall structures. *Optics and Lasers in Engineering* **50** 1779 – 1784.

[16] Bazaleeva K O, Tsvetkova E V, Balakirev E V 2016 Processes of recrystallization in austenitic alloy obtained by methods of selective laser melting *Vestnik Bauman Moscow State Technical University. Seriya: Mashinostroenie* **5** 117–127

[17] Shvartsev S L 2017 Do additive technologies have a future? *Herald of the Russian Academy of Sciences* **87** 267-275

[18] Dvirniy V V, Leonenkov A D 2017 Prospects of applying additive technologies in the aerospace industry, *Reshetnevskie chteniya* 632-633

[19] Kovalev D S, Kovalenkov P A 2017 Prospects of introduction of additive technologies in industry, *Actual problems of aviation and cosmonautics* **1** 398-400.

[20] Sovetnikov Ye I 2015 Evaluation of development additive technologies *Technology of Light Alloys* **№3** 17-31

IOP Conf. Series: Materials Science and Engineering **683** (2019) 012051 doi:10.1088/1757-899X/683/1/012051

Polymer composites based on halogen-containing oligoethers

R Ch Bazheva [1], A S Borodulin[2], A N Kalinnikov[2] and A M Charaev[1]

[1]Kabardino-Balkaria State University named after Kh.M. Berbekov, Nalchik, Chernyshevsky street, 173, Russian

[2]Intersectoral Engeneering Center "Composites of Russia", Bauman Moscow State Technical University, Moscow

Abstract. Halogen-containing oligosulfones based on 1,1-dichloro-2,2-di (4-hydroxyphenyl) ethylene and 4,4'-dichlorodiphenylsulfone on various degrees of condensation were synthesized in the solution by high-temperature polycondensation. Conducted a physical modification of indutrial polycarbonate bisphenol A oligosulfones obtained in a wide concentration range. Physical and mechanical properties of composites are investigated. The compatibility of the obtained oligosulfones with polycarbonate was studied using viscometry, differential scanning calorimetry, and probe microscopy. It was shown that the introduction of oligosulfones into the polycarbonate matrix promotes an increase in the glass transition temperature, found by differential scanning calorimetry, at 3 to 20 ° C, depending on the composition, the composites have good dielectric and technological properties. The temperature dependences of the dielectric properties of PCs and PC-based composites with different oligosulfone content are characterized by the presence of a single loss on the dielectric loss tangent dependences on the temperature.

Keywords: oligosulfone, polycarbonate, viscometry, differential scanning calorimetry, compatibility.

Introduction

Recently, a noticeable expansion of scopes of rigid chain glassy polymers characterized by low resistance to cracking is largely due to the development of compositions based on them.

Bisphenol A (PC) polycarbonate, along with a complex of valuable properties, has a number of disadvantages, which significantly limit its fields of application. In particular, high internal (residual) stresses, which lead to cracking of products during operation, slow-going relaxation processes, low adhesion, low resistance to basic media, high melt viscosity, and hence the processing difficulty associated with this. In addition, the use of PC in construction also determines the increased requirements for flammability [1-12].

Despite numerous scientific studies on the modification of polycarbonate, the range of hand-made is small. Numerous attempts to modify polycarbonate with low-molecular compounds did not give positive results. Over 50 compounds of various classes have been tested. However, to date, modification of polycarbonate in order to improve a number of performance characteristics is relevant. In this direction, the most promising way is the physical modification, i.e. development of compositions based on polycarbonate and various modifiers. In addition, it is known that the introduction of halogen atoms into a polymer contributes to an increase in fire resistance [13–37].

In order to create polymer composites based on polycarbonate, oligosulfones based on 1,1-dichloro-2,2-di (4-hydroxyphenyl) ethylene and 4,4'-dichlorobiphenyl sulfone were synthesized and some properties of the composites were studied.

Experimental

In a three-necked flask equipped with a stirrer, a reflux condenser with a Dean-Stark trap, a gas supply bubbler and a thermometer, 5.62281 g (0.02 mol) 1,1-dichloro-2,2-di (4-hydroxyphenyl) ethylene are introduced, 40 ml of dimethyl sulfoxide (DMSO) and 30 ml of toluene. When stirring, nitrogen is passed in and the temperature is raised to 70 ° C. After complete dissolution of 1,1-dichloro-2,2-di (4-hydroxyphenyl) ethylene, 3.98406 ml of 10.04 n is added. sodium hydroxide solution. The temperature is raised to 140-145 ° C and the azeotropic toluol-water mixture is distilled off until the water is completely removed. The reaction mass is cooled to 40-50 ° C and 2.87294 g (0.01 mol) of 4.4'-dichlorodiphenylsulfone is added. The reaction is carried out at 140-145 ° C for 2 hours. The resulting mass is diluted with 10 ml of DMSO and precipitated into acidified distilled water. The precipitate formed is filtered off and washed with distilled water until the filtrate is negatively reacted with chlorine ion. The resulting oligoether sulfone is dried at 100 ° C under vacuum for 24 hours.

The yield of light yellow product 97.5%; softening temperature 120-1220C; the content of the basic substance is not less than 99.9%; sodium chloride in the amount of not more than 0.05-0.1% is allowed as an impurity; they say m. = 776.52084; elemental composition, %: C = 61.87 / 61.75; H = 3.37 / 3.40; O = 12.36 / 12.31 (calculated in the numerator, found in the denominator). The content of hydroxyl groups - 4,38 / 4,431 (in the numerator is calculated, in the denominator - found). In the IR spectra, the absorption bands in the region, cm – 1: 940–920 (ether bond), 1100, 1150, and 1290 (sulfonyl group), 980 (> C = CCl2 group), 3600–3300 (hydroxyl Group).

Similarly, by changing the ratio of 1,1-dichloro-2,2-di (4-hydroxyphenyl) ethylene and 4,4'-dichloro-diphenylsulfone, oligosulfones of various degrees of condensation were synthesized. In all cases, the amount of 1,1-dichloro-2,2-di (4-hydroxyphenyl) ethylene is 1 mol more than the amount of 4,4'-dichlorobiphenylsulfone.

Ubbelohde viscometer was used to determine the intrinsic viscosity. Solutions with relative viscosity = 1.1 were diluted for subsequent measurements of viscosity in such a way that the intervals of the obtained concentrations were approximately the same, and the η of the most diluted solution should not be less than 1.1. These limits of measurement specific viscosity η = 1.1-1.5 correspond to concentrations of polymer solutions for which the dependence of [η] on the velocity gradient of fluid outflow from the capillary is not observed and with dilution there is no deviation from the linear dependence of ηsp / C or ln ηrel. / C from S.

Next, 0.75 is successively added to the viscometer; 1.00; 1.25; 2.00 ml and 4.00 ml of the solvent, and after each dilution the solution expiration time was measured. For the solvent and each polymer solution, the average value of the expiration time from three measurements was determined.

The concentration of the solutions was determined by the formula:

$$C = \frac{aV_1 \cdot 100}{V(V_1 + V_2)}, (1)$$

where: a is the polymer sample, g;

V, V1, V2 - respectively, the volumes of the solvent consumed to prepare the solution, the initial solution placed in the viscometer, and the solvent added to the viscometer upon dilution, ml.

On the basis of the data obtained, η / η0 = ti / t0, ηud = (ti-t0) / t0 (where ti is the expiration time of the polymer solution, s; t0 is the expiration time of the solvent, s), ηud were calculated. / C and t0 (η / η0 / C). Then built a graphical dependence ηud. / C or t0 (η / η0 / C) from C. Extrapolation of straight lines to the zero concentration of the polymer cuts an [η] segment on the ordinate axis.

Study of the surface of the samples on a scanning probe microscope. The sample is mounted on a polycore substrate 20x25x10 mm in size and then attached to the scanner in the horizontal position. A measuring head with removable probes is installed above the sample. Rapid supply of the probe is carried out by a stepper motor in about 1-3 minutes. The measurement time depends on the speed and field of the scan and is approximately 0.5-5 minutes.

Melt indices of polymers were measured by capillary viscometry on an IITT-M cargo capillary viscometer with a constant piston load of 21.19 N and a temperature of 220-280 ° C. Every 5 min. the extruded melt was cut from the capillary with a knife and weighed. The test result was taken as the arithmetic average of two determinations on three pieces of material, the difference in mass, between which did not exceed 5%. The melt flow rate (melt index) was calculated by the formula:

$i = \frac{10 \cdot Q}{t}$ (2)

where: Q is the mass of the polymer, g;
t - time extrusion, min.

Results and discussions

Oligosulfones based on 1,1-dichloro-2,2-di (4-hydroxyphenyl) ethylene and 4,4'-dichlorodiphenylsulfone were synthesized by nucleophilic substitution in an aprotic dipolar solvent dimethyl sulfoxide at a temperature of 140-145 ° C for 2 hours. Toluene was used as azeotropically. The resulting halogen-containing oligosulfone had the formula:

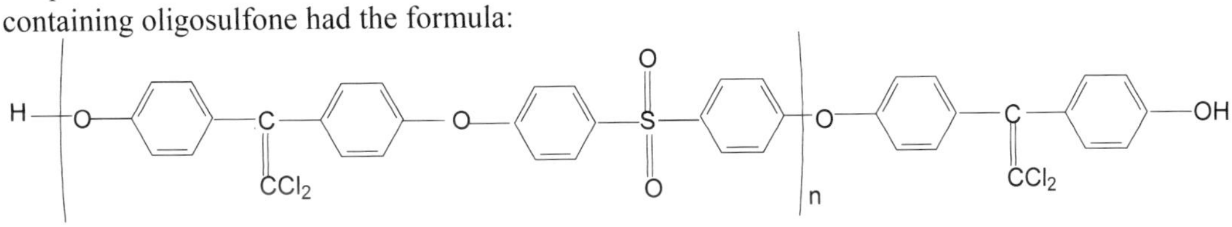

where n = 1-20.

The main properties of these oligosulfones are described in detail in [29-36].

Unambiguous methods for studying compatibility in polymer-plasticizer systems, polymer-polymer is not easy to find. In the block state, compatible polymers form transparent films and fibers, which in a phase-contrast microscope with high magnifications or in an electron microscope do not show a heterogeneous structure under any methods of contrasting. In addition, blends of compatible polymers should have the same glass transition temperature, regardless of research methods. These criteria are, in principle, unambiguous criteria for compatibility, but in practice there may be some difficulties in using them. In particular, incompatible polymers form transparent films if the refractive indices of both polymers are the same or if the refractive indices, although different, but the polymers can form a two-layer film that appears transparent when it is obtained by evaporating the solvent from the polymer solution. Such cases are not particularly difficult for analysis, since transparent films from blends of incompatible polymers are characterized by two glass transition temperatures corresponding to the glass transition temperatures of the components, if the latter differ sufficiently and can be determined using the method of investigation used.

The above makes it critical to refer to any measurements on polymer blends, since it is very difficult to determine if the mixture is in an equilibrium state. Usually, without evidence, it is assumed that a film from a mixture of polymers obtained from a solution is in a more equilibrium state than samples obtained by mixing polymers in a block state. Everything said predetermined the need to determine the compatibility of polycarbonate with the obtained oligosulfones.

To determine the compatibility, as well as to determine the nature of the distribution of the synthesized oligosulfones in the PC matrix, we used the method of scanning probe microscopy.

Studies have shown that the nature of the surface composition depends on the composition. However, in samples with a modifier content of up to 10% by weight. there is good compatibility between the components. Oligosulfones are evenly distributed in the polycarbonate matrix. The particle size of the modifier depends on the composition of the composition and is for the sample containing from 0.31-3.13 microns.

Later, oligosulfone with a degree of polycondensation n = 5 (OC-5C-2) was used to modify PC.

All this is well illustrated in Fig. 1.

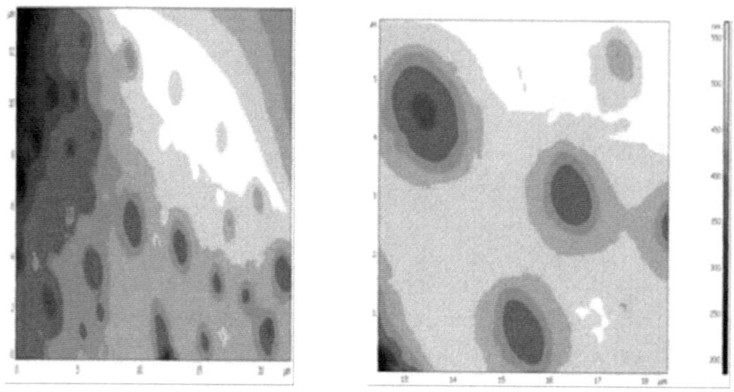

a **b**

Fig.1. SPM photographs of a PC composition with OC-5C-2, containing 5 (a) and 10 (b)% of the mass.

X-ray phase analysis showed that all compositions of composites are amorphous.

To determine the compatibility of the oligosulfones obtained, we also used the viscometric method. The studies were carried out in the Ubbelohde viscometer at a temperature of 25 ° C. At a given temperature, polymer-solvent systems (methylene chloride) were studied for composites containing various mass% oligosulfone, i.e. PC samples were injected from 1 to 10% of the mass. oligomer.

It turned out that for these systems the maximum compatibility is observed at a ratio of 95% by weight. PC and 5% of the mass. OS-5C-2. The data are presented in Fig.2.

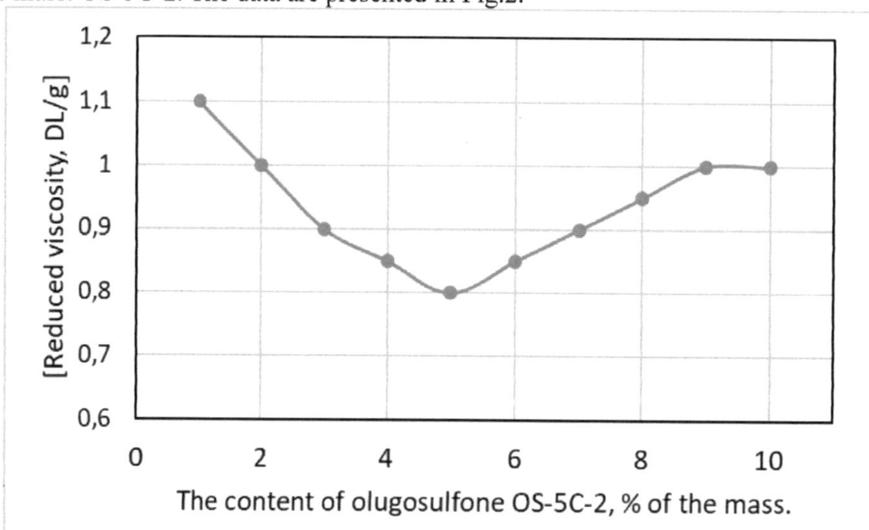

Fig.2. The dependence of the reduced viscosity of the solutions of composites from the content of OC-5C-2

The glass transition temperatures of composites found by differential scanning calorimetry are shown in Fig. 3. As can be seen from the figure, with an increase in the content of oligosulfone in the composition, Tg increases. An increase in the glass transition temperature probably indicates the interaction of the filler with the matrix.

IOP Conf. Series: Materials Science and Engineering **683** (2019) 012051 doi:10.1088/1757-899X/683/1/012051

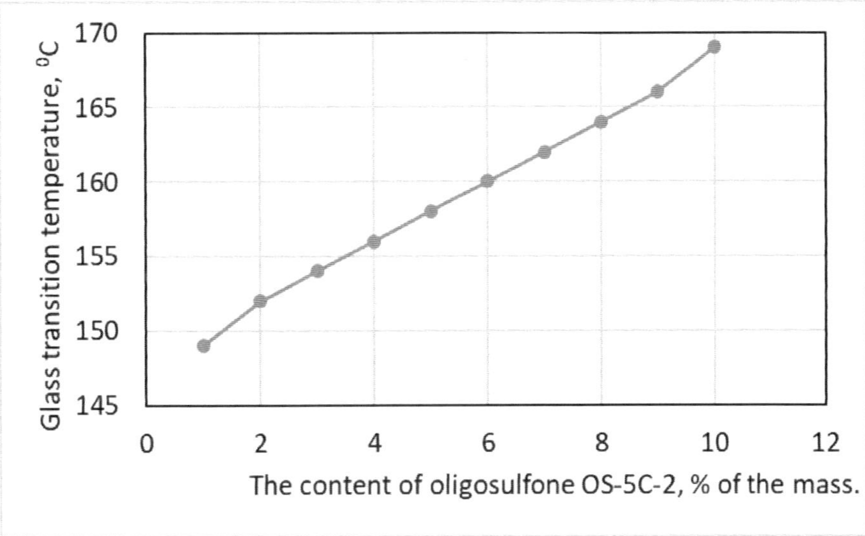

Fig.3. The dependence of the glass transition temperature of composites on the content of OS-5C-2

For the compositions obtained, the temperature dependences of the dielectric constant (ε ') and the tangent of the dielectric loss angle were studied at a frequency of 104 Hz. The values of the dielectric constant of all the investigated samples of composites are ~ 3–3.6 and are stable in the temperature range from 20 to 2000 ° C. The temperature dependences of the dielectric properties of PCs and PC-based composites with different oligosulfone content are characterized by the presence of a single loss on the dielectric loss tangent dependences on the temperature. The values of ε ' for PC and all composites do not depend on the composition and, within the limits of error, coincide and correspond to values of 2.4-2.5.

Comparison of melt indices of the original PC and composites based on them (Fig. 4) showed that the introduction of small amounts of OC-5C-2 in PC significantly affects the melt flow rate (MFR) of these composites. Thus, the introduction of only 1-3% of the mass. OS increases the PTR PC 1.5-2 times. Moreover, the introduction of small amounts is more effective - up to 5% by weight. OS.

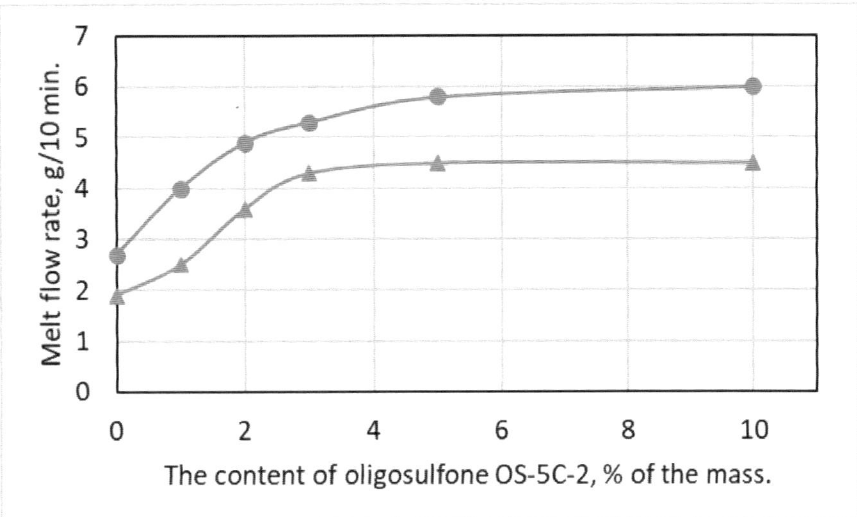

Fig.4. The dependence of the melt flow index on the content of OC-5C-2 at a processing temperature of 260⁰ C (▲) and 280⁰ C (•)

All obtained composites easily dissolve in such easily volatile and relatively non-expensive solvents, such as chloroform, methylene chloride, dioxane, dimethylformamide, etc. This will allow their processing into film products from solutions, for example, in available and volatile methylene chloride.

Conclusion

Composite polymeric materials based on aromatic polycarbonate bisphenol A and synthesized on the basis of 1,1-dichloro-2,2-di (4-hydroxyphenyl) ethylene and 4,4'-dichloro diphenyl sulfone oligosulfones of various molecular weights were developed. The study of some properties of PC compositions with oligosulfone showed that these OS can be used to improve some of the performance properties of polycarbonate.

The compatibility of obtained OS with industrial poly-carbonate has been studied by various methods. It was shown that the introduction of oligosulfones into the polycarbonate matrix promotes an increase in the glass transition temperature by 3 to 200 ° C, depending on the composition. The resulting composites have good dielectric properties.

Acknowledgments

The reported study was funded by RFBR according to the research project № 18-29-18063

References

[1] Amerik V V, Radzinsky S A, Zolkina I Yu, Andreeva T I , Simonov-Emelyanov I D , Fedotova T I, Levchuk A V 2013 Polycarbonate - market analysis and development prospects, *Plastic materials* **11** 10-13.

[2] Ponomareva V T, Likhacheva N N 2000 State and prospects of structural thermoplastics market. *Plast. Masses* **6**

[3] Bolotina L M, Chebotarev V P 2003 Development of research in the field of chemistry and technology of aromatic polysulfones, *Plast. Masses* **11** 3-7

[4] Alperstein D, Knani D, Borchmann N, Spekowius M, Hopmann C 2014 Prediction of environmental stress cracking in polycarbonate by molecular modeling *Polymers for Advanced Technologies* **25**

[5] Ye Liu, Hui Zhou, Jia-Zhi Guo, Wei-Min Ren, Xiao-Bing Lu 2017 Completely Recyclable Monomers and Polycarbonate: Approach to Sustainable Polymers, *Angewandte Chemie* **129**

[6] Wang J, McKenna G B 2015 Viscoelastic properties of ultrathin polycarbonate films by liquid dewetting, *Journal of Polymer Science Part B: Polymer Physics* **53**

[7] Bagotia N, Choudhary V, Sharma D K 2018 A review on the mechanical, electrical and EMI shielding properties of carbon nanotubes and graphene reinforced polycarbonate nanocomposites, *Polymers for Advanced Technologies* **29**

[8] Lee B Y, Park J Y, Kim Y C 2015 Effect of polycarbonate structure and reduction time on graphene oxide dispersion, *Polymers for Advanced Technologies* **26**

[9] Militskova E A, Artemov S V, Andreeva T I, Kolerov A S 1991 Properties, processing and application of a number of complex aromatic polyesters, *Production and processing of plastics and synthetic resins*, m Niitekhim 65

[10] 2010 Polycarbonate, *Application in modern construction* (SPb: NOT) 200

[11] Wu X, Yang B, Zhuang S 2018 A novel polycarbonate composite for waveguides, *Journal of Applied Polymer Science* **135**

[12] Sacher E 1995 Secondary structural motions in polycarbonate. II. Identification of the motions and their relation to impact strength, *J. Macromol. Sci.* **11** 403-410.

[13] Bazheva R Ch, Kharayev A M, Mikitayev A K 2006 Temperature transitions in polycarbonate-polytetramethylenoxide block copolymer resins, *Polymers, Polymer Blends, Polymer Composites and Filled Polymers. Syntesis, Properties, Application*, Nova Science Publishers pp. 115-120.

[14] Kharaev A M, Bazheva R Ch, Chaika A A, Barokova E B 2006 Chemical modification of polycarbonate (review), *Plastics* **9** 25-30

[15] Bazhev R C, Kharaev A M, Storozhuk I P 2007 Synthesis and some properties of polycarbonate-polyalkylene oxide block copolymers, *News of universities. North Caucasus region. Natural* Sciences **3** pp.42-44.

[16] Bazheva R Ch, Bashorov M T, Kharaev A M, Mikitaev A K 2007 Study of temperature transitions in polycarbonate and polycarbonate-polytetramethylene oxide block copolymers, *Chemical industry today* **8** p. 35-40.

[17] Bazheva R Ch, Kharaev A M, Bazhev A Z, Inarkieva Z I, Beslaneeva Z L 2016 Fire-resistant copolycarbonates, *Plastics 5-6* p. 26-30.

[18] Kharaev A M, Bazheva R Ch, Lukozhev R V 2014 Synthesis of polyesters based on oligosu-lfones containing a dichlorethylene group, *News of the Kabardino-Balkarian State University* **6** 62-68

[19] Bazheva R Ch, Kharaev A M, Inarkieva Z I, Beslaneeva Z L 2017 Copolycarbonates containing dichloroethylene groups in the main chain, *Plastics* **3-4** Pp. 32-35.

[20] Kharayev A M, Bazheva R C, Chayka A A 2006 *Polymers, Polymer Blends, Polymer Composites and Filled Polymers: Synthesis, Properties and Applications* 115-120.

[21] Kharaev A M, Mikitaev A K, Bazheva R Ch and others 2006 Modified flame retardant aromatic polyesters, *Chemical industry today* **8** 24-28.

[22] Kharaev A M, Bazheva R Ch, Chaika A A 2006 Composite materials based on polycarbonate *Plastics* **8** 26-31.

[23] Kyarov A A, Bazhev R Ch, Kharaev A M 2010 Compatibility study of polycarbonate-based composite materials. *In the collection: New polymer composite materials* (Nalchik) p. 34-38.

[24] Zapornikov V A, Osipchik V S, Redkina A A 2014 Influence of modifying additives on manufacturability and physicomechanical properties of polycarbonate // *News of higher educational institutions. Series: Chemistry and Chemical Technology* **57** 65-67.

[25] Petrova G N, Rumyantseva T V, Beyder E Ya 2013 The effect of modifying additives on the flame-retardant properties and manufacturability of polycarbonate, *Trudy VIAM* **6** 6

[26] Simonov-Emelyanov I D, Yurkin A A, Surikov P V, Shembel N N, Andreeva T I, Radzinsky S A, Zolkina I Yu, Amerik V V 2015 Evaluation of the effectiveness of the action of rheological additives in the processing of polycarbonate, *Plastics* **7-8** pp. 37-40.

[27] Zubkova O A, Lapova T V, Gorlenko N P, Sarkisov Yu S, Konoplyansky D A, Reznikov I V, Smirnov A P 2015 Improving the sustainability of polycarbonate products to the action of ultraviolet radiation, *Tomsk State University of Architecture and Civil Engineering Bulletin* **6** (53) Pp. 135-140.

[28] Kharaev A M, Bazheva R Ch, Chaika A A 2007 Composite materials based on polycarbonate (review), *International Polymer Science and Technology*, 34 **10** 27-33

[29] Kharaev A M, Bazheva R C, Barokova E B 2008 Fireproof aromatic block copolymer resin on the basis of 1,1-dichlor-2,2 di(n-oxyphenyl) ethylene, *Modern Tendencies in Organic and Bioorganic Chemistry: Today and Tomorrow* (New York) 219-222.

[30] Barokova E B, Haraev A M, Bazheva R C, Haraeva R A 2010 Oligosulfons on the basis of 1,1-dichlor-2,2- di(4-oxyphenyl)ethylene and 4,4'-dichlorodiphenylsulfone obtained by high temperature polycondensation, *Handbook of Condensed Phase Chemistry* 233-236.

[31] Nelyub V A 2016 A study of the microstructure of dressed glass fibers, *Polymer Science - Series D* **9** pp. 96-100

[32] Nelyub V A 2015 Determination of adhesion interaction between carbon fiber and epoxy binder, *Polymer Science - Series D* **8** (1) pp. 6-8.

[33] Nelyub V A, Borodulin A S, Kobets L P, Malysheva G V 2016 Capillary hydrodynamic of oligomer binder, *Polymer Science – Series D* **9**(3) pp.322-325

IOP Conf. Series: Materials Science and Engineering **683** (2019) 012051 doi:10.1088/1757-899X/683/1/012051

[34] Nelyub V A, Borodulin A S, Kobets L P, Malysheva G V 2016 A study of structure formation in a binder depending on the surface microrelief of carbon fiber, *Polymer Science – Series D* **9**(3) pp.286-289.
[35] Borodulin A S, Kalinnikov A N, Bazheva R C 2018 Receipt and investigation of performance characteristics of super constructions polyesters, *International Journal of Mechanical Engineering and Technology* **9** (13) pp. 1117-1127.
[36] Borodulin A S, Kalinnikov A N, Bazheva R C 2018 Synthesis and properties of aromatic polyethersulfones, *IJMET* **9** pp. 1109–1116.
[37] Borodulin A S, Kalinnikov A N, Kharaev A M 2018 New Polymeric Binders for the Production of Composit, *Materials today:proceedings* **11** pp. 1107–1111

Improving the fire resistance of PVC plastic the introduction ammonium octamolybdate

T A Borukaev[1], A M Kharaev[1], A Kh Shaov[1], A S Borodulin[2] and A N Kalinnikov[2]

[1]Kabardino-Balkarian State University, 360004, Nalchik, Chernishevskiy St., 173, Russian

[2] Intersectoral Engeneering Center "Composites of Russia", Bauman Moscow State Technical University, Moscow, Russian

E-mail: boruk-chemical@mail.ru

Abstract. An efficient method for the synthesis of ammonium octamolybdate with a quantitative yield has been developed. Compounds based on PVC plastic and octamolybdate are obtained. It was found that the introduction of ammonium octamolybdate in PVC plastic leads to an increase in the fire resistance of the compound and to a decrease in smoke formation during its combustion. At the same time, the mechanism for reducing smoke emission during PVC plastic combustion in the presence of ammonium octamolybdate consists in its catalytic action in the conversion of soot particles into various oxidized forms. It is shown that PVC plastics containing ammonium octamolybdate have elevated oxygen index and coke residue.

Keywords: PVC plastic, ammonium octamolybdate, synthesis, compounds, fire resistance, smoke suppression.

Introduction

One of the large polymer materials that is widely used in industry is polyvinyl chloride (PVC) [1, 2]. In particular, the volume of production of PVC and composites based on it in our country is more than 140 thousand tons per year [3]. At the same time, the fields of application of PVC and composites based on it are extremely extensive. Thus, PVC in significant amounts is used in the cable industry, where the plasticized polymer is used as insulation, filling and sheathing of electrical wires and cables [3, 4]. In the absence of flammable ingredients such as combustible plasticizers, PVC with a limiting oxygen index of about 37 does not ignite in air under normal conditions [2]. However, the majority of soft materials based on PVC, which are cable plastics, use different plastics, which leads to an increase in the flammability of these materials [5]. At the same time, the burning and smoldering of cable PVC plastic compounds is accompanied by the formation of a significant amount of smoke, which has a toxic effect on all living things in the environment [6]. In this regard, developers who deal with the problem of increasing fire resistance and reducing smoke generation during the burning of PVC plastic compounds face the problem of searching for and creating effective additives - flame retardants [7-11]. In this regard, molybdenum compounds can be quite promising and effective dyes suppressors [12-21].
The purpose of this work was to increase the fire resistance of PVC plastic, the introduction of ammonium octamolybdate, as well as to evaluate its effect on smoke formation during the combustion of plastic. The choice of ammonium octamolybdate rather than ammonium heptamolybdate is due to the

fact that thermal decomposition of ammonium octamolybdate occurs at elevated temperature, accompanied by the formation of a significant amount of coke.

Experimental

In this work, ammonium octamolybdate (OMA) was used as a flame retardant to increase the fire resistance and reduce the smoke produced during PVC burning. OMA was obtained according to the following procedure. Thus, ammonium heptamolybdate (GMA), purified from impurities, was dissolved in distilled water. Then the solution was heated to a temperature of 90-95 °C and an aqueous suspension MoO_3 was added to It, also heated to a temperature of 90-95 °C. At this temperature, the reaction mixture was stirred for 3–5 h until all of the molybdenum trioxide dissolved. The reaction scheme can be represented as follows: $(NH_4)_6Mo_7O_{24} + MoO_3 \rightarrow (NH_4)_4Mo_8O_{26} + 2NH_3\uparrow + H_2O$.

The resulting solution was evaporated to a volume of 300-400 ml using a rotary evaporator. Then the solution was left at room temperature until precipitation. After that, the precipitated white crystals were filtered off, washed with a small amount of water, and dried in a vacuum cabinet at a temperature of 70 ° C to constant weight.

Synthesized OMA in PVC-plastic was introduced by dispersing it at the stage of hot mixing (mixer DIOSNAD-49074 Germany) of the main components in the preparation of plastic [15]. Then the mixture was extruded using a twin-screw extruder with a screw diameter of 20 mm at a temperature of 140 - 165 °C. The content of OMA in plastic was 3÷10 mass. %. Then, the ex-laborers were granulated and used to make the appropriate test specimens.

The duration of burning compounds was determined according to GOST 21207-81 and UL-94. The test specimens were bars of 100×10×4 mm.

The oxygen index (OI) was determined by the GTS 12.1.044-89 on the instrument of the company Noselab ast. Fire testing model Oxygen Index (Italy). The samples were bars with a length of 100×10×4 mm.

Coke residue (CR) was determined as follows. Samples based on the original and OMA-containing PVC plastic were kept at a temperature of 800 °C in a muffle furnace for one hour. Then, the difference in the mass of sample weights was found before and after keeping in a muffle furnace, and the coke residue (%) was calculated.

To determine the amount of heat released and the specific volume of smoke produced during combustion of plasti cates, studies were performed on a calorimeter (made in Italy) according to ISO 5660-1: 2002. Compound samples (plates with dimensions of 150 × 150 × 2.0 mm) based on the original I40-13A, p.8/2 PVC compound and modified 10 wt.% OMA were exposed to a heat flux of 35 kW/m^2. The flame was at a distance of 6,4 mm from the sample.

Results and discussion

In the work to compare the efficacy of OMA used PVC plastic stamps 40-13A, p. 8/2 and a compound based on it and OMA. Studies have shown that PVC plastic I-40-13A, p.8/2 burns while in the flame zone (OI = 26%). During removal from the flame zone, the sample stopped burning after 10 seconds. At the same time, in the process of burning PVC plastic of the brand I40-13A, p.8/2, a significant amount of smoke is emitted. In turn, PVC plastic, containing OMA, when burning out of the flame zone, stopped burning almost immediately.

The increase in fire resistance of PVC plastic with the introduction of OMA is also confirmed by the determination of the OI values of the samples (see table). So it can be seen from the table that in PVC KI plastic compound increases markedly with the introduction of OMA. This behavior of the compounds is due to the decomposition of OMA and the release of non-combustible products into the flame zone [23].

It is important to note that in the process of burning in the fiery zone of PVC plastic compounds containing OMA, the amount of smoke is much less produced than in the case of burning PVC plastic of the I40-13A, p.8/2 type. The increase in fire resistance and the slight formation of smoke during the burning of PVC plastic compounds containing OMA are associated with changes in the process and mechanism of combustion of the material. It is known [22, 24], that molybdenum-containing compounds effectively influence the combustion processes of polymeric materials.

It should be noted that in the obtained composites, due to the decrease in the thermal effect of the combustion process and the catalysis of the coking process, the amount of coke residue after the destruction of PVC plastic compounds should be greater. As shown by the results of thermal studies of PVC plastic compounds containing OMA, the additive contributes to an increase in the coke residue (CO) in comparison with PVC plastic from the I40-13A brand, p.8/2 (see table).

The formation of a larger number of CR during combustion of samples was confirmed by photos taken from the surface of PVC plastics exposed to flame. The photos showed that during the burning of PVC plastic with the addition of OMA, the surface was covered with a dense coke crust. The resulting dense crust will prevent the flow of combustible products of destruction of plastic in the flame zone.

Table The influence of OMA on CR and OI PVC plastic

Indicator	The content of OMA in PVC plastic, mass. %				
	0	3	5	8	10
Coke residue, %	3	4	6	7	8
Oxygen index (OI), %	26	28	29	30	31

As noted above, a significant amount of plastic-fixer is used in PVC plastics [2, 4], which leads to an increase in the flammability of the material. So, when burning standard I40-13A PVC compound, a significant amount of smoke is formed.

Con-calorimetric studies have shown that the introduction of OMA in PVC plastic leads to a significant change in the entire process of burning material. In particular, to determine the effect of OMA on the resistance to burning of PVC plastic, a calculation was made of the rate of heat generation using the results of a calorimetric analysis. The results of research are shown in Fig. 1 and 2.

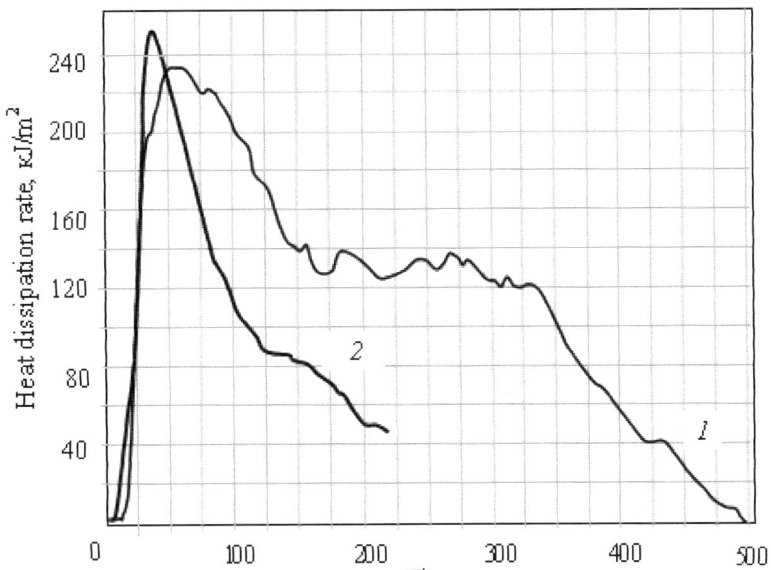

Fig. 1. The dependence of the rate of heat generation on the time of combustion of the samples: 1 - PVC plastic compound I40-13A; 2 - PVC plastic + 10% OMA.

From figure 1 it can be seen that the introduction of OMA in PVC plastic leads to a significant change in the rate of heat generation during the combustion of the material. In particular, at the initial moment of the combustion process of the initial PVC compound and the sample containing OMA, the heat-fission rate is almost the same. However, then the heat release rate for PVC plastic containing OMA is significantly reduced. Such behavior of PVC compound with OMA is explained by the fact that in the initial stage of material burning a solid coke crust forms on its surface, which effectively prevents the process of heat generation.

It is important to note that when samples are burned, the total amount of heat released for PVC plastic, containing OMA, is significantly lower than for standard plastic. This effective reduction of heat generation is obviously due to the fact that OMA is working mostly in the solid phase of plastic, catalyzing various processes. Firstly, the structural fragments of the polymer with the trans-configuration of macromolecules are formed [25]. The formation and presence of such sufficiently stable structures in PVC macromolecules inhibits the further process of polymer dehydrohalogenation, reducing the possibility of the formation of aromatic elements [26, 27]. Secondly, the processes of coking and charring are activated, leading to the formation of a crust-resistant crust. This is evidenced by the presence of a single peak on the heat release rate curve for PVC plastic with OHM (Fig. 1). Such a stable crust will effectively interfere with the process of heat generation and the output of various products of plastic decomposition. As a result, under equal conditions of combustion of two PVC plastic compounds, the initial sample (PVC plastic of the I40-13A brand, p.8/2), unlike the plastic compound containing OMA, will contribute to the further spread of the burning zone.

The results of an estimate of the amount of smoke produced during the burning of I40-13A, p.8/2 PVC and OMA plastic compound (Fig. 2) also indicate a reduction in smoke. As can be seen from Figure 2, the amount of smoke produced when burning plastic compound containing OMA is significantly lower than when burning PVC plastic of the brand I40-13A, p. 8/2. Such a decrease in the amount of smoke emitted during the burning of PVC plastic containing OMA confirms the effectiveness of the use of OMA. In particular, the reduction of smoke can be explained by the oxidative effect of OMA on soot particles [28], which are formed during the decomposition of the material in its combustion zone, i.e. OMA effectively demonstrates its ability to eliminate carbon particles of smoke by the reaction: $C + H_2O \rightarrow CO + H_2$.

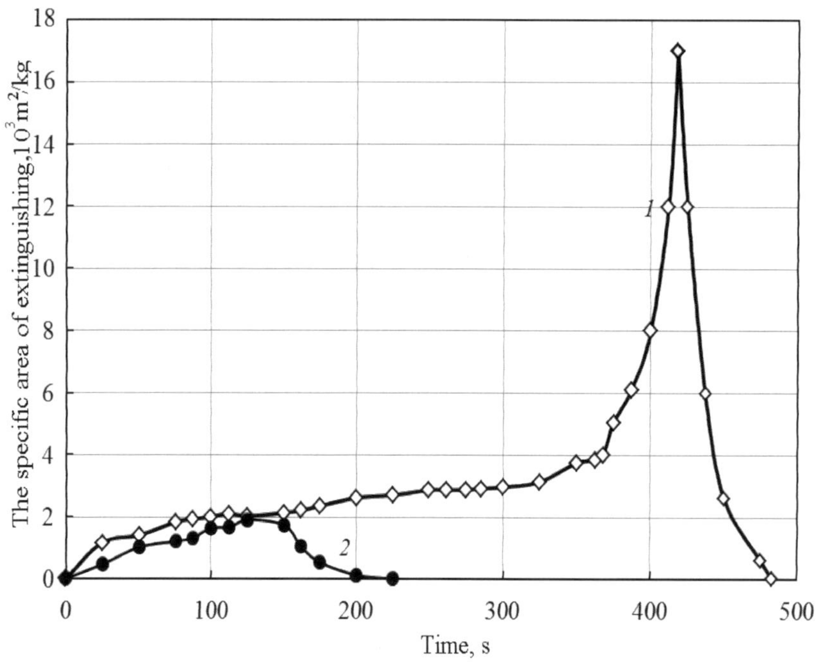

Fig. 2. The dependence of the rate of smoke emission on the time of combustion of the samples:

1 - PVC plastic board brand I40-13A; 2 - PVC plastic + 10% OMA the specific area of extinguishing

In this case, the oxidation processes, obviously, proceed through the metal carbonyl intermediate compounds. It is for this purpose that molybdenum is used in the form of OMA, which reacts at the required temperatures.

It should be noted that the insignificant amount of smoke that is formed during the burning of PVC plastic compound containing OMA (Fig. 2) has a significantly lower temperature by 15 - 20 °C, in comparison with the standard material.

Conclusions

Thus, the obtained with quantitative yield ammonium octamolybdate. In this case, the introduction of ammonium octamolybdate in PVC plastic compound leads to an increase in fire resistance, as well as to a decrease in the speed and amount of heat released. These changes in the properties of PVC plasticate are due to the fact that ammonium octamolibdate has a catalytic effect on the formation of structures of macromolecules resistant to dehydrohalogenation and a strong coke crust on the surface of the compound, which prevent the flow of combustible products of destruction of the material into the flame zone. n addition, octamolybdate ammonium can effectively oxidize soot is produced when burning PVC plastic, preventing the process of smoke generation.

Acknowledgments

This article is executed within the framework of the agreement No. 14.577.21.0241 from 26 September 2017 "Development of a new high-performance technology for producing heat-resistant dielectric superconstructive polymers of long service life", a unique identifier of works (project) RFMEFI57717X0241, funded by the Ministry of education and science of the Russian Federation

References

[1] Kiljachkov A 2014 PVC production in Russia: state and prospects, *Plastics* **134** pp. 42-47.

[2] Grossman R F and Guzeeva V V 2009 Guide to the development of compositions based on PVC (SPb: Scientific fundamentals and technologies) 614

[3] Mitkevich A S, Paverman N G and Elagina A N 2007 Cable compositions based on polyethylene and polyvinyl chloride *Development trends in Russia. Science and technology* **302** pp.3-7.

[4] Guzeev V V 2012 Strukture and Properties fille Polyvinylkhloride (SPb: Scientific basis and technology) 284

[5] Edgeriey P C and Oldland S R 1980 *HCL-Bildung beim Verbrenung von PVC Kunststoffe* **70** 217 - 221.

[6] Cherkasov O A 2014 The study of the physical and biophysical characteristics of the polyvinyl chloride in the combustion process, Bulletin of Saratov state technical University **75** 68-71.

[7] Ableev R I 2009 Actual problems in development and production of non-combustible polymer com pounds for the cable industry *Cable-news* **6-7** pp. 64 - 69.

[8] Sapaev Kh Kh, Musov I V, Mikitaev A K, Vindusheva A S and Khaschirova S Yu 2012 Polyvinyl chloride plastic compound with high fire resistance *High technology* **1** pp. 27-30.

[9] Gukepsheva L M, Tkhakakhov R B, Begretov M M and Tkhakakhov E R 2006 The effect of concentration and degree of decomposition of antipersonnel on the physical properties of PVC compounds *International Polymer Science and Technology* **6** pp. 13-14.

[10] Fomin D L and Deberdeev R Ya 2012 Influence of aluminum and magnesium hydroxides on the properties of PVC compounds *International Polymer Science and Technology* **12** pp. 47-50.

[11] Khaschirova S Yu, Borukaev T A and Sapaev Kh Kh 2012 Layered silicate nanocomposites based on polyvinyl chloride plasticate Mat. VIII International scientific- prakt. conf. *New polymer composite materials* (Nalchik) pp. 218-225.

[12] 1984 Fire resistance of PVC compositions with reduced smoke and gas emission: overview information M NITAI Khimprom,. 39

IOP Conf. Series: Materials Science and Engineering **683** (2019) 012052 doi:10.1088/1757-899X/683/1/012052

[13] Moor F and Ference R 2005 Molybdenum compounds as smoko suppressants AMAX Materials Research Center, pp.129-145.
[14] Borukaev T A and Alakaeva Z T 2018 The effect of zinc molybdate flame retardant and physico-mechanical properties of PVC Promising material **9** pp. 60-67.
[15] Nelyub V A 2016 A study of the microstructure of dressed glass fibers, *Polymer Science: Series D* **9** pp. 96-100
[16] Nelyub V A 2015 Determination of adhesion interaction between carbon fiber and epoxy binder, *Polymer Science - Series D* **8** pp. 6-8.
[17] Nelyub V A, Borodulin A S, Kobets L P, Malysheva G V 2016 Capillary hydrodynamic of oligomer binder, *Polymer Science – Series D* **9**(3) pp.322-325
[18] Nelyub V A, Borodulin A S, Kobets L P, Malysheva G V 2016 A study of structure formation in a binder depending on the surface microrelief of carbon fiber, *Polymer Science – Series D* **9** pp.286-289.
[19] Borodulin A S, Kalinnikov A N, Bazheva R C 2018 Receipt and investigation of performance characteristics of super constructions polyesters, *International Journal of Mechanical Engineering and Technology* **9** pp. 1117-1127.
[20] Borodulin, A S, Kalinnikov A N, Bazheva R C 2018 Synthesis and properties of aromatic polyethersulfones, *International Journal of Mechanical Engineering and Technology* **9** pp. 1109–1116.
[21] Borodulin A S, Kalinnikov A N, Kharaev A M at all 2019 New Polymeric Binders for the Production of Composit, *International Journal of Materialstoday:proceedings* **11**. pp. 1107–1111
[22] Borukaev T A, Shaov A Kh, Khasanov V V, Midova Z Kh, Evazova Z E and Shakova A A 2017 Increase of fire resistance of PVC-plastic by introduction of various flame retardants *Izvestiya Kabardino-Balkarian state University* **VII** pp. 15-19.
[23] Guzeeva T I, Krasilnikov V A, Andreev G G, Levashov A S, Voroshilov F A and Makarov F V 2004 Recovery of ammonium paramolybdate in hydrogen from the fluorine electrolyzer Izvestiya Tomsk Polytechnic University **307** pp.108-113.
[24] Borukaev T A, Shaov A Kh and Malkanduev Yu A 2017 Influence of molybdenum compounds on fire resistance and physical and mechanical properties of PVC-plasticate, *International Polymer Science and Technology* **11-12** pp. 35-40.
[25] Yanborisov V M and Borisevich S S 2005 The mechanism of initiation and growth of polyene se quences during thermal degradation of polyvinylchloride, *Polymer Science* **47** pp. 1478 - 1490.
[26] Fisch M H and Bacaloglu R J 1994 Degradation and stabilization of poly(vinyl chloride). II. Simulation of the poly(vinyl chloride) degradation processes initiated in the polymer backbone *Polym. Degrad. Stab.* **45** pp. 315-324.
[27] Bacaloglu R and Fisch M H 1995 Degradation and stabilization of poly (vinyl chloride). V. Reaction mechanism of poly(vinyl chloride) degradation, *Polym. Degrad. Stab.* **47** pp. 33-57.
[28] Chygrin P G, Lebuhova N V and Ustinov A Yu 2013 Structural transformations of $CuMoO_4$ in the process of catalytic oxidation of carbon, *Kinetics and catalysis* **54** pp.1-6.

IOP Conf. Series: Materials Science and Engineering **683** (2019) 012053 doi:10.1088/1757-899X/683/1/012053

Investigation of the scale factor and deformation features during compression of a spatially reinforced carbon composite material

K A Ponomarev[1] and I V Magnitskiy[1]

[1]Department of Carbon Materials, JSC "Kompozit", Korolev, Russia

Email: kirill.a.ponomarev@gmail.com

Abstract. An experimental analysis of the scale effects of the strength and rigidity of CCCM with a 4-DL reinforcement scheme under compression has been carried out. The influence of the size and shape of the sample on the observed tensile strength and modulus of elasticity of the material is described. The features of nonlinear deformation of the material, the effect of multiple loading are considered. The X-ray computed tomography method analyzed the processes occurring in the structural components of the material.

Keywords: CCCM, tensile strength, modulus of elasticity, pre-fracture, scale effect

1. Introduction

When designing structural elements for aviation and space technology products, an important task is to simulate the behavior of the materials used under thermal load conditions. Often, carbon-carbon composite materials (CCCM) are used for heat-loaded elements of such structures. Such composites have a heterogeneous structure on a scale comparable to the characteristic dimensions of the nodes of products. As a consequence, there are pronounced scale effects of strength and rigidity. An interest in modeling is also taking into account the nonlinear behavior of such materials and structural changes that occur during the loading.

The object of the analysis was a high-density CCCM with 4DL reinforcement pattern on a core frame with a characteristic rod size of 0.7 mm with a matrix of coke of coal tar pitch, past multiple high-temperature processing.

The effect of uniaxial compression stresses in the direction perpendicular to the plane of transversal isotropy of the material was considered.

The objectives of the paper were to assess the influence of scale factors on strength and elasticity characteristics of the CCCM during compression, as well as the study of structural changes of CCCM during loading, including cyclic loads.

2. Formulation of the problem

In assessing the influence of the shape and size of samples on the observed properties of the material with compression their various types were considered, a list of which is given in Table 1. The basic type of the sample was considered to be Group A, as the closest to those used in industry standards.

The study determined the change in strength and modulus of elasticity of the samples compared to the base type.

Table 1. Considered options for samples.

	Group Designation	Dimensions of the working part (mm)	
		cross-section	length
Prism 14 × 14 × 45 mm, tips filled with epoxy resin	Д	14 × 14	45
Prism 14 × 14 × 40 mm	Г	14 × 14	40
Prism 14 × 14 × 28 mm	А	14 × 14	28
Prism 14 × 14 × 20 mm	Б	14 × 14	20
Cylinder Ø 12 × 20 mm	В	Ø12	20
Prism 10 × 10 × 20 mm	Е	10 × 10	20
Cylinder Ø 8 × 14 mm	И	Ø8	14
Prism 7 × 7 × 14 mm	Ж	7 × 7	14
Paddle 20 × 20 × 56 mm	К	10 × 10	20
Paddle 20 × 20 × 40 mm	Л	10 × 10	0
Prism 20 × 20 × 40 mm w/hole	М	5 × 20 (2 parts)	0
Prism 14 × 14 × 28 mm	Н	14 × 28	14

Since the typical type of loading diagram for the material under consideration, as well as for a number of carbon composites [1, 2], is with a linear section, a pseudo-fluidity area and a strengthening section (see Figure 1), to identify the mechanisms of material destruction on the base-type samples 4 loading groups were performed: A1 - before fragmentation; A2 - prior to the start of the pseudo-fluidity platform (point I); A3 - before the start of the hardening section after the pseudo-fluidity site (point II); A4 - to point II for 15 cycles. After reaching the end point, the samples were unloaded, the values of apparent density, open porosity and thermal conductivity in the longitudinal direction were measured, if possible, and the tomogram was measured using an XT H 320 LC device. After that, the samples were re-loaded to failure, with measurement of tensile strength.

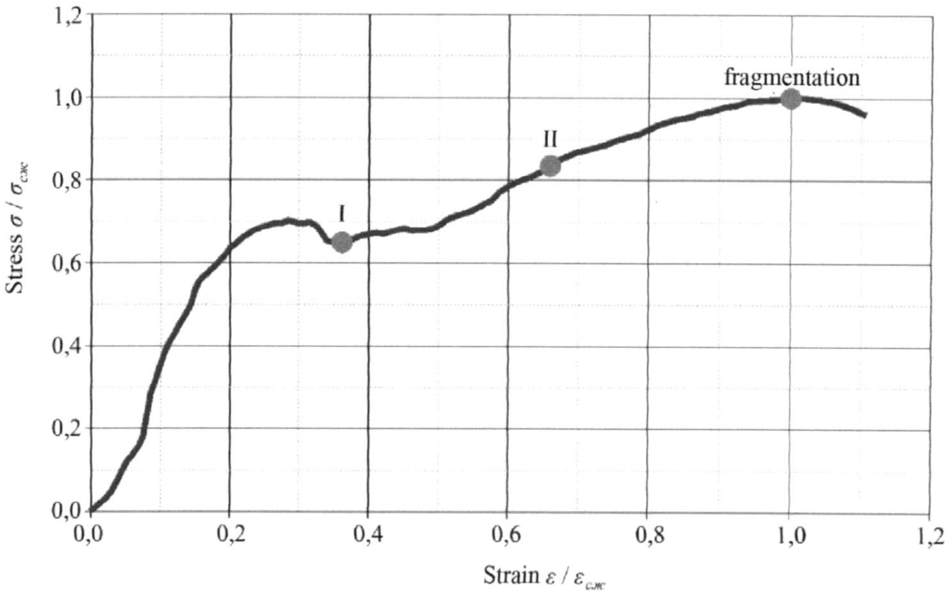

Figure 1. Characteristic compression strain pattern.

3. Results

The measured values of tensile strength and modulus of elasticity, referred to those for group A, are shown in Table 2. The indicated confidence intervals of the values are calculated for a confidence level of 95%.

Table 2. The results of measurement of tensile strength and modulus of elasticity.

Sample group	Area of working section (mm^2)	Minimum linear size (mm)	Length of the working part (mm)	Relative tensile strength, σ_i / σ_A	Relative elasticity modulus, E_i / E_A
Д	196	14	45	0.97	1.00
Г	196	14	40	0.99 ± 0.03	1.03 ± 0.12
А	196	14	**28**	**1.00** ± 0.02	**1.00** ± 0.04
Б	196	14	20	1.04 ± 0.04	1.04 ± 0.09
В	113	12	20	0.94 ± 0.03	1.01 ± 0.05
Е	100	10	20	0.89 ± 0.09	0.98 ± 0.14
И	50	8	14	0.80 ± 0.03	–
Ж	49	7	14	0.76 ± 0.03	–
К	100	10	20	0.91 ± 0.03	1.13 ± 0.07
Л	100	10	0	0.92 ± 0.02	–
М	200	5	0	0.74 ± 0.02	–
Н	392	14	14	0.75 ± 0.02	–

The results showed that the values of the ultimate strength do not depend on the method of transferring the load to the working part (through the supporting platforms / special cages with the sample ends filled with resin / directly to the ends of the working part), but only on the dimensions of its cross section and length. Thus, samples of groups E and K, having the same shape and dimensions of the working part, showed the same values of tensile strength. This fact is largely due to the high strength of the material under study when bent.

For groups of samples having the same shape and size of the cross section, the effect of reducing the tensile strength with increasing length of the working part is observed, typical of fibrous composites [1, 3–5].

The material is characterized by the dependences of the observed tensile strength both on the cross-sectional area of the sample and on its minimum linear size. The first is satisfactorily approximated by a logarithmic function, and the second - by linear.

However, the results for samples of group M (prisms with a hole) are determined only by the minimum linear size of the cross section and do not depend on its cross-sectional area. The priority of the minimum linear size of the section is indicated by the results on samples with its round and square shape. With comparable areas, square specimens with a smaller transverse linear dimension have smaller strength limits than circular specimens. On the contrary, the results for samples of group H (prisms with a working cross section 28 mm × 14 mm) are determined by the value of the working area, and not the minimum linear size. This fact may be a manifestation of a change in the mode of edge effects in the transition from small dimensions of the cross section to sizes that can be considered representative of the volume of the material [6, 7]. It should be noted that the dependence on the characteristics of the cross section of the working part is more significant than the dependence on the length of the working part.

No significant dependences of the elastic modulus and ultimate strain on the size of the working part were found. There is a slight increase in elasticity module values for the sample blades (with support sites), which may also be associated with manifestation of the edge effect.

The results of the study groups of samples A1-A4 are shown in Table 3.

Table 3. Characteristics of structural changes at different stages of deformation.

Sample group	Seeming flatness ρ_i / ρ_A	Open porosity Π_i / Π_A	Ratio of thermal conductivity at 50 °C λ_i / λ_A	Strength at reloading σ_i / σ_A
A (original)	1.00	1.00	1.00	1.00
A2 (point I)	1.00	1.06	1.01	1.04
A3 (point II)	1.00	1.05	0.95	1.07
A4 (point II, 15 cycles)	1.00	1.09	–	0.97
A1 (fragmentation)	1.00	0.96	–	0.55

After a single loading of the samples above the proportionality limit with repeated loading, a change in the shape of the deformation diagram and a decrease in the elastic modulus was observed, which is shown in Fig. 1 for samples of group A4.

For samples of group A2, a detachment or fragmentation of individual parts of the matrix was observed on the surface. In terms of volume, the number of visible delaminations between the longitudinal rods and the matrix increased, while the open porosity increased.

The samples retained their integrity and exhibited the original form of the strain diagram upon repeated loading.

Samples of group A3 were characterized by an increase in the number of bundles compared to the A2 group and the presence of individual longitudinal rods destroyed by shear cracks. At the same time, the open porosity remained at the level of the A2 group, thermal conductivity fell in the longitudinal direction, which is associated with the failure of a certain fraction of longitudinal rods. With repeated loading on the samples, the diagrams shown in Figure 2 were implemented.

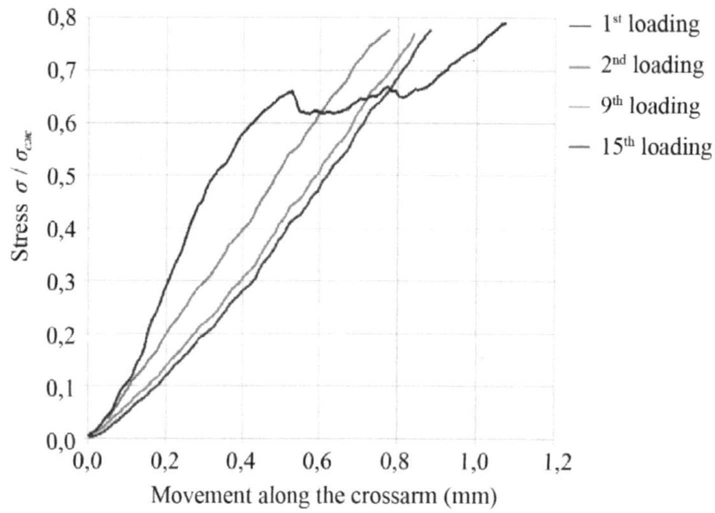

Figure 2. Multiple loading diagrams beyond the limit of proportionality.

The samples of group A1 were characterized by the presence of main shear cracks along the body of the sample, which were also observed separately on the rods upon their detailed examination. At the same time, the open porosity decreased in comparison with the A2 and A3 groups. Upon repeated loading, the samples showed lower values of the limiting stresses.

Compared with group A3, samples of group A4 were characterized by an increase in the number of locally broken rods and bundles. Open porosity is also increased. In general, it can be assumed that, initially, during compression, the mechanical connections between the longitudinal rods and the surrounding rods of transverse directions and a matrix that in the deformation diagram corresponds to a portion of the pseudo-fluidity area. In this case, the effective stiffness of the material is significantly reduced. This is followed by the destruction of the rods, apparently due to tangential stresses.

4. Conclusion

In general, for a spatially reinforced CCCM, the observed strength values do not depend on the method of transferring the load on the working part, but only on its cross-section and length. In this case, dependences both on the cross-sectional area of the sample and on its minimum linear size, and the dependence for the cross section turns out to be more significant than the dependence on the length of the working part.

When loading beyond the limits of elasticity in the material, there is a violation of the mechanical connections between the longitudinal rods and the rods surrounding them of transverse directions and the matrix, the rigidity is significantly reduced, followed by the destruction of the rods from tangential stresses.

References

1. Hatta H, Taniguchi K и Kogo Y 2005 Compressive strength of three-dimensionally reinforced carbon/carbon composite *Carbon* **43** pp 351–358
2. Suo T, Fan X, Hu G, Li Y, Tang Z и Xue P 2013 Compressive behavior of C/SiC composites over a wide range of strain rates and temperatures *Carbon* **62** pp 481–492
3. Pagano N J 1989 Interlaminar Response of Composite Materials *Composite Materials Series* **5** (Amsterdam: Elsevier)
4. Aubard X, Cluzel C, Guitard L и Ladeveze P 1999 Damage modelling of a 4D carbon/carbon composite for high temperature application *Ceramics International* **26** pp 631–637
5. Wang B L, Liang J и Liu Y 2008 Study on tensile and compressive properties and size effect of 3D fine woven punctured C/C composites Guti *Huojian Jishu/Journal of Solid Rocket Technology* **31** pp 184-187
6. Sutherland L S, Shenoi R A и Lewis S M 1999 Size and scale effects in composites: I. Literature review *Composites Science and Technology* **59** pp 209–220
7. Liu T, Sun B и Gu B 2018 Size effects on compressive behaviors of three-dimensional braided composites under high strain rates *Journal of Composite Materials* **52** pp 3895–3908

IOP Conf. Series: Materials Science and Engineering **683** (2019) 012054 doi:10.1088/1757-899X/683/1/012054

Technology of applying antifriction coatings using an electric arc pulsating power

I N Kravchenko[1], D I Petrovsky[1] and M A Glinsky[1]

[1] Russian State Agrarian University - K.A. Timiryazev Moscow Agricultural Academy (FGBOU VO RGAU-MSHA named after K.A. Timiryazev), Moscow, Russia

Abstract. The coating of a mechanical mixture of metal powders and solid lubricant material is due to difficulties due to the low adhesive and cohesive strength of the resulting coatings. This problem can be avoided when using in the process of the pulsating power electric arc (PPEA). The object of research in the paper is an anti-friction coating based on ПГ-ХН80СР3 powder with the addition of 20% by volume of graphite ПГ-50, of which plasma coating using PPEA gave a coating with friction coefficients *fmp min* = 0.038 and *fmp max* = 0.062 under contact load of 300 MPa. The test results of a friction pair, carried out for 500 hours on a path of friction of 55,000 m, showed the stability of the friction coefficient within *fmp* = 0.056 ... 0.061. The application of the coating process using the surface activation of the PPEA allowed us to obtain high-strength anti-friction coatings with the content of a solid lubricant element in the coating composition up to 20% of the total coating volume.

Keywords Antifriction coatings, plasma spraying, pulsating power electric arc (PPEA), graphite, clad materials.

1. Introduction

The process of applying coatings from a mechanical mixture of metal powders and a solid lubricant material is associated with certain difficulties. Such fine solid lubricants, such as graphite, molybdenite, boron nitride, copper fluorocyanide, etc., envelop the metal particles and prevent their solid connection between themselves in the coating composition and with the surface of the part.

The content of powder solid lubricant usually does not exceed 3-8% of the volume of the coating material. With a higher content of solid lubricant powder in the coating volume, its adhesion and cohesive strength sharply decrease, up to shedding the coating when it contains 12-15% solid lubricant. However, only at 15-30% of the content of solid lubricants unique anti-friction properties of solid lubricants appear [1-4].

Excellent material for creating antifriction coatings - graphite, has the ability at a temperature exceeding 10,000 K in an oxidizing atmosphere to interact with such materials as: nickel, cobalt, molybdenum, silicon, boron, titanium, niobium, or most carbides [5].

As carbon dissolves to form carbide compounds, the solubility of the elements decreases and some of the unbound graphite remains, firmly combined with materials [3].

Due to the high activation energy (Ea = 7.5 eV), chemical interaction of metals with graphite, with the formation of a sufficiently strong compound, can occur only in areas with the highest energy, causing thermal fluctuation of chemical bonds. The temperature of interaction of materials such as Ni, Cr, B, Ti with graphite is 1200-16000K [6].

2. Materials and methods

ПГ-ХН80СР3 powder was taken as the initial composition of the antifriction coating with dispersibility (63-80)·10-6 m. In the coating 20% by volume of graphite ПГ-50 was introduced with dispersibility (40-63)·10-6 m.

The heating temperature of the coating at the local points of attachment of the PPEA and the contact time of the phases were limited by the condition of the formation of a minimum thickness of the Ni – Ti intermetallic compound layer. The adopted maximum heating temperature is Tmax = 13,000 K.

The contact time of the phases determines the formation of a durable metallurgical compound of materials in the event that they have time to go through the processes of developing physical contact between the liquid and solid phases of materials and the chemical interaction of the phases. The duration of the contact phase tk is determined by the following equation (1):

$$t_к = t_p + t_д + t_{оэ} + t_{pxc} \quad (1)$$

where: t_p, $t_д$, $t_{оэ}$, t_{pxc} – respectively, the time of interphase surface energy retardation (the delay time of diffuse transitions); the time of development of heterodiffusion processes to the formation of supersaturated solid solutions; the time required for the formation of new chemical compounds; time of further growth of chemical compounds.

By limiting the contact time $t_к$ to only the first two terms of equation (1), the condition of joining materials is satisfied. The boundary of the molten volume of the coating material,

as shown by metallographic studies, is a spherical surface with a center in the spot of the binding of an electric arc on the surface [7].

The thickness of the coating *h* is much less than the thickness of the part *H* $(h << H)$. In this case, the entire thermophysics of the cooling process is determined by the material of the part.

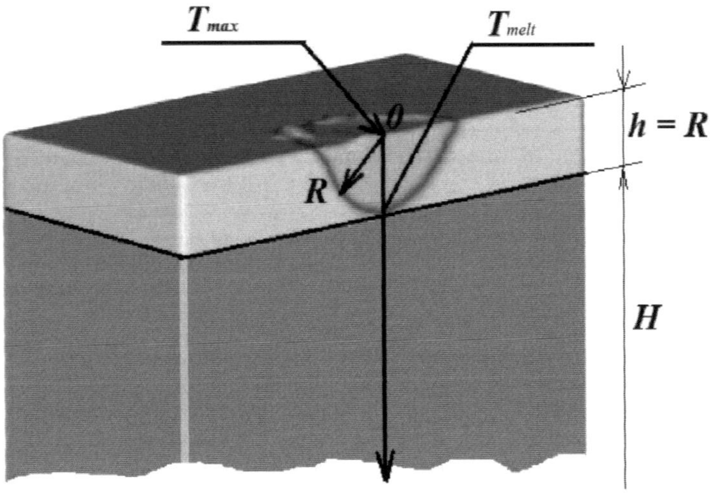

Figure 1. Melt bath cooling model

Since the molten bath of the coating material is a finite volume (Figure 1) bordering on a cold metal, we use the model of cooling a rod of finite length, the more so that the thermal conductivity of air $\lambda_{возд}$ is much less than the thermal conductivity of metal parts $\lambda_д$ ($\lambda_{возд} << \lambda_д$) and the losses to the surrounding the atmosphere can be neglected [8]. In this case, the heat distribution is isotropic; thermal processes in the coating can be neglected and the thermophysical parameters of the coating material at the melting point can be taken.

Thus, the task becomes one-dimensional, and on the basis of [5], we write the heat equation in the form (2):

$$\frac{c\rho}{\lambda}\frac{dT}{dt}=\frac{dT^2}{dx^2} \tag{2}$$

where: c, ρ, λ are the specific heat capacity, density, and thermal conductivity, respectively.
The initial and boundary conditions are determined by the following equations (3-6):

$$T\big|_{t=0}=T(X) \tag{3}$$

$$T\big|_{x=0}=T_{\max} \tag{4}$$

$$T\big|_{x=r}=T_{пл} \tag{5}$$

$$T\big|_{x=H}=0 \tag{6}$$

The final solution of equation (1) is presented in the form (7):

$$T(x,t)=T_{\max}\cos\frac{\pi d}{6r}x\exp\left(-\frac{D^2\pi^2\lambda}{36r^2c\rho}\right)t \tag{7}$$

From the obtained expression (7), it is possible to estimate the contact time $t_к$ of the phases of the material, provided that at the center of the molten coating volume (x = 0) the temperature ($T(0, t_к) = T_{пл}$) of melting is reached during cooling. From here we get the equation (8):

$$t_к=-\frac{36h^2c\rho}{\pi^2D^2\lambda}\ln\frac{T_{пл}}{T_{\max}} \tag{8}$$

The formation of a strong phase connection and the alternation of melted inclusions in a certain order given by the speed of movement of the plasma torch relative to the part and the frequency of current pulses of the PPEA contributes to the formation of a coating possessing the properties of a composite material.
Substituting in the expression (8) the values of thermophysical constants for the Ni material, the coating thickness is $5\cdot10^{-4}$ m (applied in one pass of the plasma torch), the value of the coefficient D= 0.2, we find that the contact time is $t_к = 10^{-2}$s.
The power of the pulse PPEA is calculated by the formula (9):

$$N=\frac{9,7T_{пл}\lambda h^2}{\sqrt{a\tau}\left[0,56\exp-(R/\sqrt{a\tau}\,2)^2-R/\left(2\sqrt{a\tau}\right)\left(1-erf\left(R/2\sqrt{a\tau}\right)\right)\right]} \tag{9}$$

and is N2 = 600 W at a frequency of 100 Hz. This value of power is allocated in PPEA at a current of I2 = 30 A at a distance from the substrate to the plasma torch nozzle section $(42\text{-}44)\cdot10^{-3}$ m. The relative velocity of the plasma torch in this case was taken equal to $v = 10^{-2}$ m/s [9, 10].

3. Results and discussion

The structure of the ПГ-ХН80СР3 coating with 20% ПГ-50 graphite is shown in Figure 2. The coating contains large inclusions of free graphite in a dense nickel matrix containing phases CrC, BC, SiC, hardness 192.4-212.8 MPa and a small number of nitrides.
The diffuse transition region from the coating to the titanium alloy was $(2\text{-}3)\cdot10^{-5}$ m, and the hardness of the transition zone is 131.4-152.4 MPa. The structure of the titanium alloy BT14-16 below the diffuse

transition zone did not change as compared with the exemplary structure, which confirms the conclusion about the allowable impact on the sample. The hardness of the titanium alloy was 36.7-44.7 MPa.

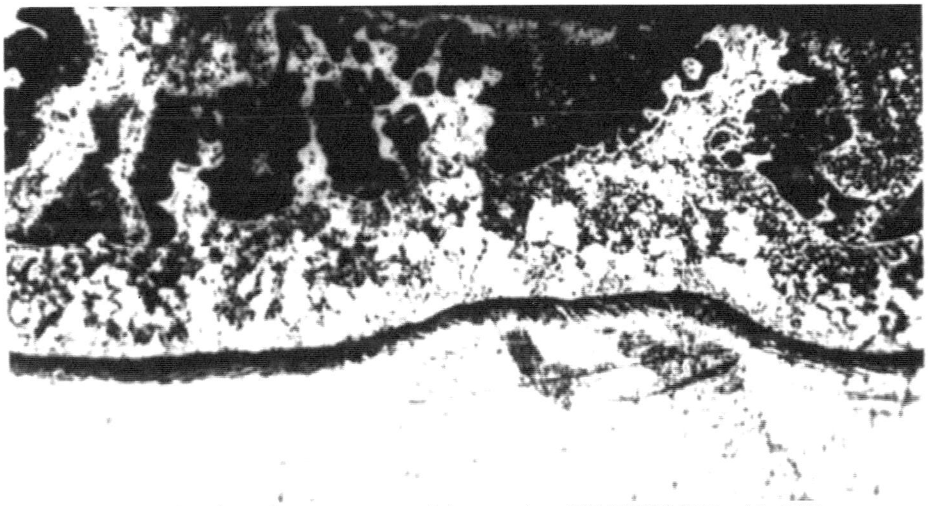

Figure 2. The microstructure of the coating ПГ-ХН80СР3 with 20% graphite applied using PPEA

The coating test was carried out at a contact load of 300 MPa on a MACT-1 friction machine. The following values of the friction coefficients are recorded: $f_{mp\ min} = 0.038$ and $f_{mp\ max} = 0.062$.
For comparison, a coating applied by the method of traditional plasma spraying without the use of PPEA was tested. A coating containing up to 10% by volume of graphite
when loaded with an indenter on a MACT-1 friction machine to a load of 0.17 MPa, was completely destroyed.

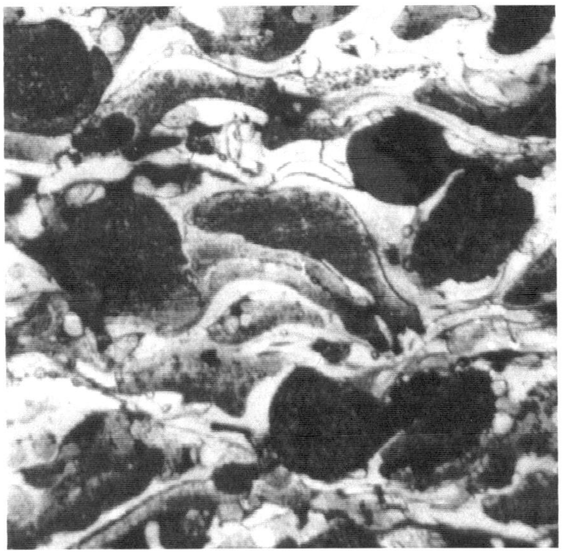

Figure 3. Coating microstructure Ni(C) + 20%Co (x 200)

From the analysis of the friction coefficient variation curves depending on time or on the friction path, selected for further testing of friction pair, in which the antifriction coating is Ni (C) + 20% Co material and the counter surface is Al_2O_3 + 50% ZrO_2 ceramics. The microstructure of the Ni (C) + 20% Co coating deposited using PPEA is shown in Figure 3.

The test results of this pair of friction, carried out for 500 h on the path of friction 55,000 m, showed the stability of the coefficient of friction within f_{rp} = 0.056-0.061. The wear of samples of a friction pair was determined on a profilogram — a model 362 profilometer and amounted to $\delta = 1.6 \cdot 10^{-6}$ m for an antifriction coating. At the same time ceramic coating wear is not detected.

The microstructure and characteristic macrostructure of the surface of the coating Ni (C) + 10% Co, applied using PPEA are shown in Figures 4 and 5.

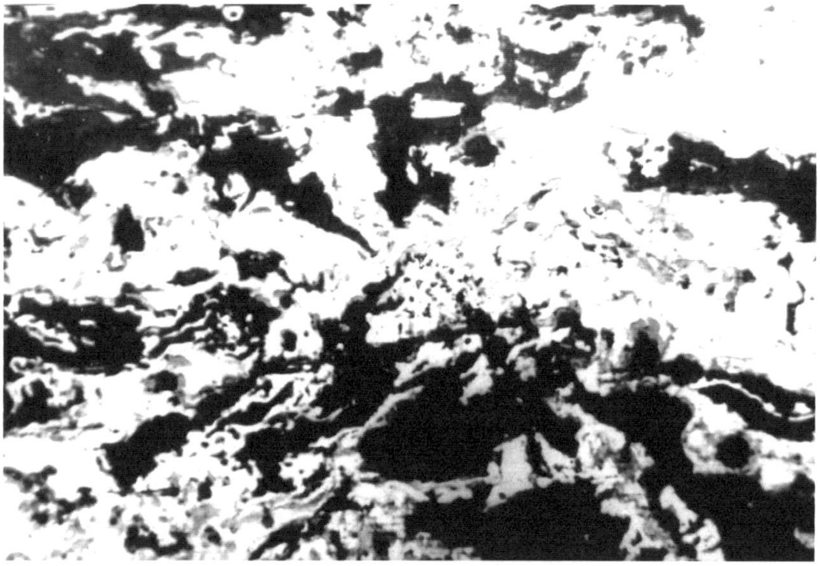

Figure 4. The microstructure of the coating Ni (C) + 10% Co, deposited using PPEA (x 200)

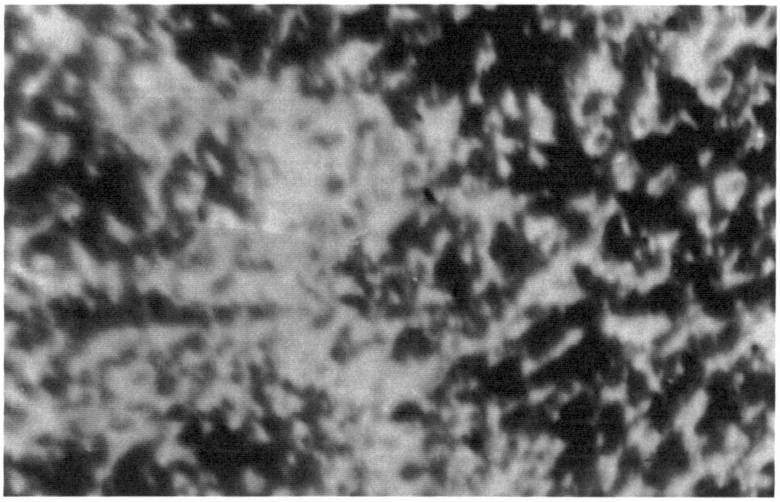

Figure 5. Characteristic surface macrostructure Ni (C) + 10% Co, applied using PPEA

IOP Conf. Series: Materials Science and Engineering **683** (2019) 012054 doi:10.1088/1757-899X/683/1/012054

The results of the analysis of the conducted studies show that the inclusions of graphite are fixed in the composition of the coating by dense metallic formations of Ni and Co. At the same time ceramic coating wear is not detected.

4. Conclusions.

1. Theoretical prerequisites for the application of high-strength antifriction coatings with a friction coefficient f = 0.038–0.062 were created.

2. The application of the coating process using the surface activation of the PPEA allowed us to obtain high-strength anti-friction coatings with the content of a solid lubricant element in the coating composition up to 20% of the total volume.

References

[1] Kutkov A A 1976 Wear-resistant and antifriction coatings (Moscow: Mashinostroyeniye) 152 p

[2] Kragelskiy I V and Alisin V V 1979 *Friction, wear and lubrication. Directory 2 books Book 1* (Moscow: Mashinostroyeniye) 358 p

[3] Matveevsky V M 1971 Temperature resistance of boundary lubricating layers and solid lubricating coatings during friction of metals and alloys (Moscow: Science) 228 p

[4] Marchenko E A et al. 1986 On the structural features and lubricating action of molybdenum diselenide.Coll. *Friction and wear in machines. Reports of the All-Union Conference,* (Chelyabinsk) 3–10.

[5] Puzryakov A F, Kravchenko I N, Sokolov I K et al. 2013 Technologies of applying protective and wear-resistant coatings of increased strength: monograph (Moscow: Publishing house "Eco-Press") 300 p.

[6] Rykalin N N 1985 High-temperature technological processes: Thermophysical basics (Moscow: Science) 171 p

[7] Zinovyeva T Yu 1999 Formation of plasma coatings when the surface is activated by an electric arc of pulsating power: Dis. ... Cand. Tech. Sciences (Samara) 154 p.

[8] Kravchenko I N, Puzryakov A F, Gladkov V. Yu. and Moskal. O.Ya. Method of impact of an electric arc of pulsating power on the adhesion strength of coatings International Journal of Applied and Basic Research **4** 14–19 URL: https://applied-research.ru/ru/article/view?id=3524.

9] Buryakin A V, Mikhailov A A, Burmistrova E E, Puzryakov A A and Puzryakov A F 2014 Patent No. 2503739 RU, IPC C23C 4/12. The method of applying coatings using an arc of pulsating power 2011142961/02 Appl. 25.10.2011 Publ. **1**

10] Kravchenko I N, Puzryakov A F, Kolomeichenko A A et al. Patent No. 150716 RU, IPC C23C14/34. Device for plasma spraying // no.2014136225/02 Appl. 08.09.2014 Publ. 20.02.2015 **5**

Engineering and technological methods of ultra-jet diagnostics of complex-profiled products made of composite materials

A A Barzov[1], V A Belov[2], A L Galinovsky[2] and I V Mazaeva[1]

[1]Lomonosov Moscow State University, Moscow, Russia

[2]FSEI HPE "Bauman Moscow State Technical University" (National Research University), Moscow, 105005, RF

Abstract An engineering technique for ultra-jet diagnostics of the state parameters of complex profiles surface layer of polymer composite materials which are obtained by winding or laying is proposed. The possibility of increasing the accuracy of obtaining express-diagnostics results under varying conditions of interaction of the hydro-ultra jet with the test surface is shown.

Keywords: ultra-jet diagnostics, composite material, mode change, express method, error model

Comparative analysis of diagnostic technologies

A distinctive feature of the formation of complex profiles of polymer composite materials (PCM) by winding or laying is the technologically-problematic provision of identical quality characteristics of the finished structure, especially the parameters of the surface layer in places of significant change in its geometric curvature. The analysis showed that the effective technology for obtaining comprehensive information about the parameters of the state of the surface layer of products from the PCM is ultra-jet hydrophysical diagnostics (UJD) developed in the relevant departments of "Bauman Moscow State Technical University". However, the use of the traditional scheme for the implementation of the UJD of complex profiles of the PCM is extremely difficult geometrically due to the change in the modes of diagnosis: variations in the angle of interaction of the ultrajet (UJ) with the surface under study and the variable distance between this surface and the cut of the formative hydraulic nozzle. These objective circumstances lead to the instability of the ultra-jet action mode (UJA) and, as a result, to the presence of significant "technological" noise in the results of the rapid diagnosis of PCM quality [1, 2].

Advantages of ultra-jet diagnostic technology

To eliminate the above considered functional deficiency of the UJD of products from PCM, an engineering diagnostic technique was developed, the distinguishing feature of which is the use of the model of the influence of deviations in the UJA modes on their information performance. The proposed algorithm for the implementation of the methodology and its appropriate software and mathematical software allow, at the first stage of experiments, to quantify the degree of information technology impact of various deviations of the UJA on their physical and technological performance. At the second stage, the data obtained are used for analytical correction of experimental UJ-diagnostics of a complex-shaped surface of a part or element of a PCM product [3, 4]. The complex of direct experimental studies showed the functional and diagnostic efficiency of the developed rapid methods, which, in general, significantly increases the effectiveness of the scientific and applied application of the UJD methods in ensuring the operational and technological quality of various composite products made of PCM, obtained primarily by winding or laying.

Acknowledgments

The studies were carried out within the framework of the RFBR grant 18-29-18081 "Evaluation of the properties of the interfacial zone of composite materials for rocket and space technology using the ultra-jet method" and a grant of the President of the Russian Federation for state support of the leading scientific schools of the Russian Federation НШ-3778.2018.8

References

[1] Barzov A A, Galinovsky A L 2009 Technologies of ultrajet processing and diagnostics of materials – ed MSTU. N.E. Bauman 246
[2] Abashin M I, Barzov A A, Galinovsky A L, Mazaeva I V 2015 Ultra-jet hydrodynamics, ed Lomonosov MSU faculty of physics 308
[3] Badanina J V, Barzov A A, Galinovsky A L, Sysoev N N 2018 Diagnostics of the physics and technology of heredity, ed Lomonosov MSU faculty of physics, 304
[4] Barzov A A, Belov V A, Sysoev N N, Fedyanin A A 2018 Hybrid physical and technological diagnostics of composite and structurally inhomogeneous materials, Preprint of the physics faculty of Lomonosov MSU **6** 44

IOP Conf. Series: Materials Science and Engineering **683** (2019) 012056 doi:10.1088/1757-899X/683/1/012056

Metallurgy of thixotropic materials: the experience of organizing the processing of structural materials in engineering Thixo and MIM methods

A B Semenov[1], A A Kutsbakh[1], A N Muranov[1] and B I Semenov[1]

[1]Interdepartmental laboratory «New methods and technologies of casting» of Bauman Moscow State Technical University, Moscow, Russia

Abstract. A brief review of the modern development of thixotropic metallurgy as a new direction in the field of science and technology of metal processing is offered.

Keywords: processing in the solid-liquid state, rheology, the thixotropy effect, microstructure and properties, the potential of industrial alloys.

1. Introduction

Starting from the 70s of the last century, aluminum alloys and features of the production of mold castings from them are the object of intensive research for materials scientists, metallurgists, and technologists, leading in a fundamentally new direction to the birth of metallurgy of thixotropic materials. Science, called rheology, has suddenly come to the fore - "a science that studies the fluidity of bodies, based on the characteristics of the medium viscosity and the rate of deformation" (P.A. Rebinder). Research by specialists of the prof. M. Flemings school (MIT, USA) of a specific metallurgical problem - hot cracks in castings, turning into a study of viscosity and the thixotropy effect in crystallizing metals, opened the way to the development of a whole group of technologies that received the generic name "thixoforming". Since 1990, this interest has been expressed in specially organized International Conferences "Processing of Metals and Composites in Solid-Liquid State" (S2P index).

2. Main part

In Thixo and MIM technologies, numerous studies by practitioners and theorists have so far focused on the transformation of empiricism and trial and error methods into the science of how to give a portion of the metal of the required, but not previously known additional quality — thixotropic properties. Prior to forming, the parts to be processed are given additional (functional) properties that correspond to the intended purpose: relaxation ability and fluidity under the influence of shear deformation, which is realized at the casting stage. The imparting of thixotropic properties to the material and the corresponding fluidity are achieved (Figure 1) by various methods and are characterized by a number of quantitatively regulated criteria. Formability is achieved by pre-relaxation of the prepared portion of solid-liquid material at the stage of filling the form.

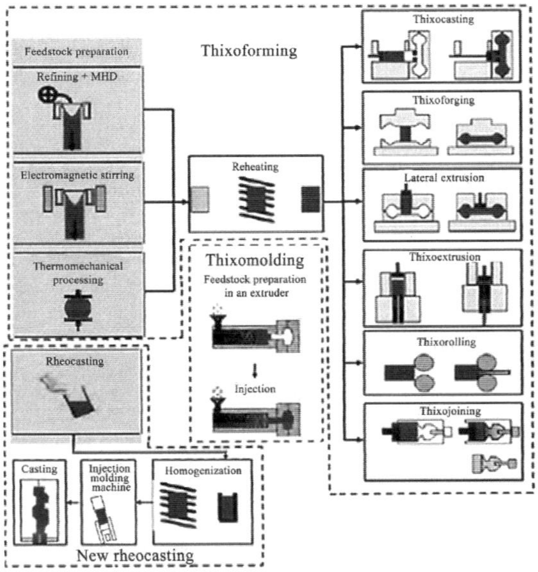

Requirement (determination method)	Parameter	Formula	Target value
Small grain size (metallography)	Mean grain diameter	$d_m = f(\dot{T}, c)$	dm < 100 μm
Globulitic structure (metallography)	Form factor	$f = \frac{4\pi A}{U^2}$	f > 0.6
Small cross-linking of the dendrites (metallography)	Contiguity	$C^{fS} = f^S \frac{2S^{SS}}{2S^{SS} + S^{SL}}$	$0.4 < C^{fS} < 0.6$
Small cross-linking volume of dendrites (metallography)	Contiguity volume	$V_C = V_S C_S$	$0.1 < V_C < 0.3$
Low viscosity of the melt under shear load (rheometer tests)	Viscosity under shear load	$\eta = k.\dot{\gamma}^{n-1}$	$\eta \approx 1-10\ cP$
Defined solid-liquid phase ratio (DTA, simulation, metallography)	F_{liq}	$f_S = 1-\left(\frac{T_M - T_L}{T_M - T}\right)^{\frac{1}{1-r}}$	40 - 60 mass. %
Large melting interval (DTA, simulation)	$dT = (T_{liq}-T_{sol})$	$dT = f(c, \dot{T}, p)$	max. 130 K min. 70 K
Temperature sensitivity (DTA, simulation)	Stop at 50% liquid portion	df_L / dT	≤ 0.015
Temperature sensitivity	Temperature sensitivity	$Df_S = 0.01\ (df_S / dT)\ T_{SS}$	≤ 0.06
Temperature sensitivity (DTA)	Temperature sensitivity	dT_{40-60}	> 10 K

Figure 1. Schemes for the implementation of thixoforming, thixomolding, new rheocasting and the criteria for the suitability of the prepared metal alloy.

Extensive experience in the organization of the processing of structural materials, and above all aluminum alloys by the methods of thixocasting and new rheocasting, has been accumulated in the global automotive industry with casting silumin Al-Si-Mg, Al-Si-Cu, Al-Si-Cu-Mg systems, where the leading positions were taken by analogs of the Russian AK7 alloy: ALTHIX A356 and ALTHIX A357 from Aluminum Pechinee and presented in figure 2 under the brand THIXALLOY five widely used in the global automotive industry analogs of this alloy offered by the company SAG (Salzburger Aluminum AG). The four added letters "THIX" means that all analogous variants of the Russian AK7 alloy, in the production of which primary Al is used and control over the Mg content is tightened, are prepared for processing in the solid-liquid state [1].

The required state of semi-finished products (feedstocks) ready for further processing is provided by the peculiarities of their structure, which must be controlled prior to the start of part molding. The main controlled factors of the structure of a similar preform are the morphological parameters of the solid α–Al phase (average grain diameter Dα, grain shape factor Fα, degree of skeletonization Cα).

The main condition for the choice of the working window of the molding process is the low sensitivity of the retained fraction of the structurally free liquid phase to temperature changes at the molding stage of the part.

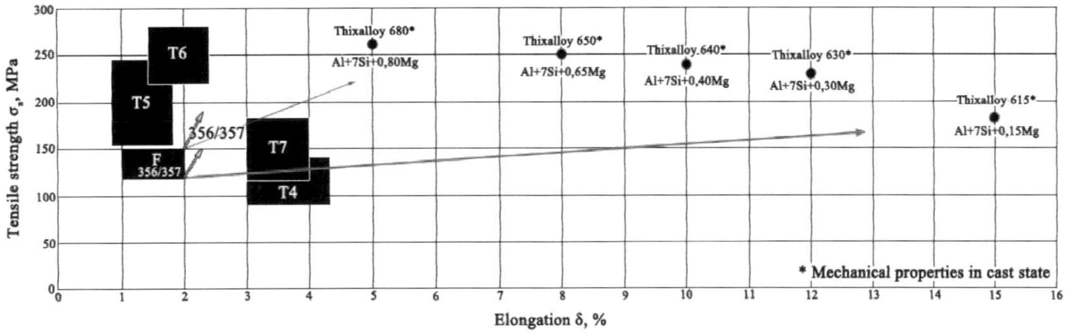

Figure 2. Analog-variants of the Russian AK7 alloy and their guaranteed mechanical characteristics in a cast state.

As a result of reliance on a new casting paradigm, a new quality of the final product is achieved. Below are the results of one of the latest publications by Chinese experts [2], in which a plate of the same size 30x100x300 mm from AZ91D alloy (Mg-8.8Al-0.6Zn-0.21Mn-0.04Si-0.002Fe) was made of the material of one heat, prepared for processing in three variants of the product molding process: liquid stamping, rheostamping, and thixostamping. Figure 3 shows the microstructures of the molded parts, and Table 1 shows the mechanical characteristics of the manufactured castings.

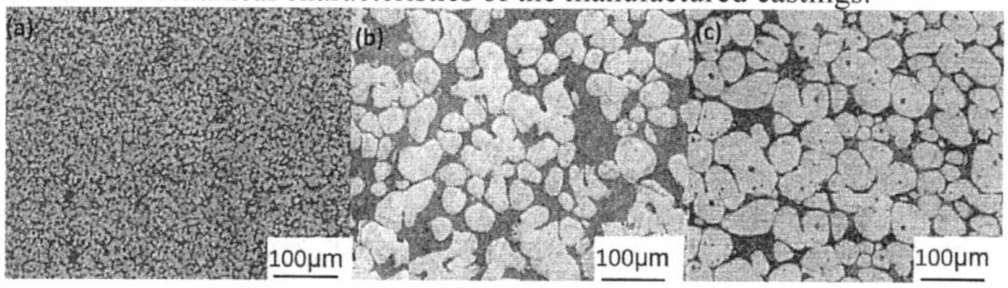

Figure 3. Microstructures of AZ91D Magnesium Alloy Parts of the same batch processed in various ways [2]: (a) liquid stamping; (b) rheostamping; (c) thixostamping.

In all cases, the microstructure of the part (Figure 3) consists of α-Mg and brittle β-Mg17Al12 of the eutectic phase, but the sizes, shape, quantities, and distribution of phases differ, which significantly affects the mechanical characteristics of the alloy in the part. Table 1 is supplemented by the properties of specimens from an alloy of the same brand AZ91D, made by thixomolding from metal powder, established in [3].

Table 1. Mechanical properties of AZ91D magnesium alloy castings of the same batch, processed by various methods [2, 3].

Process	$\sigma_{0.2}$, MPa	σ_B, MPa	δ, %
Liquid stamping	149	238	4,5
Rheostamping	152	269	4,7
Thixostamping	171	282	7,9
Thixomolding	155-165	230-235	3-6

Guaranteed mechanical properties change the proposed routes and solutions to the scientific and engineering problems of designing and producing castings. The successive stages of development in the preparation of industrial technology for the production of parts by the method of casting a solid-liquid metal [4] are preliminarily carried out with a virtual model of the product and look as follows (Figure 4).

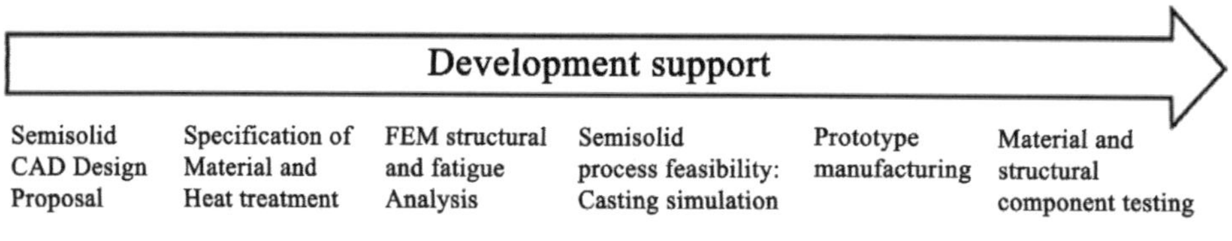

Figure 4. Stages of preparation for the introduction of the technology of mass production of parts by casting solid-liquid metal.

These requirements are the same for Thixo and for MIM technologies, in which the casting processes are based on the use of a general theoretical model of the flow of non-Newtonian fluids - the Herschel-

Bulkley model [5-10]. The most important factor in casting technology is the rheological properties of the workpiece; the main scientific task of the theory is to build a complete curve of the flow of the material of the workpiece for different cooling conditions; the main engineering problem of the theory is to establish such dependencies for industrially important casting systems (alloys) and test the suitability of flow models for specific conditions of a mold. The correctness of the choice of the casting method is established when testing the prototype parts

The technical and technological backwardness that emerged in Russia over the past decades is characterized by the following figures [11]: in industrialized countries, the production of castings from aluminum alloys exceeds the production of steel castings: in Italy - 10.7 times, in Japan - 7.6 times, in Germany - 4.2 times, in France - 3.6 times. In Russia, on the contrary, the production of steel castings exceeds the production of aluminum castings 2 times. The advanced foundries existing in Russia, such as Ford Sollers (Rosalit plant), Nemak (Ulyanovsk), Motordetal (Kostroma), Federal Mogul (Naberezhnye Chelny), work in narrowly specialized areas, satisfying a small fraction of the demand for cast automotive components for the Russian market.

A set of interdisciplinary knowledge to improve the quality of castings from standardized Al-alloys to the level required in the cast automotive components, the means-tested and recommended today for the industrial production of castings, according to Russian and foreign experts, they look different (Figure 5).

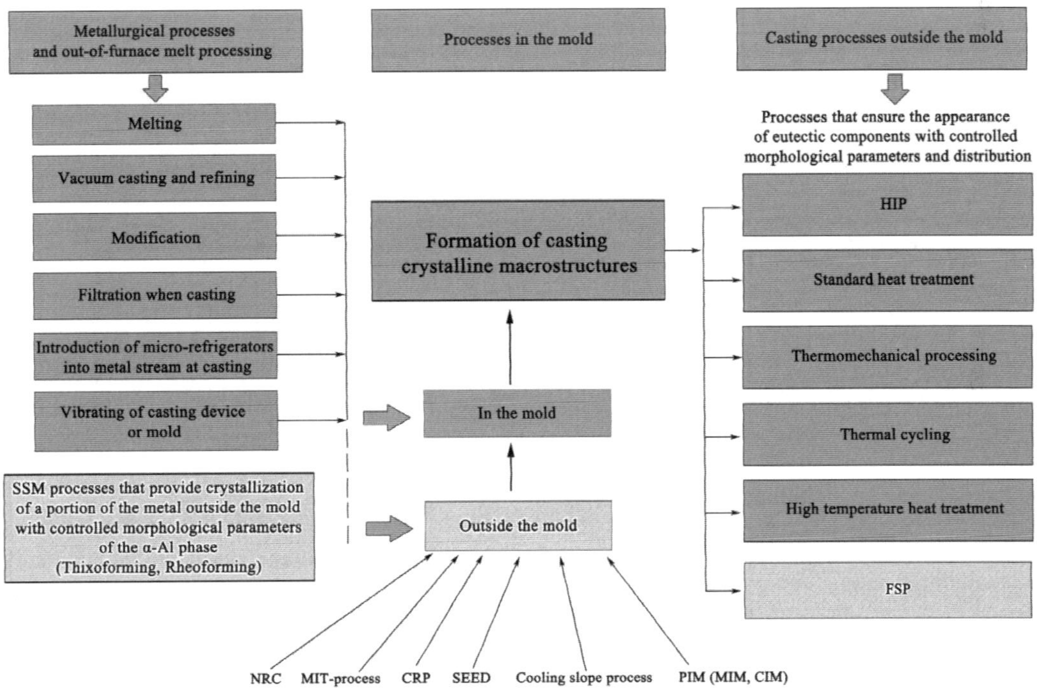

Figure 5. Modern industrial processes recommended for producing castings with desired properties from standardized Al-alloys.

According to the current curricula and reference books of Russian machine builders and foundry specialists, they are prepared only for the implementation of the processes highlighted in gray in the figure, with a complete lack of knowledge about SSM processes. Modern technologies are precision forming of metals in the solid-liquid state, i.e. in the range of solidification or melting temperatures. During molding, the material that remains in this state does not spontaneously spread, but, possessing thixotropic properties, quickly dilutes and flows, like a liquid, during shear deformation. Figure 6 shows most fully how the formation of the crystal structure of castings begins in the considered technologies. Numerous studies have shown [12–14] that when molding exact shaped parts from multiphase solid-liquid materials, the decisive condition for their processing is not the spheroidization of solid particles, but a quantity of structurally free liquid sufficient for yield. The structure of the material of the product is formed in advance almost completely (positions (b) - (d) in Figure 6) and can only evolve. The fluidity of such highly filled casting systems can be provided by its own liquid phase (eutectic) or melted low-temperature polymer mixture (PIM, MIM, CIM methods).

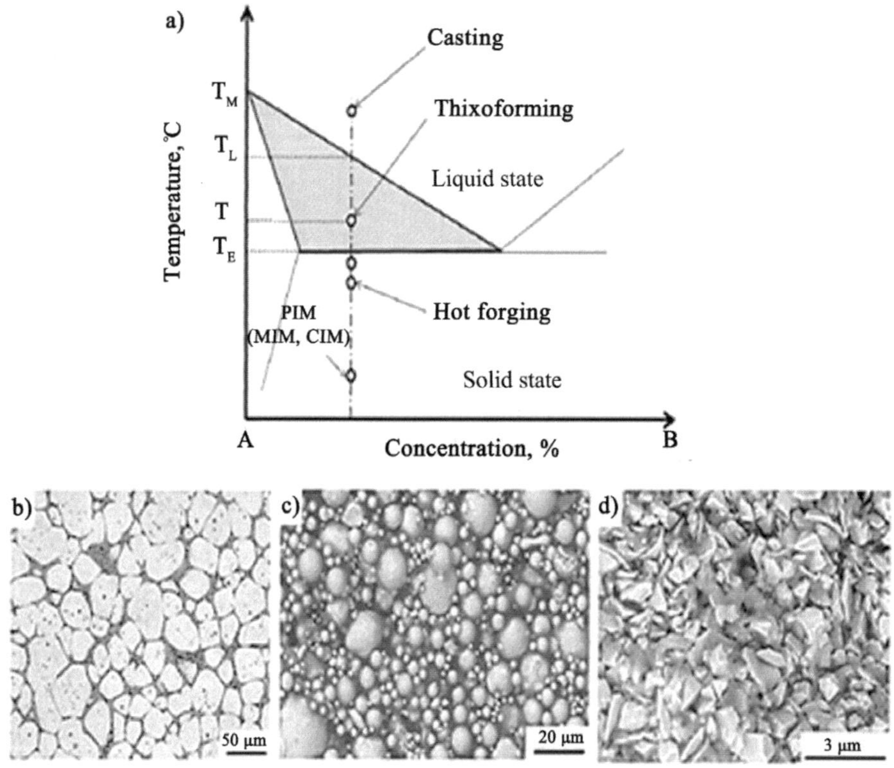

Figure 6. Casting conditions and the structure of thixotropic metals (a, b) and PIM workpieces – feedstock from small metal (c) and ceramic (d) powders

3. Conclusion

The development of competitive engineering foundry technologies in Russia, intended for subsequent commercialization in the production of metal products, composites, and ceramics in the solid-liquid state, is the imperative of the time. The accumulated backlog is manageable, but considerable efforts must be made in the field of training and retraining of engineering personnel. It is necessary to create modern competitive industries. The solution of such tasks is possible only within the framework of the creation of the Federal Target Program "Metallurgy of thixotropic materials"

References

[1] Semenov B I, Kushtarov K M 2010 *Manufacture of metal products in solid-liquid state. New industrial technologies: studies. manual for universities*. M .: Bauman Moscow State Technical University Publish House 223

[2] Tong W T, Yang H 2018 Microstructure and Mechanical properties of AZ91D Alloy by Three Types of Forging Process, *Proc. of 15th Int. Conf. on Semi-Solid Processings of Alloys and Composites (S2P 2018)* 22-24.10.2018 (China) 87-91.

[3] Zoqui E J 2014 Alloys for Semisolid Processing / *Comprehensive Materials Processing* ed by S Hashmi (Amsterdam: Elsevier Science & Technology) **5** pp 163-190.

[4] Winklhofer J 2018 Semi-Solid Casting of Aluminium from an Industrial Point of View // *Proc. of 15th Int. Conf. on Semi-Solid Processings of Alloys and Composites (S2P 2018)* 22-24.10.2018, Shenzhen, (China) pp 24-30.

[5] Semenov A B, Gavrilenko A E and Semenov B I 2016 Foundry technologies of the new generation, their exploration and development in Russia. Part 1. At the origins of the new technological paradigm *Metal technology* **4** 13-25.

[6] Semenov A B, Muranov A N, Semenov B I 2016 Next-generation casting technologies and their adaptation and development in Russia. Part 2-1. Physical nature and features of models of materials with thixotropic properties *Metal technology* **8** 8-17.

[7] Semenov A B, Gavrilenko A E and Semenov B I 2016 Next-generation casting technologies and their adaptation and development in Russia: I. at the beginning of a new technological paradigm / *Metally* **13** 1231–40. DOI: 10.1134/S0036029516130152.

[8] Semenov A B, Kutsbakh A A, Muranov A N, Semenov B I 2019 The development of methods for mathematical modeling of the process of injection molding polymers filled with metal powders / *Metal technology* **1** 9-16. DOI: 10.31044 / 1684-2499-2019-1-0-9-16.

[9] Favier V 2014 Modelling of Semisolid Flow / *Comprehensive Materials Processing* ed by S Hashmi Amsterdam: Elsevier Science & Technology **5** pp 191-209.

[10] Semenov A B, Semenov B I 2017 Two Paradigms of Technologies for Casting Metal Products, *News of universities. Engineering* **5** Pp. 79-98.

[11] Dibrov I A 2018 World production of aluminum casting. The state and prospects of the market for aluminum castings. *In the materials of All-Russia. conf. "Import substitution and export development of cast products from aluminum alloys"* (Moscow: RSPP) 13-14

[12] Patent DE 4314694 C1 Prepn. of sinter mouldings from a mixt. of a sinterable powder and an oxymethylene binding agent - involving removal of binding agent by treatment with a gaseous acid which is solid at room temp.

[13] Semenov A B, Semenov B I 2018 Current requirements and possibilities for preparing the structure of specialized blanks from Al-alloys, suitable for processing by Thixo- and MIM methods *In the materials of All-Russia. conf. "Import substitution and export development of cast products from aluminum alloys"* (Moscow: RSPP) 13-14

[14] Medvedovski E, Peltsman M 2012 Low pressure injection moulding mass production technology of complex shape advanced ceramic components, *Advances in Applied Ceramics* **111** 333-44. DOI: 10.1179/1743676112Y.0000000025.

Development of medical material based on nickel-titanium

R S Fakhurtdinov[1], I M Fedyuk[2] and E O Nasakina[3]

[1] Department of "Materials Science", Bauman MSTU, 5, 2nd Baumanskaya Str., 1, Moscow, 105005

[2] Department of "Materials Science", Bauman MSTU, 5, 2nd Baumanskaya Str., 1, Moscow, 105005

[3] Laboratory of Rare and Nonferrous Metals of IMET RAS; 49, Leninsky Prospect Str., Moscow, Russia, 119334

Abstract. The paper discusses methods for improving the biomechanical properties of titanium nickelide for its further use as a medical material. The aim of the paper was to study the processes of formation of oxide and nitride surface layers on titanium nickelide by the methods of oxidation and nitration, as well as the effect of these treatments on the mechanical properties and biocompatibility of the alloy.

Keywords:Nickel titanium, nitinol, composite materials, cava filters, stents, oxidation, nitriding.

1. Introduction

The most famous material with shape memory effect (SME) is titanium nickelide [1–3]. This alloy has proven itself as a medical material because of its physical and mechanical properties, close to the tissues of the body, as well as due to good biocompatibility.

Titanium nickelide is an equiatomic intermetallic compound with 55 at% Ni. The ratio of nickel and titanium can be varied to achieve the desired transition temperature. A nonstoichiometric oxide shell, which arises due to the reaction of titanium and a large number of oxygen atoms at the grain boundaries of the alloy, protects the material from general corrosion.

However, the high content of nickel that is toxic to a living organism in the alloy and the probability of its release from it as a result of corrosion require an increase in the corrosion resistance of titanium nickelide and its biocompatibility [4–11]. Studies show that the chemical and phase composition of the alloy surface largely determines the corrosion resistance of titanium nickelide.

2. Materials and methods for their preparation

There are various methods for solving the problem of biological compatibility and corrosion resistance of titanium nickelide: creation of organic and inorganic coatings, thermal and chemical-thermal treatment, alloying [12-17]. Among them, the formation of natural surface layers, titanium oxide and nitride, can be emphasized.

The purpose of this paper is to study the processes of formation of oxide and nitride surface layers on titanium nickelide and their influence on the mechanical properties of the alloy and its biocompatibility.

For the study, a wire with a diameter of 280 μm was obtained from nanostructured titanium nickelide of composition 55.91 wt% Ni - 44.03 wt% Ti. The wire was obtained by the method of multiple stepped hot drawing through a synthetic diamond die. Intermediate heat treatments were performed between hot drawing operations to stabilize the structure of the material and eliminate mechanical stresses [18–20]. The preparation of the wire surface was carried out mechanically - grinding and subsequent polishing with GOI paste to a mirror surface.

An oxide surface layer was obtained by annealing in an air environment of a LOIP LF series muffle furnace, model 7/13-G2. The wire was subjected to annealing in the initial state after drawing and after machining. The test samples were placed in an oven heated to an annealing temperature of 450°C, held for 15 minutes, and then cooled in air.

Nitriding of titanium nickelide was carried out in a nitrogen atmosphere of a graphite hot pressing furnace Thermal Technology Inc. high temperature experts, model HP20-3560-20. The samples were placed in a cold oven, which was heated to a nitriding temperature of 1150°C at a rate of 10 K/min. One sample was cooled with the furnace without exposure, and the other - before cooling, kept for five minutes.

In the initial state, after drawing, the surface of the wire is rough and non-uniform, the light and dark areas alternate. It was determined that a large amount of carbon is present in the dark area, which is explained by the use of carbon lubricant in drawing, and oxide and titanium nitride with various impurities are present in the light area.

The layer-by-layer chemical analysis of the samples after oxidation and nitriding was performed by Auger electron spectroscopy on a JAMP-9500F Auger spectrometer by JEOL.

After grinding and annealing, the surface of the wire becomes homogeneous. According to chemical analysis data, an increased (up to 40%) oxygen concentration and a reduced nickel concentration are observed in the near-surface layer. At a distance of 25–50 nm from the surface, nickel is practically absent (Figure 1). The change in the concentration of oxygen and nickel extends to a thickness of 100-150 nm. There is a slight increase in the concentration of carbon, which fell into the composition of graphite grease during hot wire drawing. It can be assumed that as a result of oxidation, a layer with a thickness of 100-150 nm is formed on the surface of the sample based on dense titanium oxide.

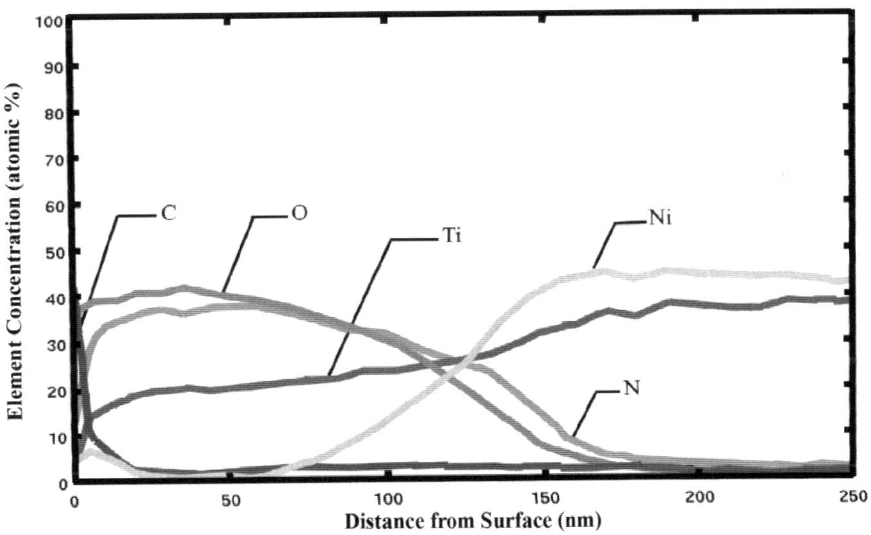

Figure 1. Layered composition of titanium nickelide after grinding and annealing.

IOP Conf. Series: Materials Science and Engineering **683** (2019) 012057 doi:10.1088/1757-899X/683/1/012057

After nitriding, there is a zone with a high nitrogen concentration, which reaches a thickness of 1500 nm (Figure 2). The maximum concentration of nitrogen on the surface is 50%, and at a distance of 200-800 nm - it reaches 60-70%. Nickel is absent on the surface and up to a distance of 750 nm from the surface and oxygen is practically absent. Thus, during nitriding of titanium nickelide without exposure, the formation of a nitride layer consisting of titanium nitride with a thickness of 1500 nm is observed, nickel is pushed to a depth of 750 nm. After nitriding with a five-minute exposure, an increase in the nickel layer of an additional 50 to 100 nm is noted.

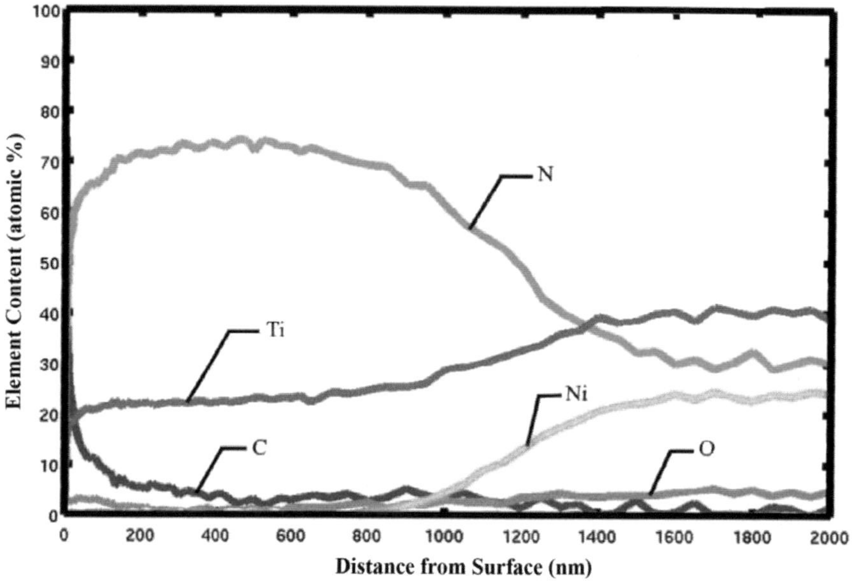

Figure 2. Layered composition of titanium nickelide after nitriding without exposure.

Mechanical properties were determined by standard methods on an Instron 3382 Instron testing machine. Table 1 shows the main mechanical properties of samples of nickel-titanium after various technological operations.

Table 1. Mechanical properties of samples after different technological processing.

Processing type	Strength limit σ_B, MPa	Yield point $\sigma_{0.2}$, MPa	Elongation δ,%
After drawing	1301 ± 10	640 ± 5	10± 0,05
Grinding	1358 ± 10	742 ± 5	20.3 ± 0.05
Grinding + oxidation	1773 ± 10	345 ± 5	13.03 ± 0.05
Nitriding without curing	1533 ± 10	313 ± 5	10.52 ± 0.05
Nitriding + curing	1533 ± 10	438 ± 5	10.17 ± 0.05

After oxidation and nitration, as compared with the sample after grinding, the tensile strength increases by 30.5% and 12.8%, respectively, the yield strength decreases by 53% and 40.9%, respectively, and the elongation decreases by 35.8% and 48% respectively. However, the mechanical properties of oxidized and nitrided samples meet the requirements for materials for the manufacture of medical devices ($\sigma_B \geq 900$ MPa; $\sigma 0.2 \geq 300$ MPa; $\delta \geq 10\%$).

3. Conclusions

Thus, during oxidation and nitriding, a diffusion layer is formed containing titanium oxides and nitrides. It was established that at the selected modes an oxide layer of 150 nm is formed during oxidation and a nitride layer with a thickness of 1500 nm during nitriding. Nickel is pushed into the depth of (100 - 150) nm with oxidation and to a depth (750 - 850) nm at nitriding. The absence of nickel on the surface of oxidized and nitrated titanium nickelide suggests that this will contribute to a better biocompatibility between a medical device made from such a material and the body.
It has been established that the mechanical properties of titanium nickelide wire after surface treatments are at the level of material requirements used for the manufacture of medical devices.

References

[1] Gunter V O, Khodorenko V N, Yasenchuk Yu F and Chekalkin T L 2006 Nickel Titanium. New generation medical material (Tomsk: Publishing House MIC) p 296

[2] Zabolotnyi V T, Belousov O K, Palii N A, Goncharenko B A, Armaderova E A and Sevost'yanov M A 2011 Materials science aspects of the production, treatment, and properties of titanium nickelide for application in endovascular surgery *Russian metallurgy (Metally)* **2011** № 5 pp. 437-448

[3] Nasakina E O, Baikin A S, Sevost'yanov M A, Kolmakov A G, Zabolotnyi V T and Solntsev K A 2014 Properties of nanostructured titanium nickelide and composite based on it *Theoretical Foundations of Chemical Engineering* **48** №4 pp. 477-486

[4] Taylor and Francis 2009 *CRC Handbook of Chemistry and Physics* ed D R Lide 90th edition (CRC Press) p 2828

[5] Lazarev N V and Gadaskina I D 1977 Harmful substances in industry: *Directory* (L.: Chemistry Volume 3) 608 p.

[6] Umarova O Zl and Kozuyutov A V The study of the resistance to crevice corrosion of alloys

based on titanium and titanium nickelide *FSBEI HPE "MATI - Russian State Technological University named after K E Tsiolkovsky*, p. 275-277

[7] Yu S Zhukova 2012 Kinetic laws, the mechanism of formation of nanoscale passive films on titanium alloys for medical purposes and their electrochemical behavior in model physiological media, ed. Zhukova, Yu S, Pustov, Yu A and Filonov M R (Physicochemistry of the surface and protection of materials t 48) No. 3 with. 267–273

[8] Muslov S A, Yarema I V and Savchenko A A 2007 Corrosive behavior of titanium nickelide in bile. *Fundamental research* No. 10 P. 85–86

[9] Starosvetsky D and Gotman I 2001Corrosion behavior of titanium nitride coated Ni–Ti shape memory surgical alloy *Biomaterials* **22** №13 pp. 1853-59

[10] S Volchkov, I Shishkovsky and Bayrikov I M 2013 Effect of porous three-dimensional implants from nitinol on the culture of multipotent mesenchymal stromal cells *Cell transplantation and tissue engineering* 8 No. 1 p. 51–56

[11] Bystrov S V 2007 A new seamless plastic technique of the main veins by a porous nitinol implant in the experiment. *Siberian Medical Journal* 22 No. 2 p. 40–44

[12] Sun Z L Effects of metal ions on osteoblast-like cell metabolism and differentia-tion Z L Sun, J C Wataha and C T Hanks 1997 *J. Biomed. Mater. Res. B: Appl. Biomater.* **34** №1 pp. 29-37

[13] Zheng C Y Enhanced corrosion resistance and cellular behavior of ultrafi-negrained biomedical NiTi alloy with a novel SrO–SiO2–TiO2 sol–gel coating Zheng C Y, Niea F L, Zheng Y L et al. 2011 *Appl. Surf. Sci.* **257** №13 pp. 5913-18

[14] Li H, Yuan B, Gao Y, Chung C Y and Zhu M Remarkable biocompatibility enhancement of porous NiTi alloys by a new surface modification approach: In-situ nitriding and in vitro and in vivo evaluation 2011 *Journal of biomedical materials re-search* **99A** №4 pp. 544-553.

[15] Neelakantan L, Swaminathan S, Spiegel M, Eggeler G and Walter Hassel A. Selective surface oxidation and nitridation of NiTi shape memory alloys by reduction annealing 2009 *Corrosion Science* **51** pp. 635-641.

[16] Sevostyanov M A et al. Kinetics of release of antibiotics from biodegradable chitosan-based biopolymer membranes 2015 *Reports of the Academy of Sciences* 465 No. 2 p. 194 - 197. [17] Nasakina, E O, Sevostyanov, MA, Demin, K.Yu., Mikhailova, AB, Goldberg, MA, Kolmakov, AG and Zabolotny V T Using the method of ion-atomic deposition to create one-dimensional composites 2015 *Materials Science* №1. 47-51

[18] Zabolotny V T, Goncharenko B A, Kolmakov A G and Sevostyanov M A
Development of technology for obtaining promising medical nanomaterials with a memory effect of the shape and products from them 2009 *Fundamental sciences in medicine: Abstracts of reports at conferences and seminars on the scientific areas of the Program in 2009* (Moscow: Slovo Company) p. 212-214

[19] Sevostyanov M A, Kolmakov A G, Goncharenko B A, Zabolotnyy V T, Shkurin S A and Asmolova Yu.O. 2010 *Study of the mechanical properties of TiNi with nanophase structure used in medicine Physico-chemistry of ultrafine (nano-) systems* IX All-Russian Conference, Izhevsk ,

IOP Conf. Series: Materials Science and Engineering **683** (2019) 012057 doi:10.1088/1757-899X/683/1/012057

November 22–26, 2010 (Rostov-on-Don: Publishing House of the Southern Federal University) p. 191 - 192.

[20] Goncharenko B A, Sevostyanov M A, Kolmakov A G et al. 2011 Getting Nitinol with Nanostructure for Medical Products NANO 2011: Collection of materials of the IV All-Russian Conference on Nanomaterials, Moscow, March 01-04, 2011 (Moscow: IMET, RAS) p. 437

Design and process for production of spacecraft structures using radial braiding and transfer molding

V I Khaliulin[1], V V Batrakov[1], V V Savitskiy[2]

[1]Kazan National Research Technical University named after A.N. Tupolev – KAI (KNRTU-KAI), 15, Tolstoy St., Kazan, 420111, Russia

[2]ISS-Reshetnev Company, 52, Lenin St., Zheleznogorsk, Krasnoyarsk region, 662972, Russia

Abstract. The paper presents the results of research dedicated to design and development of spacecraft structures using radial braiding and transfer molding. Experience in development, optimization and manufacturing of structures with closed section and biaxial and triaxial reinforcement is described for bearing rod, X-fitting, support platform, transformable lattice structure and 'capillary' tube.

1. Introduction

Generally, new material processing technologies offer a promising potential for improvement of the quality of structural design. As for composite parts, it is achieved via efficient reinforcement, promotion of integrity of the structure and sometimes via new structural design. As a result, there is a mass-efficient structure.

These days radial braiding is state of the art trend in composite industry [1, 2, 3]. Experience of its application in composite structures production is rather small, and still there is a wide range of capabilities to explore.

The aim of the study, presented in this paper, is to explore the potential of radial braiding and transfer molding (Figure 1) to improve spacecraft structures.

Radial braiding is used to produce preforms from dry reinforcement material that later will be impregnated using one of transfer molding methods: infusion, RTM, Light RTM or RFI [4].

First of all, it is efficient to use braiding for production of parts with closed section. And from this point of view, braiding is an alternative to winding with additional range of options.

2. Thermally stable tubular structures

Main advantage of radial braiding is, first of all, a possibility to produce preforms with biaxial and triaxial reinforcement (Figure 1 (b), (c)). An angle of reinforcement about the x-axis of the part may vary in a wide range. Besides, various weave patterns in the nodes may be organized (Figure 1 (d)).

Complex design and processing challenges may be dealt with owing to these processing capabilities, such as production of thermally stable structures. Figure 2 (a) shows tubular rod of a subreflector that has to meet stringent requirements when it comes to mechanical properties and minimization of deformation in a wide temperature range.

According to strength, stiffness and thermal stability requirements a set of rods was designed and produced (Figure 2 (b)). Essentially, these rods are thin-walled tubes with integrated flanges.

Mechanical properties were achieved via adjustment of components of the composite material and mainly by selected reinforcement pattern. Loading simulation and simulation of the structure under temperature were performed analytically and using finite element method to determine required properties [5, 6].

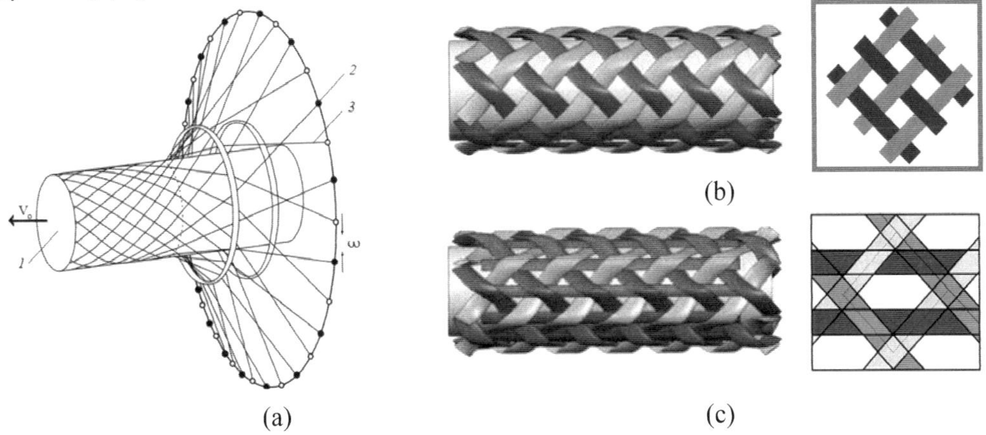

Figure 1. Radial braiding scheme (a), biaxial (b) and triaxial (c) braiding patterns: *1* – mandrel; *2* – spindle; *3* – reinforcement material (roving).

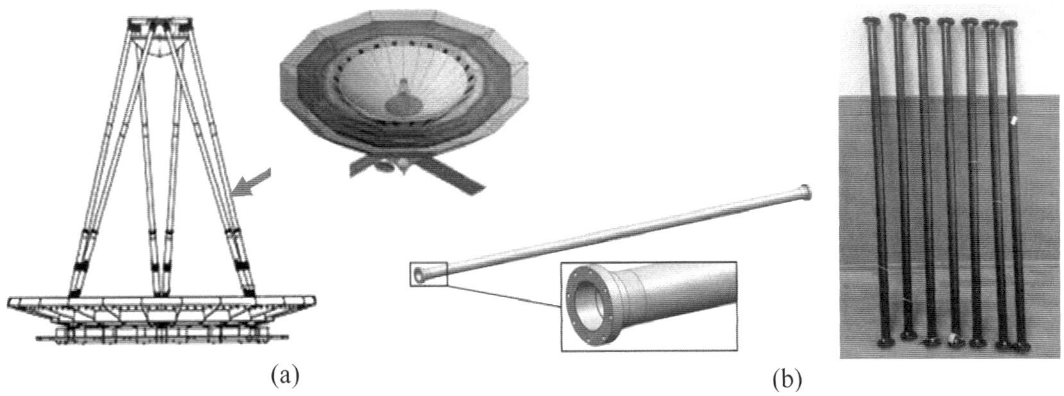

Figure 2. Thermally stable rods: (a) subreflector support (b) bearing rods.

Coefficient of thermal expansion (α_x) and mechanical properties (E_x, G_{xy}, E_y, M_{xy}) as functions of braiding pattern (biaxial or triaxial), longitudinal to spiral fibers volume ratio as well as braiding angles ω were determined (Figure 3).

As a result of a compromise between strength, stiffness and thermal stability requirements an efficient reinforcement pattern was used: M46J 0° (96.3 %) / 16x1K T300 ±60 ° (3.7 %) (called quasi-unidirectional).

Next step was process simulation to determine equipment adjustment parameters. In particular [7, 8, 9] it is necessary to calculate spindles rotation velocity, axial advance velocity of the mandrel and actual reinforcement angle. Besides, it was necessary to predict and eliminate porosity [10].

The rods were molded using RTM, so molding process was simulated in PAM-RTM (ESI Group software) first, and the tool was manufactured (Figure 4).

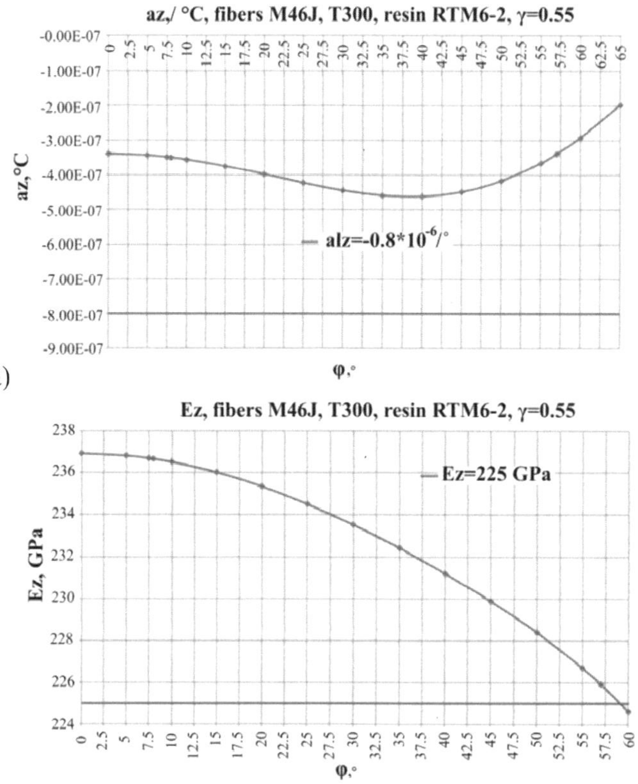

(a)

(b)

Figure 3. Lengthwise coefficient of thermal expansion (CTE) (a) and lengthwise Young's modulus (b) as a function of braiding angle.

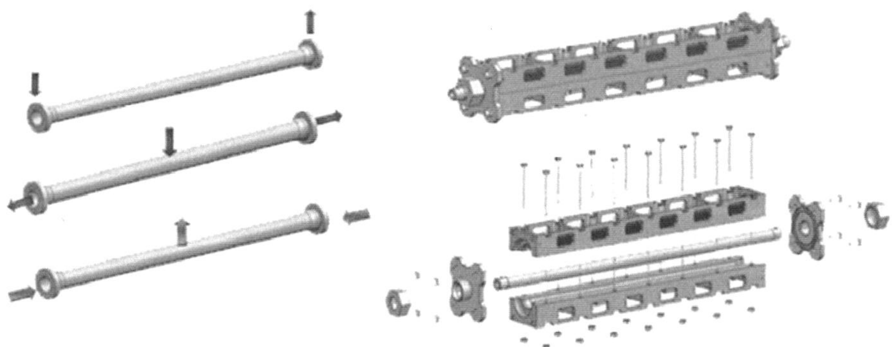

(a) (b)

Figure 4. Simulation of resin transfer (a) and design of the mold (b).

3. X-fitting

The other variant of support structure assumes intersection of the rods that is implemented using the X-fitting (Figure 5).

If a preform of the X-fitting is manufactured by lay-up, the reinforcement template turns out to be chaotic. In this case ply book contains a large amount of patterns that leads to discontinuities in reinforcement fibers.

As an alternative, kinematics of movement of the braiding equipment as well as reinforcement pattern for X-fitting radial braiding followed by RTM were developed.

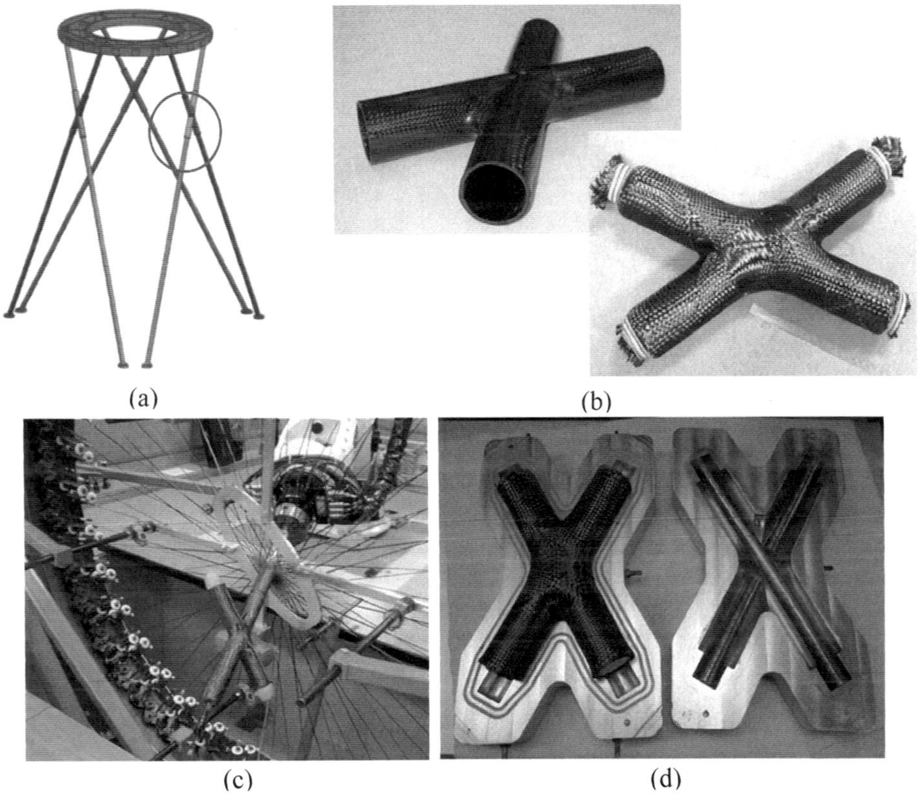

Figure 5. Support structure with intersecting rods: (a) assembly; (b) X-fitting and the preform; (c) braiding process; (d) RTM mold.

4. Support platform

Based on the requirements of reinforcement pattern regularity and a possibility to simulate mechanical properties and thermal stability, a technology of spacecraft support platform manufacturing was developed (Figure 6).

The preform was braided using Rohacell® inserts. After the inserts were assembled in the mold, the whole structure was impregnated in a single-shot RTM. Figure 6 (e) demonstrates a fragment of circular support demonstrator. Considering the fact that the mandrels of the subpreforms have rectangular section and curvature, it is needed to develop technique for processing parameters calculation, and reinforcement angles in particular (equation (1)).

$$\alpha_1' = \arctan\left(\frac{a}{\frac{(a+c)\phi_1}{\tan\alpha}\frac{l}{\pi} - \frac{\frac{c}{2}\tan\left(\frac{\gamma}{2}\right)}{l}}\right). \quad (1)$$

Here, γ is mandrel curvature angle, l is the length of curved mandrel with rectangular cross-section and a, c and φ_l are parameters of mandrel cross-section (Figure 7).

IOP Conf. Series: Materials Science and Engineering **683** (2019) 012058 doi:10.1088/1757-899X/683/1/012058

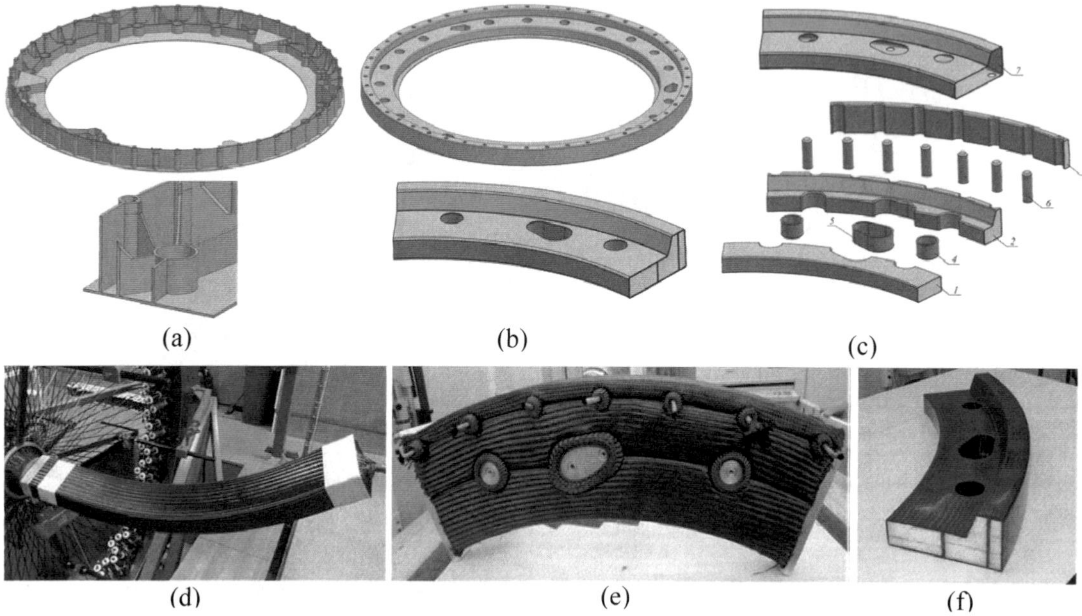

Figure 6. Circular support: (a) ribbed structure; (b) integral braided structure; (c) spaced view; (d) braiding of subpreform; (e) complex preform; (f) fragment of an integral panel.

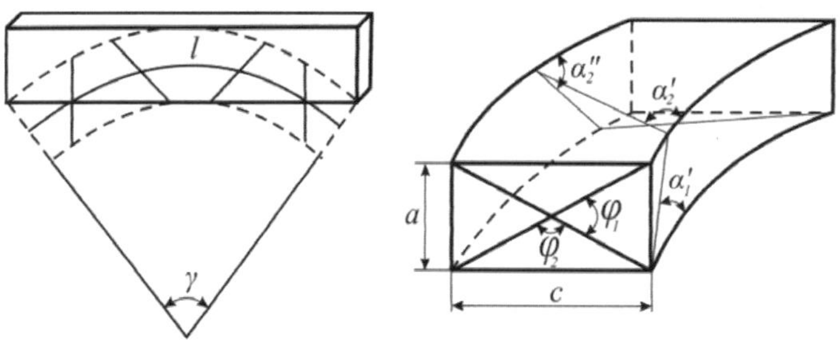

Figure 7. Position of roving on the curved mandrel.

5. Transformable lattice tubes

Another variant of ultra-light structure is a lattice tube (Figure 8) [11 - 13]. It has a braided preform that is impregnated by infusion.

One of the advantages of the said structure is an authentic weave of the fibers in the nodal areas (Figure 8 (c)) that ensures enhanced strength in these areas.

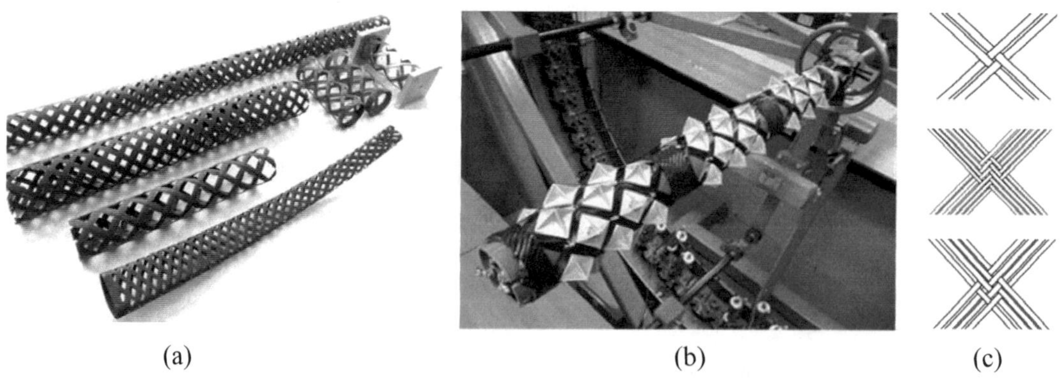

Figure 8. Lattice tube: (a) lattice structure; (b) braiding process; (c) patterns in nodal areas.

To braid lattice preforms mass properties as well as kinematic of movements of the braiding equipment were simulated. The main calculated parameters are as follows:

Elastic modulus of braided fivers (equation (2)) that provides a possibility to determine elastic properties of nodal areas depending on the types of fiber and resin and braiding angle:

$$E_x = \frac{\mathrm{fl}}{A\int\left(\frac{M\partial M}{\mathrm{EI}\,\partial f}\sin(\gamma)+\frac{f\cos(\gamma)}{\mathrm{AE}}\cos(\gamma)\right)\mathrm{dx}}. \qquad (2)$$

Here, I is inertia moment of the cross-section, E is elasticity modulus of the fiber, f is the force acting in the direction of the fiber, M is bending moment, γ is the angle of fiber curvature, bf is the distance between centers of rovings, A is a constant.

Mass of the lattice structure (equation (3)):

$$m_s = 2\frac{L}{\cos\phi}t_r h_r \rho_m \frac{L}{a_r}\sin\phi + 2\pi \mathrm{rL}\rho_m \frac{t_a}{a_a}h_a = 2\pi \mathrm{rL}\rho_m\left(2\frac{t_r h_r}{a_r}+\frac{t_a h_a}{a_a}\right). \qquad (3)$$

Here, L, r are length and radius of the structure, respectively; tr(a), hr(a), ar(a) are width, thickness and distance between spiral and lengthwise ribs, respectively; φ is the angle between spiral and lengthwise ribs.

Mandrel feed rate V_i for conical surface braiding (equation (4)):

$$V_i = \frac{\omega}{\mathrm{tg}\alpha}\left(\frac{x}{L}(r_1 - r_0) + r_0\right). \qquad (4)$$

Here, r_i is a radius of cone cross-section that changes along its axis; ω is angular velocity of the spindles; α is braiding angle; x/L are parameters of conic surface.

Fiber volume fraction in nodal areas for biaxial and triaxial braiding (equation (5)):

$$v_{f_{\mathrm{triaxial}}} = \frac{\frac{8a_1b_1}{\pi b_f}\sqrt{(2a_1)^2+\left(\frac{b_f}{\pi\cos\alpha}\right)^2}\,\xi\left(\frac{\pi}{b_f}\Big|\frac{1}{1+\left(\frac{b_f}{2\pi a_1\cos\alpha}\right)^2}\right)+\frac{4a_2b_2}{\pi b_{\mathrm{axial}}}}{2a_1+a_2}. \qquad (5)$$

Here, ξ is an elliptic line integral; bf, baxial is a distance between the centers of spiral and axial rovings; b1, a1 are semi-major and semi-minor axes of spiral roving cross-section; b2, a2 are semi-major and semi-minor axes of lengthwise roving cross-section.

This lattice structure was designed as part of self-deployable system. Its kinematic state is characterized not only by lengthwise folding, but also by lateral compression till the shape of a heavily oblong ellipse (Figure 9). The structure is transformed due to its intrinsic elasticity without any external actuators.

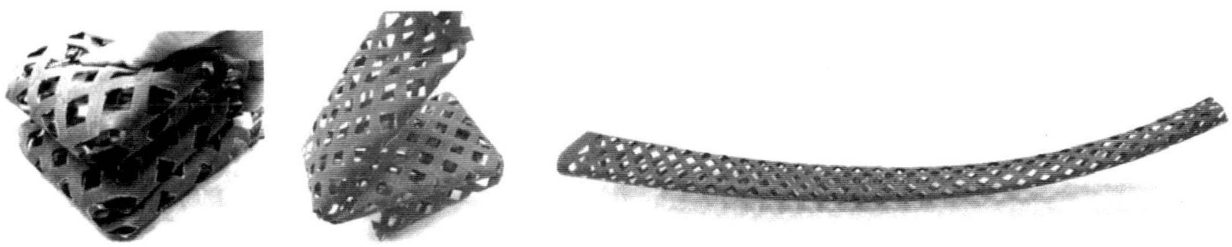

Figure 9. Deployment of the lattice tube.

6. 'Capillary' tube

Figure 10 (a) shows variants of tubes with multicell cross-section [14]. External shell has a rather large thickness compared to the tube elements inside. The structure is integral.

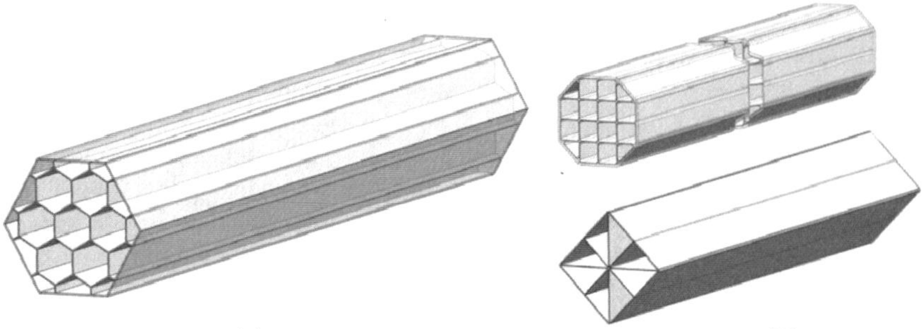

(a) (b)

Figure 10. 'Capillary' tube: (a) with hexahedral channels; (b) possible variants of cross-sections.

Manufacturing of 'capillary' tube has the following steps: braiding of preforms of the internal tubes using hexahedral mandrels; assembly of hexahedral mandrels in the single unit; braiding of the outer shell around the single unit. Then, the whole preform is impregnated using infusion or RTM technique. It is assumed that a thin-walled 'capillary' tube will be more light-weight compared to the usual tubular structure due to higher resistance to local buckling.

Conclusion

The study, including deflected mode and process simulation as well as an experiment, demonstrated that radial braiding together with transfer molding techniques provides the following possibilities:
-To generate lightweight thermally stable parts
-To design new, integral structures
-To ensure a proper level of automation and safety of production.

Acknowledgments

Authors are grateful to the Ministry of Science and Higher Education of Russian Federation for the financial support of research within the scope of the project under unique identification number RFMEFI57717X0262.

References

[1] Michael C Y Niu 2010 *Composite Airframe Structures* (Hong Kong: Hong Kong Conmilit Press) 664 p
[2] Monnot P, Lévesque J and Laberge Lebel L 2017 Automated braiding of a complex aircraft fuselage frame using a non-circular braiding model *Comp. Part A: Appl. Sci. and Man.* **102** 48–63
[3] Changliang L, Junbiao W, Chuang L, Hualin F, Binbin X. and Kaijing W 2015 Flexible tooling and local consolidation process to manufacture 1D lattice truss composite structure *Comp. Sci. and Tech.* **113** 63–70
[4] Cauchois J-P and Raynal J 2007 Closed-mould technologies in composite processing *Jec Comp. Mag.* **31** 37-39
[5] Dudchenko A A, Lur'e S A, Solyaev Yu O, Zhavoronok S I, Khaliulin V I and Batrakov V V2016 Calculation, design and production of thermally stable composite rod *Comp. Struct.* **1(141)** 3-11 (In Russian)
[6] Sidorov I N, Samipour S A and Tuktarova I V 2015 Calculation of natural frequencies and oscillation modes of the three-dimensional composite support structure of a primary radiator basing on the method of homogenization *Theory and Practice Of Technologies of Manufacturing Composite Materials And New Metal Alloys Products (TPCMM)/ Proc. of the 7-th Moscow International Conference* October 6-8 Moscow Russia 188-194 (In Russian)

[7] Samipour S A, Khaliulin V I and Batrakov V V 2018 Development of the Technology of Manufacturing Aerospace Composite Tubular Elements by Radial Braiding *Journal of Mach. Manuf. And Rel.* **47(3)** 284–289
[8] Samipour S A, Khaliulin V I and Batrakov V V 2018 A method for calculating the parameters for manufacturing preforms via radial braiding *Journal of Mach. Manuf. And Rel.* **46(3)** 302–308
[9] Samipour S A and Danilov Ya S 2016 Development and verification of an analytic technique to determine the stiffness parameters of braided tubular parts *Rus. Aero.* **59(4)** 460-465
[10] Batrakov V V, Petrushenko R Yu, Konstantinov D Yu and Khamidullin O L 2018 Influence of Autoclave Molding Modes and Stack Structure on Porosity of Composite Materials *Rus. Aero* **61(1)** 93-97
[11] Samipour S A, Batrakov V V and Khaliulin V I 2018 Patent RU 182 999 U1 A mandrel for composite lattice structures production, published September 09, 2018, report No. 25
[12] Samipour S A, Batrakov V V and Khaliulin V I 2018 Patent RU 183 464 U1 A triaxial composite lattice structure, published September 24, 2018, report No. 27
[13] Samipour S A, Batrakov V V and Khaliulin V I 2018 Patent RU 183 461 U1 A biaxial composite lattice structure, published September 24, 2018, report No. 27
[14] Endogur A I 2012 *Aircraft structure. Airframe units design* (Moscow: MAI-PRINT), 496 p

IOP Conf. Series: Materials Science and Engineering **683** (2019) 012059 doi:10.1088/1757-899X/683/1/012059

Sol-gel synthesis of oxide protective coatings of carbon fibers to increase heat resistance in aggressive oxidizing environments

A A Tselovalnikova[1], D A Trubin1, A N Karandashev[1]

[1]Peter the Great St. Petersburg Polytechnic University (SPbSTU),St. Petersburg, Russia

Abstract. Ceramic coatings of the Al2O3/SiO2 composition by the sol-gel method on carbon fibers were obtained in order to prevent oxidation and destruction of the fiber at temperatures above 450°C. Various combinations and ratios of aluminum and silicon oxides were used. The results of scanning electron microscopy and X-ray diffraction showed a change in the topographic properties of carbon fiber during the deposition of coatings thereon. Thermal analysis showed that oxidative stability improved significantly due to coverage, increased to 1100°C.

1. Introduction

To create new ceramic materials with enhanced properties and characteristics, the development of ceramic-matrix composites (CMC) using various fibers as a reinforcing component seems to be the most relevant.

The use of various fibers as a reinforcing component can significantly improve the final properties of the composite material, namely, to increase the impact strength, reduce brittleness, and preserve the carrying capacity. However, the mere use of fibers as a component of CMC does not yet guarantee automatic improvement of the properties of the material being developed.

Undoubtedly, the use of fibers leads to an increase in the strength and rigidity of the CMC, since in the direction of the fiber axis in the reinforced materials,

the ability to absorb loads increases by an order of magnitude compared with unreinforced material of the same chemical composition. However, in the case when the determining factors for the effectiveness of the application is a complex of material properties — for example, its high thermal resistance and thermal strength combined with low density — there is a need to apply special processing methods for fibrous materials to optimize their properties relative to the conditions of manufacturing processes or subsequent operation of the final product.

The most difficult problem when creating a CMC with a reinforcing component in the form of fibers is the choice of fiber. The most applicable are carbon fibers oxidized with air oxygen at temperatures above 450°C, which is a limitation for their use in a wide temperature range [1-11].

To solve this problem, it was proposed to apply thin ceramic coatings on carbon fiber, which will be resistant to oxidation, as well as be a binder between the matrix and fiber materials.

For carbon fiber as a coating material, it was decided to choose alumina and silicon oxide at the first stage, since these compounds have good heat resistance and are also fairly simple to produce. The sol-gel method was chosen as the method of preparation, since it is simple, affordable and does not require expensive equipment [12-17]

2. Materials and methods

In this paper, a carbon fiber grade URAL H-400Э was used. The mass fraction of the coupling agent on the fiber (synthetic rubber) is 0.8%. Technical ethyl alcohol, acetone and distilled water were used to prepare the carbon fiber for synthesis. Removal of the coupling agent was carried out as follows. The coupling agent was originally removed from the surface of the carbon fiber using a mixture of acetone and ethanol in a volume ratio of 1:1, followed by purification in distilled water using ultrasound. Then, for additional purification, the fiber samples were annealed in a furnace in air at a temperature of 240°C.

The following initial reagents were used in the paper: aluminum nitrate Al(NO3)3 9H2O, ammonia aqueous concentrated NH3 H2O, tetraethoxysilane (C2H5O)4Si, technical ethyl alcohol C2H5OH, hydrochloric acid HCl.

2.1. Preparation of sols of aluminum hydroxide and silicon oxide

To prepare an aluminum hydroxide sol, a concentrated solution of ammonia in water (NH4OH) was added to the aqueous solution of aluminum nitrate (Al(NO3)3) while stirring the solution on a magnetic stirrer with a speed of 120 rpm and a temperature of 50°C until a gel-like precipitate formed aluminum hydroxide (Al(OH)3). Further, the solution with the precipitate continued to stir for 20-25 minutes until the complete transition of the precipitate to the sol.

The method of preparation of the silica sol was as follows: The required amount of tetroethoxysilane was mixed with ethyl alcohol. Then, with stirring on a magnetic stirrer with a speed of 120 rpm, hydrochloric acid solution was added. When this occurred, the mixture was heated to a temperature of 50°C. The stability of the solution and its suitability for the formation of films was determined by the change in viscosity.

2.2. Carbon Coating Preparation

For the synthesis of coatings on the surface of the fiber, the fiber was passed through a bath containing a sol at a speed of 10 cm/min.

After passing through the sol bath, the fiber was subjected to 3-step heat treatment.

Stage 1 included drying the fiber at a temperature of about 22-24°C until the fiber completely dried. At the 2nd stage, the fiber passed through the first furnace and dried at the temperature of from 80°C to 150°C in an oxidizing atmosphere for 30 minutes. At the 3rd stage, the fiber passed through the second furnace, where it was annealed. The annealing of the fiber was carried out in a furnace, the temperature of the annealing was regulated in the range of 400°C to 570°C in an oxidizing atmosphere for 20-60 minutes.

Temperature selection is associated with the transition of coating materials to the required phases.

3. Results and discussion

The microstructure, elemental composition of the samples and the composition of individual phases were determined by scanning electron microscopy (SEM) and electron microprobe analysis (EMPA) using a FEI Quanta 200 scanning electron microscope with a thermal emission
cathode and an energy dispersive Si (Li) EDAX detector for microanalysis.

Using a scanning electron microscope, surface images of coated carbon fiber samples were obtained. According to the images, an analysis of the surface morphology of these samples was carried out.

Figure 1 depicts images of the surface of a carbon fiber with a single-layer (a) and four-layer (b) Al2O3 coating.

(a) (b)

Figure 1. SEM images of the carbon fiber surface with a single layer (a) and four-layer (b) Al_2O_3 coating.

As can be seen from the images, a single-layer coating of aluminum oxide forms a thin, fairly uniform coating. Separate areas of inhomogeneities can be associated with the roughness of the fiber itself or residual coupling on the fiber. Four-layer coating forms an uneven cracked coating. Many separate agglomerates are formed. This may be due to the fact that the initial irregularities of the fiber formed points of growth, on which agglomerates were subsequently formed upon repeated application of coatings.

Figure 2 shows images of the surface of a carbon fiber with a single-layer (a) and four-layer (b) SiO2 coating.

(a) (b)

Figure 2. SEM images of the carbon fiber surface with a single layer (a) and four-layer (b) SiO_2 coating.

These images show that the single-layer coating is a homogeneous structure, without pores and defects, consisting of small elongated crystals. On the surface of the carbon fiber there are agglomerates, which may be due to the uneven surface of the fiber itself, residual dressing or other contaminants. Four-layer coating forms an uneven coating with a large number of cracks. Coating morphology is heterogeneous. A large number of cracks in the coating may be due to the stress in the layer caused by the thickness of the coating.

Thermal analysis of the samples was carried out in a muffle furnace in an atmosphere of air. According to the results of thermal analysis, graphs of the dependence of the mass loss on temperature were plotted.

Fig. 3 shows a plot of mass loss (%) versus temperature in the range from 500 to 1100°C for 5 samples of carbon fiber with alumina obtained from $Al(OH)_3$ sol.

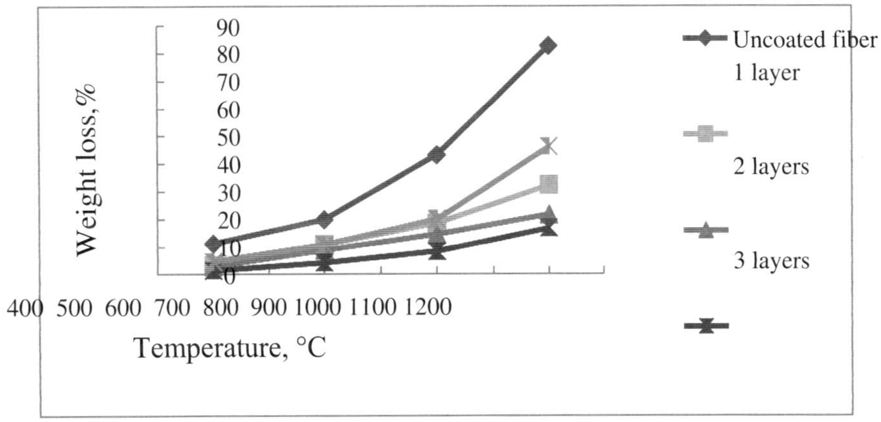

Figure 3. Graph of mass loss (%) versus temperature for 5 carbon fiber samples

In analyzing the data obtained, we can conclude about the behavior of the fiber with and without coating at elevated temperatures. At a temperature of 500°C, the weight loss is <5% for all coated fibers. Loss of fiber mass at the same temperature of 11%. As the temperature increases, the mass loss for all samples increases. However, for samples without coating and with a 4-layer coating, the temperature curve is described by an exponential relationship.

In the case of a 1,2,3-layer coating, the dependence can be approximated by a linear function. The best indicators of oxidative resistance of the fiber with a 3-layer coating. This may be due to the fact that the coating has sufficient thickness and density, and oxygen practically does not penetrate to the surface of the carbon fiber. With an increase in the thickness of the coating, a large number of defects form and the coating cracks; these defects are a source of oxygen penetration to the fiber surface. The mass loss in this case is comparable to the mass loss of the uncoated fiber.

Fig. 4 depicts a plot of mass loss (%) versus temperature in the range from 500 to 1100°C for 5 carbon fiber samples with silicon oxide obtained from SiO2 sol.

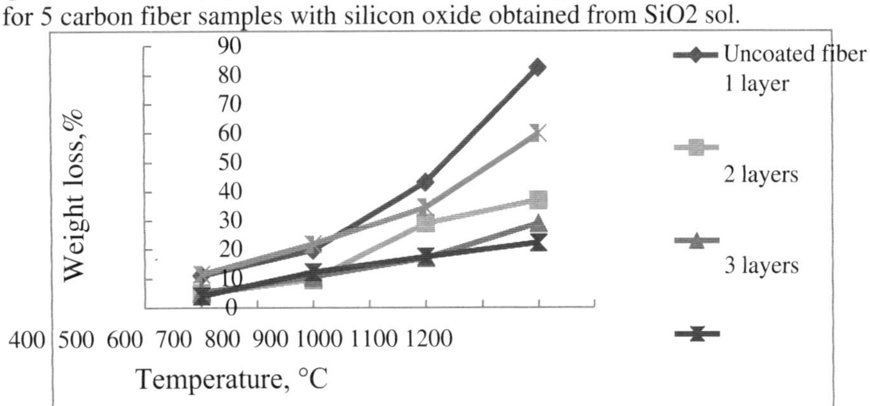

Figure 4. Graph of mass loss (%) versus temperature for 5 carbon fiber samples

For samples with silicon oxide at a temperature of 500°C, the mass loss is <5% for samples with a 1,2,3-layer coating. The loss of fiber mass without coating and with a 4-layer coating at the same temperature is 12%. With increasing temperature, fiber with 2- and 3-layer coatings shows the best properties. However, at a temperature of 1100°C, the oxidation of a fiber with a 2-layer coating occurs more strongly than with a 3-layer coating. The temperature dependence of

the sample with a 3-layer coating on this t
emperature range is linear, the remaining dependencies tend to be exponential.

4. Conclusion

Test samples were obtained and analyzed by scanning electron microscopy and thermal analysis.
Thin coatings exhibit optimal properties, with the best temperature resistance shown by coatings consisting of 2 and 3 layers. Multilayer coatings have a large number of defects, which causes the destruction of the fibers at increased temperatures. Also, multilayer coatings adversely affect the mechanical properties of the fibers, the fiber becomes less flexible, difficult to weave.
Analysis of the morphology of the coating also shows the advantage of thin, single and several layer coatings over multilayer coatings. According to the images obtained using SEM, multilayer coatings have increased defects, uneven surface and a large number of cracks.

References

[1] Kablov E, Grashchenkov D 2010 Advanced high-temperature ceramic composite materials *Russian Chem. Mag.* **5**
[2] Dham T, Bahl O 1995 Oxidation Protection of Carbon Composites *Carbon* **33** 349–59
[3] Jian H, ZengXie F 2003 Mullite-Al_2O_3-SiC oxidation protective coating for carbon/carbon composites *Carbon* **41** 2825–9
[4] Tkachenko L., Shaulov A. 2012 Protective heat-resistant coatings of carbon materials *Inorg. Mat.* **48** 261–71
[5] Baosheng X, Changqing H 2015 Preparation of a multi-composition coating for oxidation protection of modified carbon-bonded carbon fiber composites by a rapid sintering method *Surf. and Coat. Tehny* **270** 109–16
[6] Lei Z, Jianfeng H 2018 A novel design of oxidation protective β-$Y_2Si_2O_7$ nanowire toughened Y_2SiO_5/Y_2O_3-Al_2O_3-SiO_2 glass ceramic coating for SiC coated carbon/carbon composit*Corros. Scien.* **135** 233-42
[7] Zang Y 2009 AntiOxidation Behavior of Alumina Coating on Carbon Fibre by Sol–Gel *Adv. Mater. Res.* **79** 819-22
[8] Landry C, Barron A 1995 MOCVD of Alumina–Silica Oxidation Resistant Coatings on Carbon Fibers *Carbon* **33** 381–7
[9] Mohammad A, Heike H 2016 Thermal and oxidation resistant barrier on carbon fiber with Si Si–Ti based pre-ceramic coatings for high temperature application *Text. Res. J.* **3** 212-6
[10] Baklanova N.I., Zima T.M., Titov A.T. 2006 Protective coatings for carbon fibers *Inorg. Mat.* **42**
[11] Baklanova N, Zima T, Nimushina T, Kosheev S 2004 The Formation of Refractory Oxide Coatings on NicalonTM Fiber by Sol–Gel Process *J. Eur. Ceram. Soc.* **24** 3139–48
[12] Fatemeh M, Hasmaliza M, Luqman C 2011 Preparation of nano-scale α-Al2O3 powder by the sol-gel method *Silikaty* **55** 378–83
[13] Qiang F, Chuan-Bao C, He-Sun Z 1999 Preparation of alumina films from a new sol-gel route *Thin Sol. Fil.* **48** 99-102
[14] Naoufal B 2000 New Sol–Gel Route for the Preparation of Pure Alumina at 950°C *Ceram. Soc.***83** 2324-6
[15] Hu B, Yao M 2014 Preparation of Al_2O_3 dense films using sol derived from $Al(NO_3)_3 \cdot 9H_2O$ *Mat. tech.: Adv. perf. mat.* **29** 47-51
[16] Qiu X, Han K 2010 Sol–gel prepared Al_2O_3 coatings for the application as tritium permeation barrier *Fus. engine. and design* **85** 1068–72
[17] NilguKn O, John P, Yong-Jin Y, Antoni P 1999 Optical properties of sol-gel deposited Al_2O_3 *Sol. Ener. Mat.and Sol. Cells* **59** 355–66

Microporous PVDF ionic membranes for actuator applications prepared with imidazole-based poly(ionic) liquid as a pore forming material

O S Morozov [1*], S S Shachneva [1] and A V Kepman1

[1]Faculty of Chemistry, Lomonosov Moscow State University, Leninskie gory, 1-3, 119991 Moscow, Russia;

*Corresponding author

E-mail: osmorozov@gmail.com

Abstract. The microporous poly(vinylidene fluoride) (PVDF) membranes were prepared by phase inversion process using poly(1-ethyl-3-vinylimidazolium tetrafluorborate) (PIL) as a pore-forming agent. The membranes were activated by soaking in room temperature ionic liquids (EMImBF4 and BMImBF4) as a liquid electrolyte. The influence of PIL content on morphology and structure, electrolyte uptake of porous membranes and ionic conductivity of the activated membranes were investigated. The properties of the PVDF microporous membranes, such as crystallinity and porosity were characterized. The membrane with the maximum porosity demonstrates the highest electrical properties. The electrolyte uptake of EMImBF4 is 210 wt.% and the ionic conductivity is 9.2 mS/cm. Actuators with carbon nanotube bucky gel electrodes based on obtained membranes were prepared. The influence of ionic liquid and actuator thickness on electromechanical properties was discussed. The actuators with BMImBF4 as a electrolyte demonstrated high blocking force up to 9.5 mN.

Keywords: Actuator, carbon nanotubes, electroactive polymer, ionic membrane

Introduction

Polyvinylidene fluoride (PVDF) is one of the most widely studied and well-accepted polymers for membrane fabrication. PVDF-unique features include good degradation resistance against radiation, outstanding chemical and thermal resistance, and excellent mechanical properties[1–7]. The membrane technology has been widely applied to various purification process[8] such as water treatment[9,10], gas separation[11] and environmental protection[12]. PVDF membranes have wide range of applications in electronic industries including supercapacitors[13], sensors[14], actuators[15,16], and lithium ion batteries[8,17]. Polymer gel electrolytes (PGE) have been developed to replace liquid electrolytes. Considering the unique and advantageous properties of ionic liquids (IL), such as high ionic conductivity, thermal stability, wide electrochemical window, IL is a good alternative to the salts solution or solid electrolytes to prepare polymer gel electrolytes. Though gel polymer electrolytes possess suitable mechanical strength and eliminate the problem of leakage, low ionic conductivity limits their applicability. The conductivity of membranes prepared from non-ionic polymers such as PVDF is determined by both the type of electrolyte chosen and the morphology of the membrane. The simplest method of membrane production is casting from a solution or a melt of PVDF and an ionic liquid. However, in this case an essential part of the electrolyte is in closed pores, which significantly reduces the conductivity. Another approach is to create a porous membrane which is then impregnated with an electrolyte. The porosity of the membrane, its tortuosity, the degree of crystallinity and the nature of the electrolyte are determining polymer gel conductivity. Ideally, all pores should be open and well interconnected. Reducing the degree of crystallinity provides additional impregnation of the membrane by electrolyte due to the swelling of amorphous polymer fragments. The most simple way to produce PVDF membrane is conventional dry-casting process[18]. The final morphology of the obtained membranes depends on process conditions and the employed materials such as polymer, solvent, and nonsolvent, composition and temperature. Porous structure of the membrane can be obtained by adding a second polymer instead of nonsolvent as a pore-former to the casting solution. Additives such as dibutyl phthalate (DPB)[19], poly(ethylene glycol) (PEG)[20], polyvinylpyrrolidone (PVP)[21], tributyl citrate (TBC)[22] and others[23,24] are removed by extraction from the film to achieve porous membrane. Dry polymer membranes are then activated by soaking in a liquid electrolyte to obtain a polymer gel electrolyte.

In this work, microporous PVDF membranes characterized by high porosity and tortuosity were prepared by dry-cast technique using imidazolium based poly(ionic liquid) (PIL) as a pore former. The purpose of this work was to investigate the effectiveness of new high-conductivity PVDF membrane for use in the electrolyte layer of actuators.

Materials and Methods

All NMR spectra were measured at 600 MHz (1H) and 151 MHz (13C) on a Bruker Avance III Ultrashield spectrometer. All NMR spectra were recorded in deuterated dimethylsulfoxide with reference to tetramethylsilane. PVDF purchased from the Konstantinov Kirovo-Chepetsk Chemical Combine (Russia) as Fluoroplast-2. Single wall carbon nanotubes (SWCNT) purchased from the OCSiAl (Russia). 1-Methylimidazole, 1-vinylimidazole, butyl bromide, ethyl bromide, sodium tetrafluoroborate, lithium trifluoromethanesulfonimide and 2,2′-azobis(2-ethylpropionitrile) were purchased from commercial sources (Acros, Aldrich) and were used without further purification. Ionic liquids were synthesized by literature procedure[25]. Poly(ionic liquid) was prepared by literature method[26] using 1-ethyl-3-vinylimidazolium tetrafluoroborate as a starting material.

Membrane Preparation. The microporous polymer membranes were prepared by casting method as follows. The casting solutions were prepared by dissolving 2 g of PVDF and PIL in different compositions (3/7, 4/6, 5/5, 6/4, 7/3) in DMF at 100oC on a hot plate at 1400 rpm for 6 h. The mixtures were degassed in an ultrasonic bath for 30 minutes. Resulting homogeneous solutions were cast on a Petri dishes, left to dry in oven at 80°C. Then the films were stripped from the dishes and weighted. The films were immersed in a large excess of acetonitrile for 1 h at 80oC to extract the PIL. Then, the membranes were dried under vacuum. The procedure was repeated until a constant mass of the membranes was achieved.

Measurements on uptake of electrolyte solution and ionic conductivity of membranes. The PVDF membranes were soaked in ionic liquid (EMImBF4, BMImBF4, EMImTFSI or BMImTFSI) at 80oC, until the polymer membranes became uniformly transparent. The saturated membrane was taken out from electrolyte solution, the additional solution on surface was removed by filter paper. The electrolyte uptake is calculated by W (wt.%) = (ms/m0 −1)×100%, where m0 is original weight, ms is stand for saturated weight of membrane. The ionic conductivity of the polymer membranes was determined by the ac complex impedance technique over the frequency range from 0.1 Hz to 10 MHz using a Novocontrol α-S high resolution dielectric analyzer equipped and controlled by WINDETA software (Novocontrol GmbH, Germany). The samples were sandwiched between symmetrical cells containing two coin-shaped steel electrodes with area of A (0.785 cm2) at the constant potential of 5 mV to measure membrane impedance, Z (Ω). The thickness of the sample L (cm) was measured with a micrometer. The conductivity (σ, S/cm) was then calculated from the equation: σ = L/(Z×A).

Membrane Characterization. Scanning electron microscopy (SEM) was performed on an TESCAN Vega 3 instrument with an accelerating voltage of 20 kV. The samples were sputter-coated with approximately 10 nm of gold before analysis. The membranes were fractured in liquid nitrogen to observe the cross-section morphology. The differential scanning calorimetry (TA Instruments Q20

V24.11) data were obtained in a sealed aluminum pan with a heating rate of 10°C/min under N2 purge. The PVDF total crystallinity Xc was calculated by Xc (%) = ΔHm/ΔHm* × 100%, where ΔHm* = 104.5 J/g[27], is the melting enthalpy for a 100% crystalline PVDF, ΔHm is the melting enthalpy of PVDF membranes measured in DSC. The membrane porosity (P) was calculated by membrane density and determined by n-butanol uptake value. In the first case Pc = (1−ρm/ρp)×100%, where ρm is the density of membrane and ρp is the density of pure PVDF (ρp = 1.77 g/cm3). The membrane density was determined by measuring the volume and the weight of membrane. In the second set of experiments membranes were placed into 10 mL vials containing n-butanol and put in oven for 12h at 100oC. Then vials ware cooled to 25oC and left for 1 h. Weighing the mass of samples with and without n-butanol gave the porosity by the following relationship: Pb = [(mb-m0)/ ρb]/(m0/ρp +(mb-m0)/ρb)×100%, where m0 and mb are membrane weight before and after impregnation with n-butanol, ρb and ρp are densities of n-butanol (0.81 g/cm3) and PVDF (1.77 g/cm3).
Preparation of actuators. In an agate mortar, 100 mg of an ionic liquid (EMImBF4) was added to 40 mg of carbon nanotubes. The mixture was triturated and 0.5 ml of DMF was added and ground again. The operation was repeated three more times. The resulting gel was added to a solution of 60 mg of PVDF in 4 ml of DMF and the mixture was placed in an ultrasonic bath for 2 hours at 50 ° C. The dispersion process can be controlled by the resistance of the gel. The resulting thick paste was transferred into a Petri dish (r = 4.7 cm) and 12 ml of DMF was added. The mixture was evenly distributed in shape, either using an ultrasonic bath or using a spatula. The dish was placed in an oven for 3 hours (100°C). The resulting electrode was used further, without removing from the form. The electrolyte layer casting solution was prepared from 450 mg of PVDF and 1.05 g PIL in DMF (8 ml). The dissolution of the components was carried out at 100°C on a magnetic stirrer (1400 rpm). After dissolving all the components, 3 ml of acetonitrile (40 vol%) was added to the mixture and the solution was placed in an ultrasonic bath for degassing (10 min). The final solution was poured over the electrode and placed in an oven (80°C). 30 ml of acetonitrile was added to the dried film to remove it from the mold. The resulting bilayer film was washed with acetonitrile three times (or until constant weight) to remove the PIL. The two-layer film was placed in a solution of ionic liquid in acetonitrile (50% concentration) and put in oven (80°C) to evaporate acetonitrile. Samples of the desired shape were cut from the resulting activated membrane. The films were bond using ion-conductive adhesive. To remove DMF, the films were placed in oven (80°C) for 2 h.

Results and Discussion
All membranes in this work were cast from solutions in DMF, a good common solvent for both the PVDF and the poly(ionic liquid). The evaporation of the solvent leads to increase of the polymers concertation resulting in gelation of the solution by crystallization of the PVDF and to the gradual formation of a solid film. DMF is a better solvent for PIL than for PVDF, so PVDF membrane containing an additive in its pores is formed. The initial concentration of the polymers did not significantly affect the film formation process. Concentration above 0.4 g of components per 1 ml of solvent leads to the formation of a too viscous gel that could not be evenly distributed on cast form. As PVDF is a semi-crystalline polymer, a solvent heating is generally needed to weaken the interactions inside the PVDF crystallites and makes it possible to dissolve the polymer. Therefore, all casting solutions were prepared at 100oC with a total polymer concentration of 0.2 g/ml. Upon crystallization of a polymer in solution, initially polymer crystalline clusters are formed, and finally the solution is converted into a physical gel due to interconnections of polymer chains in the medium via the crystalline clusters. Therefore, the degree of crystallinity of the polymer membrane is influenced by the evaporation rate and drying temperature. To determine the effect of drying conditions on the crystallinity of PVDF, we prepared a series of membranes (PVDF/PIL, 1/1 by weight) at different temperatures (see Table 1). Before testing PIL was extracted by acetonitrile and the membranes were dried in vacuum. The PVDF crystallinity was calculated from measured by DSC ΔHm values. The membrane prepared at 80°C demonstrated minimum crystallinity of PVDF. Increased amorphous phase content provides growth of electrolyte uptake and, consequently, better electromechanical properties of PVDF membranes.

Table 1. Crystallinity of PVDF membrane prepared at different temperatures.

T, °C	Time, h	T_m, °C	ΔH_m, J/g	Crystallinity
60	12	170.8	58.19	56%
80	6	171.5	42.80	41%
90	4	171.2	49.50	47%
100	3	170.5	50.25	48%
120	1	172.6	50.92	49%

For further studies, PVDF membranes with different PIL contents (20, 30, 40, 50, 60, 70 and 80 wt%) were prepared. After drying the cast solution, the PIL remained in the membrane and may not be uniformly dispersed in the PVDF matrix. Some amount of the additive may be incapsulated by PVDF. At the extraction step, the additive is replaced by extractants in solvent extraction. After the extractants evaporated, the position of the additives in the membranes became porous structure. In cases of the membrane with PVDF content below 50 wt%, the removal of the additive was quantitative. It means that all formed pores are open. Membrane with 80 wt% of PIL lost its integrity during extraction step. During cast process, there is always one surface of the membrane facing the substrate while the other one faces the air. These two surfaces may differ in morphology. The SEM pictures both surfaces, and cross section of membrane prepared with 50 wt% PVDF content are shown in Fig. 1. The average pores size in the bottom surface is little larger than that of the upper surface. Cross section image reveals a macroscopically uniform porosity throughout the membrane. The pores are well interconnected and open to both surfaces.

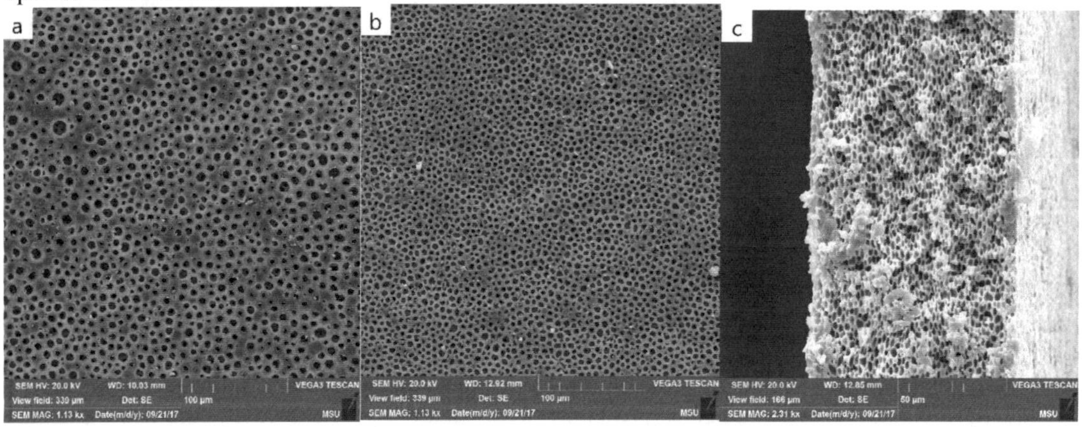

Figure 1. SEM micrographs of membranes PVDF with 50 wt% additive of PIL (a: bottom surface, b: top surface; c: cross section)

The porosities of these membranes measured by its density are very close to additive content (see Table 2). This means that the polymer and the additive have practically the same density. On the other hand, the porosities calculated by n-BuOH uptake are lower. PVDF is poor wettable with butyl alcohol, so only microporous are probably occupied. All the membranes ware soaked in ionic liquid (EMImBF4) to define electrolyte uptake and ionic conductivities. After soaking in ionic liquid, the membranes increased in thickness (~15-20 %), while the width and length decreased. The total volume of the membrane remained practically unchanged. This can be explained by the fact that the pores in the dry membrane have the shape of a flattened sphere (see Fig. 1 c) and upon impregnation the pores take more circular shape. The electrolyte uptake increased with the increase of porosity of membrane. The volume uptake was calculated using electrolyte density (d=1,294 g/cm3) to compare its values with porosities (Table 2). In all cases, the volume uptake exceeds the porosity values. The amorphous phase of PVDF facilitates the trapping of liquid electrolyte providing mobile ions more free volume and hence helping in enhancing ionic conductivity. Ion conductivities are listed in Table 2. Although the membrane with the maximum porosity exhibits a very high ion conductivity (9.2 mScm-1), there is a direct correlation between the conductivity and the membrane porosity or electrolyte uptake. Even though all the pores in the membranes are open, the average pore connectivity has a significant effect on the conductivity. The non-linear path from one side of the membrane to the other and can be expressed by the tortuosity (T)[21,28]. There are many definitions of tortuosity, with the simplest based on geometrical considerations, it is usual to define T by equation[19]: $T=\sqrt{\sigma_0/\sigma \times P}$, where $\sigma 0$ is the conductivity of pure liquid electrolyte (13.6 mScm-1), σ is the conductivity of the electrolyte membrane and P is the porosity of the membrane. The membrane formed with 70wt% of PIL content has very close to ideal tortuosity (1.02), which means there is a nearly linear path for electrolyte from one side to the other.

Table 2. Crystallinity of PVDF membrane prepared at different temperatures.

Membrane	$P_{c,\%}$	$P_{b,\%}$	Electrolyte uptake, wt%	Uptake by volume, %	σ, mS cm-1	*T*
PVDF/PIL, 70/30	22	-	47	39	0.19	3.97
PVDF/PIL, 60/40	35	17	89	55	0.39	3.49
PVDF/PIL, 50/50	51	50	116	61	1.36	2.26
PVDF/PIL, 40/60	59	58	160	69	3.98	1.42
PVDF/PIL, 30/70	72	71	210	74	9.2	1.02

There are two general approaches to prepare ionic electroactive polymer actuators with carbon nanotube bucky gel electrodes. The first one supposes hot pressing of independently prepared an electrolyte membrane and two films of electrodes. The main drawback of this approach is delamination of actuator due to low adhesion of electrodes to the membrane. The second method based on layer-by-layer cast of electrode and membrane solutions. We tested both approaches. The ionic liquid impregnated PVDF membranes were sandwiched between two carbon nanotube electrodes and hot pressed at 100°C for 30 min to fabricate the final actuators. Consequent casting was carried out at 80°C for 6 hours for each layer. In both cases the actuators delaminated during electromechanical tests. Notably, the only one electrode layer was delaminated from actuators prepared by consequent casting. Worse noting, unequal actuation to the different directions upon changing of voltage polarity was observed for all actuators. This is probably due to differences in the morphology of the membrane surfaces. To overcome all these problems, we applied another approach. The essence of this method is to stick together two two-layer films. In this case, two layers of the electrode and the ion membrane are successively casted. The device obtained in this way will be completely symmetrical, which will positively affect its effectiveness. The main difficulty in this method is to bind two films, since the external surface of the electrolyte layer has a large porosity and, therefore, a small effective surface. Attempts to apply hot pressing was failed. Therefore, it was decided to develop an adhesive material that can effectively bind the films. Conventional glue in this case is not applicable, since it could penetrate into the pores thereby, closing them and preventing the migration of ions. The adhesive material must simultaneously have good adhesion to PVDF and at the same time possess ionic conductivity not lower than the membrane itself. As such an ion-conductive adhesive, poly (ionic liquid) was tested, the same as that used in the preparation of membranes. Solution of PIL in ionic liquid at concentration below 50% is a sticky and very viscous substance. For bonding the final devices, a solution PIL in DMF with 40% additive EMImBF4 was prepared. The presence of DMF in the adhesive composition contributes to the swelling and partial dissolution of the surface layer of the PVDF membrane, which makes it possible to more effectively bind the films. A batch of samples of actuators with geometrical dimensions 2x1 cm of different thickness and with two types of ionic liquid (EMImBF4 and BMImBF4) were obtained. A new method allowed to obtain actuators with good reproducibility of electromechanical properties. The actuators with the same thickness (310±20 μm, 5 samples) generates similar blocking force of 3.1±0.2 mN at 3V DC. The actuator with increased thickness (600 μm) generates more blocking force than thinner one at the same rate at 3 V DC (Fig. 2 left diagram). The influence of voltage on generated blocking force for the actuator showed at Figure 2 (center). Despite of better electromechanical properties at 4 V DC, the back relaxation effect was observed. Using BMImBF4 as a electrolyte in spite of EMImBF4 lead to drastic increase in generated force. The maximum value of 9.5 mN was obtained in 13 minutes.

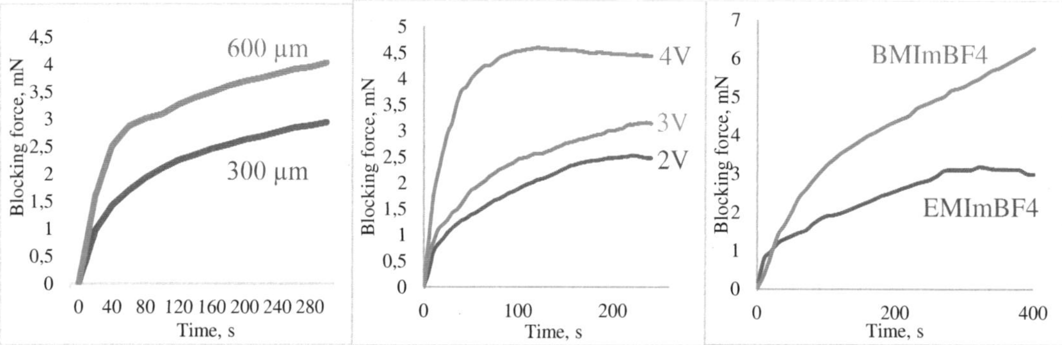

Figure 2. Time dependence of generated blocking force of the actuators.

Conclusions

The dry-cast technique appears to be a simple and attractive technique for the preparation of microporous membranes from PVDF. The main drawback of this method is the formation of an asymmetric structure of the membrane. Thus, the PVDF nearly symmetrical microporous membranes were successfully prepared by utilization of poly(ionic liquid) as a pore former. The influence of casting conditions and additive content on membrane properties were investigated. The minimum crystallinity (41%) was reached at 80oC. The polymer gel electrolytes ware prepared from obtained membranes by soaking in room temperature ionic liquid EMImBF4. Direct dependency of the porosity and electrolyte uptake microporous membranes formed from PVDF/PIL was observed. Thus, the ionic conductivity of membranes obtained from different compositions differ significantly. The membrane prepared from film with 70 wt% of PIL content demonstrate high ionic conductivity of 9.2 mScm-1 (68% of the pure ionic liquid value). The new method of preparation of ionic actuators was suggested to overcome the problem of membrane asymmetricity. The essence of this method is to stick together two two-layer films. Obtained totally symmetrical actuators demonstrated high electrotechnical properties. Namely, the actuators allowed to achieve blocking force up to 9.5 mN.

Acknowledgments

The reported study was funded by RFBR according to the research project № 18-29-18090

References

[1] Ouyang Z.-W., Chen E.-C., Wu T.-M., 2015 Thermal Stability and Magnetic Properties of Polyvinylidene Fluoride/Magnetite Nanocomposites, Materials (Basel). **8** 4553–4564. doi:10.3390/ma8074553.

[2] Jaleh B., Gavar N., Fakhri P., Muensit N., Taheri S.M., 2015 Characteristics of PVDF membranes irradiated by electron beam, Membranes (Basel). **5** 1–10. doi:10.3390/membranes5010001.

[3] Ong C.S., Lau W.J., Al-anzi B., Ismail A.F., 2017 Photodegradation stability study of PVDF- and PEI-based membranes for oily wastewater treatment process, Membr. Water Treat. **8** 211–223. doi:10.12989/mwt.2017.8.3.211.

[4] Faiz R., Fallanza M., Boributh S., Jiraratananon R., Ortiz I., Li K., 2013 Long term stability of PTFE and PVDF membrane contactors in the application of propylene/propane separation using AgNO3 solution, Chem. Eng. Sci. **94** 108–119. doi:10.1016/j.ces.2013.02.048.

[5] Nelyub V.A., Borodulin A.S., Kobets L.P., Malysheva G.V. 2016 A study of structure formation

in a binder depending on the surface microrelief of carbon fiber *Polymer Science –Series D*, , **9**(3).286-289R.

[6] Nelyub V.A. 2015 Determination of adhesion interaction between carbon fiber and epoxy binder *Polymer Science - Series D* . **8** (1) 6-8.

[7] Tiwari V.K., Kulriya P.K., Avasthi D.K., Maiti P., 2009 Radiation-resistant behavior of poly(vinylidene fluoride)/layered silicate nanocomposites, ACS Appl. Mater. Interfaces. **1** 311–318. doi:10.1021/am800040q.

[8] G. dong Kang, Y. ming Cao, 2014 Application and modification of poly(vinylidene fluoride) (PVDF) membranes - A review, J. Memb. Sci.**463**145–165.doi:10.1016/j.memsci.2014.03.055.

[9] Xie Z., Hoang M., Duong T., Ng D., Singh P., Ray P., Reddy A.V.R., Parashuram K., Maurya S., 2011 Effect of Membrane Properties on Performance of Membrane Distillation for Ammonia Removal, J. Mater. Sci. Res. **1** 37–44. doi:10.5539/jmsr.v1n1p37.

[10] Franken A.C.M., Nolten J.A.M.,. Mulder M.H.V, Bargeman D., Smolders C.A., 1987 Wetting criteria for the applicability of membrane distillation, J. Memb. Sci. **33** 315–328. doi:10.1016/S0376-7388(00)80288-4.

[11] Zhang Y., Sunarso J., Liu S., Wang R., 2013 Current status and development of membranes for CO2/CH4 separation: A review, Int. J. Greenh. Gas Control. **12** 84–107. doi:10.1016/j.ijggc.2012.10.009.

[12] Khalilpour R., Mumford K., Zhai H., Abbas A., Stevens G., Rubin E.S., 2015 Membrane-based carbon capture from flue gas: A review, J. Clean. Prod. **103** 286–300. doi:10.1016/j.jclepro.2014.10.050.

[13] Gao Y., 2017 Graphene and Polymer Composites for Supercapacitor Applications: a Review, Nanoscale Res. Lett. **12** 387. doi:10.1186/s11671-017-2150-5.

[14] Xin Y., Sun H., Tian H., Guo C., Li X., Wang S., Wang C., 2016 The use of polyvinylidene fluoride (PVDF) films as sensors for vibration measurement: A brief review, Ferroelectrics. **502** 28–42. doi:10.1080/00150193.2016.1232582.

[15] Mejri R., Dias J.C., Hentati S.B., Martins M.S., Costa C.M., Lanceros-Mendez S., 2016 Effect of anion type in the performance of ionic liquid/poly(vinylidene fluoride) electromechanical actuators, J. Non. Cryst. Solids. **453** 8–15. doi:10.1016/j.jnoncrysol.2016.09.014.

[16] HAGIWARA R., LEE J.S., 2007Ionic Liquids for Electrochemical Devices, Electrochemistry. **75** 23–34. doi:10.5796/electrochemistry.75.23.

[17] Zhang S.S., 2007 A review on the separators of liquid electrolyte Li-ion batteries, J. Power Sources. **164** 351–364. doi:10.1016/j.jpowsour.2006.10.065.

[18] Jung J.T., Kim J.F., Wang H.H., E. di Nicolo, Drioli E., Lee Y.M., 2016 Understanding the non-solvent induced phase separation (NIPS) effect during the fabrication of microporous PVDF membranes via thermally induced phase separation (TIPS), J. Memb. Sci. **514** 250–263. doi:10.1016/j.memsci.2016.04.069.

[19] Du C.H., Xu Y.Y., Zhu B.K., 2009 Plasticizer effect of dibutyl phthalate on the morphology and mechanical properties of hard elastic poly(vinylidene fluoride) fibers, J. Appl. Polym. Sci. **114** 3645–3651. doi:10.1002/app.30105.

[20] . Kim J.H, Lee K.-H., 1998 Effect of PEG additive on membrane formation by phase inversion, J. Memb. Sci. **138** 153–163. doi:10.1016/S0376-7388(97)00224-X.

[21] Cao J.H.,. Zhu B.K, Xu Y.Y., 2006 Structure and ionic conductivity of porous polymer electrolytes based on PVDF-HFP copolymer membranes, J. Memb. Sci. **281** 446–453. doi:10.1016/j.memsci.2006.04.013.

[22] Zhang Z.C., Guo C.G., Lv J.L., 2015 Tributyl citrate as diluent for preparation of PVDF porous membrane via thermally induced phase separation, Polym. Polym. Compos. **23** 175–180.

[23] H. Bin Li, Shi W.Y., Zhang Y.F., Liu D.Q.,. Liu X.F, 2014 Effects of additives on the morphology and performance of PPTA/PVDF in situ blend UF membrane, Polymers (Basel). **6** 1846–1861. doi:10.3390/polym6061846.

[24] Guo L., Liu Y., Zhang C., Chen J., 2011 Preparation of PVDF-based polymer inclusion membrane using ionic liquid plasticizer and Cyphos IL 104 carrier for Cr(VI) transport, J. Memb. Sci. **372** 314–321. doi:10.1016/J.MEMSCI.2011.02.014.

[25] . Kim D.J, Oh K.H., Park J.K., 2014 A general and direct synthesis of imidazolium ionic liquids using orthoesters, Green Chem. **16** 4098. doi:10.1039/C4GC00684D.

[26] Smith T.W., Zhao M., Yang F., Smith D., Cebe P., 2013 Imidazole Polymers Derived from Ionic Liquid 4-Vinylimidazolium Monomers: Their Synthesis and Thermal and Dielectric Properties, Macromolecules. **46** 1133–1143. doi:10.1021/ma300862t.

[27] Ma W., Zhang J., Wang X., Wang S., 2007 Effect of PMMA on crystallization behavior and hydrophilicity of poly(vinylidene fluoride)/poly(methyl methacrylate) blend prepared in semi-dilute solutions, Appl. Surf. Sci. **253** 8377–8388. doi:10.1016/j.apsusc.2007.04.001.

[28] Quartarone E, Mustarell P, Magistris A 2002 Transport properties of porous PVDF membranes, J. Phys. Chem. B. **106** 10828–10833. doi:10.1021/jp0139843.

IOP Conf. Series: Materials Science and Engineering **683** (2019) 012061 doi:10.1088/1757-899X/683/1/012061

Topology optimization of the lattice payload adapter for carrier rocket

R S Baldzhiev[1], A A Alekseyev[1] and A V Azarov[1]

[1]Bauman Moscow State Technical University, 2nd Baumanskaya str., b. 5, Moscow, 105005, Russia

E-mail: r.baldji@yandex.ru

Abstract. Current study is dedicated to the evaluation of the topology optimization method for the design of load-bearing aerospace structures. The subject of research was composite spacecraft attach fitting (adapter) which provides the interface between a spacecraft and a commercial launcher. Modelling of stress-strain state of the lattice payload adapter under operation conditions was performed. Compression strengths of structural ribs were defined together with the stiffness and buckling stability taken as the design constrains. By means of the topology optimization procedure optimal material distribution in the payload adapter was obtained. The workability of the adapter was proved by the confirmatory analysis.

1. Introduction

Currently there are several design approaches for the spacecraft payload attach fitting - adapter design, namely: sandwich panels, reinforced shells and lattice structures. Each variant has its certain pros and cons. However, lattice payload adapter with ribs manufactured from unidirectional carbon fiber reinforced plastic (CFRP) demonstrates highest weight efficiency (of all the above-mentioned variants) accompanied with high stiffness [1]-[3].

Recent advances in composite manufacturing technologies development allow switching from the traditional filament winding to the modern automated fiber placement (AFP) technology. AFP equipment combined advantages of both techniques — fiber placement and winding due to utilization of automated operational system and special software. Consequently, AFP allows obtaining products with intricate shapes [4]. Also composite parts with complex shapes can be achieved using emerging composite 3D printing technology [5]-[6]. Application of these new techniques enables to create optimal structures by varying ribs pathways (trajectory), their number and cross-section shapes.

However, a great number of variables make the design problem of lattice composite load bearing adapter quite complex and resource-consuming. Topology optimization method could overcome this issue.

Topology optimization is the design approach that allows to define reasonable material distribution in a certain volume taking into account the design constraints that depend on the operation conditions and type of the applied loads [7]-[10]. Such analysis is performed by means of the well-known calculation method – Finite Element Method (FEM)

This paper aims to develop an optimal structure of the CFRP payload adapter of the carrier rocket by utilizing topology optimization method.

2. Object of the research

The lattice payload adapter (Figure 1) consists of several types of ribs made from unidirectional CFRP. The biggest amount of the ribs belongs to spiral type - 48 pairs of main (1) and 16 additional (2).

Circumferential ribs (3) are located at the intersections of the main spiral ribs. The adapter is mounted to the neighboring structures with the upper (4) and lower (5) frames. In the lower cross-section the adapter is connected with the upper stage of a launcher using 8 groups of bolts (Figure 2). The diameter of the lower frame is 2085 mm, of the upper – 1212 mm, the adapter height is 901.5 mm. Total weight of the structure is 64 kg. The geometry parameters of the structural ribs cross-sections are presented in Table 1.

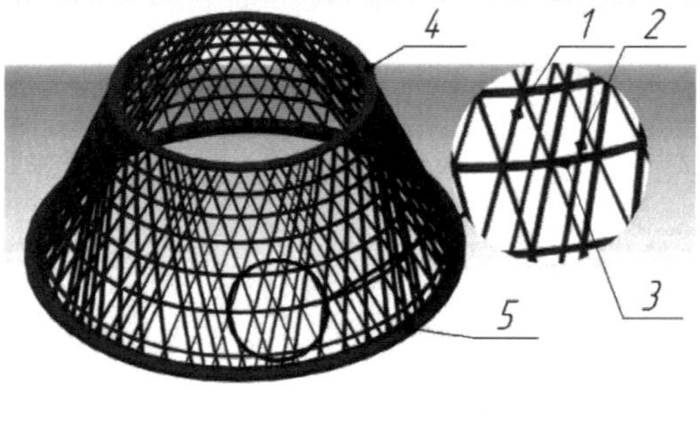

Figure 1. The CFRP adapter: main spiral ribs (1); additional spiral ribs (2); circumferential ribs (3); upper frame (4); lower frame (5).

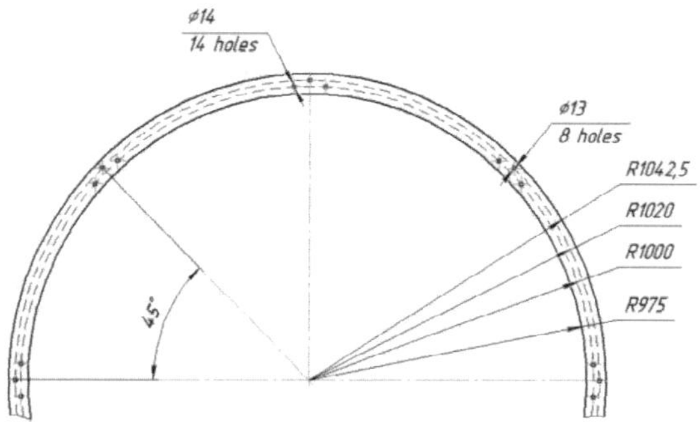

Figure 2. The bolting holes location on the lower frame scheme.

Table 1. Geometry parameters of the structural ribs.

Element	Main and additional spiral ribs	Circumferential ribs	Upper frame	Lower frame
Cross-section dimensions	8x25	4x25	25x25	30x30

3. Topology optimization of the lattice adaptor structure

3.1. Trajectory parameters determination

The input parameters for the topology optimization procedure were the adapter dimensions – upper and lower frame diameters, and positioning of the bolt holes on the lower frame. During the payload orbiting the adapter must withstand intensive compressive stresses while keep its dimensions and shape unchanged. Therefore strength and stiffness characteristics of the initial lattice structure were assumed as the constraints for optimization.

Modelling of the stress-strain condition of the adapter was conducted in order to define structure mechanical properties. Stress-strain analysis was conducted in Femap – Siemens PLM Software. The applied loads are presented in Table 2.

Table 2. The values of the applied loads.

Characteristics	Loads types		Values	
			kN	kN×m
Strength	Combined	Axial force	73	
		Bending moment		165
		Side force	77	
Stiffness	Axial force		120	
	Bending moment			70

The modelling results (Figure 3) demonstrate that maximum compression strain in the ribs is 310 MPa, buckling coefficient of the ribs – 4.4, displacement under the axial load is 0.09 mm, while under side load – 0.13 mm. All the obtained results do not exceed the ultimate values for CFRP (Table 3), that indicates the structure operation capacity.

Table 3. Stress-strain state modelling results for the initial adaptor structure.

Analysis type	Parameters	Obtained value	Ultimate value
Strength	Maximum compression stress in the ribs, MPa	310	450
Buckling	Buckling safety coefficient	4.40	> 1
Stiffness	Displacement under the axial load influence	0.09	< 0.5
	Displacement under the bending moment influence	0.13	< 0.5

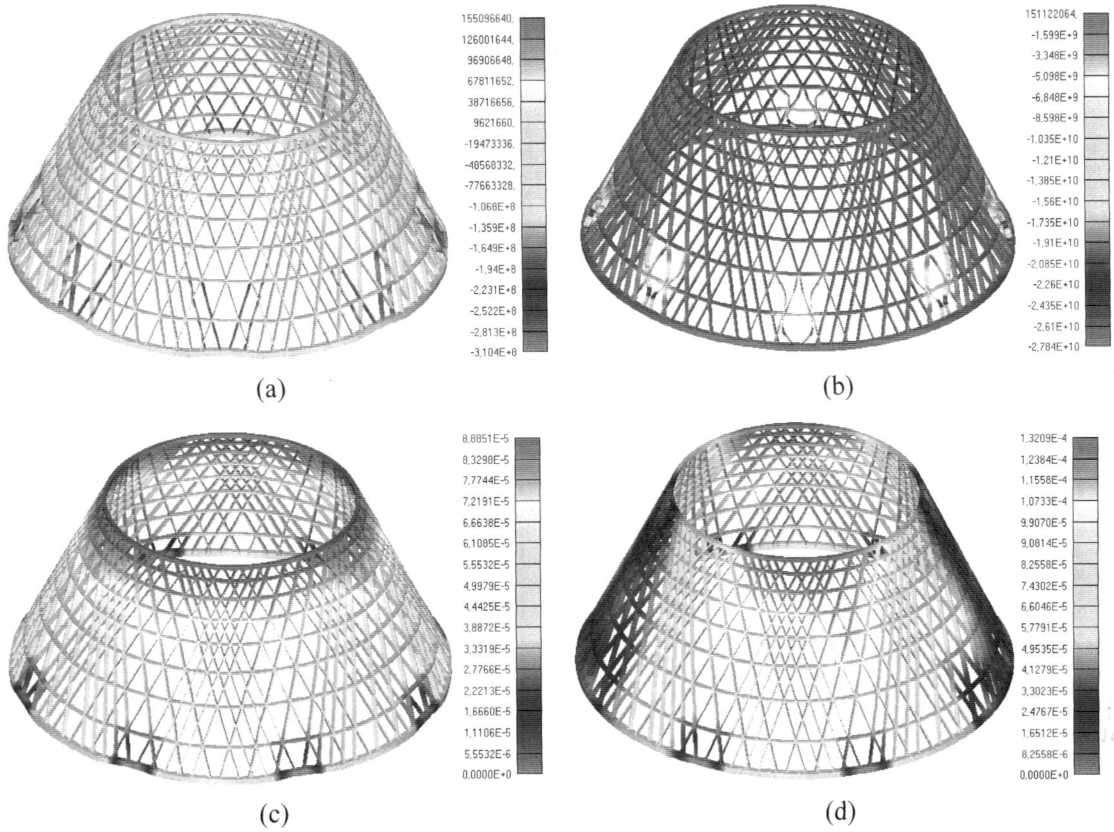

Figure 3. Initial adaptor structure tress-strain analysis results: (a) compression stress fields; (b) buckling stability; (c) ribs displacement under the axial load; (d) ribs displacement under the side load.

3.2. Topology optimization

The 3D-model for the topology optimization as well as all geometry modeling was carried out using the Shape Design module of computer-aided design system Catia V5. This adapter model had the same overall shape as the initial adaptor, but by contrast to the original one the model had a solid wall (Figure 4). Topology optimization was conducted in the Optistruct module of the FEM program Altair Hypermesh with a great toolkit for a complex optimization analysis. The resulting structure after topology optimization after cleanup is shown in Figure 5. The geometrical parameters of the resulting ribs are presented in Table 4.

Figure 4. 3D-model of the adaptor for the topology optimization.

Figure 5. Optimized adaptor geometry after cleanup.

Table 4. The geometrical parameters of the resulting ribs.

Structural element	Cross-sections dimensions, mm
Main ribs	12x30
Additional ribs	8x30
Upper frame	30x30
Lower frame	135x30

The resulting lattice adapter had a maximum ribs thickness of 12 mm, and weight of 27 kg, that is almost twice lower than the initial structure weight. However, the resulting structure before cleanup had certain imperfections. Thus, it had numerous non-joint sections together with complex ribs shapes and different lengths (Figure 6). Consequently, for the further analysis, the geometry was modified leaving the basis concept provided by the topology optimization.

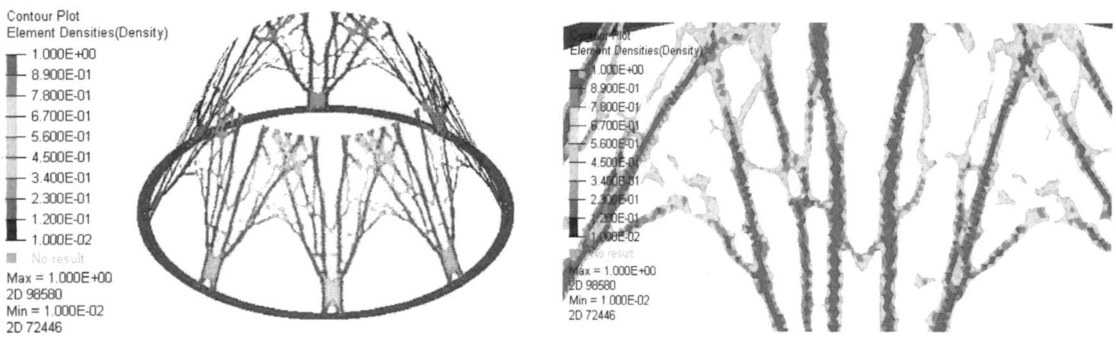

(a) (b)

Figure 6. Topology optimization results: (a) adaptor geometry before cleanup; (b) geometry imperfections – non-joint section, complex ribs shapes and different lengths.

3.3 Optimized adaptor geometry verification and post processing

After the structure modification some elements were deleted and that is why a new stress-strain analysis of the modified structure was carried out (Figure 7). The results shown in Table 5 demonstrate that the elements displacements exceed the ultimate value and the structure needs to be stiffened. For this reason additional lateral stiffness ring (8x30 mm) was added to the intersections of the additional ribs (Figure 8). This lateral stiffness ring reduces the displacement down below the allowable value (Figure 9, Table 6).

IOP Conf. Series: Materials Science and Engineering **683** (2019) 012061 doi:10.1088/1757-899X/683/1/012061

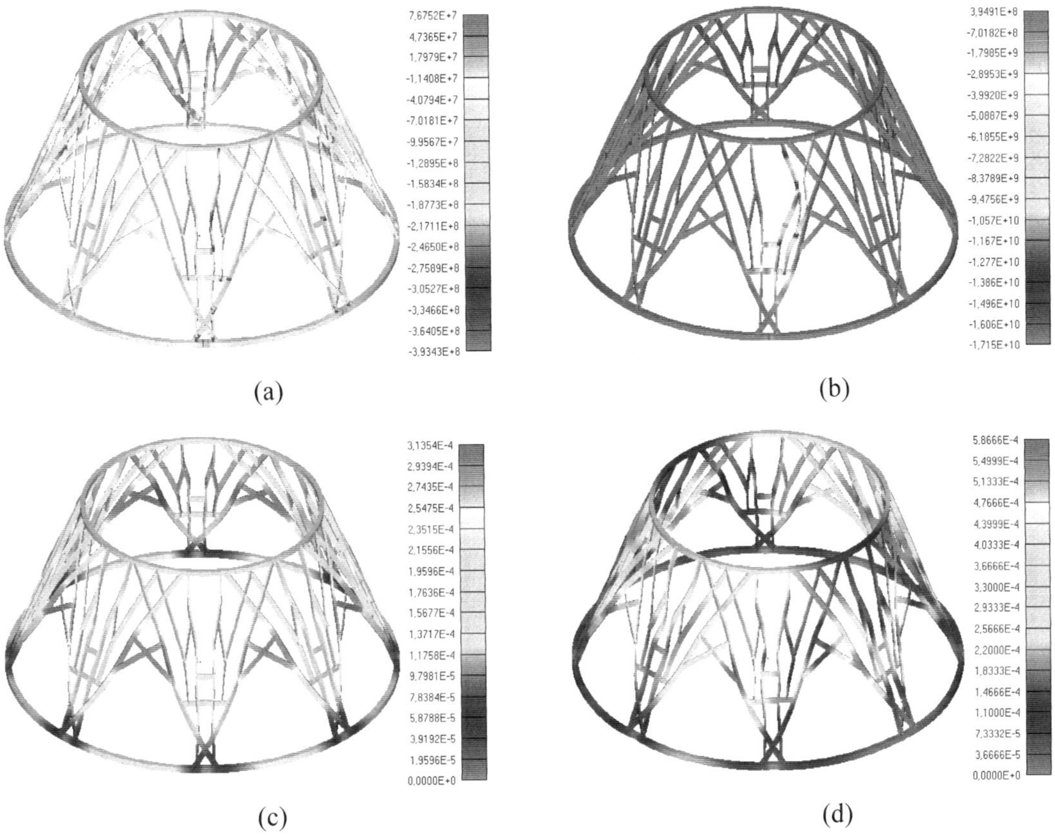

Figure 7. Modified adapter structure stress-strain analysis results: (a) compression stress fields; (b) buckling stability; (c) ribs displacement under the axial load; (d) ribs displacement under the side load.

Table 5. Stress-strain state modelling results for the optimized adapter structure after modification.

Analysis type	Parameters	Obtained value	Ultimate value
Strength	Maximum compression stress in the ribs, MPa	393.43	450
Buckling	Buckling safety coefficient	5.09	> 1
Stiffness	Displacement under the axial load influence	0.31	< 0.5
	Displacement under the bending moment influence	0.59	< 0.5

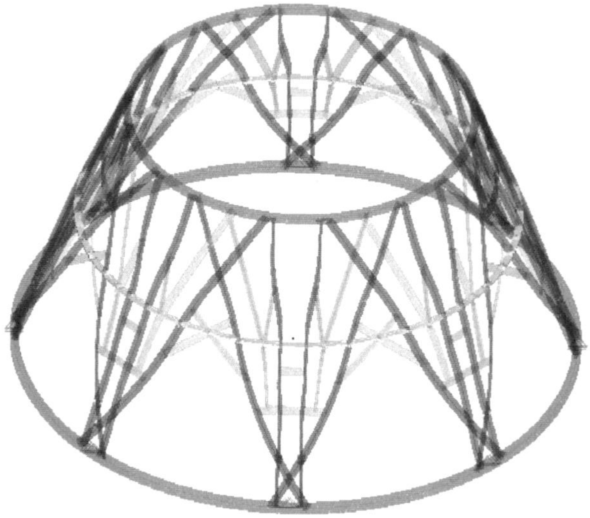

Figure 8. Optimized adapter with the additional lateral ring.

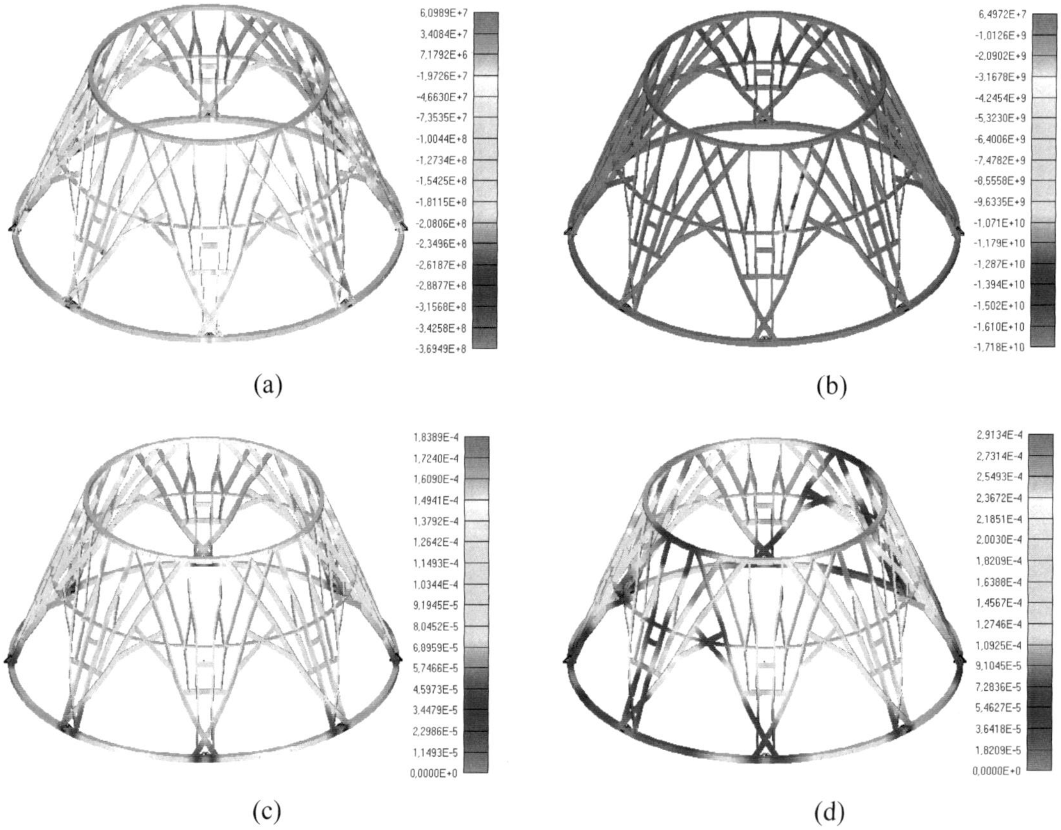

Figure 9. Stress-strain analysis of the optimized adapter with the additional lateral ring:
(a) compression stress fields; (b) buckling stability;
(c) ribs displacement under the axial load; (d) ribs displacement under the side load.

Table 6. Stress-strain state modelling results for the optimized adapter with the additional lateral ring.

Analysis type	Parameters	Obtained value	Ultimate value
Strength	Maximum compression stress in the ribs, MPa	369.49	450
Buckling	Buckling safety coefficient	6.72	> 1
Stiffness	Displacement under the axial load influence	0.18	< 0.5
	Displacement under the bending moment influence	0.29	< 0.5

The implementation of topology optimization enables to reduce the mass of the adapter for 45 %. Thus, the initial structure mass was 53 kg, mass of the adaptor without the additional lateral ring – 27 kg, mass of the adaptor with the additional lateral ring – 28.7 kg.

4. Conclusion

The research results demonstrate that topology optimization is a feasible instrument for aerospace lattice structures design. Utilization of this approach leads to the remarkable increase of weight efficiency. Nevertheless, the assumptions used during the optimal structure modification leads to the results deviations. Therefore the optimized geometry needs to be verified. However, the modified adaptor structure demonstrates 45 % weight reduction that proves high potential of the described methodology.

References

[1] Razin A F and Vasiliev V V 2004 Development of composite anisogrid spacecraft attach fitting *Proc. of 11th European Conf. on Composite Materials (May 31 – June 3,RhodesGreece)* C202 9 pё

[2] Vasiliev V V, Barynin V A and Razin A F 2012 Anisogrid composite lattice structures Development and aerospace applications *Composite structures* **94(3)** 1117-1127

[3] Azarov A V 2013 Theory of composite grid shells *Mechanics of Solids* **48** no 1 57-67

[4] Frketic J, Dickens T and Ramakrishnan S 2017 Automated manufacturing and processing of fiber-reinforced polymer (FRP) composites: An additive review of contemporary and modern techniques for advanced materials manufacturing *Additive Manufacturing* **14** 69-86

[5] Azarov A V, Antonov F K, Vasiliev V V, Golubev M V, Krasovskii D S, Razin A F, Salov V A, Stupnikov V V and Khaziev A R 2017 Development of a two-matrix composite material fabricated by 3d printing *Polymer Science* Series D vol **10** no 1 pp 87-90

[6] Xin Wang, Man Jiang and Zuowan Zhou 2017 3D printing of polymer matrix composites: A review and prospective *Composites* Part B: Engineering **110** 442-458

[7] Bendsoe M P and Sigmund O 2003 *Topology optimization: theory, methods, and applications* (Berlin, Heidelberg, New York, Barcelona, Hong Kong, London, Milan, Paris, Tokyo:Springer Science & Business Media) 370

[8] Chugunov M V and Schekin A V 2015 Practical aspects for development of add-in solid works applications on example of optimization problems *International Journal of Applied Science and Engineering Research* **4** no 4 494-499

[9] Borovikov A A and Tenenbaum S M 2016 Topology Optimization of Spacecraft Transfer Compartment *Aerospace Scientific Journal of the Bauman MSTU* **2** no 5 16-30

[10] Belova O V and Vulf M D 2017 New to Power Equipment Design Approaches with Additive Manufacturing prospects *IOP Conf. Series: Journal of Physics: Conf. Series* **891** 012211

IOP Conf. Series: Materials Science and Engineering **683** (2019) 012062 doi:10.1088/1757-899X/683/1/012062

Development of the radiant heating facilities for the testing of ceramic materials with operating temperatures up to 2000 K

R S Baldzhiev*, P V Prosuntsov and M O Zabezhailov

Bauman Moscow State Technical University, Department SM-13 "Rocket and Space Composite Structures", 105005 2-ya Baumanskaya 5-1, Moscow, Russian Federation

E-mail: r.baldji@yandex.ru

Abstract. The radiant heating facilities based on tungsten halogen lamps are frequently used for high-temperature materials experimental study and aerospace hot structures functional testing. The ease of maintenance and operation as well as the relatively low cost and high efficiency are the main advantages of halogen lamps as a heating source. However, the significant disadvantage of halogen lamps is the limitation of the facility operating temperature. It could not exceed 1500 K due to the tungsten-halogen cycle disruption. Current paper investigates the feasibility of the facility operating temperature expansion up to 2000 K by means of the lamps cooling with a compressed air flow. The mathematical simulation of a complex heat transfer in the facility operating area was performed. The design-layout schemes for several facilities were proposed. The modelling results proved the performance of the proposed design-layout schemes.

1. Introduction

One of the main problems of the aerospace engineering is the thermal protection of structures against high-enthalpy airflow. Currently, the high-temperature ceramic materials are widely used in the hot structures of reusable aerospace vehicles. The operating temperatures of such materials could reach 1800 °C [1, 2, 3]. In addition to the high operating temperatures, these materials have several advantages, such as, high chemical resistance, mechanical performance and weight efficiency by contrast to the ablative, carbon-carbon and other materials.

However, the information about ceramic materials behavior at high temperatures is required to design aerospace structures. Thus, it is necessary to conduct numerous thermal tests for the determination of material's thermophysical and mechanical properties. Moreover, these tests should be carried out with high alternating values of the heating rate in order to confirm material performance under equal to in-situ conditions. In addition, due to the requirements of the aerospace vehicles reusability, it is necessary to conduct the long-term thermal-cycling tests in order to estimate the material operating life and its resistance to the thermal-cyclic loads.

Currently, the radiant heating facilities are the most widespread among all the types of heating test facilities in Russia and abroad. Even during the development of the thermal protection of such well-known space systems as Buran and Space Shuttle, most of the experiments were carried out via thermal test facilities based on tungsten halogen lamps (THLs) [3, 4, 5]. The advantages of this type of facilities are ease of operation and maintenance, high efficiency and relatively low cost. This is due to the use of commercially available serial produced halogen lamps as a radiation source. The performance and operation principle of THL are similar to the other types of the incandescent lamps (Fig. 1). A specific

feature of THL is the utilization of halogen vapors as a buffering gas inside the lamp bulb. Due to this, a so-called tungsten-halogen cycle occurs during lamps operation. In this cycle, tungsten atoms that are evaporated from the filament could not deposit on the lamp bulb due to a chemical reaction with the halogens. This fact significantly increases the lamps operating life and prevents changes in the light flux from filament due to the loss of the bulbs transparency. The serial production provides relatively low cost heaters with high efficiency from a wide range of current leads types, lamp wattage and sizes.
However, the operation of the tungsten-halogen cycle is possible only in a limited bulbs temperature range (from 523 to 1270 K) [5, 6]. If the bulb temperature exceeds this limitation, tungsten atoms will deposit on it. This leads to the loss of the bulb transparency and, a as result, to the lamp break down (Fig. 2). Moreover, besides the destruction of the relatively low cost lamps it could also lead to a significant damage of the high cost facility systems. Because of this in existing facilities based on THL, test specimen temperature could not exceed 1500 K [5, 6]. Thus, the issue of the development of the new facilities for the thermal testing of ceramic materials with operating temperatures up to 2000 K is relevant.

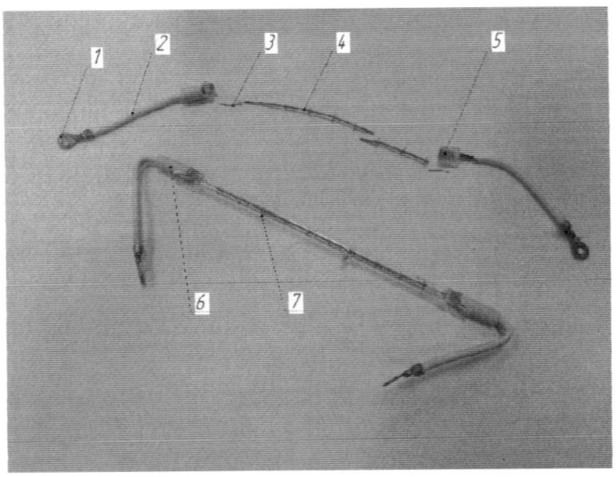

Figure 1. The halogen lamp KG-220-2000-5: (1) a clamp; (2) a connection wire; (3) a lead wire; (4) a tungsten filament; (5) a coupling plate; (6) a ceramic lamp base; (7) a quartz glass bulb

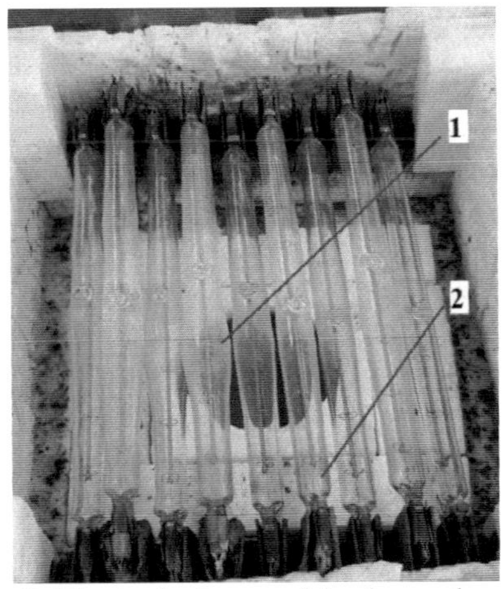

Figure 2. The typical case of the lamps break down in the thermal test facility due to bulbs overheating: (1) buckling failure of emolliated bulbs under the internal pressure load; (2) deposition of the tungsten vapor on the walls of lamps bulbs

2. Object of the research

In [7], the principal possibility of creating such a facility by means of the lamps active cooling with compressed room-temperature air was shown. Based on the results of this research, the operating area design-layout schemes for several facilities were created:

1) The basic facility design for the serial thermal stability testing of high-temperature ceramic materials.
2) The facility for the estimation of materials and structures operating life via long-term thermal-cyclic testing with cooling system for the facility metal frame and thermal insulation.
3) The modified facility for the precision research of the materials thermophysical properties with a more uniform temperature field distribution over the test specimen surface.

The creation of such facilities by a "cut-and-try" method is an extremely expensive and almost unpromising event. Therefore, it is necessary to rely on the mathematical simulation of a complex heat transfer in the facility operating area, which occurs parallel and interrelated via three heat transfer mechanisms: the heat conductivity in the test specimen, thermal insulation, heaters, bulbs and glass partitions; the convection in the flowing gas medium and the radiation in the semi-transparent bodies and the gas medium. In addition, the temperature, the gas velocity and the radiation fields are interrelated and interdependent. Moreover, the spectral and the temperature functions of the materials thermophysical and optical properties must be taken into account.

To solve this problem the mathematical modelling was conducted on an advanced computing systems (Intel Core i7-4790 processor) in the finite-element package ANSYS Fluent (License Number: 339001) – a powerful computational fluid dynamics software. Besides this, Fluent could be used for the modelling of thermophysical processes, such as convection, conductivity and radiation (even in semi-transparent bodies), which make it capable for simulation of the multi-physical processes occurred in facility.

The heating facilities should also meet a number of requirements to the operating area geometry (shape and size) as well as requirements to the heaters power and the uniformity of the incident radiation flux on the test specimen. Hence, the proposed facility design-layout scheme should satisfy all these requirements for the level of temperature, unevenness of the temperature field on the test specimen surface and the rates of heating and cooling.

3. The Design-Layout Schemes of the Proposed Facilities

3.1. Basic facility design for the serial thermal stability testing of high-temperature materials

The development of a design-layout scheme for the new radiant heating facility was based on the previous results obtained in [7]. The structure of the facility thermal insulation was subjected to significant changes. The main material for the facility thermal insulation is the lightweight high-temperature insulation made of ceramic material based on alumina (Al2O3) for thermal insulation up to 1800 °C. However, the cost of this material is rather high and the standard block thickness is 40 mm. Still, according to the modelling results for the facility thermal insulation, blocks thickness should be 50 mm. Thus, it was considered to use double layer thermal insulation – the main layer from alumina and the additional layer from even more lightweight and relatively low-cost ceramic material based on silica (SiO2). Due to this, it is possible to achieve the required total insulation thickness and to reduce its weight and cost. However, the maximum operating temperature for the silica-based ceramic material does not exceed 1100 °C. Thus, an additional thermal simulation required for checking the feasibility of this solution.

The geometry model of the new facility operating area (Fig. 3) comprises the following components: a silica-based thermal insulation blocks (10 mm thickness each); an alumina-based thermal insulation blocks (40 mm thickness each); a reflective screen from the polished heat-resistant steel (1 mm thickness); nine THLs (11 mm outer diameter and 1.5 mm bulb wall thickness), each of 2.0 kW in capacity; a test specimen (100×100×10 mm overall dimensions); a system of two quartz glass partitions (2 mm thickness each and at the 4 mm distance between each other) for the separation of the THL zone from the test specimen zone. The overall dimensions of the facility operating area are 200×150×250 mm. The proposed facility design is shown in Fig. 4 and it is now under assembly at the Joint Stock Company «Obninsk Research and Production Enterprise «Technologiya» named after A.G. Romashin (JSC «ORPE «Technologiya» named after A.G. Romashin), Obninsk, Kaluga Region, Russian Federation.

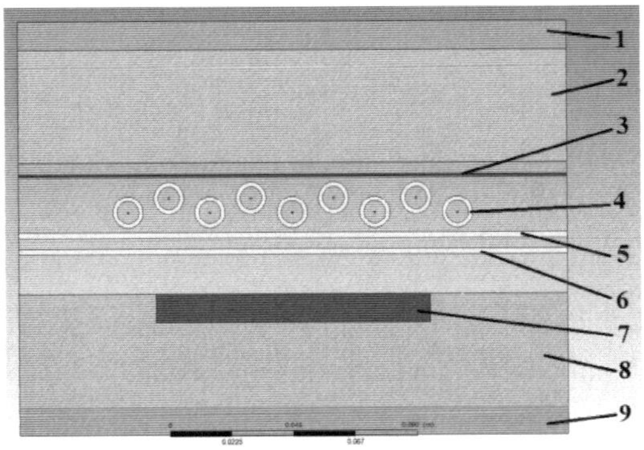

Figure 3. The design-layout scheme of the basic facility: (1, 9) a silica-based thermal insulation; (2, 8) an alumina-based thermal insulation; (3) a steel reflective screen; (4) nine THLs; (5, 6) separating glasses; (7) a test specimen

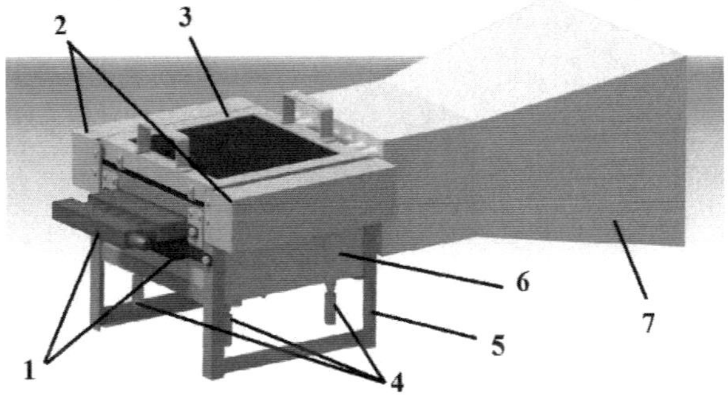

Figure 4. The geometry model of the proposed facility: (1) an air supply duct; (2) lamp housings; (3) a removable facility cover; (4) a removable base with test specimen; (5) a steel frame; (6) the facility operating area; (7) a hot air disposal duct

This design model illustrates the key ideas of the project: the air-cooling of lamps arranged in checkerboard order; dividing of the facility operating area into several zones and alternating air blowing directions in neighboring zones. Dividing of the facility operating area allows to avoid the test specimen cooling during the lamps air blowing. An airflow in the test specimen zone is switched on only after the heating stage is completed and allow to achieve the high cooling rate, which is an additional advantage of the proposed design in comparison with the traditional one. Alternating air blowing directions in neighboring zones allow to decrease the temperature field non-uniformity. Finally, the lamps arranging in checkerboard order allow to improve the heat removal from the bulbs.

Based on this geometry a finite-element model was created. The characteristic dimension of the finite-element mesh was ranged from 0.1 to 1 mm depending on the geometry and size of the facility structural components. The total amount of the mesh elements was about 31500 pcs. The time increment also ranged from the 0.01 s in the beginning and to the 0.5 s after the first several seconds of the calculation. The total process duration for this type of the heating experiment was 600 s, the duration of computations varied from 6 to 8 hours on an advanced computing systems.

The modelling results (Figs. 5, 6) demonstrate that the maximum specimen temperature reaches 2100 K, along with this, the temperature of the THL bulbs as well as the temperature of other facility structural components does not exceed the permitted values (1270 K for the THL bulbs and glass partitions, 1300 K for the heat-resistant steel, 1370 K for the silica-based thermal insulation blocks). The presence of the reflective screen increases the efficiency of the specimen heating due to the re-reflection of the radiation from the lamps to the test specimen. In addition, it could be seen that the thermal insulation blocks above the lamps are almost not heated due to the air-cooling and the presence of the reflective screen.

The maximum values of the airflow velocity (Fig. 7) are found in the clearances between the THL bulbs and the glass partitions and (or) the reflective screen, constituting 22 m/s, the mass-average velocity is about 7 m/s. The total gas flow on cooling the facility structural components is about 0.05 kg/s. The average air temperature at the operating area output (Fig. 8) is about 500 K.

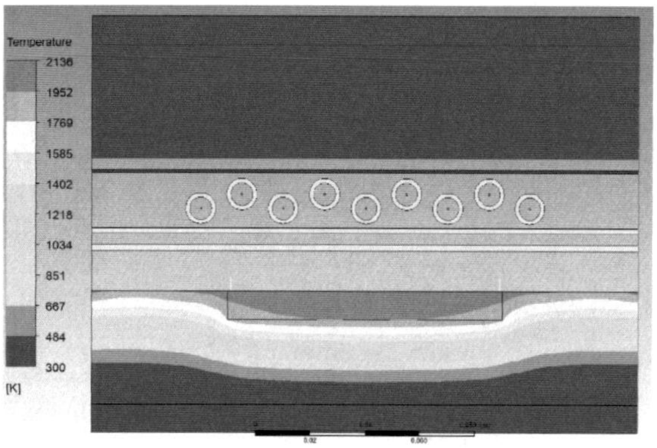

Figure 5. The temperature of the thermal insulation and the specimen under intense heating for 300 s, K

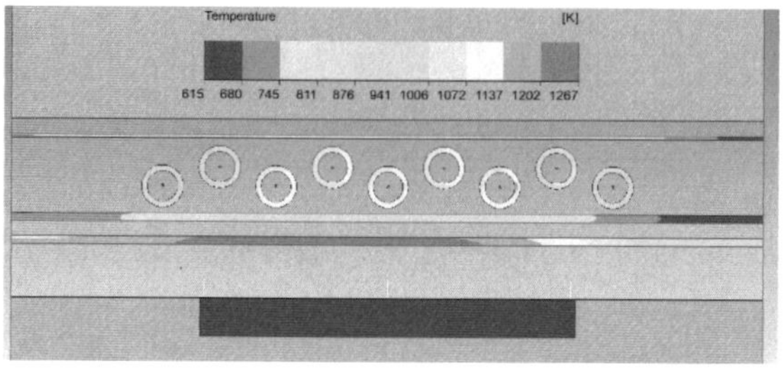

Figure 6. The temperature of THL bulbs, glass partitions and reflective screen at 300 s, K

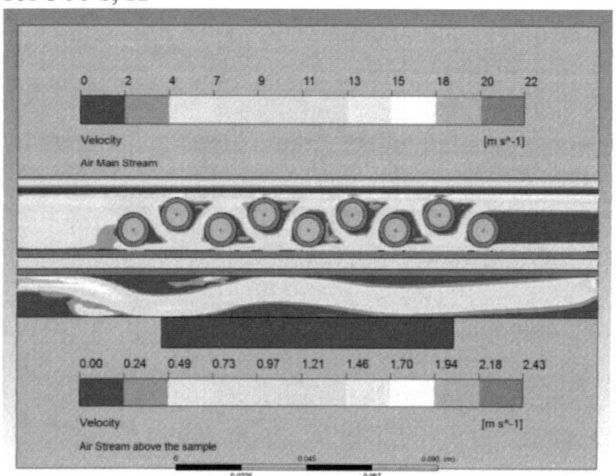

Figure 7. The air velocity field at 301 s, m/s

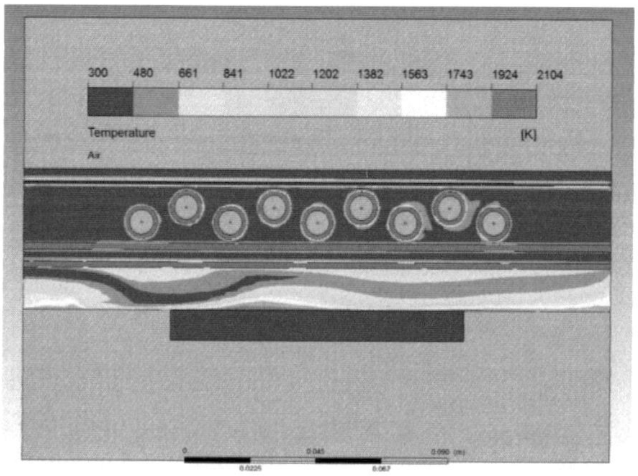

Figure 8. The air temperature field at 301 s, K

In addition to the temperature level, for ensuring the thermal stability testing, the facility should meet the requirements of the non-uniformity of the temperature distribution over the specimen surface, the temperature gradient through the specimen thickness and the rates of heating and cooling. The THL as a radiation source allow to achieve high heating rate.

Still, due to the relatively small lamp diameter, THL is a discrete radiation source. Therefore, the temperature distribution over the test specimen surface is wavy (Fig. 9). The local temperature maximums are located under the lamps. To ensure the testing of the materials thermal stability and determination of the specimen thermal conductivity through the thickness, the non-uniformity of the temperature distribution over the specimen surface should not exceed 5 %. For the traditional THL-based facility design, this is a significant problem. However, the proposed design-layout scheme meets this requirement – the maximum non-uniformity of the temperature distribution over the test specimen surface at 300 seconds is less than 93 °C (4.35 %) (Fig. 9).

Another significant problem of the traditional facility design is inability to achieve high cooling rate due to the thick thermal insulation. However, one of the unique advantages of the proposed design is the possibility to ensure an effective heat removal during the specimen cooling by air blowing of the specimen surface and thus achieving of the high cooling rate. Besides, there is a possibility to install under the specimen a special fridge that could increase the cooling rate and the temperature gradient through the specimen thickness.

In summary, there are four possible opportunities for specimen cooling: 1) just by switching off the THLs; 2) by using, in addition, the active air-cooling in the specimen zone; 3) by using the water-cooled fridge located under the specimen, instead of the air-cooling; 4) by means of combined air and water cooling.

According to the modelling results, the maximum heating and cooling rates respectively are 196 and 236 K/s (Fig. 10). Finally, the maximum temperature gradient through the specimen thickness for the proposed facility is 165 K. For comparison, the required value for the thermal stability test and for the determination of the specimen thermal conductivity through the thickness is 100 K.

Thus, modelling results prove the possibility of using the THL-based heating facilities with active air-cooling for the thermal tests of high-temperature ceramic materials.

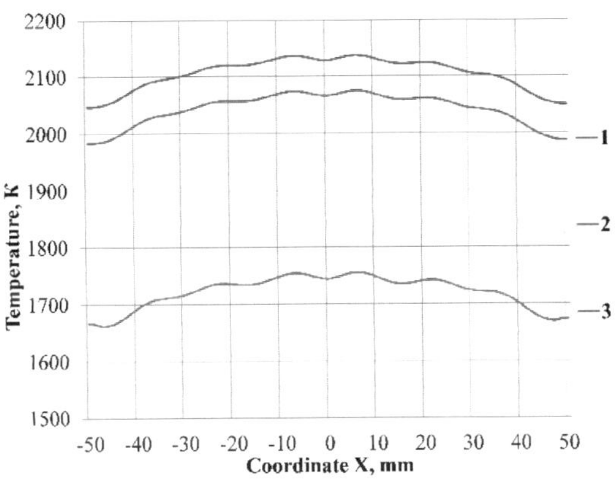

Figure 9. The temperature distribution over the test specimen surface, K:
(1) at 60 s; (2) at 180 s; (3) at 300 s

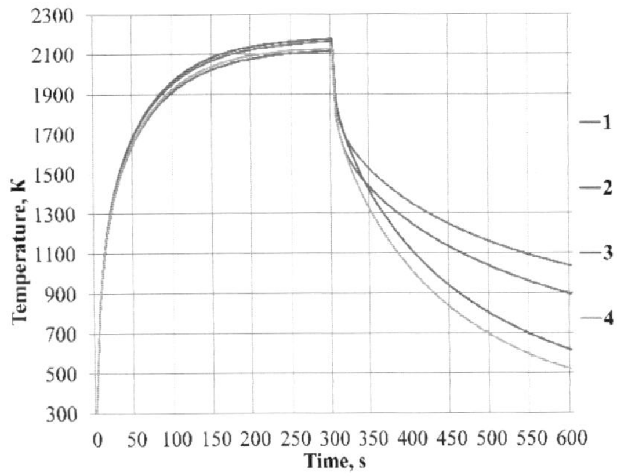

Figure 10. The temperature at a central point on the specimen surface vs. time, K: (1) free convection; (2) active air-cooling; (3) water-cooling; (4) combined air and water cooling

3.2. Facility for the estimation of materials operating life via long-term thermal-cyclic testing

The long-term thermal-cyclic testing is required to prove the possibility of materials application in reusable aerospace vehicles and to estimate its operating life. The thermal-cycling tests mean the implementation of the longstanding repeated cycles of alternate heating and cooling of the test specimen. Thus, in addition to the requirements of the lamps bulbs temperature state and high specimen heating and cooling rates, it is necessary to ensure the efficient thermal insulation of the facility metal structure during the long-term thermal-cycling test. Moreover, the temperatures of other facility structural components also must not exceed their maximum allowable values.

Therefore, the simulation of the facility operating during the thermal-cycling test is required. The total process duration for this type of the heating experiment was 1800 s (three circles of heating and cooling duration for 600 s each). The duration of computations was more than 24 hours. The modelling results (Fig. 11) demonstrate that the temperature on the border between the thermal insulation and the metal frame at the end of the experiment is close to the maximum allowable value for the structural steel – 1100 K.

Thus, it is necessary to ensure the cooling of the facility metal frame and thermal insulation. There are three possible opportunities for the metal frame and thermal insulation cooling: 1) just by free convection; 2) by active air-cooling; 3) by water-cooling. The metal frame temperature must be less than 70 °C to ensure safe facility operating.

The modelling results (Fig. 12) demonstrates that facility design will meet this requirement just in case of water-cooling. Therefore, the possibility of using THL-based facilities for the high-temperature materials long-term thermal-cycling tests has been proved via complex heat transfer simulation. The facility design-layout scheme with water-cooled thermal insulation was proposed.

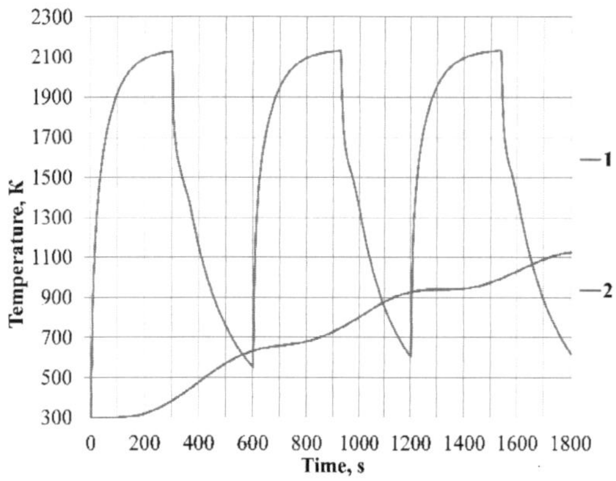

Figure 11. The temperature vs. time graph for the thermal-cyclic test, K: (1) the test specimen temperature; (2) the temperature on the border between the thermal insulation and the metal frame

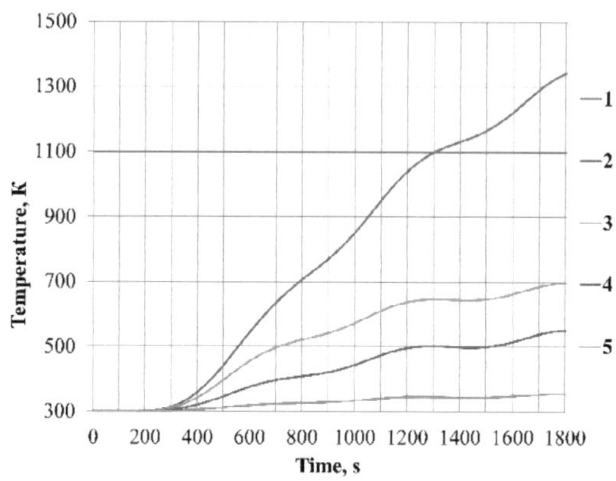

Figure 12. The temperature on the border between the thermal insulation and the metal frame vs. time, K: (1) the temperature limit; (2) isolated frame wall; (3) free convection; (4) active air-cooling; (5) water-cooling

3.3 Modified facility for the precision research of the materials thermophysical properties

The apparatus for solving one-dimensional inverse heat conduction problem is usually used for the determination of materials thermophysical properties. In this case, the facility design must meet the requirements for the uniformity of the temperature distribution over the test specimen surface and as a consequence the uniformity of the incident radiation flux. For the serial testing, the value of the non-uniformity of the temperature distribution over the test specimen surface during the experiment must be less than 5 %. However, for the precision thermophysical tests this value must be less than 3 %. There are two different possibilities to increase the uniformity of the temperature distribution over the test specimen surface.

The first option to meet this requirement is to increase a number of heating sources. The basic facility design-layout scheme allows the adding of four additional halogen lamps at the operating area. Besides this, the lamps power must be optimized. The modelling results allow to propose the dividing of the lamps into four groups and to distribute the power in the following way: first three lamps continue to work in nominal mode, power of the next three lamps was reduced by half, then for next three it was reduced to 0.6 of nominal and for the final group of four lamps it was reduced to 0.75 of nominal. With these modifications, it is possible to achieve 2.51 % non-uniformity of the temperature distribution over the specimen surface during the experiment (Fig. 13).

The second option is to reduce the test specimen size down to 70×70×10 mm and to move it at 5 mm to the right (in the blowing direction). In addition, as well as in the previous case, the lamps power must be optimized. In this case, the modelling results allow to propose dividing the lamps into three groups and to distribute the power in the following way: three left lamps continue to work in nominal mode, for the three central lamps power was reduced to 0.8 of nominal and for the three right lamps it was increased to 1.075 from the nominal. By this way, it is possible to achieve even better result – 1.73 % non-uniformity of the temperature distribution over the specimen surface (Fig. 14). However, this modification restricts the specimen size.

Therefore, the possibility to improve the accuracy of the determination of the materials thermophysical properties via THL-based heating facilities was proved. The modifications of the basic facility design-layout scheme were proposed.

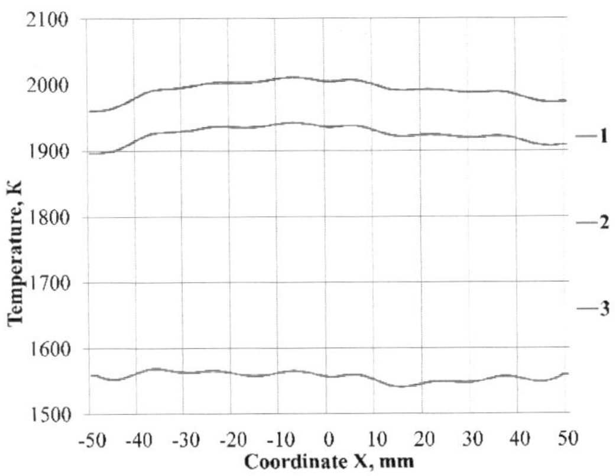

Figure 13. The temperature distribution over the test specimen surface for the 13 lamps scheme, K: (1) at 60 s; (2) at 180 s; (3) at 300 s

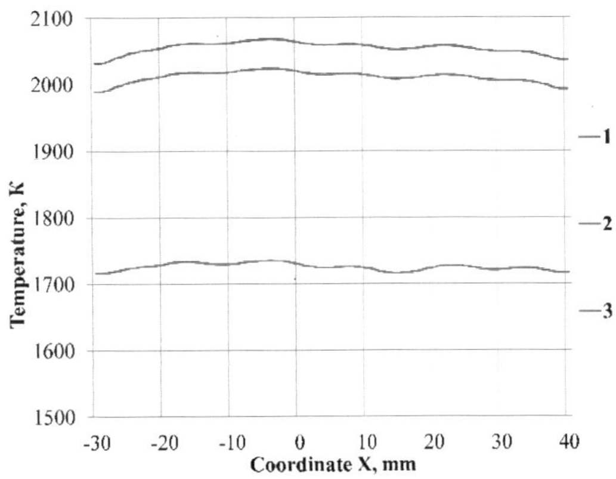

Figure 14. The temperature distribution over the test specimen surface for the 70 mm specimen scheme, K: (1) at 60 s; (2) at 180 s; (3) at 300 s

4. Conclusion

The research results demonstrate that the mathematical modelling of a complex heat transfer process is a feasible tool for the designing of the heating test facilities that are used in aerospace engineering. As a result, it is possible to significantly reduce the amount of time and material resources required for the development of the facilities. The usage of the tungsten halogen lamps in the radiant heating facilities for the thermal testing of ceramic materials with operating temperatures up to 2000 K is possible due to the lamps bulbs active air-cooling. However, the development of such facilities requires multi-physical processes simulation and study of the different facilities systems operation.

Acknowledgement

Certain scientific results were obtained within the research project № 18-38-00687 and were funded by Russian Foundation for Basic Research (RFBR).

References

[1] Bansal N P (ed.) 2005 *Handbook of ceramic composites* (Kluwer Academic Publishers. Springer Science & Business Media) 554 p. DOI: 10.1007/b104068

[2] Prosuntsov P V and Taraskin N Y 2017 Theoretical and numerical characterization of the thermal conductivity of carbon ceramic material from realistic microscale imaging *EDP Sciences – MATEC Web of Conferences* vol. **92** art. no. 01008 5 p. DOI: 10.1051/matecconf/20179201008

[3] Reznik S V 2000 Mathematical models of radiative-conductive heat exchange in heat-shield materials of shuttle space transportation systems *Journal of Engineering Physics and Thermophysics* vol. **73** no. 1 pp. 9-24 DOI: 10.1007/BF02681671

[4] Petrova A A and Reznik S V 2016 Theoretical Rationale of Heating Block for Testing Bench of Aerospace Crafts Thermal Protection Elements *EDP Sciences – EPJ Web of Conferences* vol. **110** art. no. 01048 5 p. DOI: 10.1051/epjconf/ 201611001048

[5] Zvorikin D B, Alexandrova A T and Baikaltsev B P 1985 *Reflective infrared heat ovens* (M.: Machinostroenie) 176 p. [in Russian]

[6] Yeliseyev V N and Tovstonog V A 2001 Characteristics of High Intensive Heat Radiation Sources and Systems *Vestnik MGTU. Mashinostroenie* no. **4(45)** pp. 3-32 [in Russian]

[7] Baldzhiev R S and Prosuntsov P V 2017 Modeling Combined Heat Transfer in the Operating Area of the High Power Heating Test Facility International *EDP Sciences – MATEC Web of Conferences* vol. **110** art. no. 01009 4 p. DOI: 10.1051/matecconf/ 201711001009

IOP Conf. Series: Materials Science and Engineering **683** (2019) 012063 doi:10.1088/1757-899X/683/1/012063

Study of the GTE stator vane samples made from a Al-SiC metal matrix composite reinforced by SiC.

V V Berezovskiy[1] and J A Kurganova[2]

[1] All-Russia Institute of Aviation Materials, Moscow, 105005 Russia deputy of chef of laboratory in

[2] Bauman State Technical University, Moscow, 105005, Russia

Abstract. Al-SiC system metal matrix composites application in gas turbine engine ventilator stator blade is shown. Mechanical properties of Al-SiC system metal matrix composite in comparison with aluminum and titanium alloys is shown. Mechanical properties of gas turbine engine ventilator stator blade is shown.

Key words: Metal matrix composites, Composites based on aluminum alloy, Composites reinforced by SiC particles, Gas turbine engine ventilator stator blade.

1. Introduction

Increasing the strength characteristics of metallic materials is currently an urgent problem of materials science in connection with the sharply increased requirements from the designers of advanced products of aviation and space technology. One of the effective ways to solve this problem is the development of technology and the creation of composite materials on a metal matrix with a reinforcing phase [1-3].

Abroad, they find more and more widespread use in aviation products, MMC equipment on an aluminum matrix reinforced with silicon carbide isometric particles, which allows reducing the mass of parts and structural elements by 15-50%, increasing their elastic modulus and fatigue characteristics by 1.5-2 times compared with matrix metal, while providing increased efficiency, structural reliability, and when compared to polymer composite materials - a significant reduction in operating costs [4-6].

As applied to fanjet engines, a striking example of the use of discretely reinforced Al-SiC system MMC is the manufacture of fan stator blades.

Initially, in the construction of aircraft engines used solid, and later hollow stator blades made of titanium alloy. The disadvantages of such blades were a lot of weight and high manufacturing cost. With the advent of polymeric materials, carbon fiber instead of titanium alloys began to be used, which led to a reduction in part weight. However, the blades of graphite-epoxy material had low ballistic characteristics, sensitivity to erosion from particles and raindrops during the flight, which led to expenses increase and a decrease in flight safety.

The solution to the problem was the joining (soldering) of titanium foil over the carbon-fiber leading edge of the blade, which led to an improvement in erosion resistance, but not ballistic properties and, as a consequence, a significant increase in the value of the blade. [7]

Since 1996, in the Pratt & Whitney 4084, 4090 and 4098 series engines, SiC reinforced MCM Al-SiC

(6092 / SiC / 17.5p) was used instead of carbon fiber. This MMC is manufactured by powder metallurgy by mixing the starting components, followed by consolidation into briquettes. Next, the resulting briquettes are extruded. Over 50,000 blades have been produced. Through the use of this MMC, a sevenfold decrease in the rate of erosion and a significant increase in ballistic resistance from foreign objects were found. In addition, the service life of the blades increased by 300%, which reduced the cost of maintenance and repair and increased the cost savings on the life cycle. The total savings from such implementation exceeded $ 100 million [8-10]

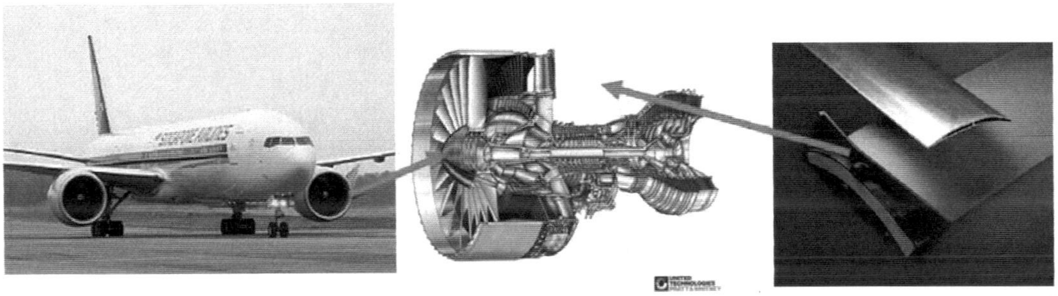

Figure 1. GT fan stator blades of MCM Al-SiC (6092 / SiC / 17.5% -T6), manufactured by DWA Aluminum Composites Inc.

The paper presents the results of studies of simulator samples of the stator blade of the GTE guide vane made from dispersion-strengthened metal composite material based on a deformable aluminum alloy D16 reinforced with silicon carbide particles MMC Al-SiC. The mechanical properties of the obtained MMC and stator blade simulator samples were determined.

2. Research methodology

The technology of the fabrication of MCM included the preparation of the initial mixture for mechanical alloying, sifting the material through the vibrating screen in order to avoid random large carbides falling into the MMC. Mechanical alloying of the charge. Briquetting of granules followed by extrusion on a hydraulic press.

The aging alloy D16 (GOST 4784. 2024 Al alloy) in the annealed state (particle size = 10 μm) was chosen as the matrix alloy. According to GOST 26327, silicon carbide powder (d = 5-10 μm) was used as a reinforcing component.

The initial components were mixed in the CмB-100 vibromixer in argon environment.

MCM samples were made by hot extrusion of pre-bricked MCM components.

The density and porosity of the obtained samples were investigated by the method of hydrostatic weighing in accordance with GOST 2160-92.

Samples for metallographic examinations were prepared on the Struers equipment complex.

Microstructure was analyzed on an Olimpus light microscope with x50 and x200 magnifications. For image processing, SIAMS software package was used.

The values of tensile strength and elasticity modulus were determined by testing samples on a tensile machine INSTRON 1195 as per GOST 4765-73, OST 6-10-411-77, and GOST 11262 - 80.

Simulator samples of stator blades were manufactured on a 1K62 turning-cutting machine with subsequent finishing.

The physicomechanical properties of the simulator samples were determined on an INSTRON 1195 tensile testing machine at FSUE "Central Institute of Aviation Motors" by Dr. Sc. Professor T.D. Karimbaev research group .

Research results and discussion.

MMC samples were made on the core of D16 (2024) aluminum alloy reinforced with silicon carbide particles with a volumetric content of a reinforcing component of 20%. The brand of the matrix alloy and the volume content of the reinforcing component were selected based on the experience of previous studies [12]

The figure shows the microstructure of MMC with a 20% by volume content of the reinforcing component. The particles of the reinforcing component are evenly distributed in the matrix alloy. Structure defects are large agglomerations of silicon carbide particles, pores are not detected visually. Most accurately, the presence or absence of pores in the structure can be judged by the difference in the values of the theoretical and experimentally measured density of the material. The theoretical density of the material is 2870 kg/m3

Porosity measured by the method of hydrostatic weighing amounts to 2820 kg/m3. Thus, the porosity of the samples does not exceed 1%.

The mechanical properties of the MMC samples of the Al-SiC system at 20°C and 300°C were investigated.

Table 1 presents the values of the physicomechanical characteristics of the investigated MCM in comparison with the matrix aluminum alloy D16 and titanium alloy BT6 (6Al-4V titanium alloy), traditionally used in the cold nodes of the gas turbine engine and airframe of a modern passenger aircraft.

Table 1. Values of the physicomechanical characteristics of the investigated MCM in comparison with the matrix aluminum alloy D16 and titanium alloy BT6

Item No.	Material	Tensile Strength at 20°C (MPa)	Density (kg/m^3)	Elasticity modulus (GPa)
1	MCM Al-SiC	600	2870	110
2	D16*	480-490	2740	72
3	BT6*	900-1100	4430	115

The table shows that the values of the physicomechanical characteristics of the developed MMC exceed the characteristics of the matrix alloy D16, and the modulus of elasticity reaches the values characteristic of titanium alloys, while maintaining the density characteristic of aluminum alloys. Compared to the matrix alloy, the strength of MMC significantly increases at elevated temperatures. This is apparently due to the fact that under the applied load, the load is transferred from the weaker matrix through the interface of the matrix / hardener to the usually more rigid hardener [11-16]. Thus, the hardening occurs by bearing the greater part of the applied load by the hardener.

Figure 1 shows the appearance of the simulator samples of the stator blade of the GTE fan guide vanes.

Figure 1. Simulator sample of the stator blade of the GTE fan guide vanes of the MMC Al-SiC system

Table 2 presents the values of the mechanical characteristics of the samples of stator blades simulators in comparison with the values of the matrix alloy D16.

Table 2. Values of the mechanical characteristics of the samples of stator blades simulators in comparison with the values of the matrix alloy D16 and titanium alloy BT6.

Item No.	Material	Tensile strength at 20°C (MPa)	Elasticity modulus (GPa)
1	Simulator samples of stator blades	505	110
2	MMC	600	110
3	D16	470-490	72

The table shows that while maintaining high values of the modulus of elasticity, the tensile strength of the simulator samples is 20% lower than the tensile strength of the MMC from which they are made. Due to the fact that all the samples collapsed in the zone of transition from the blade airfoil to the lock - the zone of stress concentration (Fig. ...), the drop in strength values is due solely to the design features of the simulator samples.

3. Conclusions

Analysis of the structure and properties demonstrates the suitability of technology for the manufacture of parts of MMC, which proves the level of mech. properties in comparison with the analogue.

MMC designed and manufactured therefrom simulator samples showed a significant increase in the physical and mechanical properties in comparison with aluminum alloys, and on the elasticity modulus reach values characteristic of the titanium alloys. Completed, together with Central Institute of Aviation Motorsv, studies of the mechanical characteristics of stator-blade simulator samples showed the possibility of using the developed MCM as a material of the stator vanes of the guide vanes of a promising gas turbine engine.

References

[1] Kablov E.N. *Materialyi novogo pokoleniya. Zaschita i bezopasnost*, 2014. No. 4. P. 28-29.

[2] Kablov E.N. 2015 Innovatsionnyie razrabotki FGUP "VIAM" GNTs RF po realizatsii «Strategicheskih napravleniy razvitiya materialov i tehnologiy ih pererabotki na period do 2030 goda». *Aviatsionnyie materialyi i tehnologii* **1** 3-33

[3] Kablov E.N., Grashchenkov D.V., Shchetanov B.V., Shavnev A.A., Nyafkin A.N., Emikh L.A.,Vdovin S.M.,Nishchev K.N., Chibirkin V.V., Eliseev V.V.*AlSiC-based metal matrix composites for power electronic devices* Composites:Mechanics,Computations,Applications,

2013.**4**(1)65-74.

[4] Graschenkov D.V. 2017 *Strategiya razvitiya nemetallicheskih materialov, metallicheskih kompozitsionnyih materialov i teplozaschityi. Aviatsionnyie materialyi i tehnologii*, No Sspecial. 264–271.

[5] Krasnov E.I., Shteynberg A.S. Shavnev, A.A., Berezovskiy V.V. 2013 Issledovanie sloistogo metallicheskogo kompozitsionnogo materiala sistemyi Ti TiAl3. *Aviatsionnyie materialyi i tehnologii*.. **3** 16–19

[6] Kosolapov D.V., Shavnev A.A., Nyafkin A.N., Grishina O.I. 2016 Issledovanie formirovaniya strukturyi kompozitsionnyih granul Al–SiC. *Aviatsionnyie materialyi i tehnologii* **3** 49-52.

[7] Shavnev A.A., Berezovskiy V.V., Kurganova Yu.A. 2015 2 Osobennosti primeneniya konstruktsionnogo metallicheskogo kompozitsionnogo materiala na osnove alyuminievogo splava, armirovannogo chastitsami SiC. Chast 2 (obzor). *Novosti materialovedeniya. Nauka i tehnika*.. **3**(15) 11-17.

[8] Chawla N. Chawla K.K.2006 *Metal Matrix Composites*,Springer Sience+Business Media, Inc, 401.

[9] Olivier B (2001) *Metal Matrix Composites (MMCs) from Space to Earth.* Eiggenossische Materialprufungs-und Forschungsanstalt, EMPA, Abt. Werkstofftechnologie, Feuerwerkerstrasse 39, CH-3602 Thun. URL:https://www.ethz.ch/content/dam/ethz/special-interest/matl/department/news/materialsday/materialsday-2001-beffort.pdf

[10] Evans, A., C.S. Marchi and A. Mortensen, 2003. Metal Matrix Composites in Industry. Kluwer Academic, New York pp: 19-38.

[11] Ganesh V.V. and Chawla N. 2003 *Effect of Reinforcement-Particle-Orientation Anisotropy on the Tensile and Fatigue Behavior of Metal-Matrix Composites.* Metallurgical and materials transactions volume **34A**, 1

[12] Berezovskiy V.V., Solyaev Yu.O., Lure S.A., Babaytsev A.V., Shavnev A.A., Kurganova Yu.A. 2014 Issledovanie mehanicheskih svoystv metallicheskogo kompozitsionnogo materiala na osnove alyuminievogo splava, armirovannogo dispersnyimi chastitsami karbida kremniya. Deformatsiya i razrushenie materialov.. **14** 12-16.

[13] Lloyd DJ. 1997 *Composite engineering handbook*. New York, NY: Marcel Dekker; 631–670.

[14] Rawal S, 2001 *Metal Matrix Composites for Space Applications* Journal of Materials, **53**14-17.

[15] Surappa M.K., 2003 *Aluminum matrix composites: challenges and opportunities* Sadhana, India. **28**, Parts 1 & 2, February/April 319–334

IOP Conf. Series: Materials Science and Engineering **683** (2019) 012064 doi:10.1088/1757-899X/683/1/012064

Estimation of the effective elastic characteristics of carbon composite materials by testing ring-shaped samples under high temperatures

K A Ponomarev[1]

[1]Carbon Material Division, JSC "Kompozit", Korolev, Russia
Email: kirill.a.ponomarev@gmail.com

Abstract. A computational and experimental method is suggested to determine indirectly the elastic properties of carbon composite materials at room temperature up to 2200°C and more. The method involves loading of ring-shaped test specimens through diametrical compression using non-cooled indenters at three points, within the elastic strain range, with ductility measurements. Based on the measured ductility values, shear moduli, as well as the longitudinal and transversal elasticity moduli in the test specimen cutting plane, can be determined by solving the identification problem. Results of experimental assessment of the permissibility of repetitive mechanical and thermal loading of the carbon-carbon composite material are presented. Method tryout with high-density needle-punched composite material is described.

Keywords: carbon-carbon composite, elasticity modulus, shear modulus, high temperature, cyclic loading

1. Introduction

Analysis of the effective mechanical characteristics, particularly, the elastic properties of high-temperature composites, within the working temperature range, is critical in terms of simulation modeling of structural components made of these materials. For this purpose, it is necessary to perform strain measurements over the gauge length of the test specimen during mechanical tests. For strain measurements, contact [1–3] as well as contactless (generally, optical [4, 5]) strain gauges are used.

Direct measurements methods have a number of hardware limitations, given the high test temperatures; therefore, of equal interest are indirect methods that involve estimation of the elastic properties based on measurements of the change in the position of the test plant crosshead.

In addition to the Young's modulus of elasticity, shear modulus measurements are also topical. For anisotropic materials, V-notched test specimens are usually used [6, 7], especially when it comes to high temperature tests. The test specimens of this sort, however, have a number of shortcomings, namely, the narrow area of pure shear strains and the difficulty of their unambiguous differentiation from bending strains. In case of high-temperature measurements, when it is impossible to use any strain-measuring means, the finite rigidity of the test fixture and, when applying local heating to the test specimen, is in a complex stress state under steep thermal gradient conditions, contributes to the measurement error.

Problem formulation

This paper describes a method for indirect estimation of the elasticity/shear moduli of carbon-based composite materials (particularly, CCCM), under uniaxial diametrical compression of ring-shaped test specimens. The method is applicable both at room temperature [8] and at high temperatures. With this method it is possible to determine elasticity moduli (longitudinal E_1 and transversal E_2) and shear modulus G_{12} for a material orthotropic in the ring cutting plane, using the same test specimen, over a wide temperature range.

The method involved measurement of the ductility values by applying compressive loads to the ring-shaped test specimen, within the elastic strain range, using square-section indenters, consecutively, first along one of the orthotropy axes (see Figure 1, a), then at 45° to this axis (see Figure 1, b) , and finally at 90° to this axis (see Figure 1, c).

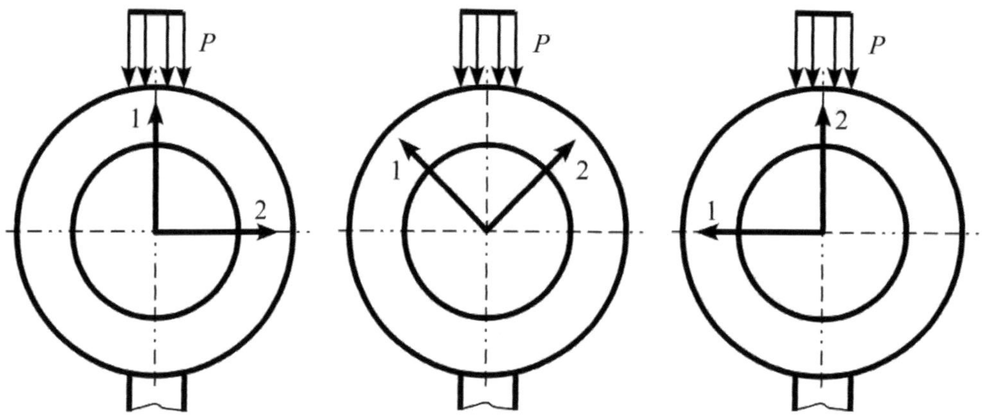

Figure 1. Load patterns (a, b, c) for consecutive loading of the ring-shaped test specimen.

Under diametrical compression, ductility values of ring-shaped test specimens are greater than those of test specimens that have no hollow space. Moreover, there are unambiguous dependences of the deflection towards the load direction on the elasticity/shear moduli. For the elastic properties within the CCCM characteristic range, influence of Poisson's ratio ν_{12} on the stress-strain state of the ring is very small.

Dependence of the ring-shaped test specimen's ductility along the load axis S^{theor} on its elastic properties, for a certain geometry, can be summarized in a table using one of the existing calculation methods, e.g. [9, 10]. In our case, we calculated the dependences using the finite-element simulation modeling. Relative ductility s of the ring-shaped test specimen of particular dimensions can be specified as a function of two relative elastic characteristics e_2 and g_{12}:

$$s = S^{theor} E_1 = f\left(e_2, g_{12}\right),$$

where

$$e_2 = \frac{E_2}{E_1},\ g_{12} = \frac{G_{12}}{E_1}.$$

Characteristic dependences of the relative ductility obtained for the ring-shaped test specimen, 50 mm outer diameter, 10 × 10 mm cross section, on e_2 and g_{12} are shown in Figure 2.

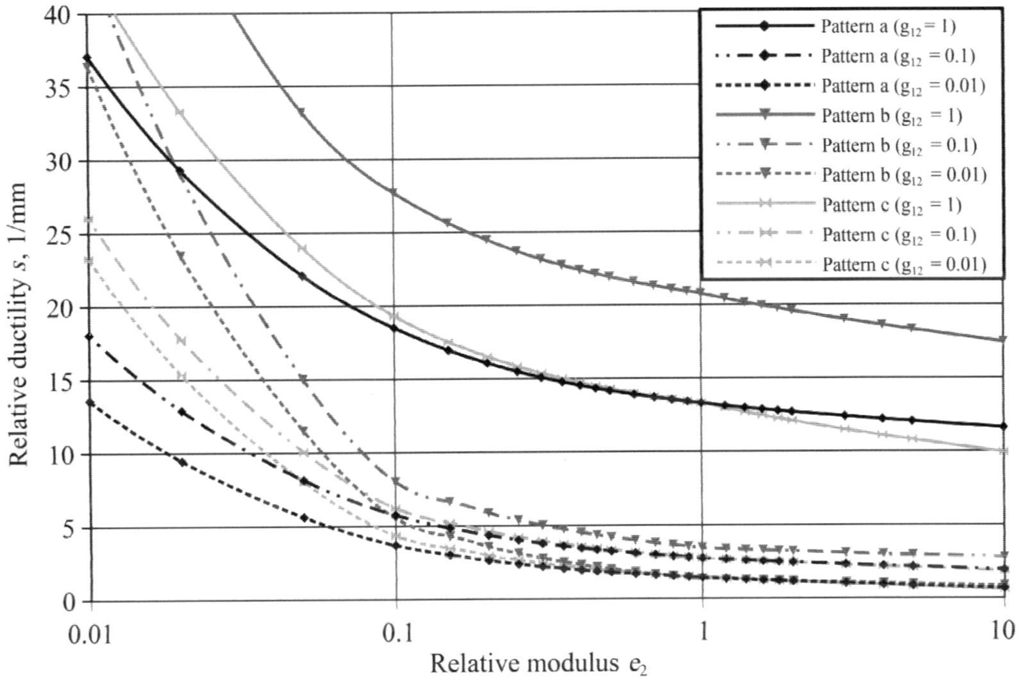

Figure 2. Characteristic dependences of ring-shaped test specimen's relative ductility on its relative elastic characteristics.

Since the suggested method does not involve loading of the test specimen beyond the elastic strain range, it is possible, given the technological features of the material being investigated, to perform a full cycle of measurements with the same test specimen over the entire temperature range, thus reducing variations in the experimental values of the elasticity modulus.

For ductility measurements under temperatures close to the room temperature, an extensometer fixed on the ring's inner surface along the load direction can be used. Under high temperatures, it is suggested to use local heating in the inert atmosphere on the UVTK-type high-temperature mechanical test plant [11].

The algorithm for determining the elastic characteristics for the ring-shaped test specimen involves the following steps:

1. Consecutive loading of the test specimen within the elastic strain range as per load patterns a, b and c (see Figure 1) at room temperature, measuring ductility S_0 of the entire powertrain, based on the crosshead displacement, and ductility S_0^{sp} of the ring by extensometer readings.
2. Evaluation of ductility contribution S' to the total ductility of the entire force loop as follows:

$$S' = S_0 - S_0^{sp}.$$

For loading, cooled test fixture is used, its temperature and ductility depending only slightly on the test specimen local heating temperature. Design and material of the non-cooled indenters are chosen so that their ductility can be deemed negligibly small as compared to the ductilities of the cooled test fixture and the heated test specimen.

3. Consecutive loading of the same test specimen at temperature T_i as per load patterns a, b and c (see Figure 1), measuring ductility S_i of the entire force loop.

4. Calculation of the test specimen ductility at temperature T_i based on the measured ductility S_i values as follows:

$$S_i^{sp} = S_i - S'.$$

Repeat steps 3 and 4 for each necessary temperature point.

5. Evaluation of moduli $E_{1\,i}$, $E_{2\,i}$ and $G_{12\,i}$ based on value sets $S_{a\,i}^{sp}$, $S_{b\,i}^{sp}$ and $S_{c\,i}^{sp}$.
Evaluation of the target moduli represents an identification problem and can be reduced to minimization of residual function Φ_i that looks as follows:

$$\Phi_i = \left(\frac{S_a^{theor}(E_1, E_2, G_{12}) - S_{a\,i}^{sp}}{S_{a\,i}^{sp}} \right)^2 + \left(\frac{S_b^{theor}(E_1, E_2, G_{12}) - S_{b\,i}^{sp}}{S_{b\,i}^{sp}} \right)^2 + \\ + \left(\frac{S_c^{theor}(E_1, E_2, G_{12}) - S_{c\,i}^{sp}}{S_{c\,i}^{sp}} \right)^2,$$

where S_a^{theor}, S_b^{theor} and S_c^{theor} – calculated ring-shaped test specimen ductility values for load patterns a, b and c, correspondingly. The target moduli values are those corresponding to the relevant calculated ductility values.

Assessment of the permissibility of repetitive mechanical and thermal loading of the test specimen

An important feature of the suggested method is a series of consecutive measurements during repetitive cycles of test specimen loading/unloading. This may result in degradation of the material properties during the experiment, which is unacceptable.

Tests of materials under investigation have shown no changes in the experimental rigidity values under uniaxial cyclic tests at room temperature, when the loads stay within the proportional elastic limit along the load direction [8].

Assessment of the permissibility of measurements under repetitive mechanical and thermal loading required additional investigations. For the assessment, we used high-density 4D-inplane rod-reinforced CCCM with a coal-tar pitch coke matrix after multiple high temperature treatment (HTT) at about 2250 °C. The tests were performed using the UVTK-type thermomechanical test plant [11], with tensile test specimens cut at 30° to the reinforcement direction, in the transversal anisotropy plane. The loading was performed according to the sinusoidal law, with the frequency of 1 Hz, from zero load up to 20% of proportional elastic limit of the material, over the target temperature range up to 2500 °C.

The results obtained over the entire temperature range (up to the HTT temperature) showed that cyclic loading, after hundreds of load cycles, does not have a significant impact on the ductility values both at the target temperature and after the test specimen has cooled down, as related to the initial ductility values. Dependence of the ductility of the powertrain on the test time including the time of test specimen heating, soaking under 2200 °C and cooling is shown in Figure 3.

Repetitive (about ten) heating cycles up to temperatures not exceeding the HTT temperature did not show any impact on the rigidity level either.

At test temperatures higher than the HTT temperature, there was a gradual change in the ductility level, both when soaked under high temperature and after the test specimen has cooled down. The change increased significantly as the temperature approached 2500°C.

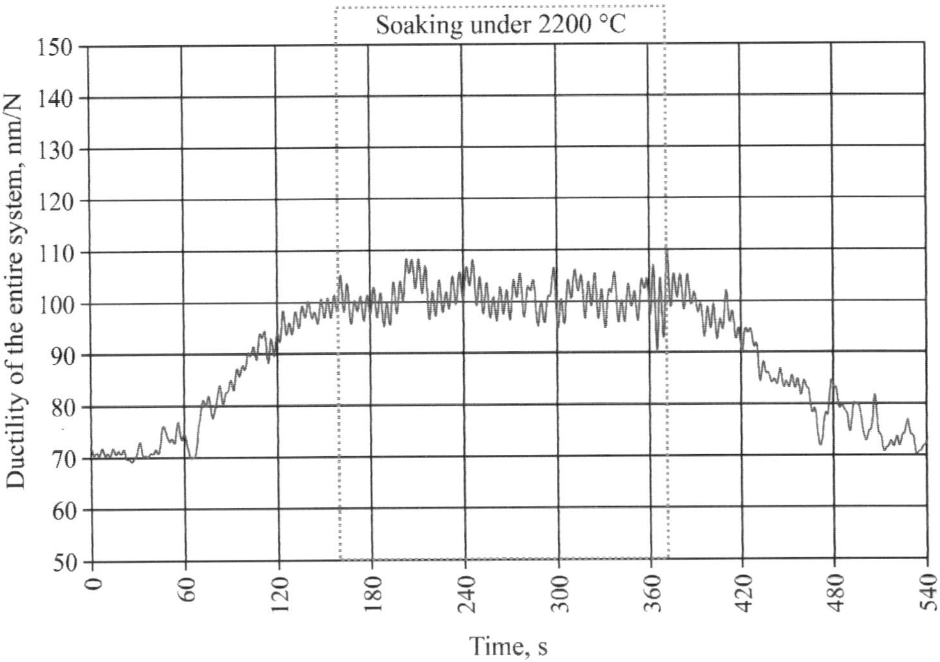

Figure 3. Time dependence of the ductility under cyclic loading during heating, soaking under high temperature and cooling (for 4D-inplane-reinforced CCCM).

Results of method tryout

For room temperature measurements, the method was tried out with several CCCM types, primarily, rod-reinforced CCCM. Results obtained for ring-shaped test specimens showed good agreement with those obtained using other direct measurement methods [8].

To try out the method over a wide temperature range, we developed a special test fixture and a special heater to provide uniform temperature distribution in the test specimen.

The method was tried out with high-density 2.5D-type needle-punched CCCM [12] based on Torayca T700S-type fiber-based carbon tape and nonwoven web with a coal-tar pitch coke matrix, after HTT at about 2250°C. During trial testing, we decided not to exceed the HTT temperature, as it would require an additional study of the influence of these temperature conditions on the elastic properties of the test specimen material.

The test specimens were cut in the XZ plane of the test material (here X – one of the carbon tape laying directions, Z – needle punching direction). Results obtained at room temperature were compared with those obtained using strain-gage sensors during tensile tests (X axis), compressive tests (X and Z axes), as well as shear tests, using V-notched shear specimens [7]. In all cases, there was a good agreement between the results obtained with the ring-shaped test specimens and those obtained using alternative methods.

High temperature tests were performed at 1000, 1500, 2000 and 2200°C. Results obtained at room and high temperatures, as well as comparison values for room temperature, are shown in Figure 4.

The temperature dependences obtained for the elastic properties during trial tests with needle-punched CCCM are in a good agreement with the dependences stated in the available publications [13], which attests to applicability of the method.

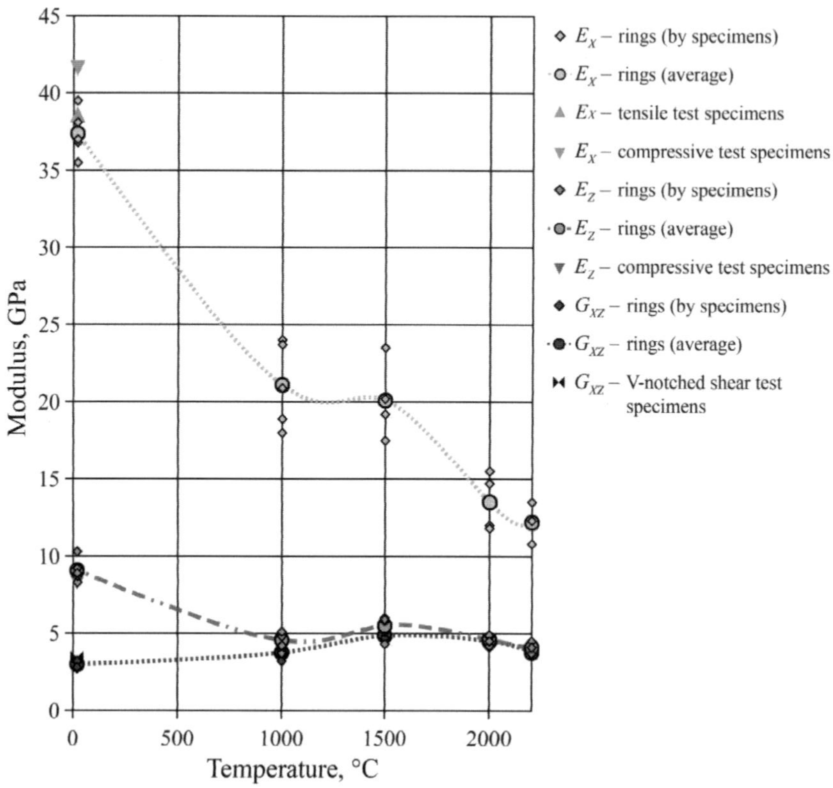

Figure 4. Results of method tryout with 2.5D-type needle-punched layered CCCM (X – one of carbon tape laying directions, Z – needle punching direction).

Conclusions

A computational and experimental method is suggested to determine indirectly the elastic properties of carbon composite materials under local heating. The method is based on diametrically-compressed test specimen ductility measurement along different directions, with concurrent measurement data processing.

Prior to high-temperature testing of CCCM, assessment of the permissibility of repetitive mechanical and thermal loading of CCCM was performed. Cyclic loading up to 20% of the proportional elastic limit, after hundreds of load cycles, was shown not to have an impact on the ductility values both at high temperature and after the test specimen has cooled down, as related to the initial ductility values.

Trial testing of the method was performed with pitch-based needle-punched CCCM at room temperature to 2200°C. Results obtained at room temperature show good agreement with the data obtained using direct measurements with strain-measuring means. The temperature dependences of the properties being investigated correspond to those described in the available publications.

The main advantages of the method are small size of the test specimen, the possibility of testing the same test specimen over a wide temperature range, as well as satisfactory evaluation of the shear modulus at high temperatures.

References

[1] Nair S V and Jakus K 1995 *High Temperature Mechanical Behavior of Ceramic Composites* (Newton: Butterworth-Heinemann)Bullock E 1989 Research *and Development of High Temperature Materials for Industry* (Barking: Elsevier Science Publishers)

[2] Klokova N P 1990 Resistance Strain Gages: Theory, Design Methods, Models (Moscow: Mashinostroenie)

[3] Krivenyuk V V 1990 *Prediction of Long-Term Strength of Refractory metals and Alloys* ed Pisarenko G S (Kiev: Naukova Dumka)

[4] Kobayashi K and Yamaguchi I 2011 High Accuracy and Wide Temperature Measurement by Non-Contact Speckle Extensometer *Strain* **47** pp e98–e103

[5] ASTM D5379M-12 *Standard Test Method for Shear Properties of Composite Materials by the V-Notched Beam Method*

[6] Belov N V, Kamyshev A B, Magnitsky I V and Taschilov S V 2011 Development of Composite Material Shear Test Methods at Normal and High Temperatures *Composite material structures* **3** pp 77–85.

[7] Bakulin A A, Magnitsky I V, Ponomarev K A and Tashchilov S V 2015 Evaluation of the Elastic Properties of Spatially-Reinforced Carbon-Carbon Composite Materials *Composite material structures* **2** pp 46–51.

[8] Tokovyy Y V and Ma C C 2009 Analytical solutions to the planar non-axisymmetric elasticity and thermoelasticity problems for homogeneous and inhomogeneous annular domains *Int. J. Eng. Sci.* **47** pp 413–437

[9] Tokovyy Y V, Hung K M and Ma C C 2010 Determination of stresses and displacements in a thin annular disc subjected to diametral compression *JOTH* **165** pp 342–354

[10] Belov N V, Vagin V P, Dvoretsky A E, Quitca E V, Magnitsky I V, Mironikhin A N, Ponomarev K A and Tashchilov S V 2013 Test Plant for Mechanical Testing of Composite Materials at High Temperatures *Proceedings of the Xth All-Russian Conference "New Technologies"* vol 2 (Miass, RAS) pp 45–51.

[11] Chen T, Liao J, Liu G, Zhang F и Gong Q 2003 Effects of needle-punched felt structure on the mechanical properties of carbon/carbon composites *Carbon* **41** pp 993–999

[12] Chen Z, Fang G, Xie J, Liang J 2016 Experimental study of high-temperature tensile mechanical properties of 3D needled C/C–SiC composites *Materials Science and Engineering: A* **654** pp 271–277

IOP Conf. Series: Materials Science and Engineering **683** (2019) 012065 doi:10.1088/1757-899X/683/1/012065

Optimization of curing mode of epoxy resin based composites

Yangyang Chen[1], G Malysheva[1]

[1]Bauman Moscow State Technical University, Moscow, Russia Federation

E-mail: yangyangchen@mail.ru, malyin@mail.ru

Abstract. In this article the techniques for determining the thermos-physical properties of epoxy binders depending on their different state of aggregation are presented. The epoxy composition based on epoxy resin and modified aliphatic polyamine was used as an object of research. Experimental studies were carried out to determine heat capacity and thermal conductivity of epoxy binders, depending on their degree of conversion. Mathematical modeling of the curing process of epoxy binder at different heating rates by the finite element method was carried out. When using the results of mathematical modeling, the kinetics of the curing process was optimized depending on different heating rates by the ideal point method. Moreover, a study was conducted on the effect of limit condition on optimization results.

1.Introduction

Epoxy oligomers are widely used in aircraft manufacturing, shipbuilding, and in the production of space technology, since they have a series of distinctive features: their good adhesive properties with respect to a large number of fillers and substrates, good water resistance and unique technological properties [1-3].

In the process of curing epoxy oligomer (and thermosetting binders of any chemical nature) there is a change in its aggregative state and the material from the liquid initially goes into a gel state and then into a solid [4,5].

A typical two-stage technological process of curing the binder consists of five sections: on the first and third, there is an increase in temperature, on the second and fourth - remained constant temperature, and on the fifth section - cooling [6]. The time and temperature of exposure in the second section are determined, as a rule, experimentally for each particular product. The time and temperature of exposure in the fourth section completely depends on the chemical structure of the binder used and is set by its developers. Thus, during the development of the curing process, the technologist must independently determine the modes in the first, second and third sections [7].

In this work, it is assumed that the second section begins after the start of the gelation process and continues until its completion. At the end of the third section is the final formation of the structure of polymer matrices.

The purpose of this study is to optimize curing modes of epoxy binders, taking into account changes in the state of aggregation during their heating.

2.Methodology

As an object of research used epoxy binder hot curing, consisting of an epoxy resin and a hardener based on aromatic amine.

2.1. Method of dynamic mechanical analysis

The holding temperature at second section corresponds to the gelation temperature. Experiments to estimate the gelation temperature depending on the heating rate were performed using dynamic mechanical analysis at heating rates: 1,3,5,7 °C / min [7]. According to the Flory-Stockmayer Theory, the degree of conversion at the gelation point does not depend on the heating rate and is constant for a particular resin [8, 9]. Therefore, in this work, we can use the Ozawa theory, and the method of approximation is used to determine the relationship between the gelling temperature and heating rates. Thus, the gelling temperature was determined at heating rates of 2, 4, 6 °C / min. The obtained values of the gelling temperature depending on the heating rate are shown in table 1.

Table 1. Gelling temperature depending on the heating rate

Heating rate(°C/min)	Gelling temperature (°C)
1	65
2	80
3	98
4	102
5	107
6	122
7	130

2.2. Differential Scanning Calorimeter (DSC) Method

In this work, the change of the thermos-physical properties of the binder at different degrees of conversion during the curing process is investigated. The values of heat capacity and thermal conductivity were evaluated directly in the curing process (for liquid, gel and solid state). The coefficients of heat capacity of the binder, depending on the conversion, are determined by the DSC method. The tests were carried out before the curing reaction start. The thermal effect was determined by the DSC method and the degree of conversion was experimentally evaluated [10].

2.3. Laser flash method

The values of thermal conductivity coefficients are determined by the laser flash method. The peculiarity of these studies is the use of a special crucible (Fig. 1) [9], since the liquid binder did not allow withstanding the required dimensions and is transparent to laser radiation.

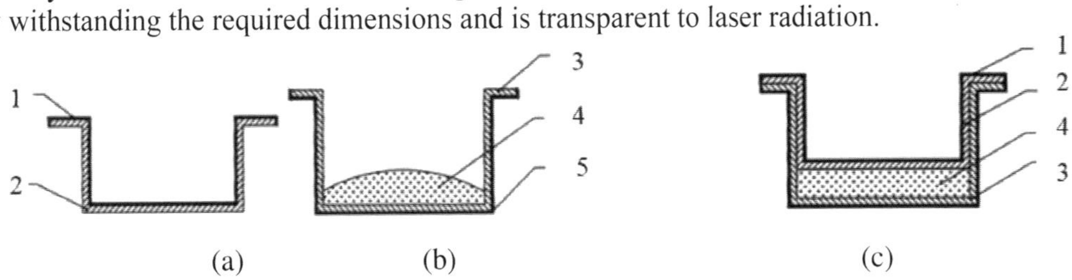

Fig. 1. Special mandrel for determining thermal conductivity. (a) Lid; (b) Crucible; (c) Crucible with lid. 1,5. Graphite grease; 2. Lid; 3. The crucible; 4. Binder.

The mandrel consists of a crucible and a lid made of a platinum-rhodium alloy. Manufacturer determined the thermos-physical and geometrical properties of the crucible material and the cover (heat capacity, density, thermal diffusivity). With the help of graphite grease in an aerosol, the external surfaces of the crucible and the lid are blackened to create the same optical properties of the surfaces. In the process of research, the effective thermal diffusivity of the binders was first determined, and then the thermal conductivity was determined. The results of the experiment are shown in Table 2.

Table 2. Dependence of heat capacity and thermal conductivity on the degree of conversion

Conversion degree, %	Cp, J/(kg/K)	λ, W/(m/K)
0	1973	0,08
5	1973	0,08
16	1921	0,14
28	1820	0,19
75	1406	0,22
87	1338	0,25

3.Modeling and optimization

3.1. Modeling the heating process

The simulation of heating process was carried out by taking into account the change in the aggregative state of the binder in the process of its curing at heating rates on the first and third sections: 1-7 °C / min. The 49 modeling options are shown in Table 3. The temperature gradients on the surface and inside the sample are presented in Fig. 2, 3. , and the duration of the heating process for different modeling options are presented in Fig. 4.

Table 3. Modeling Options

№ Modeling Options		Heating rate in the first section (°C/min)						
		1	**2**	**3**	**4**	**5**	**6**	**7**
Heating rate in the third section (°C/min)	**1**	1	8	15	22	29	36	43
	2	2	9	16	23	30	37	44
	3	3	10	17	24	31	38	45
	4	4	11	18	25	32	39	46
	5	5	12	19	26	33	40	47
	6	6	13	20	27	34	41	48
	7	7	14	21	28	35	42	49

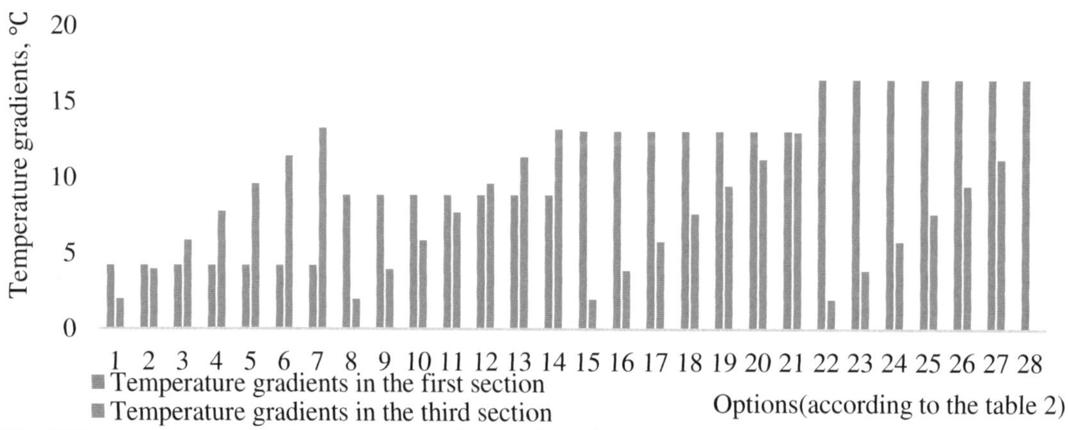

Fig. 2. The temperature gradients on the surface and inside the sample for modeling options 1-28

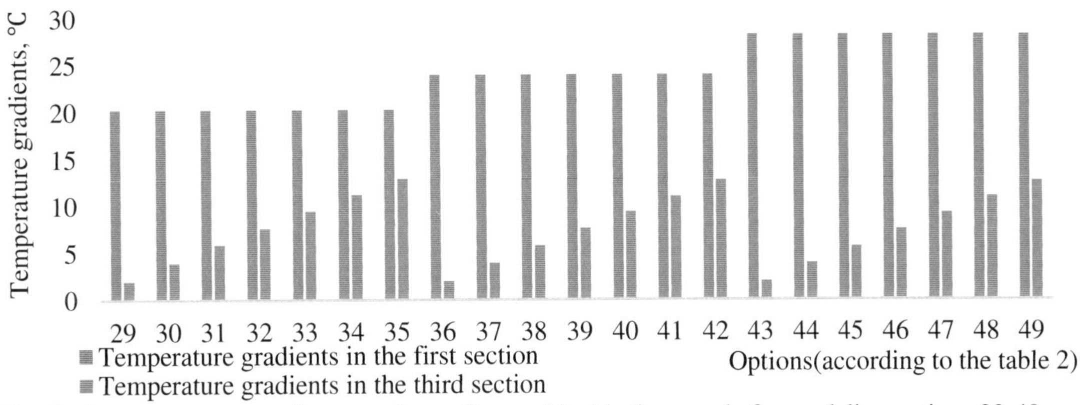

Fig. 3. The temperature gradients on the surface and inside the sample for modeling options 29-49

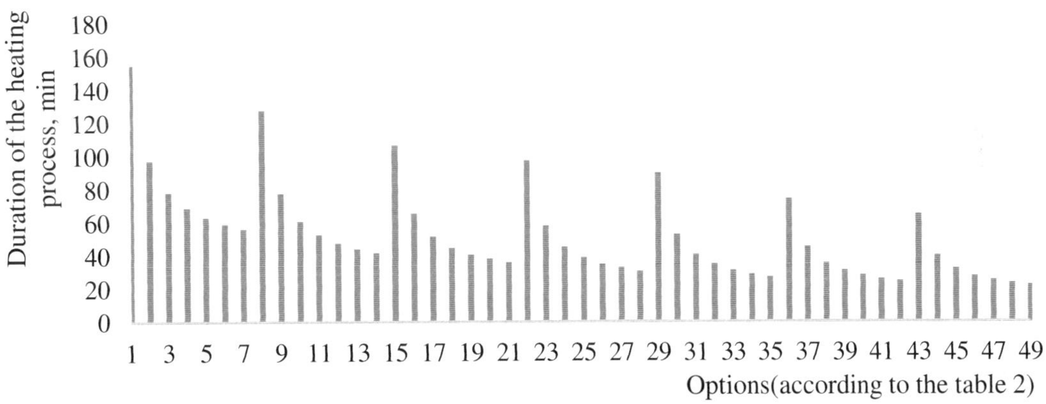

Fig. 4. Duration of the heating process for different modeling options

3.2. Method of optimization the curing process

All of the above options for technological regimes were investigated for optimization.

For the optimization criterion in this paper, it is proposed to use:

(1) The maximum temperature difference on the surface and inside the sample in the area of the first ΔTmaxI;

(2) The maximum temperature difference on the surface and inside the sample in the third section ΔTmaxIII;

(3) The total duration of the heating processτ.

All listed in table. options for technological regimes were investigated using the ideal point method [11, 12]. The most acceptable is considered an alternative, in which the distance from the "ideal point" is minimal:

$$R_{Ai} = \sqrt[N]{\sum_{i=1}^{N} (x_{id,j} - x_{ij})^2} \qquad (1)$$

where RAi – the distance of the point of the i-th alternative from the ideal point; N – number of criteria for evaluating alternatives; xid, j – idealvalue according to the j-th criterion for the ideal variant; xij – value by the j-th criterion for the i-th alternative.

3.3. Study of the influence of the restriction condition on the optimization results

If without condition of limitation:

In this case, the coordinates of the ideal point: (4.2 2.0 22.2), N = 3, i = 49. The minimum distance from the ideal point: RA28 = 6.95.
It has been established that mode 28 is optimal, in which the heating rate on the I stretch is 4 °C / min, the heating rate on the III stretch is 7 °C / min, and the holding temperature is 102 ° C.
Limit condition 1:
(1)ΔT_{maxI} <20 °C,(2) ΔT_{maxIII} <10 °C,(3) τ <60 мин.
In this case, the coordinates of the ideal point: (8.8 3.9 34.85), N = 3, i = 9. The minimum distance from the ideal point: RA19 = 6.95.
It has been established that mode 19 is optimal, in which the heating rate on the I stretch is 3 °C / min, the heating rate on the III stretch is 5 °C / min, and the holding temperature is 98 °C.
Limit condition 2:
(1)ΔT_{maxI} <22 °C,(2) ΔT_{maxIII} <10 °C,(3) τ <60 мин.
In this case, the coordinates of the ideal point: (8.8 3.9 31.2), N = 3, i = 13. The minimum distance from the ideal point: RA28 = 4.72.
It has been established that mode 26 is optimal, in which the heating rate on the I stretch is 4 °C / min, the heating rate on the III stretch is 5 °C / min, and the holding temperature is 102 ° C.
Limit condition 3:
(1) ΔT_{maxI} <20 °C, (2) ΔT_{maxIII} <15 °C,(3) τ <60 мин.
In this case, the coordinates of the ideal point: (4.2 7.6 30.25), N = 3, i = 17. The minimum distance from the ideal point: RA28 = 5.21.
It has been established that mode 21 is optimal, in which the heating rate on the I stretch is 3 °C / min, the heating rate on the III stretch is 7 °C / min, and the holding temperature is 98 ° C.
Table 4 shows the optimization results for different constraint conditions.

Table 4. The results of optimization for different constraint conditions

Limit condition	The optimal option		
	Heating rate in the first section (°C/min)	Heating rate in the third section (°C/min)	Temperature in the second section(°C)
Without limit condition	4	7	102
Limit condition 1	3	5	98
Limit condition 2	4	5	102
Limit condition 3	3	7	98

4.Conclusion

A technique has been developed for determining the heat capacity and thermal conductivity of epoxy binders depending on their different state of aggregation (in a liquid, gel-like, and solid state). It was found that with an increase in the degree of conversion, the heat capacity of epoxy binders decreases, and the thermal conductivity increases.
The mathematical modeling of the curing process was carried out for different heating rates: 1-7 °C /min, the temperature gradient and the duration of the heating process were calculated in the curing mode.
The optimization of the curing mode was carried out depending on different heating rates. It was found that limitation No. 3 is the most severe, and in this case the best option: the heating rate on the first section is 3 °C / min, the heating rate in the third section is 5 °C / min the temperature in the second section is 98 °C.

References

[1] Bachenon S., Berlin A. 2010. Polymer composite materials 352 p.

[2] Kablov E.N. The formation of domestic space materials scienceю *Bulletin of the Russian Foundation for Basic Research. 2017*. № 3 (95). P.97–105.

[3] Shimina Y.Y., Malysheva G.V. 2015 Determination of failure-free operation probability in adhesive joints at high temperature *Polymer Science – Series D* **8** 207-210.

[4] Nelyub V.A., Borodulin A.S., Kobets L.P., Malysheva G.V. 2016 Capillary hydrodynamics of oligomer binder. *Polymer Science – Series D.*. **9**(3). Pp.322-325.

[5] Baurova N.I., Zorin V.A., Prikhodko V.M 2015 Description of the process of degradation of material properties using the apparatus of catastrophe theory. *Polymer Science – Series D.* 8(2). Pp.92-95.

[6] Baurova N.I., Zorin V.A., Prikhodko V.M. 2016 Manifesrarion of a synergistic effect in technological heredity. *Polymer Science – Series D.*. **9**(2). Pp.209-211.

[7] A.N. Polilov 2015 Experimental mechanics of composites: Proc. allowance. M .: Publishing House of Moscow State Technical University. N.E. Bauman,. 375 p.

[8] Baurova, N.I. 2016 The use of polymer composite materials in the production and repair of machines: study guide / N.I. Baurova, V.A. Zorin. - M .: MADI 264 p.

[9] Nelyub V.A., Borodulin A.S., Kobets L.P., Malysheva G.V. 2017 Viscous husteresis in filled siloxane binders // Polymer Science – Series D.. **10**(1) 19-22.

[10] Petrova A.P., Malysheva G.V. 2017 Glues, adhesive binders and adhesive prepregs: Proc. allowance. Moscow: VIAM Publishing House 472

[11] Boyko AP, Ulakhovich D A. 2016 The problem of multi-criteria selection of the optimal vector quantization method for linear prediction parameters. *News of St. Petersburg State Electrotechnical University* LETI **3** 10-14.

[12] Vishnekov AV, Karpova I.P., Ferapontova E.S. 2010 Calculation of the value of alternatives for the "ideal point" method in multi-criteria decision-making tasks for situational centers. *Quality. Innovation. Education* 56-60.

Design calculation of braided high-temperature composite structures

A A Smerdov

Bauman MSTU, Moscow, Russia, asmerdov@mail.ru

Abstract. Following a review of published papers devoted to calculations of braided composite structures, basic requirements for computational models are formulated. A mathematical model intended for design calculations of braided structures of shells of revolution is presented. The dependences for calculating the stiffness, thermal deformation, and strength characteristics of a braided composite material are given depending on the braiding scheme, reinforcement angle, and the current radius of the shell. The relationship between the angle of reinforcement, the thickness of the single-pass layer and the coverage rate of the structure that determines the degree of continuity of the layer is investigated. The features of calculations of structures with a two-axis and three-axis braiding scheme are discussed. Examples of calculations of specific composite structures are given, the dependences of elastic moduli, coefficients of linear thermal expansion and ultimate strengths on the variation of the radius of a typical braided carbon-ceramic composite shell are demonstrated.

Keywords: Composite, braided structure, design calculation, stiffness, thermoelasticity, strength

1. Features of braided structures and approaches to their calculation

Braided structures are widely used in modern rocket and space technology, and their scope in the foreseeable future will expand. They serve as the basis for fairings, wing elements and the fuselage of hypersonic products, nozzles of rocket engines and other supporting structures operating at high temperatures and intensive force loading. The main bearing element of such structures is carbon yarns; however, braiding is also possible with quartz filaments to create radio-transparent structures.

Braided structures have a number of features that separate them from traditional winding and laying structures. First of all, it is the presence of bearing elements - yarns, - sizes of which cannot be considered as small sixez not only in comparison with the thickness of a single layer, but often in comparison with the thickness of the entire wall of the shell structure. Secondly, it is a variety of types of braiding (two-axis and three-axis braiding schemes, braiding with a constant coverage rate and with a constant angle of reinforcement, and so on). At the same time, the properties of the braided structure strongly depend on both the change in the design parameters and the change in perimeter of the structure.

In recent years, there has been a rapid increase in the number of publications devoted to various aspects of the development of composite structures based on braided preforms. There are hundreds of already published articles and articles continue to appear in various print and electronic publications. The main part of these publications covers various production and technological issues. However, to date, quite a lot of articles has been published on various approaches to calculation of braided composite structures. These approaches have varying accuracy and may be used for both verification

and design calculations. At the same time, both analytical methods and methods of numerical calculation, including those based on the finite element method (FEM), are considered.
Among the papers published in recent years, it is possible to emphasize the articles [1–7] devoted to the calculation of the stiffness characteristics of braided composites, the papers [8–12], in which the issues of determining their thermal deformation properties, as well as the publications [9-19], devoted to the analysis of the processes of failure of braided structures and the determination of their strength characteristics. There are also commercial software products [20-22].
A characteristic feature of almost all approaches being developed is the striving to maximally take into account all the features of the complex structure of a braided preform. It uses regular models of yarns braiding strongly idealizing real-life materials.
The main question that arises when analyzing existing computational models is where to get the necessary input data to fill them in: at the stage when design calculations are carried out, usually there are not enough experimental data to correctly evaluate the subtle effects associated with interlacing yarns. For example, the well-known DIGIMAT software [20] provides a solution to the contact problem when applying individual monofilaments; at the same time, the properties of fibers necessary for the calculation in the transverse direction, implemented in a braided structure, cannot be determined from any known experiments.

2. Basic principles for constructing a mathematical model for design calculations of braided structures

First of all, the basic principles include reasonable schematization of the structure, discarding unnecessary details. Of course, the key is the question of what particular details should be considered unnecessary. The answer to this question should be given by the experiment. Base of the experiment - the second key principle. Summarizing, we can formulate the following maxim: anything that cannot be measured in an accessible experiment should not be contained in mathematical model. Of course, in this case some loss of accuracy is inevitable. The third principle is that the purpose of the design calculations is to compare various design options, and the main thing in these calculations is to ensure comparable accuracy of the calculation of these options. Details can be specified in the check calculations after selecting the optimal structure.
When constructing a model for design calculations, the most serious is the question of considering the curvature of the yarns. When analyzing it, one should take into account that information on the properties of non-curved filaments is practically inaccessible. On the other hand, any experiment on the braided structure provides the information about the properties of curved filaments, which can be put into the calculation.
Based on all these considerations, the main provisions of the calculation model are formulated:
the wall of the shell is created by an integer number of identical passes (the material is uniform in thickness); material characteristics are given to the conditional thickness of the single pass layer;
a structural model is adopted, the components of which are yarns located in two or three directions, as well as a conditional medium filling the space between the yarns; the properties of this medium are used to describe the characteristics of the material in those directions that are not provided by the fibers;
characteristics of the impregnated yarns and characteristics of the conditional medium in the transverse direction and in shear are selected according to the results of experiments on the same type material, while the characteristics of the yarns in the transverse direction are not separately taken into account, but are included in the medium filling the space between the yarns;
failure of the braided structure is assumed to be associated with that of the yarns of at least one of the families that constitute this structure; if there are no yarns in this direction, then the matrix strength is considered.
These provisions are used in the derivation of specific calculated dependencies. In this case braided shells of revolution are considered; however, the presented approach can be easily generalized to other forms of braided structures.

For structures with a biaxial braiding scheme ±φ, formulas used for cross-reinforced materials can be used [23]. For a three-axis braiding scheme 0/±φ, it is necessary to develop a new concept of a layer reinforced in three directions. This concept proceeds from the presence of a structure ±φ, to which a set of meridional filaments is added. The coordinate system associated with the rotation shell is used: the x axis is directed along the meridian, the y axis — along the parallel.

First of all, a link should be established between the angle of reinforcement, the thickness of the single-pass layer and the coverage rate of the structure, which determines the degree of continuity of the layer.

The width of the yarn remains constant under the condition of a layer hiding less than 100%. With a decrease in diameter or an increase in the angle of braiding, the coverage rate approaches 100% and after the yarns of the braid begin to touch each other, the shape of the yarns begins to change: from a strongly flat ellipse to almost the circle. After reaching the limit form of the cross-section (circle), the yarns will no longer be retained against a surface of the molding tool, and there will be a loss of quality of the material in the design. The range of angles φ corresponding to the total coverage rate is determined by the formulas

$$\arccos\frac{nb}{2\pi R} \le \varphi \le \arccos\frac{nd}{2\pi R} \quad \text{or} \quad \arccos\frac{b}{\frac{2\pi R}{n} - \mathrm{b}} \le \varphi \le \arccos\frac{d}{\frac{2\pi R}{n} - d},$$

moreover, the first formula is used with a two-axis braiding scheme, and the second with a three-axis braiding scheme. In the last formulas the following is indicated: R is the current radius of the shell, n is the number of yarns of each family, b is the initial width of the yarn, and d is the maximum width of the yarn (with a circular shape).

To ensure constant full coverage rate, the width of the yarn must be related to the radius, the angle of reinforcement and the number of yarns by dependencies

$$b(R) = \frac{2\pi R\cos\varphi}{n} \quad \text{or} \quad b(R) = \frac{2\pi R}{n\left(1 + \frac{1}{\cos\varphi}\right)},$$

moreover, the first formula is used with a two-axis braiding scheme, and the second with a three-axis braiding scheme. The expected thickness of the yarns in a layer formed in a single pass can be approximately determined by the formula that is valid for elliptical yarns:

$$h(R) = \frac{4S}{\pi b(R)},$$

where S is the area of the filled yarn. The thickness of the layer formed in a single pass, H, can be approximated by the formula

$$H \approx 2\mathrm{h(R)}.$$

According to the results of experiments on cylindrical samples of braided structures [24] the values of the characteristics of the components of the composite structure are determined: stiffness of the impregnated yarn EF, elastic moduli of conditional medium E2 and G12, linear thermal expansion ratios (CLTE) of the α yarns and α2 medium, as well as ultimate loads under tension and compression of the impregnated N+ and N– yarns, and ultimate stresses under tension and compression of the inter-filament medium S2+ and S2– (for a given temperature).

Based on this information, the coefficients of the braided layer stiffness matrix are calculated.

$$g_{xx} = \frac{\left\{\frac{nEF}{\pi RH\cos\varphi} + E_2\left(1 - \frac{nS}{\pi RH\cos\varphi}\right) + \left[\mathrm{E}_2\left(tg^2\varphi + 2\nu_{12}\right) + 4G_{12}\right]tg^2\varphi\right\}\cos^4\varphi}{1 - \nu_{12}\nu_{21}} -$$

for the two-axis braiding scheme,

$$g_{xx} = \frac{nEF}{2\pi RH} + \frac{\left\{ \frac{nEF}{\pi RH\cos\varphi} + E_2\left(1 - \frac{nS}{\pi RH\cos\varphi}\right) + \left[E_2\left(\mathrm{tg}^2\varphi + 2\nu_{12}\right) + 4G_{12}\right]\mathrm{tg}^2\varphi \right\}\left(1 - \frac{nS}{2\pi RH}\right)\cos^4\varphi}{1\text{-}\nu_{12}\nu_{21}} -$$

for the three-axis braiding scheme,

$$g_{xy} = \left(\frac{\frac{nEF}{\pi RH\cos\varphi} + E_2\left(1 - \frac{nS}{\pi RH\cos\varphi}\right) + E_2}{1\text{-}\nu_{12}\nu_{21}} - 4G_{12} \right)\cos^2\varphi\sin^2\varphi + \frac{\nu_{12}E_2}{1\text{-}\nu_{12}\nu_{21}}\left(\cos^4\varphi + \sin^4\varphi\right),$$

$$g_{yy} = \frac{\left[\frac{nEF}{\pi RH\cos\varphi} + E_2\left(1 - \frac{nS}{\pi RH\cos\varphi}\right)\right]\sin^4\varphi + E_2\left(\cos^2\varphi + 2\nu_{12}\sin^2\varphi\right)\cos^2\varphi}{1\text{-}\nu_{12}\nu_{21}} + 4G_{12}\cos^2\varphi\sin^2\varphi\,,$$

$$g_{ss} = \left(\frac{\frac{nEF}{\pi RH\,cos\,\varphi} + E_2\left(1 - \frac{nS}{\pi RH\cos\varphi}\right) + E_2\left(1 - 2\nu_{12}\right)}{1\text{-}\nu_{12}\nu_{21}} \right)\cos^2\varphi\sin^2\varphi + G_{12}\left(\cos^2\varphi - \sin^2\varphi\right)^2.$$

In the last formulas the following is indicated: R is the current radius of the shell, H is the thickness of the layer formed in one pass, φ is the angle of reinforcement, n is the number of yarns of each family,

$$\nu_{12} = \frac{E_2}{2G_{12}} - 1 \quad (\text{if } E_2 < 2G_{12}, \nu_{12} = 0), \quad \nu_{21} = \frac{\nu_{21}E_2}{\frac{nEF}{\pi RH\cos\varphi} + E_2\left(1 - \frac{nS}{\pi RH\cos\varphi}\right)}.$$

Technical constants of elasticity [23] of the braided layer

$$E_x = g_{xx} - \frac{g_{xy}^2}{g_{yy}}, \quad E_y = g_{yy} - \frac{g_{xy}^2}{g_{xx}}, \quad G_{xy} = g_{ss}, \quad \nu_{xy} = \frac{g_{xy}}{g_{yy}}, \quad \nu_{yx} = \frac{g_{xy}}{g_{xx}}. \tag{1}$$

The coefficients of layer thermal stress with the two-axis braiding scheme

$$\beta_x = \left[\frac{\alpha nEF}{\pi RH\cos\varphi} + \frac{\alpha_2 E_2}{1 - \nu_{12}\nu_{21}}\left(1 - \frac{nS}{\pi RH\cos\varphi}\right)\right]\cos^2\varphi + \frac{\alpha_2 E_2}{1 - \nu_{12}\nu_{21}}\sin^2\varphi,$$

$$\beta_y = \frac{\alpha_2 E_2}{1 - \nu_{12}\nu_{21}}\cos^2\varphi + \left[\frac{\alpha nEF}{\pi RH\,cos\,\varphi} + \frac{\alpha_2 E_2}{1 - \nu_{12}\nu_{21}}\left(1 - \frac{nS}{\pi RH\cos\varphi}\right)\right]\sin^2\varphi. \tag{2}$$

For the three-axis braiding scheme, instead of the first formula (2), the dependence should be used

$$\beta_x = \frac{\alpha nEF}{2\pi RH} + \left\{\left[\frac{\alpha nEF}{\pi RH\cos\varphi} + \frac{\alpha_2 E_2}{1 - \nu_{12}\nu_{21}}\left(1 - \frac{nS}{\pi RH\cos\varphi}\right)\right]\cos^2\varphi + \frac{\alpha_2 E_2}{1 - \nu_{12}\nu_{21}}\sin^2\varphi\right\}\left(1 - \frac{nS}{2\pi RH}\right),$$

the second formula remains unchanged.

CLTE of the braided layer

$$\alpha_x = \frac{\beta_x - \nu_{xy}\beta_y}{E_x}, \quad \alpha_y = \frac{\beta_y - \nu_{yx}\beta_x}{E_y},$$

where technical elastic constants are determined according to (1).

When calculating the strength, deformations of the yarns of each family acting under the given loads are calculated; the ratio of the ultimate deformation of each yarn to the current one gives local strength factor. Two more local strength factors are limiting deformation ratios of the conditional medium to the magnitudes of the actual deformations, then the smallest of local factors is selected. Thus, the criterion of maximum strains [23] is used. The strength factor ξ at stresses, σx and σy, acting at the given point of the braided structure is calculated using the formulas

$$\xi = \min\left\{\frac{N_s}{EF\varepsilon_s};\frac{S_x}{E_2\varepsilon_x};\frac{S_y}{E_2\varepsilon_y}\right\} \text{ or } \xi = \min\left\{\frac{N_x}{EF\varepsilon_x};\frac{N_s}{EF\varepsilon_s};\frac{S_x}{E_2\varepsilon_x};\frac{S_y}{E_2\varepsilon_y}\right\}, \quad (3)$$

the first of the formulas (3) is used for the two-axis braiding scheme, and the second for the three-axis braiding scheme. In formulas (3) it is indicated:

$$\varepsilon_x = \frac{\sigma_x - \nu_{xy}\sigma_y}{E_x}, \quad \varepsilon_y = \frac{\sigma_y - \nu_{yx}\sigma_x}{E_y}, \quad \varepsilon_s = \frac{\sigma_x}{E_x}\left(\cos^2\varphi - \nu_{xy}\sin^2\varphi\right) + \frac{\sigma_y}{E_y}\left(\sin^2\varphi - \nu_{yx}\cos^2\varphi\right),$$

$$N_s = \begin{cases} N^+, \text{ if } \varepsilon_s > 0 \\ -N^-, \text{ if } \varepsilon_s < 0, \end{cases} \quad N_x = \begin{cases} N^+, \text{ if } \varepsilon_x > 0 \\ -N^-, \text{ if } \varepsilon_x < 0, \end{cases}$$

$$S_x = \begin{cases} S_2^+, \text{ if } \varepsilon_x > 0 \\ -S_2^-, \text{ if } \varepsilon_x < 0, \end{cases} \quad S_y = \begin{cases} S_2^+, \text{ if } \varepsilon_y > 0 \\ -S_2^-, \text{ if } \varepsilon_y < 0. \end{cases}$$

If one of the terms in formulas (3) contains a zero value in the denominator, this term is excluded from consideration.

Formulas (3) describe the so-called first failure of the composite [23]. As is known, the first failure does not always immediately lead to the exhaustion of the bearing capacity of the composite structure; in this regard, the mechanisms of the first failure can be divided into fatal and non-fatal. In the case, the fatal failure are the cases when the smallest safety factor in the first formula (3) corresponds to the first term and in the second formula (3) - to the first or second member. In this case, the reinforcing yarns are destroyed, ensuring the strength of the structure.

The situation is more complicated in cases where the first failure corresponds to one of two last terms in formulas (3), that is, the environment between the yarns at first begins to break down. In these cases, it is possible both the immediate exhaustion of the bearing capacity of the structure, and the continuation of its perception of the increasing load. Everything is determined by the specific braiding scheme, reinforcement angles and type of loading.

At present, it is not possible to create an algorithm for automatically accounting for changes in the bearing capacity of braided structures of high-temperature composites at the beginning of the binder failure, because this requires a missing today amount of experimental research data. However, we can recommend an approximate method of accounting for these effects “in the manual mode”.

The meaning of this technique is to carry out in complex cases the sequence of calculations with a gradual change (artificially high) of strength characteristics of the interlayer medium. This is done until in formulas (3) the terms determining the binder failure will not exceed one of the members associated with the failure of the yarns. Thus, an upper estimate of the strength of the structure under investigation can be obtained after obtaining its lower estimate at the first calculation.

If the braided structure is a cylinder, the angle of reinforcement is kept constant along its length; thus, the properties of the braided material are the same in any section. If the shell of rotation is braided, then with a change in the radius, the properties of the material will change. The following main options are possible:

- braiding with a constant angle of reinforcement, when the spacing between the centers of the yarns increases, the coverage rate decreases or the thickness changes with increasing radius;
- braiding with constant coverage rate, when the angle of reinforcement increases with increasing radius;
- intermediate braiding schemes, when the angle changes according to a given program from the smallest radius to the largest; this changes both the thickness and the coverage rate.

3. Calculation examples

Examples of calculations for braided structures are illustrated in Figure 1, which shows graphs of

changes in material characteristics along the axis of a braided shell of rotation, such as an uncooled nozzle of a rocket engine. The shell profile is shown in the graphs. The initial radius of the shell is 150 mm, the final radius is 540 mm, and the length is 1200 mm. Baseline data for the calculation: EF = 150 kN, S = 0.63 mm2, α = –0.3·10–6 K-1, E2 = 16 GPa, G12 = 6.4 GPa, α2 = 3.5·10–6 K–1. Ultimate tensile load of the yarns is 500 N, when compressed – 360 N, the ultimate strength of inter-yarn medium under tension and compression - 40 and 45 MPa. Different three-axis braiding scheme options are considered on 122 yarns in each direction: braiding scheme with a constant angle of 70° (line 1 on the graphs), braiding scheme with constant total coverage rate and an initial angle of 50° (line 2), and braiding scheme with a constant coverage rate of 80% and an initial angle of 20° (line 3).
It is seen that all the properties of the braided material vary greatly along the axis of the shell. By changing the braiding scheme and the initial angles of reinforcement, one can choose the most appropriate combination of characteristics in the meridional and circumferential direction, based on the operating conditions of each specific composite structure.
If the braided structure is designed to work at a variable temperature, then graphs similar to those shown in Figure 1 are plotted for two or more reference temperatures between which interpolation is carried out. When conducting a finite element calculation, the structure is divided into zones, in each of which characteristics are taken for the braided material corresponding to a given temperature.

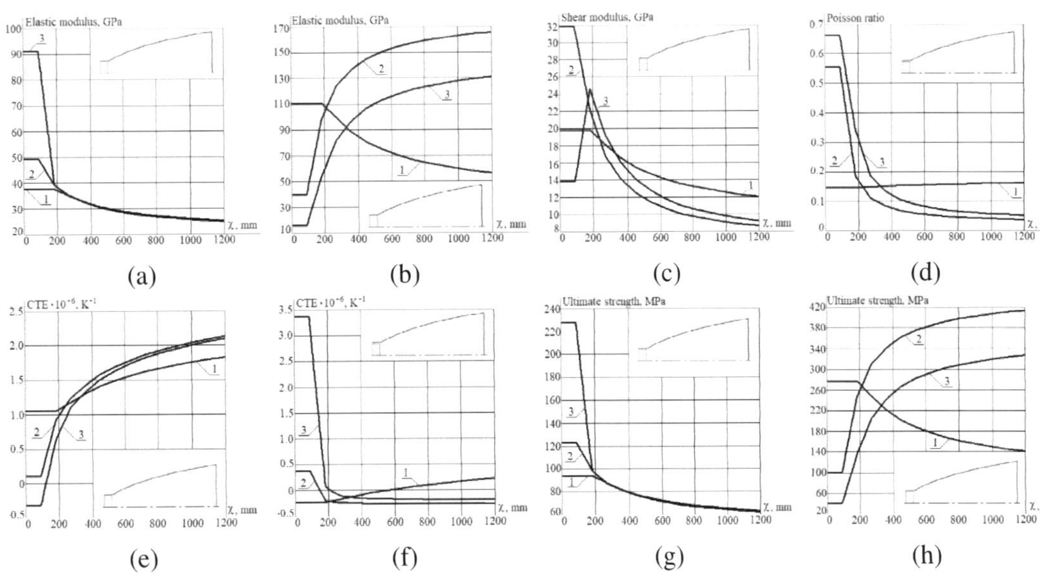

Figure 1. The dependence of the characteristics of the braided composite on the axial coordinate of the shell χ: (a) meridional modulus of elasticity, (b) circumferential modulus of elasticity, (c) shear modulus, (d) Poisson's ratio, (e) meridional CLTE, (f) circumferential CLTE, (g) strength under meridional tension, (h) circumferential tensile strength

References

[1] Shokrieh M M and Mazloomi M S 2010 An analytical method for calculating stiffness of two-dimensional tri-axial braided composites *Compos. Struct.* **92** 2901-05
[2] Xu L, Kim S J, Ong C-H and Ha S K 2012 Prediction of material properties of biaxial and triaxial braided textile composites *J. Compos. Mater.* **46** 2255-70
[3] Ji X, Khatri A M, Chia E S, Cha R K, Yeo B T, Joshi S C and Chen Z 2013 Multi-scale simulation and finite-element-assisted computation of elastic properties of braided textile reinforced composites *J. Compos. Mater.* **48** 931-49
[4] Ayranci C and Carey J P 2010 Predicting the longitudinal elastic modulus of braided tubular composites using a curved unit-cell geometry *Compos.* B *Engineering* **41** 229-35
[5] Samipur S A and Danilov Y S 2016 Development and verification of an analytic technique to

determine the stiffness parameters of braided tubular parts *Russian Aeronautics* **59(4)** 460–65

[6] Zhang C, Binienda K W and Kohlman L W 2014 Analytical model and numerical analysis of the elastic behavior of triaxial braided composites *J. Aerospace Engineering* **27** 473-83

[7] Zhang P, Gui L J and Fan Z J 2009 An analytical model for predicting the elastic properties of triaxially braided composites J. *Reinforced Plastics and Composites* **28** 1903–16

[8] Sun X K and Sun C J 2004 Mechanical properties of three-dimensional braided composites *Compos. Struct.* **65** 485-92

[9] Xu Y and Zhang W 2009 Numerical prediction of the effective coefficient of thermal expansion of 3D braided C/SiC composite Int. *J. Simul. Multidisci. Des. Optim.* **3** 443-48

[10] Jiang L-L, Xu G-D, Cheng S, Lu X-M and Zeng T 2014 Finite element analysis of thermo- mechanical properties of 3D braided composites *Applied Composite Materials* **21(2)** 325–40

[11] Zhai J, Cheng S, Zeng T, Wang Z and Jiang L 2017 Thermo-mechanical behavior analysis of 3D braided composites by multiscale finite element method *Compos. Struct.* **176** 664-72

[12] Lu Z, Wang C, Xia B and Zhou Y 2014 Effect of interfacial properties on the thermophysical properties of 3D braided composites: 3D multiscale finite element study *Polymer Composites* **35(9)** 1690-1700

[13] Wang C, Zhong Y, Adaikalaraj P F B, Ji X, Roy A, Silberschmidt V V and Chen Z 2016 Strength prediction for bi-axial braided composites by a multi-scale modelling approach *J. Materials Science* **51(12)** 6002–18

[14] Ivanov D S, Baudry F, Van Den Broucke B, Lomov S V, Xie H and Verpoest I 2009 Failure analysis of triaxial braided composite *Compos. Sci. Technol.* **69(9)** 1372–80

[15] Xiao X, Kia H G and Gong X J 2011 Strength prediction of a triaxially braided composite *Compos.* A *Appl. Sci. Manuf.* **42** 1000-06

[16] Binienda W K and Li X 2010 Mesomechanical model for numerical study of two-dimensional triaxially braided composite *J. Eng. Mech.* **136** 1366-79

[17] Zhang C, Binienda W K, Goldberg R K and Kohlman L W 2014 Meso-scale failure modeling of single layer triaxial braided composite using finite element method *Compos.* A *Appl. Sci. Manuf.* **58** 36-46

[18] Xu L, Jin C Z and Kyu H S 2014 Ultimate strength prediction of braided textile composites using a multi-scale approach *J. Compos. Mater.* **49(4)** 477-94

[19] Li X, Binienda W K and Goldberg R K 2011 Finite-element model for failure study of two-dimensional triaxially braided composite *J. Aerosp. Eng.* **24** 170-80

[20] Gontiuk A DIGIMAT software package as the basis of an integrated solution for additive technologies 2018 *CAD/CAM/CAE Observer* **121** 68-72 (in Russian)

[21] Verpoest I and .Lomov S V 2005 Virtual textile composites software WiseTex: Integration with micro-mechanical, permeability and structural analysis *Composites Science and Technology* **65(15–16)** 2563-74

[22] Richardson R S 2015 CATIA Composites braiding https://cdmhub.org/resources/761

[23] Vasiliev V V and Morozov E V 2001 *Mechanics and analysis of composite materials* (Amsterdam: Elsevier) 412 p

[24] Smerdov A A, Tairova L P and Timofeev I A 2018 Experimental study of tubular specimens based on carbon fiber braided multilayered carcasses with polymer, carbon and ceramic matrices *Composite materials constructions ("Konstrukcii iz kompozicionnyh materialov")* **2-2018** 52-59 (in Russian)

IOP Conf. Series: Materials Science and Engineering **683** (2019) 012067 doi:10.1088/1757-899X/683/1/012067

Hybrid composites in reusable space vehicles wing structures

E R Ashikhmina, T G Ageyeva and P V Prosuntsov

Bauman Moscow State Technical University, 2nd Baumanskaya str., b. 5, Moscow, 105005, Russia

E-mail: tageeva888@gmail.com

Abstract. The study is dedicated to the solution of a complex problem – development of an optimal thermal design of the tourist class reusable space vehicle (RSV TC) wing structure. The complexity of the problem resides in the necessity to consider critical factors of the suborbital flight and in a number of nesting tasks,that need to be solved. The authors implemented a multiscale modeling approach covering design of hybrid composite material thermal properties (i.e. mesoscale level), determination of thermal loads and thermal state analysis of a composite wing (i.e. macroscale and structural scale levels). Analysis of thermal fields demonstrated the necessity of thermal protection (TP) application for the wing structure. Consequently, optimal from the weight efficiency point of view TP thicknesses were determined. Summing up, the results of a RSV TC wing structure development are presented in the current paper.

1. Introduction

Space tourism has become a subject of particular interest during the last few decades. The facts that facilitate this interest reside in a vast experience in spacecraft development, a number of successful flights of non-professional astronauts to the International Space Station (ISS), and high potential interest of business in this field [1]. Nowadays two variants of space tours are technologically feasible – orbital and suborbital. The last one implies ballistic trajectory flight beyond the Karman line (100 km above sea level). Due to some technical, financial and medical reasons suborbital flights are now much more attractive for both business and customers. The potential high demand of suborbital flights is proved by the results of market research provided by IPSOS Group (France) conducted by the order of EADS Astrium [2]. Duration of the proposed suborbital tour will not exceed 80 min with the exposure to weightlessness of 3-5 min. The ticket price will lie in the range of 150-200 ths. USD. For commercial feasibility of such flights RSV development is a question of essential importance [3]. Recently, a number of private companies all over the world are involved into the space tourism activity. The examples of suborbital RSV projects could be RocketplaneXP (Rocketplane Kistler, USA, 2005), New Shepard (Blue Origin, USA, 2006), Space Plane (EADS Astrium, France, 2007), Lynx Mark (XCOR Aerospace, USA, 2008), SpaceShipTwo (Virgin Galactic, USA, 2010), Oduvanchik (Bauman MSTU, Russia, 2007), Reusable Suborbital Space Complex (Cosmocourse, Russia, 2014) and others.

Weight efficiency together with high service security are the key issues in RSV TC design. Therefore, polymer composite materials that can meet mentioned requirements and have proved their reliability for structural application in aerospace for decades are the right choice for RSV load bearing structures (fuselage, wings, spars, etc.). Due to the demand of economic feasibility of RSV, stipulated by its tourist application, hybrid polymer composites, that contain two or even more types of the reinforcement (herein after hybrid composites), have great opportunities. Combination of expensive carbon fibers

(CFs), that demonstrate high strength-to-weight performance, with cheap but lower in mechanical properties glass fibers (GFs) allows to achieve optimal design solution from technical and economical positions.

However, design of suborbital RSV TC is a complex problem, that involves a number of nesting tasks, i.e. trajectory parameters and flight loads determination, materials choice and optimal structure development, etc. Modern engineering design approach implicates a wide use of Finite Element Analysis (FEA) methods for all mentioned subtasks, thus reducing the development time and cost [4]. The type of the analysis, that literally covers the design in micro-, meso-, macro- and structural levels, is referred as a multiscale analysis. Micro-, meso- and macro-scale types belong to the modeling of material and aims to diminish numerous laboratory tests, while structural level is responsible for the global structure modelling, thus reducing full-scale trials costs.

Consequently, current work aims to develop an optimal design methodology for suborbital RSV TC load bearing structures. The implementation of the mentioned methodology is demonstrated via the design of a hybrid composite wing for the suborbital RSV, engineered in Bauman MSTU [5].

The reminder of this paper is organized as follows. Section 2 describes the design procedure; in Section 3 we report the results of the solved nesting tasks; Section 4 concludes the paper.

2. Development of the design procedure

Design of the RSV TC is a complex task that requires solving of a certain number of separate engineering problems. The general scheme of RSV design is presented at Figure 1. Our previous study covers the structural optimal design of the RSV wing structure [5], and the current study is mostly concentrated on the thermal design, that includes the following stages: trajectory parameters evaluation, materials thermal properties definition, aerodynamic thermal loads modelling, structure thermal analysis and TP optimization.

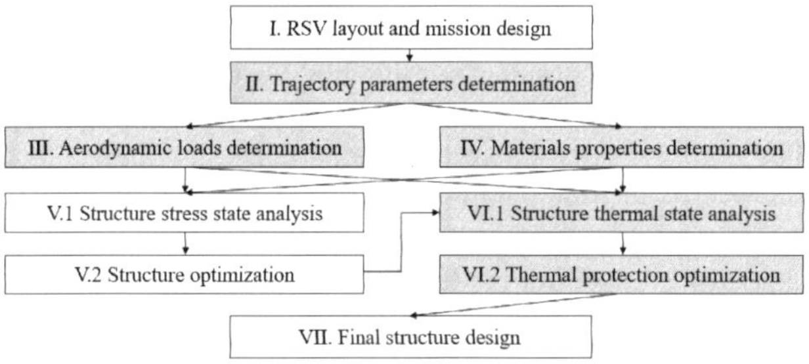

Figure 1. Principal scheme of RSV design.

3. Solving of the formulated tasks

3.1 Trajectory parameters determination

Two phases could be distinguished during the suborbital flight, namely active and passive. Active phase involves acceleration and injection of RSV either with a rocket booster or with an aircraft up to the certain height. Passive phase includes ascent of RSV to the height of 100 km, staying at zero-gravity conditions for 3-5 min following by the glide descent in the atmosphere and return to the landing point. In current study trajectory parameters (i.e. flight altitude, attack angle, RSV velocity, etc.) were defined by means of special program, developed in NPO Molniya, that simulates RSV center-of-mass motion [6]. Due to the economic requirements the velocity at the end the active phase should be minimum but enough to reach the desired height with holding-up at this point for not less than 3 min. Therefore, trajectory slope to the horizontal plane at the beginning of the passive phase is rather high. For the considered suborbital RSV, the flight passive phase starts from 46 km altitude with the velocity of 3.8M and trajectory angle of 56°. It is worse to mention that the value of the side overload is limited by 4.5g, that is acceptable for trained space tourists. Some results of the trajectory parameters determination are presented at Figure 2. This data is used as the input parameters for the next design stage.

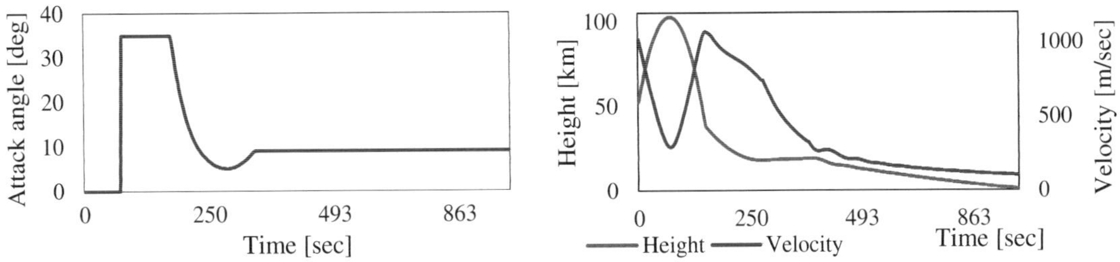

(a) (b)

Figure 2. Trajectory parameters of the suborbital RSV TC:
(a) attack angle vs. time; (b) flight height and RSV velocity vs. time.

3.2 Aerodynamic loads determination

Modelling of RSV aerodynamic flow was performed by means of Fluid Flow module of FEA-package ANSYS Workbench. 3D-model of RSV was enclosed into a gaseous medium with the total volume of almost 100 000 m3. Such volume was large enough to leave out medium disturbance bringing by the side areas. Flow direction and velocity were assigned on the hemisphere frontal area of the gaseous medium, while on the back flat area free outflow conditions were applied. Modelling was conducted for the most critically loaded trajectory interval – RSV reentry in a dense atmosphere with the high values of attack angle. The resulting heat flow from the gaseous medium to the RSV surface distributes unevenly (Figure 3) and moreover changes each second. However, critical regions of the wing surface could be distinguished, e.g. the leading edge and neighboring section in the windward side of the wing are subjected to the highest temperatures influence. Thus, the gaseous temperature close to these regions could reach 945 K (Figure 4).

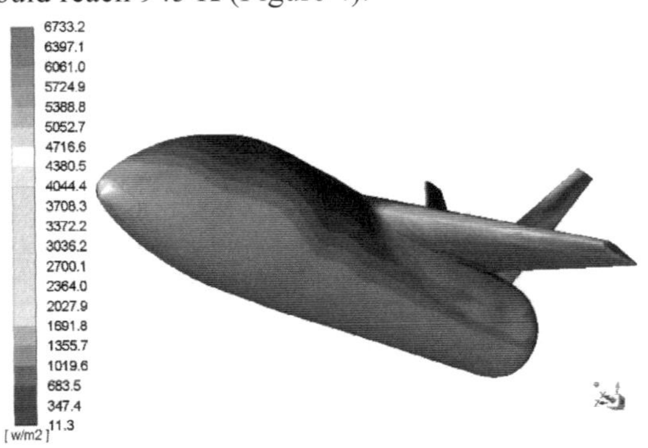

Figure 3. Total heat flux from the gaseous medium to the RSV surface [W/m^2].

Figure 4. Temperature of the gaseous medium close to the RSV surface [K].

3.3 Determination of the material thermal properties

The wing skin consists of a laminated composite combining layers reinforced with GFs and CFs, and a lightweight core made of Kevlar®. Thermal conductivity of such hybrid structure depends on numerous variables, i.e. nature and thermal properties of both reinforcement and matrix materials, reinforcement content, defects in composites, reinforcement architecture, etc. Moreover, thermal conductivity of laminated composites has a certain anisotropy and varies with the temperature. Such a big number of variables leads to a great amount of possible combinations, and as a result each material exhibits unique thermal response. Thus, it is wise to utilize FEA-analysis to determine hybrid composite thermal properties. In this study we implemented Digimat and ANSYS Workbench software to solve this task utilizing reference data about thermal properties of composite distinct constituents (fibers and matrix) [7-9].

The thermal properties of each component of a hybrid composite were defined separately. Consequently, three different models (for CF reinforced plastic (CFRP), GF reinforced plastic (GFRP) and Kevlar® honeycomb) were developed. The parameters and assumptions of each model are presented in Table 1.

IOP Conf. Series: Materials Science and Engineering **683** (2019) 012067 doi:10.1088/1757-899X/683/1/012067

Table 1. Parameters of the computational models for determination of materials thermal properties.

Parameter	CFRP	GFRP	Kevlar® honeycomb
Software	Digimat	Digimat	Ansys WB
Reinforcement architecture/type	UD tape/ AKSA A-49 24K4	Satin fabric 8/3 /T10	Satin fabric/Kevlar
Representative material volume (RMV)			
RMV dimensions [mm]	0.04 x 0.04 x 0.4	4 x 2.2 x 0.24	$h = 10$, $d = 3$, $\delta = 0.085$
Fiber diameter [µm]	6	6	-
Reinforcement volume fraction [vol.%]	60	60	60
Density [kg/m^3]: reinforcement/matrix	1 790/1 300	2 520/1 300	1 435/1 300
Heat transfer	Conductive	Conductive	Combined

The important issue of the described problem is the character of heat transfer in each material model. Whereas thermal conductivity is inherent for CFRP and GFRP, honeycomb, due to its peculiar structure, exhibits more complex heat transfer. Thus, heat transfer in honeycomb is carried out via conductivity, air convection inside the cell and radiation of the cell walls (with the emissivity factor equal to 0.9). By applying different temperatures for the opposite cell faces the aggregated heat flow coming through the cell was simulated, following by the evaluation of the effective honeycomb thermal conductivity.

The modelling results indicate, that CFRP demonstrates downward temperature dependency in transverse and longitudinal directions, while GFRP – upward (Figure 5 (a)). Note, that due to isotropic nature of GFs the thermal conductivity of the respective composite was investigated only in one direction. At the same time the effective heat transfer of the honeycomb shows non-linear rise with the temperature increase. This fact could be explained by the incremental contribution of the radiative heat transfer with the temperature growth.

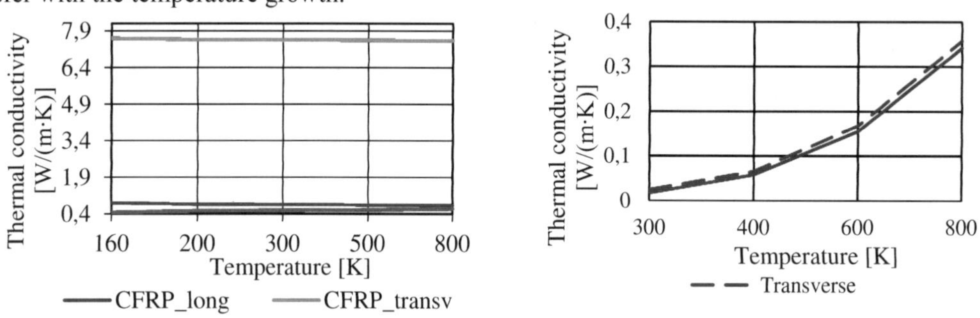

(a) (b)

Figure 5. Thermal conductivity of hybrid composite constituents:
(a) CFRP (transverse and longitudinal) and GFRP; (b) Kevlar® honeycomb.

3.4 Modeling of the wing thermal state

At a previous design step described in the study [5,10] an optimal structure of the hybrid composite wing skin was determined, so here we present only the resulting wing structure (Figure 6). Next, thermal state of the wing under the applied thermal loads was simulated in ANSYS WB. In addition to aerodynamic heat, radiative heat transfer between the inner wing walls and outer surface with the ambient medium were considered. The resulting temperature distribution on the wing skin surface demonstrates that maximum temperatures are localized in the leading edge and could exceed 600 K (Figure 7). So, in order to withstand the thermal loads, hybrid composite wing needs additional TP. However, it is not wise to apply TP to the whole wing surface, as it will increase the structural mass enormously [11]. Therefore, the optimization of the TP thickness and its positioning on the wing surface was carried out.

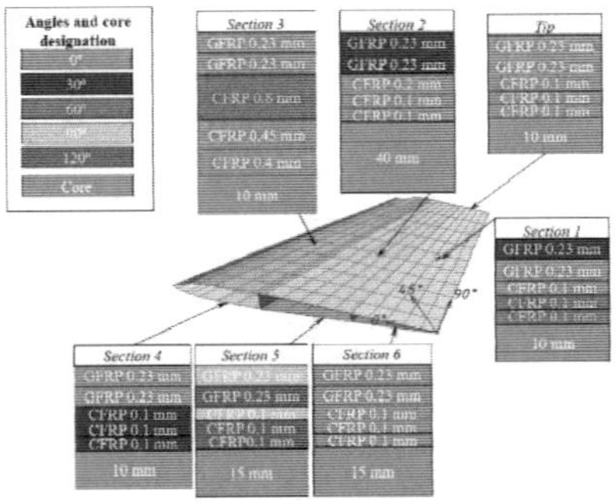

Figure 6. Optimized hybrid composite wing skin.

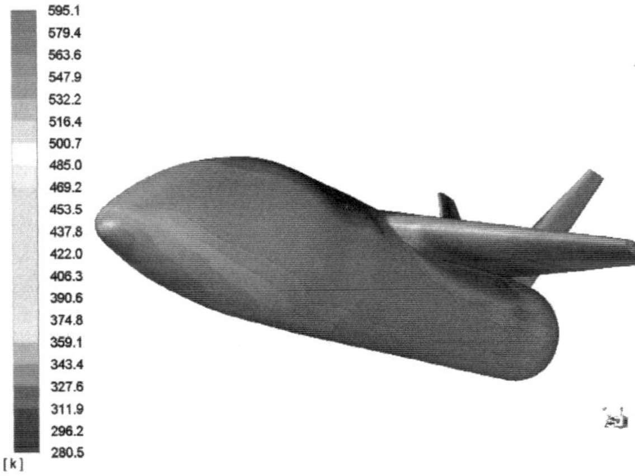

Figure 7. RSV surface temperature at 50 km altitude (210 sec of passive phase) [K].

3.5 Optimization of TP of the wing

Some wing sections are exposed to the influence of the temperatures, that exceed the allowable service temperature for polymer composites. Thus, the maximum service temperature for polyimide resin does not exceed 530 K. Therefore, the leading edge as well as parts of the windward and leeward sides of the wing needs to be thermally protected. Needless to say, that as the temperature distribution is uneven within the wing surface and thereby it is reasonable to apply different thicknesses of the TP layer on the different sections. Hence, it's necessary to find and optimum distribution of the TP within the RSV wing surface. In our previous work we have proved the feasibility of using this material for RSV wing TP [12]. However, parametric optimization of the TP thickness is necessary [13]. Such optimization could be conducted by means of introduction of the set of line-segment functions with apexes, which are coincident with the borders of the root and tip wing chords. Between this borders the TP thickness varies linearly in both directions – along the chord length and along the wing span. Therefore, the optimization problem was formulated as follows:

$$W = \int_0^{S_{wing}} \rho \cdot \delta(x,y) dS \rightarrow min, \tag{1}$$

$$T(x,y) \leq T_{max}, (x,y) \in S_{wing}, \tag{2}$$

where W is the TP mass, kg; ρ – TP density, kg/m3; δ – TP thickness, mm; S, S_{wing} – TP and wing surface area respectively, m2; T, T_{max} – calculated and maximum allowable temperature respectively, K.

Maximum TP thickness was taken to be equal to 20 mm, while discretization step – equal to 2.5 mm. The additional constraint was the equivalence of the TP thickness in the borders of the wing sections. Genetic algorithm was chosen for optimization. The resulting distribution of TP thickness is presented in Figure 8. The modelling results demonstrate that wing excluding Sections 1 and 2 need additional TP with the thicknesses from 0 up to 10 mm. The total mass of such TP will be 68 kg, that not exceed 20% of the total wing mass (380 kg).

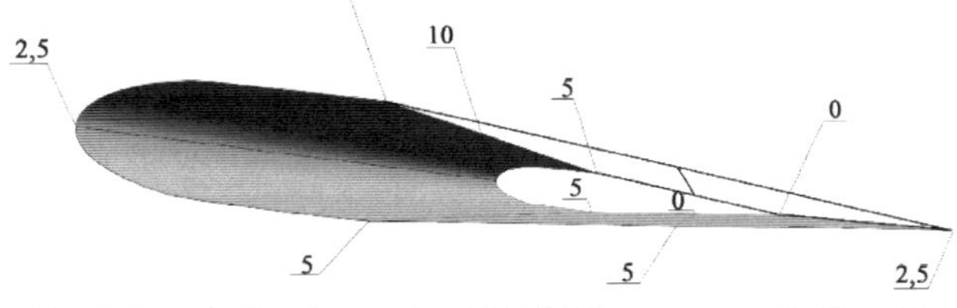

Figure 8. Results of the TP optimization: wing TP thicknesses excluding Sections 1 and 2 [mm].

4. Conclusion

An optimal design methodology for suborbital RSV TC load bearing structures was developed in this study. The feasibility of the developed methodology was illustrated with the example of a hybrid composite wing for RSV TC. The results of the solution of several distinct engineering problems were presented, e.g. trajectory parameters determination, aerodynamic flow modelling, material properties simulation, and TP optimization.

References

[1] Chang, Y.-W 2015. The first decade of commercial space tourism. *Acta Astronautica* , **108**, 79 91.

[2]. Chavagna, C. In *The suborbital space tourism project of eads astrium*, 45th AIAA/ASME/SAE/ASEE Joint Propulsion Conference & Exhibit, Denver, Colorado, 2 - 5 August, 2009; American Institute of Aeronautics and Astronautics, Inc. : Denver, Colorado, p 9.

[3]. Baiocco, P.; Bonnal, C. 2016 Technology demonstration for reusable launchers. *Acta Astronautica* **120**, 43-58.

[4] Qu, F.; Sun, D.; Zuo, G. 2018 A study of upwind schemes on the laminar hypersonic heating predictions for the reusable space vehicle. *Acta Astronautica*, **147**, 412-420.

[5] Ageyeva, T.G.; Ashikhmina, E.R.; Prosuntsov, P.V. 2018 Optimization of hybrid composite material structure for wing skin of tourist class reusable space vehicle. *Vestn. Mosk. Gos. Tekh. Univ. im. N.E. Baumana, Mashinostr. [Herald of the Bauman Moscow State Tech. Univ., Mech. Eng.]* **1**, 4-19.

[6] Ageeva T.G., Dudar E.N., Reznik S.V. **2010** Complex approach to the wing construction design for reusable spacecraft. *Aviakosmicheskaya tekhnika i tekhnologiya. [Aerospace Technic and Technology]*, **2**, 3-8.

[7]. Prosuntsov P.V., Taraskin N.Y. 2016 Theoretical and numerical characterization of the thermal conductivity of carbon ceramic material from realistic microscale imaging. *Matec Web of Conferences*, , **92**, 01008.

[8] Aleksandrov I.A., Prosuntsov P.V. **2016** Determination of the effect of carbon nanosized particles on thermophysical characteristics of polymer composite materials. *Polymer Science Series D*, **9**, 377–381.

[9]. Prosuntsov P.V., Taraskin N.Y. 2016 Theoretical and Numerical Characterization of the Thermal Physical Properties of Carbon Ceramic Material. *Matec Web of Conferences,* **, 72**, 01092.

[10]. Prosuntsov P.V., Shulyakovsky A.V., Taraskin N.Y. 2017 Numerical Simulation of a Thermal-Protection Element of a Promising Reusable Capsule-Type Lander. *Journal of Engineering Physics and Thermophysics*, , **90**, 110-116.

[11]. Kandinsky R.O., Prosuntsov P.V. 2015 Modelling combined heat exchange in the leading edge of perspective aircraft wing. *Matec Web of Conferences*, , **23**, 01019

.[12]. Ashikhmina, E.R.; Ageeva, T.G.; Prosuntsov, P.V. 2017 Thermal design of the wing skin of a tourist class reusable spacecraft *Engineering Journal: Science and Innovation,* , **12**, 1-13.

[13]. Reznik S.V., Prosuntsov P.V., Mikhailovsky K.V. 2015 Prediction of thermophysical and thermomechanical characteristics of porous carbon–ceramic composite materials of the heat shield of aerospace craft. *Journal of Engineering Physics and Thermophysics*, **88**, 594-601.

Development of branched aromatic maleimides as components of self-healing binders for polymer composites

N.V. Karelina[1], V I Solodilov[3], A.V. Polezhaev[1,2]

[1] Interbranch Engineering Center "Composites of Russia" of Bauman Moscow State Technical University, Lefortovskaya Emb. 1, Moscow, Russia, 105005

[2] A.N.Nesmeyanov Institute of Organoelement Compounds of Russian Academy of Sciences , 28, Vavilova Str., Moscow, 119334, Russia

[3] Bauman Moscow State Technical University, 105005, Russia, Moscow, 2nd Baumanskaya str., 5/1

E-mail: knv@emtc.ru

Abstract. Various modifications of epoxy resins with maleimides and polyimide fragments results in the formation of heat-resistant materials with excellent mechanical and electrical §properties together with high chemical, corrosion, and radiation resistance. Such materials can be used in the aerospace industry, electronics, instrument making, and other industries. The ability of some epoxy-maleimide resins to react with furanic groups through thermo-reversible Diels-Alder reaction open a new area in composite material science - self-healing binders. Self-healing materials can restore their structure and mechanical properties after a damage that results in an increase of the service life several times and, therefore, reduce their cost. In this paper, maleimides were modified by diglycidyl ethers to obtain not previously described polyethers, capable of self-healing of mechanical properties after damage. All compounds were characterized by ^{1}H NMR.

1. Introduction

Thermosetting resins are widely used in industries, for example, as adhesives or binders for fiber-reinforced composite materials for the aerospace industry [1]. The most widely used of thermosets are epoxy resins and polyimides. Epoxy resins are known as materials with high tensile strength, modulus of elasticity, excellent adhesive properties, low shrinkage during curing, high chemical and corrosion resistance. The simplicity of chemical modification and processing also increase the utility of epoxy resins. However, applications of epoxy resins in aerospace and some other areas are limited with their insufficient thermal stability and sometimes poor mechanics [2]. Polyimides showed a unique combination of extremely high thermal and heat resistance, excellent mechanical and electrical characteristics in together with high chemical and radiation resistance, which leads to their widespread use in the aerospace industry, electronics, electrical engineering, instrument engineering and others [3]. However, polyimides have some disadvantages, such as poor processability, fragility, and high cost. Therefore, the development of various modifications of epoxy resins with polyimides in order to obtain materials with suitable properties and an acceptable cost has attracted the attention of scientists for a long time [4-14].

Also, epoxy resins modified with maleimide fragments, are of particular interest to developers of self-healing binders based on thermo-reversible Diels-Alder reaction between furan and maleimide as a diene and dienophile, self-healing process could occur within the 100-150°C range. [15-20].
We demonstrate here a modification of maleimides with diglycidyl ethers to obtain polyethers not previously described, potentially capable of self-healing of mechanical properties after damage.

2. Experimental

2.1. Materials and methods
Maleic anhydride (p.a., Khimreaktiv, Russia), 4-aminophenol (97%, Acros Organics, Italy), 3,5-diaminobenzoic acid (98%, Acros Organics, United Kingdom), 5-aminoisophthalic acid (94 %, Aldrich Chemistry, India), sodium acetate anhydrous (99%, Acros Organics, Spain), acetic anhydride (tech., Ya.M. Sverdlov factory, Russia), acetonitrile (99%, Component-Reactiv, Russia), para-toluenesulfonic acid monohydrate (98%, Sigma-Aldrich, USA), 1,4-butanediol diglycidyl ether (Khimekslimited, Russia), diglycidyl ether of diethylene glycol DEG-1 (NIIKHIMPOLIMER, Russia), 4-dimethylaminopyridine DMAP (99%, Acros Organics, USA), tetrabutylammonium hydrogen sulfate (98%, Acros Organics, India), triphenylphosphine, TTP (99%, Acros Organics, China), hexamethyl trimethylammonium bromide, CTAB 97%, Acros Organics, USA), epichlorohydrin (≥ 99%, Fluka Analytical, Germany), sodium hydroxide (p.a., Labtech, Russia), anhydrous sodium sulfate (p.a., Labtech, Russia), phosphorus oxide (V) (p.a., Component- Reagent, Russia), acetic acid (p.a., OOO Baum-Lux , Russia), acetone (p.a., Component- Reagent, Russia), chloroform (r.g., Component-Reagent, Russia), toluene (p.a., Component- Reagent, Russia), ethyl alcohol (medical antiseptic solution, 95%, Ferein, Russia), 4-dimethylformamide DMF (99%, Acros Organics, USA). DEG-1 was purified by vacuum distillation; the remaining reagents were used without further purification. ATR FTIR spectra were recorded on a Thermo Fisher Scientific Nicolet iS10 (USA) spectrometer. All 1H-NMR spectra were recorded on a Bruker Avance 600 spectrometer (600.15 MHz). The values of chemical shifts (δ, ppm) in the spectra were determined relative to the residual signal of the solvent.

2.2. Synthesis of N-(4-hydroxyphenyl)maleamic acid 1
In a three-neck flask with reflux, 64.8 g (0.66 mol) of maleic anhydride were dissolved in acetone (525 ml) at the room temperature. Then 65.4 g (0.60 mol) of p-aminophenol was added to the solution in portions over 30 minutes and stirred for 1.5 hours. The precipitate was filtered, washed with acetone and dried in vacuum over P_2O_5. 115.8 g (yield 93%) of the target substance was collected as lemon-yellow crystalline precipitate.
Spectrum ^{1}H-NMR (DMSO-d6): 13.19 (s, 1H, -COOH), 10.50 (s, 1H, –NH), 7.42 (d, 2H, J = 8.9, Ar), 6.74 (d, 2H, J = 8.9, Ar), 6.48 (d, 2H, J = 12.4, = CH- C (O) -NH), 6.29 (d, 2H, J = 12.4, = CH-COOH).

2.3. Synthesis of N-(4-hydroxyphenyl)maleimide 2
70.40 g of acid 1 and 5.12 g of p-toluenesulfonic acid were boiled in a mixture of DMF (36 ml) and toluene (400 ml) at 110 °C with a Dean-Stark nozzle for 5 hours until the precipitate dissolved completely and the stoichiometric amount of water was collected. After evaporation of toluene, the thick solution was poured with water, the precipitate was filtered, washed with 5% NaHSO4 solution and water, dried in vacuum over P2O5. 50.40 g (yield 78%) of the target substance was received as bright yellow crystalline precipitate. ^{1}H-NMR (DMSO-d6): 7.34 (d, 2H, J = 8.8, Ar), 7.17 (d, 2H, J = 8.8, Ar), 6.84 (s, 2H, -CH = CH-maleimide ring).

2.4. Synthesis of 3,5-bis{[(2Z)-3-carboxyprop-2-enoyl]amino}benzoic acid 3
9.70 g (0.1 mol) of maleic anhydride and 5.02 g (0.033 mol) of 3,5-diaminobenzoic acid were boiled in chloroform (300 ml) for 20 hours. The precipitate was filtered, washed with chloroform and dried in vacuum over KOH. 11.31 g (98.0% yield) of the desired substance was obtained as light green crystalline

precipitate. ^{1}H-NMR (DMSO-d6): 11.05 (s, 3H, -OH), 10.62 (s, 2H, –NH), 8.38 (s, 1H, Ar), 8.02 (s, 2H, Ar), 6.52–6.24 (m, 4H, - (O) C-CH = CH-C (O) -).

2.5. Synthesis of 3,5-bis(2,5-dioxo-2,5-dihydro-1H-pyrrol-1-il)benzoic acid 4

5.45 g (0.016 mol) of acid was stirred with 0.4 eq. (0, 51 g, 0.0063 mol) of sodium acetate in acetic anhydride (120 ml) at 90-100°C for 1.5 hours. Then the solution was poured onto ice, 750 ml of cold water was added, left for 10 minutes, then vigorously stirred for 1 hour. The precipitate was filtered, washed with ~ 300 ml of cold water and dried in vacuum over KOH. 2.41 g (yield 49.2%) of the target substance was received in the form of beige crystalline precipitate. ^{1}H-NMR (DMSO-d6): 12.96 (s, 1H, -OH), 8.07-7.96 (m, 2H, Ar), 7.69 (s, 1H, Ar), 7.26 (s, 4H, -CH = CH- maleimide ring).

2.6. Synthesis of 5-{[(2Z)-3-carboxyprop-2-enoyl]amino}benzene-1,3-dicarboxylic acid 5

To the solution of 2.89 g (0.0295 mol) of maleic anhydride in glacial acetic acid (55 ml), 5.16 g (0.0285 mol) of 5-aminoisophthalic acid were added with vigorous stirring. In 5 hours, the precipitate was filtered, washed with water and dried in vacuum over KOH. 2.422 g (yield 30.4%) of the target substance was received in the form of light beige crystalline precipitate. ^{1}H-NMR (DMSO-d6): 12.89 (s, 3H, -OH), 10.72 (s, 1H, –NH), 8.50 (s, 2H, Ar), 8.23 (s, 1H, Ar), 6.55–6.36 (m, 2H, - (O) C-CH = CH-C (O) -).

2.7. Synthesis of 5-(2,5-dioxo-2,5-dihydro-1H-pyrrol-1-yl)benzene-1,3-dicarboxylic acid 6

1.46 g (0.0052 mol) of acid was mixed with 0.25 g (0.003 mol) of sodium acetate in acetic anhydride (25 ml) at 90°C for 17 hours. Then the solution was poured into ice water, extracted with diethyl ether, the organic layer was dried over sodium sulfate, filtered, evaporated, and the residue was recrystallized from ethyl alcohol. Received 0.97 g (yield 71.5%) of the target substance in the form of beige crystalline precipitate. ^{1}H-NMR (DMSO-d6): 12.92 (s, 1H, -OH), 7.98 (s, 2H, Ar), 7.65 (s, 1H, Ar), 7.23 (s, 2H, -CH = CH-maleimide ring).

3. Results and discussion

Self-healing polymers based on the Diels-Alder thermo-reversible reaction are very promising due to their ability to restore the structure and mechanical properties repeatedly. A modification of the epoxy binder with furan fragments is most often used for the introduction of diene groups, and bismaleimide or other dimaleimides are added to the binder as a dienophile [1, 21-23]. The price of dimaleimides is quite high and it greatly increases the final cost of the developed materials.

Most of the work on self-healing polymers is devoted to the synthesis of furan components, and a single diphenylmethane (bismaleimide) has been often used as the dienophile. We consider that the use of new maleimides based on inexpensive aromatic amines will significantly expand the possibilities of creating new self-healing polymers. In this regard, we developed a series of various branched aromatic maleimides. It is also should be noted that in this paper mainly simple methods of synthesis allowing to obtain high-purity compounds without additional purification were used.

The proposed synthetic scheme begins with the synthesis of simple aromatic maleimides from commercially available p-aminophenol, 3,5-diaminobenzoic acid and 5-aminoisophthalic acid. The reaction was carried out in two stages with the preparation of maleamic acid and subsequent cyclization by acetic anhydride or p-toluenesulfonic acid. Maleimides 2, 4, 6 of high purity were obtained during the work. The most likely reason of small yields in the preparation of compounds 2, 4 and 5 is their partial solubility in water (Scheme 1).

IOP Conf. Series: Materials Science and Engineering **683** (2019) 012068 doi:10.1088/1757-899X/683/1/012068

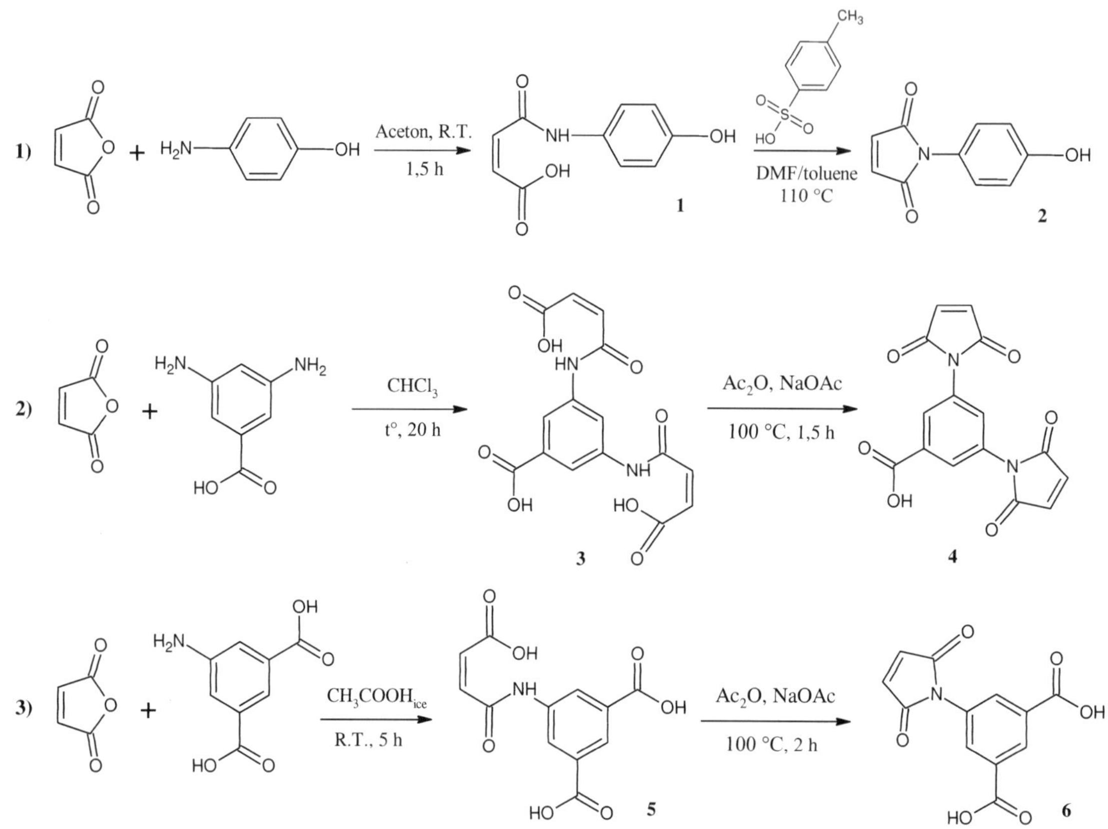

Figure 1. Synthesis of maleimides

On the next stage maleimide containing precursors bearing active functional groups were reacted with inexpensive and commercialy available epoxydes such as epichlorohydrine and diglycydyl ethers of aliphatic diols. First we used literature procedure [24] for the synthesis of compound 7, however, could not obtained the product due to the probable ring opening of the maleimide ring under basic conditions. The courses of reactions targeted compounds 8 and 9 were monitored by disappearance in IR spectra of an epoxy groups bands at 837, 912, 1275 and 1519 cm-1 for compound 8 and 855, 910, 1336 and 1479 cm^{-1} for compound 9, respectively. However, NMR analysis showed the presence of a mixture of reaction products. Thus, it was decided to abandon attempts to obtain esters of N-(4-hydroxy phenyl)maleimide. We are working now on the synthesis of compounds 10 and 11, since known methods for similar compounds described in the literature [25] require optimization.

At the next stage, we are planning to synthesize Diels-Alder adducts from these compounds, to study the process of curing the obtained materials with various amines and to determine the ability of these binders to self-healing after mechanical damage.

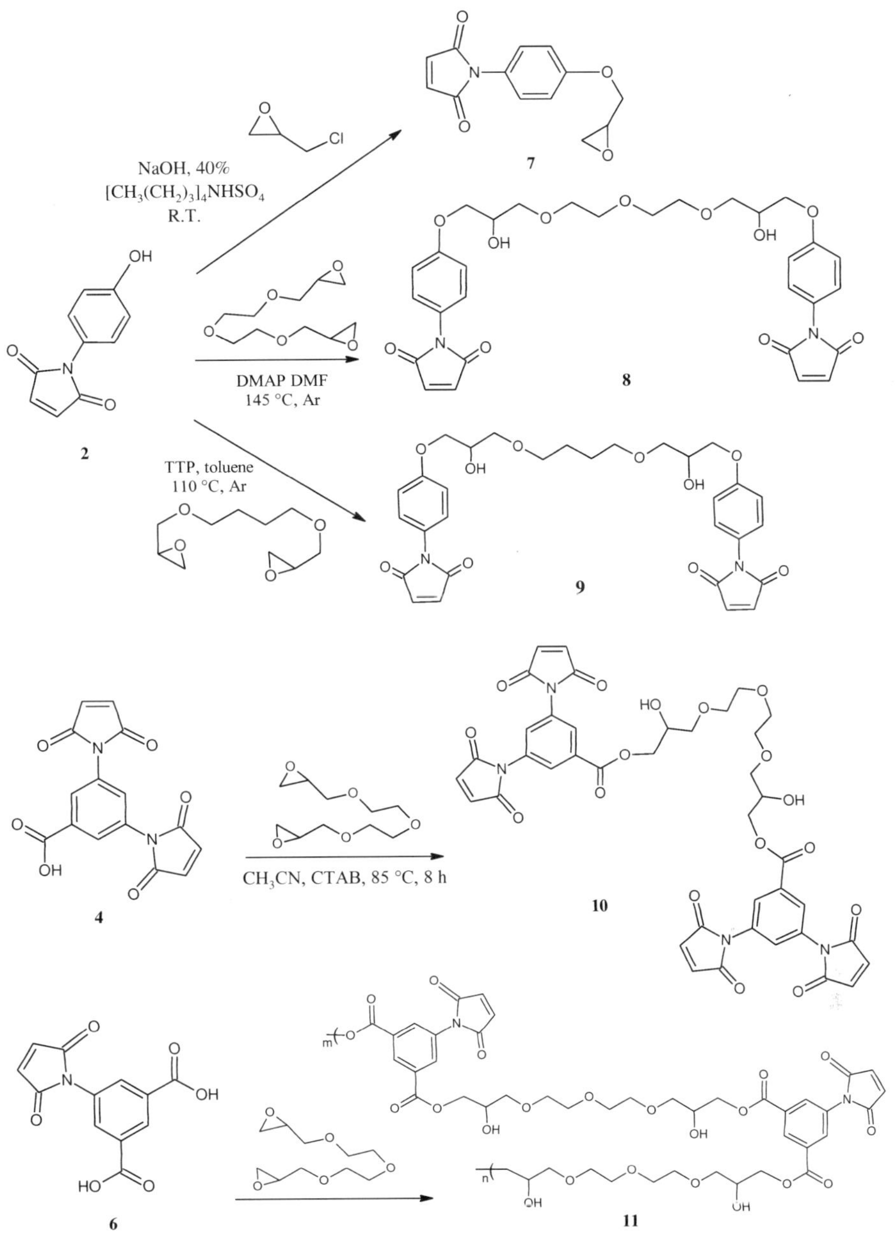

Figure 2. Schemes for producing ester precursors of resins

Acknowledgments

This work was supported by Russian Foundation of Basic Research (RFBR) grants 18-29-18036 and 18-29-18037. NMR experiments were performed with the financial support from Ministry of Science and Higher Education of the Russian Federation using the equipment of Center for molecular composition studies of INEOS RAS.

References

[1] F.R. J 1994 *Chemistry and technology of epoxy resins* vol 8 (London: Chapman & Hall)
[2] Liu Y L, Chen Y J and Wei W L 2003 Novel thermosetting resins based on 4-(N-maleimidophenyl)glycidylether I. Preparation and characterization of monomer and cured

resins *Polymer* **44** 6465-73
[3] Hoffman I.V., Abalov I.V., Yudin V.E. and Tyrants V.G. 2011 Mechanical and thermal properties of nanocomposite films based on aromatic polyimide and carbon nanocones *Phys. Solid State* **53** 1433-9
[4] Kumar A A, Alagar M and Rao R M V G K 2002 Synthesis and characterization of siliconized epoxy-1, 3-bis(maleimido)benzene intercrosslinked matrix materials *Polymer* **43** 693-702
[5] Musto P, Martuscelli E, Ragosta G, Russo P and Scarinzi G 1998 An interpenetrated system based on a tetrafunctional epoxy resin and a thermosetting bismaleimide: Structure - Properties correlation *J. Appl. Polym. Sci.* **69** 1029-42
[6] Han J L, Chern Y C, Li K Y and Hsieh K H 1998 Interpenetrating polymer networks of bismaleimide and polyurethane crosslinked epoxy *J. Appl. Polym. Sci.* **70** 529-36
[7] Kim W G and Nam T Y 1996 Curing characteristics of o-cresol novolac epoxy resin modified by bismaleimide *J. Polym. Sci. Pol. Chem.* **34** 957-62
[8] Ichino T and Hasuda Y 1987 New Epoxy-Imide Resins Cured with Bis(Hydroxyphthalimide)S *J. Appl. Polym. Sci.* **34** 1667-75
[9] Gaw K, Jikei M, Kakimoto M, Imai Y and Mochjizuki A 1997 Adhesion behaviour of polyamic acid cured epoxy *Polymer* **38** 4413-5
[10] Hodgkin J H, Simon G P and Varley R J 1998 Thermoplastic toughening of epoxy resins: a critical review *Polym. Advan. Technol.* **9** 3-10
[11] Hay J N, Woodfine B and Davies M 1996 Toughening of epoxy resins by polyimides synthesized from bisanilines *High. Perform. Polym.* **8** 35-56
[12] Mantecon A, Cadiz V, Serra A and Martinez P A 1988 Synthesis of Asymmetric Diglycidyl Ester Compounds - Reaction with Aromatic Diamines *Angew. Makromol. Chem.* **156** 37-48
[13] Adhinarayanan K, Packirisamy S, George R S, Rao V L and Ramaswamy R 1991 Epoxy- Imide Resins Based on Bis(Carboxyphthalimide)S *J. Appl. Polym. Sci* **43** 783-91
[14] Abraham G, Packirisamy S, Adhinarayanan K, Feby A G and Ramaswamy R 2000 Epoxy-imide resins from N-(4-and 3-carboxyphenyl)trimellitimides. I. Adhesive and thermal properties *J. Appl. Polym. Sci.* **78** 1729-36
[15] Mphahlele K, Ray S S and Kolesnikov A 2017 Self-Healing Polymeric Composite Material Design, Failure Analysis and Future Outlook: A Review *Polymers* **9** 535
[16] Garcia S J 2014 Effect of polymer architecture on the intrinsic self-healing character of polymers *Eur. Polym. J.* **53** 118-25
[17] Toncelli C, De Reus D C, Picchioni F and Broekhuis A A 2012 Properties of Reversible Diels-Alder Furan/Maleimide Polymer Networks as Function of Crosslink Density *Macromol. Chem. Phys.* **213** 157-65
[18] Chen X X, Dam M A, Ono K, Mal A, Shen H B, Nutt S R, Sheran K and Wudl F 2002 A thermally re-mendable cross-linked polymeric material *Science* **295** 1698-702
[19] Chen X X, Wudl F, Mal A K, Shen H B and Nutt S R 2003 New thermally remendable highly cross-linked polymeric materials *Macromolecules* **36** 1802-7
[20] Liu Y L and Chen Y W 2007 Thermally reversible cross-linked polyamides with high toughness and self-repairing ability from maleimide- and furan-functionalized aromatic polyamides *Macromol. Chem. Phys.* **208** 224-32
[21] Weizman H, Nielsen C, Weizman O S and Nemat-Nasser S 2011 Synthesis of a Self-Healing Polymer Based on Reversible Diels-Alder Reaction: An Advanced Undergraduate Laboratory at the Interface of Organic Chemistry and Materials Science *J. Chem. Educ.* **88** 1137-40
[22] Postiglione G, Turri S and Levi M 2015 Effect of the plasticizer on the self-healing properties of a polymer coating based on the thermoreversible Diels-Alder reaction *Prog. Org. Coat.* **78** 526-31
[23] Coope T S, Turkenburg D H, Fischer H R, Luterbacher R, Van Bracht H and Bond I P 2016 Novel Diels-Alder based self-healing epoxies for aerospace composites *Smart. Mater. Struct.* **25** 084010

[24] Mouzin G. C H, Rieu J-P., Duflos A. 1983 A convenient one-step synthesis of glycidyl ethers *Synthesis* 117-9
[25] Khalafi-Nezhad A, Rad M N S and Khoshnood A 2003 An efficient method for the chemoselective preparation of benzoylated 1,2-diols from epoxides *Synthesis* 2552-8
[26] Platonova E O, Vlasov E, Pavlov A A, Kireynov A, Nelyub V A and Polezhaev A V 2019 Self-healing polyurethane based on a difuranic monomerfrom biorenewable source *J. Appl. Polym. Sci.* **136** 47869
[27] Homer A and Winans C F 1939 Preparation of furfurylamines US Patent 2175585A

The use of inflatable structures for the removal of spacecraft from orbit

V V Koryanov[1], A G Toporkov[1], V P Kazakovtsev[1], A A Nedogarok[1] and E K Gnezdova[1]

[1] Dynamics and flight control of motion of rockets and spacecraft, Bauman Moscow State Technical University, 2-nd Baumanskaya, 5, b.1, 105005, Moscow, Russia

E-mail: vkoryanov@bmstu.ru

Abstract. Every year the concentration of objects of space debris is steadily growing, which significantly complicates the conduct of both modern and future space missions using automatic, and especially manned space vehicles. To date, over 15,000 artificial objects and fragments larger than 5 cm have been recorded in near-Earth space. Therefore, the issues of cleansing outer space from objects of space debris of various sizes are quite relevant. In this paper, are considered questions use inflatable structures for deorbiting spacecraft, in order to avoid the formation of new space debris. At the end of the work, conclusions are drawn about the effectiveness using of such inflatable structures.

1. Introduction

Due to the intensive development of the direction for the creation of small space vehicles and in particular on the basis of CubeSat, the issue de-orbiting of spacecraft after the end of their active existence in order to reduce the number of orbital debris in orbit is becoming very topical.
From 2000 to 2017 years, were launched into the low earth orbit (using standard launch vehicles) more than 800 spacecrafts of various types, and more than 200 spacecrafts only in 2017 year.
The majority of spacecraft are concentrated in orbits from 400 km to 700 km. At the same time at an altitude of about 400 km is the International Space Station, which since 2012 year is the platform for the mass launch of CubeSat.
Currently, the international community of scientists and engineers is in the active phase of working out space missions to clean up of space from objects of space debris. At the same time, various technologies for the reduction orbital lifetime and disposal of space debris are being developed.
As the main technologies for the deorbiting of space debris, most popular ones can be distinguished:
Installation of additional devices (Thruster de-orbiting kit, TDK) in the nozzle space debris (see in Figure 1 [1]). Functionality of such devices will allow to transfer the space debris to the disposal orbit.
Capture of space debris with the help of an active spacecraft. It is assumed that the active spacecraft will capture the space debris, transfer it to a given orbit, undock from the space debris, and continue to carry out scheduled flight operations. Within the framework of this concept, a lot of methods for capturing space debris are considered, which are subdivided into active and passive [2].
Deorbiting of a spacecraft by using a solar sail [3]. But for different shapes [4, 5] and different materials there are problem questions [6, 7].

Deorbiting of a spacecraft by increasing the aerodynamic resistance of the atmosphere after the deployment of braking devices. In the last two years only two projects have been proposed in Russia that are based on the principle of deploying braking devices [8, 9].

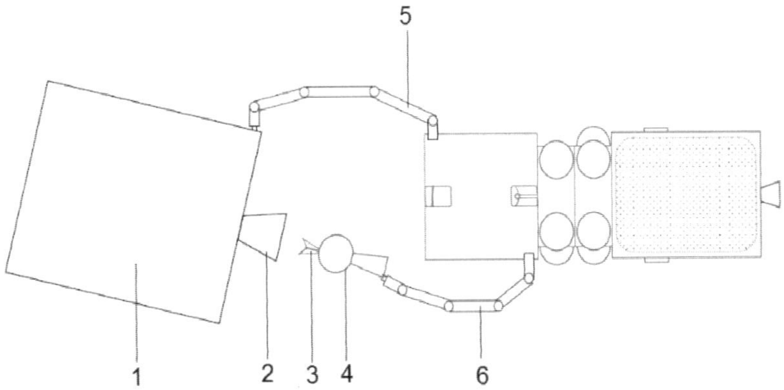

Figure 1. Installation of the TDK device in the nozzle space debris: 1- space debris, 2-nozzle space debris, 3-element fastening, 4-TDK, 5-docking manipulator, 6-manipulator.

The principle of operation of this technology is to increase the ballistic coefficient by creating a larger cross-sectional area of the space vehicle (compared to the original one) due to the deployment of braking devices. The final form of the disclosed braking device may be different.
In this case, the spacecraft can initially be equipped with brake devices during design and assembly on Earth.
Another option is when the braking device is install on the object after it has been classification like the space debris. In this case, will first need to capture the space debris and install the device. But space debris should have standardized interface for docking.

2. De-orbiting device configurations

De-orbiting devices can be designed in shapes of sphere, thorus, cone, pyramid, either dome, plane etc. The development of inflatable structures is covered in other articles [10, 11, 12]. According to the newthonian model for low-density high-velocity flow gas particles impact surface of de-orbiting device.Body get the momentum component normal to it's surface.
The result force applied to area S is equal

$$\vec{R}_a = -\vec{n}V_\infty^2\rho \cdot S\sin^2\alpha = -\vec{n}\left(\vec{V}_\infty \cdot \vec{n}\right)^2 \rho \cdot S, \qquad (1)$$

$\vec{n}$ – unit vector normal to surface S.
V_∞ – gas partickles velocity, equal to spacecraft velocity relative to moving atmosphere V_{atm}, with minus sign,
ρ – atmosphere density,
α – angle between flow velocity vector and surface.
Drag force $\vec{X}_a$ respectively is equal

$$\vec{X}_a = -\frac{\vec{V}_\infty}{V_\infty^2}\left(\vec{n} \cdot \vec{V}_\infty\right)^3 \rho \cdot S, \qquad (2)$$

or

$$X_a = \frac{1}{V_\infty}\left(\vec{V}_\infty\vec{n}\right)^3 \rho \cdot S, \qquad (3)$$

Spacecraft velocity relative to atmosphere, rotating with Earth (airspeed)

$$\vec{V}_{\text{atm}} = -\vec{V}_\infty = \vec{V} + [\vec{r} \times \vec{\omega}_3], \qquad (4)$$

$\vec{V}$ – spacecraft velocity in ECSF reference frame,
$\vec{r}$ – spacecraft position vector in ECSF reference frame,

$\vec{\omega}_3$ – Earth rotation angular velocity vector.

Drag force is usually expressed through dynamic pressure q_∞ and drag coefficient c_{xa}, or through ballistic coefficient σ, respectively

$$X_a = c_{xa} \cdot q_\infty \cdot S, \quad (5)$$

where $q_\infty = \frac{\rho V_\infty^2}{2}$, or

$$X_a = m \cdot \sigma \cdot V_\infty^2 \cdot \rho. \quad (6)$$

From (5) and (6)

$$\sigma = \frac{c_{xa} S}{2m}, \quad (7)$$

m – spacecraft mass.

From (3) and (5) c_{xa} for plane surface

$$c_{xa} = \frac{2}{V_\infty^3}\left(\vec{V}_\infty \vec{n}\right)^3 = 2\cos^3\left(\widehat{\vec{V}\vec{n}}\right), \quad (8)$$

and (7) ballistic coefficient

$$\sigma = \frac{1}{m}\cos^3\left(\widehat{\vec{V}\vec{n}}\right) \cdot S. \quad (9)$$

Calculational software was developed in appliance with newthonian flow theory, which used for estimation of drag and ballistic coefficients for different configurations of deorbiting device and their orientation in the flow. The data are listed below.

For example, *the cone design* is compact in the folded form and provides a large area in the expanded form. The conical shape facilitates the angular stabilization of the spacecraft near the zero angle of attack.

An example of the polar diagram of the dependence of c_{xa} on the direction of blowing for a cone with an half angle of 30° is shown in Figure 2.

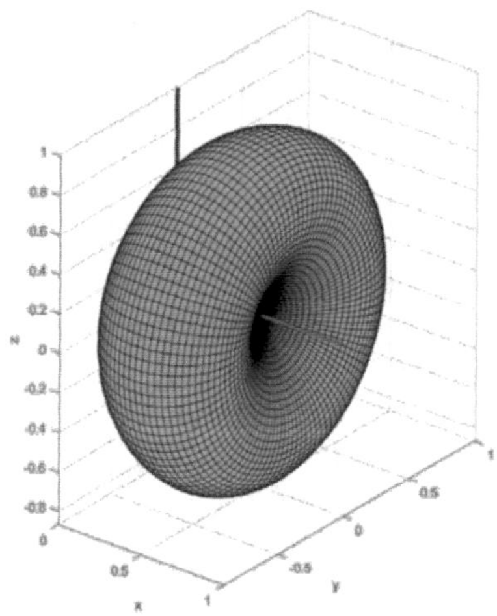

Figure 2. Polar diagram of the dependence of the drag coefficient of the cone on the direction of the speed of the oncoming flow.

The maximum drag coefficient in the front hemisphere c_{xa}= 1.05 is achieved with the direction of motion perpendicular to the cone generator, the minimum c_{xa}= 0.505, when moving with the toe in the direction of the velocity vector.

The tetrahedral shape of the braking device can be supported by deployable reinforcing elements, for example, wire or strip ribs unwound from the coil. This design does not need a pressure generator and can not be subject to leakage if the film sheath is damaged, but has less stability and requires the presence of drives for mechanical deployment. The tetrahedral shape contributes to the angular stabilization of the spacecraft near the zero angle of attack.

An example of the polar diagram of the dependence of c_{xa} on the direction of blowing for a cone with an half angle of 30° is shown in Figure 3.

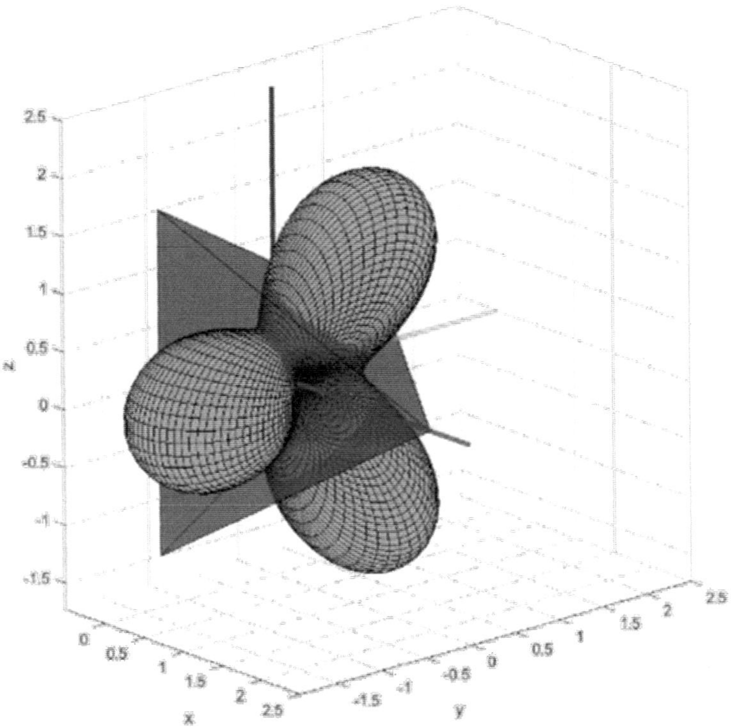

Figure 3. Polar diagram of the dependence of the coefficient of drag of a tetrahedron on the direction of the speed of the oncoming flow.

The maximum drag coefficient in the front hemisphere c_{xa} = 2 is achieved with the direction of the incident stream vector perpendicular to the lateral face of the tetrahedron, the minimum c_{xa} = 0.213 - when moving with the toe in the direction of the velocity vector.

The brake device is *a flat frame* of various shapes, on which the film is stretched. Has the greatest drag coefficient c_{xa} = 2 (see Figure 4).

The maximum drag coefficient in the front hemisphere c_{xa} = 2 is achieved with the direction of the vector of the incident flow perpendicular to the plane of the brake.

The cupola device is a parachute. The film canvas is stretched by means of unfold able wire "slings" to maintain the maximum area and radius of curvature. Unlike a flat device, does not have a reinforcing frame around the perimeter, so under the influence of the oncoming flow the shape can differ substantially from the flat one and have a radius of curvature. However, in comparison with a flat device, the structure and the deployment scheme are simplified, and the pressure of the oncoming flow naturally contributes to maintaining the shape of the cupola and stabilizing the objects.

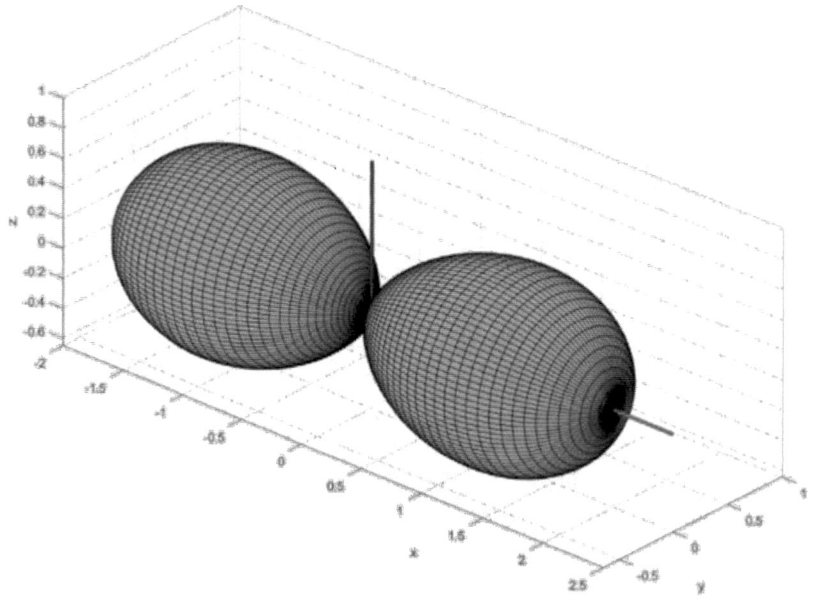

Figure 4. Polar diagram of the dependence of the coefficient of the drag of the plane on the direction of the speed of the oncoming flow.

Spherical braking device is structurally the simplest and does not require reinforcing elements. It can be a thin-film ball, supercharged by a gas generator. Has the same drag coefficient c_{xa} = 1 for any orientation of the objects.

3. Estimating the lifetime of a spacecraft on orbit

To estimate the spacecraft's orbital lifetime, a software complex for predicting the motion of the center of mass of the spacecraft was developed, which makes it possible to evaluate the possibilities of using braking devices of various configurations.
For an altitude of 200 km, the disturbing effect of atmospheric inhibition is 10^{-4} m/s^2. For an altitude of 400 km, the order of the perturbing acceleration decreases and is 10^{-6} m/s^2.
For altitudes from 800 km to 1500 km, the order of acceleration varies from 10^{-9} m/s^2 to 10^{-12} m/s^2.
However, even a slight perturbing acceleration from the atmosphere on will make a significant contribution to deceleration, since the drag force acts constantly and the direction of action is always opposite to the direction of motion of the space vehicle.
This fact makes it possible to use inflatable, unfolding and other braking devices quite efficiently for the purpose of reducing spacecraft from orbit.

4. Calculation results

As part of the research:

- calculation of the rate of fall of the orbit altitude of the spacecraft was carried out for nominal estimates of the forecast of the values of the indices *F10.7* and *ap*;
- the mass of satellite varied from 5 kg to 600 kg;
- the initial cross-sectional area of the satellite (S_0) varied from 0.05 m^2 to 1 m^2;
- considered two values of the cross-sectional area of the braking device after opening: S_1 = 4 m^2 and S_2 = 10 m^2;
- the shape of the disclosed braking device was selected as a tetrahedron.

Mass and geometric parameters of satellites with shown in table 1.

Table 1. Mass and geometric parameters of satellites

Number	Mass, kg	S_0, m²
1	5	0.05
2	10	0.10
3	20	0.20
4	30	0.30
5	40	0.40
6	50	0.50
7	100	0.65
8	150	0.75
9	200	0.80
10	300	0.85
11	400	0.95
12	500	0.95
13	600	1.00

The results of calculating the lifetime of satellite on a circular orbit of 600 km shown in Figure 5.

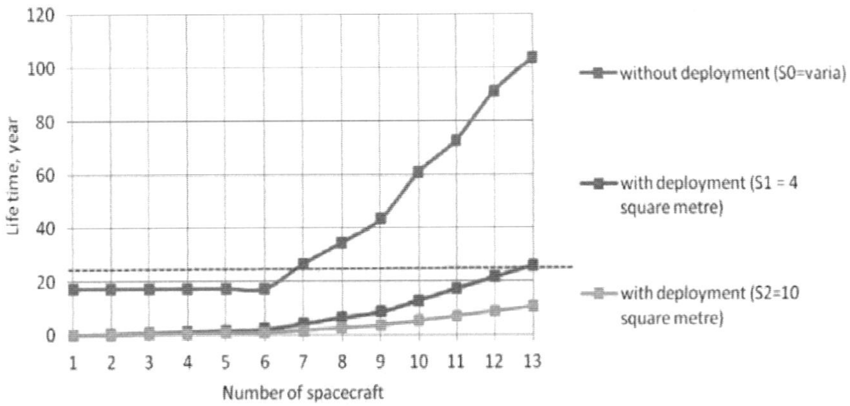

Figure 5. Changes in the height of spacecraft of various configurations for the initial altitude H_0 = 600 km.

The results of calculating the lifetime of satellite on a circular orbit of 700 km shown in Figure 6.

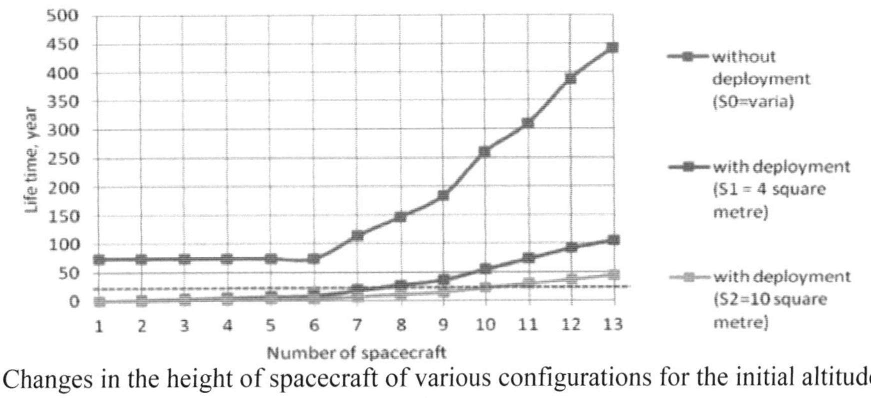

Figure 6. Changes in the height of spacecraft of various configurations for the initial altitude H_0 = 700 km.

As can be seen from the obtained results Figures 5-6, the disclosure of the inflatable braking device allows to significantly reduce the duration of passive ballistic spacecraft existence in orbit. And in most cases, the existence time does not exceed 25 years, which meets one of the requirements of the provisions of the Inter-Agency Committee on Space Debris [13].

So, according to preliminary calculations, an inflatable braking device in the form of a tetrahedron with a cross-sectional area of 4 m^2 will weigh about 1.5 kg.

5. Conclusions

Taking into account the obtained results, we can say that the use of passive braking devices in the form of inflatable structures can significantly reduce the time of existence of spacecraft in orbit, so the demand for such systems in the future will be very high.

The main advantage of inflatable structures before rigid ones is a small mass and the possibility of compact stacking in the required volume when putting into orbit. However, to date, inflatable braking devices are not sufficiently tested in outer space conditions, although ground handling has been conducted for a long time.

At the same time, there is a probability of breakdown of the inflatable structure by the fragments of the space debris and the occurrence of leaks of the supercharged gas, which can significantly reduce the efficiency of such a system. It is assumed that such leaks can be compensated by the multiple actuation of the gas generator.

Acknowledgments

The research was performed at Bauman Moscow State Technical University with the financial support of the Ministry of Education and Science of the Russian Federation under the Federal Target Program "Research and development on priority directions of scientific and technological complex of Russia for 2014-2020". Agreement №14.577.21.0247 (unique identifier RFMEFI57717X0247)

References

[1] Marco M, Castronuovo 2011 Active space debris removal—A preliminary mission analysis and design. *Acta Astronautica* **69** 848-59.

[2] Shan M, Guo J and Gill E 2015 Review and Comparison of Active Space Debris Capturing and Removal Methods. *Progress in Aerospace Sciences* **80** 18–32.

[3] Trofimov S P 2015 The removal of small spacecraft from low Earth orbits. PhD dissertation. 125.

[4] Mayorova V I, Tenenbaum S M, Rachkin D A, Nerovnyy N A and Popov A S 2013 Space cexperiment "BMSTU-SAIL". *Proceedings of the international astronautical congress,* IAC 64th pp 10157-65.

[5] Mayorova V I, Nerovnyy N A, Rachkin D A, Tenenbaum S M, Kotsur O S and Popov A S 2015 Current status of BMSTU-Sail space experiment. *66 International Astronautical Congress* **14** pp 10679-83.

[6] Nerovny N A, Lapina I E and Grigorjev A S 2017 Light radiation pressure upon an optically orthotropic surface. *J. Quant. Spectros. Radiat. Transf.* **202** 64-73 https://doi.org/10.1016/j.jqsrt.2017.07.016.

[7] Zarubin V S, Zimin V N and Kuvyrkin G N 2017 Temperature Distribution in the Spherical Shell of a Gauge-Alignment Spacecraft. *J. Appl. Mech. Tech. Phys.* **58.** 1083-90. https://doi.org/10.1134/S0021894417060141.

[8] Scientific significance. [Electronic resource]: Project "Mayak". - Access mode: http://cosmomayak.ru/about/science.

[9] Lavochkin and MAI specialists came up with a way to clean up space debris from orbit. [Electronic resource]: Joint-stock company "Scientific and Production Association. S.A. Lavochkin" - Access mode: http://www.laspace.ru/press/news/events/20171018_kak_ubrat%27_musor_s_orbity/.

[10] Koryanov V V 2013 Research of the dynamics motion of landing vehicle with inflatable braking device in the planet atmosphere. *Proceedings of the International Astronautical Congress*, IAC **8**, pp. 5831-36.

[11] Aleksashkin S N, Pichkhadze K M and Finchenko V S 2012 Design principles in planetary atmospheres reentry vehicles with inflatable braking systems. *Vestnik "NPO imeni Lavochkina"* **2** 4-11.

[12] Leonov V V 2016 The design features of inflatable large-scale mirror concentrators for space high-temperature solar power plant. *Proceedings of the international astronautical congress*, IAC 67, Making Space Accessible and Affordable to All Countries.

[13] Report of the Committee on the Peaceful Uses of Outer Space. General Assembly Official Records Sixty-second session Supplement No. 20 (A/62/20). p.50, 2007.

Furfurylglycidyl ether: a new effective active diluent for epoxy resins from bio-renewable raw materials

T V Petrova[1], V I Solodilov[1], V E Kabantseva[1], N V Karelina[1], A V Polezhaev [1,2,*]

[1]Bauman Moscow State Technical University, 105005, Russia, Moscow, 2nd Baumanskaya str., 5/1

[2]A.N. Nesmeyanov Institute of Organoelement Compounds of Russian Academy of Sciences, 119334, Vavilova street 28

E-mail: avp@emtc.ru

Abstract. New reactive diluents based on bio-renewable raw materials are widely explored as additives to traditional resins. We investigated the effect of furfurylglycidyl ether on the rheological and thermomechanical properties of a widely used epoxy binder with an anhydride hardener. It was shown that the addition of 5-15% active diluent significantly reduces the viscosity of the composition to the values optimal for RTM and vacuum infusion technologies for fiber-reinforced plastics and does not lead to a significant decrease in heat resistance and Young's modulus of the cured binder

1. Introduction

Currently, epoxy resins are used as matrices for fiber-reinforced composites, surface coatings, adhesives, due to low cost, and specific properties such as heat and chemical resistance [1]. However, the high viscosity and poor processability of some commercial epoxy resins severely limit wide applications. An effective way to eliminate these disadvantages is the use of suitable reactive diluents so that there was no substantial loss of basic properties [2, 3]. Reactive diluents are usually low molecular weight compounds containing at least one reactive functional group. Such diluents reduce a viscosity of the resin, improve its workability, and react with monomers and oligomers to form homogeneous materials [4].

In the past decade, a lot of attention was paid to the development of substitutes for petroleum-based products due to their non-renewability. The move from petroleum-based monomers and building blocks to bio renewables involves the use of natural materials for the production of different materials. Furfural and other furan containing monomers are especially valuable because of production from a wasted source such as corn cobs, wood products or cereal byproducts such as the hulls of cottonseed, oats, and rice.[5-7]. Furan monomers can be obtained from hemicellulose as one of the main components of plant biomass contained in large quantities in agricultural waste. Furfural, as the most critical first-generation furan derivatives industrially produced up to 200 000 tons per year, both in wealth and in developing countries. More than 80% of furfural has usually converted to furfuryl alcohol (FA) through a simple reduction process [8].

In a recent publication [9], the authors suggested using furfurylmetacrylate obtained by acrylation of furfuryl alcohol as a replacement for styrene in a polyester resin. Furfurylglycidyl ether (FGE) is another

FA derivative. FGE was explored as a monomer in the anionic polymerization with formation poly (FGE) homo and deblock (with propylene oxide) copolymers [10, 11], for modifying natural polymers such as chitosan [12], and for producing thermostable polymers by the Diels-Alder reaction [13, 14]. Furfurylglycidyl ether was recently reported as a very useful active diluent in epoxy compositions with aliphatic amine hardener. It was shown that FGE effectively reduced the viscosity of the composition, slightly decreases the glass transition temperature of the cured resin and does not lead to degradation of mechanical properties [2]. However, there is no information on FGE use in exopy/anhydride binders. Epoxy/anhydride resins are usually used for the production of high-performance composites with high heat resistance and excellent mechanical properties.

We decided to investigate the potential of FGE as an active diluent in anhydride curing composition and to investigate the rheological and thermomechanical properties of the composition before and after curing.

2. Experimental

2.1. Materials and methods

The epoxy resin DER331 (epoxy value 23%, DOW Chemicals) was used without further purification, furfurylglycidyl ether (FGE, epoxy value 27.92%, DOROS, Russia) was additionally distilled under vacuum. Methylhexahydrophthalic anhydride (MHHPA, 98%, Acros Organics) and 2-methylimidazole (2MI, Himex) were used as received.

The ATR-FTIR spectra were registered on a Thermo Nicolet iS10 spectrometer with a Fourier transducer using a germanium crystal in the range of 4000-650 cm^{-1}. The dynamic viscosity was measured on a rotational cone-plate viscometer Brookfield Cap 2000+ (rotational speed of 10 rpm, spindle No. 1). The study of the curing process of the samples was carried out by differential scanning calorimetry (DSC) on NETZH DSC 204 F1 Phoenix calorimeter in the temperature range from 25 to 260 ° C at the heating ate of 5 K*min^{-1} under an argon atmosphere. Dynamic mechanical analysis (DMA) of cured samples (20 mm x 6 mm x 2 mm, 3 samples for each composition) was carried out in a three-point bending mode on a Netzsch DMA 242 E Artemis instrument within the temperature range of 25-250 ° C, with a heating rate of 2 K / min, at frequency of 1 Hz.

2.2. Resin formulation (DER331/FGE/MHHPA/2MI)

We prepared 7 combinations with different FGE content (FGE: 0, 5, 10, 15, 20, 25 и 30% with respect to the DER331 resin) amount of hardener and catalyst was calculated with the following equation.

$$m = X \cdot \frac{M_0}{M}$$

M_0 - hardener molecular mass (g/mol);

M - epoxy group molecular weight (43 g/mol);

X - ammount of epoxy groups (g) per 100 g of resin (Russian standard 10587-84).

All samples were cured for 3 h at 100°C, 6 h at 140°C.

3. Results and discussion

The low processability of highly viscous binders limits their widespread use, and therefore active diluents are widely used. Unlike solvents, active diluents cure with epoxy resin and do not require subsequent removal. We investigated the viscosity of the binder with the content of FGE from 5 to 30% with an interval of 5%. Temperature dependence was shown for each composition (Table 1, Figure 1). As expected, the viscosity significantly decreases with increasing amount of active diluent, and with increasing temperature. The data show that FGE effectively reduces the viscosity of the Bisphenol A-based epoxy resin, which is consistent with previously obtained data [2, 3].

Table 1. The dependence of viscosity on temperature and FGE content.

DER331+ X% FGE		Temperature, °C					
		25	30	35	40	45	50
Viscosity, Pa*s	0%	16,00	7,50	3,80	2,00	1,10	0,60
	5%	5,00	2,60	1,40	0,80	0,40	0,20
	10%	2,00	1,00	0,50	0,30	0,13	0,04
	15%	0,80	0,40	0,20	0,09	0,02	-
	20%	0,40	0,20	0,06	0,06	-	-
	25%	0,15	0,04	0,02	-	-	-
	30%	0,05	0,03	0,02	-	-	-

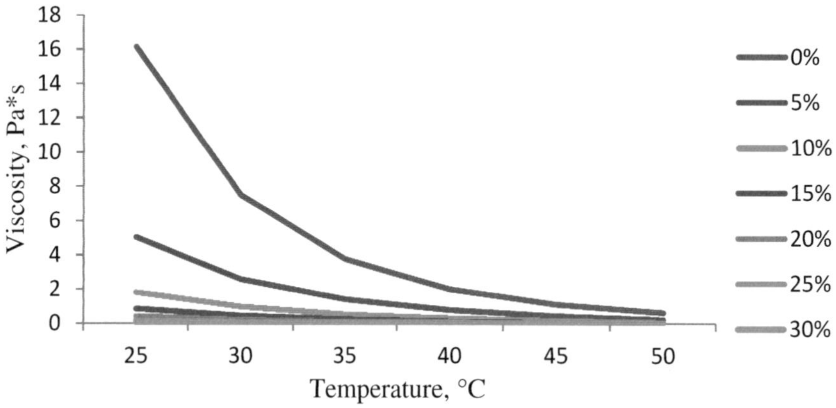

Fig. 1. The dependence of viscosity on temperature and FGE content.

To study the effect of the active diluent content on the heat and the curing temperature of the binder, the DSC method was applied with the following mode: isotherm 10 min, 25 ° C; heating, from 25-260 ° C, 5 K / min; cooling to 25 ° C, 20 K / min; isotherm 5 min. As can be seen from the results in Table 2, the FGE additive has almost no effect on the position of the maximum of the curing exotherm, which is in the around 160 ° C. The enthalpy of curing process is significantly reduced, but it is unlikely that this is due to a real decrease in the amount of heat released. The measurement was carried out in open crucibles and a sufficiently rapid heating leads to the partial evaporation of volatile active diluent from a crucible, which leads to underestimated values of enthalpy. However, the remaining parameters, such as T_{onset}. and T_{max}., which are crucial to the speed of curing, remain unchanged. Thus, we can conclude that the use of an active diluent does not require changes in the standard temperature curing regimes.

Table 2. DSC measurement results for formulations with different FGE content.

FGE content/T	T_{onset}, °C	T_{max}, °C	ΔH, J/g
0%	125,1	164,0	356,0
10%	128,4	158,5	340,9
20%	132,7	160,5	201,0
30%	142,5	159,6	170,3

To study the effect of the active diluent on the mechanical properties of the cured composition, we measured the cured samples of the binder using the three-point bending method on a DMA instrument.

The binder was cast into silicone molds in the form of bars (80 x 10 x 12 mm - length, width, thickness), cured in the following mode: 3 hours at 100 ° C and 6 hours at 140 ° C. From one bar received 3 small samples for DMA size 20x6x2mm were cut. The measurements were carried out at a heating rate of 2 K / min in the range of 25–250 ° C at a frequency of 1 Hz with a proportionality coefficient of 1.1 and an absolute amplitude of 60 μm and a maximum dynamic force of 10 N. The measurement results are shown in Table 3 and Figure 2. The glass transition temperature Tg of the studied samples decreases linearly with an increase in the amount of active diluent: for pure resin, the value is 156 ° C was obtained, and for the composition with 30% FGE - 117 ° C. This disadvantage is characteristic of any active diluents [2, 3]. It should be noted that the glass transition temperature does not drop as much as in the case of active diluents based on aliphatic alcohols or glycols. The reduction in viscosity required for such technological processes as vacuum infusion is achieved already at 5-10% of FGE. At these concentrations a decrease in heat resistance of 10–15 ° C can be considered insignificant.

Table 3. DMA results

Content	E_1' (MPaMPa) at 45°C	E_2' (MPa) at 45°C	E_3' (MPa) at 45°C	Sample 1, T_g, °C	Sample 2, T_g, °C	Sample 3, T_g, °C	Average E', MPa	Average T_g, °C
DER331	3898	3830	3936	156,4	156,6	155,4	3888	156,1
5%FGE	3377	3342	3747	149,7	149,5	147,1	3489	148,8
10%FGE	4175	3337	3120	142,9	143,2	142,2	3544	142,8
15%FGE	3885	3113	3499	136,9	135,5	134,0	3499	135,5
20%FGE	4695	4236	3678	130,2	129,6	129,7	4203	129,8
25%FGE	4478	4847	4072	124,0	121,1	124,9	4466	123,3
30%FGE	4405	4231	4308	117,6	116,5	117,7	4315	117,3

The effect of the amount of FGE on the Young's modulus is non-linear: the addition of 5-15% FGE to the epoxy-diane resin leads to a slight decrease in the modulus from 4 to 3.5 GPa, a further increase in concentration leads to an increase in the modulus to 4.5 GPa. However, it should be noted that such quantities of diluent are deliberately non-technological, so the use of this effect is not practical in real systems.

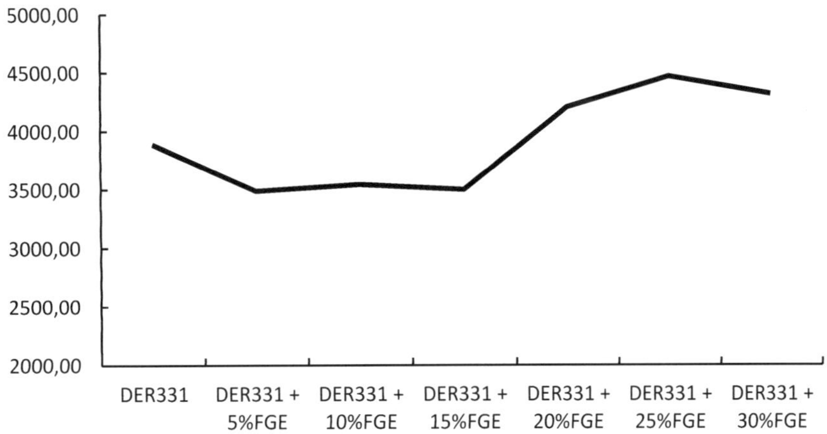

Figure 2. The dependence of elastic the modulus on the FGE content.

To confirm the completeness of the reaction, the binder was characterized by ATR-FTIR spectroscopy. As can be seen from the spectra in Figure 3, the bands of the carbonyl group of the anhydride hardener at 1785 and 1861 cm-1 completely disappear, as does the strip of the epoxy group at 889 cm-1, which confirms the complete course of curing.

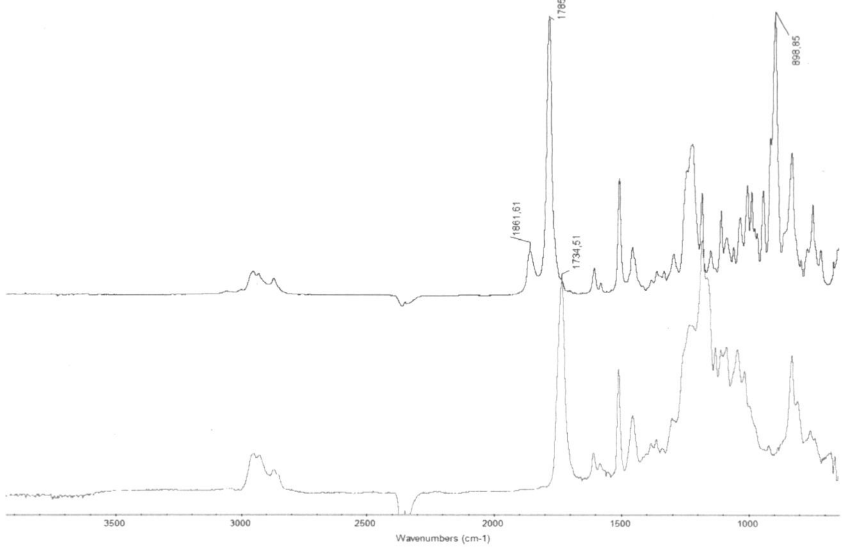

Figure 3. IR-ATR spectra of uncured (top) and cured (bottom) compositions with 20% of FGE4.

4. Conclusions

New active diluent - furfurylglycidyl ether was investigated in composition with Bisphenol A-based epoxy resin and anhydrite hardener. Rheological studies demonstrated high efficiency in reducing viscosity, a DSC study showed that the FGE additive does not affect the nature of the curing reaction, and testing of samples with the DMA method revealed a slight drop in heat resistance and the Young's modulus after addition of 10-15% FGE to the commercial epoxy.

Acknowledgments

This work was supported by Russian Foundation of Basic Research (RFBR) grants 18-29-18036 and 18-29-18037. NMR experiments were performed with the financial support from Ministry of Science and Higher Education of the Russian Federation using the equipment of Center for molecular composition studies of INEOS RAS.

References

[1] Ellis B and ebrary Inc. 1993 *Chemistry and technology of epoxy resins* (London ; New York: Blackie Academic & Professional)

[2] Karami Z, Nademi F, Zohuriaan-Mehr M J and Rostami A 2017 An efficient fully bio-based reactive diluent for epoxy thermosets: 2-[(Oxiran-2-ylmethoxy) methyl] furan versus a petroleum-based counterpart *J. Appl. Polym. Sci.* **134** 44957

[3] Sinha A, Islam Khan N, Das S, Zhang J and Halder S 2018 Effect of reactive and non-reactive diluents on thermal and mechanical properties of epoxy resin *High Perform. Polym.* **30** 1159-68

[4] Johansson K and Johansson M 2006 A model study on fatty acid methyl esters as reactive diluents in thermally cured coil coating systems *Prog. Org. Coat.* **55** 382-7

[5] Bessonov I V, Kopitsyna M N, Polezhaev A V and Nelyub V A 2016 A mechanistic study of the reaction between furfural-acetone resins and polyamines *Pol. Sci. Ser. D* **9** 17-21

[6] Bessonov I V, Polezhaev A V, Kuznetsova M N, Nelub V A, Buyanov I A, Chudnov I V and Borodulin A S 2013 Rheological and thermal analysis of low-viscosity epoxy-furan

composites *Pol. Sci. Ser. D* **6** 308-11
[7] Gandini A, Lacerda T M, Carvalho A J F and Trovatti E 2016 Progress of Polymers from Renewable Resources: Furans, Vegetable Oils, and Polysaccharides *Chem. Rev.* **116** 1637-69
[8] Suib S L 2013 New and future developments in catalysis catalysis by nanoparticles. (Amsterdam ; Waltham, MA: Elsevier) p 1
[9] Zhang Y, Li Y, Thakur V K, Wang L, Gu J, Gao Z, Fan B, Wu Q and Kessler M R 2018 Bio-based reactive diluents as sustainable replacements for styrene in MAESO resin *RSC Advances* **8** 13780-8
[10] Barthel M J, Rudolph T, Crotty S, Schacher F H and Schubert U S 2012 Homo- and diblock copolymers of poly(furfuryl glycidyl ether) by living anionic polymerization: Toward reversibly core-crosslinked micelles *J. Pol. Sci Pol. Chem.* **50** 4958-65
[11] Wagner M, Barthel M J, Freund R R A, Hoeppener S, Traeger A, Schacher F H and Schubert U S 2014 Solution self-assembly of poly(ethylene oxide)-block-poly(furfuryl glycidyl ether)-block-poly(allyl glycidyl ether) based triblock terpolymers: a field-flow fractionation study *Pol. Chem.* **5** 6943-56
[12] Na H-N, Park S-H, Kim K-I, Kim M K and Son T-I 2012 Photocurable O-carboxymethyl chitosan derivatives for biomedical applications: Synthesis, in vitro biocompatibility, and their wound healing effects *Macromol. Res.* **20** 1144-9
[13] Tian Q, Rong M Z, Zhang M Q and Yuan Y C 2010 Synthesis and characterization of epoxy with improved thermal remendability based on Diels-Alder reaction *Polym. Int.* 59 1339-45
[14] Platonova E O, Vlasov E, Pavlov A A, Kireynov A, Nelyub V A and Polezhaev A V 2019 Self-healing polyurethane based on a difuranic monomer from biorenewable source **136** 47869

IOP Conf. Series: Materials Science and Engineering **683** (2019) 012071 doi:10.1088/1757-899X/683/1/012071

The effect of triethylene glycol additive on the self-healing properties of epoxy binder

T V Petrova[a], V I Solodilov[a], V E Kabantseva[a], N V Karelina[a], A V Polezhaev[a,b,*]

[a]Bauman Moscow State Technical University, 105005, Russia, Moscow, 2nd Baumanskaya str., 5/1

[b]A.N. Nesmeyanov Institute of Organoelement Compounds of Russian Academy of Sciences, 119334, Vavilova street 28

E-mail: avp@emtc.ru

Abstract. A triethyleneglycol-modified epoxy composition based on diglycidylfurfurylamine (FuEp) and 1,1'- (methylene-di-4,1-phenylene)bismaleimide (BMI) was cured with methyl hexahydrophthalic anhydride (MHPPA). The components of the composition can reversely binds through Diels-Alder reaction between furan and maleimide moieties. We studied the effect of TEG additive on the rheological and thermophysical properties of the binder, as well as the ability through thermal treatment. All compounds were characterized with NMR and IR spectroscopy. It was shown that TEG significantly reduces the curing time and the glass transition temperature of the cured samples, which allows to speed up the process of self-healing.

1. Introduction

Self-healing materials belong to the class of "smart" materials that can partially or completely repair damage caused by external influence. [1-3]. Such materials should be able to self-heal cracks and other defects arising during the operation of the product. The development of various strategies and approaches to the creation of self-healing materials is one of the most important tasks in the creation of new composite materials. [4, 5].

There are two different approaches to the creation of such materials: intrinsic and extrinsic self-healing. In the case of extrinsic self-healing, the liquid reducing agent is placed in spherical capsules or hollow tubes. When a crack is formed, the liquid components spread out, filling the crack, and polymerize under the action of a hardener or catalyst that was previously inserted into the composite matrix. [6-11]. Thus, this type of self-healing only provides one-time recovery.

Intrinsic self-healing does not require the introduction of various healing agents. This method is based on the ability of materials to re-form covalent bonds between functional groups on the surface of a crack. To form these bonds, the monomeric components of the material must be fully or partially polymerized due to a reversible, under certain conditions, reaction. If damaged, this material may be subjected to depolymerization to release fragments that fill the crack, followed by reverse polymerization to the crosslinked state. [12-14].

Intrinsic self-healing is a more promising method for creating self-healing materials, as it allows self-healing to be controlled and repeated. Furan components of such polymers are usually synthesized from furfural - a product of processing plant raw materials, which is important from the point of view of "green" chemistry. [15, 16]. We present here a reversible self-healing approach based on reversible Diels-Alder reaction between furan and maleimide pair [17, 18, 19].

2. Experimental

2.1. Materials and methods

BMI (1,1'-(methylene-di-4,1-phenylene)bismaleimide, 95%, Sigma-Aldrich), MHHFA (hexahydro-4-methylphthalic anhydride, 98%, Acros Organics), triethylene glycol (TEG), furfurylamine (99 %, Acros Organics), epichlorohydrin (99%, Fluka), NaOH (Labtech), tetrabutylammonium hydrosulfate (98%, Acros Organics) were used without further purification. All solvents used: ethyl acetate, petroleum ether, hexane (all reagent grade) - were produced by the company "Component-Reagent". IR-ATR spectra were obtained on a Thermo Nicolet iS10 spectrometer with a Fourier transducer using a germanium crystal in the range of 4000-650 cm-1. NMR spectra were recorded on a Bruker Avance 600 spectrometer 600 (600.1 MHz) in DMSO-d6. Calorimetric measurements were performed on DSC NETZH DSC 204 F1 Phoenix calorimeter in the temperature range from 20 to 200 ° C with a heating / cooling rate of 10 K min^{-1} in argon atmosphere. Viscosity was determined on a rotational cone-plate viscometer Brookfield Cap 2000+ (rotation speed of 8 rpm, spindle No. 1)..

2.2. Synthesis of diglycidylfurfurylamine FuEp

The target substance was obtained by a modified literature method. [17]. Briefly, the synthesis was carried out in two stages. At the first stage, furfurylamine was slowly added to the epichlorohydrin, the product obtained, without isolation, was cyclized by the action of an alkali solution. Yield 33,9 g. ^1H NMR (400 MHz, DMSO-d_6, ppm): 3.1–2.4, 6.17–6.25 (=CH-CH= furan ring, 2H), 7.31 (-O-CH=C-furna ring, 1H), 3,7-4,07 (-NCH2, 2H). IR (ATR, cm^{-1}): 3121, 3049, 2995, 2922, 2831, 1739, 1601, 1442, 1350, 1257, 1151, 1076, 1011, 917, 856, 830.

2.3. Synthesis of cross-linked polymer (FuEp/BMI/MHHFA/TEG)

BMI 0.451 g (1.3 mmol) was mixed with 0.526 g (2.5 mmol) of FuEp at 70 ° C for 15 minutes. TEG was added to the mixture (1, 5 or 10% with respect to FuEp). The curing conditions were chosen in accordance with the optimum temperature for the formation of Diels-Alder bonds, so that most of the furan and maleimide groups reacted before the system was cured. Samples were cured with 0.680 g (4.0 mmol) of methyl hexahydrophthalic anhydride (MGGFA) at 70 °C for 24 h.

3. Results and discussion

Tian and coauthors showed [17] that a three-component binder consisting of FuEp, BMI and MHGFA can cure to form a cross-linked polymer that forms a double interpenetrating network: one part of the network was formed by a copolymerisation of epoxy and anhydride, and the second was formed through DA adducts between the maleimide BMI and the furan ring of FuE. Under heating to 120 ° C in a cross-linked polymer, a retro-DA reaction occurred with the destruction of the bonds between furan and maleimide with the release of BMI, the subsequent prolonged curing at 60 ° C leads to the reformation of the DA adducts, which resulted in the restoration of mechanical properties after damage (Scheme 1)

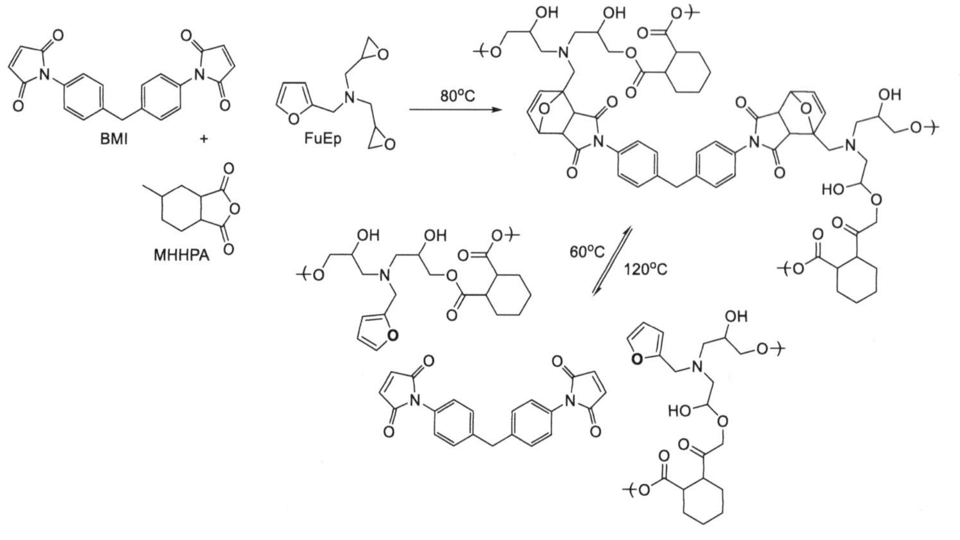

Scheme 1. Formation of a cross-linked polymer and reversible decomposition of DA adducts.

We repeated this work and noticed some shortcomings. For example, the restoration cycle proposed by the authors (20 minutes at 120 ° C and 72 hours at 70 ° C) is long and impractical for industrial use. Another significant drawback is the fragility of the resulting composition. It is known that one of the most commonly used technologies for manufacturing of fiber reinforced composite is vacuum infusion. This method requires the use of low viscosity binders to ensure good impregnation quality. We suggest that the addition of triethylene glycol could help to reduce all the above disadvantages. It is known that glycols act as a catalyst for an anhydride curable epoxy composition and also as a plasticizer. The addition of such a component can lead to a decrease in the curing time and a decrease in the glass transition temperature, and, consequently, to an acceleration of the self-healing process (Scheme 2).

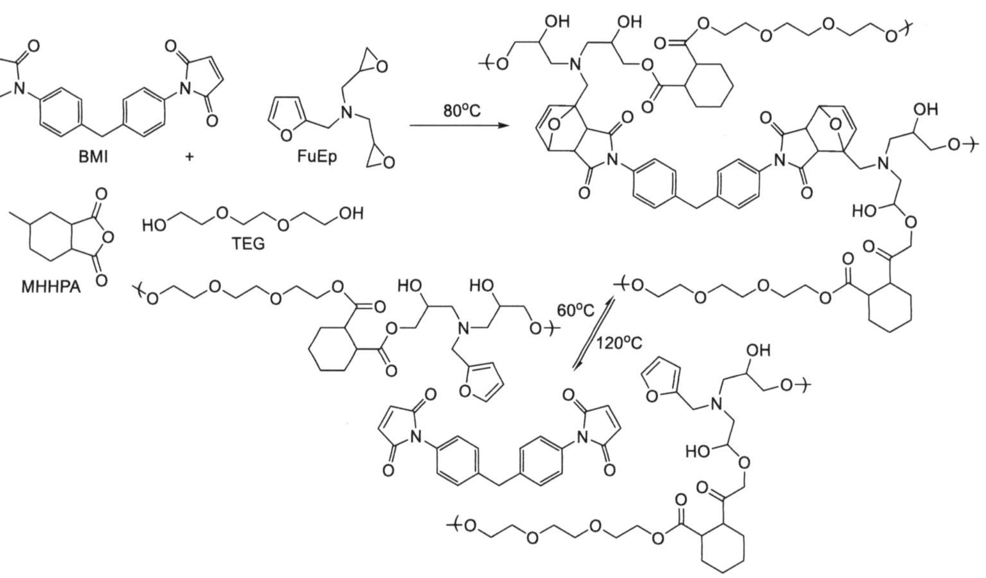

Scheme 2. Formation of a cross-linked polymer in the presence of TEG and reversible destruction of DA adducts.

3.1. Synthesis and structure of the epoxy component of the binder (FuEp)

The epoxy component of the binder was obtained in two stages, at the first stage, furfurylamine was slowly added to epichlorohydrin, controlling the temperature, then the resulting mixture was treated with 40% NaOH solution, extracted and purified by silica gel flash chromatography (Scheme 3).

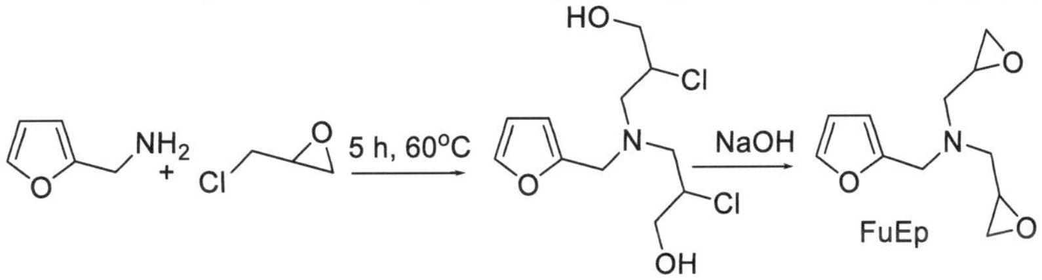

Scheme 3. Synthesis of an epoxy component.

The compound obtained was characterized with IR and NMR spectroscopies. Figure 1 shows the FTIR (ATR) spectrum of the FuEp. The bands at 3049 cm^{-1} (C–H of oxirane ring), 1257 and 857 cm^{-1} (C – O – C), and 918 cm^{-1} correspond to the oxirane cycle. Stretches of the furan ring in FuEp appeared as bands at 3122 cm^{-1} (C–H), 1508 cm^{-1} (C = C), 1351 cm^{-1} (C – N), 1152 cm^{-1} (C – O), and 1011 $^{-1}$.

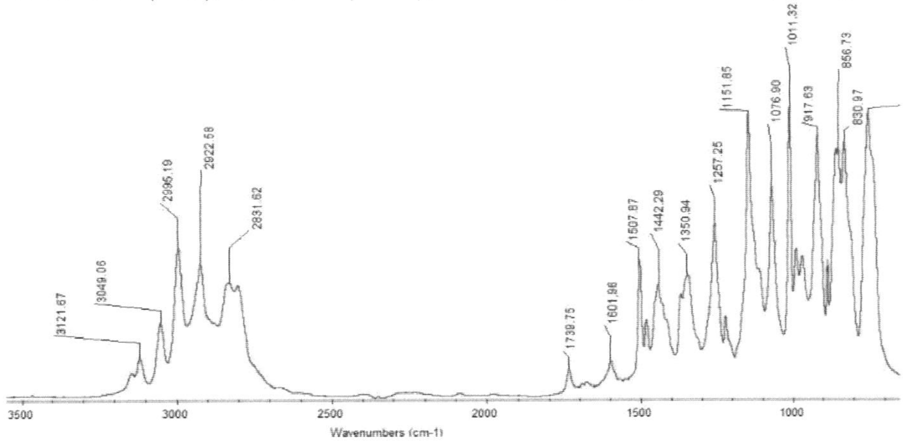

Fig. 1. FTIR spectrum of FuEp.

In the ^{1}H NMR spectrum (Figure 2), the FuEp peaks at 3.1–2.4 ppm appeared from protons of oxirane rings ("5"). The furan cycle was showed by peaks at 6.17–6.25 ppm (= CH – CH =, 2H) and 7.31 ppm (–O – CH = C–, 1H), which corresponds to protons, denoted as 2, 3, and 1 , respectively. The characteristic peak at 3.7-4.07 ppm (-NCH2-furan ring, 2H) corresponds to the methylene group in position "4". The data obtained unequivocally confirm the structure of the synthesized epoxy component.

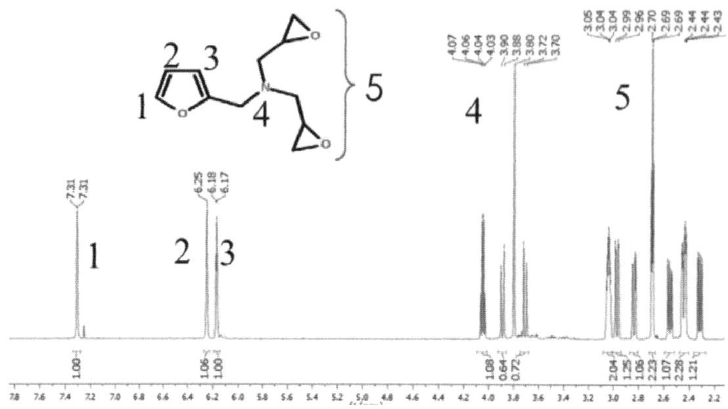

Figure 2. ^{1}H NMR spectrum of FuEp

3.2. Obtaining a binder (FuE / BMI / MHHFA / TEG) and the study of its properties

A binder was obtained by mixing of the components in a solution of dichloromethane, followed by removal of the solvent in high vacuum. The resulting viscous solution was kept at 80 ° C for 10 minutes to ensure the formation of DA adducts between the furan and maleimide components. We studied the dependence of the viscosity of the composition FuEp / BMI / MHHFA from the amount of added triethylene glycol. Viscosity measurement showed a significant dependence of the dynamic viscosity on the TEG content, as well as on temperature. Thus, the introduction of 1% TEG into the binder resulted in a decrease in viscosity from 3.21 to 0.75 Pa * s. Adding more TEG does not linearly decrease viscosity, but has only a minor effect.

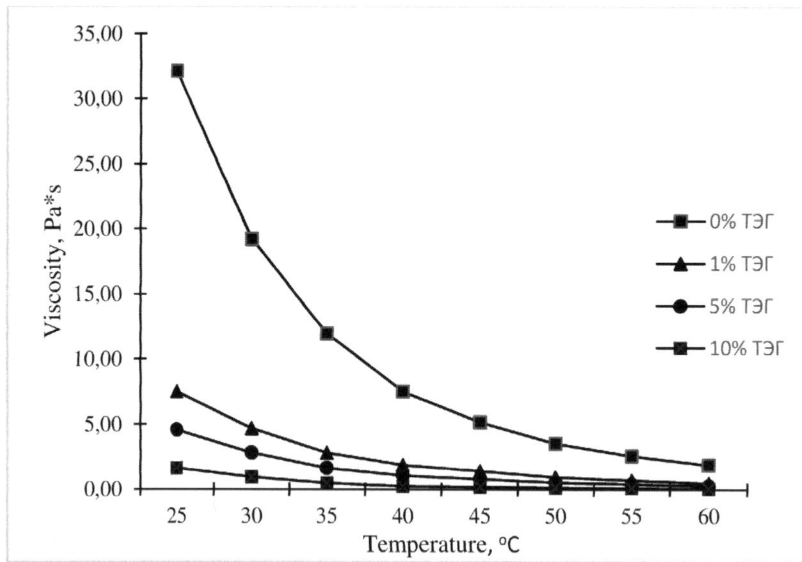

Figure 3. Dependence of viscosity from temperature for different compositions

The DSC method was used to determine the maximum curing exotherm temperature for the binder, as well as the glass transition temperature of the cured material. The results of thermal analysis are presented in Table 1. At the first heating of the composition, an exothermic peak appears on the DSC curve, corresponding to the reaction of copolymerization of epoxide and anhydrite hardener and TEG.

The decrease in the maximum of the exotherm when adding TEG shows that TEG acts as an accelerator in the curing process.

Tab. 1. DSC results.

TEG, %	T_{max} (°C)	T_c (°C)
0	148,97	138,17
1	107,17	117,80
5	105,58	115,69
10	103,79	107,28

The addition of 1% of TEG leaded to a drop of the maximum of the exotherm by 45 °C, but at the same time reduced the glass transition temperature, and hence the upper limit of the heat resistance of the polymer, by 30 °C. The increase in the concentration of triethylene glycol in this composition, in our opinion, is inexpedient, so further changes in the characteristics were insignificant. Also, DSC analysis allowed to indirectly assess the degree of self-healing. Samples of a composition with 1% TEG were subjected to thermal cycling according to the following program: 120 °C for 20 minutes to completely destroy DA adducts between furan and maleimide, and then kept at 60 ° C for various periods of time. After that, the samples were measured by the DSC method. To assess the degree of reduction of the DA adducts, the area of the endothermic peak of the Diels – Alder back reaction was used, which is proportional to the number of bonds formed, which in turn indicates the completeness of the reduction (Figure 4).

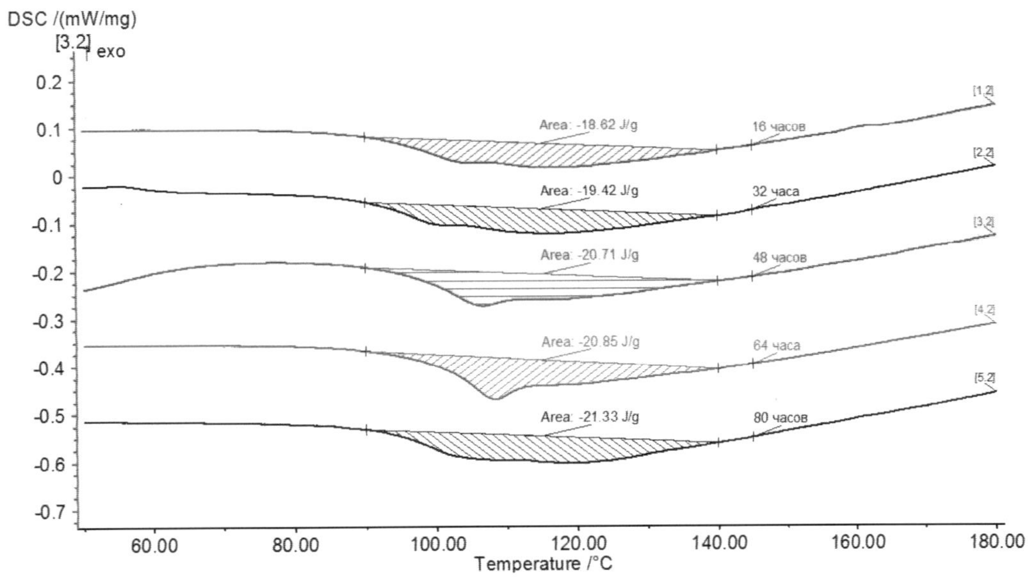

Figure. 4. DSC thermograms of samples containing 1% of TEG treated for 60 °C for 16, 32,48, 64, and 80 h.

The peak areas of the Diels-Alder reverse reaction for samples with different TEG contents are shown on Tab. 2. As can be seen, for samples with the content of 1% TEG, after 32 h, almost complete restoration of bonds is observed. An increase in the holding time does not contribute to a significant increase in the peak area and varies within 5%. For samples with higher TEG content, the glass transition temperature decreased and the glass transition curve overlaps with the signal corresponding to the Diels-Alder reverse reaction; therefore, it was not possible to separate these peaks and correctly analyze their areas.

Table 2. The results of measurements of the peak area of the Diels-Alder back reaction depending on the time of keeping the samples in the furnace and the TEG content

Time, h	16	32	48	64	80
TEG content, %	Square of the peak, J/g				
0	-9,69	-10,06	-11,79	-13,39	-12,36
1	-18,76	-19,63	-20,82	-21,02	-22,15
5	-13,69	-38,60	-18,29	-20,22	-40,71
10	-27,86	-12,98	-27,70	-28,34	-13,05

We also carried out an optimization of the composition of the binder in order to determine the optimal TEG content, providing a reduction in viscosity for values acceptable for vacuum infusion and at the same time minimal loss of heat resistance (Fig.5). It was shown that an increase in the TEG content above 1% is unnecessary.

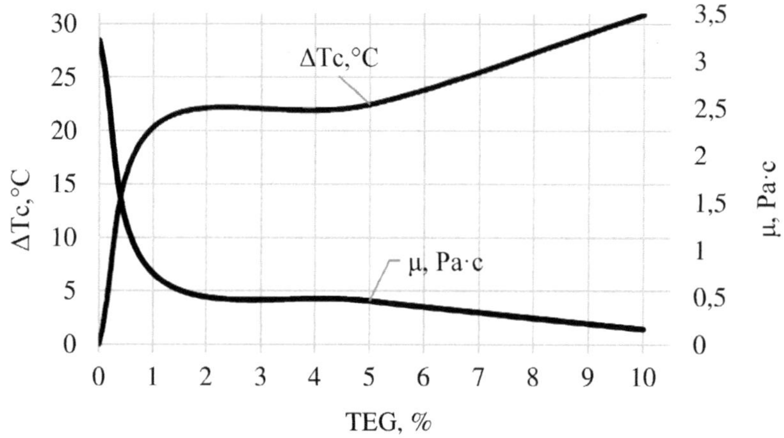

Figure 5. Dependence of viscosity from glass transition temperature and TEG content

4. Conclusions

It was found that the optimal content of triethylene glycol in the composition with FuEp/BMI/MHHFA T about 1%. the addition of TEG makes possible to achieve a reduction of the temperature of the reaction of the epoxy ring opening from 148 °C to 107 °C, as well as to reduce the viscosity by 76%.

Acknowledgments

This work was supported by Russian Foundation of Basic Research (RFBR) grants 18-29-18036 and 18-29-18037. NMR experiments were performed with the financial support from Ministry of Science and Higher Education of the Russian Federation using the equipment of Center for molecular composition studies of INEOS RAS.

References

[1] Wool R P 2008 Self-healing materials: a review *Soft Mat.* **4** 400-18
[2] Bergman S D and Wudl F 2008 Mendable polymers *J. Mater. Chem.* **18** 41-62
[3] Yuan Y, Yin T, Rong M and Zhang M 2008 Self-healing in polymers and polymer composites. Concepts, realization and outlook: A review *Express Polym. Let.* **2** 238-50
[4] Blaiszik B J, Kramer S L, Olugebefola S C, Moore J S, Sottos N R and White S R 2010 Self-

healing polymers and composites *Annual review of materials research* **40** 179-211
[5] Osswald T A and Menges G 2012 *Materials science of polymers for engineers*: Carl Hanser Verlag GmbH Co KG)
[6] White S R, Sottos N, Geubelle P, Moore J, Kessler M R, Sriram S, Brown E and Viswanathan S 2001 Autonomic healing of polymer composites *Nature* **409** 794
[7] Patrick J, Sottos N and White S 2012 Microvascular based self-healing polymeric foam *Polymer* **53** 4231-40
[8] Zhang H, Wang P and Yang J 2014 Self-healing epoxy via epoxy–amine chemistry in dual hollow glass bubbles *Compos. Sci. Tech.* **94** 23-9
[9] Nji J and Li G 2010 A biomimic shape memory polymer based self-healing particulate composite *Polymer* **51** 6021-9
[10] Toohey K S, Sottos N R, Lewis J A, Moore J S and White S R 2007 Self-healing materials with microvascular networks *Nature mater.* **6** 581
[11] Jin H, Mangun C L, Stradley D S, Moore J S, Sottos N R and White S R 2012 Self-healing thermoset using encapsulated epoxy-amine healing chemistry *Polymer* **53** 581-7
[12] Billiet S, Hillewaere X K, Teixeira R F and Du Prez F E 2013 Chemistry of crosslinking processes for self- healing polymers *Macromol. rapid commun.* **34** 290-309
[13] Turkenburg D and Fischer H 2015 Diels-Alder based, thermo-reversible cross-linked epoxies for use in self-healing composites *Polymer* **79** 187-94
[14] Hia I L, Vahedi V and Pasbakhsh P 2016 Self-healing polymer composites: prospects, challenges, and applications *Polym. Rev.* **56** 225-61
[15] Bessonov I V , Kopitsyna M N, Polezhaev A V, Nelyub V A 2016 A mechanistic study of the reaction between furfural-acetone resins and polyamines *Pol. Sci., Ser.* D **9** 17–21
[16] Bessonov I V, Polezhaev A V , Kuznetsova M N, Nelub V A, Buyanov I A, Chudnov I V, Borodulin A S 2013 Rheological and Thermal Analysis of Low Viscosity Epoxy Furan Composites *Pol. Sci., Ser.* D **6** 308–311
[17] Platonova E O, Vlasov E, Pavlov A A, Kireynov A, Nelyub V A and Polezhaev A V 2019 Self-healing polyurethane based on a difuranic monomer from biorenewable source **136** 47869
[18] Tian Q, Yuan Y C, Rong M Z and Zhang M Q 2009 A thermally remendable epoxy resin *J. Mater. Chem.* **19** 1289-96.
[19] Gu L and Wu Q Y 2018 Recyclable bio- based crosslinked polyurethanes with self- healing ability *J. Appl. Polym. Sci.* **135** 46272

IOP Conf. Series: Materials Science and Engineering **683** (2019) 012073 doi:10.1088/1757-899X/683/1/012073

Modified Oil Bitumen

L V Fedorova[1], Z O Tretyakova[2] and M V Voronina[2]

[1]Bauman Moscow State Technical University, 5/1, st. Baumanskaya 2-ya, Moscow, 105005, Russia.

[2]Saint-Petersburg Mining University, 2, 21st Line, St Petersburg 199106, Russia.

E-mail: momd@yandex.ru

Abstract. The article describes the modification of petroleum bitumen with elastomers of different chemical nature and molecular weight. The effectiveness and expediency of such a modification is shown. The objects of research in the work were the following brands of oil bitumen: BN-70/30 and BND-90/130. The following modifiers were used: butyl rubber (BK-1675), liquid thiocol - a condensation product of aliphatic dihalo derivatives with sodium polysulfide, rubber crumb. Shown undoubted efficiency and the prospect of improving the technological and operational properties of bitumen systems by modifying the latter with selected polymers. In addition, this modification is economically feasible due to the low cost of the selected modifiers and their presence in the range of enterprises of the domestic chemical industry.

Key words: bitumen, modification, elastomers, binder, devulcanizate.

1. Introduction

Oil bitumen, thanks to a valuable set of technical properties, have long been widely used as a plastic binder in building materials. As natural asphalt, bitumen has been exploited since antiquity as a source of road paving, caulk, and mortar and is still used for these purposes in some parts of the world [1]. The large scale of industrial production of bitumen and the relatively low cost make them indispensable for the construction of highways, the production of soft roofs, the manufacture of waterproofing mastics, etc. The global production of bitumen is estimated at hundreds of millions of tons. However, the quality of many bituminous materials does not meet the modern requirements of the construction industry. The indicators of temperature stability are low and have low working capacity under alternating deformation conditions [2]. And there is a real technological problem of introducing modifying additives to overcome the forces of surface tension and the distribution of quantitatively small and relatively light additives in comparison to the base material [3]. These disadvantages lead to a decrease in the durability of materials in which bitumen serve as a binder and impregnating component.

The steadily growing demands on the quality and performance of materials based on bitumen cannot be satisfied only by choosing raw materials and improving bitumen production technology. In this case, resort to their modification. The most common techniques are the modification of fillers, surfactants, and polymeric additives.

The objects of research in the work were the following brands of oil bitumen: BN-70/30 and BND-90/130. The following modifiers were used: butyl rubber (BK-1675), liquid thiocol - a condensation product of aliphatic dihalo derivatives with sodium polysulfide, rubber crumb (TU 38-10436).

We used standard methods for assessing the properties of bitumen compositions.

It is known that high-molecular industrial rubbers such as butyl rubber (BC) are poorly compatible with bitumen. One of the ways to increase their compatibility is to reduce the molecular weight, which in this work was achieved with the help of radiation destruction of BC. The absorbed dose of γ-irradiation of rubber samples was 125 Mrad. The modifying effect of the use of BC is manifested in the improvement of the performance characteristics of the compositions (Table 1). So, at 5% concentration of BD destructive agent, softening temperature (Tp) of bitumen increases from 690C to 91 0C, frost resistance from -80C to -280C, water absorption index decreases by more than 2 times.

The technological difficulties of combining bitumen with rubber (the need for destruction of the latter) caused the need to turn to liquid oligomeric rubbers, which are usually reactive [4]. The use of the latter allows to simplify the technology of modifying bitumen with elastomers. At the same time, it is necessary to consider that the curing of oligomers is carried out in a bitumen environment, which affects the speed of curing and the formation of polymeric structures. Therefore, the modification of bitumen BN-70/30 with thiokols (liquid polysulfide rubbers) was investigated. The dissolution of thiokol in bitumen occurs quickly, and the viscosity of the composition after cooling is lower than that of the original bitumen. The positive effect of this modification is obvious (Table 1).

Another modifier was a rubber destructive agent [5]. According to statistics, the mechanical loss of rubber in the operation of automobile tires constitute 10% of their initial mass. At the same time, rubber as a structural material, undergoes slight structural changes by the time the products are out of service, which is facilitated by the presence of an inhibitor that inhibits the development of the oxidation process, which underlies the aging of rubber. Consequently, worn tires are a source of valuable polymeric materials. Along with this, they are a source of long-term and persistent environmental pollution, due to the high resistance to natural factors. Thus, recycling used tires is of great economic and environmental importance. The most profitable of the implemented directions for the processing of used tires is to obtain a regenerate. To use regenerates as modifying additives in bitumen is a very tempting way to recycle tires, since huge amounts of bitumen in construction can "absorb" all this waste.

A modification of the road bitumen of the BND-90/130 brand with a vulcanised rubber crumb was carried out, the test results of which are presented in Table 1.

Table 1. The test results of bitumen-polymer compositions

Composition	**Softening temperature, ^{0}C**	**Penetration, x0,1 мм**		**Ductility at 25^{0}C, cm**	**Flexibility on a bar Ø50мм, 0^{0}C**	**Water absorption after 42 days, %**
		at 25^{0}C	**at 0^{0}C**			
BN-70/30	69	13	12	3	-8	0,67
BN-70/30 + BK 125MRad	91	6,0	5,0	1,1	-22	0,32
BN-70/30 + 5% Thiocol	94	2,5	1,5	3,8	-28	0,3
BND-90/130	42	126	25	7	-10	0,823
BND-90/130 +5% destructive rubber crumbs	68	45	20	140	-38,5	0,007

Thus, it has been established that to achieve the required physical and mechanical parameters, it is advisable to introduce rubber destructive agents up to 5% into bitumen. The higher content of this elastomer in bitumen does not give a significant effect.

With unlimited compatibility of thiokol with bitumen, showing a plasticizer function for the period of mixing, its vulcanization in a bitumen matrix leads to an increase in viscosity and an increase in all the basic operational and technical indicators of the resulting composition, which should be recommended as a roofing material.

The use of rubber crumb devulcanizate to modify the bitumen binder also showed the promise of its use in the form of improving the technical performance of the compositions.

Shown undoubted efficiency and the prospect of improving the technological and operational properties of bitumen systems by modifying the latter with selected polymers. In addition, this modification is economically feasible due to the low cost of the selected modifiers and their presence in the range of enterprises of the domestic chemical industry.

On average, additives of polymers allow to improve the main indicators of bitumen 1.4-1.8 times. The use of polymers as modifiers of road bitumen will significantly improve the quality of road asphalt concrete in coatings, increase their shear stability, water resistance, frost and crack resistance. Polymer modifiers can increase the service life of road and waterproofing coatings by a factor of 5–10, bringing the turnaround time between 15–25 years instead of 2–5 years for coatings of traditional bituminous materials. Thereby, material, labor and energy resources are significantly saved, material consumption of coatings decreases.

References

[1] Meyer R F, Attanasi E D, and Freeman, P A 2007 Heavy oil and natural bitumen resources in geological basins of the world. *U S Geological Survey Open-File Report 2007-1084*

[2] Voronov V M, Solovyova O Yu, Nespolovskaya T N 1995 The use of crushed vulcanizates in mastic compositions based on bitumen *Rubber and rubber* №3 p. 34-39.

[3] Kurganova Y A, Scherbakov S P 2017 Influence of a Discrete Additive of Aluminum Oxide on Structure and Properties of Aluminum Alloy. *Zapiski Gornogo instituta*. **228**, p. 717-721. DOI: 10.25515/PMI.2017.6.717

[4] Khakimullin Yu N, Sungatova Z O, Murafa A V 1998 Thiocol's modified bitumen composition. (Texas. USA) p. 62.

[5] Khakimullin Yu N, Sungatova Z O, Murafa A V, Khozin V G 1998 Bitum-elastomer compositions for construction purposes *Proceedings of the ISTC"Modern problems of building materials"* Part 1. PGASA (Penza) p. 143-144.

IOP Conf. Series: Materials Science and Engineering **683** (2019) 012074 doi:10.1088/1757-899X/683/1/012074

Influence of aggressive environment on basalt fiber based microplastics

R A Korokhin [1], V I Solodilov [1], A V Kireinov [1,2] and M A Orlov [2]

[1] Semenov Institute of Chemical Physics, 119991, Russia, Moscow, Kosygina str., 4

[2] Bauman Moscow State Technical University, 105005, Russia, Moscow, 2nd Baumanskaya str., 5/1

Abstract. Physical and mechanical properties of basalt fibers and microplastics based on them have been investigated. Influence of various environments such as distilled and sea water, alkaline solutions, UV-radiation, on microplastics' characteristics have been studied. Alkaline solutions of high concentration are shown to affect the most the characteristics of studied materials.

1. Introduction

Basalt is an igneous rock constituting 30% of the earth's crust. The mechanical properties of basalt fibers are comparable with glass fibers but have a higher chemical resistance [1-5]. Thus, basalt plastics successfully compete with metal products, surpassing them in corrosion, alkali- and acid- resistance and etc. Therefore, basalt fibers and based on them composites are widely studied throughout the world [6-12].

2. Experiment

In our study the strength of basalt harnesses and monofilaments and basalt microplastics' resistance to water, alkaline solutions and UV-radiation has been investigated. Epoxy oligomer ARALDITE CY8615 + XB 5173 HARDENER (Huntsman) has been used as impregnation binder. Linear density of basalt fibers occurred to be 1200 tex.

Time of exposure of microplastics to both distilled and sea water as well as alkaline solutions (pH 12.5 – 14.1) was 0-60 days, to UV-radiation 0-21 days.

Samples of monofilaments, microplastics were tested for tensile, strength σ and elastic modulus E. Samples for shear strength τ determination have been obtained by pressing. The samples consisted of plates with asymmetric cuts (Fig. 1), which were tested by means of uniaxial tension [11], and the shear strength was calculated by formula (1)

$$\tau = \frac{P}{b \cdot w}, \qquad (1)$$

where P is breaking load, b is the width of the sample, w is distance between the cuts.

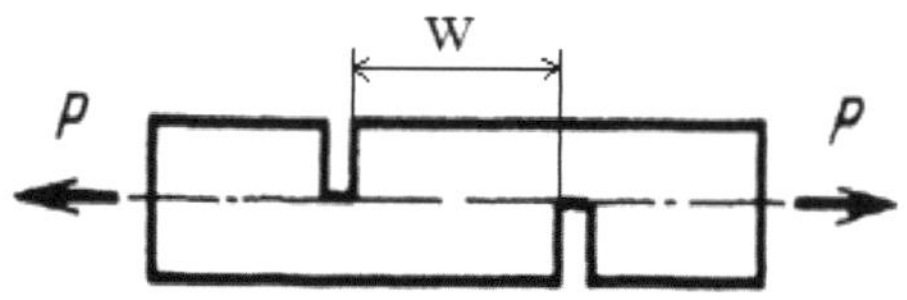

Figure 1. Scheme of test method of samples with cuts under tension or compression

3. Results and discussion

Monofilaments microphotographs and distribution by diameter are presented in figure 2.

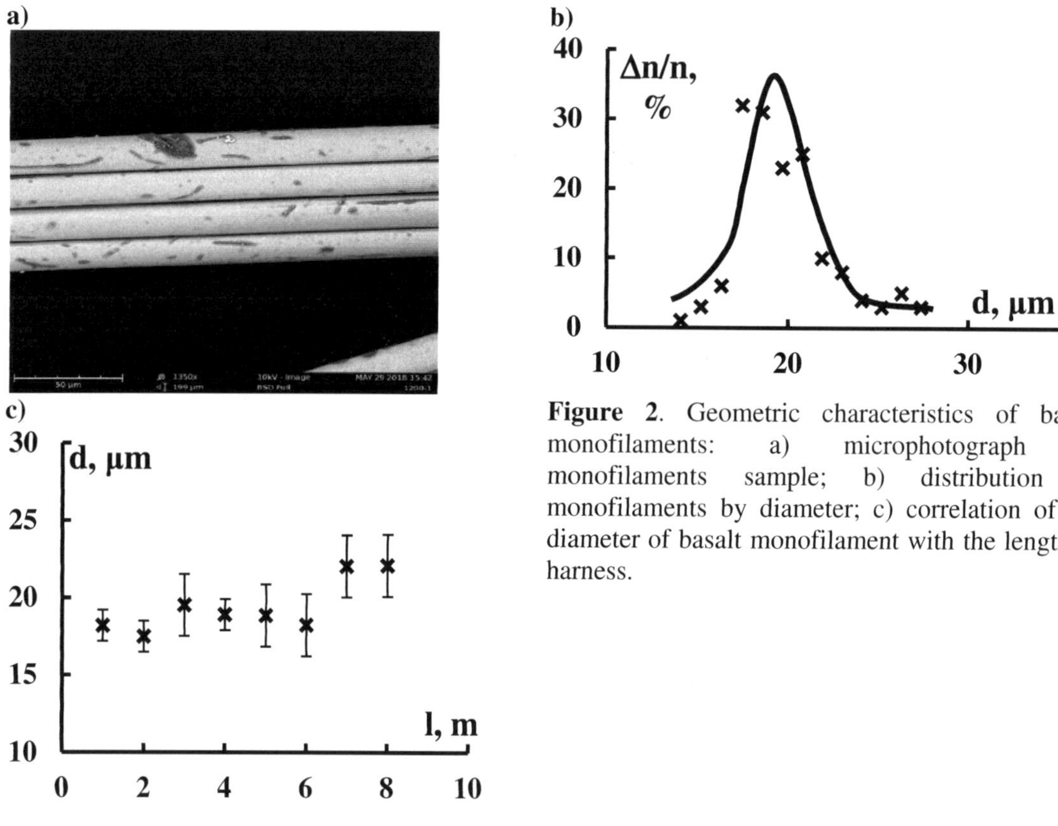

Figure 2. Geometric characteristics of basalt monofilaments: a) microphotograph of monofilaments sample; b) distribution of monofilaments by diameter; c) correlation of the diameter of basalt monofilament with the length of harness.

Based on microphotographs (an example is shown in Fig. 2a, the distribution curve of monofilaments by diameter (Fig. 2b) was constructed, showing that the average diameter of the basalt monofilament is 19 μm. Measurement of the monofilaments' diameter in respect of the length of the bundle (Fig. 3c) showed the average diameter is also 19–20 μm.

The dependence of the monofilaments' strength on the number of averaged values is shown on fig. 3a. The average strength of the monofilament varies insignificantly when considering over 30 values, i.e. you can talk about the well-established value of strength. With a smaller number of averages the strength of the monofilament can differ by 40%. Such a spread overlaps the spread of the obtained data. A similar trend is observed in determining the elastic modulus (Fig. 3b).

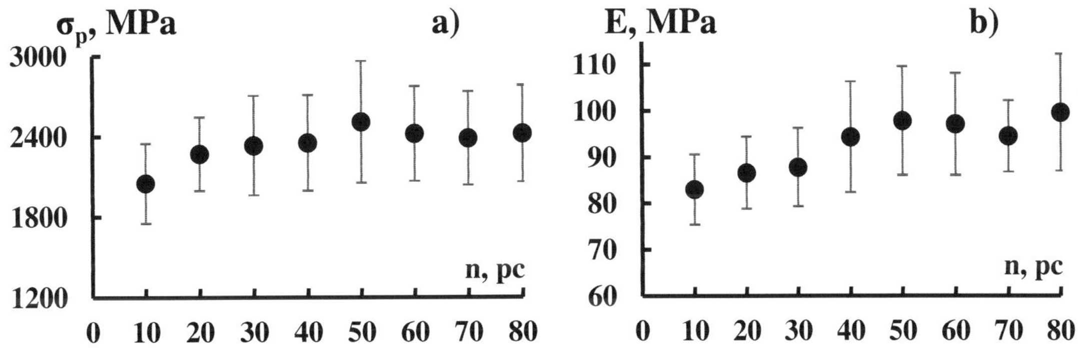

Figure 3. Correlation of monofilaments' strength (a) and elasticity module (b) with number of averaged values.

As a result, the average value of the strength of the monofilament is 2400 MPa, with the harness strength values being 1.5 times lower, 1600 MPa. Such a difference testifies to the diversity in tension and diversity in length of the fibers in the bundle.

The strength of the control samples of basalt-based microplastics is 1460 MPa, while the fiber content in the samples is 65-70 vol. %. Using the formula (2) for calculating the fiber realization coefficient of the fiber k [12], we get the value k~0.87-0.93.

$$k = \frac{\sigma_{mp}}{\sigma_f \cdot w}, \qquad (2)$$

Where σ_{mp} is strength of microplastic, σ_f is strength of monofilament, w is distance between cuts.

The results of determination the microplastics strength after exposure to various environments are shown in Fig. 4

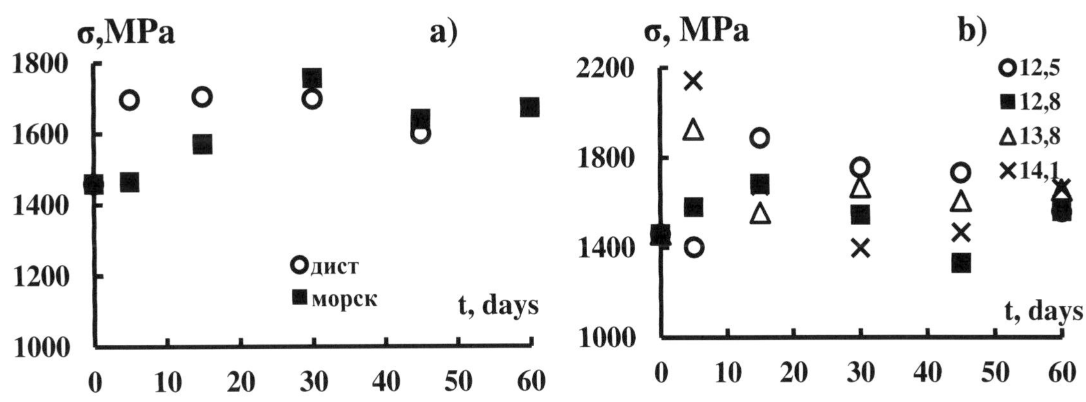

IOP Conf. Series: Materials Science and Engineering **683** (2019) 012074 doi:10.1088/1757-899X/683/1/012074

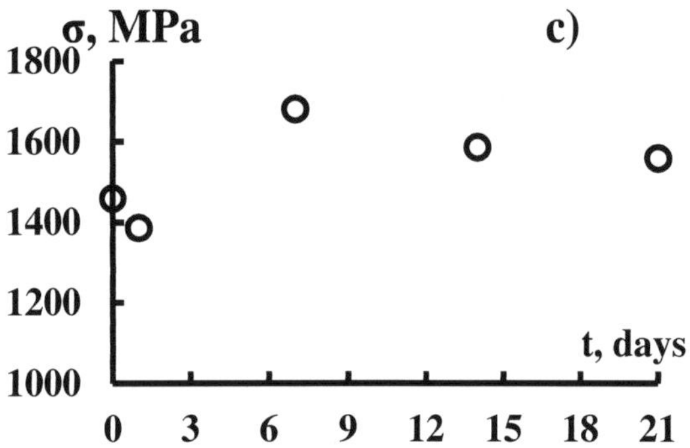

Figure 4. The strength of microplastics after exposure to: a) distilled and sea water; b) alkaline solutions (pH is shown in the figure); c) ultraviolet radiation.

As can be seen from the figures, the strength of microplastic depends on the time of exposure to various environments. The increase in strength after exposure to water (Fig. 4a) is associated with leaching of residues of the uncured resin and improvement of contact at the matrix – fiber interface.

In case of alkali solutions (Fig. 4b) it is associated with healing of edge defects, which occurs more efficiently at high concentrations of alkali in solution with short exposure times (up to 5-10 days). Longer exposure to the solutions has a negative impact on the matrix and fiber.

Exposure to UV radiation (Fig. 4c) leads to the curing of the epoxy matrix, i.e. hardening of the interphase boundary, and hence the growth of strength characteristics of loaded sample.

When determining the modulus of elasticity (Fig. 5), the tendency of strength characteristic remains the same.

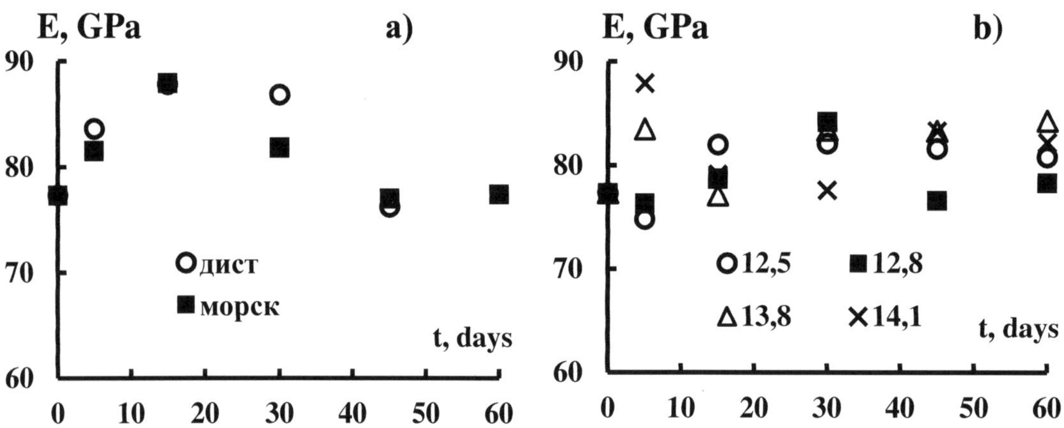

IOP Conf. Series: Materials Science and Engineering **683** (2019) 012074 doi:10.1088/1757-899X/683/1/012074

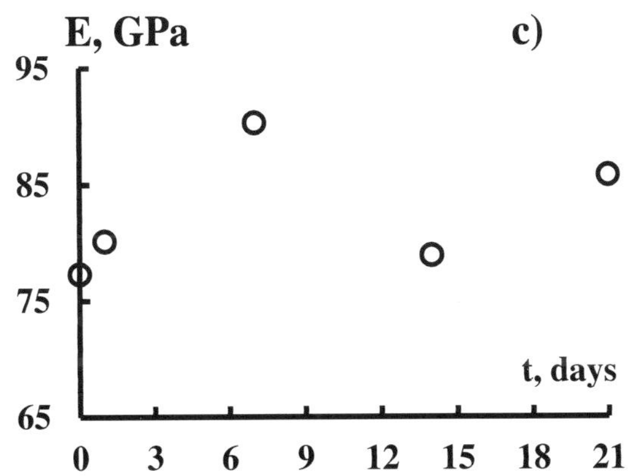

Figure 5. Elasticity module of microplastics after exposure to: a) distilled and sea water; b) alkaline solutions (pH is shown in the figure); c) ultraviolet radiation.

The mass loss of microplastics during their exposure to different conditions was measured. It was found that a noticeable change in mass occurs only when the microplastics are exposed to alkaline solutions (Fig. 6). There is a sharp decrease in the mass of the sample to 0.5-0.8% in first 5 days, then the mass loss rate reaches plateau and after a 60-day exposure is 1.3-1.8%.

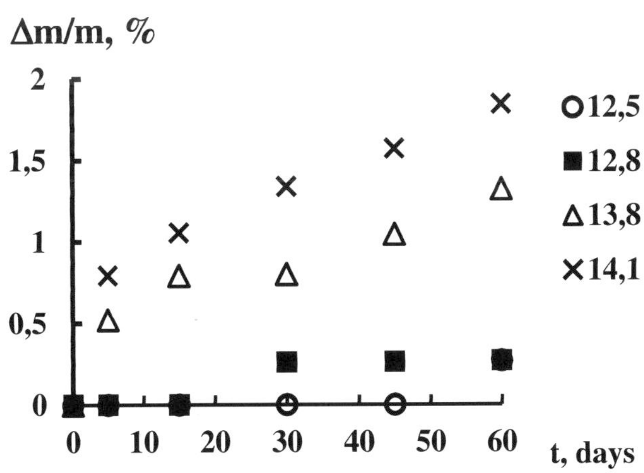

Figure 6. Mass loss of microplastics of alkalis in alkali solutions (pH is shown in the figure)

Analysis of shear strength (Fig. 7) showed its significant dependence on the area of the sample's working region.

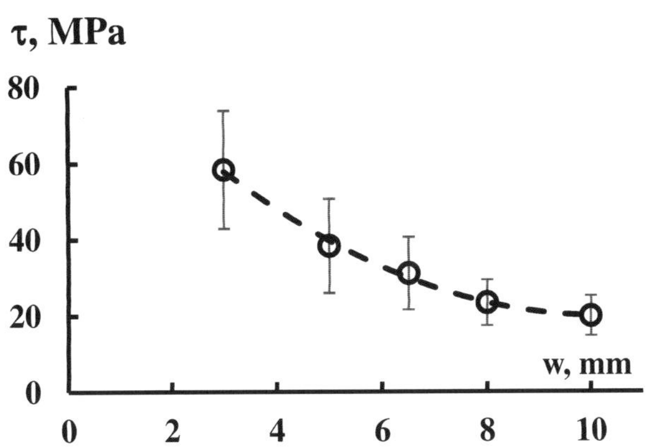

Figure 7. Basalt-plastics sherar strength.

The figure shows that as the distance between cuts increases, a monotonic drop in shear strength occurs., The values of τ fall from 58 MPa to 20 MPa when w increases from 3 mm to 10 mm, which is associated with an increase in the flexural stress arising in the process of application of axial tensile forces with increasing distance between the cuts. The values obtained at w = 3 mm correspond to the previously obtained experimental data [11].

4. Conclusions

Thus, the work showed the promising use of basalt plastics, since they have physical and mechanical characteristics comparable to those of glass-reinforced plastics, but at the same time have a higher resistance to the action of the aqueous medium and alkali solution, as well as resistance to the action of ultraviolet radiation. At the same time, in some cases, the influence of aggressive media leads to an improvement in the characteristics of materials.

Acknowledgments

The work was supported by the Ministry of Education and Science of the Russian Federation: Contract No. 03.G25.31.0264

References

[1] Basaltic Fiber Composite Materials and Structures [in Russian] (Kiev) 1980.
[2] Trostyanskaya E B, Kutyrev U V 1976 Bazal'toplastiki. Plasticheskiye massy, 1144-1146.
[3] Nelyub 2015 *Polym. Sci. Ser. D* **8**(1) 6-8
[4] V A Nelyub 2016 *Polym. Sci. Ser. D* **9**(1) 96-100.
[5] Dalinkevich A A, Gumargaliyeva K Z, Marakhovskiy A V, Sukhanov S S 2010 Konstruktsii iz kompozitsionnykh materialov **3** 35-54.
[6] KupermanA M, GorbatkinaY A, GorenbergA Y, Ivanova-MumzhievaV G, ZakharovaT Y, Lebedeva O V, Solodilov V I, Petrosyan G K 2008 Physical-mechanical properties of andesite-dacite fibers and their modifications. Glass And Ceramics **65**(1-2) 11-15.
[7] Kuperman A M, Gorbatkina Y A, Gorenberg A Y, Ivanova-Mumzhieva V G, Zakharova T Y, Lebedeva O V, Solodilov V I, Petrosyan G K 2009 Physical-mechanical properties of fibers made from scoria and materials based on them. *Glass And Ceramics* **66**(6-7) 245-248.
[8] Ross A 2006 *Compos Technol* **12** 44–48.
[9] Colombo C, Vergani L, Burman M 2012 *Compos. Struct.* **94**(3) 1165–1174.
[10] Pavlovski D, Mislavsky B, Antonov A 2012 *Reinf. Plast.* **51**(4) 36–37.
[11] Tarnapol'skiy U M, Kintsis T Ya 1975 Metody staticheskikh ispytaniy armirovannykh plastikov [in Russian] M.: Khimiya 264 s.
[12] Gunyayev G M 1981 Struktura i svoystva polimernykh voloknistykh kompozitov [in Russian] M.: Khimiya, 232 s.

Newer approaches to the creation of the thermostable space platforms for the Earth remote sounding

M A Gorodetsky[1], **K V Mikhaylovsky**[1] **and S V Reznik**[1]

[1]BMSTU, Moscow, Russia

E-mail: mikgord@gmail.com

Abstract. A method for the integrated analysis and the determination of the composite construction parameters of a platform of the Earth remote sounding spacecraft, is proposed. The conditions of thermal load for a flight along a sun-synchronous orbit were analysed. The mathematical simulation of the operating conditions ensuring the effective operation of such satellites was performed. The method was tested with account of the input parameters of the «BelKA» satellite orbit. This method will be useful when selecting a working orbit for such satellites at the stage of technical proposals, since it includes determination of their orbital characteristics, as well as the heat flows onto the elements of their construction. The results of simulation of thermal regime of the composite structure options are presented.

Key words: the Earth remote sounding spacecrafts, sun-synchronous orbit, polymer composite materials, nonwoven material.

1. Introduction

For rocket and space equipmet, three-layer constructures with the covers made of a composite material (CM) and a honeycomb core are commonly used, which have increased bending stiffness with relatively low weight. Such constructions are often used for the Earth remote sensing (ERS) spacecraft (SC). The desired physicomechanical and thermophysical characteristics of the constructions can be achieved by combining the materials of the bearing courses and fillers. For the improvement of economic feasibility and reliabilty and for the reduction of three-layer constructions mass with strength and rigidity maintenance, the development of new types of covers and fillers, based on new materials, is required.

Applied to the SC platforms with stable dimensions, the important indicators are strength, aging and damping capacity. The CMs of high modulus carbon fibers meet these requirements. Along with the common CMs, the hybrid CMs of nonwovens are increasingly used. The nonwovens actually also belong to the composites, since they consist of two elements, one of which functions as a filler, the other - as a binder element [1]. Recently, new types of nonwovens have been created, the structure of which is similar to the honeycomb cores. However, in the literature there is no information on the characteristics of multi-layered panels in which nonwovens are used instead of the honeycomb core [2].

Creating of a ERS SC is a complex interdisciplinary problem. An integral part of their designing is the determination of the most effective orbit parameters from the point of view of information content, power supply and the duration of active spacecraft existence [3–6]. For the ERS SC with the platforms with stable dimensions and reflectors of space antennas [7–14], the development of simulation methods and construction optimization in view of the requirements of operation and technological limitations are essential.

IOP Conf. Series: Materials Science and Engineering **683** (2019) 012075 doi:10.1088/1757-899X/683/1/012075

The purpose of this work is to improve the mass-dimensional characteristics of three-layer panels of the ERS SC platforms with the use of hybrid CMs of carbon fiber and organic nonwovens, as well as the reduction of the time spent on the designing in the framework of an integrated methodology for the ERS SC exploitation conditions..

2. The ERS SC dynamics simulation

A part of the design studies is the simulation of the orbital motion of a SCs, corresponding to their intended purpose. A set of requirements is imposed to the aims of ballistic design of the ERS SC: the Earth global view provision; the periodicity of the Earth selected areas review with a given frequency; SCs flight performing in a circular orbit, with constant height, as the SC equipment is usually adjusted to predetermined height before launching; the Earth's territories survey on illuminated ascending or descending branches of the orbit; minimal changes in the conditions of the SC track illumination during the entire life of the SC; correction of the orbit height decreasing due to atmospheric deceleration.

As is known, due to the non-centrality of the Earth gravitational field, the orbital plane of the SC precesses noticeably around the earth's axis with a low speed of the orbit node displacement. This makes it possible to choose the height and inclination of the orbit, which ensures the equality of the precession velocity and the average angular velocity of the Earth's annual motion around the Sun. Such orbits are called sun-synchronous orbits (SSO), or the Sun "tracking" ones, i.e. changing their position synchronously with the direction of the Sun-Earth. Since, unlike the diurnal, the annual motion of the Sun along the celestial sphere occurs from west to east, the SSO should also precess in this direction, therefore, the SSO inclination should exceed 90°. One of the important features of the SSO is the almost unchanged solar lapse of time of an SC passing over the same points of the route from turn to turn.

The main disturbing factor during the flight is the decelerating impact of the atmosphere, which reduces altitude and leads to a flight track shift of an SC. The correction maintains stability of the altitude profile, the properties of solar synchronicity and the characteristics of the tracks grid. The local time of the ascending node should be chosen so as to provide the required illumination conditions of the sub-satellite point for the surveys.

In this paper, we have constructed a mathematical model of the conditions for the operation of an ERS SC for the analysis of illumination and full coverage of the Earth's surface by flight tracks. The paper takes into account the effect of the atmosphere onto the orbital motion and the corrections for maintainance of the altitude of the orbit. The methodology has been tested for an ERS SC, similar to «BelKA». [15].

With the help of mathematical models implemented as algorithms in the Matlab software package, the ERS SC working orbit parameters were determined: orbit height h=514 km, orbit inclination i=97,4°, local ascending node transit time $m\Omega$=6 h. The orbit type is near-circular– (with eccentricity e=0,00135), sun-synchronous, multiple (the multiplicity period is 14 days).

3. Simulation of the temperature field and the stress-strain state of the ERS SC platform.

The SC design is a complex procedure, the implementation of which is associated with the various physical processes. The ones of the most significant are heat exchange processes. They are determined by heating from the direct solar heat flow and the heat radiation flow from the Earth, and also by the onboard heat radiation.

In order to assess changes in the temperature field and its effect on the stress-strain state of the ERS SC platform, the special conditions were established that correspond to the motion in the orbit calculated above.

The study platform of an ERS SC had a box-like parallelepiped shape, the walls of which were three-layer panels formed by two carbon-fiber platings 2 mm thick and aluminum honeycomb core 12 mm thick. Simulation of the temperature state was carried out for a representative element of the construction with dimensions of 500x500x16 mm. Data on thermophysical characteristics necessary for simulation was borrowed from [16], and data on optical characteristics from [17]. The simulation was carried out

IOP Conf. Series: Materials Science and Engineering **683** (2019) 012075 doi:10.1088/1757-899X/683/1/012075

in the NX / Nastran software package. For all elements of the construction, the three-dimensional type of grid was set, with the element size of 10 mm.

The calculations were performed with a use of the "NX/Space Systems Thermal" module that implements the Monte Carlo method. For this method the orbit parameters, the SC orientation in space, the position of the Sun relative to the orbit during the simulation period, were set. The calculation was performed for one orbit pass of 45 s step. To study the impact of the redistribution of radiation from the elements of the construction, the object of modeling Radiation was used: setting of the radiant heat exchange between the model surfaces, as well as the influence of the surrounding space. Initial dynamic data of the SC was set in the object of modeling "Orbital heating". It helped to determine the conditions of thermal load, caused by the heat radiation flows of the Sun and the Earth. The simulation showed that temperature varies from minus 70° to plus 50° during the year (Fig. 1).

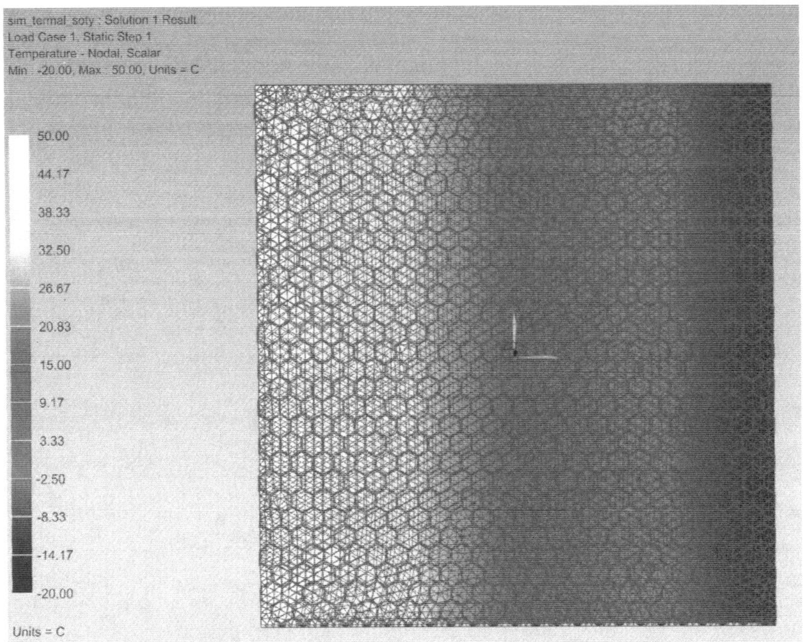

Fig 1. Temperature field on the cover surface

Operating stress in the longitudinal and transverse directions of a three-layer panel under extension/compression and bending ranges from minus 70 to plus 80 MPa. The displacements in the cover of a three-layer panel range within 0.1 mm in the longitudinal, and within 0.001 mm in the transverse direction (Fig. 2).

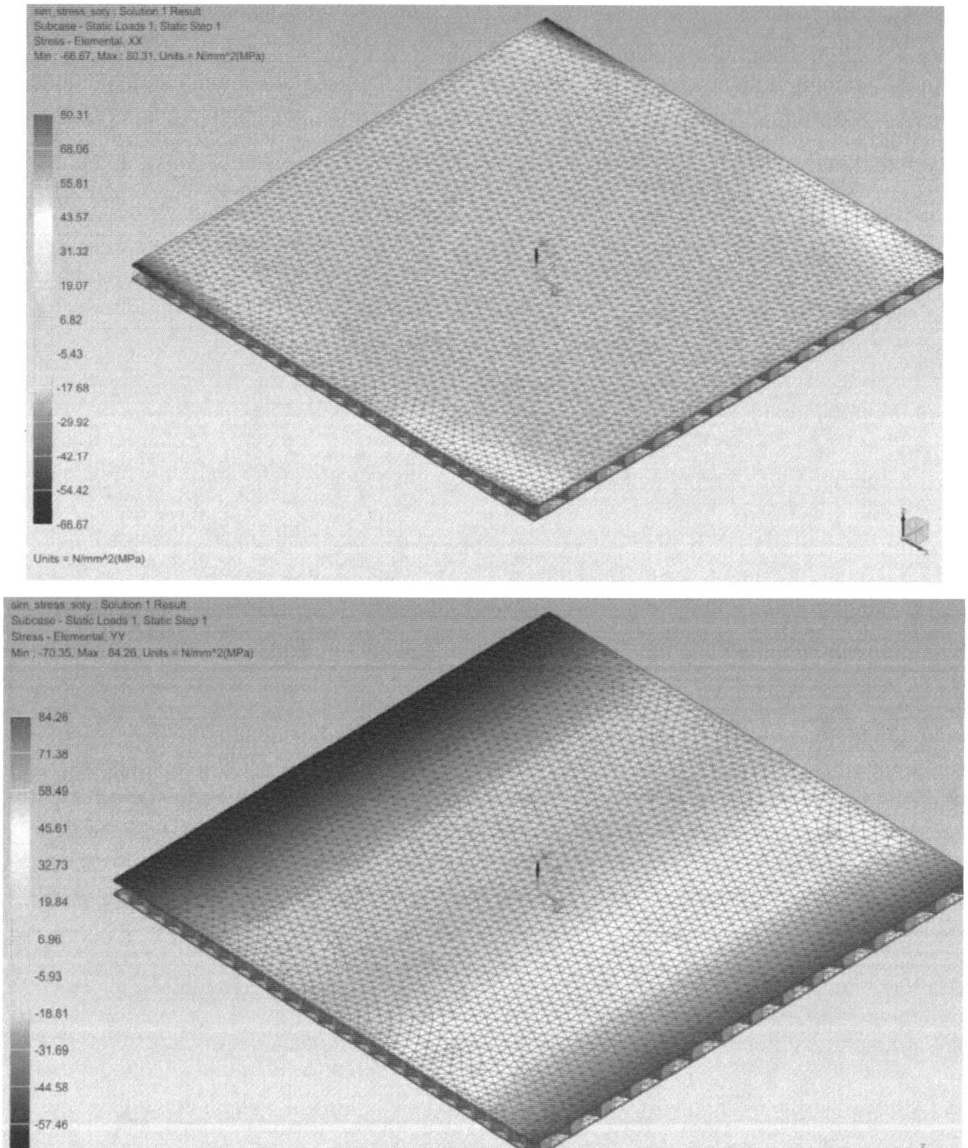

(a) (b)

Fig 2. Stress-strain state in the longitudinal (a) and transverse (b) directions

4. Experimental study of SC samples

Different types of reinforcing fillers are used in the three-layer constructions: metal, paper honeycombs, honeycombs from CM. The ways to perfection of such constructions are unlimited. Promising ones are the new types of fillers in the form of hybrid CM based on nonwovens.

In this work, the characteristics of the hybrid CMs with basis of polyester nonwovens with an ordered honeycomb structure were studied. Between themselves, the cells are separated by the channels that contain microspheres (Fig. 3).

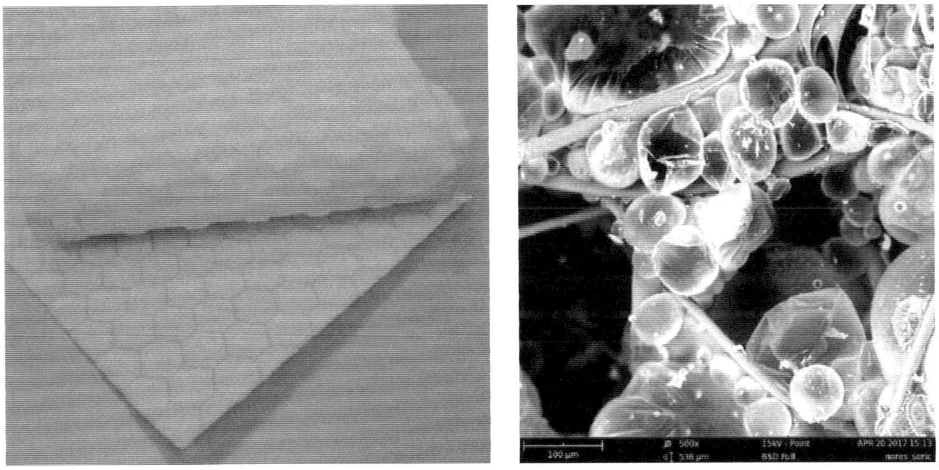

(a) (b)

Fig 3. Photo of nonwoven material in the initial state (a); at 500 times magnification (b)

Glass fiber of the brand “Ortex 560” and epoxy binder based on resin “ED-20” were used as the fibrous filler, as a hardener - isomethyltetrahydrophthalic anhydride and as an active solvent - diethylene glycol of the brand “DEG-1”. The mechanical characteristics of the materials used are given in Table 1.

Table 1. The characteristics of the materials used

№	Material	Characteristics	
1	Polyester nonwoven	Specific weight, g/m^2	130
		Strength, MPa at:	
		- bending	16;
		- compression	4.
		Flexural modulus, MPa	1000
2	Fiberglass cloth	Linear density (tex):	
		- basis	1200;
		- weft	1200.
		Breaking load, N	
		- basis …	6500;
		- weft	5700.
3	Epoxy binder	Strength, MPa at:	
		- extension	82
		- bending	105
		Glass transition temperature, °C	110

Three types of samples were prepared and studied: without nonwoven, and with one and two layers of nonwoven material. Sample #1 was made of six layers of fiberglass, #2 - of seven layers (three layers

of fiberglass + one nonwoven layer + three layers of fiberglass), and #3 - of eight layers (two layers of fiberglass + one nonwoven layer + two layers of fiberglass + 1 layer of nonwoven fabric + two layers of fiberglass). For all samples and all layers, the angle of calculation was 0°.
The mechanical characteristics of the produced CMs were studied with a use of a Netzsch 242E Artemis dynamo-mechanical analyzer (Table 2).

Table 2. Values of the modulus of elasticity of the multilayered polymer CMs

Test temperature, °C	Sample №		
	1	2	3
30	11024	10664	13305
50	9060	10575	13286
100	1439	10183	12745
150	1292	8367	9285
200	1396	4867	4551

Analysis of the results showed that modulus of elasticity of a CM, consisting of two layers of nonwoven (sample #3) increases by almost 25% at a temperature of 30°C compared with the sample consisting of one layer of nonwoven (sample #2). Samples #2 and #3, compared with the original sample of fiberglass (sample #1) have at an average 3 times greater values of modulus of elasticity at temperatures up to 200°C [2]. Modulus of elasticity of a CM, containing a non-woven material, compared with the same but without it, starts to increase at a temperature of 50°C and further, as the temperature increases, these differences increase. Thus, the use of non-woven material in the structure of a CM allows to increase the values of modulus of elasticity.

5. Conclusion

A system of mathematical models was proposed. It was implemented as algorithms in the MatLab software package. The parameters of the working orbit of an ERS SC, similar to the BelKA SC meeting the requirements for the ERS SC orbits, were estimated.

With the help of numerical simulation, it was shown that the operating stresses in the longitudinal and transverse directions of a three-layer platform panel of an ERS SC flying through a solar-synchronous orbit, at extension/compression and bending, range from -70 to +80 MPa, which corresponds to the limit of allowable values of strength limit for carbon fiber. However, the main stress is taken by the cover, the material and structure of which were studied in this work. Displacements in the cover of a three-layer panel are expected to be within 0.1 mm in the longitudinal direction, and within 0.001 mm in the transverse direction.
The developed methodological techniques can be useful at the stage of technical proposals for the materials selection for the ERS SC constructions with stable form and dimensions.

Acknowledgments

The authors are grateful to the Inter-Sectoral Engineering Center of Composites of BMSTU for providing equipment and Tun Lin Htett for providing samples of a hybrid CM based on nonwoven material.

References

[1] Gorodetsky M A, Htet T L and Malysheva G V 2017 Forming technology of hybrid composites based on organic nonwoven material *10th All-Russian conference for Young scientists and*

specialists (with international participation) "The Future of Russian Engineering": report collection (Moscow: BMSTU Publishing House), 545–547.

[2] Nelyub V A, Gorodetsky MA, Tun L X and Malysheva GV 2017 Properties of multilayer polymer composites based on non-woven material made of polyethylene terephthalate *Bulletin of Kazan Technological University* **20,** 24, 74–77

[3] Mikhailovsky KV and Gorodetsky MA 2017 Development of methods for determining and correcting the parameters of the working orbit of the Earth remote sensing spacecraft *Bulletin of RUDN. Engineering research* **18,** 3, 361–372

[4] Montenbruck O and Gill E 2000 *Satellite orbits: models, methods, applications* (Berlin: Heidelberg Springer – Verlag), 369 p.

[5] Hastings D and Garrett H 2004 *Spacecraft-Environment interactions* (Cambridge: Cambridge University press), 320 p.

[6] Schaub H and Junkins J L 2009 *Analytical mechanics of space systems. 2-nd edition* (Reston: AIAA), 819 p.

[7] Reznik S V, Prosuntsov P V, Mikhailovsky K V and Shafikova I R 2016 Material science problems of building space antennas with a transformable reflector 100 m in diameter *IOP Conference Series: Materials Science and Engineering* **153**, 1. 012012 10 p. DOI: 10.1088/1757-899X/153/1/012001

[8] Sayapin S N and Shkapov P M 2016 Kinematics of deployment of petal type large space antenna reflectors with axisymmetric petal packaging *J. of Machinery Manufacture and Reliability* **45**, 5, 387-397

[9] Reznik S V, Prosuntsov P V and Azarov A V 2015 Substantiation of the structural-layout scheme of the mirror-space-antenna reflector with a high shape stability and a low density per unit length *J. Eng. Phys. Thermophy* **88**, 3, 699-705 DOI:10.1007/s10891-015-1239-x

[10] Reznik S V, Prosuntsov P V and Azarov A V 2015 Modeling of the temperature and stressed-strained states of the reflector of a mirror space antenna *J. Eng. Phys. Thermophy* **88**, 4, 978–983 DOI:10.1007/s10891-015-1273-8

[11] Golovatov D, Mikhaylov M and Bosov A 2016 Optimization of technological parameters of impregnation of load-bearing rod elements of reflector made of polymer composite materials by transfer molding method *Indian J. of Science and Technology* **9**, 46 No 107492 DOI: 10.17485/ijst/2016/v9i46/107492

[13] Prosuntsov P V, Reznik S V, Mikhailovsky K V, Novikov A D and Zaw Ye Aung 2016 Study variants of hard CFRP reflector for intersatellite communication *IOP Conf. Series: Materials Science and Engineering* **Vol 153** No 1. 012012. 7 p. DOI: 10.1088/1757-899X/153/1/012012

[14] Reznik S V and Novikov A D 2016 Comparative analysis of the honeycomb and thin-shell space antenna reflectors *MATEC Web of Conferences 01012 (2017). Thermophysical Basis ofEnergy Technologies,* 5 p. DOI: 10.1051/matecconf/2017920101292

[15] Website: https://www.energia.ru/rus/news/news-2006/press_release-07-26.html.

[16] Bolton W 2007 *Construction materials: metals, alloys, polymers, ceramics, composites. Pocket Guide* translation from English (Moscow: Metallurgy) 336 p.

[17] Latyev L N, Petrov V Ya, Chekhovskaya E N, Shestakov E N 1974 *Radiative properties of solid materials* ed. A. E. Sheindlin. (Moscow: Energy) 472 p.

IOP Conf. Series: Materials Science and Engineering **683** (2019) 012076 doi:10.1088/1757-899X/683/1/012076

Active thermography as a contemporary method for ensuring the quality of composite material products

I O Kotovshchikov[1,2]

[1] Limited Company «LOCUS», Saint Petersburg, Russia.

[2] Department of Control Systems and Robotics, University of Information Technologies, Mechanics and Optics, Saint Petersburg, Russia.

Abstract. This paper examines the current trends in the quality control of composites and the perspectives of application of active thermography method to ensure the quality of products made of composite materials. Using NDTherm™ – system of active thermography, the research has been carried out on composite samples and products with inbuilt artificial defects. Based on research results the author highlights the most promising applications of the active thermography method.

1. Introduction

The long-term service of aviation equipment implies high workload and strict requirements for non-destructive testing (NDT) of parts and assembly units in the aerospace industry. In modern aircraft, as of 2015, the volume share of products made of composite materials reaches 50% [3]. Due to well-known features of physical and mechanical properties of composite materials, there is a tendency of constant increase their share by volume in the aircraft industry which makes it relevant to develop, improve and implement new methods to ensure the quality of products made of composite materials. Because of geometric features of composite material products, preference is given to high-efficiency NDT methods that can be used without special preliminary treatment of the inspected object surface and with easy test accessibility.

Active thermography is one of the promising NDT methods that can be used for products made of composite materials (Figure 1).

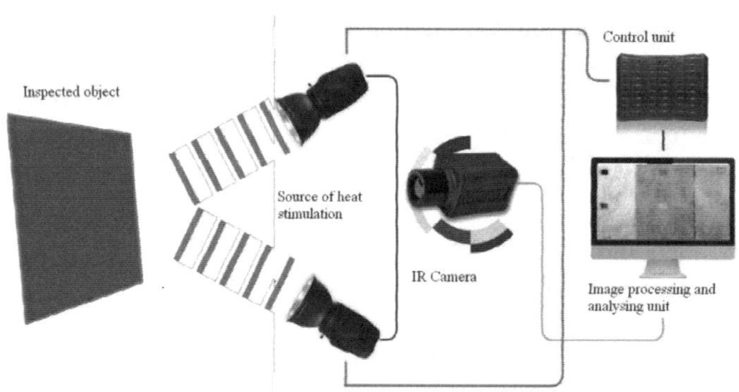

Figure 1. Scheme of inspection by means of active thermography

Unlike passive thermography, the active thermography (AT) method requires heat load monitoring to the inspected object. At the same time the AT method does not require physical contact with the inspected object for the load to the inspection system or for obtaining the response.

At present this method has been successfully used by enterprises manufacturing or operating products and structures made of composite materials. Such large aerospace enterprises like Boeing and Airbus pay special attention to the AT method, include it in their specifications that describe the inspection procedure of the fuselage, cellular panels, etc. [4].

The AT method is included in the international NDT standards (NAS 410, EN 4179, etc.) and in the Russian State standards (GOST R 17359-2015, GOST R 18434-1-2013, etc.). The national standard of the Russian Federation GOST R 56787-2015 [5] concerned with NDT polymer composites recommends the AT method to detect such defects as:

- lamination;
- change of density;
- breaking of connections;
- breaking of connections between fibers;
- breaking of fiber function;
- breaks;
- inclusion;
- moisture;
- porosity;
- change of the thickness;
- emptiness.

In addition, advantages of the AT method are also as follows:

- non-contact application;
- high performance (in certain cases, 5 times faster than ultrasonic or radiographic inspection methods);
- possibility of structure inspection from one side only;
- safety;
- a wide range of inspectable composite materials;
- a wide variety of problems solved by this method.

Like any other method, AT has its limitations:

- it is necessary to ensure uniform heating of entire inspected surface of the part;
- a heavily soiled or glossy surface causes interference and noise and makes it difficult to interpret results;
- going deeper the heat wave attenuates dramatically, so the thickness of the tested layer is usually limited up to 10 mm.

The great opportunities of the AT method are connected to the variety of its technical implementations. There are various approaches to the generation of heat loading (heat stimulation) in the inspected object. According to the physics principle, they can be divided into: optical (flash lamps, halogen lamps, lasers), induction (eddy currents), microwave, electric current, mechanical (ultrasound or vibration), hot or cold gases or liquids. Depending on the problem being solved, different type of the heating profile can be used: apportioned, local, band-pass, "grid". Modern pieces of equipment that implement the AT method also allows to manage the shape of the heat stimulation pulse: Dirac pulse, rectangular, Gaussian pulse or a sequence of pulses. In addition, the effective NDT technology that uses the AT method is impossible without special software (SW) processing infrared (IR) images of the inspected object [6].

In order to evaluate the perspective of the AT method in practice, the active thermography system NDTherm™ by Opgal is used in this study. It is a high-performance NDT system that implements the AT method. NDTherm™ consists of the following basic elements:

- Uncooled high-sensitive infrared microbolometer camera;
- Halogen lamps with a total power of 3.2 kW;
- Electronic control unit;
- Special image processing Opgal algorithms.

An overview of the system with the identification of main units is shown in Figure 2. The source of heat stimulation in NDTherm™ active thermography system is 4 halogen lamps (pos. 5) which ensure uniform heating of the inspected object surface. By the use of an electronic control unit (pos. 1) it is possible to manage the type of heat stimulation (the shape and duration of the thermal pulse). Heat stimulation is generated by light pulse with duration from 0.1 to 20 seconds. Special SW included in the unit of the image processing and analyzing (pos. 2) allows implementation of special algorithms for processing IR images of the inspected object and the use of tools for analyzing the obtained thermographic defectograms (thermograms).

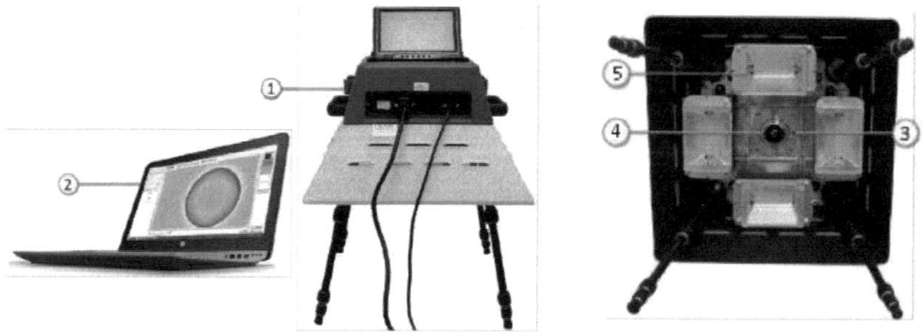

Figure 2. Overview of the active thermography system NDTherm™:
1 – electronic control unit; 2 – image processing and analyzing unit; 3 – camera for visible light; 4 – IR camera; 5 – sources of heat stimulation (halogen lamps)

The results of inspection by NDTherm™ system are presented in the form of time sequences of thermograms after processing by special algorithms. Thus, the output parameters are pixel functions of the dependence of temperature (or its derivative) on time and on coordinates. In general, the algorithms used for processing are determined by the task being solved (type and size of the defects, product material, inspection process conditions, etc.), but NDTherm™ system typically contains a set of the most universal IR image processing algorithms. The purpose of using these algorithms is to increase the ability of flaw detection. In this study, the following algorithms have been used:

- Reconstruction of an image with the use of polynomial fitting;
- First time derivative;
- The second time derivative;
- Fourier transform.

A more detailed description of the practical applicability of the above methods for processing IR images can be found in [7].

2. Experimental studies

In order to evaluate the applicability of the AT method implemented in the NDTherm™ system, experimental research of products and samples made of composite materials has been carried out.

2.1 Flaw detection in composites

Inspected objects with inbuilt artificial defects have been investigated. The results of these studies give positive results for the detection of the following types of defects: laminations in products made of carbon fiber-reinforced plastic (Figure 3), laminations in sandwich panels, lamination in aluminum composites, non-gluing in honeycomb panels (Figure 4), inclusions in glass fiber-reinforced plastic (GFRP), etc.

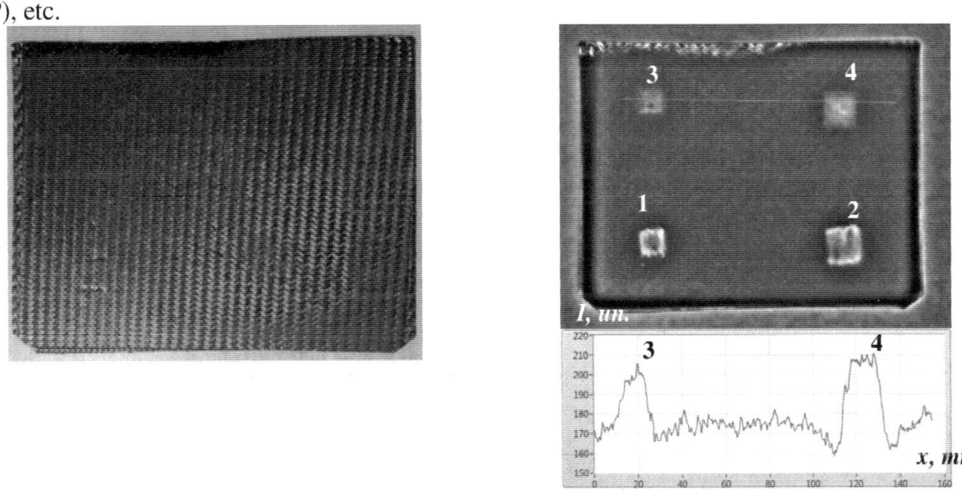

(a) (b)

Figure 3. Detection of laminations in carbon fiber-reinforced plastic (CFRP) sample using AT method: (a) visible image of the sample; (b) thermogram of the sample with detected defects

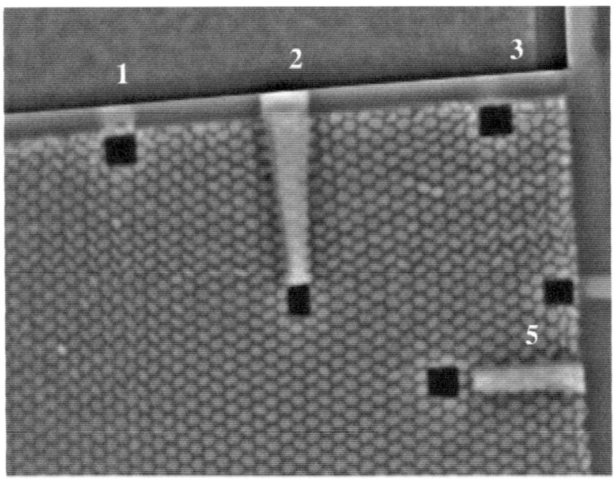

Figure 4. Thermogram of a honeycomb panels of aerospaces structure with detected artificial defects

Software tools allow building a profile of temperature (brightness of the thermogram) change along the line drawn by the user, as shown in Figure 3b. After selecting the analyzed region of the thermogram, a graph of the brightness distribution along the profile appears. The value on the abscissa axis on this graph correspond the coordinate of a point on the profile (x), and the value on the ordinate axis is a value of brightness in system units (I). This tool allows using a quantitative estimate of defects for categorizing identified heterogeneities.

The detection of defects in complex structures of alternating layers of CFRP and GFRP using the NDTherm^{TM} active thermography system is described in detail in [8].

2.2 Thermal tomography

At ensuring one-side uniform heating of the inspected object surface, it becomes possible to reconstruct the structure of the object or the defect. Thermal tomography (dynamic thermal tomography) uses time sequential of the thermograms showing the dynamics of the thermal response of the inspected object surface (Figure 5) versus classical computed tomography which uses the multi-view images [4].

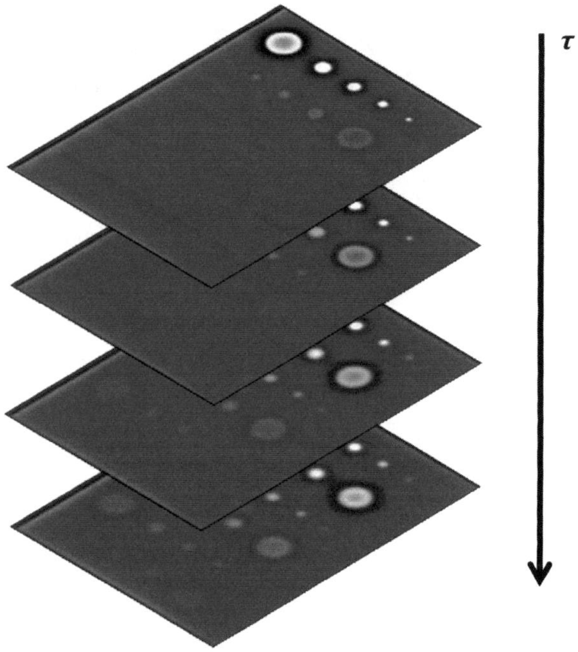

Figure 5. Time sequence of thermograms of the reference sample obtained using the NDTherm™ AT system. The images are arranged in an ascending order of thermal response time and accordingly the depth of defects accordingly.

As noted above, the NDTherm™ system provides results in the form of time sequence of thermograms obtained at different times which makes it possible to use thermal tomography in this equipment.

The practical application of thermal tomography is the ability to evaluate the depth of the defect and its length in the depth (Z axis) of the sample. The speed of propagation of heat waves inside one material is constant (under normal conditions) and is determined by the physical properties of the material. Deeper defects appear on thermograms later than less deep ones. Thus, using a reference sample with artificial defects at different depths, like in example in Figure 5, made of the same material as the inspected object it is possible to measure reference values of the response time for specific depths of defects. Further, comparing the maximum amplitude parameter of the thermal response of the detected defect with the similar parameter for the reference defect, it is possible to estimate the depth of the defect, which is detected in the object.

2.3 Material engineering tasks

The properties of composite material are determined by the properties of its components and their configuration. Since the manufacture of composite materials is a technologically complex process, NDT methods for material engineering tasks are of a particular interest. The study of the properties of the composite material components allows improving the technology of its manufacture.

The use of highly-sensitive IR cameras in the equipment that implements the AT method allows registering structural elements with different thermal characteristics. For example, the high contrast of the thermogram shown in Figure 4 allows identifying and analyzing the structure of the product: evaluating the cell size of the honeycomb structure, analyzing their shape, etc. In case of monitoring reinforced composite materials, including CFRP, resolution of thermograms obtained by the

NDTherm™ system helps to identify individual fibers (Figure 6), which makes it possible to detect such defects as breaking of connections and fiber function.

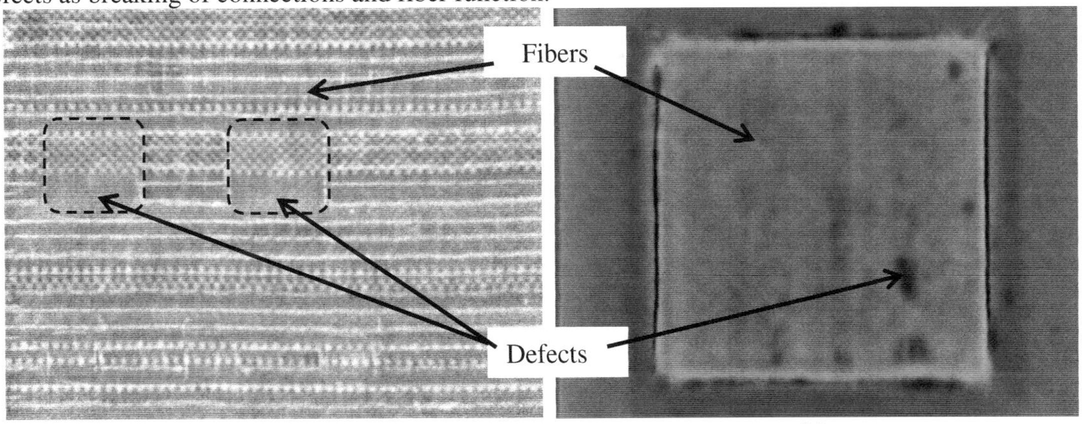

(a) (b)

Figure 6. The examples of the structure imaging in samples made of composite materials: (a) layered carbon-carbon composite; (b) CFRP

In [9], technique of pixel function matrices factorization application (functions of the dependence of temperature on time and on coordinates) – "principal component analysis" with the use of equipment for AT is described. Since the main components of these matrices contain information about the thermos-physical characteristics (thermal diffusivity, heat capacity and thermal conductivity) of the material being inspected, this approach provides an opportunity to evaluate the porosity of the CFRP structure.

In this study, the AT method was used for porosity non-contact evaluation in a composite material (Figure 7). The specimens made of CFRP with different porosities and different thicknesses have been studied:

- Specimen №1 – h = 2 mm; porosity = 1.6 %;
- Specimen №2 – h = 2 mm; porosity = 4.4 %;
- Specimen №3 – h = 3 mm; porosity = 4.8 %.

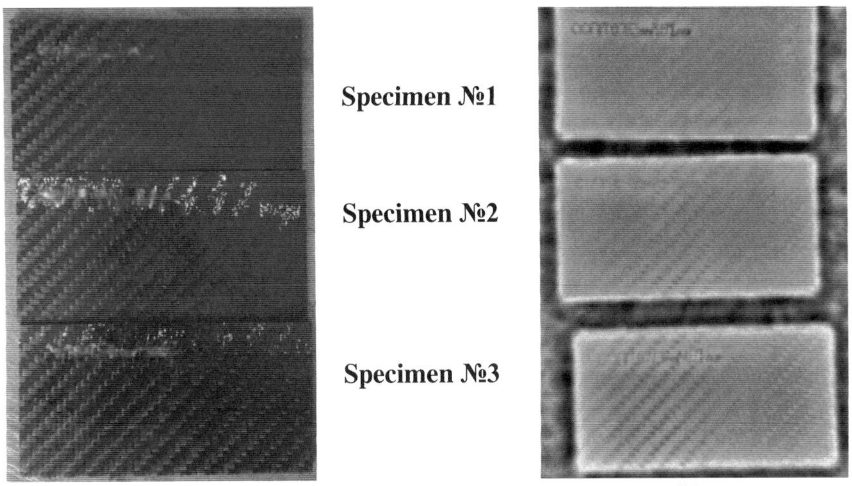

(a) (b)

Figure 7. Porosity evaluation task: the AT method for inspection of the CFRP: (a) visible image of the specimens; (b) thermograms of specimens

The use of samples with different thicknesses permits choosing the optimal operation mode of the equipment – the one in which thermal response is more affected by porosity than the difference in thickness. As a result, it was concluded that the porosity of the material affects its thermos-physical characteristics, changing the characteristics of heat propagation inside the sample: on the thermogram in Figure 7, it can be seen that with the same thermal stimulation the IR images of specimens №1 and №2 have a different contrast structural elements of the composite. Thus, the AT method can be used to evaluate the porosity of a composite material by using reference samples.

3. Conclusions

Based on the results of the study, the following promising applications of the AT method can be highlighted:

- Non-destructive inspection of a wide range of composite material products.
- Detection of the most common defects in composite materials – laminations, voids, inclusions, moisture, non-glue, etc.
- Carrying out dynamic thermal tomography of products with the capability to identify the characteristics of the detected defects (linear dimensions and depth).
- Study of the composite materials and components, that they are composed of.
- Evaluation of the thermos-physical characteristics of the material (thermal diffusivity, heat capacity and thermal conductivity, etc.).
- Evaluation of porosity and other physical characteristics of the material affecting its thermal response.

Thus, the use of the AT method allows not only carrying out the flaw detection of products, but also analyzing the structure and thermos-physical properties of the material. Based on the aerospace industrial sector world leaders' experience at successful application of the AT method in the production and operation of products made of composite materials, this NDT method should be introduced to Russian enterprises. The use of AT method can improve the technology of composite material manufacture by the ensuring of its quality.

References

[1] Smotrova S A, Naumov S M and Smotrov A V 2015 *Manufacturing technology of power units of aircraft structures made of polymer composite materials* (Moscow: TECHNOSPHERA)
[2] Wandelt M, Kroger T and Johannes M 2016 *Active Thermography – a Powerful Non-Destructive Method for Large-Sized Composite Product Testing (NDT world)* **19** no 1 pp 8–12
[3] GOST R 56787-2015 *Polymer composites. Nondestructive testing*
[4] Vavilov V P 2009 *Infrared thermography and heat control* (Moscow: Spektr)
[5] Kotovshchikov I O 2018 *Proc. Russian. Conf. "TestMat"* (Moscow) pp 119–129
[6] Poplawski J M 2016 *NDT Study of Complex CFRP/GFRP Structure by Means of Active Thermography (NDT world)* **19** no 1 pp 18–20
[7] Ciliberto A, Cavaccini G, Salvetti O et al. 2002 *Porosity detection in composite aeronautical structures (Infrared Physics & Technology)* **43** pp 139–143

IOP Conf. Series: Materials Science and Engineering **683** (2019) 012077 doi:10.1088/1757-899X/683/1/012077

Metallic coating technologies for carbon fabrics and methods for assessing their quality

V A Nelyub

Bauman MSTU, Moscow, Russia

E-mail: admin@emtc.ru

Abstract. The results of experimental analyses of the properties of carbon tapes (and filaments) after the application of metal coatings on them are presented. Various methods have been investigated for pre-cleaning the surface of carbon tapes and fabrics before applying a metallic coating thereon. A magnetron sputtering method was used to apply the metallic coating. An algorithm has been developed that allows determining the method of pre-cleaning carbon tape before its metallization. An adhesive test method was used as a criterion, allowing to determine the adhesion strength of the metallic coating to the carbon tape. An algorithm has been developed that allows determining the modes of applying metallic coatings on a carbon tape. Modes of applying metallic coatings on carbon tapes have been developed.

Keywords: carbon tape, elementary thread, metal coating, sputtering modes.

Introduction

Carbon fabrics, tapes and fibers are widely used as reinforcing materials in the manufacture of rocket and space technology products, aircraft industry, automotive industry and many other industries, due to their high specific characteristics, unique thermal and operational properties [1, 2]. In recent years, new binders have been created that allow creation of carbon plastics with higher mechanical properties, including increased strength for interlayer shear [3,4], new theoretical [5-7] and experimental methods for evaluating their properties are proposed [8-10]

To impart functional properties to CFRPs, technologies are used to apply thin metallic coatings to the surface of carbon fabrics [11, 12]. The main purpose of metallization is to change the microstructure and physicochemical properties of the surface of carbon fabrics, which allows to regulate their reactivity, wear resistance, thermal and electrophysical properties, biocompatibility, etc. [13].

Metallization occurs as a result of chemical (or electrochemical) reactions from aqueous solutions. A typical technological process of metallization of fabrics consists of the operations of degreasing the surface, etching, sensitization, activation and direct deposition of the coating. The thickness of metal coatings using electrochemical metallization can vary from tenths of a micrometer to several micrometers [14]. If instead of the electrochemical method, a chemical method is used, then it is possible to obtain coatings from units of nanometers to several microns [15]. Another method of metallization of fabrics is sputtering technology, in which the metal is initially converted into a gaseous or liquid state, and then transferred to the surface of the fabric, where it turns into a solid material, forming a metal coating.

The main methods of sputtering are: vacuum, plasma and gas plasma [15, 16]. During vacuum

sputtering, the material is heated in any inert medium, it passes into a gaseous state and then condenses on the surface of the fabric. When using this technology, coatings of any composition can be applied. When using plasma or gas-plasma sputtering, an electric arc arises between a pair of electrodes. An inert gas is injected into its combustion zone that can ionize to form a plasma. The powder of the applied metal enters the plasma jet, which melts and transfers to the substrate, which is fabric. The method of magnetron sputtering is widespread, and its modern modification - reactive magnetron sputtering. Using modern sputtering plants, coatings of various thickness and composition can be obtained.
Before applying metallic coatings on the surface of metal fabrics, it is necessary to carry out their preliminary cleaning, the main purpose of which is to remove various types of contaminants from the surface, which according to a number of specialists [14] will increase the adhesive strength of adhesion of the metal coating to the fabric.
The purpose of this paper is to develop a technology for activating the surface of a carbon tape and a technology for applying a metal coating on its surface.

Objects and Methodology
The research objects are carbon fabric ТВЧ 2/2 312-10 and carbon tape LUP and FibArm Tape-230|300, on which metal coatings of stainless steel 12Х18Н10Т and titanium were applied by way of magnetron sputtering.
We used three methods of activating the surface of the carbon tape:
1) processing with xenon lamp INP7 / 120;
2) processing on a special unit, plasma activation of the УПМ-500;
3) processing directly in the chamber of the magnetron sputtering unit.
For each of these three activation methods, the magnitude of the discharge current, the pressure, and the duration of the treatment were varied as treatment modes. The processing of all samples of carbon tape was performed from two sides.
The duration of treatment in the air plasma ranged from 10 seconds to 3 minutes. The pressure was from 10 to 100 MPa. The total complexity of processing (along with the process of installing the sample and creating a vacuum) was 15 minutes.
When using the "Betta-One-2" unit, only the duration of exposure was varied from 1 to 10 min, which corresponded to doses of 1 min - 240 mJ / cm2; 3 min - 720 mJ / cm2, etc.
As the main criterion for the quality of the pretreatment, the strength characteristic of the filament after processing carbon fabrics and ribbons was used, the value of which should not be lower than for the initial one.
To carry out mechanical tests, the reinforcing filler was divided into filaments, which were tested on a Favimat + Textechno tensile testing machine. A feature of this equipment was its high accuracy and the possibility, along with the magnitude of the breaking stress under tension, elongation and modulus, also to determine the diameter of the filaments and the linear density. Only those samples that had no external damage to the carbon reinforcing fillers were subjected to mechanical testing.
The thickness of the coating was determined directly on the filament using an atomic force microscope of the brand SOLVER47PRO.

The deposition of thin films of metallic coatings of titanium and stainless on samples of carbon tapes was carried out in a laboratory magnetron sputtering unit MIR-2. The diameter of all targets was 130 mm, the size of the chamber made it possible to process samples of carbon tapes 210 × 297 mm in size.
The quality of the adhesion of the metallic coating to the carbon tape was assessed by adhesive tape test (ASTMD3359).

Results and discussion
The algorithm for choosing a technique for treating carbon fabric before applying a metal coating thereon is shown in Fig.1. Table 1 shows the average values of the breaking stress (σ) and the diameter of the filaments for all used processing modes.

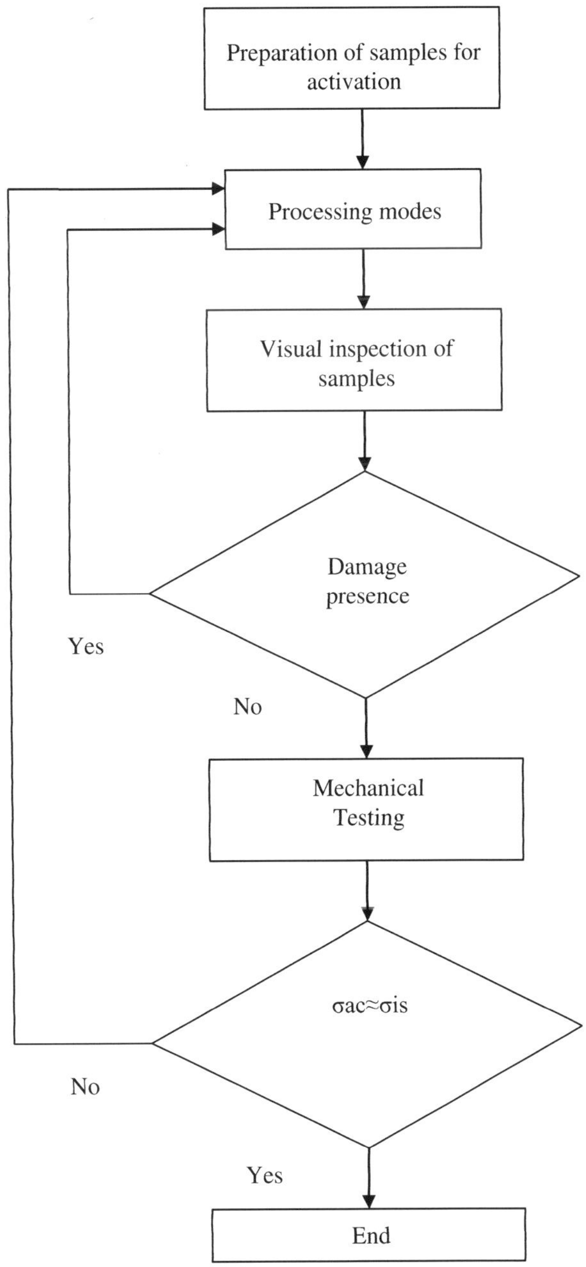

Fig.1. Algorithm for selecting carbon tape activation modes before plating

Table 1. Properties of filaments before and after processing the ТВЧ 2/2 312-10 carbon fabric

Item No.	Modes of processing	Destructive stress, σ (MPa)	Diameter of the elementary thread (μ)
1	Original	1850	5.98
	"Betta-One-2" unit, lamp irradiation for		
2	10 min	1750	5.96
3	5 min	1800	5.97
4	3 min	1850	5.97
5	1 min	1850	5.97
	УПМ-500 unit, irradiation in the air plasma		
6	5 min	1400	5.24
7	4 min	1600	5.33
8	3 min	1750	5.95
9	1 min	1750	6.01
10	30 sec	1800	5.95
11	10 sec	1800	5.97
	Bulat-6 unit, processing in argon		
12	5 min	950	4.82
13	3 min	1200	5.22
14	1 min	1400	5.89

An analysis of the results showed that processing on Bulat-6 units for more than 1 minute and on the УПМ-500 units for more than 3 minutes results in a significant decrease in the strength of the filament and a decrease in its diameter. However, the fact of a change in the strength characteristics indicates a relationship between the type and duration of treatment and the properties of the carbon tape. Surface treatment of carbon fabric ТВЧ 2/2 312-10 using "Bette-One-2" unit does not change the mechanical characteristics of the filament, which is an indirect confirmation that this type of treatment is not at all effective, because it has no effect on the surface of the carbon fabric.
Following the results of the analyses, for the activation of carbon fabric surface, the best results were shown by modes No. 8-11 (using the УПМ-500 unit) and mode 14 (using the Bulat unit), which were used in further work to prepare the surface of all the used adhesive tapes before applying of metallic coatings on them.
Using the activation mode found, metal coatings were applied to carbon tapes. Sputtering was performed under the following modes: Plasma gas pressure (Ar) 1.3x10-3-1.8x10-3tor, discharge current 2-4 A, sputtering time 2-3 min. Argon was used as a plasma gas. The algorithm for determining the metallization modes is shown in Fig. 2

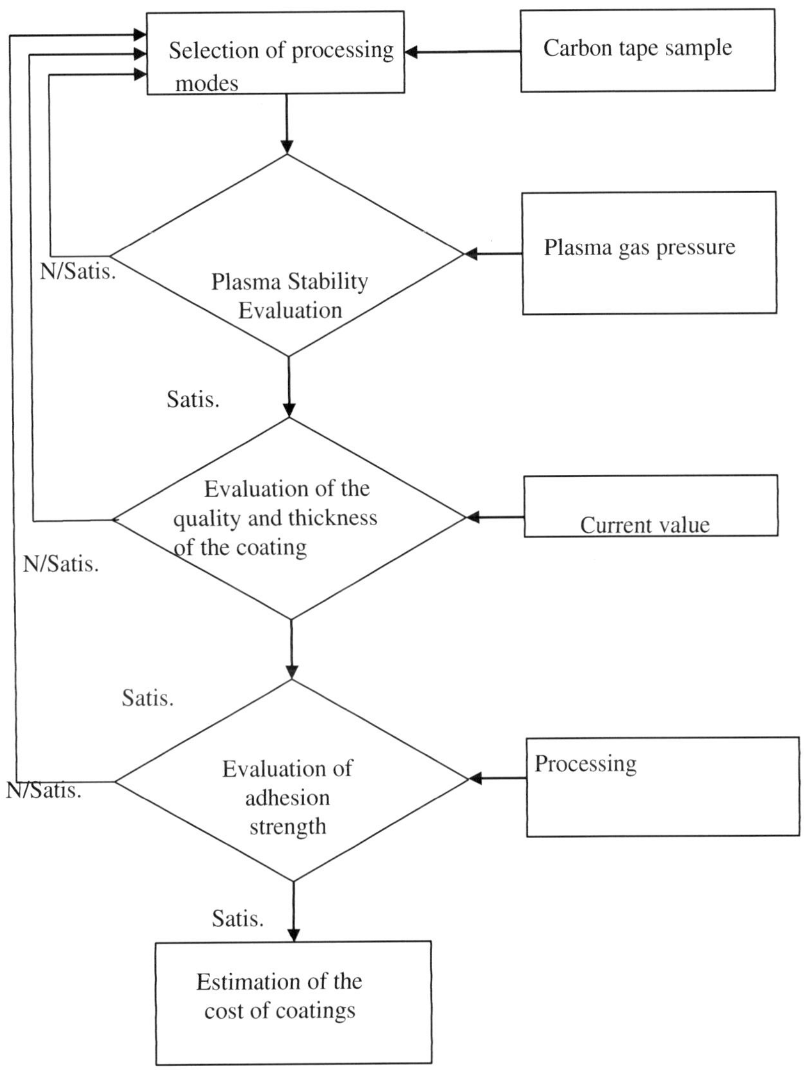

Fig. 2. Algorithm for determining the modes of metallization

IOP Conf. Series: Materials Science and Engineering **683** (2019) 012077 doi:10.1088/1757-899X/683/1/012077

The pressure range of the plasma gas (argon) was determined from the following considerations: at pressures below 1.3x10-3tor, the plasma around the target cathode becomes unstable, which adversely affects the quality of the coatings. When argon pressures are higher than 1.8x10 -3tor, the deposited metal films become loose, which significantly impairs their adhesion and the quality of metal coatings. As a result of the tests, it was found that the discharge current should not be higher than 2-4 amperes, since at a lower current, the adhesion of the coating to the substrate is deteriorated (assessed by the method of adhesive tape test), and the deposition rate also decreases, the value of which is directly proportional to the discharge current. An increase in current leads to overheating of the substrate, which leads to its deterioration or to a significant increase in gas evolution, which significantly degrades the quality of coatings. Spray time varied in the range of 2-3 minutes. Reducing the sputtering time led to the formation of a non-continuous metal film, and with increasing duration, the thickness of the coatings increased.

The total duration of metallization of one sample from one side on the unit MIR -2 was 2.5 hours. Most of the time (2 hours) was required on the creation of a vacuum, not more than 1 minute, was spent on the flow of argon, the duration of deposition depended on the mode and ranged from 5 seconds to 5 minutes.

The results of the adhesive tape test are given in table. 2. The test was performed using a graphite plate, the surface of which was prepared in various ways before titanium was sprayed onto it, as well as directly on carbon tapes LUP and FibArm Tape-230|300 and carbon fabric ТВЧ 2/2 312-10

Table 2. Adhesive Tape Test Results

Surface preparation technology and type of coating	% removed area of the total coverage area
Graphite plate, titanium coating	
Without prior preparation	30-35
Cleaning in the unit УПМ-500 –1 min	5-10
Cleaning in the unit УПМ-500 –2 min	3-5
Cleaning in the unit УПМ-500 –10 min	< 1
Graphite plate, stainless steel coating	
Without prior preparation	10-20
Cleaning in the unit УПМ-500 –2 min	1-3
Carbon fabric ТВЧ 2/2 312-10	
Without prior preparation, coating of titanium	70-90
Cleaning in the unit УПМ-500 –3 min, coating of titanium	10-20
Without prior preparation, coating of stainless steel	30-50
Cleaning in the unit УПМ-500 –3 min, coating of stainless steel	10-20
Carbon tape LUP	
Without prior preparation, coating of titanium	60-80
Cleaning in the unit УПМ-500 –3 min, coating of titanium	15-20
Without prior preparation, coating of stainless steel	40-60

Cleaning in the unit УПМ-500 –3 min, coating of stainless steel	< 10
Carbon tape FibArm Tape-230 I 300	
Without prior preparation, coating of titanium	70-80
Cleaning in the unit УПМ-500 –3 min, coating of titanium	30-40
Without prior preparation, coating of stainless steel	60-75
Cleaning in the unit УПМ-500 –3 min, coating of stainless steel	10-30

The results of the adhesion test using the adhesive tape test showed that the adhesive strength of the stainless steel coating is somewhat higher than for titanium. Comparison of two carbon tapes with each other showed that the adhesion strength of metal coatings with carbon tape of LUP is higher than with Fibra tape, but lower than with carbon fiber fabric 2/2 312-10. For all investigated coatings and carbon fabrics and tapes, surface activation before metallization

Conclusions

An algorithm and a pretreatment mode for carbon tape before metallization, as well as an algorithm and modes for applying metallic coatings, have been developed. As a result of the research it was found that to activate the surface of the carbon tape, it is advisable to process them on the УПМ-500 unit (irradiation in air plasma) for 1-3 minutes immediately before applying metal coatings.

When working out the coating modes on the installation of a magnetron sputtering MIR, it was found that the discharge current should not be higher - 4 amperes, and the sputtering time should not exceed 2-3 minutes. When the current decreases below 2 amps, the adhesion strength of the metal coating to the carbon tape deteriorates, and the deposition rate also decreases, the magnitude of which is directly proportional to the discharge current. An increase in current leads to overheating of the substrate and to a decrease in the uniformity of the metallic coating. A decrease in the duration of the deposition of coatings below 2 min of sputtering led to the formation of a non-continuous metal film, and with an increase in the duration, the thickness of the coatings increased.

Acknowledgments

The results of this paper were obtained within the framework of the project on the topic "Scientific research on the development of composite materials with the structure of controlled chaos and their use in high-tech production" on the assignment number 11.7291.2017 / CU.

References

[1] Bazhenov S L 2010 Polymer composite materials. *Dolgoprudny: Publishing House Intellect*. 352 p.
[2] Nelyub V A, Borodulin A S, Kobets L P, Malysheva G V 2017 Viscous husteresis in filled siloxane binders. *Polymer Science – Series D* **10**(1) Pp.19-22.
[3] Shimina Y Y, Malysheva G V 2015 Determination of failure-free operation probability in adhesive joints at high temperature. *Polymer Science – Series D*. **8**(3) Pp.207-210.
[4] Nelyub V A, Borodulin A S, Kobets L P, Malysheva G V 2016 Capillary hydrodynamics of oligomer binder. *Polymer Science – Series D* **9**(3) Pp.322-325.
[5] Baurova N I, Zorin V A, Prikhodko V M 2015 Description of the process of degradation of material properties using the apparatus of catastrophe theory. *Polymer Science – Series D* **8**(2) Pp.92-95.
[6] Baurova N I, Zorin V A, Prikhodko V M 2016 Manifesrarion of a synergistic effect in technological heredity *Polymer Science – Series D* **9**(2). Pp.209-211.
[7] Polilov A N 2015 *Experimental mechanics of composites: Tutorial.* M: Publishing house of MSTU named after N E Bauman 375 p.
[8] Baurova, N I 2016 *Use of polymer composite materials in the manufacture and repair of machines: Tutorial* M: MADI 264 p.
[9] Dmitriev O S, Kirillov V N, Zuev A V, Cherepakhina A A 2011 Influence of the type of filler

on the optimal curing regimes of thick-walled PCM. *Adhesives. Sealants. Technology*. no.11. p.27-36.
[10] Petrova A P, Malysheva G V 2017 G*lues, adhesive binders and adhesive prepregs: Tutorial.* M: Publishing house VIAM 472 p.
[11] Nelyub V A 2018 Metallization technology of carbon fabrics and properties of carbon plastics based on them. *Metal technology* №3 Pp.7-10
[12] Nelyub V A 2018 Evaluation of the effect of the metallization of the surface of carbon tapes on the mechanical characteristics of the filaments. *All materials. Encyclopedic reference* no.4. p.2-6.
[13] Berlin E V, Seidman L A 2010 Ion-plasma processes in thin-film technology. (Moscow) Technosphere 528p.
[14] Berlin E V and Seidman L A 2014 Production of thin films by reactive magnetron sputtering. Technosphere 256 p.
[15] Babin S V, Fursov A A Yegorov E N 2017 Study of the effect of a plasma-sprayed intermediate layer on the adhesive strength of a fiberglass-metal joint. *Bulletin of the Moscow Aviation Institute* **24** № 3 p. 195 -201.

IOP Conf. Series: Materials Science and Engineering **683** (2019) 012078 doi:10.1088/1757-899X/683/1/012078

Metal-coated carbon fiber model

V A Nelyub[1], A A Bocharov[1]

[1]Bauman MSTU, Moscow, Russia

E-mail: admin@emtc.ru

Abstract. The results of the modeling of the stress-strain state of the filaments from which the carbon unidirectional tape LUP is made, are provided for the latter's initial state and after applying a grade 12X18H10T stainless steel coating thereon. The results of calculations were compared to experimental data. Magnetron sputtering was used to apply a metallic coating, with pre-cleaning the carbon tape. These models allow to optimize the thickness and choose the type of metallic coating.

Keywords: carbon tape, filament, metal coating, modeling.

Introduction

Carbon fabrics, tapes and fibers are widely used in aerospace manufacturing, aircraft manufacturing, automotive, due to their high specific characteristics [1-3]. The values of specific strength and specific stiffness for titanium alloys are 30 and 200, respectively, whereas for CFRP these characteristics are at least 90 and 1050. The benefits of carbon plastics also include unique thermal characteristics. Modern technologies for the molding of parts and products from carbon plastics, for example, vacuum infusion, ensure the specified quality while significantly reducing the cost of production [4-6].

To impart functional properties to CFRPs, technologies are used to apply thin metallic coatings to the surface of carbon fabrics [7]. The main purpose of metallization is to change the microstructure and physicochemical properties of the surface of carbon fabrics, which allows to regulate their reactivity, wear resistance, thermal and electrophysical properties, biocompatibility, etc.

The properties of metallized carbon fabrics and carbon plastics based on them are investigated by experimental methods that are laborious and expensive [8, 9]. The task of creating a mathematical model of an elementary carbon filament, on the surface of which a thin metallic coating is applied, is relevant because it will allow modeling the behavior under load of various systems. When conducting a simulation of the stress-strain state of a metal-coated filament, we used the method of solving inverse problems. A distinctive feature of solving inverse problems (as compared with direct problems) was the difference in cause-effect relationships. Creating adequate mathematical models will significantly reduce the path from the idea to the experimental design sample. In this paper, the task was to determine the values of displacements at known values of breaking load. This formulation of the problem was related to the fact that earlier [7], the values of breaking loads and displacements for given loads were experimentally found.

The purpose of this paper is to develop mathematical models that allow predicting the stress-strain state of filaments with a metallic coating.

Objects and Methodology

The research object is a domestic unidirectional carbon tape LUP, to which a metal coating of stainless steel 12X18H10T was applied by way of magnetron sputtering. Before applying the metal coating, plasma-chemical treatment of the carbon tape was performed. The thickness of the coating was determined directly on the filament using an atomic force microscope of the brand SOLVER47PRO.

To carry out mechanical tests, the carbon tape was divided into filaments, which were tested on a Favimat + Textechno tensile testing machine. A feature of this equipment was its high accuracy and the possibility, along with the magnitude of the breaking stress under tension, elongation and modulus, also to determine the diameter of the filaments and the linear density.

The model of the filament was built using the Siemens NX program; all calculations were performed using the NX / Nastran software package. When building a geometric model, the diameter of the filament was assumed to be 6 μm, its length varied from 10 to 100 microns. Calculations of the stress-strain state of a filament with a change in its length were carried out in order to compare the results of theoretical and experimental analyses. In tensile tests, especially when using Textechno with the Favimat + prefix, the length of each filament was not measured and, in fact, was a random value. A load of 200-5000 cN was applied to the model in the form of a distributed tensile force. On the opposite side of the sample the embedment was used.

Results and Discussion

The model of an elementary thread with a metallic coating is shown in Figures. Calculations of the stress-strain state of the filament (without a metallic coating and with a coating) were carried out for filaments of various lengths. These calculations were performed in order to compare the results of theoretical and experimental studies. In experimental studies, the length of each filament was not measured and, in fact, was a random value. The calculation showed that with an increase in the length of the filament, the values of displacements in their absolute value practically do not change, but are redistributed along its length. When conducting mechanical tests, the magnitude of the maximum strain at which the filament is destroyed was determined. For the convenience of comparing the theoretical and experimental results, the displacement values obtained by calculation were recalculated into a strain.

The results of experimental and theoretical studies are given in the Table.

Table 1. Results of theoretical and experimental studies

Coating thickness (nm)	Strain (%)	Stress (MPa)
Calculation results		
Without coating	0.78	1900
50	0.67	2050
100	0.56	2210
200	0.40	2562
Results of experimental studies		
Without coating	0.78	1773
40-80	0.73	1893
140-210	0.48	1912

The calculation showed that with an increase in the length of the filament, the values of displacements in their absolute value practically do not change but are redistributed along its length. The magnitude of the elongation at which the rupture of the filament occurs is 0.78%, which is fully consistent with the results of experimental studies. Thus, as a result of a comparison of theoretical and experimental data, it has been established that the created filament model is adequate.

From the obtained research results it also follows that with an increase in the thickness of the coating, the magnitude of the strain decreases, and the stress values, on the contrary, increase. The nature of the theoretical laws obtained corresponds to the general laws obtained as a result of experimental studies, but they differ in their quantitative estimates.

(a)

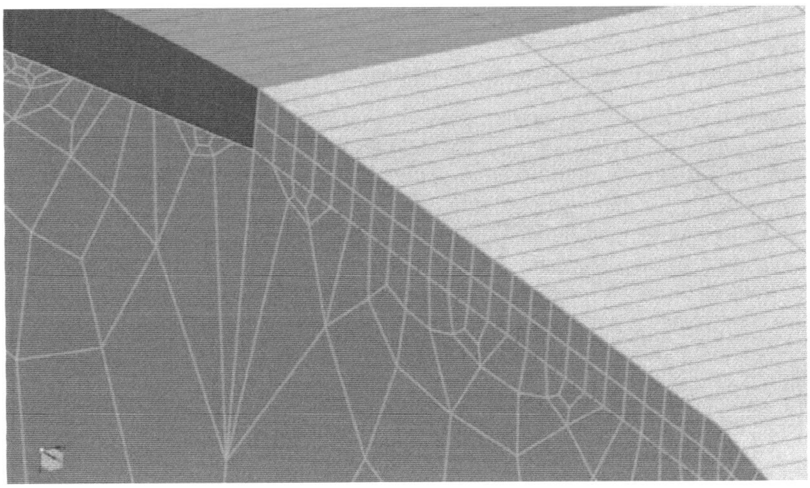

(б)

Fig.1. Model of an elementary thread with a metal coating at magnification x100 (a) and x500 (b)

This difference is probably due to the lack of methods for determining the exact values of the thickness of metallic coatings directly on the filament. The study of the structure of the coatings, carried out using an atomic force microscope, made it possible to establish that the stainless-steel coating represents granules ranging in size from 20 to 200 nm. Such a large variation in the size of the granules leads to a significant difference in the thickness of the metallic coating.

The equipment used for mechanical testing allows determine the value of the linear density of filaments. The results of the analyses show that the average linear density of the filament without a metallic coating is 4.9 tex. With an increase in the thickness of the metal coating, the value of the linear density also increases for a coating with a thickness of 50 nm is 5.0 tex, and with a coating thickness of 100 nm it is 6.5 tex. Thus, the greater the thickness of the metal coating is, the lower the weight efficiency of such materials is and, therefore, it is not advisable to increase the thickness of the metal coating more than 100 nm.

Conclusions

The mathematical model of a metal-coated filament has been developed, which makes it possible to estimate the stress-strain state. The experimental studies carried out confirmed the adequacy of the theoretical model created. As a result of the calculations, it was established that with an increase in the thickness of the coatings from 50 nm to 200 nm, the strain characteristics of the filaments decrease, the stresses increase and the linear density increase. Using the methods of atomic force microscopy, we could not accurately estimate the thickness of the metal coating directly on the filaments, due to the inhomogeneity of the coating. The results obtained can be used in predicting the properties of reinforced fabrics with various metallic coatings, including not only carbon, but glass or basalt ones.

Acknowledgments

The results of this work were obtained in the framework of the project on the topic "Scientific research on the development of composite materials with the structure of controlled chaos and their use in high-tech production" on the assignment number 11.7291.2017 / CU.

References

[1] Bazhenov S L Polymer composite materials
[2] Berlin A A, Kulkov V G, Oshmyan 2010 Dolgoprudny: Publishing House Intellect - 352 p.
[3] Baurova N I 2016 Use of polymer composite materials in the manufacture and repair of machines: Tutorial MADI - 264 p.
[4] Nelyub V A, Borodulin A S, Kobets L P, Malysheva G V 2017 Viscous husteresis in filled siloxane binders *Polymer Science – Series D* **10**(1) pp.19-22.
[5] Shimina Y Y, Malysheva G V 2015 Determination of failure-free operation probability in adhesive joints at high temperature *Polymer Science – Series D.* **8**(3) pp. 207-210.
[6] Nelyub V A, Borodulin A S, Kobets L P, Malysheva G V 2016 Capillary hydrodynamics of Capillary hydrodynamics of oligomer binder *Polymer Science – Series D* **9**(3) pp. 322-325.
[7] Polilov A N 2015 Experimental mechanics of composites: Tutorial. 375 p.
[8] Nelyub V A 2018 Metallization technology of carbon fabrics and properties of carbon plastics based on them *Metal technology*. №3. P.7-10.
[9] Baurova N I, Zorin V A, Prikhodko V M 2015 Description of the process of degradation of material properties using the apparatus of catastrophe theory *Polymer Science – Series D.* **8**(2) pp. 92-95.
[10] Baurova N I, Zorin V A, Prikhodko V M 2016 Manifesrarion of a synergistic effect in technological heredity *Polymer Science – Series D.* **9**(2) pp. 209-211.

Information about authors

* Nelyub Vladimir Aleksandrovich, Candidate of Technical Sciences, Associate Professor at the Department of Rocket and Space Composite Structures, MSTU. N E Bauman
** Bocharov Artem Alekseevich, Master of the Department of Rocket and Space Composite Structures, Bauman MSTU.

About the mechanisms of fibre reinforced plastic behaviour under cycling loading

V N Paimushin[1,2], R A Kayumov[1,3] and S A Kholmogorov[1]

[1] Kazan National Research Technical University named after A.N. Tupolev, Kazan, K.Marks str. 10, 420111, Russia

[2] Kazan Federal University, Kazan, Kremlyovskay str. 18, 420008, Russia

[3] Kazan State University of Architecture and Building Construction, Kazan, Zelenaya str. 1, 420043, Russia

E-mail: vpajmushin@mail.ru

Abstract. In order to study the peculiarities of the behavior of fiber reinforced plastics under cyclic loading, tests of specimens made of ELUR-P carbon tape and cold curing binder XT-118 with a stacking sequence $[+45^\circ/-45^\circ]_{2s}$ (s - the number of laminas) for cyclic tension were carried out. It is established that stabilization of the secant modulus of elasticity is observed as the number of cycles increases. The experimental data were compared with the results of numerical modeling of the deformation processes under study. Based on numerical calculations and analytic solutions found, for a number of types of loads and creep kernels, the value of the secant modulus of elasticity does not depend on the load amplitudes, when using relations with the Abel creep kernel, the ratio of viscous parts of strain shifted by values, does not depend on the period of cyclic load, the amplitude of the load and the parameter that determines the degree of viscosity of the material, but depends only on the parameter that determines the degree of attenuation of the creeping process. By using the method of decomposing the load in a series of trigonometric functions with the same cyclic periods, it is shown that the effect is manifested in the general case of cyclic loading using relations with the Abel creep kernel. There is established the independence of secant and tangential moduli of elasticity from the absolute values of maximum and minimum stresses for any kinds of loads and creep kernels using the linear theory of hereditarily elastic body. A method has been developed for identifying creep kernel parameters, relations for viscoplastic and elastic strain based on the analysis of experimental data on cyclic loading. For the separation of hereditarily elastic and viscoplastic strain, the hypothesis was introduced that their rates at large times are very different. This made it possible, on the basis of theoretical results obtained for the Abel kernel, to obtain ratios from which, independently of other mechanical characteristics, one can find the attenuation parameter from the strain values, measured in the experiment in times shifted by an integer number of periods. The method is applied to processing the data of real experiments conducted by the authors.

1. The results of experimental studies

Currently, in the scientific literature there are a large number of works devoted to the study of the mechanical properties of fiber reinforced plastics (FRP) under cyclic loading. Most of them are devoted to studying the residual strength and stiffness of FRP under cycling loading by comparing the elastic modulus and tensile strength after a certain number of loading cycles with their values obtained in the first loading cycle. On the basis of experimental data, as a rule, a relationship is built between the current modulus of elasticity of the composite and the number of cycles. For example, the results of such experiments for a cantilever specimen of a FRP can be found in [1]. A review of various models characterizing the residual stiffness of composites can be found, in particular, in [2]. The work [3] is devoted to the problems of building models describing the decreasing of the elastic modulus in the process of multiple loading for a FRP based on carbon fiber and an epoxy binder. It assumes that the stiffness at failure is proportional to the actual stress. In [4], a study was carried out for a unidirectional composite based on glass fibers under shear conditions for repeated cyclic loading of the specimens with a stacking sequence $[+45^{\circ}/-45^{\circ}]_{2s}$. From the experimental stress-strain diagrams, the secant shear modulus G_{12} was determined and it was concluded that it decreases after loading the specimen with ten cycles.

The authors of this work carried out a series of experimental studies of a unidirectional FRP based on carbon tape ELUR-P and a binder HT-118 under multiple loading by testing specimens with a stacking sequence $[+45^{\circ}/-45^{\circ}]_{2s}$ for tension and compression. The tests were carried out according to the method described in the article [5], on specimens aged eight months after their manufacture. Measurement of axial strain was carried out using a contact extensometer with a base of measurement 50 mm. In Figure 1 they are presented for the case of $\sigma_x^{\max} = 55$ MPa for the purpose of illustration.

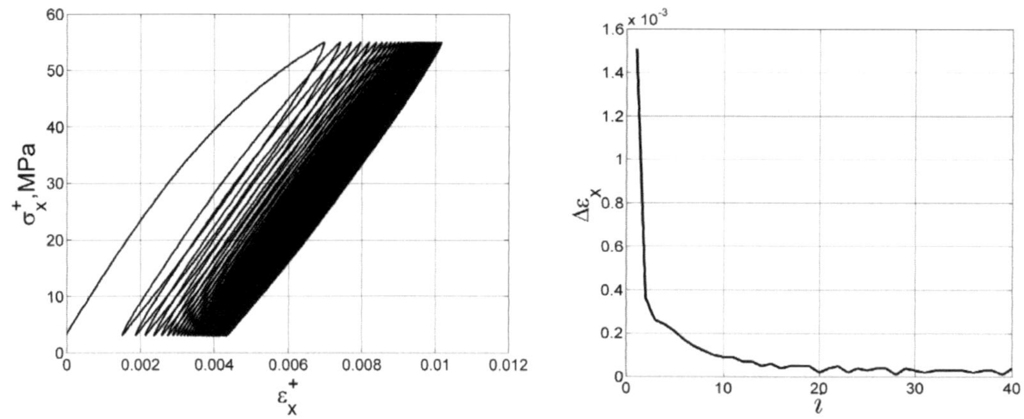

(a) (b)

Figure 1. Stress-strain curve (a) and dependence of residual strain increment on number of cycles (b)

It was revealed that the loading and unloading curves form a hysteresis loop, and the axial strain ε_x on each subsequent cycle increases by the value $\Delta\varepsilon_x^{(i)}$ (Fig. 1), where i - is the cycle number. With the increase in the number of cycles, the value $\Delta\varepsilon_x^{(i)}$ decreases. Figure 2 shows the dependences of the secant moduli of elasticity under loading and unloading at three stress levels (hereinafter, $E_+^{(i)}$ - the secant modulus under loading, $E_-^{(i)}$ - during unloading).

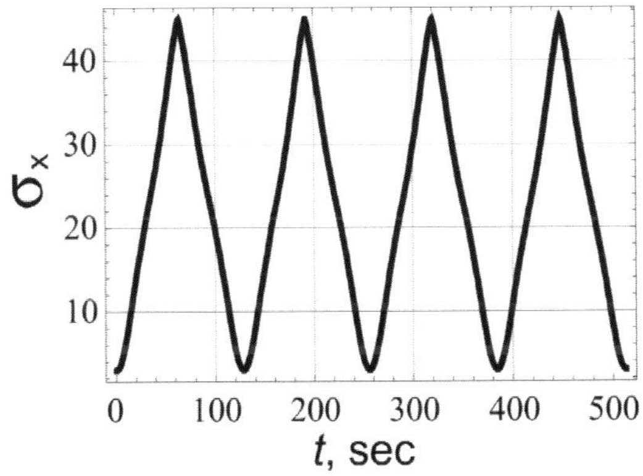

(a)

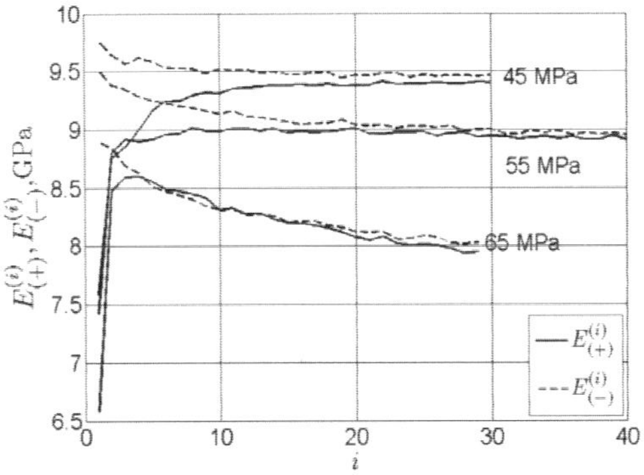

(b)

Figure 2. Dependence of stress on time (a) and moduli of elasticity on cycle number for displacement rate 0.5 mm/min

Analysis of the results of the above and other experiments, as well as a review of the scientific literature indicates the presence of a number of features of the behavior of specimens from materials such as carbon fiber reinforced plastic under cyclic loading [6, 7]. They consist in the fact that, in the general case, on each cycle, the mechanical characteristics change, which determine elastic strain, creep strain, and also strain caused by microrearrengement of the composite structure. In this connection, the task of a theoretical description of the stress-strain state being formed in a spesimen is rather complicated, since it is necessary to construct the dependences of the rheological and elastic characteristics of the material not only on the level of average stresses, but also on the number of cycles, asymmetry coefficients.

To clarify the structure of strain, the authors carried out a series of experimental studies not only on cyclic loading, but also on long-term tension with long-continued exposure under stress of specimens from cross-ply FRP [8]. Based on the analysis of the obtained results, it was found that the total axial strain can be represented as a sum of the following components:

$$\varepsilon = \frac{\sigma}{E_0} + \varepsilon^r(\sigma,t) + \varepsilon^v(\sigma,t) + \varepsilon^{nel}(\sigma) + \varepsilon^R(\sigma), \qquad (1)$$

$$\varepsilon^v = \int_0^t f(\sigma)\, H(\sigma, t-\tau)\, d\tau, \qquad (2)$$

$$d\varepsilon^r / dt = F(\sigma, \varepsilon^r). \qquad (3)$$

Here $\varepsilon^{nel} = \varphi(\sigma)$ – is a nonlinearly reversible part, $\varepsilon^R(\sigma)$ – is an irreversible instantaneous part, $\varepsilon^r(\sigma,t)$ – is an irreversible creep strain, ε^v – is a hereditarily elastic (viscoelastic) reversible component, H – is the creep kernel

In this paper, we will consider in what follows a fair generalization of Kachanov's hypothesis [9] that the strain ε^v, ε^R, ε^{nel}, ε^r develop independently of each other.

The choice of the physical ratio for ε^v must be determined in accordance with the results of experiments. As such a relation, you can use the view representation (the model with the Abel creep kernel)

$$\varepsilon^v = B\int_0^t \frac{\sigma(\tau)}{(t-\tau)^\alpha} d\tau, \quad \alpha < 1, \quad B > 0 \qquad (4)$$

which, as shown by an analysis of the literature and the processing results of the author's experiments, is well confirmed for FRP. Equation (4) also has the advantage of having the least amount of mechanical characteristics.

Next, let us accept the experimentally confirmed hypothesis that the damping of the creep strain rate is much faster than the damping of the strain rate of hereditary elasticity. This suggests that the increment of instantaneous residual strain after a sufficiently large number of cycles represents only an increment of instantly reversible and hereditarily elastic strain.

2. Theoretical studies of hereditarily elastic strain

Assuming that the stresses in the relation (4) vary according to an arbitrary cyclic law with a period T

$$\sigma = \sigma(t) = \sigma(t+T) \qquad (5)$$

we will conduct a theoretical study of the problem of changes in hereditarily elastic strain. Since $\sigma(\tau)$ a periodic function, it can be decomposed into the following trigonometric series

$$\sigma(\tau) = \sigma_0\left(1 + \left[\sum_{m=1}^{\infty} s_m \sin\frac{2\pi m\tau}{T} + c_m \cos\frac{2\pi m\tau}{T}\right]\right). \qquad (6)$$

When introducing a change of variables

$$\tau^* = \tau/T,\ t^* = t/T\ , \qquad (7)$$

relation (6) will be the following

$$\sigma(\tau) = \sigma_0\left(1 + \sum_{m=1}^{\infty}\left[s_m \sin(2\pi m\tau^*) + c_m \cos(2\pi m\tau^*)\right]\right) = \sigma_0 \cdot f(\tau^*), \qquad (8)$$

which allow relation (4) represent in the form of

$$\varepsilon^v(t) = BT^{1-\alpha}\sigma_0 \int_0^{t^*} \frac{f(\tau^*)}{(t^* - \tau^*)^{\alpha}} d\tau^* \qquad (9)$$

We now write the ratio of strain ε^v calculated with a shift, when the time in question differs from some time on a loading cycle by an amount multiple of the cycle period. Then we get

$$\varepsilon^v(t+Tn)/\varepsilon^v(t+Tk) \;=\; \int_0^{t^*+n} \frac{f(\tau^*)}{(t^* + n - \tau^*)^{\alpha}} d\tau^* \Bigg/ \int_0^{t^*+k} \frac{f(\tau^*)}{(t^* + k - \tau^*)^{\alpha}} d\tau^* \qquad (10)$$

It follows that the ratio of viscous parts of the strain at such times will not depend on T , B, σ_0 .

3. Inverse problems of identification of mechanical properties

We now consider the inverse problem of cyclic loading performed, for example, by a multilinear law (Fig. 2). The issue in the problem of identification of rheological characteristics is the issue of separation in the experiment of viscoelastic strain ε^v and creep strain ε^r . For such a separation, we accept the experimentally confirmed hypothesis that the decay of the rate of irreversible creep strain occurs much earlier than the decay of the rate of strain of hereditary elasticity. Therefore, provided that $t_i, t_j \gg t_1$ (t_1 - the time of the first cycle), it is possible with some error (which we will assume to be small) to write the relation

$$\Delta_{ij}\varepsilon^{\exp} = \varepsilon_i^{\exp} - \varepsilon_j^{\exp} = \varepsilon_i^v - \varepsilon_j^v \qquad (11)$$

Writing them for different times t_i, t_j, but with the same k , we obtain from (5) a system of equations for the parameter α . By minimizing the quadratic residual of this system, a constant α can be found that is included in the approximating expressions (4) for the creep kernel H . Then, from (11) we again obtain the system of equations, but already containing the parameter B . The minimization of the quadratic residual of this second system can be found B .

Next, it is necessary to identify the parameters of relation (3) for the theory of aging. In the future, it was chosen in its simplest form with respect to stress

$$d\varepsilon^r/dt = \chi_0\sigma/(1 + \chi_1(\varepsilon^r)^m), \qquad (12)$$

where m, χ_0, χ_1 – unknown constants.

At the last stage, one can find the secant modulus of elasticity E, considering the strain-stress curve at the initial part of the first loading cycle, but taking into account the laws of viscoelastic and viscoplastic strain already found.

As a result of processing the data of the experiments, the following values were obtained

$B = 4.0110^{-6}$ sec1 $^{\alpha-1}$/MPa, $\alpha = 0.7101$, $E = 12400$ MPa,

$\chi_0 = 4.015 \cdot 10^{-8}$ (sec·MPa)$^{-1}$, $\chi_1 = 1.913 \cdot 10^3$, $m = 22.1$

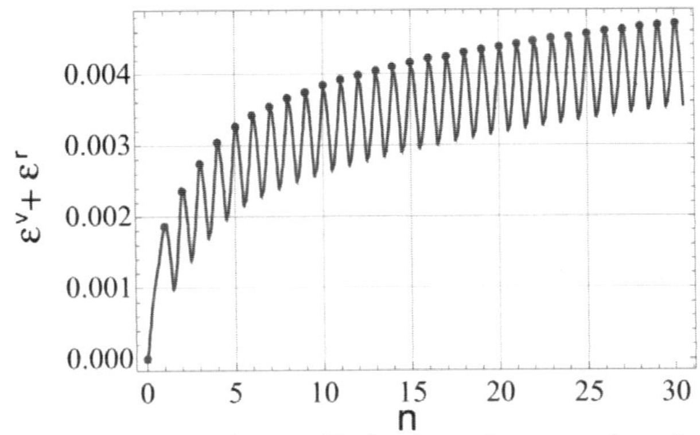

Figure 3. Dependence of inelastic strain on number of cycle (markers denote the experimental data)

Figure 3 shows for clarity only the strain values minus the elastic parts at maximum stresses. Strictly speaking, maximum strain are achieved with some delays compared to stresses. However, this delay for the considered case is almost indistinctly graphically.

Acknowledgments
The research results were obtained in the framework of the fulfillment of the state task of the Ministry of Education and Science of Russia No. 9.5762.2017/VU (project No. 9.1395.2017/PCh), due to subsidies from the federal budget for the purpose of implementing a comprehensive project to create high-tech production (project No. «2017-218-09-015», development and manufacture of specimens) and the Russian Science Foundation (project No. 19-19-00059 - formulation and solution of the inverse problem).

References

[1] Van Paepegem W, Degrieck J 2002 A new coupled approach of residual stiffness and strength for fatigue of fibre-reinforced composites *International Journal of Fatigue* **24** pp 747-762.
[2] Van Paepegem W, Fatigue 2010 Damage modelling of composite materials with the phenomenological residual stiffness approach *Fatigue Life Prediction of Composites and Composite Structures* **1** P pp 102-138.
[3] Whitworth H A 1998 A stiffness degradation model for composite laminates under fatigue loading *Composite Structures* **40** no. 2 pp 95-101.
[4] Van Paepegem W, De Baere I., Degrieck J 2006 Modelling the nonlinear shear stress-strain response of glass fiber-reinforced composites Part I: Experimental results *Composite Science and Technology* **66** pp 1455–1464.
[5] Paimushin V N, Kholmogorov S A, Makarov M V, Tarlakovskii D V, Lukaszewicz A 2018 Mechanics of fiber composites: Forms of loss of stability and fracture of test specimens resulting from three-point bending tests *ZAMM Zeitschrift fur Angewandte Mathematik und Mechanik* (Article in Press) DOI: 10.1002/zamm.201800063.
[6] Rabotnov Yu N 1966 *Creep of structural elements* (Moscow: Nauka Press) p 752. (In Russian)

IOP Conf. Series: Materials Science and Engineering **683** (2019) 012079 doi:10.1088/1757-899X/683/1/012079

[7] Yudin V E, Volodin V P, Gubanov G N 1997 Features of the viscoelastic behavior of carbon plastics on the basis of the polymer matrix: a model study and calculation *Mechanics of composite materials* **5** pp 656-669. (In Russian)

[8] Kyumov R A, Lukankin S A, Paimushin V N, Kholmogorov S A 2015 Identification of FRP mechanical properties *UCHENYE ZAPISKI KAZANSKOGO UNIVERSITETA-SERIYA FIZIKO-MATEMATICHESKIE NAUKI* **157** no.4 pp.112-132.

[9] Kachanov L M. 1958 About the time of failure in creep conditions *Izv. AN SSSR. OTN* **8** (In Russian)

Layered nanostructured composites with self-healing effect based on plastic borosiloxane matrices

N N Sitnikov [1, 2], I A Khabibullina [1] and V I Mashchenko [3]

[1] Federal State Unitary Enterprise Keldysh Research Center, 8, Onezhskaya Street, Moscow,125438, Russia

[2] National Research Nuclear University MEPhI (Moscow Engineering Physics Institute), 31, Kashirskoe shosse, Moscow, 115409, Russia

[3] Moscow Region State University, 24, Very Voloshinoy street, Mytishi, 141014, Russia

E-mail: sitnikov_nikolay@mail.ru

Abstract. This paper presents a layered composite material with an internal self-healing matrix based on borosiloxane compounds. The effects of self-healing in materials of various chemical nature (polymers, cements, ceramics, metals and composite materials) are considered and evaluated from the point of view of their practical application and the creation on their basis of self-healing composites. The properties and characteristics of borosiloxane are described. The prospects of using of borosiloxane as an internal self-healing matrix of layered or reinforced composites are shown. The effect of self-healing of the developed layered composite at breakdown and cut is demonstrated.

Keywords: self-healing materials, borsiloxane (borosiloxane), multilayer composites, matrix.

1. Introduction

The concept of self-healing man-made materials capable of recovering their initial characteristics after damage has been appeared relatively recently - several decades ago. This area of materials science attracts the scientific community actively due to the present development of technologies of materials science as well as the opening up prospects for the application of such materials. Artificially created substances or systems capable of partially or fully restoring the initial characteristics after the caused damage would open up tremendous opportunities, especially in cases when it is necessary to support operations of materials in hard-to-reach areas for as long as possible. Ideally, recovery in such materials has to occur without any external intervention, especially of a human. There are a few successful applications of such technologies in scientific publications. Most ideas for the implementation of artificial self-healing of non-biological materials are presented as the laboratory research models [1-17].

The following definition based on the analyzed literature and authors' own experimental data are proposed: the effect of self-healing (self-repairing) in man-made materials - a full or partial reduction of the surface area of material's damage due to mass transfer and consolidation of boundaries (bonds) with full or partial recovery of the functional characteristics of the material. This effect is observed in materials after their damage is usually accompanied by a discontinuity with increasing surface area. In such materials, the consolidation of the boundaries of damage (responsible for the concept of "healing") occurs after their drawing together, realized through mass transfer (responsible for the concept of "self"). The processes of mass transfer and consolidation of boundaries can occur autonomic (for example, due to the flow of material) or nonautonomic, i.e. healing is stimulated by external influence. An example of external influence is temperature increase or ultraviolet radiation. The self-repair mechanisms of man-made materials are divided into “external” and “internal” according to the method of organizing the processes of self-healing. “External” self-healing mechanisms are based on recovering components specially embedded in the matrix of the base material, for example, microcapsules with healing substances. The “internal” self-healing mechanisms are not required any additional repairng compositions [1].

The ability of artificial materials to self-repair of any properties can allow to increase their service life, reduce the cost of maintaining them in working condition and repair, as well as increase the level of safety of the structure or product as a whole. Based on the opening opportunities more of academic and industrial research organizations support the development of new self-healing materials and composite systems based on them.

The aim of this work is to determine the possibility of using borosiloxanes as a healing matrix for the preparation of self-healing layered composites.

2. Evaluation of self-healing effects in materials of various kinds and technologies based on them

Self-healing materials are a wide class of substances and can be subdivided into conditionally “neat” materials (polymers and semipolymers, ceramics, cements and metals) and composite materials and systems, which are presented in various combinations (reinforced materials, encapsulated materials, systems with hollow and filled fibers, vascular systems, laminates, sandwich panels with liquid reagents, etc.) [1-8].

Today, self-healing in “neat” materials is most successfully implemented in oligomers (semipolymers) and polymers due to relatively high diffusion rates because of having transverse molecular bonds which are able to restore bonds by crosslinking of polymer chains under specific conditions. Some low molecular weight polymers have reversible covalent bonds allowing partial or complete self-healing when reconnected with additional heating or without it [2, 3]. Consolidation of damage boundaries, reinstatement of bonds, directly responsible for the healing effect, can be achieved both by the use of covalent bonds and by non-covalent interactions. In the first case, various crosslinking reactions are used, for example, Diels-Alder reaction. Non-covalent healing can be achieved through the formation of hydrogen bonds and complex compounds, aromatic interaction (π-π-interaction), ionic interaction, Van der Waals forces, etc. Materials with covalent bonds have hidh strength, but at the same time they need the introduction of micro-inclusions of initiating substances and monomers, which provide healing. Because of this, the resource of such materials is limited, because over time there is a shortage of substance for the implementation of mass transfer. The bonds of the second type of polymers are more easily broken and restored, which undoubtedly affects the possibility of multiple healing and, accordingly, the durability of the material, but at the same time they withstand less stress and temperature [2-5].

Many of the developed polymers demonstrating self-healing properties are elastic materials (rubbers, etc.) or viscous substances (resins, gels), most of which are oriented to use at temperatures from 0 to 150 °C. The type and size of the healed defects have a wide range, self-healing of disconnected surfaces require their connection. The recovery time of properties is ranging from a few seconds to hours. Sometimes, an increase of the temperature (relative to the exploitation's temperature) or exposure to radiation, e.g. ultraviolet, is required to realize the self-healing of the initial parameters [1–8].

In ceramic materials to obtain the effect of self-healing processes are usually used oxidation of the constituent parts of the ceramic matrix or additional substances introduced into it at high temperatures (about 1000 ° C). For such processes, accompanied by an increase in mass in places of defects, usually use ceramics phase Mn+1AXn, where M is the transition metal, A is the IIIA element or IVA subgroup of the periodic system, X is carbon or nitrogen (for example, the Ti2AlN phase) for processes accompanied by mass gain in the defect areas. As a result, defects caused by wear or thermal stress are obtained by filled bonding oxides of the A-element. Products of oxidation reactions are used to fill small cracks (up to about 100 microns). The healing time is tens of hours and more [9, 10].

Self-healing of cement materials is also accompanied by mass of material gain in the defect areas, mainly due to mineral components, which provide healing of cracks in consequence of hydration, calcination, oxidation or other chemical processes. The sizes of the recovered defects are in the range from several microns to mm. These processes occur in air upon contact with water, and are characterized by a long time (tens of days - months) in a wide range of temperatures [7, 11].

Self-healing of metallic materials, due to their special properties, is achieved more difficult than in most other classes of materials. One of the problems is the nature of the bond between atoms and their low mobility at working temperatures. In general, defects in metals are healed by more fusible and plastic phases introduced into the main matrix, or by the accelerated formation of agglomerates from phases droping out of the basic material at the sites of defects under certain conditions. For example, autonomous damage recovery upon heating to 550 °C can be achieved in iron containing a small amount of gold due to creep of the precipitated phases[12, 13].

Improving self-healing characteristics of various materials is the development of composite systems based on them with introduced various fibers, materials with special properties or chemical components, allowing autonomic or non-autonomic reducing the destruction of the initial material and contributing to the faster and complete healing of the defect. For example, some elastic fibers are introduced into the polymer matrix, bring together after deformation the boundaries of the polymer's destroyed area,then this surfaces form bonds and the defect is healed by the previously described methods (autonomous action). Non-autonomous self-healing is realized by introducing into the matrix materials with special properties that are able to additionally influence under external impact, e.g. to expand and thereby reduce the size of the destroyed area. As such materials there can be various entangled fibers and materials with a "shape memory effect" (SME), expanding or contracting with increasing temperature, as well as various substances, e.g. gels, capable at several times to increase in size at a certain external impact [6, 14, 15].

There are lot of works that study self-healing composite materials, in which inert thin-walled breakable capsule with a healing substance are introduced in the basic matrix. The capsule breaks when a defect occurs, such as a crack, and the healing agent get out and spreads into the crack. At the same time, it either interacts with the matrix or the external environment or mixes with a catalyst — a solidifier pre-embedded in the material (separately from the capsules), hardens and seals the crack [1-8].

Layered composite materials (sandwich panels) are considered one of the most promising areas for creating self-repairing systems, having in its composition a layer or several layers with some kind of self-healing mechanism. In this scheme, each layer performs its specific function, and in the general system, the layered composite material is able to minimize damage and restore its original macrocharacteristics. Such sandwich panels can include various solid, viscous and liquid fillers, which react together to form a solid phase in case of material defect [6–8, 16–18]. In most cases, the healing of composite encapsulated materials is represented as the filling of the discontinuities appearing in the material with some substance other than the base material, sometimes with completely different properties from the matrix material. In fact, this is not the recovery of the original characteristics of the material, but the formation of a new material with a different structure and properties. However, self-healing in most cases means the reduction of the bulk or surface integrity of the product with the simultaneous partial or full recovery of important performance properties, such as leaktightness, strength characteristics, electrical conductivity, exterior etc.

The concept of self-healing laminated composite materials is broad and may include various self-healing mechanisms in one system, allowing to obtain for unique self-healing effects that are unattainable in other materials.

3. Properties and characteristics of borosiloxane oligomeric and polymeric compounds

Based on the analytical evaluations and characteristics of self-healing materials, borosiloxane-based materials (BS) were chosen as the main self-healing matrix of the layered composite [8, 19-22]. BS are oligomers or polymers based on elemental organosiloxanes in which the Si-O-B group is present. They have the properties of silicone oils and polymer materials, combining such characteristics as fluidity under static load and elasticity under short-term or shock load. It is assumed that these properties determine due to intermolecular donor-acceptor interactions of boron and oxygen atoms. Thus, BS is a dilatant (non-Newtonian) fluid, which behaves like a chewing gum or elastomer when stretched, and turns into a viscous fluid, spreading over the underlayer surface in the absence of external stress [23, 24]. Mass transfer to the damaged area in borsiloxane composites occurs due to the ability to spread at low loads (Figure 1).

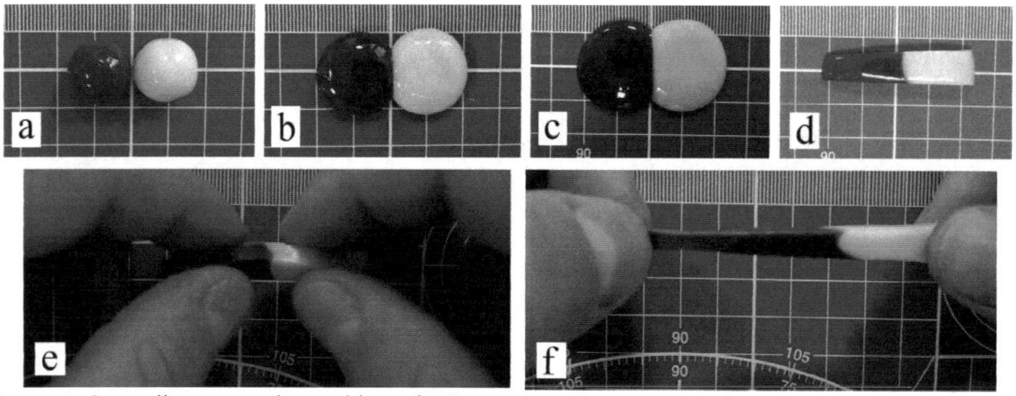

Figure 1. Spreading out and stretching of BS samples after consolidation of surfaces under normal conditions. (a) - (c) Samples of BS in the form of balls with the addition of 0.1% carbon nanotubes (left) and without: (a) time t = 0, (b) t = 1.5 hours, (c) t = 3 hours (d) - (f) Behavior of the area of the healed material with load.

Figure 1 shows an experiment demonstrating the process of self-healing of disconnected boundary of BS. Two balls of the same size were made of BS, then the balls were brought into a point touch. Without external intervention the BS balls spread out under the action of gravity. The contact area of brought together balls gradually increases, in which the consolidation of boundaries is implemented. The subsequent stretching of the plate cut out from the consolidated sample shows that the place of consolidation has the same mechanical properties as the whole sample. The material is stretched, and then breaks not at the place of consolidation of the boundary (when stresses greater than the strength of the sample are applied). The experiment demonstrates the phenomenon of mass transfer with the integration of boundaries, leading to a decrease of the total surface area, and the phenomenon of interdiffusion during the consolidation of surfaces.

Thus, BS is a promising material for use as an internal self-healing matrix of a layered composite material.

4. Layered self-healing composite material with internal borsiloxane oligomeric matrix

The developed composite with internal borosiloxane matrix has a multilayer structure. The multilayer scheme has the advantage of simplicity of obtaining and is most convenient for checking the effectiveness of the selected materials for each layer of the laminated composite material. As an internal self-healing matrix, one or several BS layers with various fillers were used. Monolithic vacuum-tight silicone sheet with a thickness of 0.5–1 mm and wicker glass-fiber cloth with a thickness of 0.3–0.4 mm impregnated with silicone were chosen as materials for the outer layers. These materials showed sufficient tensile strength and satisfactory functional properties. The choice of outer layers of silicone-based materials is justified by the fact that these materials exhibit good adhesion in contact with borosiloxane caused by similar bonds.

Samples of BS used in the work were obtained by mixing polymethylsiloxane with boric acid and Lewis acid, which was a catalyst. Boric acid was taken in the amount of 4-40 weight part per 100 weight part polymethylsiloxane, Lewis acid was taken in the amount of 0.001-3 weight part per 100 weight part polymethylsiloxane. The obtaining blend compound was heated and introduced into it one additive at least

- a filler (aerosil, talc or kaolin) or a plasticizer (oleic acid). Heating was carried out by microwave radiation to a temperature of 150-250 °C for 10-90 minutes [24]. Carbon and polyamide fibers were added additionally to the BS to reinforce it. The obtained BS samples showed good flowability and consolidation of boundary at their bringing together both in the reinforced state and in the initial state.

The general manufacturing technology of the laminated composite was in the following steps:

- preliminary isothermal exposure of the BS at a temperature of about 100 °C;
- crumpling material to obtain a homogeneous mass, followed by vacuuming to remove air bubbles;
- rolling BS to a layer of a given thickness;
- successive lay-up of the BS layer with other layers of the composite;
- homogeneous distribution of the BS in the composite material until it is completely adjoining to other layers.

Some various layered self-healing composite material with an internal borsiloxane matrix with thickness of 0.5 mm, 1 mm and 1.5 mm, with and without fillers, were manufactured. A sample of a laminate self-healing composite material consisting of three layers with an internal borosiloxane matrix with thickness of 1mm is shown in Figure 2.

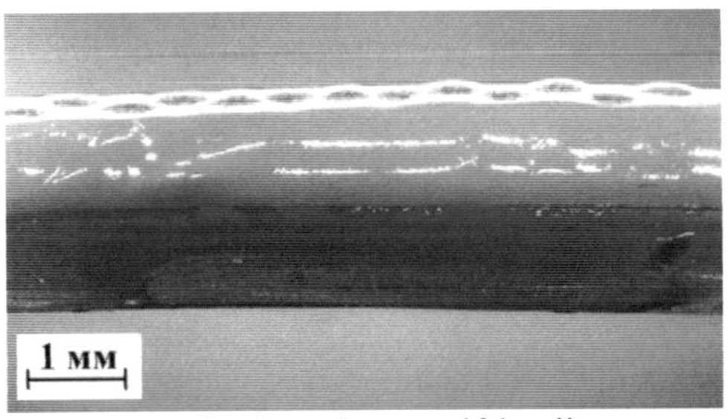

Figure 2. Typical cross-sectional view of a three-layer self-healing composite with an internal layer of BS.

The demonstration of the self-healing effect and the experimental determination of the healing characteristics of the produced layered composite during breakdown and cut were performed in a laboratory setup, which is a chamber with a flanged window (Figure 3).

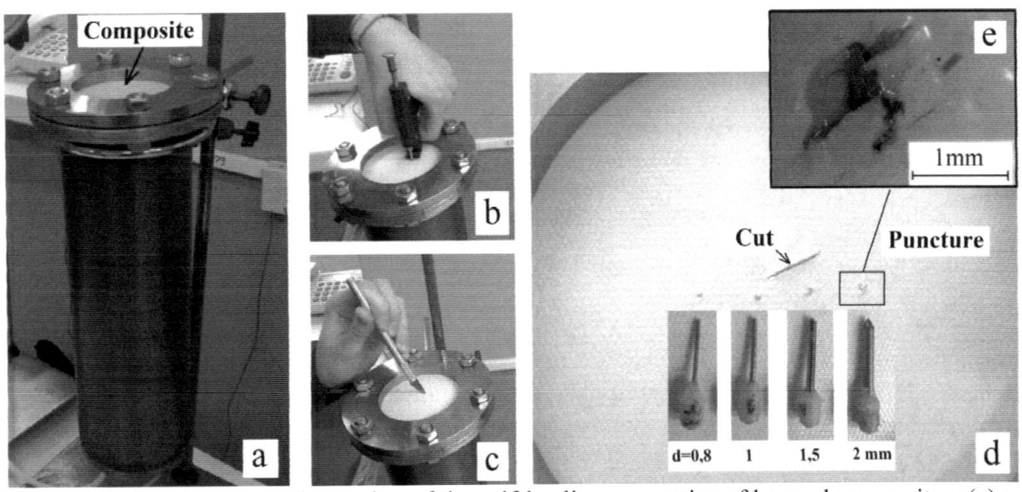

Figure 3. Laboratory setup for testing of the self-healing properties of layered composites: (a) an overpressure chamber, in the upper part of which a sample of a laminate composite with a white top layer is clamped by a flange joint; (b), (c) breakdown and cut of the composite; (d) the image of

surface defects formed by the piercers, (e) an enlarged image of the damage area on the upper layer of the composite.

Experimental laboratory setup has valves for inlet and outlet of gas creating pressure in the chamber volume, the pressure is monitored according to the installed pressure gauge. A sample of the self-healing composite material is placed in the flange window and leaktight fixed in it, an overpressure of about 0.1-0.3 atm is created in the chamber. This setup allows to study the processes of self-healing in manufactured composites after their end-to-end breakdown by any object. Evaluation of the effect of self-healing of the test sample of a layered composite and the dynamics of healing was carried out the pressure drop in the chamber, after a defect was formed in the composite with a discontinuity. The parameters of self-healing of the defect are monitored by pressure drop: when the pressure in the chamber after the breakdown decreases and then stabilizes, it is assumed that the hole from the damage was tightened and there was a consolidation of the boundaries in the BS [18]. Testing of the self-healing properties of the composite was carried out by its breakdown with pointed piercers of 0.8 mm, 1 mm, 1.5 mm and 2 mm of various diameters, as well as by a through-cut with a scalpel.

All samples of the testing layered self-healing composite with an internal borosiloxane matrix showed an effect of self-healing at the end-to-end breakdown: the holes healed almost instantly, the pressure in the chamber after the breakdown stabilized less than 1 second. Figure 3 shows the characteristic damage places on the upper layer of the composite left in the composite. A cut of about 6 mm long was formed with a through-cut of the self-healing composite by a scalpel with a blade dipping to its full length. A pressure drop of 0.03 atm was observed from the original 0.25 at after removing the scalpel blade from the composite in the chamber. The self-healing time of the testing laminate composite was 2 seconds, after that the pressure in the chamber has stabilized

5. Conclusion

The paper considers and analyzes the effects of self-healing in materials of various chemical nature (polymers, cements, ceramics, metals and composite materials) from the point of view of their practical application and the creation on their basis of layered self-healing composites. The properties and characteristics of BS compounds are described and the prospect of their using as an internal self-healing matrix of layered or reinforced composites is shown. An approach to the creation of a new layered composite using BS as a self-healing layer was developed. The effect of self-healing of the developed laminate composite with its through damage by breakdown and cut is demonstrated using the original laboratory setup. Damage formed from piercers with a diameter of 0.8 mm, 1 mm, 1.5 mm and 2 mm healed almost instantly, and the cut of about 6 mm long healed in 2 seconds.

Acknowledgments

This work is supported by the Russian Foundation for Basic Research agreement 18-29-18095\18.

References

[1] Thakur V K and Kessler M R 2015 Self-healing polymer nanocomposite materials: A review *Polymer* **69** 369–383
[2] Urdl K, Kandelbauer A, Kern W et al. 2017 Self-healing of densely crosslinked thermoset polymers – a critical review *Progress in Organic Coatings* **104** 232–249
[3] Scheiner M, Dickens T J and Okoli O 2016 Progress towards self-healing polymers for composite structural applications *Polymer* **83** 260–282
[4] Bekas D G, Tsirka K, Baltzis D et al. 2016 Self-healing materials: A review of advances in materials, evaluation, characterization and monitoring techniques *Composites Part B* **87** 92–119
[5] Zhua D Yu, Rong M Z and Zhang M Q 2015 Self-healing polymeric materials based on microencapsulatedhealing agents: From design to preparation *Progress in Polymer Science* **49-50** 175–220

[6] Hia I L, Vahedi V and Pasbakhsh P 2016 Self-Healing Polymer Composites: Prospects, Challenges, and Applications Polym*er Reviews* **56 (2)** 225-261
[7] Sitnikov N N, Khabibullina I A, Mashchenko V I and Rizakhanov R N 2018 Prospects of application of self-healing materials and technologies based on them *Inorganic Materials: Applied Research* **9 (5)** 785–793.
[8] Sitnikov N N, Khabibullina I A and Mashchenko V I 2018 Self-healing materials: an overview of self-healing mechanisms and their applications Video *Science* **1 (9)** 1–29
[9] Farle A, Kwarkernaak C, Van der Zwaag S et al. 2015 A conceptual study into the potential of Mn+1AXn-phase ceramics for self-healing of crack damage *Journal of the European Ceramic Society* **35** 37–45
[10] Shibo L, Guiming S, Kwakernaak K et al. 2012 Multiple crack healing of a Ti_2AlC ceramic *Journal of the European Ceramic Society* **32 (8)** 1813–20
[11] Ahn T H and Kishi T 2010 Crack self-healing behavior of cementitious composites incorporating various miner admixtures *J Adv. Concr. Technol.* **8 (2)** 171–186
[12] Zhang S et al. 2015 Self-healing of creep damage by gold precipitation in iron alloys *Advanced Engineering Materials* **17(5)** 21–26
[13] Zhang S et al. 2013 Defect-induced Au precipitation in Fe–Au and Fe–Au–B–N alloys studied by in situ small-angle neutron scattering *Acta Materialia* **61(18)** 7009–19
[14] Luo X and Mather P T 2013 Shape memory assisted self-healing coating *ACS Macro Lett.* **2 (2)** 152–156
[15] Rivero G, Nguyen L-T T, Hillewaere X K D et al. 2014 One-pot thermoremendable shape memory polyurethanes *Macromolecules* **47 (6)** 2010–18
[16] Zavada S R, McHardy N R et al. 2015 Rapid, Puncture-Initiated Healing via Oxygen-Mediated Polymerization *ACS Macro Let*t. **4** 819–824
[17] Aissa B, Tagziria K and Haddad E 2012 The Self-healing capability of carbon fibre composite structures subjectedto hypervelocity impacts simulating orbital space debris *ISRN Nanomaterials* 16
[18] Sitnikov N N, Khabibullina I A, Mashchenko V I and Vysotina E A 2018 Self-healing mechanisms in materials of various kinds and prospects for their use *Collection of abstracts of the Int.. Conf. on Composite Materials "Key Composite Trends: Science and Technology"* (Russia, Moscow: Bauman Moscow State Technical University,) pp. 75.
[19] Wang J, Xu W, Li Z and Zhou Z 2018 Damping and Self-Healing Properties of Polyborosiloxane Composites *Polymeric Materials Science and Engineering* **34(10)** 84-90
[20] Wu T and Chen B 2016 Synthesis of Multiwalled Carbon Nanotube-Reinforced Polyborosiloxane Nanocomposites with Mechanically Adaptive and Self-Healing Capabilities for Flexible Conductors *ACS Applied Materials & Interfaces* **8 (36)** 24071–78
[21] D'Elia E, Barg S, Ni N, Rocha V G and Saiz E 2015 Self-Healing Graphene-Based Composites with Sensing Capabilities *Adv. Mater* **27** 4788-94
[22] Shariaty D A, Qian D, Cheng Y-T and Odom S A 2018 Application of Cross-Linked Polyborosiloxanes and Organically Modified Boron Silicate Binders in Silicon-Containing Anodes for Lithium-Ion Batteries *Journal of The Electrochemical Society* **165 (5)** 731–735
[23] Mashchenko V I, Shashkova Yu O, Solomatin A C, et al. 2017 Features of the formation of the microstructure of borosiloxane-based liquid crystal composites *Bulletin of MGOU. Series: Physics-Mathematics* **2** 34–45
[24] Mashchenko V I, Alekseev A N et al. 2011 The method of obtaining of masses for modeling with biocidal properties *Patent 2473216 (RU)* date of registr. 27.05.2011

Numerical calculation of the stress-strain state of piezoelectric layered polymer composite materials equipped with a control piezoactuator

A N Anoshkin[1], P V Pisarev[1], A A Pan'kov[1] and V A Ashikhmin[1]

[1]Perm National Research Polytechnic University (PNRPU), Komsomolsky Prospect, 29, Perm, Russia

Abstract. A mathematical model is formulated for calculating the stress-strain state (SSS) of piezoelectric layered polymer composite materials (PCM) equipped with a control piezo-actuator. The model implements the principle of thermo-piezoelectric analogy, which allows to move from the coupled boundary value problem ofelectroelasticityto the boundary value problem of thermoelasticity. A comparison of the deformation fields for model layered samples from PCM with a different arrangement of piezoactuatorswas carried out. Verification of the developed mathematical models was carried out. The main patterns and mechanisms are revealedfor controlling the geometry of a layered PCM using embedded piezo-actuators.

1. Introduction

There is a recent trend of using SMART structures with controlled geometry in aviation technologies. SMART structures make it possibleto significantly improve the aircraft performance by reducing the vibroacoustic loading of highly loaded elements, such as an airplane wing or helicopter blades [1]. The development of SMART structures made of PCM equipped with piezoelements on the one hand will reduce the weight of the structure, and on the other hand will eliminate mechanical drives and skew automatons, which will significantly reduce the cost of maintaining structures with controlled geometry and fuel consumption of an aircraft.

SMART designs can be most efficient while developing new aviation wings with variable geometry, which will allow optimal adaptation to the aerodynamic parameters of the air flow [2]. The application of SMART technologies to damping oscillations of helicopter blades is now being considered as well. The possibility of using SMART-technologies in the creation of fan blades and an air engine rectifier apparatus is also under consideration [3].

There are some scientific works describing the theoretical foundations of creating SMART structures from PCM with controlled geometry [4,5]. Numerical models of the blade cross section have been developed, and to identify its optimal configuration, parametric studies of the relative rigidity of the controlled structure equipped with AFC actuators have been carried out.

In [6], numerical results are presented for a four-bladed, bearingless model of a rotor with self-adjusting piezoelectric actuators located at an angle of 45° on the upper and lower surfaces of the composite blade. To analyze the dynamic loads on the main rotor with composite blades, a complex model was developed. To simulate the behavior of a layered structure from a smart composite, an approach based on the theory of high-order deformation was applied [7].

In [8–10], a series of works was carried out related to the analysis of the structural and aerodynamic parameters of rotor blades in order to better understand the effect of active twisting. The influence of the active twisting of the blade, the required power, the load on the blade and the hub were evaluated in relation to changes in stiffness (torsional and flexural in the swing and rotation planes), sectional mass and torsion inertia, center of gravity and location of the elastic axis of the blade.

However, the scientific problem limiting the creation of SMART structures (introduction into the practice of engineering design) is the lack of scientifically based approaches and proven methods for their design. To solve this problem, it is necessary to carry out a complex of simulation and experimental studies aimed at developing new mathematical models of composite materials with piezoactive structural elements (piezocomposites), and developing methods for solving related boundary problems of electro-magnetic elasticity for inhomogeneous mediums with piezoactive structure elements. There is a need to develop methods for calculating structures, choosing the control elements, developing concepts, calculation methods, creating methods for designing structures, testing models and methods using prototypes for SMART structures, developing experimental technologies for their manufacture from modern materials, testing these prototypes.

In this study, new mathematical models of composite materials with piezoactive structural elements (piezo composites) have been developed. In the simulation experiments, the dependences of the effective electromagnetic coupling factors of piezocomposites on the structural geometric and physicomechanical parameters of the composite and the parameters of the electric field (control voltage) are investigated.

2. Numerical model

To carry out a numerical simulation of the SSS of a layered sample of PCM, equipped with piezoactuators, a three-dimensional computer model was developed with an explicit description of the layered structure. The model consists of 8 anisotropic layers and a rectangular piezo actuator. Geometrical characteristics of layers: 400×45×0.21 mm, piezo-actuator: 102x35x0.3 mm. The solution of the problem was carried out by the finite element method (FEM) using the ANSYS Workbench software package.

The sample under study was made of carbon-fiber plastic based on equal-strength woven filler with a reinforcement scheme [0°/90°/45°/-45°/-45°/45°/90°/0°]. Technical elastic constants of CFRP and micro fiber piezoactuator (MFC) are taken from [13, 14]):

GFRP – E_x=E_y= 63,9 GPa, E_z=20 GPa, G_{xy}=19,5 GPa, G_{yz} = G_{xz} = 2,7 GPa, ν_{xy} = 0,04, ν_{yz} = ν_{xz} = 0,3; MFC – E_x = 30 GPa, E_y = E_z = 15,5 GPa, G_{xy} = G_{xz} = 10,7 GPa, G_{yz} = 5,7 GPa, ν_{xy} = ν_{xz} = 0,44, ν_{yz} = 0,35, d_{31} = d_{32} = -1,98*10 m/V, d_{33} = 4,18*10^{-10} m/V.

Similarly, in the course of numerical simulation, convergence was studied. For better convergence of the solution and to reduce the errors in the results obtained, a computational grid was generated, the cells of which have a prismatic shape. The maximum element size was set to 0.9 mm., the minimum – to 0.21 mm.

Also, in the numerical simulation the mechanical behavior of the samples was investigated at various levels of the control voltage applied to the piezo-actuator. To describe the inverse piezoelectric effect in the framework of the developed mathematical model, it is proposed to use a thermo-piezoelectric analogy between piezoelectric and thermoinduced deformations.

The analogy makes it possible to move from the need to solve a coupled boundary value problem of electroelasticity to the solution of a substantially simpler unrelated boundary value problem of thermoelasticity and is based on the hypothesis of invariance of an electrically inhomogeneous field in a piezoelectric element (arising from the action of a control electrical voltage on its electrodes) to deformation fields inside the volume of the piezoelectric element and, as a result , to the deformation of the structure, on the surface of which the piezoelectric element (piezoactuator) is fixed. That is to say, the hypothesis suggests the presence of the "inverse piezoelectric effect" in the piezoelectric element in the absence of the "direct piezoelectric effect". Such an assumption is admissible for the cases when the "controlling" component of the electric field in the piezoelectric element, caused by the action of the control electric voltage $U_{упр}$ on the electrodes of the piezoelectric element, is significantly greater than the "deformation" component of the electric field caused by the "direct piezoelectric effect" from the influence of the essentially non-uniform deformation that appear and initially unknown fields in the piezoelectric element. Note that the non-uniformity of the deformation fields in the volume of the piezoelectric element is caused by many factors, in particular, the shape and location of the electrodes on the surfaces of the piezoelectric element, the characteristics of the control electrical voltage on the electrodes and the attachment of the piezoelectric element to the surface of the structure [11, 12].

The formulation of a coupled boundary problem of electroelasticity for a structurally inhomogeneous body with an outer boundary, in particular, for a fragment of an elastic composite structure with a piezo-actuator fixed on the surface in the form of a composite plate with piezoactive and piezopassive phases and electrodes consists of several equilibrium equations

$$\sigma_{ij,j}(\mathbf{r}) = 0, \tag{1}$$

continuity equation

$$\widehat{D}_{i,i}(\mathbf{r}) = 0, \tag{2}$$

equilibrium equation of defining relations

$$\begin{aligned} \sigma_{ij}(\mathbf{r}) &= C_{ijmn}(\mathbf{r})\varepsilon_{mn}(\mathbf{r}) - e_{nij}(\mathbf{r})\widehat{E}_n(\mathbf{r}) \\ \widehat{D}_i(\mathbf{r}) &= e_{imn}(\mathbf{r})\varepsilon_{mn}(\mathbf{r}) + \lambda_{in}(\mathbf{r})\widehat{E}_n(\mathbf{r}) \end{aligned} \tag{3}$$

equilibrium equation for Cauchy relations of small elastic deformations

$$\varepsilon_{ij}(\mathbf{r}) = [u_{i,j}(\mathbf{r}) + u_{j,i}(\mathbf{r})]/2, \tag{4}$$

equilibrium equation for coupling the stress

$$\widehat{E}_i(\mathbf{r}) = -\phi_{,i}(\mathbf{r}) \tag{5}$$

with the field of electric field potentials $\varphi(\mathbf{r})$ and boundary conditions, for example, in the form of given non-zero values of electric potentials on the electrodes of the piezo-actuator, mechanical fixation conditions (for displacement vector) or loading (for voltage vector) and, in addition, potentials or inductions (normal components to the border areas a) electric field for all parts of the outer boundary $\mathbf{\Gamma}$ of the area $\mathbf{V}$ with the fulfillment of the conditions of ideal contact on the interphase surfaces of continuity of the vectors of displacements, voltages and inductions the electric field in the area $\mathbf{V}$ where $\boldsymbol{\sigma}(\mathbf{r})$, $\boldsymbol{\varepsilon}(\mathbf{r})$ and $\mathbf{u}(\mathbf{r})$ are the fields of stresses, strains and displacements. $\widehat{\mathrm{D}}(\mathbf{r})$, $\widehat{\mathrm{E}}(\mathbf{r})$, $\varphi(\mathbf{r})$ are the fields of inductions, strengths and potentials of the electric field. As a result, the statement of the coupled boundary value problem of electroelasticity in the area $\mathbf{V}$ with an external boundary $\mathbf{\Gamma}$ will take the form:

$$\begin{aligned} &[C_{ijmn}(\mathbf{r})u_{m,n}(\mathbf{r})]_{,j} + [e_{nij}(\mathbf{r})\phi_{,n}(\mathbf{r})]_{,j} = 0 \\ &[e_{jmn}(\mathbf{r})u_{m,n}(\mathbf{r})]_{,j} - [\lambda_{jn}(\mathbf{r})\phi_{,n}(\mathbf{r})]_{,j} = 0 \end{aligned} \tag{6}$$

for finding the displacement fields $\mathbf{u}(\mathbf{r})$ and electric field potentials $\varphi(\mathbf{r})$ taking into account the given boundary conditions, the continuity conditions of the electroelastic fields at the internal interphase boundaries and the conditions for loading the area $\mathbf{V}$ with an electric voltage between the actuator electrodes.

For the case of the composite areas $\mathbf{V}_{1,2}$ of the structure fragment and the piezo-actuator in the area $\mathbf{V}$ with the corresponding homogeneous regions with effective, in general, anisotropic electroelastic properties, we have

$$\begin{Bmatrix} \mathbf{C}(\mathbf{r}) \\ \mathbf{e}(\mathbf{r}) \\ \boldsymbol{\lambda}(\mathbf{r}) \end{Bmatrix} = \sum_{f=1}^{2} \begin{Bmatrix} \mathbf{C}_f^* \\ \mathbf{e}_f^* \\ \boldsymbol{\lambda}_f^* \end{Bmatrix} i_f(\mathbf{r}), \quad i_f(\mathbf{r}) = \begin{cases} 1, & \mathbf{r} \in \mathbf{V}_f \\ 0, & \mathbf{r} \notin \mathbf{V}_f \end{cases} \tag{7}$$

for the coefficients of differential operators (6), where the indicator functions $i_{1,2}(\mathbf{r})$ for the areas $\mathbf{V}_1$ of the actuator and the fragment of the structure $\mathbf{V}_2 = \mathbf{V}/\mathbf{V}_1$, the tensors of the effective elastic $\mathbf{C}^*_{1,2}$, piezoelectric $\mathbf{e}^*_{1,2}$ (in particular $\mathbf{e}^*_{1,2} = 0$) and dielectric $\lambda^*_{1,2}$ properties of the areas $\mathbf{V}_{1,2}$ [12].

The use of "thermo-piezoelectric analogy" (2), (3) allows you to move from the system of differential equations (6) to a simpler

$$[\overset{*}{C}_{ijmn}(\mathbf{r})u_{m,n}(\mathbf{r})]_{,j} - \widehat{\beta}^*_{ij,j}(\mathbf{r})\Delta\widehat{T} = 0, \tag{8}$$

where the field of temperature coefficients

$$\widehat{\beta}^*_{ij}(\mathbf{r}) = \begin{cases} \overset{*}{e}_{(1)nij}(\mathbf{r})\zeta_{(1)n}(\mathbf{r}), & \mathbf{r} \in \mathbf{V}_1 \\ 0, & \mathbf{r} \notin \mathbf{V}_1 \end{cases} \tag{9}$$

calculated through a known field $\boldsymbol{\zeta}(\mathbf{r})$ with components

$$\zeta_{(1)n}(\mathbf{r}) = \widehat{E}_{(1)n}(\mathbf{r}) / U_{\text{упр}}, \tag{10}$$

identifying the magnitude of the control voltage U_{cont} on the electrodes of the piezoelectric element with the reduced heating temperature $\Delta\widehat{T} = U_{cont}$. Note that in some approximation, the inhomogeneous field $\zeta_n(\mathbf{r})$ in the area $\mathbf{V}_1$ can be calculated on the basis of solving a model boundary value problem of electroelasticity for an unrelated piezo-actuator design. While the tensor of the coefficients of the linear thermal expansion is determined by the following ratio:

$$\widehat{\boldsymbol{\alpha}}^*(\mathbf{r}) = \begin{cases} \widehat{\boldsymbol{\alpha}}^*_1(\mathbf{r}), & \mathbf{r} \in \mathbf{V}_1 \\ 0, & \mathbf{r} \notin \mathbf{V}_1 \end{cases}, \tag{11}$$

where the components $\widehat{\alpha}^*_{(1)ij} = \overset{*-1}{C}_{(1)ijmn}\overset{*}{e}_{(1)pmn}(\mathbf{r})\zeta_{(1)p}(\mathbf{r})$ are calculated through the known tensors of the effective elastic C^*_1 and piezoelectric $\mathbf{e}^*_1$ properties of the actuator and the field of coefficients $\zeta_1(\mathbf{r})$ that take into account the peculiarities of the geometric shape and arrangement of the electrodes of the piezo actuator.

When implementing the connection $\Delta T = \Delta U$ in the framework of the PC finite element analysis, you can use the simplified relation of the following form:

$$\alpha_{3i} = \frac{d_{3i}}{\Delta_{el}} (i = 1,2,3), \tag{12}$$

where Δ_{el} is the effective distance between the electrodes, the piezoelectric strain factor d_{3i} characterizes the electromechanical properties of the piezoelectric material and describes the deformation of the material depending on the orientation of the electric field. The first index indicates the direction of the electric field, and the second index indicates the direction of deformation.

Knowing the values of the three piezoelectric strain coefficients d_{31}, d_{32} and d_{33} is enough to fully characterize the electromechanical properties of the piezoelectric material.

3. Results of numerical simulation

According to the results of model layered PCM samples mechanical behavior numerical simulation, stress and displacement fields (Fig. 1,2) were obtained for samples with different locations of piezoactuators

An analysis of the axial displacements fields U_z revealed that when the bending piezoactuator is moved away from the fixing point, the normal displacements U_z decrease. When comparing the results obtained for a sample with a bottom and upper arrangement of a piezoactuator, a decrease in normal displacements of U_z by 5.07 times was registered.

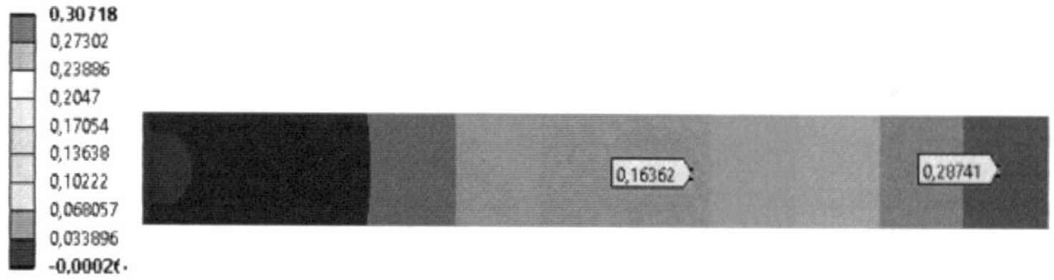

a

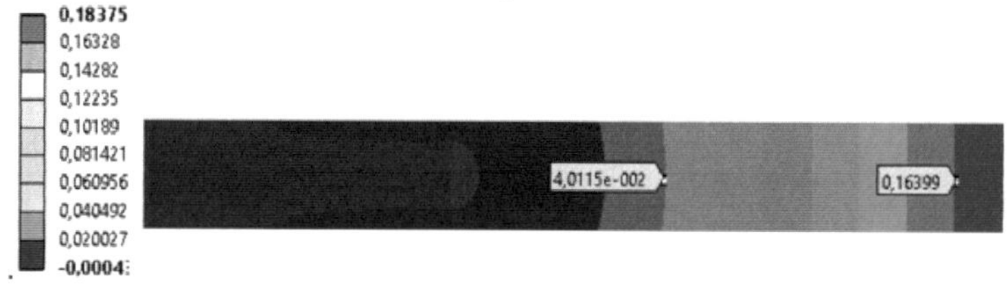

b

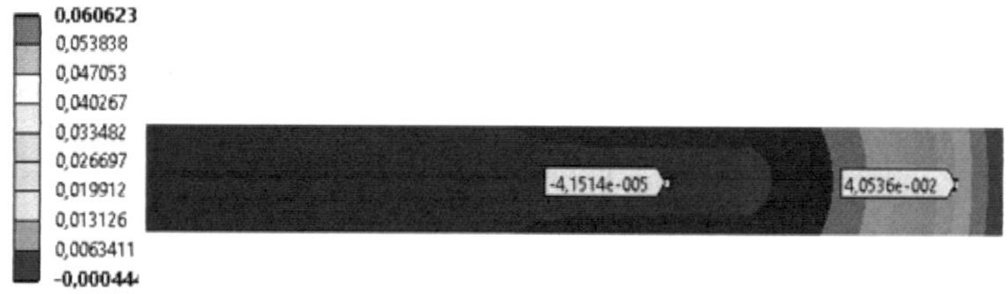

c

Figure 1. Fields of distribution of normal displacements U_z at the applied control voltage 120V: a - bottom location; b - middle location; c - upper location.

When comparing the results obtained for samples equipped with two (bottom and middle location) and three (bottom, middle and upper location) flexural piezoactuators with results obtained for a sample with one (bottom location) piezoactuator (at the embedment), an increase in normal displacements U_z was 1.66 and 2 times respectively.

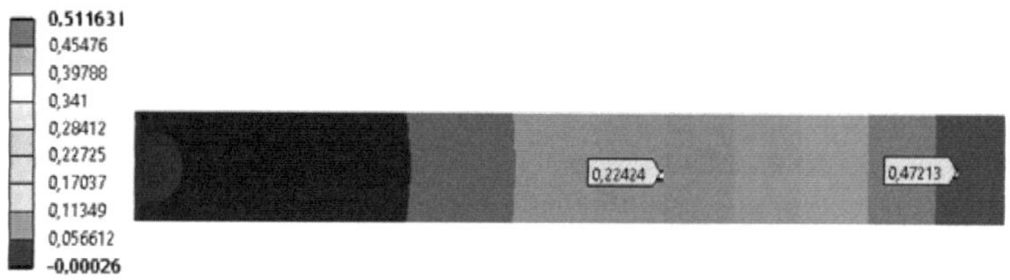

a

b

Figure 2. Fields of distribution of normal displacements U_z at the applied control voltage 120V: a - bottom/middle location; b - bottom/middle/upper location.

According to the results of the research, we can formulate the following recommendations: when designing controlled structures made of PCM that work in bending, when the piezoactuator location is distant from sealing, the bending effect decreases, therefore, it is recommended that the piezoactuators should be placed as close as possible to the mounting location of the structure.

4. Comparison of numerical simulation results and laboratory tests data

As part of the verification of the developed mathematical models, a comparison was made of the results of a numerical simulation of the model layered samples of PCM equipped with piezoactuators mechanical behavior with the results of laboratory tests. A general view of a laboratory setup with a sample of carbon fiber is presented in Figure 3. According to the results of laboratory bending tests for samples with a reinforcement scheme: [0°/90°/45°/-45°/-45°/45°/90°/0°], the values of normal displacements in the control design points at various levels of electrical control voltage were determined.

On the basis of the obtained results, graphs of the sample displacement at the control points from the control voltage applied to the piezoactuator are plotted and are shown in Figure 4.

Figure 3. General view of the laboratory setup with a sample of carbon fiber.

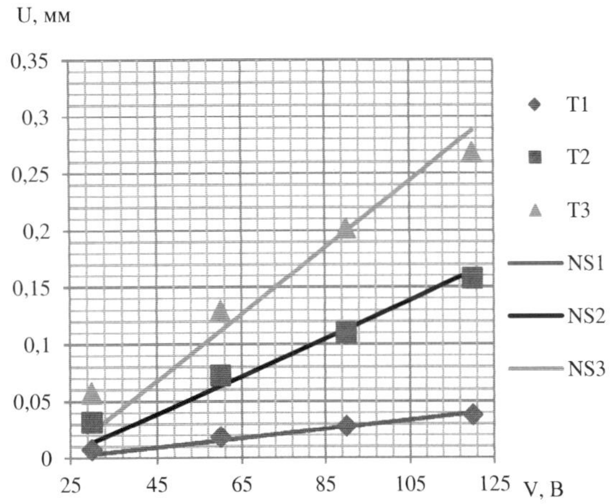

Figure 4. Graph of sample movement at the control points depending on the control voltage applied to the piezoactuator.

The analysis of dependencies revealed that the results of numerical simulation are within the range of the bend test results. The error in comparison with the average values for numerical models with the obtained results for laboratory tests on bending does not exceed 8.32%.

5. Conclusion

Thus, according to the results of the research, a mathematical model was developed for calculating the SSS of piezoelectric layered PCMs equipped with a control piezoactuator. The model implements the principle of thermo-piezoelectric analogy, which allows moving from the coupled boundary problem of electroelasticity to the boundary problem of thermoelasticity. A study of the convergence of finite element models of the samples was carried out. The mechanical behavior of layered PCM samples equipped with piezoactuators at different levels of control voltage was simulated. The main regularities and mechanisms of geometry management (deformation) of layered PCM with the built-in piezo actuators are revealed. Recommendations for the location of the bending piezo-actuators in the design of structures SMART structures of PCM were formulated. As part of the verification of the developed mathematical models, a comparison was made of the mechanical behavior of samples equipped with piezoactuators numerical simulation results with the results of laboratory tests. The error in comparing the results of numerical experiments with laboratory tests was no more than 9%.

Acknowledgments
The research was carried out at Perm National Research Polytechnic University with the support of the Russian Science Foundation (project № 18-19-00722).

References

[1] Kovalovs A, Barkanov E and Gluhihs S 2007 Active control of structures using Macro-Fiber Composite (MFC) *Journal of Physics: Conference Series* **93**(1) 3-9

[2] Brudanov A M 2015 Practical Application of Smart-Materials and Smart-Designs Young *Scientist* **104**(24) 101 - 104

[3] Chen P and Chopra I 1996 Induced strain actuation of composite beams and rotor blades with embedded piezoceramic elements *Smart Mate. Struct.* **5**(1) 35-48

[4] Cesnik C E S and Shin S J 2001 On the twist performance of a multiple-cell active helicopter blade *Smart Mater. Struct.* **10**(1) 53—61

[5] Cesnik C E S and Shin S J 2001 On the modelling of integrally actuated helicopter blades Int. *J. Solids Struct.* **38**(10) 1765-89

[6] Chattopadhyay A, Lin Q and Gu H 2000 Vibration reduction in rotor blades using active composite box beam *AIAA J.* **38**(7) 1125-31

[7] Chattopadhyay A, Lin Q and Gu H 1999 Modelling of smart composite box beams with nonlinear induced strain Composites*: Pt B, Eng.* **30**(6) 603-12

[8] Sekula M K, Wilbur M L and Yeager W T 1998 Aerodynamic design study of an advanced active twist rotor *Proc. Amer. Helicopter Soc. 4th Decennial Specialist Conf. Aeromech.* p 12.

[9] Sekula M K, Wilbur M L and Yeager W T 2005 A parametric study of the structural design of an advanced active twist rotor *Proc. 61st Annual Forum Am. Helicopter Soc.* p14

[10] Wilbur M L and Sekula M K 2005 The effect of tip geometry on active-twist rotor response *Proc. 61st Annual Forum Am. HelicopterSoc.* p 14

[11] Makarova I S 2012 Solution of a disconnected thermoelasticity problem with boundary conditions of the first kind *Vestn. sam.gos tech. un-that. Ser. Phys.-mat. Science* **3**(28) 191-195

[12] Pan'kov A A 2009 *Statistical mechanics of piezocomposites* Perm: Publishing house of the Perm state technical university p 480

[13] Kovalev A, Barkanov E, Ruchevsky S and Wesolovsky M 2017 Modeling and design of a full-scale rotor blade with embedded piezo-composite actuators *Mechanics of composite materials* **53**(2) 1-20

[14] Grinev M A, Anoshkin A N, Pisarev P V, Zuiko V Yu and Shipunov G S 2015 CAD/CAE modelling of mechanical behavior of composite outlet guide vane for aircraft jet engine PNRPU *Mechanics Bulletin* **3** 38-51

IOP Conf. Series: Materials Science and Engineering **683** (2019) 012082 doi:10.1088/1757-899X/683/1/012082

Study of the effect of permeability coefficients in a three-dimensional formulation for fabrics various types of fabric structures

Pyi Phyo Maung[1], **G. Malysheva** [2]

[1,2] Bauman Moscow State Technical University, Moscow, Russia Federation

E-mail: pyaephyo@mail.ru, malyin@mail.ru

Abstract. The results of theoretical and experimental studies of the kinetics of the impregnation process for 4 types of woven structures depending on the thickness of the molded parts are presented. The calculations were carried out with and without the permeability coefficient in the z plane. All calculations were performed in the PAM-RTM program. It has been established that considering the permeability coefficients along the x, y, z axes allows to increase simulation accuracy. As a result of the research it was found that with increasing laminate thickness more than 3 mm, it is necessary to consider the influence of the coefficient of permeability in the z-plane. Simulation error without taking into account the through thickness permeability coefficient increases from 6% to 22%.

1. Introduction

Polymer composite materials (PCM), made of glass, carbon fabrics and epoxy binders, are now widely used in the manufacturing of different products, which is associated with a unique set of their technological, mechanical and thermal properties [1, 2]. One of the main disadvantages, which limiting the application of PCM in the industry, is their high cost. Recently, more and more attention has been paid to direct molding methods, where dry fabric is used, without prepregs. The most widely used technologies in the fiber reinforced products are vacuum infusion and resin transferred molding. The main technological operations of these technologies are: the manufacture of a binder, fabric pattern, laying up, impregnation and curing. Quality losses can occur at each of these technological operations, but the most complex is the impregnation operation [3, 4]. To optimize the impregnation technology, the PAM-RTM software is used, which allows to determine the binder supply channels, the duration of the impregnation process, etc. These mechanical characteristics of final products depend on the fabric structure, including the type of weaving. When laying up the fabrics on the mold with double curvature surface, then in the process, all structures of the unit cell are deformed, which leads to a change in the values of permeability coefficients [9].

Currently, numerous studies have been carried out in which methods are proposed for determining the values of permeability coefficients in a two-dimensional formulation, i.e. along the X and Y axes. However, the permeability coefficient through the thickness of the fabric laminate also has impact on the kinetics of the impregnation processes.

The purpose of this work is to determine the values of permeability coefficients in a three-dimensional formulation, which will improve the accuracy of calculations when simulating the impregnation process.

2. Methodology

The flow rate of the impregnation process was determined in accordance with the law of Darcy [5].

$$V = -\frac{K}{\mu}\frac{\Delta P}{L} \qquad (1)$$

where V – flow rate of resin; ΔP – pressure drop; μ – resin viscosity, K – fabric permeability; L– length of specimen.

In a three-dimensional model, the permeability coefficient is a symmetric tensor with components Kxy = Kyx; Kxz = Kzx; Kyz = Kzy

$$[\mathrm{K}] = \begin{bmatrix} \mathrm{K_{xx}} & 0 & 0 \\ 0 & \mathrm{K_{yy}} & 0 \\ 0 & 0 & \mathrm{K_{zz}} \end{bmatrix} \qquad (2)$$

Fabrics, as a rule, were considered as two-layer porous [1], which have different permeabilities in the x and y planes. Permeability in the plane of the laminate with S number of layers was investigated by the method of Mogavero and Advani [2]:

$$\mathrm{K_{xx\ или\ yy}} = \frac{1}{\mathrm{L}}\sum_{\mathrm{i=1}}^{\mathrm{S}} \mathrm{t_i K_i} \qquad (3)$$

where t_i- thickness i-layer, K_i – permeability coefficient i- layer, S – number of layers.

In this work, it is assumed that the fibrous thread consists of parallel impermeable filaments. It is assumed that the impermeable filaments, in turn, are located in a periodic pattern. The pattern considered are a quadratic array and a hexagonal array. For two cases of quadratic and hexagonal arrangement of fibrous filaments, Gehbart [4] describes permeability as

$$\mathrm{K_{i_{xx}}} = \frac{8\mathrm{R_f^2}}{\mathrm{c_1}}\frac{(1-\mathrm{V_f})^3}{\mathrm{V_f}^2} \qquad (4)$$

$$\mathrm{K_{i_{yy}}} = \mathrm{c_2}\left(\sqrt{\frac{\mathrm{V_{fmax}}}{\mathrm{V_f}}}-1\right)^{5/2}\mathrm{R_f}^2 \qquad (5)$$

where V_f – fiber volume fraction in the thread, R_f – filament radius, c_1, c_2 и V_{fmax} –constants.

For these two arrays, constants are defined as according to the recommendation of Gebart [8].

For the case of quadratic arrangement of filaments in the thread

$$\mathrm{c_1} = \frac{16}{9\pi\sqrt{2}}$$

$$\mathrm{c_2} = 57$$

$$\mathrm{V_{fmax}} = \frac{\pi}{4}$$

For the case of hexagonal arrangement of filaments in the thread

$$\mathrm{c_1} = \frac{16}{9\pi\sqrt{6}}$$

$$\mathrm{c_2} = 53$$

$$\mathrm{V_{fmax}} = \frac{\pi}{2\sqrt{3}}$$

To determine the value of the permeability coefficient in the z plane, the authors of [6, 7] proposed equation (6)

$$K_{zz} = \frac{N\pi r^4 \sin\beta}{8A} \qquad (6)$$

In this case,

$$\pi r^2 = \pi n_f R_f^4$$
$$K_{zz} = \frac{N\pi n_f R_f^4 \sin\beta}{8A} \qquad (7)$$

where N – number of pores consisting in the fabric surface area (N = 1 for unit cell), r – radius of warp and weft, n_f – number of filaments in the warp and weft, R_f – radius of filament, β – angle between the warp and fabric axis, A – area of unit cell structures.

2.1. Objects

The object of the research is an aircraft fairing model made of fiberglass with various types of weaving and epoxy binder. The geometrical parameters of the aircraft fairing are shown in figure 1. Four types of fiberglass produced by Interglas, differing in the type of weaving (Table 1) were used in this study.

Table 1. Geometric parameters of different fabrics

Type of Interglas	Weaving types	V_f, %	Layer thickness, mm	R_f, μm	Liner density, tex	Unit Cell	
						Length, mm	Width, mm
92110	Plain	55	0,23	4,3	400	6,6	6,6
92125	Twill (2/1)	52	0,3	5,9	200	7,4	7,4
92140	Twill (2/2)	50	0,17	5,7	200	8,5	8,5
92145	Satin	58	0,36	6,1	400	4,9	4,9

In this work, the permeability coefficient values were calculated (table 2) using equations (4-7). All calculations were made on the assumption that the fiber content of fabrics in polymer composite is from 50 to 58%. The permeability coefficients in the x and y axes are equal.

Table 2. Permeability coefficients values of fiberglass

Type of Interglas	Permeability coefficients in the direction of the axes, m^2 ($\times 10^{-11}$)	
	K_{xx}, K_{yy}	K_{zz}
92110	1,4	0,4
92125	1,62	0,67
92140	1,9	0,88
92145	1,12	0,15

In this research, the simulation of the duration of the impregnation process depending on the number of layers of fabric (thickness) was carried out (figure 2). Theoretical studies were performed on two-dimensional and three-dimensional models, i.e. without taking into account the permeability coefficient in the z plane. To assess the adequacy of theoretical calculations, experimental studies were carried out, as a result of which the total duration of the impregnation process was also evaluated depending on the structure of the fabric used and the thickness of the sample (figure 3). The comparison between the modelling and experimental results are presented in figure 4.

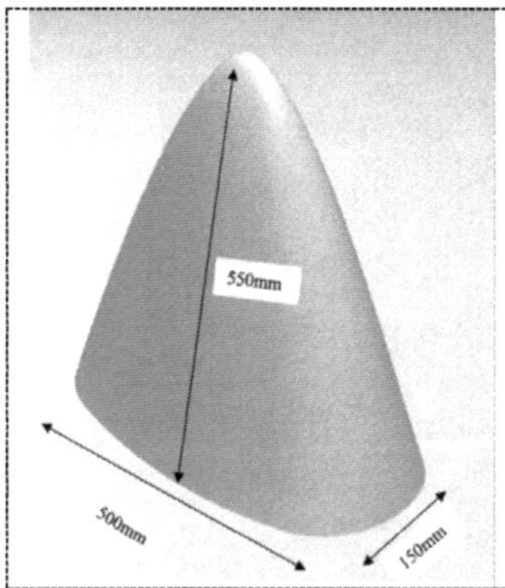

Figure 1. Geometric parameters of aircraft fairing

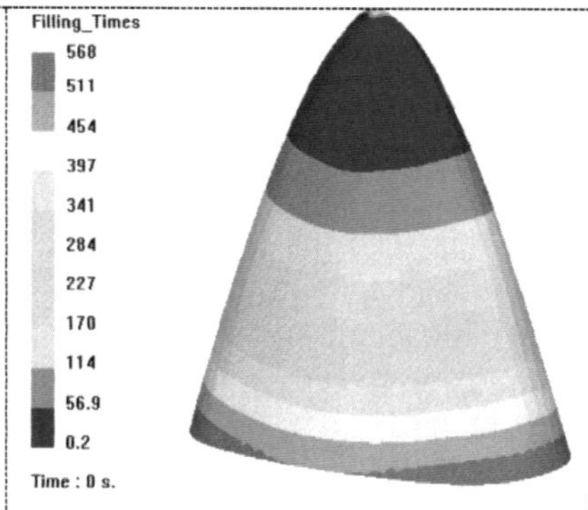

Figure 2. Modelling of impregnation process

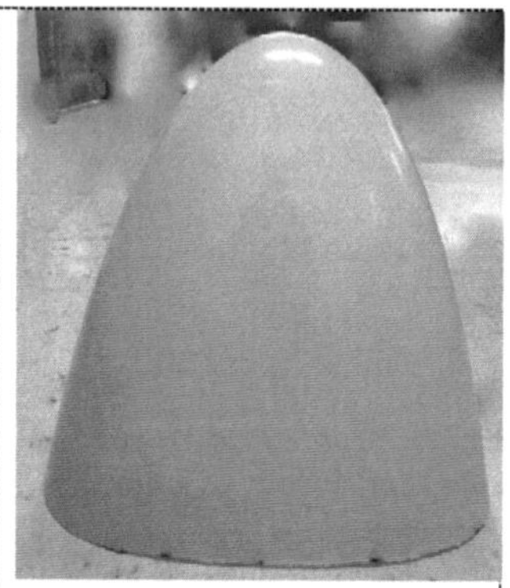

Figure 3. Experimental model of aircraft fairing

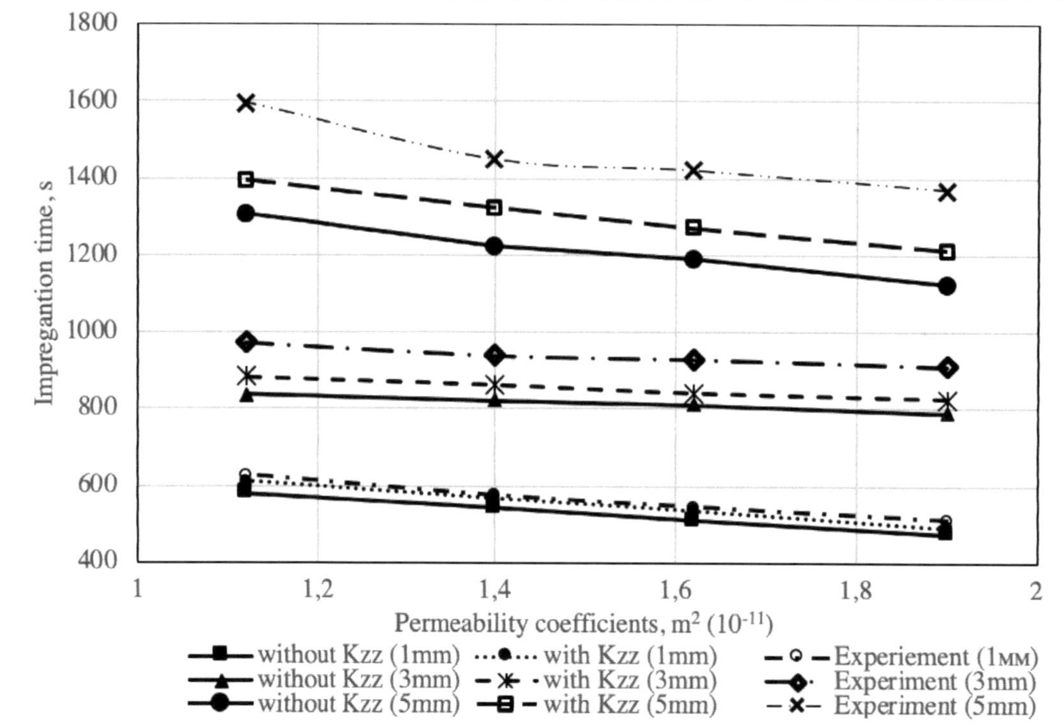

Figure 4. Comparative analysis of the results for different sample thicknesses

2.2. Results

As a result of the research it was found that with an increase in the number of layers (thickness of the laminate), the influence of the kinetics of the impregnation process in the z plane increases. It is shown that for a thickness of 1 mm the error between the theoretical and the experiment is not very visible, for 5 mm thickness it increases significantly. Considering the permeability coefficient in the z-plane, the error is much less than without taking into account K_{zz} in the comparison with the experiment.

3. Conclusion

The permeability coefficients in the directions x, y, z of the fabric with different weaving pattern are determined. Then the analytical values are used to investigate the effects of permeability coefficient in the z direction when simulating the impregnation process of the aircraft nose cone by vacuum infusion method. As a result of the research it was found that with increasing laminate thickness more than 3 mm, it is necessary to consider the influence of the coefficient of permeability in the z-plane. Simulation error without taking into account the through thickness permeability coefficient increases from 6% to 22%.

4. References

[1] Baurova N I, Zorin V A, Prikhodko V M 2015 Description of the process of degradation of material properties using the apparatus of catastrophe theory *Polymer Science – Series D*. **8**(2). Pp.92-95.

[2] Baurova N I, Zorin V A, Prikhodko V M 2016 Manifesrarion of a synergistic effect in technological heredity *Polymer Science – Series D*. **9**(2). Pp.209-211.

[3] Nelyub V A, Borodulin A S, Kobets L P, Malysheva G V 2017 Viscous husteresis in filled siloxane binders *Polymer Science – Series D*. **10**(1). Pp.19-22.

[4] Nelyub V A, Borodulin A S, Kobets L P, Malysheva G V 2016 Capillary hydrodynamics of oligomer binder *Polymer Science – Series D*. **9**(3). Pp.322-325.

[5] Sorokin A 2012 Theoretical modelling of permeability coefficient for filtration of incompressible liquids News of higher educational institutions. *Geology and exploration* **6** pp 47–54.

[6] Endruweit A, Ermanni P 2013 The in-plane permeability of sheared textiles. Experimental observations and a predictive conversion model *Composites Part A: applied science and manufacturing* p 264.

[7] Li Qiu, Xiao-Dong Chen, Rui Wang, De Peng Wan 2016 Macro fluid analysis of laminated fabric permeability *Thermal Science* **20**(3) pp 835-838.

[8] Gebart BR 1992 Permeability of unidirectional reinforcements for RTM J. of Composite Materials **26**(8) pp 1100–33.

[9] Maung PP, Malysheva G and Gusev S 2016 A study of the effect of network angle of fabrics on kinetics of impregnation upon molding of articles made from carbon plastics. *Polymer Science - Series D* **9**(4) pp. 407-10.

IOP Conf. Series: Materials Science and Engineering **683** (2019) 012083 doi:10.1088/1757-899X/683/1/012083

Related problems of designing heat-loaded structures from ceramic matrix

A D Ezhov[1], **L V Bykov**[1], **N V Artemchuk**[1]

[1] Department of aviation and space heat engineering, Moscow Aviation Institute

(National Research University), Volokolamskoe shosse, 4, Moscow, A-80, GSP-3, 125993, Russian Federation.

e-mail: bykov@mai.ru

Abstract. The paper presents a method of mathematical modeling of the mechanical and thermal state of the contact pairs on the example of contact between two composites. The algorithm of the study including the preparation of profilograms roughness of the contacting surfaces of the samples (identification roughness parameters), construction of a three-dimensional model of the surface roughness, mechanical interaction analysis microroughness contacting surfaces, determining the thermal contact resistance and the calculation of thermal fields in the contact zone. The technique was tested on a contact pair of ceramic-matrix composites for various values of mechanical and thermal loading. The obtained calculated data are verified according to the experimental study.

Key words: *composite materials, contact pair, surface roughness, mathematical modeling*

1. Introduction

The development of technological capabilities of modern materials has led to a significant amount of composite materials with radically new properties. The rapid growth of the spectrum of such materials leads to the need to develop methods for the rapid and reliable analysis of the contact properties of pairs of composite materials under their mechanical and thermal loading.

The performance characteristics of the contacting parts are directly determined by the contact properties of the mating surfaces. The processes of friction and wear occur on the actual contact area and depend not only on the material properties, but also on the pressure on this area, since the actual pressure determines the destruction of surface films and the occurrence of adhesive bonds in the contact. The flow of high-intensity heat fluxes through the contact of two materials is a rather complicated process that requires reliable determination of contact properties in a wide range of temperatures and pressures.

The study of the properties of new materials is carried out in the form of complex tests, on the basis of which it can be concluded that different materials are compatible in contact pairs. The program of complex tests is quite time consuming and is associated with known material costs. Conducting preliminary numerical studies based on a mathematical model of contact using the minimum set of known characteristics of the materials constituting the contact pair should allow a significant reduction in material and time costs by optimizing the test program. At the same time, to obtain the necessary contact parameters as a result of a numerical study, it is possible to formulate requirements for both the contacting materials themselves and the state of their surface.

The study of the contact properties of materials is devoted to a variety of works [1-10]. It should be noted that most authors conduct research in fairly local ranges of mechanical and temperature loads and a limited range of materials under study. As a rule, such an approach is caused by the need to solve an urgent, but narrowly specialized task. The tasks of contact interaction are often focused on specific aspects of contact interaction: separately on tribology issues, separately on modeling contact interaction of rough surfaces, separately on calculating the actual touch area and contact volume, separately on contact thermal resistance and thermal interaction of surfaces. The most common methods of calculating contact properties today are based on various analytical dependencies obtained as a result of generalization of experimental data. Unfortunately, such approaches, when used for a wide range of contacting materials, have extremely low accuracy, i.e. existing analytical dependencies are applicable only for a certain range of materials and fixed ranges of variation of mechanical and thermal loading of the contact zone.

In this paper, the authors attempted to calculate the comprehensive thermomechanical characteristics of contact surfaces, based on the influence of roughness on the contact parameters.

2. Algorithm of the study

To solve the problems of thermal interaction in the contact zone, the authors developed a technique [8, 9, 11], which allows, based on microgeometry of surface roughness, to simulate contact interaction of irregularities for subsequent thermal analysis. The variation of the geometrical parameters of the surface roughness profile made it possible to optimize the surface microgeometry in accordance with the necessary conditions of contact thermal interaction. This technique also makes it possible to simulate the introduction of a film into the contact zone and to evaluate its influence on the contact properties.

The algorithm for conducting research according to the existing methodology is as follows:

2.1. Obtaining profilogram of surface roughness of specific samples or determination of surface roughness parameters by known profile parameters from existing databases.

To determine the location and contact area of the surfaces of real connections, it is necessary to remove the profilogram of the surface under study. Then, as a result of processing the profilogram, its three-

IOP Conf. Series: Materials Science and Engineering **683** (2019) 012083 doi:10.1088/1757-899X/683/1/012083

dimensional model can be built, from which the characteristic contact spots can be distinguished. To implement this item, it is necessary to obtain the surface data as an array of points (x, z) in electronic format. This can be done using any modern profilometer.

If it is impossible to obtain an array of profilogram points in electronic form, you can use the base of the surface roughness profiles, from which you can select the most suitable profile based on the values of the average height of the irregularities and the step of the microprojections.

2.2. Construction of a three-dimensional model of surface roughness.

Building a three-dimensional model is possible only if there is a three-dimensional array of points with coordinates (x, y, z). A three-dimensional array of points can be obtained using the Fortran program [12], which allows, based on one surface roughness profile, to obtain an array of data with coordinates of points (x, y, z) of the surface asperities.

The resulting array of points is processed in the SolidWorks three-dimensional design environment, where the surface roughness geometry is modeled using the method of constructing a spline surface using points.

2.3. Mechanical analysis of interaction between asperities of contacting surfaces.

To perform a mechanical analysis of the interaction of asperities, a computational mesh model is constructed. As a result of the mechanical analysis of the asperity of asperities, the amount of convergence of the contacting surfaces, the area of the actual contact and the contact volume are determined.

2.4. Thermal analysis of the interaction of the contacting surfaces.

On the basis of the deformed model obtained, a thermal calculation is made of the contacting parts under given thermal loads. The temperature field of the contacting surfaces obtained as a result of thermal analysis allows determining the thermal state of the entire contact (intensity of heat fluxes and contact thermal resistance).

The obtained values of the thermal state of the contact allow you to make the necessary adjustments to the roughness of the contacting surfaces due to changes or improvements in the technological process of processing the contacting parts.

The data on the roughness parameters obtained as a result of such an analysis can later be used to develop design and technological documentation, and the data itself is recorded in the database of contact pairs of materials.

3. Testing methods

Testing of the technique was carried out on a contact pair of samples of composite materials KMK-MS - KMK-MS with an arithmetic average deviation of the profile R_a= 12.27 μm. The prototypes were cylinders with a length of 50 mm and a diameter of 30 mm. Samples were in contact with their ends and placed one above the other. The upper sample was heated from the side of the free end, heat was removed from the free end of the second sample. Samples were insulated from the surrounding space. It was possible to apply and control the clamping pressure.

At the first stage, profilograms of the contacting surfaces were determined using a Surftest SJ-210 profilometer. According to the results of roughness measurement, an average value was obtained for the three measurements made. The profilogram of one of the contacting surfaces is shown in Figure 1.

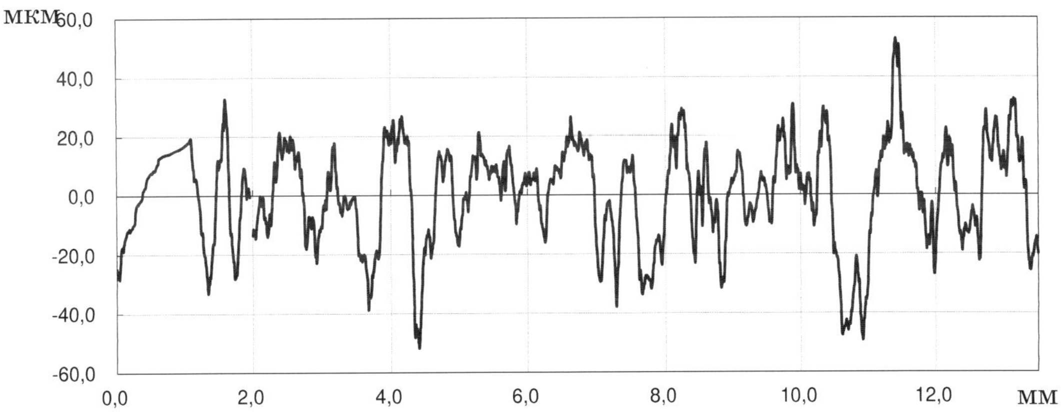

Fig.1. Surface roughness of KMK-MS.

R_a = 12.27 μm at the base length L = 14 mm.

On the basis of an array of surface roughness profile points obtained using a profilometer using the program [12], a three-dimensional array of surface microrelief points was obtained (Fig.2).

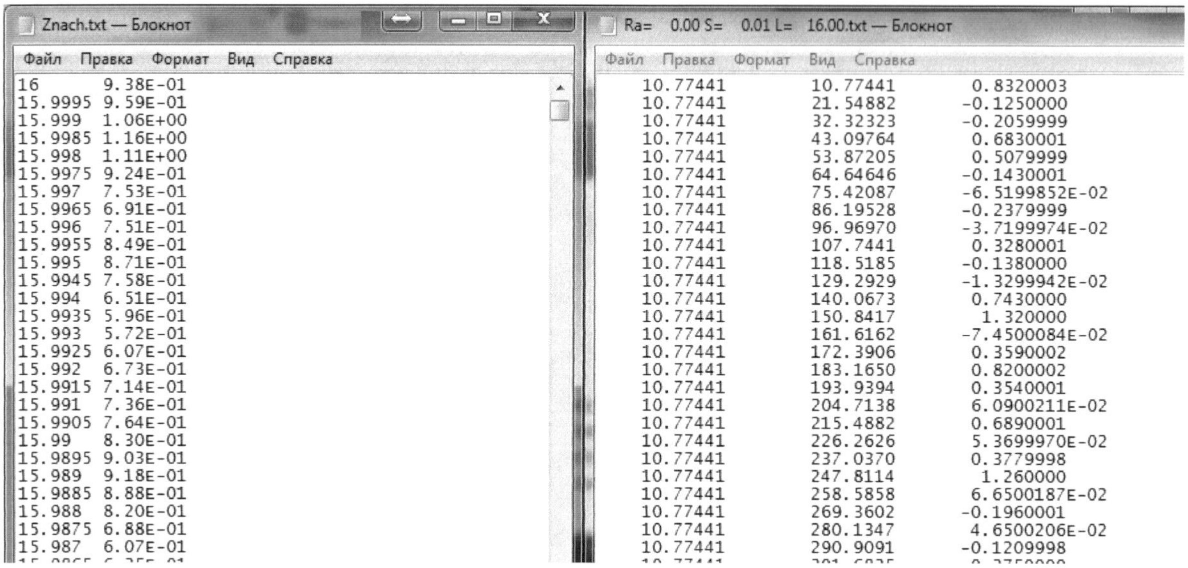

Fig.2. An array of profilogram points (left) and an array of surface microrelief points (right).

Based on the array of points of the surface microrelief, a three-dimensional model of the surface and a calculated finite element mesh model were constructed. The result of building a mesh model based on a three-dimensional model of surface roughness is presented in Figure 3.

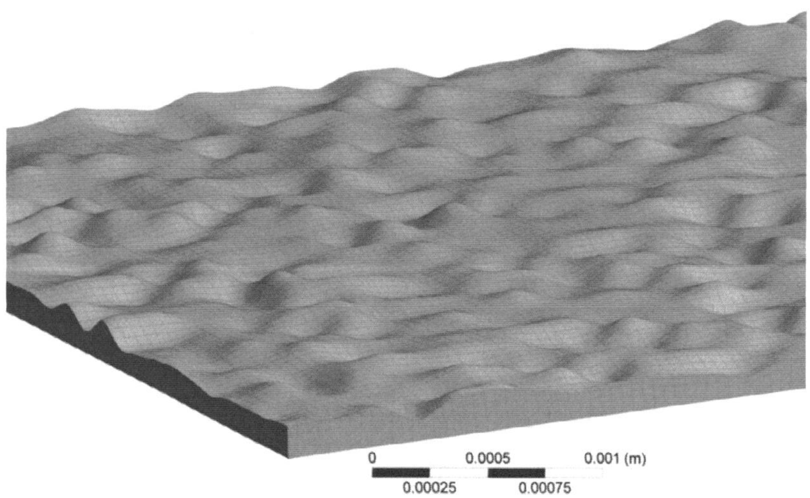

Fig. 3. Mesh model of surface roughness.

R_a = 12.27 microns (isometric).

For the mechanical analysis, the following boundary conditions were specified:

- pressure P (load A) is applied to the upper boundary, which varies in the range from 10 to 100 atm.,

- boundary conditions of the sliding seal without friction (load C, D) are applied to the lateral borders of the samples,
- to the lower boundary the boundary conditions of rigid fitting are applied (load B) (Fig. 4).

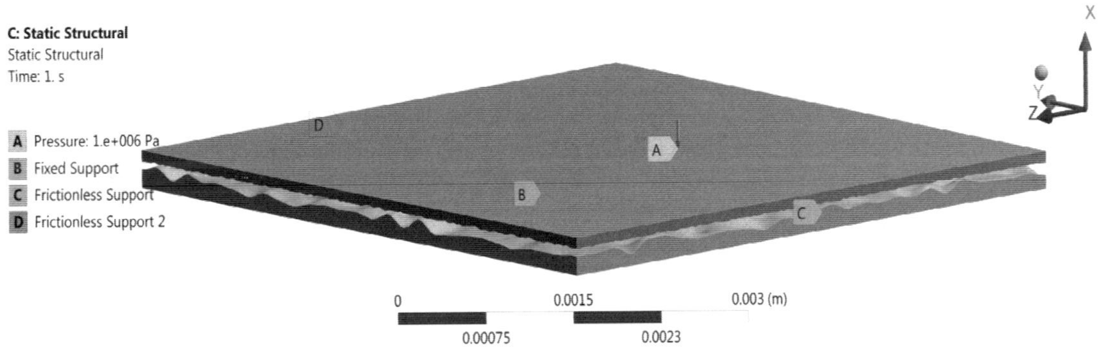

Fig. 4. Boundary conditions.

According to the results of mechanical analysis, the values of equivalent stresses and inter-contact pressure, the magnitude of the penetration of the surface of one part into the surface of another part were obtained. The field of equivalent stresses on the surface of the KMK-MS sample for an applied pressure of 10^6 Pa is presented in Figure 5.

IOP Conf. Series: Materials Science and Engineering **683** (2019) 012083 doi:10.1088/1757-899X/683/1/012083

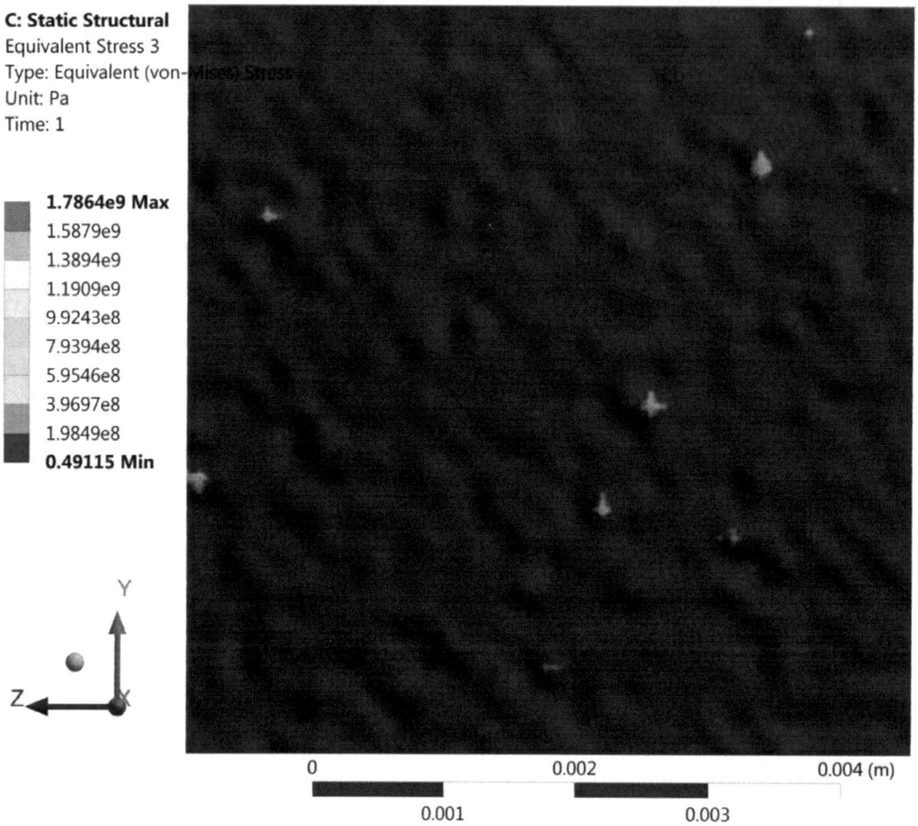

Fig. 5. The field of equivalent stresses on the surface of the KMK-MS sample for an applied pressure of 10^6 Pa.

For each value of the pressure applied to the sample (10, 20, 30 ...), at least 10 calculation sub-steps were performed, based on the results of which, by moving the pressure of the upper surface of one sample, the displacement coordinates of the upper surface of one sample were calculated to find the value of its penetration into the surface of the other sample.

The field of equivalent stresses on the surface of the KMK-MS for a pressure of 10^7 Pa is presented in Figure 6.

IOP Conf. Series: Materials Science and Engineering **683** (2019) 012083 doi:10.1088/1757-899X/683/1/012083

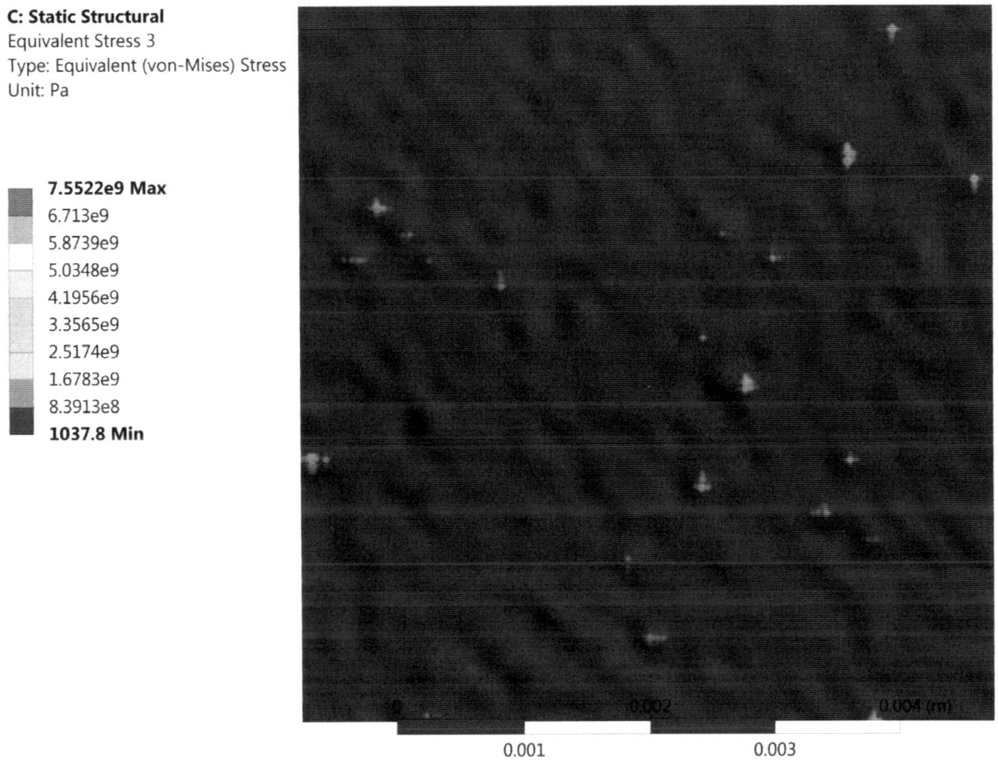

Fig. 6. The field of equivalent stresses on the surface KMK-MS for an applied pressure of 10^7 Pa.

The presented figures make it possible to notice that the number of contact points and their area depend on the applied pressure: the number of contact areas and their area increase with increasing pressure applied.

As a result of the performed strength calculation, the magnitude of the displacement of the surface of the upper sample relative to the surface of the lower sample was obtained, which is necessary to determine the actual area of contact in the contact pair of two rough surfaces. The magnitude of the displacement for the considered contact pair, depending on the applied pressure are given in table 1.

Table 1. Dependence of the displacement of the pressure applied to the pair of materials.

Applied pressure, MPa	Displacement, mm
1.00	$7.21 \cdot 10^{-6}$
2.00	$8.60 \cdot 10^{-6}$
3.00	$9.77 \cdot 10^{-6}$
4.00	$1.06 \cdot 10^{-5}$
5.00	$1.13 \cdot 10^{-5}$
6.00	$1.21 \cdot 10^{-5}$
7.00	$1.27 \cdot 10^{-5}$
8.00	$1.34 \cdot 10^{-5}$
9.00	$1.40 \cdot 10^{-5}$
10.00	$1.46 \cdot 10^{-5}$

The determination of the parameters of the temperature field, depending on the magnitude of the heat flux, was made by dividing one calculation step into 10 substeps. As a result of the thermal analysis, the temperature field, the direction and magnitude of the heat flux in the contacting surfaces were obtained. Figure 7 shows the temperature field distribution on the contacting surface.

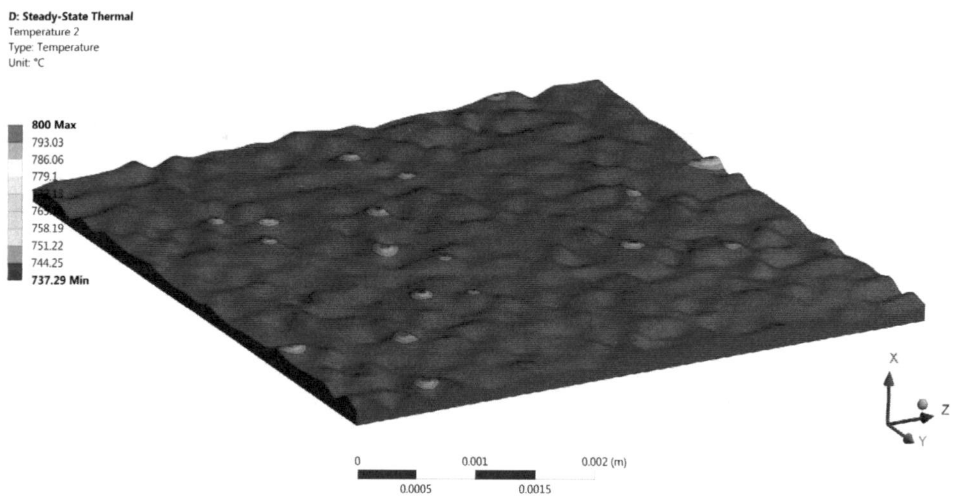

Fig. 7. Distribution of the temperature field on the surface of the KMK-MS at an intercontact pressure of $5 \cdot 10^6$ Pa.

The values of the calculated contact thermal resistance R_K and the surface temperature of the samples for the considered contact pair, depending on the applied pressure, are given in Table 2.

Table 2. Values of contact thermal resistance R_K and temperature in the contact pair KMK-MS - KMK-MS

Applied pressure, MPa	$R_K \cdot 10^4$, **K·m²/W**	**Top sample temperature, C°**	**Bottom sample temperature, C°**
1.00	187.03	625.78	438.75
2.00	148.88	673.43	524.55
3.00	131.65	719.14	587.49
4.00	119.09	728.34	609.25
5.00	111.05	735.13	624.08
6.00	105.09	739.56	634.46
7.00	100.48	742.99	642.51
8.00	96.40	745.31	648.92
9.00	93.60	748.50	654.90
10.00	90.52	751.95	675.57

The conducted test calculation showed good (within 6%) agreement with the results of an experimental study of the properties of a KMK-MS-KMK-MS contact pair under power and thermal loading.

4. Conclusions

The use of the presented technique allows determining by the method of mathematical modeling contact thermal resistances for any contact pair in the permissible pressure range. And having the value of contact thermal resistance for the considered pairs of materials in a wide range of pressures, it is possible with certain reliability to calculate the thermal state of the structure under actual operating conditions of contact pairs of materials.

References

[1] Popov V M 1971 Heat transfer in the contact zone of detachable and one-piece connections. *Energiya, Moscow* (in Russian)

[2] Ezhov A D, Bykov L V, Mesnyankin S Yu, Bogachev E A, Razina A S 2015 *Teplovyye protsessy v tekhnike*, no 21 pp.510-516

[3] Thompson M K 2007 *A multi-scale iterative approach for finite element modeling of thermal contact resistance*. PhD Thesis, Massachusetts Institute of Technology, Boston, Massachusetts, USA

[4] Mesnyankin S Yu, Ezhov A D, Basov A A 2014 Izvestiya RAN. *Energetika,* no 5 pp. 65-74

[5] Murashov M V, Panin S D, Klimov S M 2015 *Nauka i obrazovaniye: nauchnoye izdaniye MGTU im. N E Baumana* no.1 pp.189-200

IOP Conf. Series: Materials Science and Engineering **683** (2019) 012083 doi:10.1088/1757-899X/683/1/012083

[6] Popov V M, Lushnikova E N, Chernoukhov P A 2012 *Lesotekhnicheskiy zhurnal,* no. 1 pp. 7-12

[7] Bolotov A N, Sutyagin O V, Rachishkin A A 2014 Mekhanika i fizika protsessov na poverkhnosti i v kontakte tverdykh tel, detaley tekhnologicheskogo i energeticheskogo oborudovaniya, no. 7 pp. 29-41

[8] Ezhov A D, Mesnyankin S Yu, Bykov L V 2017 V sbornike: Materialy KHKH Yubileynoy Mezhdunarodnoy konferentsii po vychislitel'noy mekhanike i sovremennym prikladnym sistemam *(VMSPPS'2017)* pp. 236-238

[9] N B Demkin, E V Ryzhov 1981 Surface Quality and Contact of Machine Components *Mashinostroenie*, Moscow (in Russian).

[10] Bykov L V, Ezhov A D 2018 V sbornike: Materialy XII Mezhdunarodnoy konferentsii po prikladnoy matematike i mekhanike v aerokosmicheskoy otrasli *(NPNJ'2018)*, Moscow pp. 559-560

[11] Ezhov A D, Bykov L V, Mesnyankin S Yu 2018 Numerical Method for Determining the Real Contact Area of Contacting Bodies. *Journal of Surface Investigation: X-ray, Synchrotron and Neutron Techniques.* **12** Issue 5 pp 914–917

[12] Certificate of state registration of computer programs №2015618021 «Determination of coordinates of points on a rough flat surface» from 28.06.2015 (in Russian)

IOP Conf. Series: Materials Science and Engineering **683** (2019) 012084 doi:10.1088/1757-899X/683/1/012084

Creation and analysis of hybrid nanocomposites based on dioxidine and nanoparticles of bioactive metals embedded in cryostructurated gelatin matrices

A V Nuzhdina[1,2], T I Shabatina[1,2], O I Vernaya[2] and V P Shabatin[2]

[1] Bauman Moscow State Technical University, Faculty "Fundamental Sciences", Moscow, Russia

[2] Lomonosov Moscow State University, Faculty of Chemistry, Moscow, Russia

Email: anastasia.nuzhdina@emtc.ru

Abstract. This paper is devoted to the creation of hybrid nanocomposites based on the antibacterial drug dioxidine and bioactive metals - silver and copper. The obtained materials were analyzed using the methods of NMR, UV and IR spectroscopy, low-temperature adsorption of argon, SEM and TEM. It is shown that they are silver nanoparticles with a size of 2-30 nm or copper nanoparticles with a size of 10-40 nm, incorporated into dioxidine particles with a size of 50-350 nm. The obtained drug nanocomposites were embedded into wide-porous biopolymer cryostructurated gelatin-based matrices. The possibility of the release of drug components from a biopolymer carrier and manifestation of its antibacterial activity has been shown. At the same time, hybrid nanocomposites based on metals and antibacterial drugs showed increased activity to suppress the growth of microbial cells of *Escherichia coli 52, Staphylococcus aureus 144, Mycobacterium cyaneum 98,* then their components separately.

1. Introduction

It is known that the efficacy of medicinal substances is largely influenced by their solubility. When reducing the particle size to 20-200 nm not only increases the solubility of the substance, but also improves the permeability of biological barriers, which leads to an overall increase in bioavailability [1]. The effectiveness of pharmacopoeial drugs can be improved, including through the creation of new hybrid forms containing known medicinal substances together with nanosized bioactive metal nanoparticles [2-4]. In this case, mixing of two or more components on a nanoscale scale can contribute to the change in existing and the acquisition of new properties by the resulting hybrid systems [5].

In addition, the current task is to create drugs with controlled release of medicinal substances, including prolonged dosage forms, in which the release of the drug substance takes longer than for the usual dosage form [6-9]. An important advantage of the dosage forms of prolonged action is to reduce the peak concentration and the absence of temporary fluctuations in the concentration of the active substance in the blood and tissues, concomitant therapy with conventional medicines. One of the methods for obtaining such dosage forms is the binding of the pharmacologically active component to the biopolymer matrix, which provides a gradual release of the drug substance from it and a long-term maintenance of its required concentration in the body or locally in a specific target organ. As biocompatible carriers for the so-called "depot forms" of medicinal substances, wide porous gelatinous matrices formed by cryostructuring techniques are of interest [10–13].

The purpose of this paper was the synthesis of hybrid nanocomposites based on the antibacterial drug dioxidine and silver and copper metal nanoparticles, the inclusion of the obtained nanocomposites in gelatinous cryostructurated matrices to ensure their prolonged release, as well as the determination of the antibacterial activity of the resulting hybrid drug nanoforms with respect to *E. coli bacterial cells 52 , S. aureus 144 and M. cyaneum 98.*

2. Experimental part

2.1. Sampling

Dioxidine substance corresponding to Pharmacopoeia Article 42-2308-97, КНД-С-К colloidal silver (TU 9154-024-74107096–2008), copper chloride II, hydrazine hydrate of qualification "analytical grade" were used without further purification. Copper nanoparticles were obtained by reducing an aqueous solution of copper chloride to hydrazine hydrate [14]. Low-temperature synthesis of gelatin cryostructurated matrices was performed according to previously published methods [15, 16].

Highly dispersed dioxidine powder and hybrid dioxidine nanocomposites with silver nanoparticles (Ag-dioxidine) or copper (Cu-dioxidine) were prepared by spraying an aqueous solution containing dioxidine and metal nanoparticles (1 wt% Dioxidine, 0.005 wt% Ag or 0.02 wt% Cu) through a pneumatic nozzle into liquid nitrogen, then the frozen solutions were subjected to cryosublimation drying for 24 hours [17–20].

Dioxidine, as well as Ag/dioxidine, Cu/dioxidine nanocomposites were included in the disks of cryostructurated matrices as follows: sponge gelatin disks were immersed for 30 minutes in aqueous solutions of dioxidine and metal nanoparticles (1 wt% dioxidine, 0.02 wt% Ag or 0.02 wt% Cu), then the discs were removed from the solution, frozen with liquid nitrogen and subjected to cryosublimation drying for 24 hours. The obtained samples of Ag-dioxidine-gelatin and Cu-dioxidine-gelatin contained 11.5 wt% dioxidine, 0.46 wt% metal.

2.2. Determination of physico-chemical properties

The ^{1}H NMR spectra were recorded in a saturated solution in deuterated water (D_2O) on a VXR 400 high resolution NMR spectrometer (Varian, USA). The IR spectra of the samples were obtained in the range of 4000–400 cm^{-1} on a Bruker Tensor II spectrometer (Germany) with the prefix ATR platinum. The IR spectra of powdered samples were recorded by the method of diffuse reflection. UV spectra of 0.01 wt% aqueous solutions of samples were taken on a Jasco V-770 spectrophotometer (Jasco, Japan) in the range of 200–700 nm. The release of dioxidine from cryostructurated matrices was monitored spectrophotometrically at 360 nm.

The microstructure of the samples was studied by transmission electron microscopy (TEM) on a LEO 912 AB Omega electron microscope (Zeiss, Germany) at magnifications of 1000–20000 times and scanning electron microscopy (SEM) on a Phenom scanning electron microscope (FEI Company, Germany) at magnifications of 20 to 4000 times.

The determination of the specific surface area (S_{sp}) of the samples was performed by the method of low-temperature adsorption — thermal desorption of argon on a laboratory setup based on a Chrom 5 chromatograph. Pre-adsorbed gases were removed from the surface of the samples in a vacuum unit. The average particle size of dioxidine was calculated by the formula: $a = 6 /(\rho S_{ud})$, where ρ is the dioxidine density.

2.3. Determination of antimicrobial activity

The antibacterial activity of hybrid nanocomposites Ag-dioxidine and Cu-dioxidine in comparison with dioxidine and solutions of colloidal silver and copper was determined by a disk-diffusion method [21] using Krasnaya Lenta filter paper disks (5 mm diameter) and gelatin cryostructurated matrices disks

(4 mm dia and 2 mm high). As test cultures the bacteria were used obtained from a cultures collection from the Department of Microbiology, Biological Faculty, Moscow State University, named after M. V. Lomonosov (*E. coli 52*, *S. aureus 144* and *M. cyaneum 98*). The experiments were performed in Petri dishes containing 20 ml of agar nutrient medium, dried for 24 hours (thickness of the medium layer 4 mm). Measurements of the zone of inhibition of growth of test cultures were performed after 16 hours of incubation. Statistically reliable results were obtained by repeating measurements of the zone of inhibition of growth three times for each series of samples.

3. Results and discussion

The determination of the chemical composition of the obtained samples was carried out by various methods of physico-chemical analysis. Comparison of the 1H NMR spectra of the starting dioxidine and its hybrid nanocomposites showed that, as part of the hybrid nanocomposites Ag-dioxidine and

Cu-dioxidine, the studied drug compound remained unchanged in chemical form (see below).

Initial dioxidine.

Spectrum ^{1}H NMR (D2O, δ, ppm, 600 Hz): 4.93-5.21 (m, 4 H, CH_2); 7.85-8.05 (m, 2 H, Ar); 8.38-8.52 (m, 2 H, Ar).

Hybrid dioxidine nanocomposites.

Spectrum ^{1}H NMR (D2O, δ, ppm, 600 Hz): 4.92-5.25 (m, 4 H, CH_2); 7.84-8.03 (m, 2 H, Ar); 8.35-8.51 (m, 2 H, Ar).

The IR spectra of the source dioxidine and the resulting hybrid composites Ag-dioxidine and Cu-dioxidine were identical (Figure 1). The UV spectra of dioxidine aqueous solutions and Ag-dioxidine and Cu-dioxidine compositions also completely coincided (λmax = 241, 259, 360 and 375 nm) and corresponded to chemically unchanged dioxidine [22, 23]. In addition, absorption spectra corresponding to dioxidine were also present in the UV spectra of aqueous extracts of samples of the Ag-dioxidine-gelatin matrix and Cu-dioxidine-gelatin matrix. Thus, on the basis of NMR, IR and UV spectroscopy data, we confirmed the presence of chemically unchanged dioxidine in the composition of hybrid nanosystems Ag-dioxidine, Cu-dioxidine, Ag-dioxidine-gelatin matrix and Cu-dioxidine-gelatin matrix.

The kinetic curve of dioxidine release from the gelatin matrix is shown in Figure 2. The release of dioxidine from the biopolymer matrix occurs within 60 minutes, that is, relatively quickly. To achieve a more prolonged effect of controlled isolation of the drug component from the carrier matrix, it is necessary to change the structure of the matrix by varying the conditions of its synthesis.

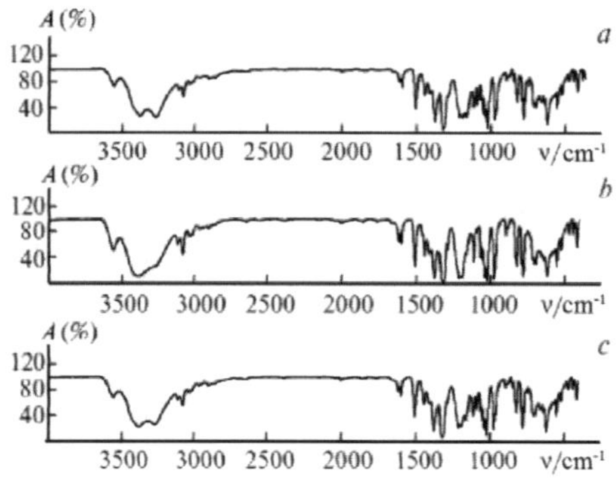

Figure 1. IR spectra of pharmacopoeial dioxidine (a) and hybrid compositions of Ag-dioxidine (b) and Cu-dioxidine (c).

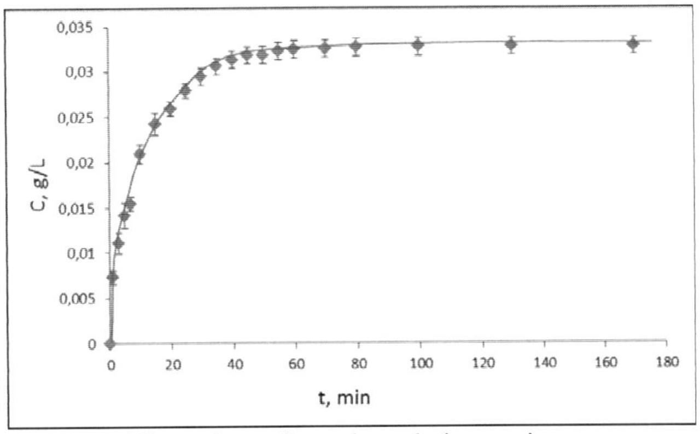

Figure 2. Kinetic curve of dioxidine release from the gelatin matrix

The TEM data indicate that Ag-dioxidine and Cu-dioxidine hybrid nanocomposites consist of organic particles 50–350 nm in size, inside of which silver particles of 2–30 nm in size or copper 10–40 nm in size are enclosed. Examples of micrographs of the Ag-dioxidine and Cu-dioxidine systems are shown in Figure 3.

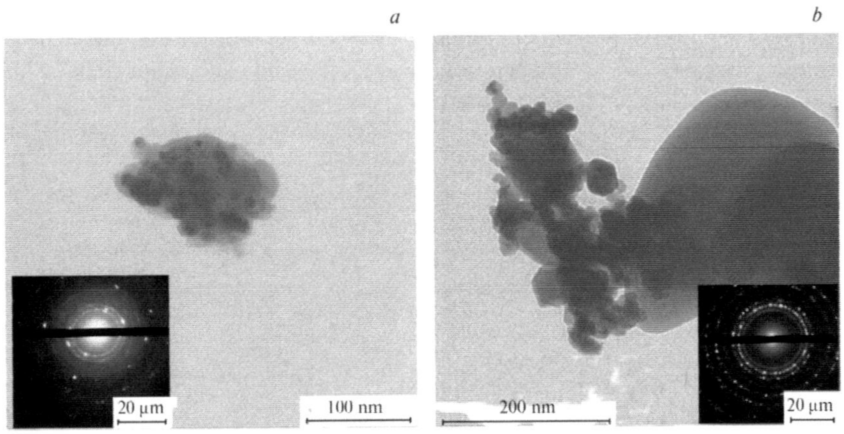

Figure 3. TEM micrographs of hybrid composites Ag-dioxidine (a) and Cu-dioxidine (b).

The specific surface values and the average particle size calculated for these values for Ag-dioxidine nanocomposites were 31 $m^2 \times g^{-1}$ and 130 nm, and for Cu-dioxidine nanocomposites - 24 $m^2 \times g^{-1}$ and 166 nm, respectively. Thus, the average particle size obtained by the method of low-temperature adsorption of argon is consistent with the data obtained by the TEM method.

SEM photomicrographs (Figure 4) indicate that gelatin cryostructurated matrices are wide porous matrices with a pore size of tens of microns. The inclusion of dioxidine and metal nanoparticles does not significantly affect their size.

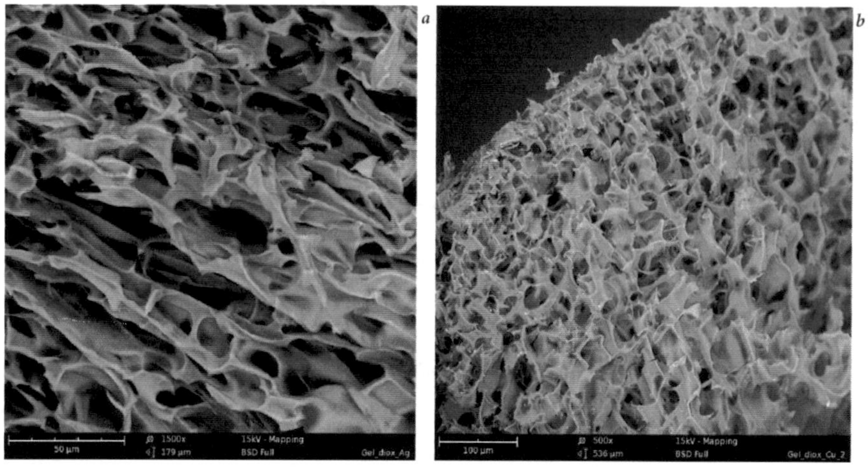

Figure 4. SEM micrographs of hybrid composites Ag-dioxidine (a) and Cu-dioxidine (b) included in gelatin cryostructurated matrices.

The antibacterial activity of various hybrid dioxidine-based nanosystems has been determined for bacterial strains *E. coli 52*, *S. aureus 144* and *M. cyaneum 98*. The data obtained are summarized in Tables 1 and 2. The antibacterial activity of hybrid nanocomposites Ag-dioxidine and Cu-dioxidine was higher than that of individual nanoparticles of metals and dioxidine. The same trend was observed when testing nanocomposites included in gelatin cryostructured matrices. Thus, a synergistic increase in the antibacterial activity of the resulting hybrid systems with respect to a number of bacterial cells was demonstrated in comparison with the effect of individual medicinal components.

Table 1. Growth Inhibition Zone for *E. coli 52*, *S. aureus 144* and *M. cyaneum 98* around filter paper discs soaked with sample solutions.

Bacterial strain	Growth Inhibition Zone, Ag, mm	Growth Inhibition Zone, Cu, mm	Growth Inhibition Zone, dioxidine, mm	Growth Inhibition Zone, Ag/dioxidine, mm	Growth Inhibition Zone, Cu/dioxidine, mm
E. coli 52	0	0	26 ± 1.2	36.7 ± 0.6	32.1 ± 1.2
S. aureus 144	0	0	30.3 ± 0.6	37.4 ± 0.6	33.0 ± 1.2
M. cyaneum 98	0	0	26.1 ± 0.8	36.3 ± 0.6	31.5 ± 0.8

[a]Dioxidine - 0.3 wt%, Ag – 0.0015 wt%, Cu — 0.006 wt%

Table 2. The growth inhibition zone of E. coli 52 around dioxidine, Cu and Ag nanoparticles, and dioxidine hybrid nanocomposites with metal nanoparticles included in gelatinous cryostructurated matrices.

Growth Inhibition Zone Cu, mm	Growth Inhibition Zone Ag, mm	Growth Inhibition Zone Dioxidine, mm	Growth Inhibition Zone Cu/Dioxidine, mm	Growth Inhibition Zone Ag/Dioxidine, mm
0	4.0 ± 1.2	35.0 ± 0.6	37.3 ± 1.2	38.1 ± 1.2

4. Conclusion

In the research we obtained hybrid nanocomposites consisting of particles of dioxidine 50–350 nm in size, inside of which silver nanoparticles with a size of 2–30 nm or copper 10–40 nm in size are enclosed. These nanocomposites were incorporated into gelatinous cryostructurated matrices, which allowed us to achieve a gradual release of the drug within 60 minutes. Tests of antimicrobial activity against strains of E. *coli 52, S. aureus 144, M. cyaneum 98* showed that the compositions have a higher antibacterial activity against cells compared to the original dioxidine and metal nanoparticles.

The resulting bioactive nanomaterials exhibit significantly greater antibacterial activity against pathogenic strains (S. *aureus 144, M.* cyaneum 98), compared with E. *coli* cells that are part of the natural bacterial composition of the human body. This eliminates the possible increased toxicity of the resulting hybrid nanocomposites and biopolymer nanomaterials based on them and opens up prospects for the creation of new dosage forms for targeted delivery and controlled release of drug components.

Acknowledgments
The team of authors is grateful to the Russian Science Foundation for funding this work within the framework of the grant No. 16-13-10365.

References

[1] Savjani K T, Gajjar A K and Savjani J K 2012 Drug Solubility: Importance and Enhancement Techniques *ISRN Pharmaceutics* **2012** 1-10.
[2] Gupta A, Saleh N M, Das R, Landis R F, Bigdeli A, Motamedchaboki K, Campos A R, Pomeroy K, Mahmoudi M and Rotello V M 2017 Synergistic antimicrobial therapy using nanoparticles and antibiotics for the treatment of multidrug-resistant bacterial infection *Nano Futures* **1** 015004
[3] Li P, Li J, Wu C and Wu Q 2005 Synergistic antibacterial effects of β-lactam antibiotic combined with silver nanoparticles *Nanotechnol.* **16** 1912-17
[4] Dong X, Awak M Al, Tomlinson N, Tang Y, Sun Y-P and Yang L 2017 Antibacterial effects of carbon dots in combination with other antimicrobial reagents *PLoS One* **12** e0185324
[5] Rao S and Prestidge C A 2016 Polymer-lipid hybrid systems: merging the benefits of polymeric and lipid-based nanocarriers to improve oral drug delivery *Expert Opinion on Drug Delivery* **13** 691-707
[6] Mitran R A, Băjenaru L and Moisescu M G 2018 Controlling drug release from mesoporous silica through an amorphous, nanoconfined 1-tetradecanol layer *European J. Pharm. Biopharm.* **127** 318-325
[7] Ding C and Li Z 2017 A review of drug release mechanisms from nanocarrier systems *Mater. Sci. Eng.* C **76** 1440-53
[8] Sami A J, Khalid M, Jamil T, Aftab S, Mangat S A, Shakoori A R and Iqbal S 2018 Formulation of novel chitosan guargum based hydrogels for sustained drug release of paracetamol *Int. J. Biological Macromol.* **108** 324-332
[9] Ito T, Takami T, Uchida Y and Murakami Y 2018 Chitosan gel sheet containing drug carriers with controllable drug-release properties *Coll. Surf. B: Biointerfaces* **163** 257-265
[10] Hasnain M S, Nayak A K, Singh M., Tabish M and Ara J 2016 Alginate-based bipolymeric-nanobioceramic composite matrices for sustained drug release *Int. J. Biological Macromolecules* **83** 71–77
[11] Jalababu R, Veni S, Reddy K V N S 2018 Synthesis and characterization of dual responsive sodium alginate-g-acryloyl phenylalanine-poly N-isopropyl acrylamide smart hydrogels for the controlled release of anticancer drug *J. Drug Deliv. Sci. Technol.* **44** 190-204
[12] Bini R A, Silva M F, Varanda L C, da Silva M A and Dreiss C A 2017 Soft nanocomposites of gelatin and poly(3-hydroxybutyrate) nanoparticles for dual drug release *Coll. Surf. B: Biointerfaces* **157** 191-198

[13] Talebian A and Mansourian A 2017 Release of Vancomycin from electrospun gelatin/chitosan nanofibers *Mater. Today: Proceed.* **4** 7065-7069
[14] Saykova S.V., Vorobyev S.A., Nikolaev. R. B and Mikhlin Yu. L. 2010 Determining the conditions for the formation of copper nanoparticles during the reduction of Cu2 + ions with hydrazine hydrate solutions Journal of General Chemistry 80 952-957
[15] Lozinsky V I, Kulakova V K, Ivanov R V, Petrenko A Yu, Rogulska O Yu and Petrenko Yu A 2018 Cryostructuring of polymer systems. 47. Preparation of wide porous gelatin-based cryostructurates in sterilizing organic media and assessment of the suitability of thus formed matrices as spongy scaffolds for 3D cell culturing *E-Polymers* **18** 172-176
[16] Rodionov I A, Grinberg N V, Burova T V, Grinberg V Ya, Shabatina T I and Lozinsky V I Cryostructuring of polymer systems. 44. Freeze-dried and then chemically cross-linked wide porous cryostructurates based on serum albumin *E-Polymers* **17** 263-274
[17] Vernaya O.I., Shabatin V.P., Semenov A.M. and Shabatina T.I. 2016 Production of cryochemically modified ultrafine dioxidine powder and determination of its antibacterial activity *MSU Bulletin Ser. 2.* Chemistry **57** 315-320
[18] Vernaya OI, Shabatin V.P., Shabatina T.I., Khvatov D.I., Semenov A.M., Yudina T.P. and Danilov V.S. 2017 Cryochemical modification of dioxidine, its activity and toxicity. Journal of Physical Chemistry 91 230-233
[19] Vernaya OI, Khvatov D.I., Nuzhdina A.V., Fedorov V.V., Shabatin V.P., Semenov A.M. and Shabatina T.I. 2017 Hybrid nanocomposites Cu / dioxidine: cryochemical synthesis and antibacterial activity *MSU Bulletin Ser. 2.* Chemistry **58** 271-272
[20] Vernaya O., Shabatin V.P., Semenov A.M. and Shabatina T.I. 2017 Cryochemical synthesis and antibacterial activity of silver nanocomposites with dioxidine *MSU Bulletin Ser.2.* Chemistry **57** 388-391
[21] Onishchenko G.G. 2004 Determination of the sensitivity of microorganisms to antibacterial drugs. Guidelines of MUK 4.2.1890-04 Moscow 40 p
[22] Vernaya O I, Shabatin V P, Semenov A M and Shabatina T I 2016 Obtaining ultradispersed dioxidine powder modified via cryochemical synthesis and determining its antibacterial activity Moscow University Chemistry Bulletin **71** 295-298
[23] Vernaya O I, Shabatin V P, Shabatina T I, Khvatov D I, Semenov A M, Yudina T P and
[24] Danilov V S 2017 Cryochemical Modification, Activity, and Toxicity of Dioxidine *Russ. J. Phys. Chem.* A **91** 229-232

IOP Conf. Series: Materials Science and Engineering **683** (2019) 012085 doi:10.1088/1757-899X/683/1/012085

Simulation of particle size distribution of mechanochemically activated charges of Al-SiC composite materials

S N Sergeenko

Federal State Budget Educational Institution of Higher Education Platov South-Russian State Polytechnic University (NPI), Novocherkassk, Russia

E-mail: sergeenko@gmail.com

Annotation. In this work, 3D Spline models of the influence of silicon carbide content on the particle size distribution of the powder mechanochemical activated mixture Al-SiC were built. The formation of agglomerates in the process of mechanochemical activation (MCA) of a multicomponent Al-SiC mixture in a liquid medium of a saturated aqueous solution of boric acid has been established. The increase in the content of SiC in the charge increases the degree of agglomeration of the composite particles of the powder mixture in the process of MCA. Subsequent manual processing leads to the grinding of agglomerates and a change in the distribution of particle sizes from bimodal to unimodal distribution. The parameters of the constructed nonlinear model of the particle size distribution of mechanochemically activated Al – SiC powder blends (5–20 wt.%) Are determined. At the optimum content of silicon carbide (10% wt.), The relationship between the extreme values of the model parameters and the maximum hardness of the hot-compacted composite Al-SiC powder was established.

1. Introduction

Promising technologies for the production of powder and cast composite materials Al-SiC provide a solution to the actual problem of manufacturing valve seats of internal combustion engines [1]. An increase in the content (9% by weight) and a decrease in the size (115 μm) of SiC particles leads to an increase in the tensile strength values in tensile tests of the Al-SiC composite material obtained by injection molding technology [2]. Studies of the effect of the content (5-10% wt.) and the size of SiC particles on the mechanical properties of the powder composite hot-compacted material showed that a decrease in the particle size of SiC leads to an increase in hardness and compressive strength [3].

The SRSPU (NPI) has developed a technology for producing hot-compacted powder composite materials Al-SiC (0-20% wt.) based on mechanochemically activated charges in a liquid medium of a saturated aqueous solution of boric acid. The increased efficiency of mechanochemical activation has been shown, as compared with the mixing technology when producing hot-composit Al-SiC composite materials [4]. As a result of the research, the hereditary effect of the dispersion-agglomeration processes during the mechano-mechanical activation of the charge on the patterns of formation of hot-compacted low-porous aluminum-based powder materials was established [5-17].

Mechanochemical activation of powder blends in high-energy mills is accompanied by the formation of a new surface, an increase in the concentration of defects and chemical transformations. The accumulation of energy in a solid leads to a decrease in the activation energy of the subsequent chemical transformation. The speed of a solid-phase reaction depends on the number and area of contacts between the interacting particles, and is also determined by the size distribution of the powder batch particles [18-21].

2. Purpose of the work

Construction of a particle size distribution model of mechanochemically activated Al-SiC powder blends.

3. Research methodology

Mechanochemical activation (MCA) of the powder mixture was carried out for 3.24 ks in a SAND – 1 ball planetary mill at a rotor speed of 4.8 s^{-1} and a mass ratio of balls (d_b = 10 mm) and charge equal to 20. To form a protective layer on the surface of the powder particles, which prevents oxidation in the process of short-term heating of the molding in an air atmosphere, activated in a liquid medium a saturated aqueous solution of boric acid (20% of the mass of the charge) [4,22]. Powders of aluminum PA-4 and silicon carbide (black) SiC (0–20 wt.%) with irregular fragmentation of particles were used as starting materials [4]. The granulometric composition of the PA-4 powder (Figure 1) was investigated on a Microtrac Bluewave S3500 laser analyzer.

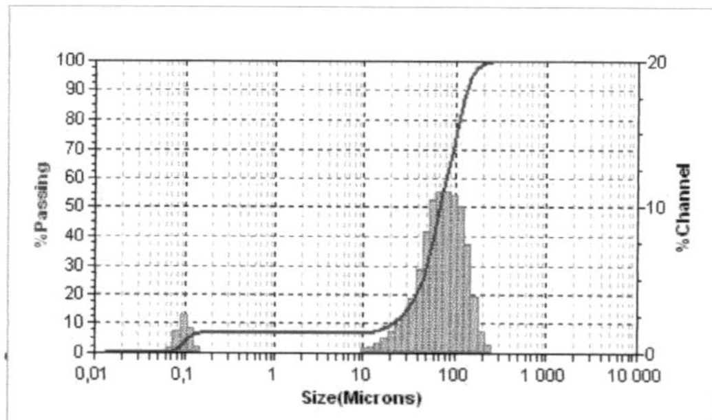

Figure 1. The particle size of the powder PA-4

The elemental analysis of the material of the mechanochemically activated charge particles was carried out using the EDAX Genesis 2000 XMS 30 microanalysis system coupled with the SEM FEI Quanta 200 (Collective Use Center "Nanotechnologies" of the SRSPU (NPI).

To assess the degree of agglomeration of the charge in the MCA process, the average particle size of the charge d0 was calculated after the mechanochemical activation of the Al-SiC powder batch and the subsequent manual pestle processing in the mortar d1 based on the results of the sieve analysis of the particle size distribution of the charge [4]. The construction of 3D Spline models of particle size distribution of the powder mixture was carried out in Statistica.

4. Results of experimental studies

Analysis of the 3D Spline model of the influence of the silicon carbide content C_{SiC} on the particle size distribution of the powder mechanochemical activated mixture Al-SiC (Figure 2) showed that agglomerates (d_0> d_1) are formed during the MCA multicomponent mixture Al-SiC. Elemental analysis (Figure 3) of the activated Al-SiC mixture (10% wt.) After manual processing showed that composite particles containing Al, C, O, Si are formed during MCA. The formation of agglomerates is accompanied by a continuous-sequential process of their dispersion-agglomeration-dispersion - ..., leading to a change in the average size of the powder particles d0. Agglomeration of composite charge particles can be represented as a process of formation of compounds by topochemical reactions proceeding along the front of particle interaction, setting of particles on the contact surfaces, caricaturing the material of soft particles with solid particles and cladding of solid particles with a layer of soft particle material [4].

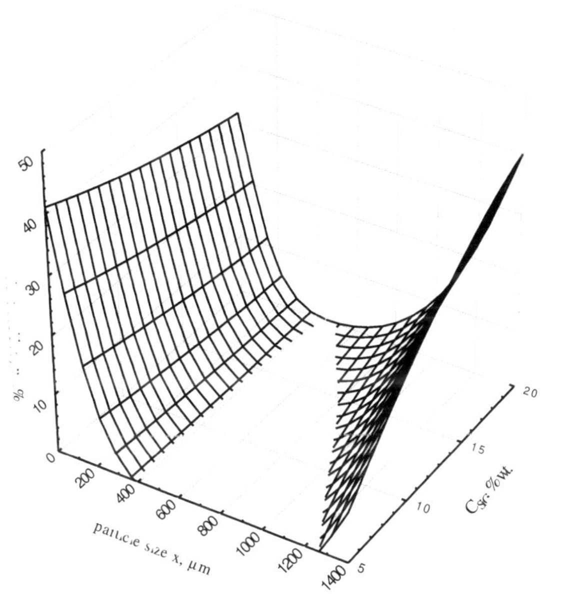

a)

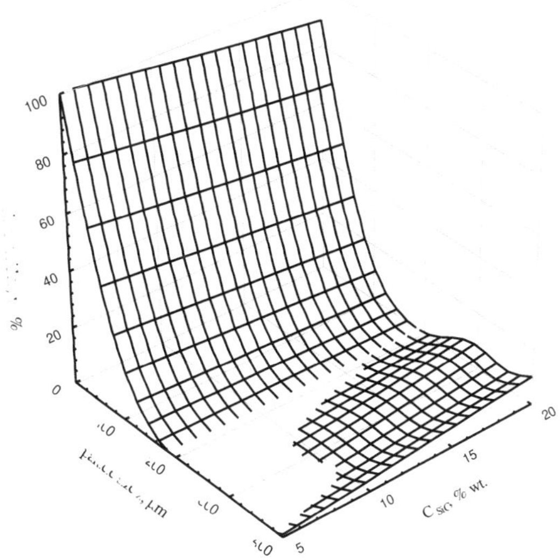

b)

Figure 2. 3D Spline model of the effect of SiC content on the particle size distribution of the powder mechanochemical activated mixture Al-SiC
a) after MCA; b) after the MCA and subsequent manual processing in a mortar

Figure 3. Elemental analysis of composite particles of mechanochemically activated mixture Al-SiC (10% wt.)

The increase in the content of SiC in the charge increases the proportion of agglomerated composite particles of the powder mixture in the process of MCA. Subsequent manual processing leads to the grinding of agglomerates and a change in the distribution of particle sizes from bimodal to unimodal distribution. The single-modal size distribution of agglomerated Al-SiC particles can be described by the Rosin-Rammler equation in a double logarithmic scale, reduced to a linear form [4, 23] and taking into account the passage values (p) of the powder mixture with dimensions greater than X

$$\ln\left(\ln\left(\frac{1}{p}\right)\right) = \ln\alpha + \beta\cdot\ln X\,, \quad (1)$$

$$\boldsymbol{y=a+b\cdot x}, \quad (2)$$

where

$$y = \ln\left(\ln\left(\frac{1}{p}\right)\right),\ x = \ln X\,,\ a = \ln\alpha\,,\ b = \beta\,,$$

α, β – parameters of the Rosin-Rammler equation

$$F(X) = \alpha\cdot\beta\cdot X^{\beta-1}\cdot exp(-\alpha\cdot X^{\beta}), \quad (3)$$

The results of studies of the mechanochemical activation of multicomponent mixtures conducted at the SRSPU (NPI) showed that dependence $y(x)$ can be described by a logistic function [24]

$$y = a + b/\left(1+\left(\frac{x}{c}\right)^{d}\right). \quad (4)$$

The values of the (a, b) parameters of the linear and (a, b, c, d) nonlinear model (4) were calculated using the standard Table Curve 2D package. The assessment of the degree of compliance of the constructed models with the results of experimental studies was assessed using the coefficient of determination r^2 (Table 1.2).

Analysis of the calculated values of the coefficient of determination r^2 showed an increase in the degree of compliance with the results of experimental studies in the transition from linear (2) to nonlinear logistic (4) models $y(x)$ for the entire range of C_{SiC} values studied. Linear models (2) cannot be used to describe the particle size distribution of mechanochemically activated powder blends of Al-SiC with

Table 1. The values of the parameters of the linear (2) and non-linear (4) models for the charge after the MCA.

C_{SiC}, % wt.	Values of parameters of a linear equation $y = a + b \cdot x$			Values of nonlinear equation parameters $y = a + b / \left(1 + \left(\frac{x}{c}\right)^d\right)$				
	a	b	r^2	a	b	c	d	r^2
5	-2,62	0,61	0,960	2,79	-9,575	2,714	1,878	0,976
10	-1,970	0,428	0,938	-8,988	10,1	2,259	-2,985	0,985
15	-1,804	0,317	0,969	-3,838	4,92	2,794	-1,804	0,983
20	-1,244	0,224	0,938	-2,509	4,156	3,828	-1,162	0,943

Table 2. The values of the parameters of the linear (2) and non-linear (4) models for the charge after the MCA and the subsequent manual processing in the mortar.

C_{SiC}, % wt.	Values of parameters of a linear equation $y = a + b \cdot x$			Values of nonlinear equation parameters $y = a + b / \left(1 + \left(\frac{x}{c}\right)^d\right)$				
	a	b	r^2	a	b	c	d	r^2
5	-2,229	0,695	0,866	-1,59	3,748	4,088	-6,94	0,976
10	-3,607	0,905	0,849	0,039	103,43	11,51	-5,877	0,891
15	-3,203	0,778	0,581	0,385	18,135	6,119	-38,56	0,917
20	-2,699	0,624	0,524	0,188	10,52	6,0	-47,11	0,943

an increased SiC content (15-20% wt.). Dependences (5) of the effect of C_{SiC} on non-linear parameter values

$$a, b, c, d = f(C_{SiC}) \quad (5)$$

models (4) $y(x)$ are extreme. It has been established that with an optimal SiC content of 10% by weight, ensuring the production of hot-composite Al-SiC composite material with a maximum hardness (371 HV) [4], extreme values of parameters a, b, c, d are observed (table 1.2) for a nonlinear particle size distribution model powder mechanochemically activated charges.

5. Conclusions

Based on the studies, a nonlinear model of the particle size distribution of mechanochemically activated Al – SiC powder blends was constructed. The hereditary effect of mechanochemical activation on the formation of the mechanical properties of the hot-compacted Al-SiC composite material has been established.

References

[1] Patel N S, Patel A D and Ranjan R K 2018 Studies on Properties of Al–SiC Metal Matrix Composite Material for Making IC Engine Valves *Proceedings of the World Congress on Engineering* 2018 **II** WCE July 4-6 2018 London U K www.iaeng.org/publication/.../WCE2018_pp667-70.pdf

[2] Youssef Y and El-Sayed M 2016 Effect of Reinforcement Particle Size and Weight Fraction on the Mechanical Properties of SiC Particle Reinforced Al Metal Matrix Composites / *International Review of Mechanical Engineering* **10** No 4 pp 261-265

[3] El-Kady O and Fathy A 2014 Effect of SiC particle size on the physical and mechanical properties of extruded Al matrix nanocomposites *Materials & Design* **54** pp 348-53

[4] Sergeenko S N and, Alabid N S 2016 Hot-deformed powder materials based on mechanochemically activated charges Al – SiC *Tsvetnye Metally* No 9 (885) pp 68-77

[5] Slabkii D V and Sergeenko S N 2016 Hot-deformed Al–Ni powder materials based on alloy D-16 mechanically-activated turnings. *Metallurgist* **59** 11 pp 1228-33

[6] Dyuzhechkin M K, Sergeenko S N and Popov Y V 2016 Features of structure and property formation for hot-deformed materials of the Al–Si and Al–Si–C systems based on mechanochemically activated charges *Metallurgist* **59** 9-10 pp 835-42

[7] Sergeenko S N, Juraev F T, Al Hilfi A X and Postnikov A A 2016 Features of mechanochemical activation and compaction of the powder mixture Al-Si-Ni-Fe-C and Al-Cu-Fe *Fizika i khimiya obrabotki materialov* [Physics and chemistry of materials treatment] **4** pp 63-72

[8] Fedoseeva M A and Sergeenko S N 2015 Structure and properties of Al–FeTi powder material based on mechanochemically activated alloy D-16 turnings *Metallurgist* **59** 5-6 pp 535-9

[9] Dorofeev Yu G, Bezborodov E N and Sergeenko S N 2003 Special features of formation of compacted material from mechanochemically activated fining of aluminum alloy D16 *Metal Science and Heat Treatment* **45** 1-2 pp 73-5

[10] Dorofeev Yu G, Bezborodov E N and Sergeenko S N 2003 Hot-strained powder materials on the base of mechanically alloyed Al-C mixture *Fizika i khimiya obrabotki materialov* [Physics and chemistry of materials treatment] **3** pp 64-72

[11] Dorofeev Yu G, Bezborodov E N and Sergeenko S N 2003 Hot wrought powders on the base of D16 chips mechanochemically activated *Tsvetnye Metally* **1** pp 81-5

[12] Dorofeev Yu G, Bezborodov E N and Sergeenko S N 2002 Development of technology for producing hot-deformed powder material based on mechanochemically activated chips of D-16 aluminum alloy *Materialovedeniye* **9** pp 40-5

[13] Dorofeev Yu G, Bezborodov E N and Sergeenko S N Kinetics 2002 of mechano-chemical activation of aluminium-based powder charge in saturated solution of orthoboric acid *Fizika i khimiya obrabotki materialov* [Physics and chemistry of materials treatment] **3** pp 51-4

[14] Dorofeev Yu G, Bezborodov E N and Sergeenko S N 2002 Effect of kinetics of mechanochemical activation of al powders on the processes of finish hot compacting *Fizika i khimiya obrabotki materialov* [Physics and chemistry of materials treatment] **4** pp 79-81

[15] Dorofeyev Yu G, Bezborodov E N and Sergeyenko S N 2002 Special features of formation of compacted material from mechanochemically activated chips of aluminum alloy D-16 *Tekhnologiya legkikh splavov* [The Technology of Light Alloys] **2** pp 23-8

[16] Dorofeyev Yu G, Bezborodov E N and Sergeyenko S N 2002 Features of the formation on the dynamic hot compacted material from mechanochemically activated chips of aluminum alloy D-16 *Tekhnologiya legkikh splavov* [The Technology of Light Alloys] **3** pp 37-41

[17] Dorofeyev Yu G, Bezborodov Ye N and Sergeyenko S N 2001 Features of high-density hot-deformed material based on "chip" powder of aluminum alloy D-16 *Izvestiya Vysshikh Uchebnykh Zavedenij. Tsvetnaya Metallurgiya* **10** pp 28-31

[18] Boldyrev V V 2006 Mechanochemistry and mechanical activation of solids Russian Chemical Reviews **75** 3 pp 203-16

[19] Sergeyenko S N 2012 Technologies for producing the powder materials on the base of mechanically alloyed mixture (review) *Tekhnologiya metallov* [Metals technology] **1** pp 46-56

[20] Sergeyenko S N 2012 Technologies for producing the powder materials on the base of mechanically alloyed mixture (review) *Tekhnologiya metallov* [Metals technology] **5** pp 46-55

[21] Sergeyenko S N Technologies for producing the powder materials on the base of mechanically alloyed mixture (review)]. *Tekhnologiya metallov* [Metals technology] 2012 **6** pp 47-56

IOP Conf. Series: Materials Science and Engineering **683** (2019) 012085 doi:10.1088/1757-899X/683/1/012085

[22] Kotiyeva L U, Galetova N L, Pavlova V V 1986 Adding strengthening phases in aluminum powder *Tsvetnaya metallurgiya* [Nonferrous metallurgy] **4**, pp. 35-7
[23] Khodakov G S 1972 *Physics of grinding* Moscow Nauka 307 p
[24] Dorofeyev Yu G, Sergeyenko S N and Kolomiyets R V 2007 Effect of the composition of a powder charges Ni-Fe-NaCl on the kinetics mechanical activation in liquid media *Fizika i khimiya obrabotki materialov* [Physics and chemistry of materials treatment] **4** pp 57-61

IOP Conf. Series: Materials Science and Engineering **683** (2019) 012086 doi:10.1088/1757-899X/683/1/012086

The effect of curing temperature on phase separation, morphology, fracture toughness and impact resistance of epoxy-polysulfone blends and fiber reinforced plastics based on them

V I Solodilov[1, 2], **A V Shapagin**[3], **R A Korokhin**[1] and **Yu A Gorbatkina**[1]

[1] Semenov Institute of Chemical Physics Russian Academy of Sciences, Russia

[2] Bauman Moscow State Technical University, Russia

[3] A.N. Frumkin Institute of Physical Chemistry and Electrochemistry Russian Academy of Sciences, Russia

e-mail: vital-yo@yandex.ru

Abstract. The influence of the curing temperature of epoxy - polysulfone compositions on crack and impact resistance of matrices and unidirectional reinforced plastics was studied. It was shown that the structure of cured material depends on the content of PSK-1 and the curing temperature. Stable effect of increasing fracture toughness, as a rule, is observed for matrices with an extended structure of thermoplastic, which is typical for the phase structures of the type "interpenetrating phases". Energy of destruction of fiber reinforced plastics is determined by the fracture toughness of the matrix and its resistance to impact, which depend on the type of structure formed in the curing process.

1. Introduction

Majority of composite materials based on epoxy matrices does not meet the increasing requirements for crack, impact and chemical resistance. At present time complex polymer composition, including modifying additives, are used to create composite materials with increased thermal and shock resistance [1-11]. Active diluents, elastomers, particles of various shapes, sizes and nature, thermoplastic polymers are used as modifying additives. The most effective as modifiers proved rigid-chain thermoplastics (polyetherimides, polyaryletherketones, polyphenylenoxide, polysulfone) [12-16].With such types of modification it is possible to increase the physical, mechanical and operational properties of reinforced plastics without reduction of heat resistance. The effect of such modification largely depends on the structure of the cured complex compositions [12-16]. At the same time, the design of optimal phase structures remains a very urgent task. To solve this problem, complex studies of phase separation of polymer mixtures, morphology and physical and mechanical properties of cured systems, as well as the establishment of the relationship of strength characteristics and phase structure are required.

In this work the physical and mechanical properties of epoxy-polysulfone matrices and unidirectional winding glass, carbon and organoplastics are investigated and possible mechanisms of their destruction are discussed.

2. Experimental

The study was performed based on epoxy oligomer ED-20 modified with polysulfone PSK-1 (Mn = 35000). The content of thermoplastic phase varied from 5 to 20 wt. % (from the mass of the oligomer). Triethanolaminatonate (TEAT) (10 wt. %), 4,4`-diaminodiphenylsulfone (DDS) (30 wt. %), iso-methyltetrahydrophthalic anhydride (MTHPA) (90 wt. %) + methylimidazole accelerator (0.5 wt. %) were used as curing agents. The curing proramms: TEAT – 8 h at 160 °C; DDS – 6 h at 180 °C; MTHPA – 2 h at 90 °C, then 12 h at 120°C.

The fracture toughness of the matrix G_{IR}^{M} was determined by splitting a double-cantilever beam [17]. The energy of destruction of epoxy-polysulfone matrices was determined at a three-point bending mode using a spring impact machine at a loading rate of 4 m/s [17]. The morphology of the fracture surface after splitting of the samples was investigated by scanning electron microscope (PSEM-500). To confirm the existence of continuous phases of polysulfone, cured samples were soaked in dichloromethane for 7 days. After being in the solvent we controlled their appearance.

RVMPN10-400 glass fibers (RPA “Stekloplastik”), Rusar-S aramid fibers (PLC “Khimvolokno”), and carbon Sigrafil C Type fibers (SGL Carbon Group) were used for composite manufacture. Heat treatment of composites were performed in the same modes as the matrix.

For winding composites resins containing a large (10 - 20 wt. %) amount of polysulfone were used We developed an original technological scheme for composite production, which combines the methods of thermoplastics and thermosets processing. Composites based on the non-modified epoxy matrix were obtained by common method [18]. We were able to produce materials with the high content of fiber (60 - 70 vol %) and low pore content (1-3 vol. %) when the viscosity of the binder does not exceed 6 Pa·s.

On the wound rings and segment cuts, the energy of destruction (E_F^{CM}) at the shear quasi-static and impact loading was determined [17], as well as interlaminar fracture toughness (G_{IR}^{CM}) [19].

3. Results and Discussion

The modification of epoxy binder with polysulfone can be very effective (Fig. 1) for reduction of crack resistance. A significant increase in G_{IR}^{M} begins only at a concentration of PSK-1 above 10 wt. % (Fig. 1a). The increase of fracture toughness of epoxy matrices begins with the contents of the PSK-1 greater than 5 wt. %. When 15 – 20 wt. % of PSK -1 was applied the G_{IR}^{M} values increased from 3 to 10 times depends on the type of the hardener. The increase in the energy of destruction at impact of matrices cured with TEAT was 20 %; for DDS cured sample – 90 %; for MTHPA cured one – 58 %.

Maintenance of the effect of increasing crack and impact resistance of epoxy-polysulfone matrices in wound reinforced plastics is shown in Fig. 2. The total energy of destruction of the E_F^{CM} was largely determined by the impact resistance of the matrix. It should be noted that E_F^{CM} glass fiber reinforced plastics grows faster than carbon fiber composites. Given that the specimens of composites were destructed in the same way (from the shear stresses), it can be assumed that the fracture energy of reinforced plastics significantly depends on the resistance of the matrix to impact and strength at the polymer-fiber interface. Any modification of polymer matrices affects not only their bulk properties, but also the surface ones. It was previously shown [20] that the adhesion properties of modifiers can be much lower than those of the initial matrices.

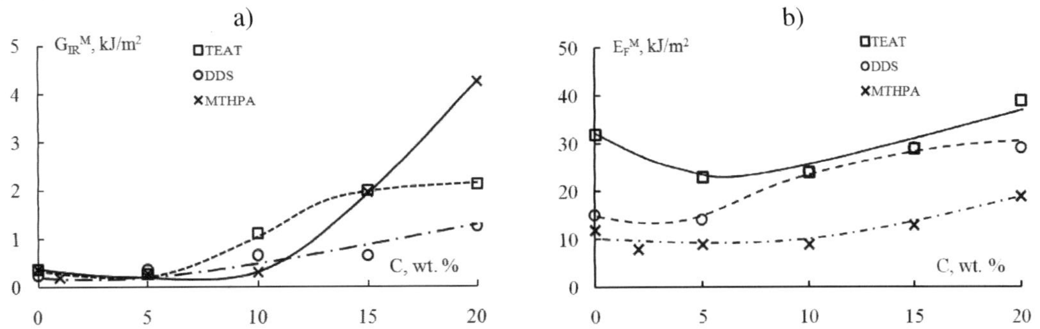

Figure1. Fracture toughness G_{IR}^{M} (a) and total fracture energy at impact E_F^{M} (b) of cured epoxy-polysulfone matrices. The type of hardener is shown on the diagram

However, the adhesive strength in the polymer-fiber system can vary along curves with the maximum or increase monotonically in case of epoxy resins modified with thermoplastics. The adhesive strength of the modified systems was 10 – 30% higher than that of the thermoplastic-free ones.

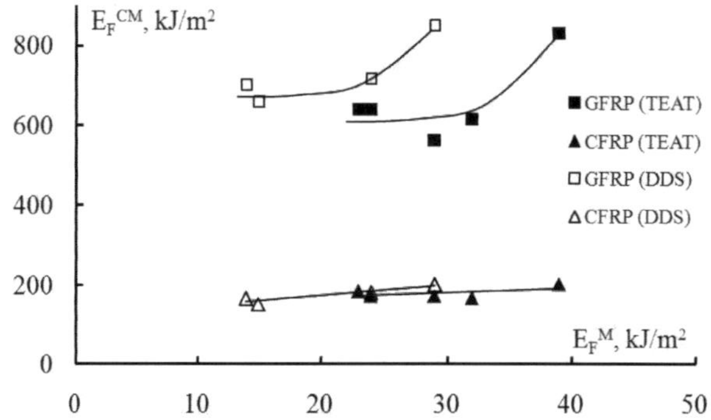

Figure2. Comparison of the total energy of destruction of composites E_F^{CM} and full energy destruction of epoxy-polysulfone matrices E_F^{M}. The diagram shows the type of material and the hardener

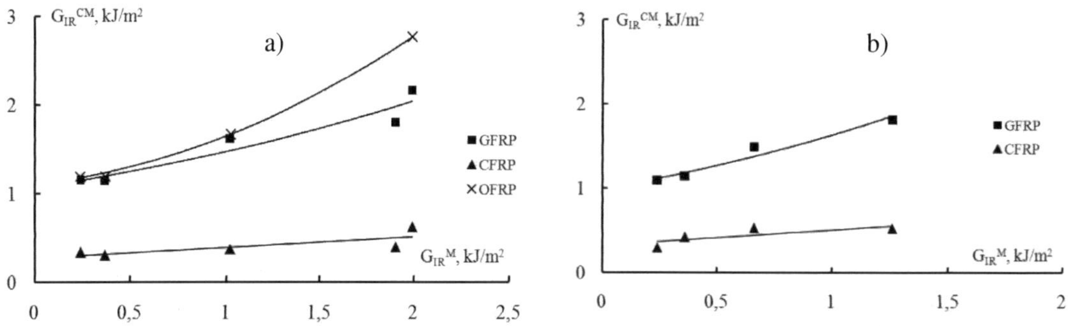

Figure3. The correlation of the fracture toughness of unidirectional composites G_{IR}^{CM} and epoxy-polysulfone matrices G_{IR}^{M}, cured with TEAT (a) or DDS (b)

Dependence of fracture toughness of reinforced plastics G_{IR}^{CM} of crack resistance of the matrix G_{IR}^{M} is presented on Fig. 3. In all cases, there is a correlation of crack resistance of the studied matrices and unidirectional composites on based on them. The growth of energy G_{IR}^{CM} for glass fiber and organic fiber reinforced plastics is faster than for carbon fiber reinforced plastics.

The points of the studied systems were plotted on the temperature-concentration field (Fig. 4). Schematically, each isotherm built curves of the process of the curing reaction. We have previously shown that heterogeneous region expanded on the scale of concentrations and temperatures with increase of conversion rate (α) [16]. It was established using microscopic studies and x-ray microanalysis that all systems were characterized with three types of structures depends on their place in one of the three areas of the phase diagram: I (○) and III (●) were the phase structures of the data type "matrix-dispersion", where the composition of the dispersed particles were close to the PSK-1 and epoxy, respectively, II (ж) – were structures of "interpenetrating phases" type.

It was found that at low concentrations of PSK-1 the system was an epoxy matrix with fine phases enriched with polysulfone according to microscopy and elemental analysis. Further increase in the concentration of thermoplastics leads to the formation of a inverted phase structure (which is an extended interpenetrating phase), and at concentrations above 15 wt. % for TEAT cured samples and 20 wt. % for DDS and MTHPA cured ones the formation of a structure where the phase enriched with polysulfone matrix and the disperse phase epoxy oligomer were observed.

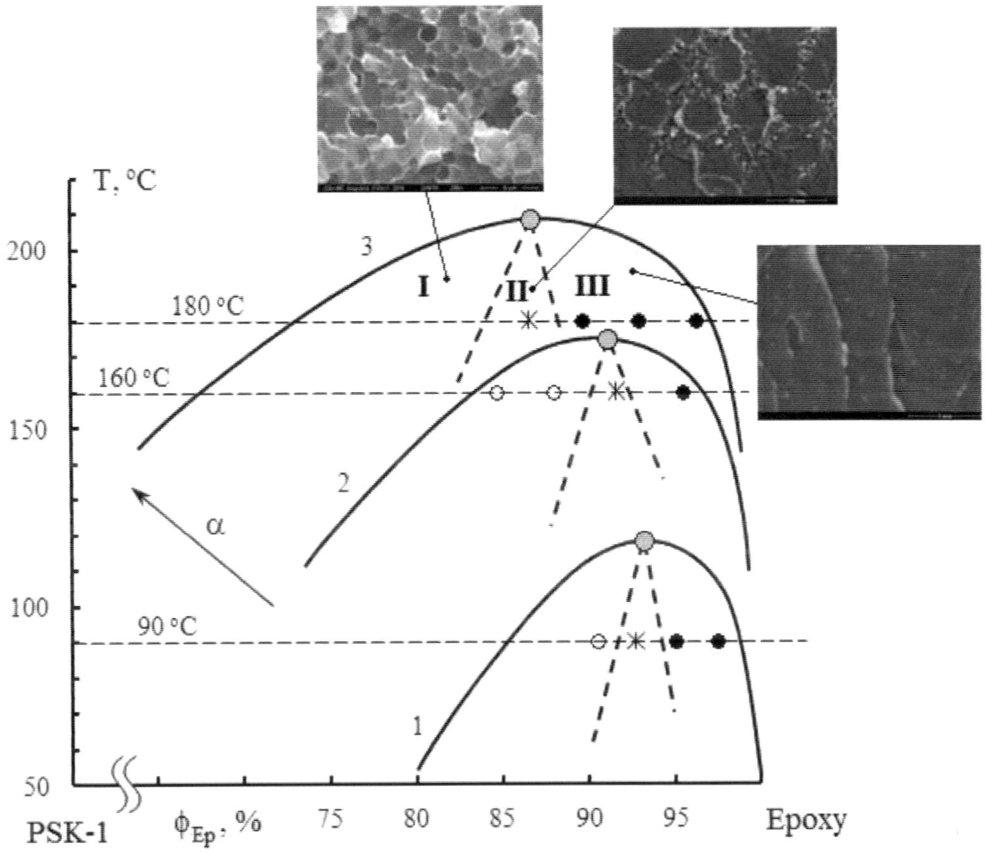

Figure4. Model of evolution of phase diagrams (1, 2, 3) of epoxy-polysulfone mixtures during curing. The points corresponding to the concentrations of the studied systems are shown on the curing isotherms (90, 160 and 180 °C). The arrow indicates the direction of displacement of the upper critical mixing point with increasing conversion rate. For further explanations see the main text.

The scanning electron microscopy and elemental analysis of the cured chips of epoxypolyurethane systems the sizes of the dispersed polysulfone and polyepoxy phases built and size distribution of particles. For all the compositions studied the particle diameter distributions are close to normal (Fig. 5). From the resulting distributions of the particles diameters determined the most frequent diameters (Tab. 1). It can be seen that for epoxy matrices cured by TEAT or DDS with the dispersed phase of PSK-1 the average size of the polysulfone particles was about 0.5 μm. In the case of the epoxy binder cured by MTHPA, the average size of the dispersed phase of PSK-1 was greater than for the above compositions (0.8 – 1 μm). For the compositions ED-20 + TEAT + 10 wt. % PSK-1, ED-20 + DDS + 20 wt. % PSK-1 and ED-20 + MTHPA + 15 wt. % PSK-1 two types of structures were observed. The first was the dispersed phase enriched with PSK-1, the second was the inverted structure, where the dispersed phase has been enriched with ED – 20 and the continuous phase was enriched with PSK-1. The size of the inverted phases was more than 100 μm. The size of the dispersed phase of PSK-1 varies from 0.5 to 1.78 μm depending on the type of hardener. In the inverted plots of the particle size of the polyepoxy varies from 1.4 to 21 μm. It should be noted that the bimodal phase distribution enriched with polyepoxy was observed for the ED-20 + MTHPA + 15 wt. % PSK-1. For this system polyepoxy particles with an average diameter of 5 μm and 21 μm were found in inverted regions. This distribution probably indicates several stages of phase separation. The characteristic size of the phase enriched with polyepoxy was in the range from 3 to 17 μm for a fully inverted structures.

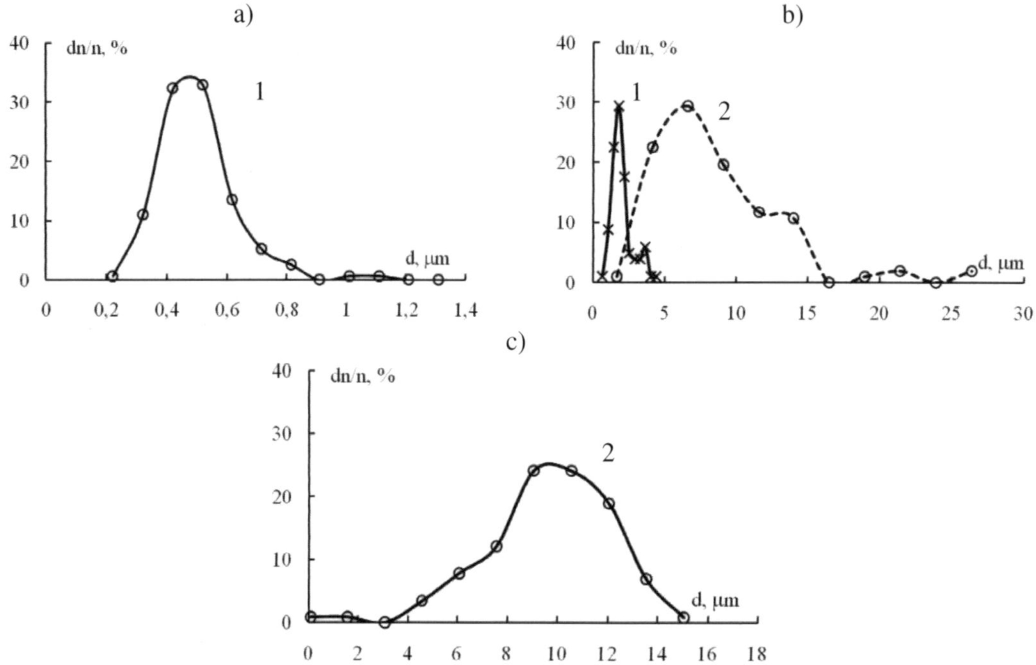

Figure5. The distribution of polysulfone (curve 1) and polyepoxy (curve 2) phase diameters for systems cured by TEAT and containing 5 wt. % (a), 10 wt. % (b) and 20 wt. % PSK-1 (c)

We should notice that a significant improvement in mechanical properties was observed for the systems with a concentration of PSK-1 resulted in inverted phase and for structures with large dispersed phases enriched with an epoxy oligomer. Thus, fine (less than 0.5 μm) phases enriched with polysulfone were structures that do not dissipate the crack energy, as the crack propagates through the epoxy matrix bypassing the dispersed phase. Macro (>10 μm) phase significantly improves the physico-mechanical characteristics of the matrices. This is mainly due to the fact that in such epoxy – polysulfone compositions the matrix is a plastic phase (enriched with thermoplastics) and consequently the crack propagation should inevitably be associated with the destruction of the plastic phase.

Table 1. Particle sizes distribution of phases enriched with polysulfone or polyepoxy.

Hardener	Concentration of PSK-1, wt. %							
	5		10		15		20	
	Phase PSK-1	Phase ED-20	Phase PSK-1	Phase ED-20	Phase PSK-1	Phase ED-20	Phase PSK-1	Phase ED-20
TEAT	0,47	-	1,78	6,58	-	17,24	-	10,55
DDS	0,38	-	0,42	-	0,56	-	0,47	1,41
MTHPA	0,81	-	1,01	-	1,10	5,0/21,10	-	2,90

To confirm the existence of a continuous phase enriched with polysulfone the cured samples were kept in dichloromethane. It was assumed that the polysulfone would dissolve and the samples would lose their initial shape. The result of the exposure of epoxy-polysulfone matrices in a solvent is shown on Fig.6. It can be seen that unmodified epoxy matrices were destroyed due to residual stresses regardless of the type of a hardener, and therefore the curing temperature.

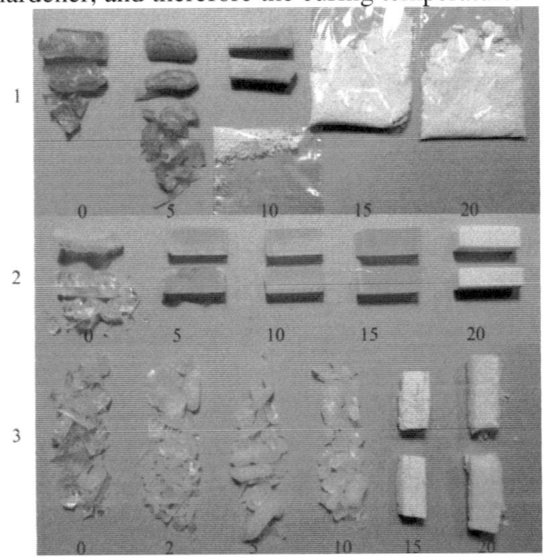

Picture 6. Appearance of the epoxypolysulfone binder, cured with TEAT (1), DDS (2) and MTHPA (3), after soaking in dichloromethane. The numbers on the figure are concentrations of PSK-1

For epoxy-polysulfone samples corresponding to region I on temperature-concentration phase diagrams (Fig. 4) were completely destroyed. The dispersed phase of polyepoxy was found on the surface of the solvent according to x-ray spectral analysis. In the case of the dispersed phase of the polysulfone (region III), the samples retain their initial shape and cavities on their surface were visible where the dispersed phase enriched with PSK-1 has been washed. The samples containing interpenetrating phases (region II) were swollen (no complete destruction was observed). The only exception was the composition of the ED-22+DDS modified with 5, 10 and 15 wt.% of PSK-1. These samples has fully retained their shape and have not been destroyed, thus showing excellent resistance to organic solvent.

4. Conclusion

We showed the effectiveness of the use of epoxy-polysulfone matrices to increase crack resistance and shear strength at low-speed impact of unidirectional winding composites based on fibers of different chemical nature. The study of the morphology of the fracture surface showed that the final form of the structure depends on the content of PSK-1 and the curing temperature. It was established that a stable effect of crack resistance increase is usually observed for matrices with extended thermoplastics structure, which is typical for phase structures of the "interpenetrating phases" type (region II in Fig. 4 The fracture energy of reinforced plastics is largely determined by the crack resistance of the matrix and its resistance to impact which depends on the type of structure formed during the curing process. With data on the evolution of phase diagrams in the curing process, Phase diagram presented allows to predict some physical and mechanical properties of the specific systems cured at specified conditions.

References

[1] Mohammadali A, Lomov S V, Gorbatikh L 2015 Morphology and fracture behavior of POM modified epoxy matrices and their carbon fiber composites *Composites Science and Technology* **110** 8–16.

[2] Kamar N T, Drzal L T 2016 Micron and Nanostructured Rubber Toughened Epoxy: A Direct Comparison of Mechanical, Thermomechanical and Fracture Properties *Polymer* doi: 10.1016/j.polymer.2016.03.084.
[3] Wang, F, Drzal L T, Qin Y, Huang Z 2016 Enhancement of fracture toughness, mechanical and thermal properties of rubber/epoxy composites by incorporation of graphene nanoplatelets *Composites: Part A* doi: 10.1016/j.compositesa.2016.04.009.
[4] Halawani N, Augé J L, Morel H, Gain O, Pruvost S 2016 Electrical, thermal and mechanical properties of poly-etherimide epoxy-diamine blend *Composites Part B* doi: 10.1016/ j.compositesb.2016.11.022.
[5] Brantseva T V, Solodilov V I, Antonov S V, Gorbunova I Y, Korohin R A, Shapagin A V, Smirnova N M 2016 Epoxy modification with poly(vinyl acetate) and poly(vinyl butyral). I. Structure, thermal, and mechanical characteristics *J. Appl. Polym. Sci.* doi: 10.1002/APP.44081.
[6] Kopitsyna M N, Bessonov I V, Gusev S A, Kireinov A V, Kopitsyn D S, Solodilov V I, Kotomin S V 2018 The properties of epoxy binders modified by furan resin and polysulfone *Polymer Science, Series B* **60** (3) 299-306.
[7] Solodilov V I, Korokhin R A, Gorbatkina Y A Kuperman A M 2012 Organic fiber reinforced plastics based on complex hybrid matrices including polysulfone and carbon nanotubes as modifiers of epoxy resins *Russian Journal of Physical Chemistry B* **6** (3) 425-432.
[8] Solodilov V I, Bessonov I V, Kireinov A V, Taraskin N Yu, Kuperman A M 2016 glass fiber reinforced plastic with epoxy matrix modified by furfural-acetone resin and polysulfone *Composites and nanostructures* **8** (2) 77-87.
[9] Borodulin, A S, Kalinnikov, A N, Bazheva, R C , Beshtoev, B Z Synthesis and properties of aromatic polyethersulfones (2018) *International Journal of Mechanical Engineering and Technology* **9** (13), pp. 1109-1116.
[10] Nelyub V A , Borodulin A S, Kobets L P and Malysheva G V A study of structure formation in a binder depending on the surface microrelief of carbon fiber *Polymer Science –Series D*, 2016, **9**(3), pp.286-289.
[11] Zheng N, Huang Y, Liu H-Y, Gao J, Mai Y-W 2017 Improvement of interlaminar fracture toughness in carbon fiber/epoxy composites with carbon nanotubes/polysulfone interleaves *Composites Science and Technology* doi: 10.1016/j.compscitech.2016.12.017.
[12] Hu Jin, Baoqing Yang, Fan-Long Jin, Soo-Jin Park 2015 Fracture toughness and surface morphology of polysulfone-modified epoxy resin *Journal of Industrial and Engineering Chemistry* **25** 9-11 doi: 10.1016/j.jiec.2014.10.032.
[13] Mimura K, Ito H, Fujioka H 2000 Improvement of thermal and mechanical properties by control of morphologies in PES-modifid epoxy resins *Polymer* **41** 4451-4459.
[14] Solodilov V I, Gorbatkina Yu A, Korokhin R A, Kuperman A M 2018 Properties of Filament-Wound Organoplastics Based on Epoxy Polysulfone Matrices and Armos and Rusar Aramid Fibers Polymer Science, Series D **11** (3) 247–251.
[15] Korokhin R A, Solodilov V I, Gorbatkina Yu A, Shapagin A V 2015 Rheological and physicomechanical properties of epoxy-polyetherimide *Mechanics of Composite Materials* **51** (3) 313-320.
[16] Chalykh A E, Gerasimov V K, Bukhteev A E, Shapagin AV, Kudryakova G K, Brantseva T V, Gorbatkina Y uA, Kerber M L 2003 Compatibility and evolution of the phase structure of blends of polysulfone with setting epoxy oligomers *Vysokomol. Soed.* **45A** (7) 1148-1159.
[17] Solodilov V I, Korokhin R A, Gorbatkina Yu A, Kuperman AM 2015 Comparison of fracture energies of epoxy-polysulfone matrices and unidirectional composites based on them Mechanics of Composite Materials **51** (2) 177-190.
[18] Solodilov V I, Gorbatkina Yu A 2006 Properties of unidirectional GFRPs based on an epoxy resin modified with polysulphone or an epoxyurethane oligomer *Mechanics of Composite Materials* **42** (6) 513-526.

IOP Conf. Series: Materials Science and Engineering **683** (2019) 012086 doi:10.1088/1757-899X/683/1/012086

[19] Solodilov V I, Bazhenov S L, Gorbatkina Yu A, Kuperman A M 2003 Determination of the interlaminar fracture toughness of glass-fiber-reinforced plastics on ring segments *Mechanics of Composite Materials* **39** (5) 407-414.
[20] Gorbatkina Yu A, Ivanova-Mumjieva V G 2018 Adhesion of the modified epoxides to the fibers *Torus Press* (Moscow) doi: 10.30826/ADHESION-GI-2018.

Increase of composite structures strength by carbon fiber surface modification with spray deposited carbon nanotubes

D D Levin[1], A V Romashkin[1], Yu A Polikarpov[1], N S Struchkov[1], A V Kireynov[2] and I A Komarov[2]

[1] National Research University of Electronic Technology, Moscow, Russian Federation

[2] Bauman Moscow State Technical University, Moscow, Russian Federation

Abstract The possibility of carbon fiber and the polymer matrix interfacial interface improvement by modification carbon fiber surface with functionalized carbon nanotubes. Nanotubes were deposited to carbon fiber surface by aerosol spray deposition, to increase the tensile strength of composite structures. The relationship between the tensile strength of the composite structure with the epoxy matrix and the uniformity of carbon nanotube coating was investigated.

Keywords: CARBON FIBER, EPOXY MATRIX, CARBON NANOTUBE, COMPOSITE STRUCTURE, INTERFACIAL INRERFACE, TENSILE STRENGTH

1. Introduction

Improvement strength properties of new composite materials requires increasing its strength properties without deteriorating other parameters of the structure as a whole, and without substantially increasing the cost of production. Carbon nanotubes (CNTs), due to their mechanical, thermal and electrical properties, are considered as promising materials for modification of the polymer matrix or filler of polymer composites [1]. The main problem in nanoscale fillers containing materials production is the uniform distribution of the filler in the polymer matrix of the composite. In particular, distribution of the CNTs in the composite is often implemented by ultrasonic processing [2, 3]. In order to improve the interaction of CNT with polymer matrix CNT surface is often functionalized by amino- or carboxyl groups. However, it is not always possible to realize the optimal interaction between the filler and the matrix, which can lead to the absence of a noticeable effect of improving the characteristics [4]. The large surface area of CNTs, together with their outstanding physical and mechanical properties is probably the main reason of significant improvement of the nanocomposite mechanical properties due to interaction of CNTs with polymer macromolecules. However, to reach such improvement in the properties of the CNT modified composite, it is necessary to find an optimal way to improve CNTs interaction with the polymer matrix on the one hand and with carbon fiber at the same time. Carbon fiber plays the role of a cheaper filler material that provides sufficient strength of the composite at the micro and macro levels. The result of modification is an increase of the strength characteristics of composite material, that was the purpose of this work.

2. Materials and methods for tasks resolving, accepted assumptions

In order to investigate the effect of CNT addition to the carbon fiber surface on the tensile strength of a composite material, a series of experiments was carried out. Three types of the samples of carbon fiber/epoxy composite were manufactured and tested. Different samples containing: uncoated fibers (reference sample), carbon fiber coated with CNT (type 1), carbon fiber coated with CNT and tetraethylenetetramine (TETA) (type 2). Carbon fibers (Toho Tenax, Tenax-E HTS45 E23, linear density 800 tex, Japan) with a thread diameter of about 5.5 mkm were used as the basis for the composite material. The sample was a bundle of carbon fibers with dimensions of 5x150 mm. Carbon nanotube layer was applied to the carbon fiber surface by iterative spray-deposition with small flow of the dispersion per unit of time (no more than 0.2 ml/min, at a pressure in the compressed air line of about 2 atm, on the substrate heated to 75°C during deposition). To prepare the dispersion of CNT in N-methylpyrrolidone (NMP) and a mixture of solvents (NMP and cyclohexanone (CG)) with concentration of 62 μg/ml, carboxylated CNTs (P3-SWNT, Carbon Solutions Inc., USA) were used [5]. The solution of tetraethylenetetramine (TETA, Dow Chemical, the Netherlands) with a concentration of 12 μg/ml in a mixture of isopropyl and benzyl alcohol was used for modification of previously deposited CNT to obtain the second type of samples. The TETA solution was deposited using a method similar to that used for the deposition of CNTs, and the amount of the deposited solution was also the same and did not depend on the number of deposited CNTs. Unmodified epoxy resin with medium viscous based on bisphenol-A (Araldite LY-556, Huntsman Inc., USA) was used as the matrix. It was prepared for “hot” curing together with the hardener isomethyl tetrahydrophthalic anhydride (iso-MTHPA, Chimex Limited, China) which amount was 90% by weight based on the weight of epoxy resin used, to which an additional 2-methylimidazole (0.2 weight % from epoxy resin weight) is added, which is a curing accelerator agent in the composition of the hardeners "hot" curing mixture. The glass transition temperature of the binder was about 140 °C. Samples were made by the wet impregnation method. Carbon fibers were soaked in a bath with a binder. Than the samples were cured on the holder, providing the tension of the samples, that was strips of fibrous material that were soaked of the binder. The curing mode was 4 hours at 110 °C. Lining pads for gripping specimens in tensile testing machine was made of basalt plastics.

The tensile strength was measured on a Zwick Z 100 tensile testing machine at a loading rate of 10 mm/min. Distance between clamping jaws was 60 mm. The tests were carried out in accordance with ASTM D 2343-09. During the tests, the load, the modulus of elasticity and the elongation at break were determined by stretching the specimen with the help of a testing machine until the specimen failed. During the test, a stress-strain diagram was recorded.

3. Results

It was found that using a mixture of solvents NMP and CG allows to obtain uniformly distributed CNTs on the carbon fiber surface (Fig. 1a) while the deposition of CNTs from the solution of pure NMP leads to the formation of conglomerates (Fig. 1b).

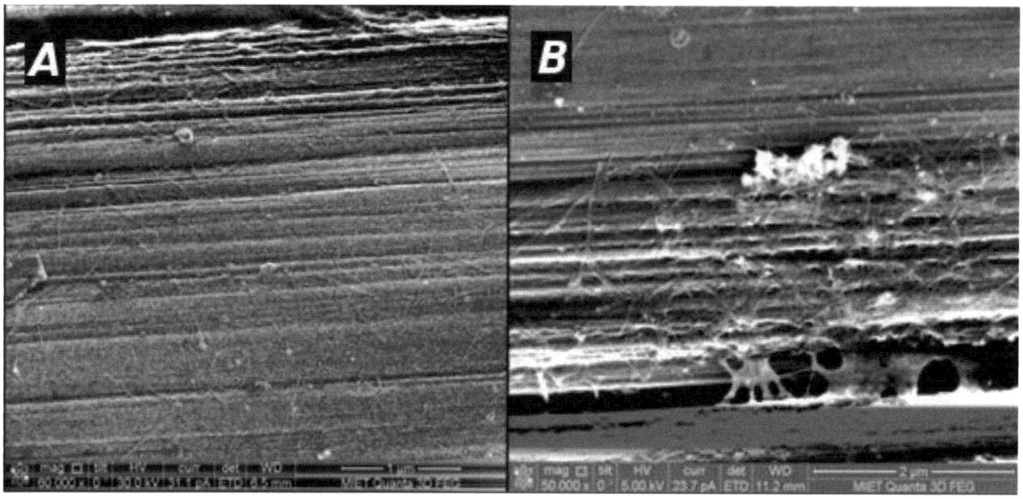

Fig. 1 - SEM image of a CNT mesh on carbon fiber surface with CNT deposited from a dispersion with NMP and CG (A) and NMP (B).

An important aspect of the formation of a composite structure with increased strength characteristics is the preliminary processing of the initial carbon fiber [6]. This effect presumably was reached by presence of carboxyl- (type 1 samples) and amino groups (type 2 samples) on the surface of CNTs. Static tensile strength measurements for samples with a cross section of 0.44 mm^2 for all specimens types showed the following results of the average values for the Young's modulus, ultimate strength, ultimate elongation respectively: 167.8 GPa, 3355.3 MPa, 0.97 mm (reference samples, averaging over 5 samples); 164.5 GPa, 2352.0 MPa, 0.65 mm (type 1 samples, averaging over 2 samples); 157.0 GPa, 3947.0 MPa, 1.38 mm (type 2 samples, averaging over 3 samples); and 166.5 GPa, 3815.0 MPa, 1.18 mm (type 2 samples with a 2-fold increase in the amount of CNTs, averaging over 2 samples).

4. Discussion of the obtained results and comparing with previously known works

Based on a comparison of the CNTs distribution uniformity on fibers surface the following conclusions were made. The CNTs distribution uniformity on the carbon fiber surface significantly affects the nature of the interfacial interaction of the polymer matrix with carbon fiber. Modification of the surface with CNTS leads to an increase in the 14-18% ultimate strength, as well as 20-40% in the ultimate elongation of the sample, and without a significant change in the modulus of elasticity as opposed to samples without modification of carbon fibers. This result correlates with previously known results on increasing the strength of composite materials containing carbon nanomaterials, where the introduction of nanoparticles in the whole volume was carried out however, with increase in the strength of the composite by more than 2 times [7]. At the same time, even a small amount of carbon nanotubes (twice reduced for simply type 2 samples) applied to the surface has a significant effect on the ultimate tensile strength and maximum elongation increment. But the average value of the elastic modulus in this case was slightly reduced relative to reference samples. This decrease can be explained by the excessive amount of TETA molecules applied on top of the CNT in the case of a smaller amount of CNTs for the 2nd specimens type, that led to greater ultimate deformation before rupture and the maximum possible elongation in combination with the minimum elastic modulus. The average number of CNT fibers distributed over the carbon fiber surface was up to 1% of the area in the case of the deposition of CNT from a dispersion with NMP and CG, while deposition from pure NMP gave up to 5% filling of the surface area of carbon fiber. Taking into account the obtained data on tensile strength measurement, excessive addition of CNT (type 1 samples) between the carbon fibers in combination with the lack of necessary functional groups (amino groups in this case) for effective binding with the epoxy matrix leads to a significant decrease in ultimate strength and elongation at break. However, the modulus of elasticity is on the same level with unmodified carbon fibers. In addition, the formation of CNT conglomerates during depositing from a dispersion with pure NMP associated with insufficient temperature of carbon fibers upon the process of CNT deposition. As a result, agglomeration of microdroplets as they are sprayed leads to uneven deposition of CNTs as microdroplets dry out and can also cause a decrease in tensile strength [8, 9]. The addition of a solvent with a low boiling point leads to a decrease in the size of the droplets and promote an increase in the residual solvent evaporation rate, contributing to a more uniform coating of carbon fibers.

5. Conclusion

The surface of carbon fibers can be uniformly covered with a CNT mesh by the iterative spray-coating that leads to reinforcement of final composite materials. The surface containing CNT mesh can be also modified with TETA molecules and these processes leads to increase in tensile strength of 14-18% and maximum elongation of 20-40% while maintaining the elasticity modulus at the level of values characteristic of unmodified samples. At the same time, it is necessary to maintain an optimal ratio between the amount of CNT and TETA deposited to carbon fiber surface, as well as the uniformity of the distribution of CNTs, the role of which appears to be quite significant when filling up to 1% of the surface area. This last fact makes it possible to implement a low-cost modification of carbon fibers with the formation of new generation composite materials with improved characteristics.

Acknowledgments
This work was supported by the Ministry of Science and Higher Education of the Russian Federation, Agreement No. 14.574.21.0184 (unique ID RFMEFI57417X0184)

References

[1] Thostenson E T, Chou T W 2006 Processing-structure-multi-functional property relationship in carbon nanotube/epoxy composites. *Carbon* **44** № 14 P. 3022-3029.
[2] Heimann M et al. 2008 Investigations of carbon nanotubes epoxy composites for electronics packaging. *Electronic Components and Technology Conference* 2008. ECTC 2008. **58**th. – IEEE, – P. 1731-1736.
[3] Bauhofer W, Kovacs J Z 2009 A review and analysis of electrical percolation in carbon nanotube polymer composites. *Composites Science and Technology*. **69** № 10 P. 1486-1498.
[4] Loos M R et al. 2008 Effect of carbon nanotubes addition on the mechanical and thermal properties of epoxy matrices. *Materials Research*. **11** № 3 P. 347-352.
[5] Itkis M E et al. 2003 Purity evaluation of as-prepared single-walled carbon nanotube soot by use of solution-phase near-IR spectroscopy. *Nano Letters*. **3** № 3 P. 309-314.
[6] Dai Z et al. 2011 Effect of heat treatment on carbon fiber surface properties and fibers/epoxy interfacial adhesion. *Applied Surface Science*. **257** № 20 P. 8457-8461.
[7] Guo H et al. 2014 Preparation and mechanical properties of epoxy/diamond nanocomposites. *Polymer Composites*. **35** № 11 P. 2144-2149.
[8] Singh B P et al. 2015 Solvent free, efficient, industrially viable, fast dispersion process based amine modified MWCNT reinforced epoxy composites of superior mechanical properties. *Adv Mater Lett*. **6** № 2 P. 104-113.
[9] Nelyub V A et.al. 2016 Capillary Hydrodynamics of Oligomer Binders. *Polymer Science, Series D*, **9** № 3 P. 322–325

Development of carbon/epoxy composite bracket for housing sensitive element of spacecraft

V A Komarov [1], **A A Pavlov** [1], **S A Pavlova** [1]

[1] Aircraft Design Department, Samara University, 34, Moskovskoye shosse, Samara, Russian Federation

Abstract. The article considers the challenge of using layered carbon/epoxy composite materials in a high-loaded airframe structures. It is demonstrated by a redesign of the bracket for housing a sensitive element of the spacecraft. The new design approach considering simultaneously structural and technological features of composite constructions. It can also be applied to other design tasks to develop a structure of minimum weight.

Key words: composite structure, layered carbon/epoxy composite material, design procedure, structural and technological features, topology optimization, test procedure.

1. Introduction.

The high-performance requirements of the aerospace industry to make structures of minimum weight and sufficient strength lead to an increase of volume fraction of high-performance composite materials in the primary structures. The replacement of traditional metal alloys for composite materials requires a marked transition in design philosophy. Conventional design practices are based on empirical and finite element approaches, therefore they do not always allow to achieve a technologically feasible product. Thus, the transfer of the technological feature in a usable form for composites application is the major challenge. The integration of conceptual design method and technological feasibility of layered composite materials is a complex task.

2. General considerations.

The design procedure described is one by which spacecraft bracket, initially designed from titanium alloy, was designed to be fabricated from carbon/epoxy layered composite materials. This research was intended to provide an example of using an integrated methodology for the design of composite structure considering manufacturing and technological features to satisfy the same performance requirements as those of the original structure.

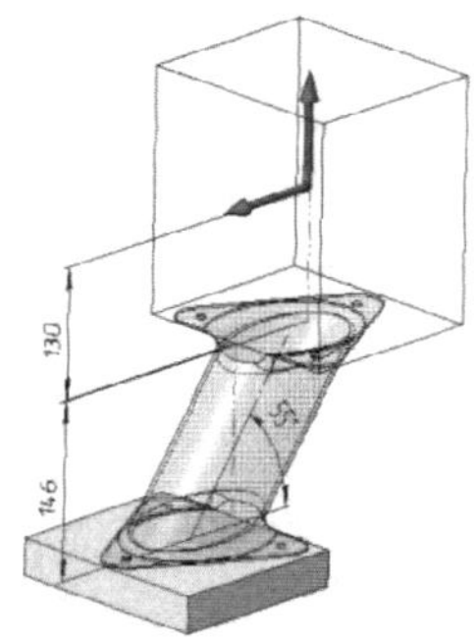

Figure 1. Construction. Load-carrying scheme.

The design chosen for the study was fairly simple. The structure, shown schematically in Figure 1, serves primary to house a sensitive element on the spacecraft body. The bracket is loaded inertial with two axial loads about both axes of its base plane. The initial metal structure consists of a center cylinder body, located at an angle to the spacecraft structure, and two (upper and bottom) parallel flat basements. Connection of the bracket to the spacecraft body and a sensitive element is provided by bolt joint by three point to both upper and bottom supports. The unit weight about 1400 gram and is made for titanium alloy. In considering replacement of traditional metal material with an advanced carbon/epoxy composite material, a weight saving of about 50 percent was desired. This transfer to composite materials sets the following requirements:

- strength of composite structure should be equal or exceed that of the existing titanium structure$
- overall dimensions and attachment point should correspond the initial structure.

Carbon/epoxy composite material, developed on the base of bi-axial carbon fabric CC201 and epoxy system SR8100/SD8824, was chosen for the present study. The mechanical properties of this material, obtained experimentally, are shown in Table 1. In addition, some material properties of titanium alloy are provided for comparison.

Table 1. Material properties

Material	*Density* $kg\backslash m^3$	*Modulus, GPa*		*Strength, MPa*				
		Elasticity		*Tensile*		*Compressive*		*Shear*
		E_{11}	E_{22}	$[\sigma_{11}]^+$	$[\sigma_{22}]^+$	$[\sigma_{11}]^-$	$[\sigma_{22}]^-$	τ_{12}
CC201[SR8100/SD8824]	1451	56,7	56,7	640	640	423	423	120
Titanium alloy	4620	96	96	930	930	930	930	168

The ultimate tensile strength of high-modulus composite material exceed those of the titanium by a factor of 2, the ultimate compressive strength of composite – by a factor 1,5, meanwhile the density of composite material is three times lower those of titanium, which shows a weight saving when carbon/epoxy composite is substituted for titanium.

3. Design practice.

The intended design approach consists of several staged. In conceptual design stage, a rational load-carrying structure was found. Based on the existing design experience, it was decided that the bracket should be an integrally fabricated shell structure, which consists of a set sequence of layers. Three design concepts (Figure 2) were considered within the overall constraint that the composite structure must the interchangeable with the titanium structure. They were differ in the cross-sectional shape. The basic analyze for these structures was stress-strain state. Shell finite-element models were created for each type of design concepts. The material of the analyzed models was set isotropic. This assumption was introduced in order to simplify the modelling process, but in allows determining the basic load-carrying scheme at the same time. The constraint and loads of the models corresponds to those on the Figure 1. The results of linear static analysis for each conceptual design are shown below.

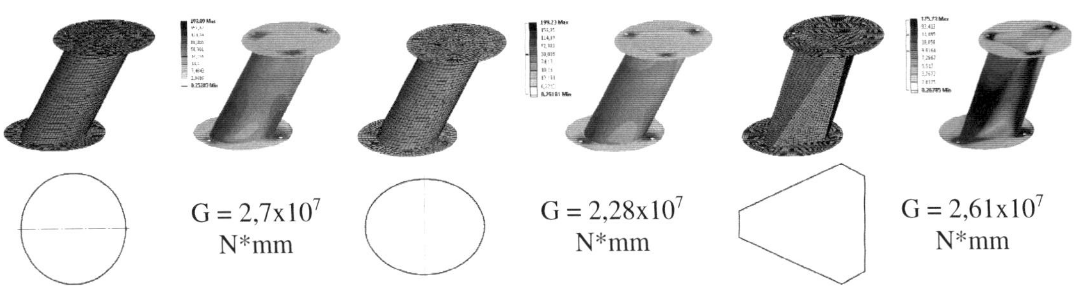

Figure 2. Conceptual design analysis.

The search of the optimal load-carrying structure was executed by means of special criterion called "load-carrying factor" G. It shows simultaneously the magnitude and the path of internal loads and is calculated by the following equation []:

$$G = \int_V \sigma_{\max}^{eqv} dV \cong \sum_{i=1}^{n} \sigma_i^{eqv} V \tag{1}$$

where $\sigma_{экв}$ – equal stresses in the stricture; i–number of the finite element; V – material volume fraction; V_i– volume of the finite element. The smaller the value of load-carrying factor, the more advantageous the design concept. The results of load-carrying factor calculation for each type of design concepts are shown on Figure 2. The second variant of a load-carrying bracket structure was chosen for the further study considering its load-carrying efficiency.

The next stage – technological design – suggested structural design considering perspective manufacturing procedure – vacuum infusion. Technological features of the designed carbon/epoxy composite bracket were taken into account by setting a lay-out scheme. It consists of unit structural and technological ply-groups. The group union occurred on the basis of equality or similarity of layer cutting. The overall lay-out scheme was made of four principal ply-groups (Figure 3), which ensures the minimum amount of the groups, the ability for stepwise layout of the preform and the compatibility of the load-carrying capacity of the contact groups. In considering strength and technological constraints, a stacking sequence $[0;45]\times n_{group}$ was assigned for each lay-out group. Figure 3 schematically shows the mutual arrangement of the groups.

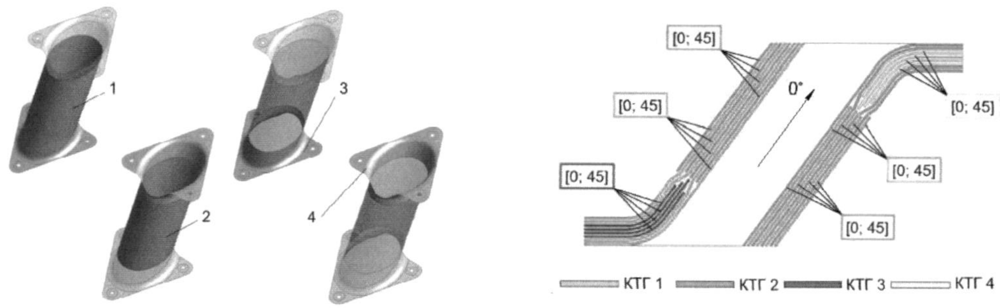

Figure 3. Lay-out scheme with stacking sequence.

At the optimal design stage, the search was made for the number of layers in each lay-out group considering the constraint of minimum mass. The finite-elements model of the bracket with anisotropic properties of carbon/epoxy composite material, was developed using shell geometric model, lay-out scheme and the chosen stacking sequence. The strength condition was set by a widely used maximum stress criterion:

$$k = \max\left(\left|\frac{\sigma_{11}}{[\sigma_{11}]}\right|, \left|\frac{\sigma_{22}}{[\sigma_{22}]}\right|, \left|\frac{\sigma_{33}}{[\sigma_{33}]}\right|, \left|\frac{\tau_{12}}{[\tau_{12}]}\right|, \left|\frac{\tau_{13}}{[\tau_{13}]}\right|, \left|\frac{\tau_{23}}{[\tau_{23}]}\right| \right) \le 1 \tag{2}$$

where σ_{11}, σ_{22}, σ_{33}, τ_{12}, τ_{13}, τ_{23}- actual stress values in a finite element, $[\sigma_{11}]$, $[\sigma_{22}]$, $[\sigma_{33}]$, $[\tau_{12}]$, $[\tau_{13}]$, $[\tau_{23}]$ – strength component of carbon/epoxy composite material.

The distribution of maximum stress criterion and thicknesses over the optimal bracket construction is shown on Figure 4. Theoretical mass of the optimal design was345 gram.

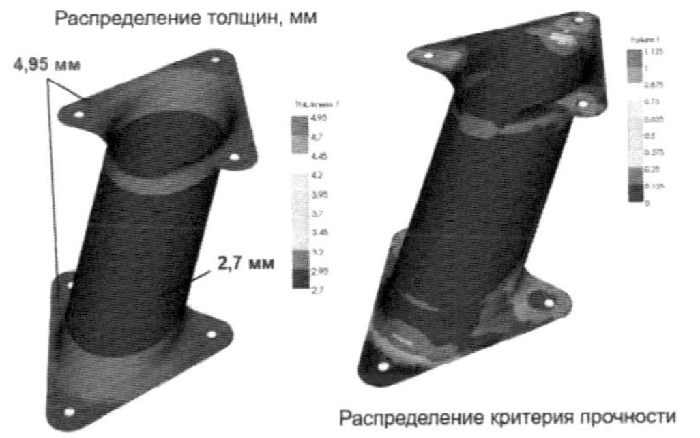

Figure 4. Distribution of stress criterion and thickness

4. Experimental study.

The composite bracket was fabricated according to the chosen vacuum infusion technology and the stacking sequence of lay-up. The overall duration of the technological cycle was 12 hours. Machining operation and drilling holes for bolt joints were made after matrix polymerization. The general view of the bracket is shown on Figure 5. The total mass of this bracket was 350 gram.

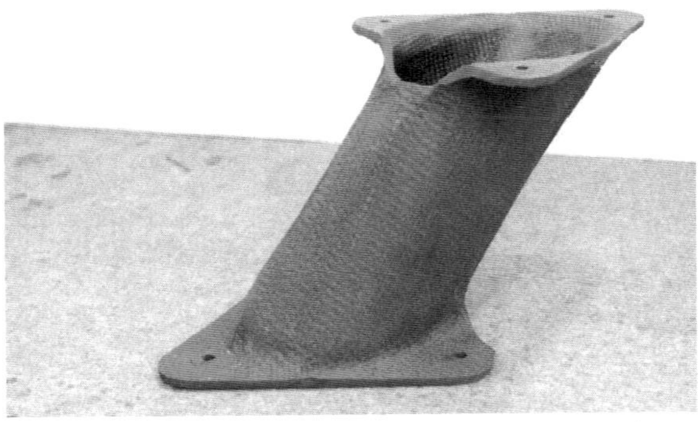

Figure 5. General view of carbon/epoxy bracket

The results of integrated design and manufacturing of the composite bracket were certificated by static strength tests. The tests were carried out using servo-hydraulic test machine MTS 322.2 with a specially designed and manufactured test equipment. The general view of the test equipment and the bracket position for initial types of loading are shown on the Figure 6(a). The first load case was chosen for the tests as the most critical one. The structure was loaded at a speed of 2 mm/min until the force value will not exceed two times the value of design load (Figure 6(b)).

IOP Conf. Series: Materials Science and Engineering **683** (2019) 012088 doi:10.1088/1757-899X/683/1/012088

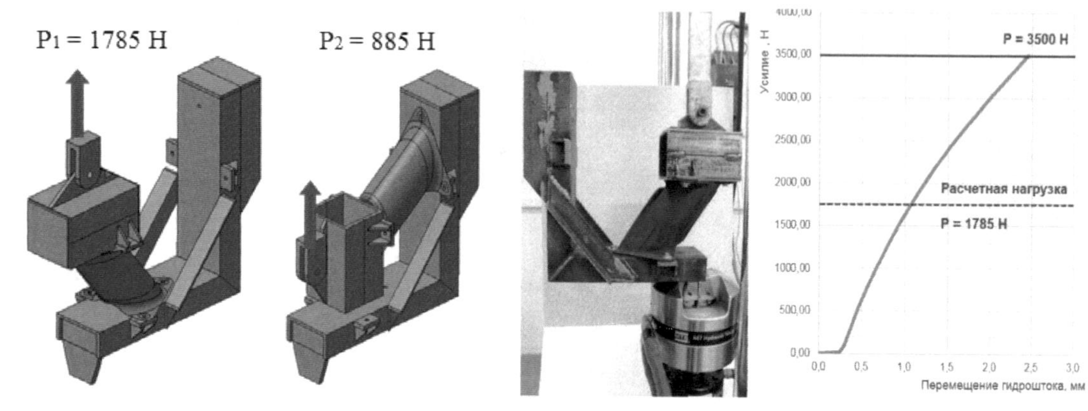

Figure 6. Static strength test of the bracket

Test results showed good agreement with the results of strength-state state analysis of the designed structure.

5. Conclusion.

In addition to presenting a case study of carbon/epoxy composite use in a spacecraft, this study supports the following conclusions:

1. The replacement of traditional metal alloys with composite materials requires consideration of the technological features at the early design stages.
2. The use of layered carbon/epoxy plastic in a case study reduced the structural mass by 80 percent providing at least double safety margin.
3. The described integral design procedure has shown good efficiency and can be recommended for solving more complex problems.

References.

[1] Komarov V A 2017 Interlaminar Strength and Stiffness of Layered Composite Materials – *Procedia Engineering* **185** pp 168 – 172.

IOP Conf. Series: Materials Science and Engineering **683** (2019) 012089 doi:10.1088/1757-899X/683/1/012089

Fabrication of samples and study on mechanical properties of metal matrix composites with system "Al-nanoAl_2O_3"

Y Chen, Yu A Kurganova, S P Shcherbakov and V K Gaaze

Department of Engineering Technology, Bauman Moscow State Technical University,

5/1, ul.Baumanskaya 2-ya, Moscow, 105005, The Russian Federation,

E-mail: chenyijin@yandex.ru

Abstract. Experimental samples of aluminum matrix composite materials filled with discrete Al_2O_3 fibers with a diameter of 10–20 nm were obtained by liquid-phase mechanical mixing. Copper powder with a diameter of 10 μm used as a transport component for the introduction of a light nanosized fraction. A comparative analysis of the hardness and microhardness of the experimental samples demonstrated the presence of a modifying effect from the introduction of discrete Al_2O_3 fibers.

Keywords: metal matrix composites, nanofibers of aluminum oxide, aluminum, liquid-phase method of mechanical mixing, mechanical properties

1. Introduction

Aluminum-composite materials (ACM) have a number of outstanding mechanical and functional properties, which allows them to be widely used both in Russia and in the world [1,2]. Recently, due to the development of nanotechnologies and technologies for producing ACM, Al_2O_3 nanofibers with a high level of elastic modulus and strength are considered one of the most promising materials that can be effectively used as fillers for ACM [3,4].

The main technological problems of introducing a discrete Al_2O_3 nanofiber into aluminum is overcoming the surface tension forces and ensuring the uniformity of their distribution. Analysis of scientific and technical literature [3-6] showed that an effective way to improve wettability is doping a matrix alloy or applying special coatings on the reinforcing phase. As the hypothesis of the experimental part of the research the use of copper as an effective component was accepted, which on the one hand can act as an alloying element, the effect of which on aluminum is well known, and on the other, an element that provides transport functions when introducing nanofiber and enhancing the assimilation of the reinforcing phase by the matrix [3- 4].

2. Experimental procedure

In this research, technical aluminum АД0 (GOST 4784-97), corresponding to 1050 ISO, which has the following chemical composition, % (by weight) is selected as a matrix:

Al	Si	Fe	Cu	Mn	Zn	Ti
>99,50	Up to 0.25	Up to 0.25	Up to 0.25	Up to 0.25	Up to 0.25	Up to 0.25

And as the filler there were used discrete nanofibers Al2O3 of the trademark "Nafen" with a diameter of 10-20 nm. Experimental samples of the composite material were fabricated using a liquid-phase method of mechanical mixing. To solve the problem of the bad wettability of Al2O3 aluminum fibers, micron powders of copper were used, which in turn improved the absorption and distribution of Al2O3 nanofibers in the matrix.

For comparative analysis, four compositions were melted under identical conditions: matrix (АД0), matrix with the addition of Al2O3 nanofibers (АД0+1.14%), matrix with transport powder (АД0+1.198% Cu) and matrix with Al2O3 on a copper transport powder (АД0+1.26% (Cu+Al2O3)).

3. Results and discussion

The microstructures of samples АД0, АД0+1.14% Al2O3, АД0+1.198%Cu, АД0+1.26% (Al2O3 +Cu powders) after etching, carried out on an Olympus GX51 microscope at 50x magnification, are shown in Fig.1. It is seen that after the introduction of Al2O3 nanofibers, the grain is decreasing. And using copper, decreasing is more efficient.

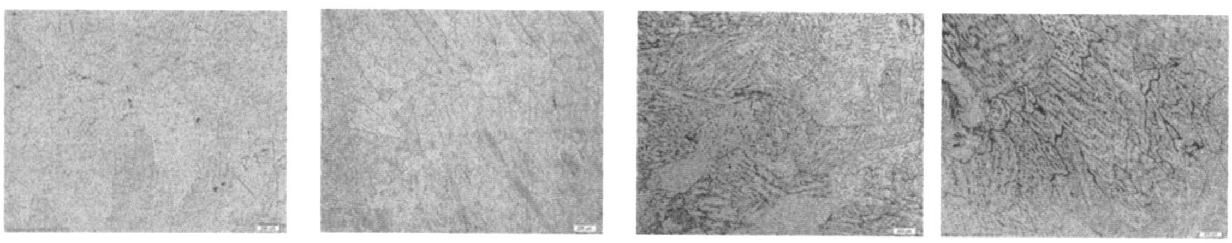

(a) (b) (c) (d)

Fig. 1 The microstructure of experimental samples after etching of АД0(a), АД0+1.14%Al_2O_3 (b), АД0+1.198%Cu (c), АД0+1.26% (Al_2O_3+Cu powders) (d)

All samples were measured for Brinell hardness (GOST 9012-59) on a Struers automatic hardness tester with a load of 62.5 kg and a ball diameter of 5 mm. The analysis of the results (Table 1) demonstrates an increase in the hardness of samples with copper powders, which is evidently due to the formation of the hardening phase CuAl2. The introduction of Al2O3 nanofibre is not significant, but it reduces hardness. Compared with the hardness of the sample АД0 and the sample АД0+1.198% Cu, the hardness value of the samples АД0+1.26% (Cu+Al2O3) increases by 1.41 times and 1.20 times, respectively.

Table 1. Hardness and microhardness of experimental samples

Samples	Brinell Hardness (HBW)	Microhardness (HV)
1. АД0	19.07±1.36	26.11±2.91
2.АД0+1,14% Al_2O_3	16.58±2.66	24.44±3.70

3.АД0+1,198%Cu	22.42±1,64	29.94 ±2.26
4. АД0+1,26%(Al_2O_3+Cu)	26.87±1.54	31.86±3.16

The results of measurement of microhardness of samples of АД0, АД0+1.14%Al2O3, АД0+1.198%Cu, АД0+1.26% (Al2O3 + Cu powders) on the automatic microhardness meter DURASCAN 70 with a load of 100 g were constructed microhardness distributions of the samples and given in Table 1. Compared to the microhardness of АД0, the value of the microhardness of the sample АД0+ 1.26% (Al2O3 + Cu powders) increases 1.22 times. The reduction in microhardness of АД0+ 1.14% Al2O3 is explained by the poor wettability of Al2O3 nanofibers with aluminum.

4. Conclusion

Thus, in the course of the work done, a method was developed and implemented for the effective introduction of discrete fibers of aluminum oxide with a diameter of 10–20 nm into the aluminum melt. The general laws of change in hardness and microhardness with a change in component composition are established. Thanks to the use of transporting copper powder with a diameter of 10 μm, it was possible not only to introduce nanofibers into the aluminum melt, but also to increase the hardness of the material.

References

[1] Kurganova Yu A 2015 Structural metal matrix composite materials: *a textbook Yu A Kurganova, A G Kolmakov*. - M.: Publishing House of Bauman MSTU 141 p.
[2] Chen Yijin 2017 Metal matrix composite materials. Prospects for effective use and production (review) *Metal Technology*. **10** p. 25 -30.
[3] Chen Y, Qin H, Kurganova Yu A and Gaaze V K 2018 A method for introduction of Al_2O_3 nanofiber into aluminum alloy. *IOP Conference Series: Materials Science and Engineering*. **347** 012050.
[4] Kurganova Yu A, Scherbakov S P 2017 Influence of discrete addition of aluminum oxide on the structure and properties of aluminum alloy. *Notes of the Mining Institute*. **228**. p. 717-721.5.
[5] Serpova V M, Shavnev A A, Grishina O I, Krasnov E I, Solyaev U O 2014 Wettability and interfacial interaction in MCM on an aluminum matrix reinforced with Al_2O_3. *Materials Science*. **12** p.29-35.
[6] Leon C A, Mendoza-Suarez G, Drew R A L 2006 Wettability and spreading kinetics of molten aluminum on copper-coated ceramics. *Journal of Material Science*. **41** P.5081-5087

Chen Yijin - Postgraduate Student, Assistant, Department of Materials Science, Bauman Moscow State Technical University (National Research University), Moscow, Russia, 2597-4708, 57202284866, 0000-0001-5587-2015, N-7839-2018.
Kurganova Yuliya Anatolyevna — Doctor of Engineering Sci., Professor, Department of Materials Science, Bauman Moscow State Technical University (National Research University), Moscow, Russia, 1015-5848, 6504373743,
Shcherbakov Svyatoslav Pavlovich - Assistant Professor, Department of Materials Science, Bauman Moscow State Technical University (National Research University), Moscow, Russia, 9914-0967, 55944400900,
Gaaze Vladislav Kirillovich - Master, Department of Materials Science, Bauman Moscow State Technical University (National Research University), Moscow, Russia, 57202288839.

Studying the Structure of Products Made of Titanium Carbide Hardened by the Impregnation Method

V Yu Kulikov[1] and Sv S Kvon[1]

[1]Department of Nanotechnologies and Metallurgy, Karaganda State Technical University, Nursultan Nazarbayev avenue, 56, Karaganda, Kazakhstan
E-mail: mlpikm@mail.ru

Abstract. The paper presents the results of studying the structure of products made of titanium carbide hardened by the impregnation method. As a bonding alloy a high-temperature alloy of the Fe-Cr-Ni-Co system was used. Samples of different porosity were impregnated with that alloy, then their structure was investigated by the methods of mercury porosimetry and electron microscopy. The data of the parameters of porosity and some mechanical properties are given. It was found that there is a correlation between the average size of open pores and the tensile strength and the ultimate strength limit. It is shown that this relationship is logical if to consider the resulting structure as a composite material: the matrix of titanium carbide which pores are filled with a metal bond. It is shown at what parameters of the porous structure the impregnation process proceeds most fully.

1. Introduction

The technology of obtaining products based on titanium carbide is well developed and consists of pressing the titanium carbide powders and a metal binder and their sintering. During the sintering process the metal binder melts and in such a way binds the particles of titanium carbide. As a result, a dense, almost nonporous product is formed that can be considered as a composite material: the matrix of titanium carbide which pores are filled with a metal binder.

One of the main problems of producing high-quality products of this type is the difficulty of combining at least two dissimilar phases. Metal-titanium carbide systems are characterized by weak binding and outflow of metal from the composition during sintering, which proceeds with the formation of a liquid phase [1-4]. It is considered that the condition for the formation of a strong bond between the cementitious metal and the nonmetallic phase is mutual complete or partial solubility.

Another relative problem of products made of titanium carbide is their fragility, low impact strength and, as a consequence, the difficulty of processing.

Meanwhile, in a number of works [5-8] it is noted that the impregnation method allows avoiding these problems and obtaining products based on titanium carbide possessing satisfactory impact toughness and new operational properties that are formed by the alloy-binder. It should be noted that the method of impregnation is used less frequently than liquid-phase sintering. This is due to the fact that in most cases they try to get a structure in which each particle of titanium carbide is surrounded by a metal binder. This structure is characterized by a greater impact strength and ductility. In order to get it easier there is used liquid-phase sintering. However, the impregnation method has other advantages. It can be used to obtain products of a complex shape, because very small pressures can be used for pressing. The products are practically non-porous, in addition, the contact time of the impregnating alloy and pressing can be adjusted. A small amount of pressure applied during pressing increases the

operating time of the mold. The method can be used for batch production.
In this paper we studied the possibility of obtaining products based on titanium carbide by the impregnation method that have enhanced heat-resistant properties. To achieve this goal it was necessary to solve several tasks:
- to determine the optimal amount of binder for impregnation;
- to define the optimal parameters of the porous structure that provide the impregnation process (the volume of open pores and the average size of open pores);
- to define the modes of pressing and sintering for developing a frame based on titanium carbide that has the predetermined parameters of porosity.

2. **Experimental** researches

As the alloy for impregnation there was used a heat-resistant alloy based on the Fe-Cr-Ni-Co system. The chemical composition of the alloy is given in Table 1. This alloy has sufficiently good heat-resistant and mechanical properties (Table 1).

Table 1. Chemical composition of the alloy for impregnation and some mechanical properties

C	Ni	Si	Mn	Cr	Co	Al	Fe	100-hour durability, MPa	
								800^0	1000^0
0.1	24	0.6	0.3	7	9	4.5	rem.	320	80

The studies carried out [9-13] have shown that the optimal amount of the impregnation binder is 37-40 %, i.e. the frame of titanium carbide after pressing and sintering should have about 40 % of open pores. To select the optimal porous structure for impregnation experiments were carried out with samples of different porosity. Samples of different porosity were impregnated with this alloy-binder and then studied with the VEGA\\TESCAN electron microscope and by mercury porosimetry using the Pascal system. Table 2 shows the results of the studies.

Table 2. Interconnection between the porous structure parameters and mechanical properties

Sample No.	Share of pores with the 4-50 μm radius, %	Share of pores with the 50-100 μm, %	Share of pores with the 100 – 116 μm, %	Share of pores with radius larger than 120 μm, %	Sample impregnated with the Cr-Ni-Co system alloy	
					Ultimate tension, MPa	Long-term strength, MPa, σ_{50}^{600}
4	15	21	35	29	820	360
5	15	36	35	14	940	485
6	17	34	34	15	980	520

It is seen from the data in Table 2, sample No. 4 has the minimum strength characteristics both at the room and elevated temperatures. If compared with the parameters of porosity, it differs from samples No. 5 and 6 by an increased content of large pores (radius larger than 120 μm). At the end of the infiltration and crystallization process, these filled pores are isolated metal sections which boundaries are incoherent with titanium carbide particles.
The carried out comparative analysis of the samples microstructures (Figure 1) showed the correctness of this assumption.

a

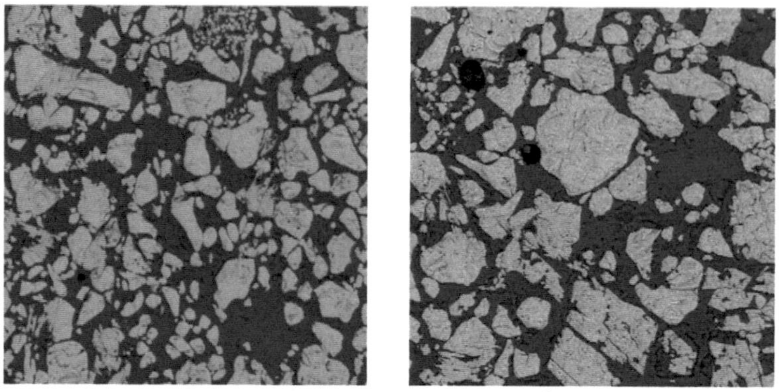

a b

Figure 1. Microstructure of coarse- (a) and fine- (b) porous impregnated cermet (×10 000).

3. Discussion of the results

The analysis of the porous structure (Table 2) showed the following. Large pores of sample 4 (Figure 1, a) are filled with the metallic melt, and the boundaries between the titanium carbide particle and the matrix are clearly visible. The structure can be considered as a two-phase structure, and the interphase boundary according to the generally accepted classification will be high-angle and, therefore, can serve as a source of destruction. In sample No. 5 the pore content of the radius that is larger than 120 μm is only 14 %. The structure in Figure 1, b) looks more homogeneous, with the same content of the metal bond, its presence in this case is manifested only in the form of interlayers between the particles of titanium carbide. With the same granulometric initial composition of titanium carbide, it is obvious that the extent of the "titanium carbide-alloy" boundaries in sample 5 is significantly higher than in sample 4, but the long-term strength of sample 5 is higher than in sample 4.

If we consider the structure of the impregnated cermet in the first approximation as the two-phase one, then titanium carbide presents one phase, and the pores filled with the alloy-binder the other. Thus, the size of the phases is determined by the titanium carbide particle size and the size of the filled pore. For clarity of the dependence a criterion for the average pore size was calculated according to the method of mercury porosimetry [14, 15]:

$$r_{cp} = 2\frac{V_t}{S}$$

where V_t is the general volume of the pores, cm^3/g;
S is the pores surface, cm^2/g.

Table 3. Long-term durability dependence on the porous structure parameters

Sample No.	General volume of the pores, cm^3/g	The pores surface, cm^2/g	Average radius of the pores, μm	Long-term strength, MPa σ_{50}^{600}
4	3.179	7. 4	0.859	360
5	3.038	7. 2	0.843	485
6	2.685	7. 3	0.736	520

IOP Conf. Series: Materials Science and Engineering **683** (2019) 012090 doi:10.1088/1757-899X/683/1/012090

From the data in Table 3 and Figure 2 it follows that the smaller (up to a certain size) alloyed pores, "grains", the higher the limit of long-term strength. For the calculation the previously obtained porosity data using the mercury porosimetry method were used.

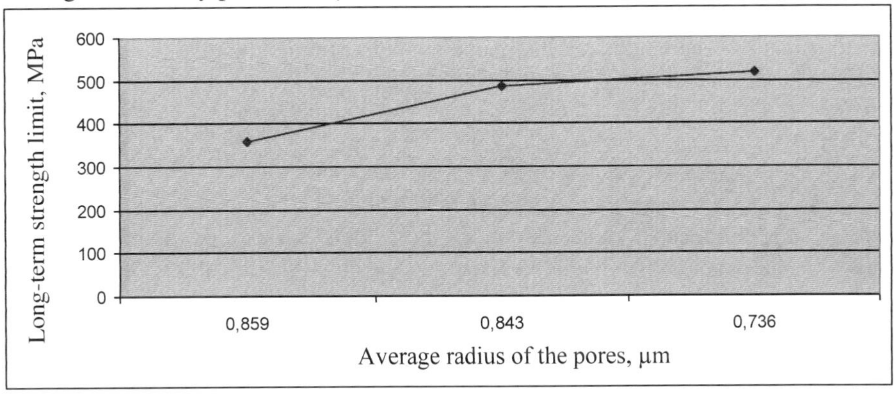

Figure 2. Long-term strength limit dependence on the average radius of the pores

This fact contradicts one of the main provisions of the theory of heat resistance. As it is known, the development of creep is directly related to the extent of grain boundaries: the higher it is, the easier the creep process develops. That is why in heat-resistant alloys they strive to obtain a fairly large grain (4-6 grades). However, in this case the opposite phenomenon is obtained experimentally.

The mechanism of strengthening the "impregnated" cermet can be explained in this case by the following circumstance. There are two main mechanisms of hardening high-temperature alloys [16-18]: direct and indirect. According to the direct mechanism in strengthening at elevated temperatures the main role is played by the composition and structure of the phases themselves and the mechanism of their interaction, i.e. in fact, heat resistance of the alloy is determined by the phase diagram.

With the indirect mechanism of hardening the level of mechanical properties is mainly determined by the grain size and the boundary morphology. The grain size as such does not directly affect the mechanical properties but determines the extent of the boundary, the dispersion, the distribution of intermetallic phases in the matrix and along the grain boundaries.

That is why in many studies the grain size is a certain criterion determining the level of mechanical properties. For two grain sizes there is the exvigosity temperature below which the fine grain is optimal, and vice versa. The temperature of exvigosity is understood as the temperature at which the strength of the matrix and the boundaries are equal. Accordingly, when the structure is formed below the temperature of exsigosity, the optimal level of properties in high-temperature alloys [19] can be achieved in fairly fine-grained structures stabilized by γ-phase precipitates.

Forming the boundary of a complex shape also contributes to hardening at elevated temperatures [20]. In pressure treatment (in particular, pressing), dislocation substructures are formed that also contribute to hardening.

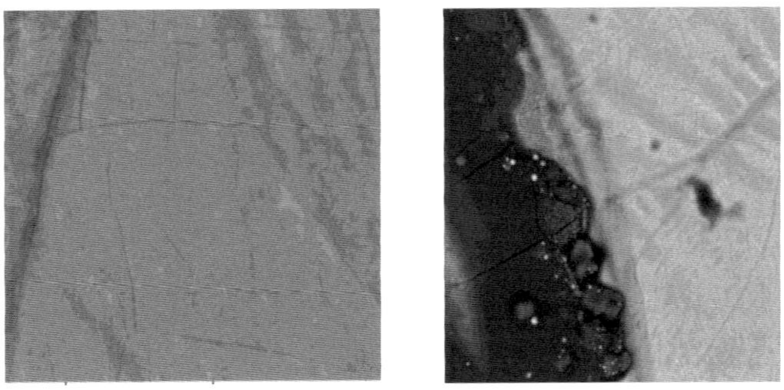

a b

Figure 3. The boundary between the matrix and the pore: a – coarse-porous sample; b – fine-porous sample

As it is seen in Figure 3, the boundary between the matrix and the large filled pore has a relatively smoothed appearance (Figure 3, a). At the same time, the boundary between the matrix and the fine pore sometimes has a sinuous, even "sawtooth" structure.

4. Conclusions

Summing up, it is possible to propose the following mechanism for strengthening the "impregnated" cermet. The sufficiently fine-porous structure of the matrix forms conditions for the formation of small-sized metallic inclusions. Accordingly, the boundary between the matrix and the metal phase also depends on the pore size: the smaller the pore, the more tortuous and complex the boundary is formed.

Thus, in order to obtain a quality product made of titanium carbide, the impregnation method requires a porous structure with an open porosity of about 40 % and an average pore size of about 0.7 μm after sintering. In this case it is possible to obtain a substantially nonporous product having sufficiently high heat-resistant properties due to the properties of the alloy-binder.

Based on the results of the studies carried out, the technology was developed and a batch of products was produced (sealing rings of stop valves for high pressure pipes) that were successfully tested in real operating conditions.

References

[1] Kiparissov S S and Libenson G A 1995 *Powder metallurgy* (M.: Metallurgia) pp 78-129.

[2] Andrievsky P A 2011 *Powder Material Science* (Moscow: Metallurgia) p. 36-84.

[3] Geguzin Y E 1994 *Physics of sintering* (Moscow: Nauka) 312

[4] Bazyleva O A, Arginbaeva E G and Turenko E Yu 2012 Heat-resistant foundry intermetallic alloys *Aviation materials and technologies* **5** 57-60.

[5] Lomberg B S, Hovsepyan S V, Bakradze M M and Mazalov I S 2012 High-Temperature Heat Resistant Nickel Alloys for Gas Turbine Engine Parts, *Aviation materials and technologies* **7** 52-57.

[6] Ramakrishnan, P 2010 *Powder metallurgy and high-temperature materials* (Chelyabinsk: Metallurgy) 352.

[7] Ejaz N, Mansoor M and Salam I 2010 Effect of high temperature exposure on the microstructure of Udimet-500 super alloy, *Journal of Materials Science* **45** 4507

[8] Ballard D L and Pilchakm A L 2010 The use of precios metal nicrelbased superalloys for thin gage application, *JOM* **62** 10.

[9] Issagulov A Z, Kvon Sv S, Filippova T A and Kulikov V Yu 2016, Determination of composition and fluidity of an alloys for imprenating hard alloys on the basis of titanium carbide, *Metallurgija,* **55** 414-416.

[10] Issagulov A Z, Kvon Sv S, Kulikov V Yu and Sakbossynova A A 2014 Cr-Ni system alloys composition impact on durability value, *Metalurgija* **53** 621 – 623.

[11] Issagulov A Z, Kvon Sv S & Kulikov V Yu 2016 Studying microstructure of heat resistant steel deoxidized by barium ferrosilicon *Metalurgija* **55** 388-390.

[12] Kovalev P V, Ryaboshuk S V, Issagulov A Z, Kvon Sv S and Kulikov V Yu 2016 Improving production technology of tube steel grades in converter process, *Metalurgija* **55** 715-718

[13] Gazaliyev A M, Kvon Sv S, Filippova, T S, Yegorov V V 2015 Using alloys of Cr-Ni-Co system as metallic bond in powder metallurgy products. *Metalurgija,* **54,** 643-646.

[14] Plachenov T G 1984 *Mercury porosimetry and its use for describing porous structure,* M.: Metallurgia 162

[15] Matus, K. (2016). Characterzation of nanometric-sized carbides formed during tempering of carbige steel cermets. *Op.Access. J (Sweden).* 2016-06-01.

[16] Gessinger, G. H. (1984). *Powder Metallurgy of superalloys* (BMM, Cambridge), p 342.
[17] Sabol, G. P. and Stickler, R. (2006). Microstructure of Nickel-Based Superalloys *J.PSS,* **3,** 344.
[18] Yiping Lu, Yong Dong, Sheng Guo A. (2014). Promising New Class of High-Temperature Alloys: Eutectic High-Entropy Alloys. *Scientific Reports,* **4,** 6200, 29.
[19] Lindblad, R. (2009). A review of the behavior of aluminide-coated superalloys. *UTUMN,* **1,** 143-170.
[20] Jun Ma, Aijun Li, Huiping Tang (2016). Study on Sintering Mechanism of Stainless Steel Fiber Felts by X-ray Computed Tomography. *Metals,* **6,** 18.

IOP Conf. Series: Materials Science and Engineering **683** (2019) 012091 doi:10.1088/1757-899X/683/1/012091

Transport properties of ceramic composites based on bismuth and indium oxides

P E Dergacheva[1], I V Kulbakin[2] and S V Fedorov[2]

[1] Bauman Moscow State Technical University (BMSTU), 5, 2nd Bauman Street, Moscow, 105005, Russian Federation
[2] Baikov Institute of Metallurgy & Materials Science (IMET RAS), 49, Leninsky Prospekt, Moscow, 119334, Russian Federation

E-mail: ivkulbakin@mail.ru

Abstract. Solid solutions based on indium oxide $In_{2-x}Zr_xO_{3+0.5x}$ (x = 0.005, 0.01, 0.015, 0.02, 0.025, 0.05, 0.075, 0.1) were synthesized and temperature dependences of their conductivity at 25-800 °C were investigated. It was found that $In_{1.985}Zr_{0.015}O_{3.0075}$ solid solution possessed the highest conductivity of about 100 S/cm at 800 °C. Dense $In_{1.985}Zr_{0.015}O_{3.0075}$ – 50 vol.% Bi_2O_3 ceramic composite was obtained and investigated, its microstructure was studied, and the formation of fine-grained structure with homogeneous distribution of components by volume was demonstrated. The total conductivity of composite was measured, and oxygen fluxes for composite 1.7 mm thick membrane at 750-800 °C was obtained. The oxygen flux was found to be 2.4×10^{-7} mol·cm^{-2}·s^{-1} at 800 °C when oxygen partial pressures on both sides of membrane was 0.21 atm/0.017 atm, while the oxygen selectivity was more than 20. This indicates the promising of investigated composite to use as an intermediate-temperature (750-800 °C) ion transport membrane for obtaining of pure oxygen from air.

1. Introduction

Nowadays the membrane technologies are actively used to produce oxygen of high purity from air and other oxygen-containing gas mixtures [1-2]. The process of separation the air components can be carried out using dense ceramic ion transport membranes (ITM) capable of selectively passing of pure oxygen under given temperatures (> 700 °C) being by the gradient of its partial pressures. However, many of known ceramic ITM fabricated by single-phase multicomponent oxides (e.g. by BSCF or LSCF) with high mixed oxygen-ionic and electronic conductivity may degrade during long-term operation due to their insufficient chemical or thermal stability [3-4], while their oxygen transport productivity of work is lowering. The solution of this problem is the creation of more stable composite membranes for long-term use, in which one of the components is electronic conductor and the other is an oxygen-ionic conductor. For example, ceramic composites based on In_2O_3 - Bi_2O_3, in which indium oxide (an electron conductor) is chemically and thermally compatible with bismuth oxide having high oxygen-ionic conductivity, have been proposed and investigated earlier [5-7]. For further application of these composites as materials of ion transport membranes, their electronic conductivity needs to be increased, which is possible by replacing pure In_2O_3 with solid solutions based on it. As it is known from the literature when doping of In_2O_3 by oxides of tetravalent cations (for example, by SnO_2), the electronic conductivity can be increased by several orders of magnitude [8-9], what is currently principally used in microelectronics to create conductive coatings, LEDs, sensors, etc.

The main goals of this work are synthesis and study the conductivity of $In_{2-x}Zr_xO_{3+0.5x}$ (x = 0.005-0.1) solid solutions, the detection of optimum zirconia additive when maximum for their conductivity is achieved; as well as obtaining and investigating the transport properties of ceramic composites "indium oxide doped with zirconia" - "bismuth oxide" to create new oxygen-permeable ITM materials.

2. Experimental

Solid solutions of $In_{2-x}Zr_xO_{3+0.5x}$ (x = 0.005, 0.01, 0.015, 0.02, 0.025, 0.05, 0.075, 0.1) were obtained from indium (III) and zirconium (IV) oxides (99.999 % of purity), which were taken in the required proportions. The charges for synthesis of solid solutions were mixed in a planetary mill Pulverisette 5 at a speed of 200 rpm with addition of ethyl alcohol in a ratio of 1:1 to the total weight of samples. Homogenization of powder mixtures was carried out for 30 min by using of agate bowls and balls (mass ratio of charges to balls was 1 to 10). Next the charges were held in a drying oven for 5 hours at 120 °C to remove alcohol and moisture, then were cold pressed at 200 MPa into parallelepipeds (a = 5 mm, b = 5 mm, l = 40 mm) with addition of a binder – 5 % aqueous solution of polyvinyl alcohol (5 % to the total mass of charges). The obtained green bodies were held in a drying oven for 5 hours at 200 °C to remove moisture, then were placed in a muffle furnace and calcined on platinum substrates in air atmosphere at 1200 °C for 100 hours, the heating and cooling rate of the furnace was 2 °C/min.

To synthesize a ceramic composite of $In_{1.985}Zr_{0.015}O_{3.0075}$ – 50 vol.% Bi_2O_3, the sintered samples of $In_{1.985}Zr_{0.015}O_{3.0075}$ were ground in an agate bowl to the powder size of 2-3 μm. This powder then was mixed with 50 vol.% Bi_2O_3 (99.999 % of purity), taking the densities of $In_{1.985}Zr_{0.015}O_{3.0075}$ and Bi_2O_3 equal to 7.15 g/cm^3 and 8.9 g/cm^3, respectively. The powder mixture was cold pressed at 200 MPa into tablets (Ø = 30 mm, h = 2.5 mm and Ø = 6 mm, h = 3 mm) for measuring the oxygen fluxes, porosity and linear thermal expansion, and into parallelepiped (a = 5 mm, b = 5 mm, l = 40 mm) for measuring of the conductivity. Next the samples were held for 5 hours at 200 °C in a drying oven, and then were calcined at 800 °C for 10 hours on platinum substrates in a muffle furnace; the heating and cooling rate of the furnace was 2 °C/min. The shrinkage of all samples during sintering was about 20 %.

X-ray phase analysis of both $In_{2-x}Zr_xO_{3+0.5x}$ (x = 0.005-0.1) solid solutions and $In_{1.985}Zr_{0.015}O_{3.0075}$ – 50 vol.% Bi_2O_3 ceramic composite was investigated by Shimadzu XRD-6000 diffractometer (CuKα - radiation, 2θ range of 10 - 70°, step for scan of 0.1°). The decoding of X-ray diffraction patterns was performed using ICDD PDF-2 database. Microstructure of polished cross section for the composite, which had been prepared by STRUERS equipment, was investigated by Olympus GX-51 optical microscope. Open, closed and total porosity of the composite was measured by hydrostatic weighing.

The conductivity of both $In_{2-x}Zr_xO_{3+0.5x}$ (x = 0.005-0.1) solid solutions and $In_{1.985}Zr_{0.015}O_{3.0075}$ – 50 vol.% Bi_2O_3 ceramic composite was measured by four-probe method in ambient air in potentiostatic mode (0.1 V) at 20-800 °C. The measurements were carried out on a Solartron 1285A potentiostate in steps of 50 °C and holding for 30 min at each temperature point to reach a steady state.

The measurements of oxygen fluxes for $In_{1.985}Zr_{0.015}O_{3.0075}$ – 50 vol.% Bi_2O_3 composite (tablet of Ø = 25 mm, h = 1.7 mm) were performed by gas chromatography technique at 750 °C, 775 °C, 800 °C on Crystallux-4000M gas chromatograph. To do this, the polished tablet was tightly connected to the end of the measuring cell, which was placed into a tube furnace and heated to the temperature of 750-800 °C. From the outside face, the tablet was in contact with air, and from the inner face it was connected to a gas chromatograph, while the sweep gas (helium) was blown through the measuring cell to remove the gas mixture penetrated through the sample. The selectivity of composite membrane was evaluated by the ratio of concentrations of oxygen over nitrogen in the analyzing gas mixture (consisted of oxygen, nitrogen, and helium). The rate of sweep gas flow was 5-20 ml/min.

The linear thermal expansion of the $In_{1.985}Zr_{0.015}O_{3.0075}$ – 50 vol.% Bi_2O_3 composite was measured under argon atmosphere on DIL 402 C dilatometer in a Netzsch tube furnace at 20-800 °C. Composite (tablet of Ø = 5 mm, h = 2.5 mm) was placed in a holder with high-precision displacement converter. The rate of argon flow was 70 ml/min, the rate of heating and cooling of the furnace was 10 °C/min.

3. Results and discussion

X-ray phase analysis of $In_{2-x}Zr_xO_{3+0.5x}$ (x = 0.005, 0.01, 0.015, 0.02, 0.025, 0.05, 0.075, 0.1) solid solutions indicates that In_2O_3 with bixbyite crystal structure (JCPDS No. 71-2194, Ia3 space group) is the main phase, and its lattice parameter a = 10.119 Å does not vary in all samples within the error of determination (± 0.005 Å). The presence of insignificant traces of second phase - ZrO_2 of tetragonal modification (JCPDS No. 81-1544, P42/nmc space group) is also detected in $In_{2-x}Zr_xO_{3+0.5x}$ (x = 0.025, 0.05, 0.075, 0.1) samples. Thus, it can be concluded that the solubility limit of zirconia in In_2O_3 bixbyite is in the range of $0 < x < 0.025$. This is confirmed by temperature dependences of total electrical conductivity for $In_{2-x}Zr_xO_{3+0.5x}$ (x = 0.005-0.1) solid solutions, which are presented in Figure 1a. The conductivity of all samples increases at an increase in the temperature in both ranges of 20 - 300 °C and 500 - 800 °C, but, on the contrary, decreases in the range of 300 - 500 °C. This is due to the features of temperature dependence of conductivity for the impurity semiconductor. At a low temperature of 20-300 °C, the concentration of free charge carriers on donor sublevel rises at an increase in the temperature, resulting to increase in total conductivity. At the second temperature range of 300 - 500 °C, an impurity donor sublevel becomes depleted, while the charge carriers pass into the conduction band, resulting to metallic conductivity. When increasing the temperature to the range of 500-800 °C, the intrinsic conductivity of the semiconductor becomes activate, so the total conductivity of all samples increases. The maximum of conductivity as for room temperature of 20 °C as for 800 °C is detected for $In_{1.985}Zr_{0.015}O_{3.0075}$ solid solution (Figure 1b), so it can be concluded that solubility limit of zirconia in In_2O_3 bixbyite is reached at x = 0.015. The conductivity of $In_{1.985}Zr_{0.015}O_{3.0075}$ solid solution at 800 °C is almost two orders of magnitude higher than that for pure In_2O_3, therefore this solid solution was chosen for preparing the composite with Bi_2O_3.

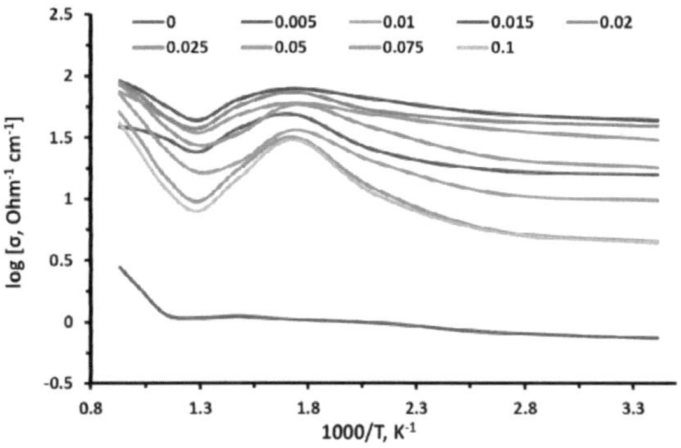

Figure 1a. Temperature dependences of the conductivity (σ) of $In_{2-x}Zr_xO_{3+0.5x}$ (x = 0, 0.005, 0.01, 0.015, 0.02, 0.025, 0.05, 0.075, 0.1)

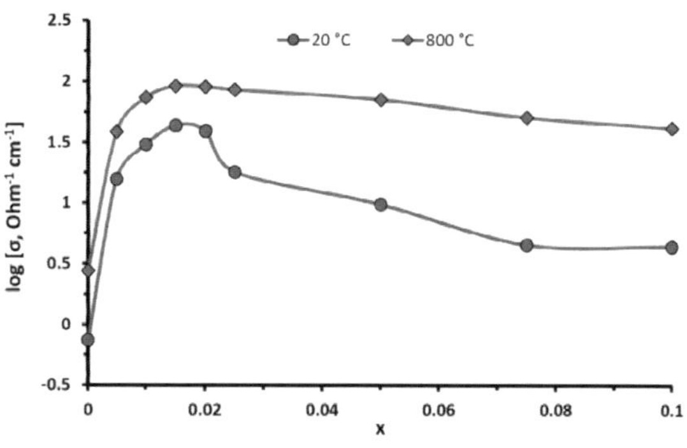

Figure 1b. Zirconia dopant concentration (x) dependences of the conductivity (σ) of $In_{2-x}Zr_xO_{3+0.5x}$ (x = 0-0.1) at 20 °C and at 800 °C

Figure 2a illustrates the micrograph of polished section of $In_{1.985}Zr_{0.015}O_{3.0075}$ – 50 vol.% Bi_2O_3 composite prepared by sintering at 800 °C in ambient air atmosphere for 10 hours. Ceramic composite with fine-grained structure is formed, where the dark and the light components ($In_{1.985}Zr_{0.015}O_{3.0075}$ and γ-Bi_2O_3, respectively, as it has been confirmed by X-ray phase analysis data) are evenly distributed. Total porosity of the ceramic composite is about 3 %, while its open porosity is of 0.15 %, which indicates the formation of high-density material by sintering. It is notable that X-ray diffraction pattern of the composite contains no peaks belonging to α-Bi_2O_3, since in the binary In_2O_3 - Bi_2O_3 system the metastable δ-$Bi_2O_3 \rightarrow \gamma$-$Bi_2O_3$ transformation is realized upon cooling of the material after sintering. According to the known literature data [10], this is because γ-Bi_2O_3 syllenite (space group I23, lattice parameter a = 10.264 Å) crystal structure is relevantly close to In_2O_3 bixbyite (space group Ia3, lattice parameter a = 10.119 Å). From this point of view, the presence of metastable state in this system, due to exhibiting of Dankov-Konobeevsky effect of the orientation-dimensional structural accordance, is favorable for further application of this composite as the material of ion transport membrane. So, this composite is allowed to be mechanically stable and free from cracks formation during thermal cycling. As it is known ceramics based on pure bismuth oxide has low mechanical stability due to dramatic polymorphous transformation of α-Bi_2O_3 → δ-Bi_2O_3 leading to a significant change in volume (6-8 %), which is accompanied by cracking [11]. Indeed, in this composite material this mechanical drawback has been overcome, that is confirmed by dilatometric studying of its linear thermal expansion, where the length change of the sample at γ-Bi_2O_3 → δ-Bi_2O_3 polymorphic transformation at 670-700 °C is less than 0.5 % (Figure 2b). This polymorphous transformation has been found to be reversible, while the linear thermal expansion coefficient for $In_{1.985}Zr_{0.015}O_{3.0075}$ – 50 vol.% Bi_2O_3 composite is observed to be 3.2×10^{-5} K^{-1} at the temperature range of 400-670 °C.

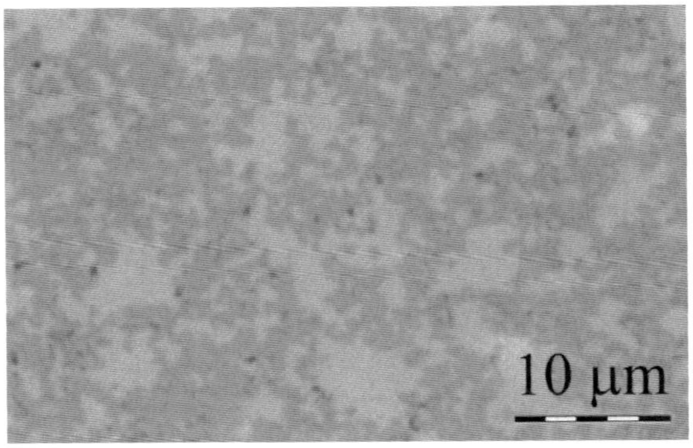

Figure 2a. Microstructure of $In_{1.985}Zr_{0.015}O_{3.0075}$ – 50 vol.% Bi_2O_3 composite, calcined at 800 °C during 10 hours

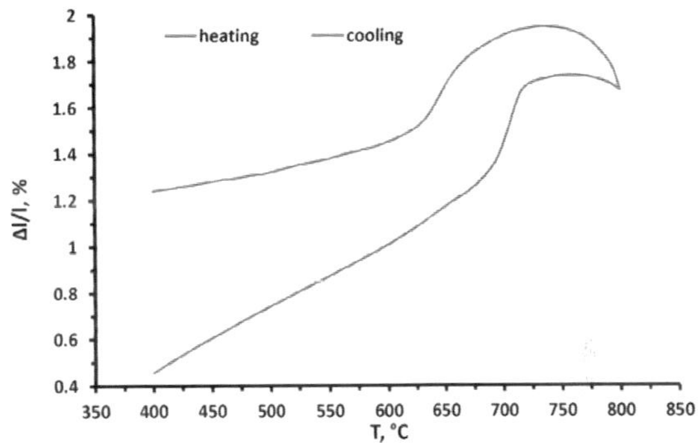

Figure 2b. Temperature dependences for linear thermal expansion (Δl/l) of $In_{1.985}Zr_{0.015}O_{3.0075}$ – 50 vol.% Bi_2O_3 composite

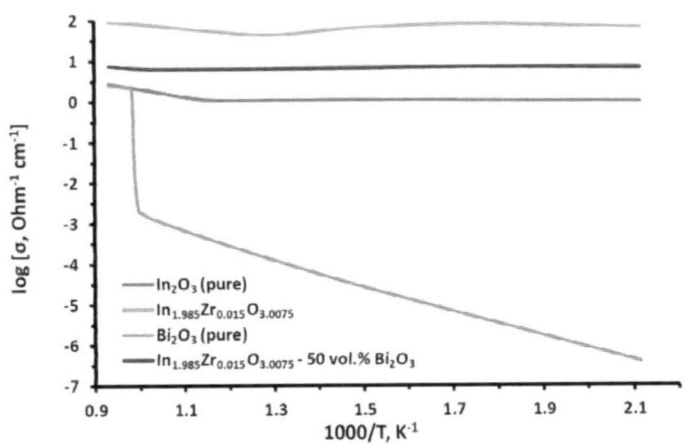

Figure 3a. Temperature dependences of the conductivity (σ) of $In_{1.985}Zr_{0.015}O_{3.0075}$ – 50 vol.% Bi_2O_3 composite, of $In_{1.985}Zr_{0.015}O_{3.0075}$ solid solution, of pure In_2O_3, and of pure Bi_2O_3

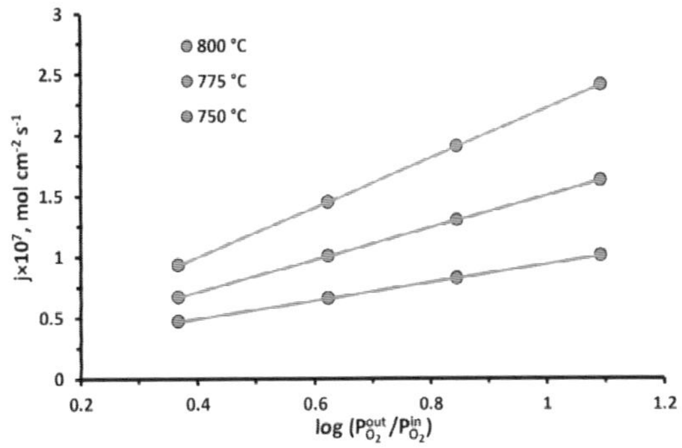

Figure 3b. Dependences of the oxygen flux (j) of $In_{1.985}Zr_{0.015}O_{3.0075}$ – 50 vol.% Bi_2O_3 composite 1.7 mm thick membrane on its partial pressure difference (log P^{out}/P^{in}) at 750-800 °C

Figure 3a shows the temperature dependences for the conductivity of $In_{1.985}Zr_{0.015}O_{3.0075}$ – 50 vol.% Bi_2O_3 composite, of $In_{1.985}Zr_{0.015}O_{3.0075}$ solid solution, of pure In_2O_3, and of pure Bi_2O_3. Up to 800 °C, the conductivity of composite varies insignificantly. Maximum of conductivity for the composite reaches the value of 8 S/cm at 800 °C, which exceeds the conductivity of pure δ-Bi_2O_3 and In_2O_3 at this temperature. Consequently, it is because of the presence of a percolated net of $In_{1.985}Zr_{0.015}O_{3.0075}$ solid solution in the composite (the conductivity of $In_{1.985}Zr_{0.015}O_{3.0075}$ solid solution is about 100 S/cm at 800 °C), which is also confirmed by its microstructure (Figure 2a). Figure 3b presents the oxygen fluxes for the 1.7 mm thick composite membrane at 750 °C, 775 °C and 800 °C. As it can be seen the oxygen flux increases at an increase in both the temperature and the partial pressure of oxygen on the outer/inner sides of the membrane, which responds to Wagner's theory of oxygen transfer in diffusion regime [12]. The maximum of oxygen flux at a difference of its partial pressures 0.21 atm/0.017 atm on both sides of membrane reaches a value of 2.4×10^{-7} mol·cm^{-2}·s^{-1} at 800 °C, while the selectivity of membrane (the flux of oxygen vs the flux of nitrogen) has been more than 20, which indicates the promise of using this composite material as an ion transport membrane to produce of pure oxygen from air. It should also be noted that this membrane material is allowed to be chemically stable during long-term operation in the presence of "acidic" (i.e., CO_2 and SO_2) gases in ambient air atmosphere, because indium and bismuth oxides are tolerant to them. As it is known this condition are not completely withstand by BSCF membranes [3-4], containing of alkaline earth barium and strontium. A further increase in the selectivity of membrane can be reached by lowering the total porosity of material, which may be achieved by controlling the particle size distribution of the initial powder charge together with applying hot pressing for consolidation of the green bodies.

4. Conclusions

A new membrane material has been synthesized that has a great potential to use in the tasks for concentrating of pure oxygen from air. The main disadvantage of this material is permanently increasing the cost of indium oxide, which is an order of magnitude higher than the cost of bismuth oxide. Apparently, a further subject of research will be the replacement of solid solutions based on indium oxide to cheaper oxides with comparable electronic conductivity, but also well compatible to bismuth oxide having an outstanding ionic conductivity.

Acknowledgements

This work was supported by the State Assignment of IMET RAS (No. 007-00129-18-00).

References

[1] Kerry F G 2007 *Industrial gas handbook: gas separation and purification* (Boca Raton, FL: CRS Press, Taylor & Francis Group) 550
[2] Zhu X and Yang W 2017 *Mixed conducting ceramic membranes: fundamentals, materials and applications* (Berlin: Springer) 367
[3] Liu Y, Zhu X, Li M, Liu H, Cong Y and Yang W 2013 *Angewandte Chemie International Edition* **52** 3232
[4] Partovi K, Liang F, Ravkina O and Caro J 2014 *ACS Applied Materials & Interfaces* **6** 10274
[5] Belousov V V, Schelkunov V A, Fedorov S V, Kul'bakin I V and Vorobiev A V 2012 *Electrochemistry Communications* **20** 60
[6] Belousov V V, Fedorov S V and Kul'bakin I V 2015 *Journal of Membrane Science & Research* **1** 96
[7] Belousov V V 2013 *MRS Communications* **3** 225
[8] Bérardan D, Guilmeau E, Maignan A and Raveau B 2008 *Journal of Applied Physics* **104** 064918
[9] Mason T O, González G B, Hwang J-H and Kammler D R 2003 *Physical Chemistry Chemical Physics* **5** 2183
[10] Kargin Yu F 2000 *Russian Journal of Inorganic Chemistry* **45** 1419
[11] Levin E M and Roth E S 1964 *Journal of Research of the National Bureau of Standards-A. Physics and Chemistry* **68A** 189
[12] Wagner C 1933 *Zeitschrift für Physikalische Chemie* **21** 25

IOP Conf. Series: Materials Science and Engineering **683** (2019) 012092 doi:10.1088/1757-899X/683/1/012092

Multiscale modeling of deformation and damage of elasticplastic particle reinforced composites

Yu I Dimitrienko, Yu V Yurin and S V Sborschikov

Department of Computational Mathematics and Mathematical Physics, Bauman Moscow State Technical University, 5 Baumanskaya 2-ya, Moscow, 105005, Russia

E-mail: dimit.bmstu@gmail.com, biofamily_7394@mail.ru

Abstract. A model is proposed for the deformation of elastic-plastic particle reinforced composite materials of a periodic structure with allowance for the damage. The model is based on a variant of the deformation theory of plasticity under active loading, in which the damage parameter is introduced, taking into account the difference in the accumulation of damage in phases during shear, tension and compression. The method of asymptotic homogenization of periodic structures was used to model the effective characteristics of elastic-plastic composites. Examples of numerical calculation are carried out for particle-reinforced metal composite - aluminum matrix filled with SiC particles.

Introduction

Methods for modeling the elastic-plastic characteristics of composites began to develop quite a long time ago [1], but until now these methods are mainly based on an approximate analysis of the mechanical behavior of composites based on certain hypotheses regarding joint deformation of composite components [2, 3]. At present, the most promising method for predicting the effective properties of composites and simulating micromechanical processes in composites is the asymptotic averaging method (homogenization method) [4–11 The purpose of this paper is to further develop the research begun in [9–11] for the case of elastic-plastic deformations of composites when taking into account the effects of detectability.

State of the problem for the mechanics of elastic-plastic composites with micro-damage

Consider a composite, to which in space R^3 there corresponds a region V with a surface Σ. A composite consists of N components V_α, $\alpha = 1...N$, components with indices $\alpha = 1,...,N-1$ are inclusions of various types, and a component with index α = N is a matrix. We denote Σ_α - the surfaces of the regions, V_α and $\Sigma_{\alpha N}$ the surfaces of the contact of the matrix and inclusions. The inclusions and the matrix are assumed to be isotropic elastic-plastic, corresponding to the deformation theory of plasticity with allowance for micro-damage [12]. Then in each area $V_\alpha, \alpha = 1...N$ we have the following problem of mechanics of elastic-plastic media with micro-damage

$$\begin{cases} \nabla_j \sigma_{ij}^{\alpha} = 0, & в V_{\alpha} \\ \sigma_{ij}^{\alpha} = \mathcal{F}_{ij}^{\alpha}\left(z^{\alpha}, \varepsilon_{kl}^{\alpha}\right), & в V_{\alpha} \cup \Sigma_{\alpha} \\ \varepsilon_{ij}^{\alpha} = \frac{1}{2}\left(\nabla_j u_i^{\alpha} + \nabla_i u_j^{\alpha}\right), & в V_{\alpha} \\ u_i^{\alpha} = u_i^{N}, \quad \left.\left(\sigma_{ij}^{\alpha} - \sigma_{ij}^{N}\right) n_j = 0\right\} & на \Sigma_{\alpha N} \\ u_i^{\alpha} = u_{ie}, на \Sigma_{\alpha e}^{1} \quad \sigma_{ij}^{\alpha} n_j = S_{ie}, на \Sigma_{\alpha e}^{2} \end{cases} \tag{1}$$

here $u_i^{\alpha}, \sigma_{ij}^{\alpha}, \varepsilon_{ij}^{\alpha}$ - displacements, stresses and deformations in the α -th component, $\nabla_j u_i^{\alpha} = \partial u_i^{\alpha} / \partial \overline{x}^{j}$ - are partial derivatives with respect to Cartesian coordinates $\overline{x}^{j}$, the last two conditions in system (1) are the conditions of ideal contact of the matrix and inclusions, and the conditions on the external surfaces, $z^{\alpha} = z^{\alpha}(\sigma_{ij}^{\alpha})$ - are micro-damage parameters that depend on stress σ_{ij}^{α}, and $\mathcal{F}_{ij}^{\alpha}\left(z^{\alpha}, \varepsilon_{kl}^{\alpha}\right)$ - nonlinear tensor function of the defining relations of elastic-plasticity with active loading.

Model of the of elastic-plasticity for media with damage

Let us choose the tensor functions $\mathcal{F}_{ij}^{\alpha}\left(z^{\alpha}, \varepsilon_{kl}^{\alpha}\right)$ corresponding to an isotropic medium obeying the modified A.Il'yushin theory of small elastoplastic deformations for active loading [12], which is modified by introducing the dependence of the elastic constants on the micro-damage parameter $z^{\alpha} = z^{\alpha}(\sigma_{ij}^{\alpha})$. In this case, the function $\mathcal{F}_{ij}^{\alpha}\left(z^{\alpha}, \varepsilon_{kl}^{\alpha}\right)$ (index α is omitted) has the following form

$$\sigma_{ij} = \mathcal{F}_{ij}\left(z, \varepsilon_{kl}\right) = (K(z) - \frac{2}{3}G(z)(1-\omega))\varepsilon_{kk}\delta_{ij} + 2G(z)(1-\omega)\varepsilon_{ij}, \tag{2}$$

where $\omega(\varepsilon_u))$ is the plasticity function of A.A. Il'yushin, which we choose in power form [23]

$$\omega(\varepsilon_u)) = \begin{cases} 0, & \varepsilon_u \le \varepsilon_T \\ 1 - (\varepsilon_T / \varepsilon_u)^{1-n}, & \varepsilon_u > \varepsilon_T \end{cases} \tag{3}$$

where n - is the hardening rate, ε_T - is the strain on the beginning of the material's fluidity, ε_u -is the second invariant (intensity) of the strain tensor [24], $K(z)$ - is the bulk modulus of the material, $G(z)$ - is the shear modulus depending on the micro-damage parameter z: $K(z) = K_0 a(z)$, $G(z) = G_0 a(z)$, where K_0 and G_0 are elastic constants, $a(z)$ - the function describing the micro-damage of the material, choose it as follows:

$$a(z) = \frac{1}{2}(\left|1 - (1-a_z)z^{m_z}\right| + 1 - (1-a_z)z^{m_z}) + \frac{a_z}{2}(1 - z - |1-z|), \tag{4}$$

where m_z is a constant. For the damage parameter z, we take the following dependence on the stress tensor invariants

$$z = \frac{\sigma_u^2}{3\sigma_S^2(1+B_1V(\sigma))} + \left(\frac{1}{\sigma_T^2} - \frac{1}{3\sigma_S^2}\right)\sigma_+^2 \qquad (5)$$

here are the invariants of the stress tensor: σ - the first, σ_u - the second and $\sigma_\pm = (|\sigma| \pm \sigma)/2$ - the sign-constant, σ_T σ_C σ_S - strength in tension, in compression and in shear. The constant B_1 is expressed through the limits of static strength under compression and shear $B = \left(\sigma_C^2 / 3\sigma_S^2 - 1\right)/\sigma_C$. The function $V(\sigma)$ in (7) describes a smooth transition of the accumulation of micro-damage between areas of tension and compression:

$$V(\sigma) = \begin{cases} 0, & if \quad \sigma > 0 \\ -\sigma, & if\ -\sigma_C < \sigma < 0 \\ \sigma_C & if \quad \sigma < -\sigma_C \end{cases} \qquad (6)$$

Model (5) with regard to (6) allows to take into account the effect of the difference in the accumulation of micro-damages in the area of tension and compression, which is characteristic of most materials.

Application of the asymptotic averaging method for elastic-plastic composites with damage

Let the composite material have a periodic structure, introduce a small parameter $\kappa = l / L << 1$ for it, as the ratio of the characteristic size of the periodicity cell (PC) to the characteristic size of the entire composite, and also introduce dimensionless global $x^k = \bar{x}^k / L$ and local $\xi^k = x^k / \kappa$ coordinates. Then we solve the solution of the problem (1),(2) in the form of asymptotic expansions of the form:

$$u_i^\alpha = u_i^{(0)}(x^k) + \kappa u_i^{\alpha(1)}(x^k, \xi^l) + \kappa^2 \ldots , \qquad (7)$$

Substituting asymptotic expansions (7) into system (1), and collecting terms at the same powers of k, we obtain the so-called local problem L_{pq} of the zero approximation over PC V_ξ :

$$\begin{aligned} &\sigma_{ij/j}^{\alpha(0)} = 0, \qquad in V_\xi \\ &\sigma_{ij}^{\alpha(0)} = \mathcal{F}_{ij}^\alpha\left(z^{\alpha(0)}, \varepsilon_{kl}^{\alpha(0)}\right), \qquad in V_\xi \cup \Sigma_s \\ &\varepsilon_{kl}^{\alpha(0)} = \delta_{kp}\delta_{lq}\bar{\varepsilon}_{pq} + \tilde{\varepsilon}_{kl}^{\alpha(0)}, \qquad \tilde{\varepsilon}_{kl}^{\alpha(0)} = \frac{1}{2}\left(u_{k/l}^{\alpha(1)} + u_{l/k}^{\alpha(1)}\right) \quad in V_\xi \\ &\left.\begin{aligned} &u_i^{\alpha(1)} = u_i^{N(1)} \\ &\left(\sigma_{ij}^{\alpha(1)} - \sigma_{ij}^{N(1)}\right)n_j = 0 \end{aligned}\right\} at \Sigma_{\xi\alpha N} \\ &\left\langle u_i^{\alpha(1)} \right\rangle = 0, \left[\!\left[u_i^{\alpha(1)} \right]\!\right] = 0. \end{aligned} \qquad (8)$$

Here are denoted $\sigma_{ij/j}^{\alpha(0)} = \partial\sigma_{ij}^{\alpha(0)} / \partial\xi_j$, $\sigma_{ij,j}^{\alpha(0)} = \partial\sigma_{ij}^{\alpha(0)} / \partial x_j$, $\left\langle u_i^\alpha \right\rangle$ - is the averaging operator over V_ξ , $\left[\!\left[u_i^\alpha \right]\!\right] = 0$ - is the condition of periodicity on the boundary of the PC. Due to the periodicity of the functions $u_i^{\alpha(1)}$, the following relationship holds : $\bar{\varepsilon}_{ij} = \left\langle \varepsilon_{ij}^{\alpha(0)} \right\rangle = \frac{1}{2}\left(u_{i,j}^{(0)} + u_{j,i}^{(0)}\right)$.

IOP Conf. Series: Materials Science and Engineering **683** (2019) 012092 doi:10.1088/1757-899X/683/1/012092

A numerical method for solving the local problem of deforming an elastic-plastic material with damage is described in [8, 9]. The software implementation of the developed numerical method was carried out using the SMCM software environment developed in the Scientific and Educational Center "Supercomputer engineering modeling and development of software systems" ("SIMPLEX") of the Bauman Moscow State Technical University and designed to solve the problems of mechanics of a deformable solid body by finite element methods.

Calculation of the effective elastic-plastic characteristics of the composite

After solving a series of problems Lpq (8) for all pq, we integrate the resulting stresses $\sigma^{(\alpha)}_{\beta\gamma(pq)}$ over the areas occupied by the fillings and the matrix

$$\bar{\sigma}_{ij} = \left\langle \sigma^{(\alpha)}_{ij} \right\rangle = \left\langle \mathcal{F}^{\alpha}_{ij}\left(z^{\alpha(0)}, \delta_{kp}\delta_{lq}\bar{\varepsilon}_{pq} + \tilde{\varepsilon}^{\alpha(0)}_{kl}\right)\right\rangle = \bar{\mathcal{F}}_{ij}\left(\bar{\varepsilon}_{pq}\right) \qquad (9)$$

as a result, we obtain the averaged stresses in the composite, which, with active loading, are functions of the averaged deformations $\bar{\varepsilon}_{pq}$. Relations (9) are non-linear defining relations of a composite. To find these constitutive relations, we apply the method of tensor-invariant approximation, according to which for the geometrically-isotropic structure of the PL, relations (9) are also non-linear-isotropic and can be represented as tensor-nonlinear isotropic functions [12]

$$\bar{\sigma}_{ij} = (K(\bar{z}) - \frac{2}{3}G(\bar{z})(1-\bar{\omega}))\bar{\varepsilon}_{kk}\delta_{ij} + 2G(\bar{z})(1-\bar{\omega})\bar{\varepsilon}_{ij} \qquad (10)$$

where $\bar{\omega}(\bar{\varepsilon}_u, \bar{\varepsilon})$ is the averaged plasticity function for a composite, generally speaking, depending on the 1st and 2nd invariant of the averaged strain tensor $\bar{\varepsilon} = \bar{\varepsilon}_{kk}, \bar{\varepsilon}_u$, and $\bar{z}$ is a measure of the composite damageability, for which, due to the isotropy of the composite, it is also possible to adopt model (4) - (6). Relations (10) can be written as the ratio between the first and second invariants of the averaged stress tensor $\bar{\sigma} = \bar{\sigma}_{kk}, \bar{\sigma}_u$ and the averaged strain tensor $\bar{\varepsilon}_u, \bar{\varepsilon}$

$$\bar{\sigma} = K(\bar{z})\bar{\varepsilon}, \qquad \bar{\sigma}_u = 2G(\bar{z})(1-\bar{\omega})\bar{\varepsilon}_u \qquad (11)$$

Modeling the processes of deformation and damage of composite materials based on the aluminum matrix and SiC particles

As an example, we performed numerical calculations of micromechanical stress and strain fields in PL in a composite material based on an aluminum alloy filled with SiC particles (Fig. 1). Figure 2 shows the deformation diagrams of the Al + SiC composite under tension along the axis direction up to separation into parts.

Figure 3 shows the dependences of the first invariant of the averaged stress tensor $\bar{\sigma} = \bar{\sigma}(\bar{\varepsilon}_u, \bar{\varepsilon})$ on the first invariant of the averaged strain tensor $\bar{\varepsilon}$ for fixed values of the 2nd invariant $\bar{\varepsilon}_u$. Calculations show that this dependence can indeed be considered linear, in accordance with (12). Figure 4 shows the dependences of the 2nd invariant of the averaged stress tensor $\bar{\sigma}_u = \bar{\sigma}_u(\bar{\varepsilon}_u, \bar{\varepsilon})$ on the 2-nd invariant of the averaged strain tensor $\bar{\varepsilon}_u$ for fixed values of the 1st invariant $\bar{\varepsilon}$. Calculations show that this dependence is indeed non-linear, but can be approximated with sufficient accuracy as a function of one variable (12), since the influence of the 1st invariant is not significant

IOP Conf. Series: Materials Science and Engineering **683** (2019) 012092 doi:10.1088/1757-899X/683/1/012092

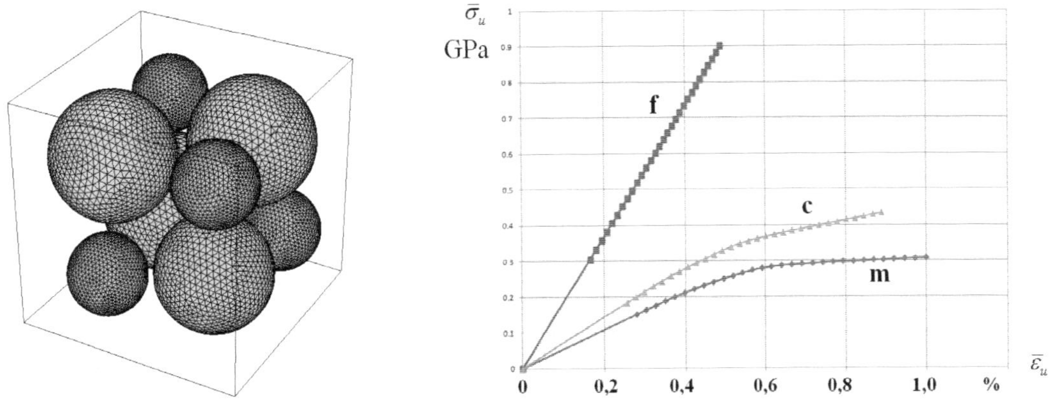

(a) (b)

Figure 1. PC of particle reinforced Al + SiC composite with a high concentration of particles (a) and calculated diagrams of the deformation of the Al matrix /m/, SiC particle /f/ and composite /c/ (b).

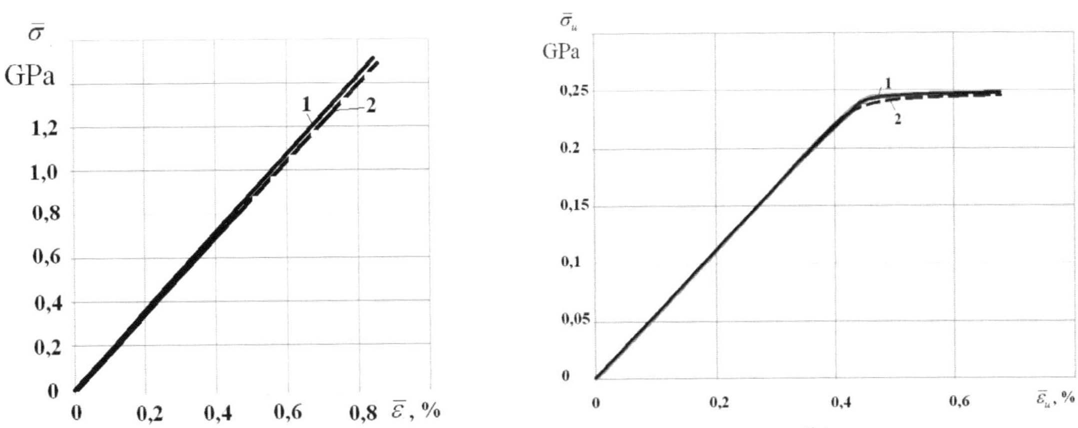

(a) (b)

Figure 2. (a) Calculated effective relation $\bar{\sigma} = \bar{\sigma}(\bar{\varepsilon}_u, \bar{\varepsilon})$ for fixed values $\bar{\varepsilon}_u$: 0 (1), 0.6% (2); (b) effective relation $\bar{\sigma}_u = \bar{\sigma}_u(\bar{\varepsilon}_u, \bar{\varepsilon})$ for fixed values $\bar{\varepsilon}$: 0 (1), 0.6% (2) for particle reinforced Al-SiC composite.

Conclusion

A model of multiscale deformation of elastic-plastic composite materials with regard to micro-damage damage is proposed. The model is based on a variant of the deformation theory of plasticity under active loading, taking into account micro-damage, which differs in tension and compression. The method of asymptotic homogenization of periodic structures is used to simulate the deformation of elastic-plastic composites. An example of numerical simulation of the process of deformation and damage for a dispersion-reinforced composite based on an aluminum matrix filled with SiC particles

IOP Conf. Series: Materials Science and Engineering **683** (2019) 012092 doi:10.1088/1757-899X/683/1/012092

is given. It is shown that the averaged defining relations of the composite are also quite well described by the same deformation theory of plasticity as the elastic-plastic matrix.

Acknowledgments
The work was carried out with support of state task of Ministry for Science and Education of the RF No. 9.5387.2017/БЧ

References

[1] Adams D F 1978 Uprugo-plasticheskoe povedenie kompozitov *Kompozicionnye-materialy. T.2. Mekhanika kompozicionnyh materialov.* M. Mir. pp 196-241.
[2] Vildeman V E, Sokolkin Yu V, Tashkinov A A 1997 *Mekhanika neuprugogo-deformirovaniya i razrusheniya kompozicionnyh materialov.* M. Nauka. 288 p.
[3] Nguen B N, Bapanapalli S K, Kunc V, Phelps J H, 2009 C.L.Tucker Prediction of the Elastic–Plastic Stress/Strain Response for Injection-Molded Long-Fiber Thermoplastics. *Journal of Composite materials*, Vol. **43**, № **3**. 217-246 p.
[4] Bensoussan A, Lions J L, Papanicolaou G 1978 *Asympotic analysis for periodic structures.* NorthHolland.
[5] Bakhvalov N S, Panasenko G P *Homogenisation: Averaging Processes in Periodic Media. Mathematical Problems in the Mechanics of Composite Materials.* Springer. 352 p.
[6] Pobedrya B.E. 1984 *Mekhanika kompozicionnyh materialov.* M. Moscow: Moscow State University Press. 352 p.
[7] Dimitrienko Yu I 2015 *Thermomechanics of Composites Structures under High Temperatures.* Springer. 367 p.
[8] Dimitrienko Yu I, Sborschikov S V, Belenovskaya Yu V, Aniskovich V A, Perevislov S N 2013 Modeling microstructural destruction and strength of ceramic composites based on the reactionbonded SiC. *Science and Education. Electronic Scientific and Technical Journal.* **11**, DOI: 10.7463/1113.0659438
[9] Dimitrienko Yu I, Gubareva E A, Sborschikov S V 2014 Finite element modulation of effective viscoelastic properties of unilateral composite materials. *Mathematical modeling and Computational Methods.* **2**. pp.28-49.
[10] Dimitrienko Yu I 2012 Asymptotic theory of multilayer thin plates. *Herald of the Bauman Moscow State Technical University Series Natural Sciences* **3** 86-100
[11] Dimitrienko Yu I, Dimitrienko I D and Sborschikov S V 2015 Multiscale Hierarchical Modeling of Fiber Reinforced Composites by Asymptotic Homogenization Method. *Applied Mathematical Sciences,* **9** 145 7211-7220
[12] Dimitrienko Yu I 2010 *Nonlinear Continuum Mechanics and Large Inelastic Deformations.* Springer. 722 p.

Analysis of anisotropy of time-dependent and nonlinear properties of unidirectional CFRP

A M Dumansky[1,2] **and Hao Liu**[2]

[1] Mechanical Engineering Research Institute of Russian Academy of Sciences, 4 Maly Kharitonievsky per., 101990, Moscow, Russia

[2] Bauman Moscow State Technical University, 5/1, Baumanskaya 2-nd Street, 105005, Moscow, Russia

Abstract Analysis of unidirectional carbon fiber reinforced plastic AS4/3501-6 regularities under strain-rate compression was given. Constitutive equations based on nonlinear hereditary mechanics relationships under shear in plane of the layer were elaborated. In order to obtain the constitutive equations, which are allowing describing the stress-strain curves of unidirectional carbon-fiber reinforced specimens under compression at different angles to the direction of reinforcement in the linear region, the algebra of resolvent operators and the relations of the elasticity theory of an anisotropic body were used. The possibility of correct using of the elaborated approach was shown.

1. Introduction

Carbon fiber reinforced plastics (CFRP) are widely used in different structures. It should be noted that most of the papers devoted to behavior under quasi-static loading [1]. In recent years, an increase number of works related to the study of the time-dependent properties of polymer-based composite materials have been observing. It is assumed that the main problem of the use of composite materials in structures is associated with the anisotropy of the mechanical properties causing the accumulation of damage during operation and the assessment of the residual resource. This is due to the presence of a polymeric binder leading to a noticeable time dependence of the properties, manifestation of the creep effects, relaxation, and strain-rate effect. In particular, in some of publications the effect of strain-rate loading on the stress-strain curves of CFRP was investigated [2-3]. Obviously, time-dependent properties most pronounced when loading in a direction, which is not coincide with the direction of reinforcement.

2. Constitutive equations

A constitutive equation of unidirectional layer takes a form depending on the direction of loading application. Clearly, in the reinforcement direction the time-dependent properties can be neglected, i.e.

$$\varepsilon_1 = \frac{1}{E_1}\sigma_1 . \qquad (1)$$

In the direction of transverse to reinforcement time-dependent properties revealed and hereditary constitutive equation can be taken in linear hereditary elastic form

$$\varepsilon_2 = \frac{1}{E_2^0}\left[1 + F_2^*\right]\sigma_2 , \qquad (2)$$

where F_2^* is a hereditary operator, which is defined by the following relationship $F_2^*\sigma = \int_0^t F_2(t-\tau)\sigma(\tau)d\tau$, F_2 is a kernel of the operator.

The most complex form of the constitutive equation is for shear in plane of the layer because of the time-dependent and nonlinear properties. Rabotnov nonlinear constitutive equation shows the following form [4]

$$\varphi(\gamma_{12}) = \left[1 + F_{12}^*\right]\tau_{12}, \qquad (3)$$

where $\varphi(\gamma_{12})$ is a function of instantaneous stress-strain curve which is along with time-dependent properties allows one to describe nonlinear properties.

3. Model description

The anisotropy of the elasticity of a unidirectional material describing by a known relationship can be described by equation of the following form

$$\frac{1}{E_\theta} = \frac{c^4}{E_1} + \left(\frac{1}{G_{12}} - \frac{2\nu_{12}}{E_1}\right)c^2 s^2 + \frac{s^4}{E_2}, \qquad (4)$$

Where E_1, E_2, G_{12} are the elastic moduli of the layer, ν_{12} is Poisson ratio, $c = \cos\theta$ and $s = \sin\theta$ are the trigonometric functions of the angle between the direction of the loading and the reinforcement.

Using the Volterra correspondence principle [4], which consists of replacing the elastic modulus with the corresponding operator expression, one can describe the deformation under a loading variable in time. In particularly, substituting in the expression for the elastic modulus (4), we can obtain an operator expression for the elastic modulus at an angle θ to the direction of reinforcement

$$\frac{1}{E_\theta^*} = \frac{c^4}{E_1} + \left(\frac{1}{G_{12}^*} - \frac{2\nu_{12}}{E_1}\right)c^2 s^2 + \frac{s^4}{E_2^*}. \qquad (5)$$

The expressions for the elastic modulus, expressed by the hereditary operators, will generally have the following form [4]:

$$\frac{1}{E^*} = \frac{1}{E^0}\left(1 + kF^*\right)$$

Substituting the Volterra representation in the expression for the elastic modulus (5), we obtain the operator expression for the elastic modulus

$$\frac{1}{E_\theta^*} = \frac{c^4}{E_1} + \left(\frac{1+F_{12}^*}{G_{12}^0} - \frac{2\nu_{12}}{E_1}\right)c^2 s^2 + \frac{s^4}{E_2^0}\left(1 + F_2^*\right) \qquad (6)$$

Expression (6) can be divided into two parts. The first one, which allows describing the instantaneous component of strain that does not depend on time and the second one, which allows describing the time-dependent properties

$$\frac{1}{E_\theta^*} = \frac{c^4}{E_1} + \left(\frac{1}{G_{12}^0} - \frac{2\nu_{12}}{E_1}\right)c^2 s^2 + \frac{s^4}{E_2^0} + \frac{1}{G_{12}^0}c^2 s^2 F_{12}^* + \frac{1}{E_2^0}s^4 F_2^*. \qquad (7)$$

Further, the problem is to identify the type and values of the hereditary operator parameters. The problem can be significantly simplified if we choose the resolvent type of the operators [4]. Linear combination of exponential functions pertains to such type of the operators presented by Prony series, as well as Rabotnov fractional exponential function [4].

In this paper, we take the condition of the hereditary operators in (2) and (3) that are similar and are Abel's operators where power function is used as a kernel. It means that F_2^* and F_{12}^* have the following kernel of operator: $I_\alpha = \frac{1}{\Gamma(1+\alpha)} t^\alpha$, $-1<\alpha<0$, where α is singularity parameter, $\Gamma(\ )$ is gamma-function. Substituting $F_2^* = k_2 I_\alpha^*$ and $F_{12}^* = k_{12} I_\alpha^*$ in (7) we get the operator expression of the modulus

$$\frac{1}{E_\theta^*} = \frac{1}{E_\theta^0} + \lambda_\theta I_\alpha^*, \tag{8}$$

where $\lambda_\theta = \frac{k_{12} c^2 s^2}{G_{12}^0} + \frac{k_2 s^4}{E_2^0}$ and $\frac{1}{E_\theta^0} = \frac{c^4}{E_1} + \left(\frac{1}{G_{12}^0} - \frac{2\nu_{12}}{E_1}\right) c^2 s^2 + \frac{s^4}{E_2^0}$.

Taking into account the assumption made earlier, we can write the following form of the constitutive equation

$$\varepsilon_\theta = \frac{1}{E_\theta^0}\left(1 + k_\theta I_\alpha^*\right)\sigma_\theta, \tag{9}$$

where $k_\theta = \lambda_\theta E_\theta^0$. It should be noted that from (9) we can obtain the operator expression for modulus at angle θ to the reinforcement direction.

Further, the values of the hereditary operators' parameters must be identified. This have been accomplished taking the experimental data obtained in [3]. Necessary to inverse constitutive equation (9), which can be done with the help of the algebra relationship for the resolvent operator [4]

$$\left(1 + k_\theta I_\alpha^*\right)^{-1} = 1 - k_\theta \Omega_\alpha^*(-k_\theta), \tag{10}$$

where $\Omega_\alpha^*(-k_\theta)\cdot 1 = t^{1+\alpha}\sum_{n=0}^{\infty}\frac{\left(-k_\theta t^{1+\alpha}\right)^n}{\Gamma\left[1+(1+\alpha)(1+n)\right]}$ is Rabotnov fractional-exponential function with the kernel expression: $\Omega_\alpha(-k_\theta, t) = t^\alpha \sum_{n=0}^{\infty}\frac{\left(-k_\theta t^{1+\alpha}\right)^n}{\Gamma\left[(1+\alpha)(1+n)\right]}$. Relationship (10) means that Rabotnov fractional-exponential function is the resolvent of Abel's operator.

Then, constitutive equation (9) can be inversed in the following form

$$\sigma_\theta = E_\theta^0\left(1 - k_\theta \Omega_\alpha^*(-k_\theta)\right)\varepsilon_\theta. \tag{11}$$

Interrelated constitutive equations (9) and (11) describe anisotropy of mechanical properties in linear elastic area. The experimental data digitized in [3] show noticeable effect of sensitivity to strain-rate and manifest physical nonlinearity at all values of the angles to the loading direction. To a less extent, they appear at angles equaling to $15^0, 90^0$ and to a more extent at angles equaling to 30°, 45° and 60°.

In the area of linear strain it is necessary to determine the following parameters: $E_1, \nu_{12}, E_2^0, k_2, G_{12}^0, k_{12}, \alpha$. The modulus of elasticity and Poisson ratio were taken from the data [1] $E_1 = 126$ GPa, $\nu_{12} = 0.28$. The singularity parameter α of the kernel was taken from previous analysis as known: $\alpha = -0.9$. With the help of the experimental data the remaining parameters E_2^0, k_2 and G_{12}^0, k_{12} can be determined. The first couple of the parameters were determined on the compression of the specimens at the angle 90°.

4. Compressive stress-strain curves

For further practical use, we give the following formula

$$\int_0^t (t-\xi)^\eta \xi^\beta d\xi = t^{1+\eta+\beta} B(1+\eta, 1+\beta) = t^{1+\eta+\beta}\frac{\Gamma(1+\eta)\Gamma(1+\beta)}{\Gamma(2+\eta+\beta)}, \tag{12}$$

where $B(\ )$ is beta function. Then, using (12) the expression for fractional exponential function acting on a power function is specified by

$$\left(1-\mathrm{Э}_{\alpha}^{*}\left(-k_{12}\right)\right)\cdot t^{b}=t^{b}-t^{1+\alpha+b}\sum_{n=1}^{\infty}\frac{\left(-k_{12}t^{1+\alpha}\right)^{n}\Gamma\left(1+b\right)}{\Gamma\left[1+b+\left(1+\alpha\right)\left(1+n\right)\right]}. \tag{13}$$

The constitutive equation describing strain-rate loading can be derived from relationship (13) when $b=1$ and may be written in the following form

$$\sigma_2=E_2^0\left[1-k_2\left(\frac{\varepsilon_2}{\dot{\varepsilon}}\right)^{1+\alpha}\sum_{n=0}^{\infty}\frac{\left(-k_2\left(\frac{\varepsilon_2}{\dot{\varepsilon}}\right)^{1+\alpha}\right)^{n}}{\Gamma\left[2+\left(1+\alpha\right)\left(1+n\right)\right]}\right]\varepsilon_2. \tag{14}$$

From (14) we define $E_2^0=15{,}5$ GPa, $k_2=0{,}2018$ $s^{-(1+\alpha)}$. Comparison experimental data and calculation presented in Figure 1.

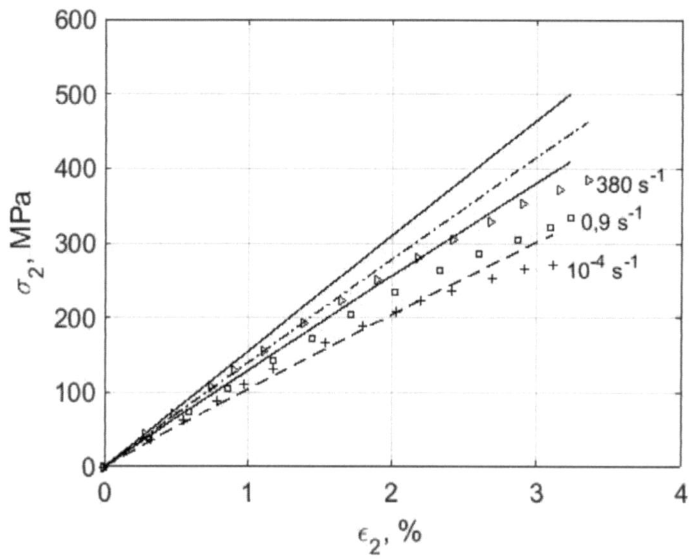

Figure 1. Stress-strain curves of unidirectional CFRP AS4/3501-6 under transverse strain-rate compression. (Upper line is instantaneous strain curve: $\sigma_2=E_2^0\varepsilon_2$)

It should be noted that a strain value of about 1.3% is a relatively small deviation from the linear law. At higher values of the strain, it is necessary to establish nonlinear constitutive equation.

The parameters characterizing the deformation under in-plane shear of the layer were determined from the testing of specimens compressed at angle 45° with taking into account the previously determined numerical values. When calculating the characteristics of in-plane shear of the layer, it was assumed that the modulus of elasticity in the direction of reinforcement weakly depends on time and can be taken equal to a constant value.

The values G_{12}^0 and k_{12} were determined by using the CFRP deformation curves at angle of 45° and strain-rate of 380, 0,8 and 10^{-4} s^{-1} with the equation similar to (14)

$$\sigma_{45}=E_{45}^0\left[1-k_{45}\left(\frac{\varepsilon_{45}}{\dot{\varepsilon}}\right)^{1+\alpha}\sum_{n=0}^{\infty}\frac{\left(-k_{45}\left(\frac{\varepsilon_{45}}{\dot{\varepsilon}}\right)^{1+\alpha}\right)^{n}}{\Gamma\left[2+\left(1+\alpha\right)\left(1+n\right)\right]}\right]\varepsilon_{45} \tag{15}$$

The parameters values can be derived as $E_{45}^{0} = 25,15$ GPa, $k_{45} = 0,2564$ $s^{-(1+\alpha)}$. Comparisons of the calculated data and the experimental ones in area of linear strain are in satisfactory agreement.

The numerical values of parameters for in-plane shear in the plane of the layer were obtained by using the relationships for E_{θ}^{0} and k_{θ} proved to be $G_{12}^{0} = 10,1$ GPa, $k_{12} = 0,301$ $s^{-(1+\alpha)}$. It should be emphasized that the above-mentioned values of the parameters were determined from the experimental data in the linear part of the stress-strain diagrams.

In Figure 2 it is shown the dependence of the parameters characterizing anisotropy of instantaneous modulus and time-dependent properties. The dependence of the instantaneous modulus is similar to the corresponding to anisotropic theory of elasticity.

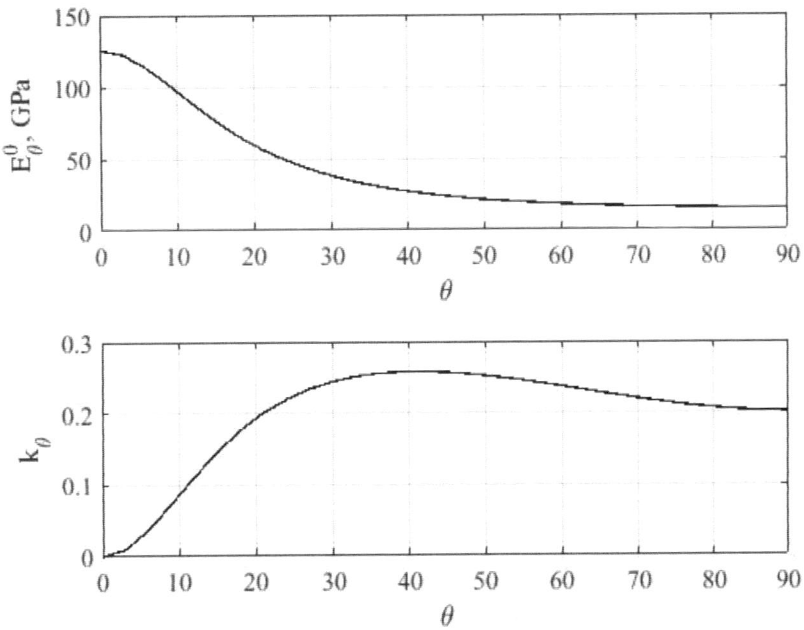

Figure 2. Dependencies of the constitutive equation anisotropy parameters.

The obtained constitutive equations allow one describing the mechanical behavior of CFRP under arbitrary loading conditions. For example, under creep, relaxation, etc.

However, the experimental data given in [3] were not enough to obtain all the necessary parameters of nonlinear constitutive equation. The value of threshold strain of the linear area can be determined as $\gamma_* = 0,75\,\%$. The nonlinear constitutive equation for in-plane shear of the layer with Abel's hereditary operator can be written as

$$\varphi(\gamma_{12}) = \left(1 + k_{12} I_{\alpha}^{*}\right)\tau_{12}. \qquad (16)$$

By using the algebra of resolvent operator equation (16), can be inversed

$$\tau_{12} = \left(1 - k_{12}\Omega_{\alpha}^{*}\left(-k_{12}\right)\right)\varphi(\gamma_{12}). \qquad (17)$$

The instantaneous stress-strain curve should be a piecewise smooth function being composed of a linear section and a non-linear section, which is the best way to describe nonlinear strain. With the help of the power approximation in the nonlinear strain area, the instantaneous deformation curve equation is given as following

$$\varphi(\gamma_{12}) = G_{12}^{0}\gamma_{12}\left[H(\gamma_{12}) - H(\gamma_{12} - \gamma_*)\right] + \left[\tau_* + a(\gamma_{12} - \gamma_*)^{b}\right]H(\gamma_{12} - \gamma_*), \qquad (18)$$

where $\tau_* = G_{12}^0 \gamma_*$, parameters a,b define nonlinear strain under shear, $H(\ )$ is Heaviside unit function of the corresponding argument.

The shear stress-strain curve is determined from the instant strain curve by the time dependence of the strain

$$\tau_{12} = \varphi(\gamma_{12}) - k_{12}\Omega_\alpha^*(-k_{12})\varphi(\gamma_{12}) \qquad (19)$$

Using expression (13) and denoting: $\psi_b(t) = (1 - k_{12}\Omega_\alpha^*(-k_{12}))\cdot t^b$, we can write the constitutive equation, which allows one to describe the deformation at a strain-rate loading: $\gamma_{12} = \dot{\gamma}\cdot t$. Substituting the last expression into (18) and then into (17) we get

$$\tau_{12} = G_{12}^0 \psi_1(t)\dot{\gamma}\left[H(\dot{\gamma}t) - H(\dot{\gamma}(t - t_*))\right] + \left[\tau_* + \psi_b(t - t_*)\dot{\gamma}^b\right] H(\dot{\gamma}(t - t_*)), \qquad (20)$$

where $\tau_* = G_{12}^0 \psi(t_*)\cdot\dot{\gamma}$, $\dot{\gamma}t_* = \gamma_*$. The following parameters of the instantaneous stress-strain curve were obtained: $a = 30$ GPa, $b = 0{,}8$. A comparison of experimental data and the calculated strain curve is shown in Figure 3.

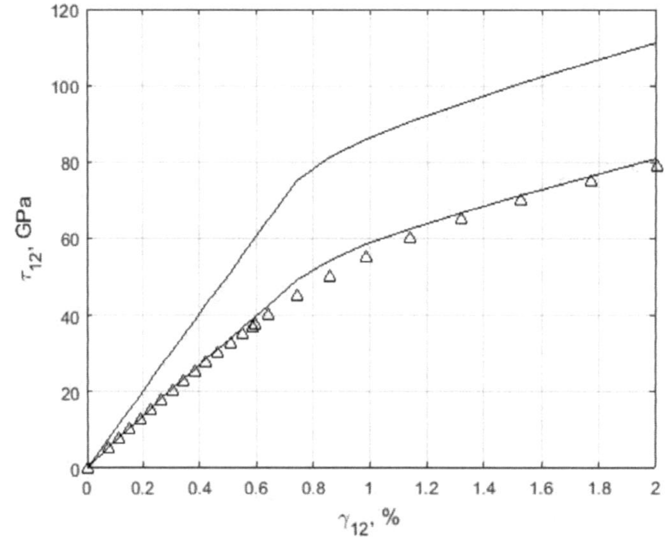

Figure 3. Stress-strain curves under shear strain-rate loading. The upper solid line is an instantaneous stress-strain curve.

5. Conclusions

The model based on constitutive equations of hereditary mechanics of solids has been derived. The method allows one to predict anisotropy of mechanical properties of unidirectional CFRP under time-variable loading has been suggested. To derive constitutive equations, the relations of anisotropic theory of elasticity, Volterra correspondence principle and relationships of resolvent operators' algebra have been used. The offered approach has been applied to experimental readings of unidirectional CFRP AS4/3501-6. The developed approach can serve as a basis for the description of the mechanical behavior regularities of CFRP.

References

[1] Hinton M J, Kaddour A S and Soden P D 2004 *Failure criteria in fibre reinforced polymer composites: The World-Wide Failure Exercise* (Amsterdam: Elsevier)

[2] Kuhn P, Ploeckl M and Koerber H 2015 Experimental investigation of the failure envelope of unidirectional carbon-epoxy composite under high strain rate transverse and off-axis tensile loading *EPJ Web of Conference 94*, 01040 (DOI: 10.1051/epjconf/20159401040)
[3] Daniel I M, Werner B T and Fenner J S 2011 Strain-rate dependent failure criteria for composites *Composite Sci. and Technology* **71** N 9 pp 357-364
[4] Rabotnov Yu N 1980 *Elements of hereditary mechanics of solids* (Mir Publishers)

Investigation of mechanical characteristics of a carbonceramic composite material based on a braided preform

A N Baryshev[1], G G Kulish[1], A A Smerdov[1], I A Timofeev[2] and S V Tsvetkov[1]

[1]Bauman MSTU, Russia, 2nd Baumanskaya Str., 5, 105005, Moscow, asmerdov@mail.ru

[2]JSC "Kompozit", Pionerskaya Str., 4, Korolev, Moscow Region, 141070, Russia

Abstract. An experimental analysis of characteristics of a carbon-ceramic composite material based on a braided preform in a plane stress state was carried out. The technical limits of strength and elastic characteristics of the material are determined. Tests of tubular specimens in a complex stress state allowed us to obtain the data necessary for the constrction of the failure criterion. The possibilities of using various failure criteria of the composite are investigated. The contours of the surface of the strength of the material in the space of stresses at the intersection of its coordinate planes are built.

Keywords: Experiment, tubular specimen, carbon-ceramic composite, failure criterion

Introduction

Carbon-ceramic composite materials (CCrCM) for temperature resistance and mechanical characteristics are close to carbon-carbon composite materials (CCbCM), but they are able to work for a long time at high temperatures in oxidizing environments [1, 2]. An important feature of hightemperature composite materials based on carbon fibers is a weak change in the strength and elastic characteristics in a wide temperature range [3]. This allows to significantly simplify the process of experimental analysis and conduct tests without heating.

The starting material for the manufacture of CCrCM is usually a billet of CCbCM, the matrix of which is saturated with silicon carbide during processing. Form, basic geometrical dimensions of the product and the structure of the filler are stipulated at the stageof making preforms and carbon-carbon blanks. The conventional method for the production of carbon-fiber reinforced plastics is the creation of shell parts –"winding" is limited for creating a reinforcing base of high-temperature composite materials in connection withthe difficulty of forming carbon and ceramic matrices in dense non-intertwined filament structures. A promising and high-performance method for manufacturing shell-type parts from high-temperature composite materials is the contour weaving on the forming tool [4]. Carbon-silicon carbide matrix can form by both a gas-phase method, providing materials with low porosity and a high content of SiC, but having a low performance, and a liquid-phase method, which features higher performance and lower costs [1, 2, 5-7].

Purpose and methodology of the experiment

The purpose of the experimental analysis was to determine the elastic and strength characteristics of the CCrCM braided frame, as well as the selection of the failure criterion that correctly describes the process of failure in a plane stress state. Due to the high sensitivity of the characteristics of composites to manufacturing technology for the production of the specimens manufacturing techniques of standard products should be applied. It is also necessary that the stress-strain state of the working part of the specimen is uniquely determined by the applied force factors, and the failure occurs within the working area.

These requirements are fully satisfied with thin-walled tubular specimens, used in this study. The sample drawing is shown in Figure 1.

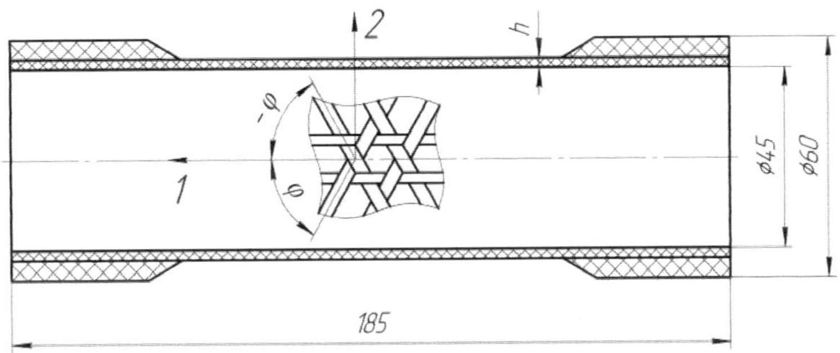

Figure 1. Thin-walled tubular sample and the direction of the axes

Tubular specimens of CCrCM type Cf / (C-SiC) m are made in JSC "Kompozit", Korolev. The reinforcing filler: carbon fiber УКН-М-3К. Volume fraction of carbon fiber:≈52%. Matrix: pyrolytic carbon, pyrolytic silicon carbide. Reinforcement pattern is 0/±70°. Silicon carbide content: ≈32% of the mass. Sizes of experimental samples: internal diameter 45 mm, height 185 mm, wall thickness: ≈ 1.9 mm. The braided preform of the specimen is formed by a system of skeletal filaments located along the generatrix and braid filaments located at angles $\varphi = \pm 70°$ to the generatrix. Figure 1 shows the coordinate system in which all the results are presented. Axis 1 is directed along the generatrix, axis 2 - in the circumferential (transverse) direction. The edges of the sample are supported by a winding made of a polymer composite material.

Tubular specimens were loaded with tensile and compressive force in the axial direction, external and internal pressure, as well as torque. These types of tests allow to determine the full set of technical characteristics of strength and elasticity of the material for the case of a plane stress state. The method and results of tests of tubular specimens from CCrCM manufactured at JSC "Kompozit" are given in [8].

When testing tubular specimens in a complex stress state, the data necessary to construct a failure criterion were obtained. During the study, biaxial compression, biaxial stretching, joint stretching in the axial direction and compression in the circumferential direction, joint compression in the axial direction and stretching in the circumferential direction, joint stretching and shear in the axial direction were realized.

To measure strains, strain gauges were used, glued to the lateral surface of the specimen in the axial direction, in the circumferential direction and at an angle of 45° tothe generatrix.When during testing the sample was subjected to external pressure, the sensors were placed on the inner surface of the specimen.

In tensile tests along the axis and in torsion, normal stresses in the direction of axis 1 and tangential stresses in the plane 12 were created in the material respectively. For fixing the sample the specialized equipment was used for the testing machine.The transfer of the axial load on the sample was carried out using self-tightening collet grips, while the torque was transmitted using pins.

During compression in the axial direction, the specimen was freely mounted on the plates of the testing machine, and in the material in the direction of the axis 1, compressive stresses were created. To ensure uniformity of the stress field in the circumferential direction, a load was applied through a spherical support.

When testing for internal and external pressure, the specimen material was loaded for tension and compression in the direction of the axis 2, respectively. Since the design of the specimen does not provide for the presence of a sealing layer, the loading was carried out through an elastic sealing shell. Devices are made in such a way that the axial component of the load does not affect the specimen.

For the tensile and compression tests in the axial direction, combined with the compression in the circumferential direction, an experimental device, shown in Figure 2, has been developed and manufactured.

Figure 2. Device for the joint loading of tubular specimens with axial force and external pressure

The external pressure acting on the specimen in the chamber of the developed device can reach 40 MPa, and the maximum force transmitted to the sample is 150 kN. The axial force is controlled independently of external pressure.A structurally similar device is described in [9], but the developed version allows for a wider range of possible load paths.

Experimental results

Strength and elastic characteristics were determined as the average value of the results for tests of at least three specimens. Characteristics are shown in Table 1.

IOP Conf. Series: Materials Science and Engineering **683** (2019) 012094 doi:10.1088/1757-899X/683/1/012094

Table 1. Characteristics of CCrCM.

Characteristics	Value
Ultimate tensile strength along the axis X + 1, MPa	174.9
Modulus of tensile elasticity along the axis E1, GPa	38.7
Poisson's ratio in tension along the ν12 axis	0.137
Compressive strength along the axis X-1, MPa	140.3
Modulus of elasticity in compression along the axis E1, GPa	41.0
Poisson's ratio in compression along the ν12 axis	0.157
Tensile strength in tension in the circumferential direction X + 2, MPa	414.7
Modulus of tensile elasticity in the circumferential direction E2, GPa	87.3
Poisson's ratio in tension in the circumferential direction ν21	0.170
Compressive strength in the circumferential direction X-2, MPa	234.2
Modulus of elasticity in compression in the circumferential direction E2, GPa	80.7
Poisson's ratio in compression in the circumferential direction ν21	0.277
Strength at shear X12, MPa	90.4
Shear modulus G12, GPa	16.4

From Table 1it follows that the tensile strength of the material significantly exceeds the compressive strength.The discrepancy between the values of elastic characteristics, determined under tension and under compression, is not more than 8%.
Figure 3 shows a typical stress-strain diagram in the longitudinal direction.Diagrams for other types of stress haveclose to linear character until the moment offailure.
The characteristic appearance of the failed specimens is shown in Figure 4.

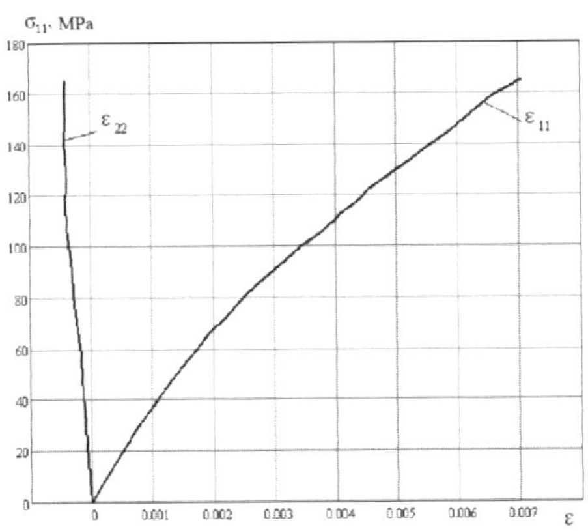

Figure 3. The characteristic pattern of the strain under longitudinal tension.

а) б) в) г)

Figure 4. View of failed specimens: a) under loading with external pressure; б) in torsion; в) under loading with internal pressure; г) when stretching along a generatrix

The results of strength tests under complex stress conditions are given in Table 2.

Table 2. Test results for complex stress state

Item No.	Test Type	Destructive stress values		
		σ_{11}, MPa	σ_{22}, MPa	σ_{12}, MPa
1.	Biaxial stretching	220.7	335.0	–
2.	Biaxial stretching	196.3	498.4	–
3.	Biaxial compression	-131.25	-262.5	–
4.	Biaxial compression	-99.8	-262.5	–
5.	Circumferential compression and tension along the axis	158.6	-199.8	–
6.	Circumferential tension and compression along the axis	-119.6	351.1	–
7.	Torsion and stretching along the axis	100.2	–	73.5
8	Torsion and stretching along the axis	186.4	–	11.8

Construction of material strength criterion

When studying the failure process of a composite material, two approaches are possible micromechanical [11, 12, 13] and phenomenological [14]. To apply the micromechanical approach, it is necessary to study the properties of the matrix and the filler. Given the fact that the CCrCM matrix is formed directly during the manufacture of the material, separate study of its properties is difficult. The development of strength criteria for composite materials for various types of stress state is the subject of a large number of works [15, 16, 17]. In practice, the criteria of the Mises-Hill, Tsai-Wu, Tsai-Hill, Hashin, maximum stresses and maximum strains and others are widely used [16, 18].

The strength of the material studied can be described using various criteria. Let us consider some of them:

- criterion of maximum stress;
- modified Mises-Hill criterion; - Tsai-Wu criterion.

The maximum stress criterion is formulated as:

$$\begin{cases} X_{-1} \le \sigma_{11} \le X_{+1} \\ X_{-2} \le \sigma_{22} \le X_{+2} \\ |\sigma_{12}| \le X_{12} \end{cases} \tag{1}$$

The expression for the modified Mises-Hill criterion is:

$$\frac{\sigma_{11}^2}{\sigma_{1B}^2} - \frac{\sigma_{11}\sigma_{22}}{\sigma_{1B}\sigma_{2B}} + \frac{\sigma_{22}^2}{\sigma_{2B}^2} + \frac{\sigma_{12}^2}{\sigma_{12B}^2} = 1 \tag{2}$$

where σ_{1B}, σ_{2B} are strength limits in direction 1 and 2, chosen according to the type of stress state ($\sigma_{1B}=X_{+1}$ at $\sigma_{11}>0$, $\sigma_{1B}=X_{-1}$ at $\sigma_{11}<0$, $\sigma_{2B}=X_{+2}$ at $\sigma_{22}>0$, $\sigma_{2B}=X_{-2}$ at $\sigma_{22}<0$), $\sigma_{12B}=X_{12}$. Tsai-Wu criterion:

$$F_1\sigma_{11} + F_{11}\sigma_{11}^2 + F_2\sigma_{22} + F_{22}\sigma_{22}^2 + F_{12}\sigma_{11}\sigma_{22} + F_{66}\sigma_{12}^2 = 1 \quad (3)$$

where F1, F11, F2, F22, F12, F66 are constants determined by test results, and F1, F11, F2, F22, F66 are uniquely determined through the tech nical limits of material strength, and F12 is selected from the condition of the best description of the experimental data obtained complex stress state [18].

Figure 5 shows the contours of the surface strength in the planes σ_{11}-σ_{22}, σ_{11}-$\sigma_{12}$and σ_{22}-σ_{12}.

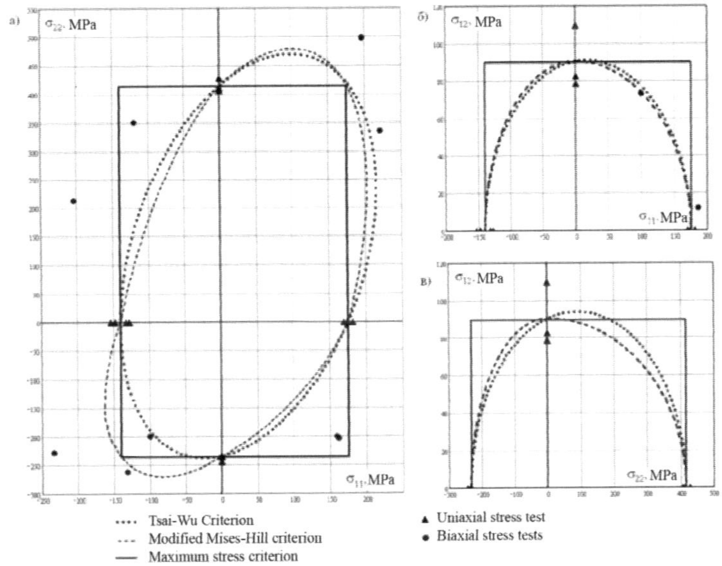

Figure 5. Contours of the surface strength.

The resulting contours and surface strengths meet the requirements: are simply connected, closed, and intersect the coordinate axes at points corresponding to technical strengths [18]. All considered criteria qualitatively correctly describe the behavior of the material in a series of experiments. To quantify the quality of the criteria, we introduce a measure of the deviation from the experimental points obtained under a complex stressed state:

$$\Delta = \sum_k \frac{\left(s_i^* - s_i\right)^2}{s_i^2} \quad (4)$$

where k is the number of experimental points obtained under a complex stress state, si is the distance from the origin of coordinates O in the stress space σ11-σ22-σ12 to the experimental point Pk, si* is the distance from the origin to the point of intersection of the OPk beam with the surface strength. For the Tsai-Wu criterion, the calculated deviation was 0.717, for the modified Mises-Hill criterion it was 0.735, and the minimum deviation value achieved using the maximum stress criterion was 0.472. Thus, among the options considered, the maximum stress criterion the bestdescribes the strength of the material studied.

Conclusions

The characteristics of the CCrCM with a reinforcement pattern of ± 70°/0° in the plane stress state are investigated. It is established that the strength of the material in the plane stress state is satisfactorily described by the maximum stress criterion. In the presence of additional experimental data, it becomes possible to carry out a procedure for identifying the characteristics of reinforcing elements and a matrix for performing calculations for products with an arbitrary reinforcement pattern.

References

[1] Kostikov V I and Varenkov A N 2003 *Superhigh-temperature composite materials* (Moscow: Intermet Engineering) 560 s.
[2] Ed NP Bansal, J Lamon 2015 *Ceramic matrix composites*: *materials, modeling and technology* (Wiley-VCH) 694 p. ISBN 978-1-118-23116-6
[3] Goto K, Hatta H, OeMand Koizumi T 2003 Tensile Strength and Deformation of a Two-Dimensional Carbon-Carbon Composite at Elevated Temperature *J. Am. Ceram. Soc.* **86 (12)** 2129-35.
[4] Bogomolov P I, Kozlov I A and Birulya M A 2017 Review of modern technologies for the manufacture of volume-reinforcing preforms for promising composite materials *Technical and technological problems of service* **1 (39)** 22-27.
[5] Timofeev A N, Mikhailovskiy V V, Timofeev I A, Bogachev E A, Koshugug E E and Shiryaev AV 2012 Silicon nitride as a precursor of silicon carbide during siliconization of carbon carbon composite materials *Structures of composite materials* **1** 11-16
[6] Timofeev P A 2017 *Formation of matrixes of composite materials of silicon carbides, nitrides and borides by the method of pyrolysis of polymer precursors* Diss. papers of Tech.Sc.Cand. (Moscow:MSTU named afterN EBauman) 126 p. (reg.ICD number:AAAA-B18- 418070290051-1)
[7] Ed W Krenkel 2008 *Ceramic Matrix Composites. Fiber Reinforced Ceramics and their Applications* (Wiley-VCH) 418 p. ISBN: 978-3-527-31361-7
[8] Smerdov A A, Tairova L P and Timofeev I A 2018 Experimental analyses of characteristics of tubular specimens of composite materials based on whole-braided multilayer carbon frames and polymer, carbon and ceramic matrices *Structures of composite materials* **2** 52-59
[9] Kaiser C, Weihs H, Zuknik K-H and Obst A 2005 Failure criteria for FRP and CMC: theory, experiments and guidelines *Proc. of the European Conference on Spacecraft Structures, Materials and Mechanical Testing* 10 - 12
[10] Dimitrienko Yu I, Sokolov A P, Shpakova Yu V and Yurin Yu V 2012 Modeling the strength surfaces of composites based on a microstructural finite elements analysis *Science and education* **11** 487-96
[11] Magnitsky I V, Ponomarev K A, Mironikhin A N and Tashchilov S V 2012 On the issue of numerical modeling of the elastic properties of composite materials with regard to reinforcement pattern *Engineering magazine: science and innovation***8 (8).**
[12] Yang Q Sand Q in QH 2004 Micro-mechanical analysis of composite materials by BEM *Engineering Analysis with Boundary Elements* **28** 919–26.
[13] Kopnov V A and Voronin V N 2002 Phenomenological criteria for the strength of a continuous anisotropic medium at different loading rates *Vestnik RUDN. Series: Engineering studies* **1** 22-26.
[14] Rotem A 2012 The Rotem failure criterion for fibrous laminated composite materials: three dimensional loading case *J. of Composite Materials* **46 (6-7)** 2379–88
[15] Soden P D, Kaddour A S and Hinton M J 2004 Recommendations for designers and researchers resulting from the world-wide failure exercise *Composites Science and Technology* **64** 589– 604.
[16] Kaddour A S, Hinton M J, LiSandSmith P A 2013 Recent progress on benchmarking cracking and damage models for fibre reinforced polymer composites. The Third World-Wide Failure Exercise (WWFE-III) *Proc. Of the 19th International conference on composite materials*, Jul 28 – Aug 2, Montreal, Canada.
[17] Vu E M, Ed. J. Sendetsky 1978 Phenomenological criteria for the destruction of anisotropic media *Mechanics of composite materials* **2**. (Moscow: Mir) 401-91

IOP Conf. Series: Materials Science and Engineering **683** (2019) 012095 doi:10.1088/1757-899X/683/1/012095

Structure and properties of the composite material based on heat-resisting VKNA-4U alloy with oxide stiffening fillers, designed for additive manufacturing technologies of power fasteners for gas turbine engines (GTE)

M V Unchikova[1], **A I Rodionov**[2] **and N A Golovlev**[1,2]

[1]Bauman MSTU, Moscow, Russian Federation

[2]VIAM FSUE, Moscow, Russian Federation

Abstract. The influence of the structure and concentration of the stiffening component on the physicomechanical properties of the heat-resisting intermetallic alloy VKNA-4U was investigated. The dependence of the strength properties of the alloy on the composition and ratio of the hardeners α-Al_2O_3, Y_2O_3, HfO_2, ZrO_2 was experimentally determined. Using scanning electron microscopy (SEM), it was found that when the alloy is reinforced with hafnium, yttrium and aluminum oxides, the dispersed hardener does not dissolve in the matrix. These compositions are considered as the most promising for the stiffening of VKNA-4U alloy.

It is shown in this paper that the required content of the oxide reinforcing filler in the VKNA-4U matrix is 2%, at which a balance is reached between the maximum value of strength and ductility. In the future, VKNA-4U powder composition mechanically alloyed with α-Al_2O_3, Y_2O_3, HfO_2 oxides can be used as a building material for additive synthesis technologies for power fasteners of gas turbine engines (GTE).

1. Introduction

One of the most important tasks of modern materials science is the creation of materials with high operating temperature while maintaining a low density, due to the need to increase the temperature of the gas at the turbine inlet. This can be achieved by developing a new class of high-temperature materials or applying new principles of hardening metal heat-resistant matrix, which will allow to obtain working temperatures higher than that of existing nickel alloys.

Stiffening and strengthening mechanisms, which are based on a multi-component doping, inevitably lead to a decrease in the melting temperature (Tm), and therefore limit the upper limit of operating temperatures. One of these principles is implemented in metal-ceramic composite materials, in which the hardening phase consists of ultrafine particles artificially introduced into the matrix, high-strength, thermodynamically stable, not soluble and not interacting with the matrix material throughout the temperature range up to the melting point of the matrix alloy. The hardening effect of the introduction of ultrafine particles is summed up with the effect of using traditional hardening mechanisms, but, unlike the latter, it is maintained at extremely high temperatures when other hardening mechanisms cease to operate [1, 2].

For power fasteners of gas turbine engines, it is advisable to use alloys with a working temperature equal to the maximum temperature of their installation. For combustion chambers, high-pressure boiler and turbine the alloys are needed with operating temperatures up to 1200°C. In most serial domestic engines power fasteners are made of serial deformable nickel alloys such as ЭИ437БУ, ЭИ698, ЭП742.

The maximum operating temperature of these alloys is 750°C, above which the material softens and its strength properties sharply decrease [3, 4].

FSUE "VIAM" has developed a number of high-strength and heat-resistant alloys that can be used to manufacture parts of an aircraft. For example, (Ni3Al) has a number of important advantages over industrial nickel alloys - operating temperature up to 1200°C for a long time (short-term temperature rises up to 1250°C are possible), the maximum heat resistance among alloys of its class at the operating temperature [5]. The disadvantage of all nickel alloys is a decrease in the strength characteristics with increasing operating temperature. This problem can be solved through adding the cermet reinforcing filler to them. It is believed that the enlargement of particles at high temperature occurs as a result of the dissociation of small particles of the following diffusion of components through the matrix and their deposition on larger particles. It follows that the fine phase should have the highest possible thermodynamic stability, and oxides meet these requirements the best [6, 7].

The addition of the oxide filler to the VKNA-4U alloy should increase the heat resistance of the finished material and eliminate the high-temperature creep of the matrix alloy, which will allow to maintain the strength and geometry of the power fasteners of GTE made from this alloy.

The purpose of this paper is to study the effect of oxide reinforcing filler on strength characteristics of matrix material of VKNA-4U alloy at room temperature.

2. Materials and methods of experiment

As a material for studying the effect of reinforcement with nanoscale powders of oxides α-Al2O3, Y2O3+Al2O3, HfO2+Y2O3+Al2O3, HfO2+Y2O3, HfO2+ZrO2+Y2O3, Al2O3+ZrO2+Y2O3, ZrO2+Y2O3 the heat-resisting intermetallic alloy VKNA-4U was selected. The starting components for the manufacture of composite powder based on VKNA-4U alloy were domestic powders.

The initial charge billet of the matrix alloy was obtained in an induction furnace UNK-1. Next, the atomization technology was applied (obtaining a micro-sized powder by spraying the charge stock) on the sixth generation HERMIGA 10/100 unit.

After gas atomization, the powder of the VKNA-4U matrix alloy was divided into fractions 0-56 μm and 56-120 μm on the Analysette 3 Spartan vibrating screen and the sieve (mesh according to GOST 6613) with a mesh size of 56 microns.

To obtain nanoscale oxide granules α-Al2O3, Y2O3+Al2O3, HfO2+Y2O3+Al2O3, HfO2+Y2O3, HfO2+ZrO2+Y2O3, Al2O3+ZrO2+Y2O3, ZrO2+Y2O3 the plasma-chemical method was applied. The fraction was synthesized on the installation of the plasma torch, the design of which was developed at the Institute of Metallurgy and Materials Science. A. A. Baykova RAS [8]. The principle of synthesis of nanoscale oxide particles is based on the interaction of dispersed raw materials with a jet of thermal plasma containing oxygen.

High-temperature chemical reactions occur in the reactor volume, leading to vaporization of the target product, which is then condensed into nanoparticles. Formed nanoparticles are deposited on the walls of the reactor and on the filter, from where they are removed in the receiving collections of the resulting product. Plasma-chemical unit is designed to obtain nanopowders as a result of oxidative processes, as well as processes of thermal decomposition of compounds. Particles of refractory compounds produced in a plasma-chemical installation, are predominantly spherical. The average particle size is no more than 300 nm.

To assess the interaction of the VKNA-4U matrix and oxide fillers, experimental samples of the following compositions were made: VKNA-4U + ZrO2 – Y2O3 – Al2O3, VKNA-4U + ZrO2 – Y2O3, VKNA-4U + HfO2 – Y2O3, VKNA-4U + HfO2 – Y2O3 – Al2O3, VKNA-4U + Y2O3 – Al2O3, VKNA-4U + HfO2 – Y2O3 – ZrO2. Experimental samples were obtained according to the following scheme: powder composite billet was obtained by the method of mechanical alloying in a planetary mill

PM400, then the powder material was compacted by cold pressing on a hydraulic press. After mechanical alloying, the average size of the composite granules was 60 μm. Compaction granules were compacted on a hydraulic press ДБ2432 with a force of 160 tons. The compacted samples were mechanically processed and annealed in a vacuum electric furnace СНВГ 4/22 at a temperature close to the working one (1200°C) for 1 h [9, 10]. The study of the interaction of the matrix alloy with stiffening filler was carried out after annealing the samples using micro-X-ray spectral analysis using an attachment with a solid-state silicon detector X- maxN 80. The mapping of the distribution of elements, the visualization of the energydispersion spectra and the calculations of the local composition were carried out using AzTec software modules.

The physicomechanical properties of samples of materials in which the dissolution of the stiffening filler in the matrix alloy did not occur were investigated.
Impact toughness was determined on the installation InstronCEAST 9050, density – as per GOST 18898-89. All tests were carried out at room temperature.

3. Research results and discussion

Due to the fact that the VKNA-4U alloy is an intermetallic alloy, it has specific disadvantages, such as insufficient low-temperature plasticity due to poorly developed mechanism for transferring the deformation along the grain boundaries. From reference sources it is known that with increasing proportion of stiffening component hardness and modulus of elasticity increase, however, at the same time, low-temperature plasticity decreases, so the optimal proportion of stiffening filler was chosen as 2% [11].
Fig. 1.2 shows the results of X-ray spectral analysis of VKNA-4U samples from 2% content of reinforcing fillers.
Figure 1 clearly shows that in VKNA-4U + ZrO2 – Y2O3 – Al2O3, VKNA-4U samples + ZrO2 – Y2O3, VKNA-4U + HfO2 – Y2O3, VKNA-4U + HfO2 – Y2O3 – ZrO2 there is a zone of interaction between the matrix and the stiffening component. Figure 2 shows the result of chemical analysis of a sample with an interaction zone.

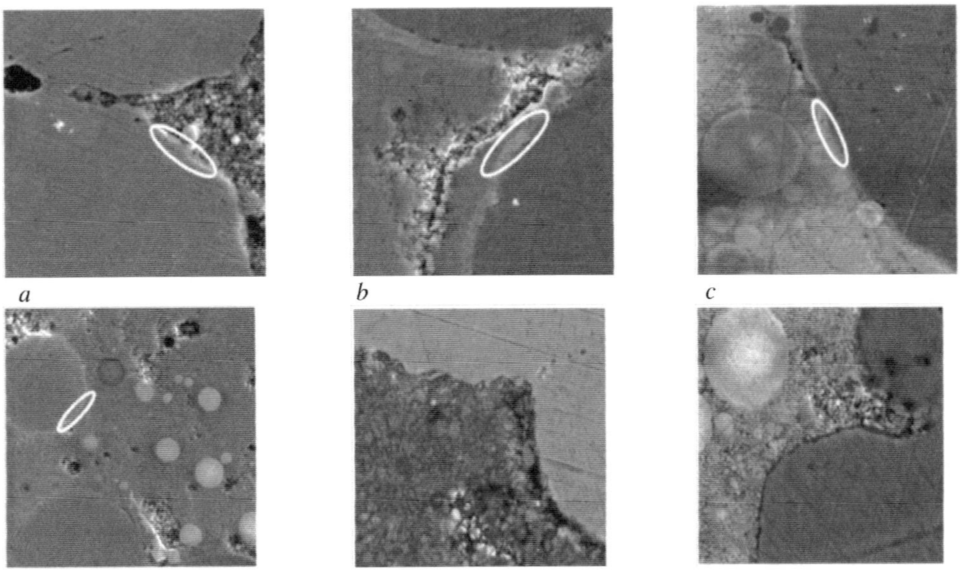

a *b* *c*

def

Figure 1. Microstructure of VKNA-4U alloy samples reinforced with various oxides: *a* - Y_2O_3

IOP Conf. Series: Materials Science and Engineering **683** (2019) 012095 doi:10.1088/1757-899X/683/1/012095

– Al_2O_3 – ZrO_2, *б* - HfO_2 – ZrO_2 – Y_2O_3, *в* - ZrO_2 – Y_2O_3, *г* - HfO_2 – Y_2O_3, *д* - Y_2O_3 – Al_2O_3, *е* - HfO_2 – Y_2O_3 – Al_2O_3. Figures a,*b*,*c*,*d* show the zone of interaction of oxide stiffening elements with the matrix, *д* and *е* - there is no interaction between the matrix and the reinforcing filler.

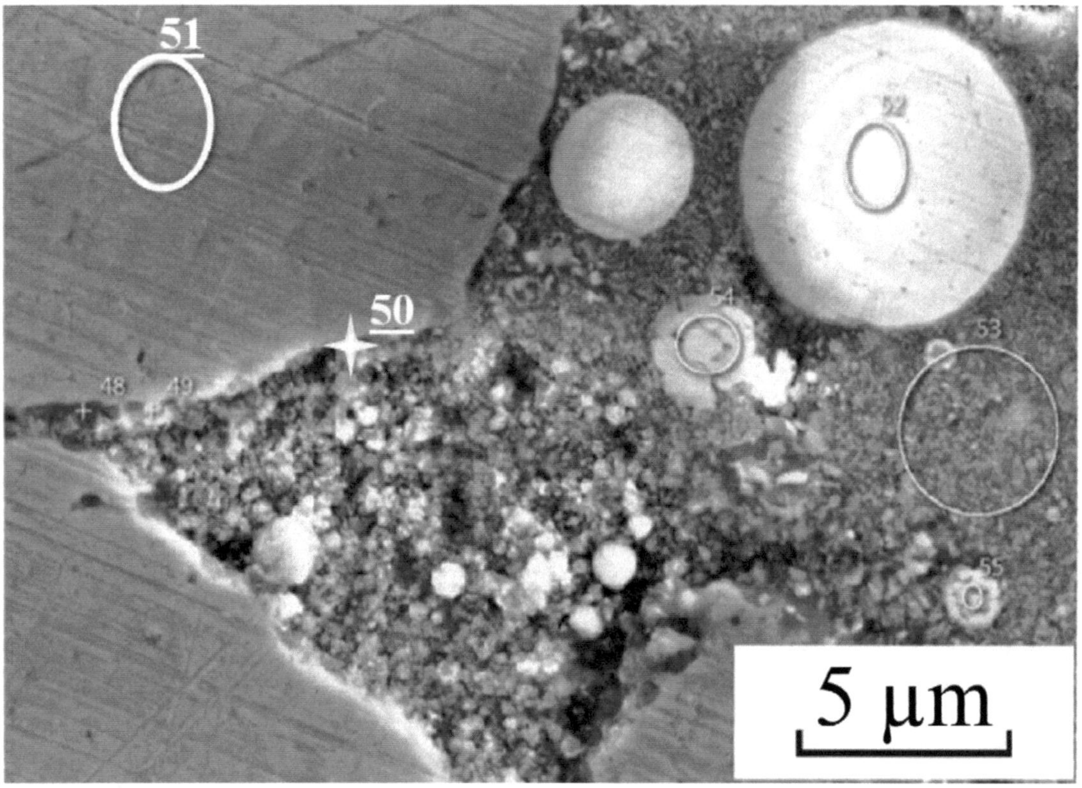

a)

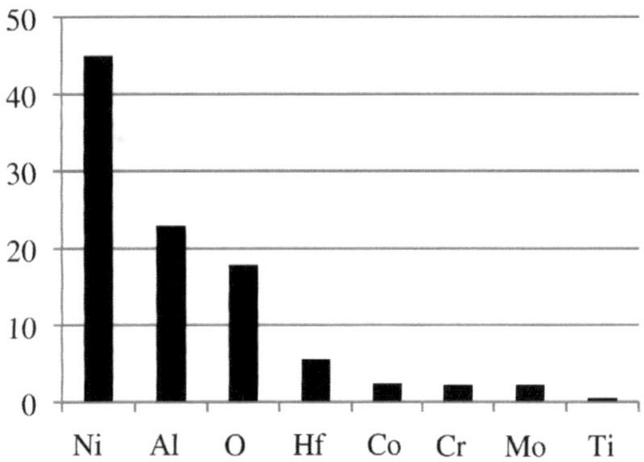

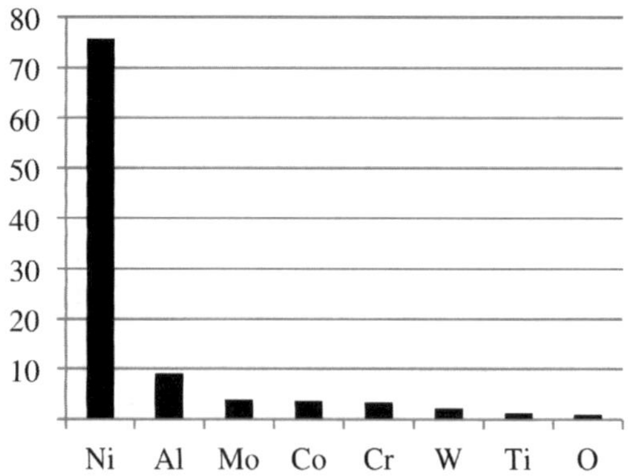

b) c)

Figure 2. The results of chemical analysis of the composite material VKNA-4U + HfO_2 – Al_2O_3 oxides,
a - general view of the research field, b - chemical analysis of the matrix (spectrum 51, wt%), c - chemical analysis of the interaction zone (spectrum 50, wt%).

The chemical reaction caused by the interaction of the stiffening element with the matrix leads to a decrease in the strength characteristics of the material [12]. In the process of interaction between the matrix and the filler, brittle metal compounds can be formed [13, 14].

After analyzing the matrix-filler interaction, for further investigation, VKNA-4U + HfO_2 – Y_2O_3 – Al_2O_3 and VKNA-4U + Y_2O_3 – Al_2O_3 alloys were selected, in which the stiffening filler is not dissolved in the matrix alloy.

Also, for all samples it can be seen that nanoscale reinforcing elements are not evenly distributed in the volume of the matrix material. This is due to the low manufacturability of the aggregate fillers, which makes them evenly distributed in the process of mechanical alloying.

To study the physicomechanical characteristics of VKNA-4U + HfO_2 – Y_2O_3 – Al_2O_3 and VKNA4U + Y_2O_3 – Al_2O_3 alloys, samples were made by mechanical alloying followed by compaction by the hybrid spark plasma sintering method [15].

When choosing the optimal content of stiffening components in the cermet powder composition for the matrix of VKNA-4U alloy, the main parameters were the density and impact strength at the room temperature. The choice of the optimal content was carried out in such a way as to ensure the preservation of plasticity at the maximum value of strength.

There were made compact samples of the material containing the reinforcing component 2 wt% and 5 wt%, a sample without a reinforcing component was also made from the matrix alloy. Adding solid stiffening components to the plastic matrix occurs according to the mechanism of plastic deformation inhibition and to a large extent depends on the number and uniform distribution of the stiffening filler by volume of the matrix material (provided there is no interaction between the matrix and the stiffening filler).

The results of studies of density and toughness, bending strength of samples VKNA-4U, VKNA4U + HfO_2 – Y_2O_3 – Al_2O_3, VKNA-4U + Y_2O_3 – Al_2O_3 are presented in table 1. The optimum content of the reinforcing component in the matrix, at which the toughness and flexural strength has a maximum value of 2 wt%, while increasing the content of the stiffening component greatly reduces the ductility of the material.

Table 1 Results of the study of density and impact strength of samples: pure VKNA-4U, VKNA-4U+HfO_2 – Y_2O_3 – Al_2O_3, VKNA-4U +HfO_2 – Al_2O_3

	VKNA-4U	VKNA-4U+ HfO_2 – Y_2O_3 – Al_2O_3		VKNA-4U + HfO_2 – Al_2O_3	
Content of reinforcing filler, wt%	0	2	5	2	5
Density, g/cm^3	7.9	7.9	7.9	7.9	7.5
Impact strength, kJ/m^2	204	260	117	149	57
Bending strength, MPa	1350	1470	1240	1400	1170

Findings

In this paper, we investigated the influence of stiffening filler for VKNA-4U alloy on it`s mechanical characteristics. The compositions of oxides that do not enter into chemical interaction with the metal matrix were selected as the most perspective for this material. Optimal content was determined for the strengthening component in the matrix, in which the strength characteristics of the material have a maximum value. An uneven distribution of the reinforcing component during mechanical alloying was established, which requires further refinement and optimization of the process. It is shown that 1-2 wt% of the oxide reinforcing filler increase bending strength by approximately 7.5%. In the future, VKNA-4U powder composition mechanically alloyed with α-Al2O3, Y2O3, HfO2 oxides can be used as a building material for additive manufacturing technologies for power fasteners of gas turbine engines (GTE).

References

[1] Kablov E N, Lomberg B S , Ospennikova O G 2012 Creation of modern heat-resistant materials and technologies for their production for aircraft engine building. *Wings of Motherland* **3 – 4** pp. 34 – 38.

[2] Kablov E N 2012 Strategic directions of the development of materials and technologies for their processing for theperiod up to 2030. *Aviation Materials and Technologies* **S** pp. 7 – 17.

[3] Kablov E N 2012 Modern materials are the basis for Russia's innovative modernization. *Metals of Eurasia* **3** pp. 10 – 15.

[4] Grashchenkov D V, Chursova L V 2012 Strategy anddevelopment of composite and functional materials. *Aviation Materials and Technologies* **S** pp. 231 – 241.

[5] Kablov E N 2015 Innovative developments FSUE VIAM SSC RF on the implementation of Strategic directions for the development of materials and technologies for their processing for the period to2030. *Aviation Materials and Technologies* **1** pp. 3 – 33.DOI: 10.18577 / 2071-9140-2015-0-1-3-33.

[6] Ospennikova O G 2012 The development strategy of heat-resistant alloys and special-purpose steels, protective and heat- protective coatings. *Aviation Materials and Technologies* **S** pp. 19 – 36.

[7] Kablov E N 2006 Structural andfunctional materials — the basis of economic andscientific and technical development of Russia. *Problems of materialsscience* **1** pp. 64 – 67.

[8] Alekseyev N V, Samokhin A V, Tsvetkov Yu V 2007 Plasma device for producingnano-powders. Patent RF 2311225 2006110838/15

[9] Grashchenkov D V, Shchetanov B V, Efimochkin I Yu 2011 Development of powder metallurgy of heatresistantw materials. *Polymer Science. Series D* **5** pp. 13 – 26.

[10] Rodionov A I , TurenkoE.Yu., Kablov E N,Bazyleva O A, Grashchenkov D V, Efimochkin I Yu, Cherenanin R N 2015 Production of high-temperature nickel-based composite. Patent RU2563084 RF 2014145770/02

[11] Kablov E N, Petrushin N V, Svetlov I L, Demonis I M 2012 Nickel casting heat-resistant alloys of a new generation. *Aviation Materials and Technologies* **S** pp. 36 – 52.

[12] Ospennikova O G, Bazyleva O A, Evgenov A G, Arginbaeva E G, Turenko E Yu 2016 Microstructural and phase transformations in anintermetallic alloy based on Ni3Al after the action ofheat treatment and hot isostatic pressing. *Aviation Materials and Technologies* **S1** pp. 36 – 43 DOI: 10.18577/ 2071-9140-2016-0-S1-36-43.

[13] Bolshakova A N, Efimochkin I Yu, Murasheva V V 2015 Mekhanicheskilegirovannyyedispersno-uprochnennyyekompozitsionnyyematerialy [Mechanicallyalloyed dispersed-hardened composite materials]. *Composite Materials Constructions*, **1** (137) pp. 36 – 40.

[14] Kablov E N, Toloraya V N 2012 VIAM is the founderof the domestic casting technology for single-crystalturbine blades of GTE and GTU. *Aviation Materials and Technologies* **S** pp. 105 – 117.

[15] Basargin O V, Kolyshev S G, Shchetanov B V, Shcheglova T M 2015 Features of high temperature tests in the bending of samples from a composite material with a matrix based on Nb. Proc. VIAM: Electron. Scientific-technical J. **5** art. 11 URL: http://www.viam-works.ru (date of circulation on March 20, 2017). DOI: 10.18577 / 2307-6046- 2015-0-5-11-11.

AUTHOR INDEX

Institute of Physics
Dirac House, Temple Back
Bristol BS1 6BE UK

ISSN: 1757-8981
ISBN 978-1-7138-0935-7

CURRAN ASSOCIATES INC.
proceedings.com